Zeta and q-Zeta Functions and Associated Series and Integrals

H. M. Srivastava

Department of Mathematics
University of Victoria
Victoria
Canada

Junesang Choi

Department of Mathematics
Dongguk University
Gyeongju
Republic of Korea

ELSEVIER

AMSTERDAM • BOSTON • HEIDELBERG • LONDON • NEW YORK • OXFORD
PARIS • SAN DIEGO • SAN FRANCISCO • SINGAPORE • SYDNEY • TOKYO

Elsevier
32 Jamestown Road, London NW1 7BY
225 Wyman Street, Waltham, MA 02451, USA

First edition 2012

British Library Cataloguing-in-Publication Data
A catalogue record for this book is available from the British Library

Library of Congress Cataloging-in-Publication Data
A catalog record for this book is available from the Library of Congress

ISBN: 978-0-323-16526-6

For information on all Elsevier publications
visit our website at *elsevierdirect.com*

This book has been manufactured using Print On Demand technology. Each copy is produced to order and is limited to black ink. The online version of this book will show color figures where appropriate.

Transferred to Digital Printing in 2016

Zeta and q-Zeta Functions and Associated Series and Integrals

Contents

Preface

This book is essentially a thoroughly revised, enlarged and updated version of the authors' work: ***Series Associated with the Zeta and Related Functions*** (Kluwer Academic Publishers, Dordrecht, Boston and London, 2001). It aims at presenting a state-of-the-art account of the theories and applications of the various methods and techniques which are used in dealing with many different families of series associated with the Riemann Zeta function and its numerous generalizations and basic (or q-) extensions. Systematic accounts of only some of these methods and techniques, which are widely scattered in journal articles and book chapters, were included in the above-mentioned book.

In recent years, there has been an increasing interest in problems involving closed-form evaluations of (and representations of the Riemann Zeta function at positive integer arguments as) various families of series associated with the Riemann Zeta function $\zeta(s)$, the Hurwitz Zeta function $\zeta(s, a)$, and their such extensions and generalizations as (for example) Lerch's transcendent (or the Hurwitz-Lerch Zeta function) $\Phi(z, s, a)$. Some of these developments have apparently stemmed from an over two-century-old theorem of Christian Goldbach (1690–1764), which was stated in a letter dated 1729 from Goldbach to Daniel Bernoulli (1700–1782), from recent rediscoveries of a fairly rapidly convergent series representation for $\zeta(3)$, which is actually contained in a 1772 paper by Leonhard Euler (1707–1783), and from another known series representation for $\zeta(3)$, which was used by Roger Apéry (1916–1994) in 1978 in his celebrated proof of the irrationality of $\zeta(3)$.

This revised, enlarged and updated version of our 2001 book is motivated essentially by the fact that the theories and applications of the various methods and techniques used in dealing with many different families of series associated with the Riemann Zeta function, its aforementioned relatives and its many different basic (or q-) extensions are to be found so far only in widely scattered journal articles published during the last decade or so. Thus, our systematic (and unified) presentation of these results on the evaluation and representation of the various families of Zeta and q-Zeta functions is expected to fill a conspicuous gap in the existing books dealing exclusively with these Zeta and q-Zeta functions.

The main objective of this revised, enlarged and updated version is to provide a systematic collection of various families of series associated with the Riemann and Hurwitz Zeta functions, as well as with many other higher transcendental functions, which are closely related to these functions (including especially the q-Zeta and related functions). It, therefore, aims at presenting a state-of-the-art account of the theory and applications of many different methods (which are available in the rather scattered literature on this subject, especially since the publication of our aforementioned 2001 book) for the derivation of the types of results considered here.

In our attempt to make this book as self-contained as possible within the obvious constraints, we include in Chapter 1 (Introduction and Preliminaries) a reasonably detailed account of such useful functions as the Gamma and Beta functions, the Polygamma and related functions, multiple Gamma functions, the Gauss hypergeometric function and its familiar generalization, the Stirling numbers of the first and second kind, the Bernoulli, Euler and Genocchi polynomials and numbers, the Apostol-Bernoulli, the Apostol-Euler and the Apostol-Genocchi polynomials and numbers, as well as some interesting inequalities for the Gamma function and the double Gamma function. In Chapter 2 (The Zeta and Related Functions), we present the definitions and various potentially useful properties (and characteristics) of the Riemann, Hurwitz and Hurwitz-Lerch Zeta functions and their generalizations, the Polylogarithm and related functions and the multiple Zeta functions, together with their analytic continuations.

In Chapter 3 (Series Involving Zeta Functions), we begin by providing a brief historical introduction to the main subject of this book. We then describe and illustrate some of the most effective methods of evaluating series associated with the Zeta and related functions. Further developments on the evaluations and (rapidly convergent) series representations of $\zeta(s)$ when $s \in \mathbb{N} \setminus \{1\}$ are presented in Chapter 4 (Evaluations and Series Representations), which also deals with various computational results on this subject.

Chapter 5 (Determinants of the Laplacians) considers the problem involving computations of the determinants of the Laplacians for the n-dimensional sphere $\mathbf{S}^n$ ($n \in \mathbb{N}$). It is here in this chapter that we show how fruitfully some of the series evaluations (which are presented in the earlier chapters) can be applied in the solution of the aforementioned problem.

In a *brand new* Chapter 6 (q-Extensions of Some Special Functions and Polynomials), we first introduce the concepts of the basic (or q-) numbers, the basic (or q-) series and the basic (or q-) polynomials. We then proceed to apply these concepts and present a reasonably detailed theory of the various basic (or q-) extensions of the Gamma and Beta functions, the derivatives, antiderivatives and integrals, the binomial theorem, the multiple Gamma functions, the Bernoulli numbers and polynomials, the Euler numbers and polynomials, the Apostol-Bernoulli polynomials, the Apostol-Euler polynomials and so on.

The last chapter (Chapter 7) contains a wide variety of miscellaneous results dealing with (for example) the analysis of several useful mathematical constants, a variety of Log-Sine integrals involving series associated with the Zeta function and Polylogarithms, applications of the Gamma and Polygamma functions involving convolutions of the Rayleigh functions, evaluations of the Bernoulli and Euler polynomials at rational arguments, and the closed-form summation of several classes of trigonometric series.

Each chapter in this book begins with a brief outline summarizing the material presented in the chapter and is then divided into a number of sections. Equations in every section are numbered separately. While referring to an equation in another section of the book, we use numbers like 3.2(18) to represent Equation (18) in Section 3.2 (that is, the second section of Chapter 3).

At the close of each chapter, we have provided a set of carefully-selected problems, which are based essentially upon the material presented in the chapter. Many of these problems are taken from recent research publications, and (in all such instances) we have chosen to include the precise references for further investigation (if necessary). Another valuable feature of this book is the extensive and up-to-date bibliography on the subject dealt with in the book.

Just as its predecessor (that is, the 2001 edition), this book is written primarily as a reference work for various seemingly diverse groups of research workers and other users of series associated with the Zeta and related functions. In particular, teachers, researchers and postgraduate students in the fields of mathematical and applied sciences will find this book especially useful, not only for its detailed and systematic presentations of the theory and applications of the various methods and techniques used in dealing with many different classes of series associated with the Zeta and related functions, or for its stimulating historical accounts of a large number of problems considered here, but also for its well-classified tables of series (and integrals) and its well-motivated presentation of many sets of closely related problems with their precise bibliographical references (if any).

Acknowledgements

Many persons have contributed rather significantly to this thoroughly revised, enlarged and updated version, just as to its predecessor (that is, the 2001 edition), both directly *and* indirectly. Contribution of subject matter is duly acknowledged throughout the text *and* in the bibliography. Indeed, we are greatly indebted to the various authors whose works we have freely consulted and who occasionally provided invaluable references and advice serving for the enrichment of the matter presented in this book. The first-named author wishes to express his deep sense of gratitude to his wife *and* colleague, Professor Rekha Srivastava, for her cooperation and support throughout the preparation of this thoroughly revised, enlarged and updated version of the 2001 book.

The collaboration of the authors on the 2001 book project was conceptualized as long ago as August 1995, and the preparation of a preliminary outline was initiated in December 1997, during the first-named author's visits to Dongguk University at Gyeongju. The first drafts of some of the chapters in this book were written during several subsequent visits of the first-named author to Dongguk University at Gyeongju. The final drafts of most of the chapters in the 2001 book were prepared during the second-named author's visit to the University of Victoria from August 1999 to August 2000, while he was on Study Leave from Dongguk University at Gyeongju. The preparation of this thoroughly revised, enlarged and updated version was carried out, in most part, during the period from January 2008 to January 2009, during the second-named author's visit to the University of Victoria, while he was on Study Leave from Dongguk University at Gyeongju for the second time. Our sincere thanks are due to the appropriate authorities of each of these universities, to the Korea Research Foundation (Support for Faculty Research Abroad under its Research Fund Program) and to the Natural Sciences and Engineering Research Council of Canada, for providing financial support and other facilities for the completion of each of the projects leading eventually to the 2001 edition and this thoroughly revised, enlarged and updated version. We especially acknowledge and appreciate the financial support that was received under the *Basic Science Research Program* through the National Research Foundation of the Republic of Korea.

We take this opportunity to express our thanks to the editorial (and technical) staff of the Elsevier Science Publishers B.V. (especially the Publisher, Ms. Lisa Tickner, for Serials and Elsevier Insights) for their continued interest in this book and for their proficient (and impeccable) handling of its publication. Springer's permission to publish this thoroughly revised, enlarged and updated edition of the 2001 book is also greatly appreciated.

Finally, we should like to record our indebtedness to the members of our respective families for their understanding, cooperation and support throughout this project.

The second-named author and his family would, especially, like to express their appreciation for the first-named author and his family's hospitality and every prudent consideration during their stay in Victoria for over one year, first from August 1999 to August 2000 and then again from January 2008 to January 2009, while the second-named author was on Study Leave from Dongguk University at Gyeongju.

H. M. Srivastava
University of Victoria
Canada

Junesang Choi
Dongguk University
Republic of Korea

February 2011

1 Introduction and Preliminaries

In this introductory chapter, we present the definitions and notations (and some of the important properties and characteristics) of the various special functions, polynomials and numbers, which are potentially useful in the remainder of the book. The special functions considered here include (for example) the Gamma, Beta and related functions, the Polygamma functions, the multiple Gamma functions, the Gaussian hypergeometric function and the generalized hypergeometric function. We also consider the Stirling numbers of the first and second kind, the Bernoulli, Euler and Genocchi polynomials and numbers and the various families of the generalized Bernoulli, Euler and Genocchi polynomials and numbers. Relevant connections of some of these functions with other special functions and polynomials, which are not listed above, are also presented here.

1.1 Gamma and Beta Functions

The Gamma Function

The origin of the *Gamma function* can be traced back to two letters from Leonhard Euler (1707–1783) to Christian Goldbach (1690–1764), just as a simple desire to extend factorials to values between the integers. The first letter (dated October 13, 1729) dealt with the interpolation problem, whereas the second letter (dated January 8, 1730) dealt with integration and tied the two together.

The Gamma function $\Gamma(z)$ developed by Euler is usually defined by

$$\Gamma(z) := \int_0^\infty e^{-t} t^{z-1}\, dt \quad (\Re(z) > 0). \tag{1}$$

We also present here several equivalent forms of the Gamma function $\Gamma(z)$, one by Weierstrass:

$$\Gamma(z) = \frac{e^{-\gamma z}}{z} \prod_{k=1}^{\infty} \left\{ \left(1 + \frac{z}{k}\right)^{-1} e^{z/k} \right\} \qquad \left(z \in \mathbb{C} \setminus \mathbb{Z}_0^-;\ \mathbb{Z}_0^- := \{0, -1, -2, \ldots\}\right), \tag{2}$$

Zeta and q-Zeta Functions and Associated Series and Integrals. DOI: 10.1016/B978-0-12-385218-2.00001-3

where γ denotes the Euler-Mascheroni constant defined by

$$\gamma := \lim_{n\to\infty}\left(\sum_{k=1}^{n}\frac{1}{k}-\log n\right)\cong 0.57721\,56649\,01532\,86060\,6512\ldots, \tag{3}$$

and the other by Gauss:

$$\begin{aligned}\Gamma(z) &= \lim_{n\to\infty}\left\{\frac{(n-1)!\,n^z}{z(z+1)\cdots(z+n-1)}\right\}\\ &= \lim_{n\to\infty}\left\{\frac{n!\,(n+1)^z}{z(z+1)\cdots(z+n)}\right\}\\ &= \lim_{n\to\infty}\left\{\frac{n!\,n^z}{z(z+1)\cdots(z+n)}\right\}\\ &\qquad (z\in\mathbb{C}\setminus\mathbb{Z}_0^-),\end{aligned} \tag{4}$$

since

$$\lim_{n\to\infty}\frac{n}{z+n}=1=\lim_{n\to\infty}\frac{n^z}{(n+1)^z}.$$

In terms of the Pochhammer symbol $(\lambda)_n$ defined (for $\lambda\in\mathbb{C}$) by

$$(\lambda)_n := \begin{cases}1 & (n=0)\\ \lambda(\lambda+1)\cdots(\lambda+n-1) & (n\in\mathbb{N}:=\{1,\,2,\,3,\,\ldots\}),\end{cases} \tag{5}$$

the definition (4) can easily be written in an equivalent form:

$$\Gamma(z)=\lim_{n\to\infty}\frac{(n-1)!\,n^z}{(z)_n}\qquad (z\in\mathbb{C}\setminus\mathbb{Z}_0^-). \tag{6}$$

By taking the reciprocal of (2) and applying the definition (3), we have

$$\begin{aligned}\frac{1}{\Gamma(z)} &= z\left[\lim_{n\to\infty}\exp\left\{\left(1+\frac{1}{2}+\cdots+\frac{1}{n}-\log n\right)z\right\}\right]\left[\lim_{n\to\infty}\prod_{k=1}^{n}\left\{\left(1+\frac{z}{k}\right)e^{-z/k}\right\}\right]\\ &= z\lim_{n\to\infty}\left[\exp\left\{\left(1+\frac{1}{2}+\cdots+\frac{1}{n}-\log n\right)z\right\}\cdot\prod_{k=1}^{n}\left\{\left(1+\frac{z}{k}\right)e^{-z/k}\right\}\right]\\ &= z\lim_{n\to\infty}\left\{n^{-z}\prod_{k=1}^{n}\left(1+\frac{z}{k}\right)\right\}\\ &= z\lim_{n\to\infty}\left[\left\{\prod_{k=1}^{n-1}\left(1+\frac{1}{k}\right)^{-z}\right\}\left\{\prod_{k=1}^{n}\left(1+\frac{z}{k}\right)\right\}\right]\\ &= z\prod_{k=1}^{\infty}\left\{\left(1+\frac{z}{k}\right)\left(1+\frac{1}{k}\right)^{-z}\right\},\end{aligned}$$

which yields Euler's product form of the Gamma function:

$$\Gamma(z) = \frac{1}{z} \prod_{k=1}^{\infty} \left\{ \left(1 + \frac{1}{k}\right)^{z} \left(1 + \frac{z}{k}\right)^{-1} \right\}. \tag{7}$$

When t in (1) is replaced by $-\log t$, (1) is also written in an equivalent form:

$$\Gamma(z) = \int_{0}^{1} \left(\log \frac{1}{t}\right)^{z-1} dt \quad (\Re(z) > 0). \tag{8}$$

This representation of the Gamma function as well as the symbol Γ are attributed to Legendre.

Integration of (1) by parts easily yields the functional relation:

$$\Gamma(z+1) = z\Gamma(z), \tag{9}$$

so that, obviously,

$$\Gamma(z) = \frac{\Gamma(z+n)}{z(z+1)\cdots(z+n-1)} \quad (n \in \mathbb{N}_0 := \mathbb{N} \cup \{0\}), \tag{10}$$

which enables us to define $\Gamma(z)$ for $\Re(z) > -n (n \in \mathbb{N}_0)$ as an analytic function except for $z = 0, -1, -2, \ldots, -n+1$. Thus, $\Gamma(z)$ can be continued analytically to the whole complex z-plane except for simple poles at $z \in \mathbb{Z}_0^-$.

The representation (2) in conjunction with the well-known product formula:

$$\sin \pi z = \pi z \prod_{n=1}^{\infty} \left(1 - \frac{z^2}{n^2}\right) \tag{11}$$

also yields the following useful relationship between the Gamma and circular functions:

$$\Gamma(z)\,\Gamma(1-z) = \frac{\pi}{\sin \pi z} \quad (z \notin \mathbb{Z} := \{0, \pm 1, \pm 2, \ldots\}), \tag{12}$$

which incidentally provides an immediate analytic continuation of $\Gamma(z)$ from right to the left half of the complex z-plane.

Several special values of $\Gamma(x)$, when x is real, are worthy of note. Indeed, from (1) we note that

$$\Gamma(1) = \int_{0}^{\infty} e^{-t}\, dt = 1, \tag{13}$$

and that $\Gamma(x) > 0$ for all x in the open interval $(0, \infty)$. Thus, (12) with $z = \frac{1}{2}$ immediately yields

$$\Gamma\left(\frac{1}{2}\right) = \sqrt{\pi}, \tag{14}$$

which, in view of (1), implies

$$\int_0^\infty \frac{e^{-t}}{\sqrt{t}}\,dt = \sqrt{\pi} \tag{15}$$

or, equivalently,

$$\int_0^\infty \exp(-t^2)\,dt = \frac{\sqrt{\pi}}{2}. \tag{16}$$

By making use of the relation (10), we obtain

$$\begin{aligned} &\Gamma(n+1) = n!; \quad \Gamma\left(n+\frac{1}{2}\right) = \frac{(2n)!\,\sqrt{\pi}}{2^{2n}\,n!}; \\ &\Gamma\left(-n+\frac{1}{2}\right) = (-1)^n \sqrt{\pi}\,\frac{2^{2n}\,n!}{(2n)!}; \quad (n \in \mathbb{N}_0). \end{aligned} \tag{17}$$

The last two results in (17) also require the use of (14). The formulas listed under (17) enable us to compute $\Gamma(x)$ when x is a positive integer and when x is half an odd integer, positive or negative.

From (2) or (4), it also follows that $\Gamma(z)$ is a meromorphic function on the whole complex z-plane with simple poles at $z = -n (n \in \mathbb{N}_0)$ with their respective residues given by

$$\operatorname*{Res}_{z=-n} \Gamma(z) = \frac{(-1)^n}{n!} \quad (n \in \mathbb{N}_0). \tag{18}$$

Since the function $\Gamma(1/w)$ has simple poles at $w = -1, -\frac{1}{2}, -\frac{1}{3}, \ldots$, which implies that $w = 0$ is an accumulation point of the poles of $\Gamma(1/w)$, the Gamma function $\Gamma(z)$ has an essential singularity at infinity. Furthermore, it follows immediately from (2) that $1/\Gamma(z)$ has no poles, and, therefore, $\Gamma(z)$ is never zero.

Pochhammer's Symbol and the Factorial Function

Since $(1)_n = n!$, the Pochhammer symbol $(\lambda)_n$ defined by (5) may be looked upon as a generalization of the elementary factorial; hence, the symbol $(\lambda)_n$ is also referred to as the *shifted* factorial.

In terms of the Gamma function, we have (*cf.* Definition (5))

$$(\lambda)_n = \frac{\Gamma(\lambda+n)}{\Gamma(\lambda)} \quad \left(\lambda \in \mathbb{C} \setminus \mathbb{Z}_0^-\right), \tag{19}$$

which can easily be verified. Furthermore, the binomial coefficient may now be expressed as

$$\binom{\lambda}{n} = \frac{\lambda(\lambda-1)\cdots(\lambda-n+1)}{n!} = \frac{(-1)^n(-\lambda)_n}{n!} \tag{20}$$

or, equivalently, as

$$\binom{\lambda}{n} = \frac{\Gamma(\lambda+1)}{n!\,\Gamma(\lambda-n+1)}. \tag{21}$$

It follows from (20) and (21) that

$$\frac{\Gamma(\lambda+1)}{\Gamma(\lambda-n+1)} = (-1)^n(-\lambda)_n,$$

which, for $\lambda = \alpha - 1$, yields

$$\frac{\Gamma(\alpha-n)}{\Gamma(\alpha)} = \frac{(-1)^n}{(1-\alpha)_n} \quad (\alpha \notin \mathbb{Z}). \tag{22}$$

Equations (19) and (22) suggest the definition:

$$(\lambda)_{-n} = \frac{(-1)^n}{(1-\lambda)_n} \quad (n \in \mathbb{N};\ \lambda \notin \mathbb{Z}). \tag{23}$$

Equation (19) also yields

$$(\lambda)_{m+n} = (\lambda)_m(\lambda+m)_n, \tag{24}$$

which, in conjunction with (23), gives

$$(\lambda)_{n-k} = \frac{(-1)^k(\lambda)_n}{(1-\lambda-n)_k} \quad (0 \leqq k \leqq n). \tag{25}$$

For $\lambda = 1$, we have

$$(n-k)! = \frac{(-1)^k n!}{(-n)_k} \quad (0 \leqq k \leqq n), \tag{26}$$

which may alternatively be written in the form:

$$(-n)_k = \begin{cases} \dfrac{(-1)^k n!}{(n-k)!} & (0 \leqq k \leqq n), \\ 0 & (k > n). \end{cases} \tag{27}$$

Multiplication Formulas of Legendre and Gauss

In view of the definition (5), it is not difficult to show that

$$(\lambda)_{2n} = 2^{2n}\left(\frac{1}{2}\lambda\right)_n\left(\frac{1}{2}\lambda+\frac{1}{2}\right)_n \quad (n \in \mathbb{N}_0), \tag{28}$$

which follows also from Legendre's duplication formula for the Gamma function, *viz*

$$\sqrt{\pi}\,\Gamma(2z) = 2^{2z-1}\,\Gamma(z)\,\Gamma\left(z+\frac{1}{2}\right) \quad \left(z \neq 0, -\frac{1}{2}, -1, -\frac{3}{2}, \ldots\right). \tag{29}$$

For every positive integer m, we have

$$(\lambda)_{mn} = m^{mn}\prod_{j=1}^{m}\left(\frac{\lambda+j-1}{m}\right)_n \quad (m \in \mathbb{N};\ n \in \mathbb{N}_0), \tag{30}$$

which reduces to (28) when $m = 2$. Starting from (30) with $\lambda = mz$, it can be proved that

$$\Gamma(mz) = (2\pi)^{\frac{1}{2}(1-m)}\,m^{mz-\frac{1}{2}}\prod_{j=1}^{m}\Gamma\left(z+\frac{j-1}{m}\right) \quad \left(z \neq 0, -\frac{1}{m}, -\frac{2}{m}, \ldots;\ m \in \mathbb{N}\right), \tag{31}$$

which is known in the literature as Gauss's multiplication theorem for the Gamma function.

Stirling's Formula for n! and its Generalizations

For a large positive integer n, it naturally becomes tedious to compute $n!$. An easy way of computing an approximate value of $n!$ for large positive integer n was initiated by Stirling in 1730 and modified subsequently by De Moivre, who showed that

$$n! \sim \left(\frac{n}{e}\right)^n\sqrt{2\pi n} \quad (n \to \infty) \tag{32}$$

or, more generally, that

$$\Gamma(x+1) \sim \left(\frac{x}{e}\right)^x\sqrt{2\pi x} \quad (x \to \infty;\ x \in \mathbb{R}), \tag{33}$$

where e is the base of the natural logarithm.

For a complex number z, we have the following asymptotic expansion:

$$\log \Gamma(z) = \left(z - \frac{1}{2}\right) \log z - z + \frac{1}{2} \log(2\pi) + \sum_{k=1}^{n} \frac{B_{2k}}{2k(2k-1)\, z^{2k-1}} + O\left(z^{-2n-1}\right)$$
$$(|z| \to \infty;\ |\arg(z)| \leqq \pi - \epsilon\, (0 < \epsilon < \pi);\ n \in \mathbb{N}_0), \tag{34}$$

which, upon taking exponentials, yields an asymptotic formula for the Gamma function:

$$\Gamma(z) = z^z e^{-z} \sqrt{\frac{2\pi}{z}} \left[1 + \frac{1}{12z} + \frac{1}{288z^2} - \frac{139}{51840z^3} - \frac{571}{2488320z^4}\right.$$
$$\left. + \frac{163879}{209018880z^5} + \frac{50043869}{75246796800z^6} + O\left(z^{-7}\right)\right] \tag{35}$$
$$(|z| \to \infty;\ |\arg(z)| \leqq \pi - \epsilon\, (0 < \epsilon < \pi)).$$

The asymptotic formula (35), in conjunction with the recurrence relation (9), is useful in computing the numerical values of $\Gamma(z)$ for large real values of z.

Some useful consequences of (34) or (35) include the asymptotic expansions:

$$\log \Gamma(z + \alpha) = \left(z + \alpha - \frac{1}{2}\right) \log z - z + \frac{1}{2} \log(2\pi) + O\left(z^{-1}\right)$$
$$(|z| \to \infty;\ |\arg(z)| \leqq \pi - \epsilon;\ |\arg(z+\alpha)| \leqq \pi - \epsilon;\ 0 < \epsilon < \pi), \tag{36}$$

and

$$\frac{\Gamma(z+\alpha)}{\Gamma(z+\beta)} = z^{\alpha-\beta} \left[1 + \frac{(\alpha-\beta)(\alpha+\beta-1)}{2z} + O\left(z^{-2}\right)\right]$$
$$(|z| \to \infty;\ |\arg(z)| \leqq \pi - \epsilon;\ |\arg(z+\alpha)| \leqq \pi - \epsilon;\ 0 < \epsilon < \pi), \tag{37}$$

where α and β are bounded complex numbers.

Yet another interesting consequence of (35) is the following asymptotic expansion of $|\Gamma(x+iy)|$:

$$|\Gamma(x+iy)| \sim \sqrt{2\pi}\, |y|^{x-\frac{1}{2}}\, e^{-\frac{1}{2}\pi |y|} \quad (|x| < \infty;\ |y| \to \infty), \tag{38}$$

where x and y take on real values.

The Beta Function

The *Beta function* $B(\alpha, \beta)$ is a function of two complex variables α and β, defined by

$$B(\alpha, \beta) := \int_0^1 t^{\alpha-1}(1-t)^{\beta-1}\, dt = B(\beta, \alpha) \quad (\Re(\alpha) > 0;\ \Re(\beta) > 0) \tag{39}$$

or, equivalently, by

$$B(\alpha, \beta) = 2 \int_0^{\pi/2} (\sin\theta)^{2\alpha-1} (\cos\theta)^{2\beta-1}\, d\theta \quad (\Re(\alpha) > 0;\ \Re(\beta) > 0), \tag{40}$$

which follows from (39) on setting $t = \sin^2\theta$.

The integrals in (39) and (1) are known as the Eulerian integrals of the first and second kind, respectively.

Putting $t = u/(1+u)$ in (39), we obtain the following representation of $B(\alpha, \beta)$ as an infinite integral:

$$B(\alpha, \beta) = \int_0^{\infty} \frac{u^{\alpha-1}}{(1+u)^{\alpha+\beta}}\, du \quad (\Re(\alpha) > 0;\ \Re(\beta) > 0). \tag{41}$$

The Beta function is closely related to the Gamma function; in fact, we have

$$B(\alpha, \beta) = \frac{\Gamma(\alpha)\,\Gamma(\beta)}{\Gamma(\alpha+\beta)} \quad \left(\alpha, \beta \notin \mathbb{Z}_0^-\right), \tag{42}$$

which not only confirms the symmetry property in (39), but also continues the Beta function analytically for all complex values of α and β, except when $\alpha, \beta \in \mathbb{Z}_0^-$. Thus, we may write

$$B(\alpha, \beta) = \begin{cases} \displaystyle\int_0^1 t^{\alpha-1}(1-t)^{\beta-1}\, dt & (\Re(\alpha) > 0;\ \Re(\beta) > 0) \\[2ex] \dfrac{\Gamma(\alpha)\,\Gamma(\beta)}{\Gamma(\alpha+\beta)} & \left(\Re(\alpha) < 0;\ \Re(\beta) < 0;\ \alpha, \beta \notin \mathbb{Z}_0^-\right). \end{cases} \tag{43}$$

Next we combine the relationship (42) with (40) and (41), and we obtain the following useful integral formulas:

$$\int_0^{\pi/2} \sin^{\mu}\theta \cos^{\nu}\theta\, d\theta = \frac{\Gamma\left(\frac{1}{2}\mu + \frac{1}{2}\right)\Gamma\left(\frac{1}{2}\nu + \frac{1}{2}\right)}{2\Gamma\left(\frac{1}{2}\mu + \frac{1}{2}\nu + 1\right)} \quad (\Re(\mu) > -1;\ \Re(\nu) > -1) \tag{44}$$

and

$$\int_0^{\infty} \frac{u^{\lambda-1}}{(1+u)^{\mu}}\, du = \frac{\Gamma(\lambda)\,\Gamma(\mu-\lambda)}{\Gamma(\mu)} \quad (0 < \Re(\lambda) < \Re(\mu)). \tag{45}$$

It should be remarked in passing that the integral (44) provides a generalization of Wallis's formula of elementary calculus and that (45), with $\mu = 1$, yields the familiar infinite integral:

$$\int_0^\infty \frac{u^{\lambda-1}}{1+u}\,du = \Gamma(\lambda)\,\Gamma(1-\lambda) = \frac{\pi}{\sin \pi\lambda} \qquad (0 < \Re(\lambda) < 1), \tag{46}$$

which is usually evaluated in the literature by contour integration (see, e.g., Copson [341, p. 139, Example 1]).

In addition to (41) and (45), by means of suitable substitutions, a number of definite integrals are expressible in terms of the Beta function:

$$\int_0^1 t^{\alpha-1}(1-t)^{\beta-1}\,(1+at)^{-\alpha-\beta}\,dt = (1+a)^{-\alpha}\,B(\alpha,\,\beta) \qquad (a > -1;\ \Re(\alpha) > 0;\ \Re(\beta) > 0); \tag{47}$$

$$\int_0^\infty t^{\beta-1}\,(1+at)^{-\alpha-\beta}\,dt = a^{-\beta}\,B(\alpha,\,\beta) \qquad (a > 0;\ \Re(\alpha) > 0;\ \Re(\beta) > 0); \tag{48}$$

$$\int_b^a (t-b)^{\alpha-1}\,(a-t)^{\beta-1}\,dt = (a-b)^{\alpha+\beta-1}\,B(\alpha,\,\beta) \qquad (a > b;\ \Re(\alpha) > 0;\ \Re(\beta) > 0); \tag{49}$$

$$\int_b^a \frac{(t-b)^{\alpha-1}\,(a-t)^{\beta-1}}{(t-c)^{\alpha+\beta}}\,dt = \frac{(a-b)^{\alpha+\beta-1}}{(a-c)^{\alpha}\,(b-c)^{\beta}}\,B(\alpha,\,\beta) \qquad (a > b > c;\ \Re(\alpha) > 0;\ \Re(\beta) > 0). \tag{50}$$

The following functional equations for the Beta function can be deduced easily from (39) and (42):

$$B(\alpha,\,\beta+1) = \frac{\beta}{\alpha}B(\alpha+1,\,\beta) = \frac{\beta}{\alpha+\beta}B(\alpha,\,\beta); \tag{51}$$

$$B(\alpha,\,\beta)\,B(\alpha+\beta,\,\gamma) = B(\beta,\,\gamma)\,B(\beta+\gamma,\,\alpha) = B(\gamma,\,\alpha)\,B(\alpha+\gamma,\,\beta); \tag{52}$$

$$B(\alpha,\,\beta)\,B(\alpha+\beta,\,\gamma)\,B(\alpha+\beta+\gamma,\,\delta) = \frac{\Gamma(\alpha)\,\Gamma(\beta)\,\Gamma(\gamma)\,\Gamma(\delta)}{\Gamma(\alpha+\beta+\gamma+\delta)}, \tag{53}$$

or, more generally,

$$\prod_{k=1}^{n} B\left(\sum_{j=1}^{k} \alpha_j,\, \alpha_{k+1}\right) = \frac{\Gamma(\alpha_1)\cdots\Gamma(\alpha_{n+1})}{\Gamma(\alpha_1+\cdots+\alpha_{n+1})} \quad (n \in \mathbb{N}); \tag{54}$$

$$\frac{1}{B(n,\, m)} = m\binom{n+m-1}{n-1} = n\binom{n+m-1}{m-1} \quad (n,\, m \in \mathbb{N}). \tag{55}$$

The Incomplete Gamma Functions

The *incomplete Gamma function* $\gamma(z, \alpha)$ and its *complement* $\Gamma(z, \alpha)$ (also known as *Prym's function*) are defined by

$$\gamma(z,\, \alpha) := \int_0^{\alpha} t^{z-1} e^{-t}\, dt \quad (\Re(z) > 0;\ |\arg(\alpha)| < \pi), \tag{56}$$

$$\Gamma(z,\, \alpha) := \int_{\alpha}^{\infty} t^{z-1} e^{-t}\, dt \quad (|\arg(\alpha)| < \pi), \tag{57}$$

so that

$$\gamma(z,\, \alpha) + \Gamma(z,\, \alpha) = \Gamma(z). \tag{58}$$

For fixed α, $\Gamma(z, \alpha)$ is an entire (integral) function of z, whereas $\gamma(z, \alpha)$ is a meromorphic function of z, with simple poles at the points $z \in \mathbb{Z}_0^-$.

The following recursion formulas are worthy of note:

$$\gamma(z+1,\, \alpha) = z\,\gamma(z,\, \alpha) - \alpha^z e^{-\alpha}, \tag{59}$$

$$\Gamma(z+1,\, \alpha) = z\,\Gamma(z,\, \alpha) + \alpha^z e^{-\alpha}. \tag{60}$$

The Incomplete Beta Functions

The *incomplete Beta function* $B_x(\alpha, \beta)$ is defined by

$$B_x(\alpha,\, \beta) := \int_0^{x} t^{\alpha-1}(1-t)^{\beta-1}\, dt \quad (\Re(\alpha) > 0). \tag{61}$$

For the associated function:

$$I_x(\alpha,\, \beta) = \frac{B_x(\alpha,\, \beta)}{B(\alpha,\, \beta)}, \tag{62}$$

we note here the following properties that are easily verifiable:

$$I_x(\alpha, \beta) = 1 - I_{1-x}(\beta, \alpha), \tag{63}$$

$$I_x(k, n-k+1) = \sum_{j=k}^{n} \binom{n}{j} x^j (1-x)^{n-j} \quad (1 \leq k \leq n), \tag{64}$$

$$I_x(\alpha, \beta) = x I_x(\alpha - 1, \beta) + (1-x) I_x(\alpha, \beta - 1), \tag{65}$$

$$(\alpha + \beta - \alpha x) I_x(\alpha, \beta) = \alpha(1-x) I_x(\alpha + 1, \beta - 1) + \beta I_x(\alpha, \beta + 1), \tag{66}$$

$$(\alpha + \beta) I_x(\alpha, \beta) = \alpha I_x(\alpha + 1, \beta) + \beta I_x(\alpha, \beta + 1). \tag{67}$$

The Error Functions

The *error function* erf(z), also known as the *probability integral* $\Phi(z)$, is defined for any complex z by

$$\operatorname{erf}(z) := \frac{2}{\sqrt{\pi}} \int_0^z \exp(-t^2)\, dt = \Phi(z), \tag{68}$$

and its *complement* by

$$\operatorname{erfc}(z) := 1 - \operatorname{erf}(z) = \frac{2}{\sqrt{\pi}} \int_z^\infty \exp(-t^2)\, dt. \tag{69}$$

Clearly, we have

$$\operatorname{erf}(0) = 0 \quad \text{and} \quad \operatorname{erfc}(0) = 1, \tag{70}$$

and, in view of the well-known result (16), we also have

$$\operatorname{erf}(\infty) = 1 \quad \text{and} \quad \operatorname{erfc}(\infty) = 0. \tag{71}$$

The following alternative notations:

$$\operatorname{Erf}(z) = \frac{\sqrt{\pi}}{2} \operatorname{erf}(z) \quad \text{and} \quad \operatorname{Erfc}(z) = \frac{\sqrt{\pi}}{2} \operatorname{erfc}(z) \tag{72}$$

are sometimes used for the error functions. Many authors use the notations Erf(z) and $\widehat{\text{Erf}}(z)$ for the error functions erf(z) and Erf(z), respectively, defined by (68) and (72), and the notations Erfc(z) and $\widehat{\text{Erfc}}(z)$ for their complements.

In terms of the incomplete Gamma functions, it is easily verified that

$$\operatorname{erf}(z) = \frac{1}{\sqrt{\pi}} \gamma\left(\frac{1}{2}, z^2\right) \quad \text{and} \quad \operatorname{erfc}(z) = \frac{1}{\sqrt{\pi}} \Gamma\left(\frac{1}{2}, z^2\right). \tag{73}$$

The Bohr-Mollerup Theorem

We have already observed that Euler's definition (1) and its such consequences as (9) and (13) enable us to compute all the real values of the Gamma function from the knowledge merely of its values in the interval $(0, 1)$, as noted in conjunction with (35). Since the solution to the interpolation problem is not determined uniquely, it makes sense to add more conditions to the problem. After various trials to find those conditions to guarantee the uniqueness of the Gamma function, in 1922, Bohr and Mollerup were able to show the remarkable fact that the Gamma function is the *only* function that satisfies the recurrence relationship and is logarithmically convex. The original proof was simplified, several years later, by Emil Artin, and the theorem, together with Artin's method of proof, now constitute the *Bohr-Mollerup-Artin theorem*:

Theorem 1.1 *Let* $f : \mathbb{R}^+ \to \mathbb{R}^+$ *satisfy each of the following properties:*

(a) $\log f(x)$ *is a convex function;*
(b) $f(x+1) = xf(x)$ *for all* $x \in \mathbb{R}^+$*;*
(c) $f(1) = 1$.

Then $f(x) = \Gamma(x)$ *for all* $x \in \mathbb{R}^+$.

Instead of giving here the proof of Theorem 1.1 (see Conway [339, p. 179] and Artin [72, p. 14]), we simply state the necessary and sufficient condition for the logarithmic convexity of a given function.

Theorem 1.2 *Let* $f : [a, b] \to \mathbb{R}$*, and suppose that* $f(x) > 0$ *for all* $x \in [a, b]$ *and that* f *has a continuous second derivative* $f''(x)$ *for* $x \in [a, b]$*. Then* f *is logarithmically convex, if and only if*

$$f''(x)f(x) - \left[f'(x)\right]^2 \geqq 0 \quad (x \in [a, b]).$$

Remmert [973] admires the following Wielandt's uniqueness theorem for the Gamma function: It is hardly known that there is also an elegant *function theoretic characterization* of $\Gamma(z)$. This uniqueness theorem was discovered by Helmut Wielandt in 1939. A function theorist ought to be as much fascinated by Wielandt's complex-analytic characterization as by the Bohr-Mollerup theorem. For further comment and applications for Wielandt's theorem, see [675, pp. 47–49], [973], and [1065]. Here, without proof, we present

Theorem 1.3 (Wielandt's Theorem) *Let* $F(z)$ *be an analytic function in the right half plane* $\mathcal{A} := \{z \in \mathbb{C} \mid \Re(z) > 0\}$ *having the following two properties:*

(a) $F(z+1) = zF(z)$ *for all* $z \in \mathcal{A}$*;*
(b) $F(z)$ *is bounded in the strip* $\mathcal{S} := \{z \in \mathbb{C} \mid 1 \leqq \Re(z) < 2\}$.

Then $F(z) = a\Gamma(z)$ *in* $\mathcal{A}$ *with* $a := F(1)$.

1.2 The Euler-Mascheroni Constant γ

The divergence of the harmonic series:

$$\sum_{k=1}^{\infty} \frac{1}{k} = 1 + \frac{1}{2} + \frac{1}{3} + \cdots = \lim_{n\to\infty} \sum_{k=1}^{n} \frac{1}{k} = \infty \tag{1}$$

was attributed by James Bernoulli to his brother (see [482]). Yet, the connection between $1 + \frac{1}{2} + \cdots + \frac{1}{n}$ and $\log n$ was first established in 1735 by Euler [427] (see Walfisz [1203]), who used the notation C for it and stated that it was *worthy of serious consideration*. We are surprised at Euler's foresight that there is a huge amount of literature on this famous mathematical constant γ among which we just refer to the book [543] and the references therein.

The Euler (or, more precisely, the Euler-Mascheroni) constant γ is defined as follows (see Eq. 1.1(3)):

$$\begin{aligned} \gamma :&= \lim_{n\to\infty} (H_n - \log n) \\ &\cong 0.57721\,56649\,01532\,86060\,65120\,90082\,40243\,1042\cdots, \end{aligned} \tag{2}$$

where H_n are called harmonic numbers defined by

$$H_n := \sum_{k=1}^{n} \frac{1}{k} \quad (n \in \mathbb{N}). \tag{3}$$

The symbol γ was first used by Mascheroni in 1790 (see [802]) and the notation C has still been used for the notation γ in (2) (see, e.g., [505]). In fact, $a_n := H_n - \log n$ $(n \in \mathbb{N})$ is a decreasing sequence and $0 < a_n < 1$ for all $n \in \mathbb{N}$. Thus, the convergence of the sequence in (2) follows by the monotone convergence theorem (see, e.g., [1202, p. 45]).

It is noted that γ *is a constant so chosen that* $\Gamma(1) = 1$ in the Weierstrass product form 1.1(2) of the Gamma function $\Gamma(z)$, and the constant γ is the very Euler constant in (2).

The relatively more familiar constants are π and e, whose transcendence was shown by Ferdinand Lindemann in 1882 and Charles Hermite in 1873, respectively. The true nature of the Euler constant γ (whether an algebraic or a transcendental number) has not yet been known. This was a part of the famous Hilbert's seventh problem. David Hilbert [557] announced 23 problems for the twentieth century in the Second International Congress of Mathematicians at Paris in 1900. We introduce here only his seventh problem: *Irrationality and Transcendence of Certain Numbers*. G. H. Hardy was alleged to have offered to give up his Savilian Chair at Oxford to anyone who proved γ to be irrational (see [543, p. 52]). The degree of possible rationality of γ has been tried (see [1222], [543, p. 97]). Appell [69] gave a proof that γ is irrational. Appell himself's finding an error, quickly he published a retraction, within a couple of weeks, of his original announcement (see Ayoub [82]).

Euler [427] gave the formula

$$1+\frac{1}{2}+\cdots+\frac{1}{x}=\gamma+\log x+\frac{1}{2x}-\frac{B_2}{2x^2}-\frac{B_4}{4x^4}-\frac{B_6}{6x^6}+\cdots,$$

B_n being Bernoulli numbers, in which, by putting $x=10$, he calculated

$$\gamma=0.57721\,56649\,01532\,5\cdots.$$

Mascheroni [802] evaluated the value of Euler constant with 32 figures as follows:

$$\gamma=0.57721\,56649\,01532\,8606\underline{1}\,811209008239\cdots$$

Soldner (see [482]) computed the value of γ as

$$\gamma=0.57721\,56649\,01532\,8606\underline{0}\,6065\cdots,$$

which differs from Mascheroni's value in the twentieth place. In fact, Mascheroni's value turned out to be incorrect. Nonetheless, since Mascheroni's error has led to eight additional calculations of this celebrated mathematical constant, so γ is often called the *Euler-Mascheroni* constant. Knuth [678] computed the first 1271 decimals. Gourdon and Demichel [503] computed a record 108 million digits of γ. Kondo [693] has computed γ to 2 billion digits, which is apparently the current world record.

The Euler (or, more popularly, the Euler-Mascheroni) constant γ is considered the third important mathematical constant next to π and e. The mathematical constants π, e and γ are often referred to as the *holy trinity*. The constant γ has been involved in a variety of mathematical formulas and results. For instance, the book [505] contains about 160 formulas involving γ. Conversely, Wilf [1228] posed as a problem the following elegant infinite product formula, which contains all of the three most important mathematical constants π, e and γ (see Eq. 3.6(19)):

$$\prod_{j=1}^{\infty}\left[e^{-\frac{1}{j}}\left(1+\frac{1}{j}+\frac{1}{2j^2}\right)\right]=\frac{e^{\frac{\pi}{2}}+e^{-\frac{\pi}{2}}}{\pi\, e^{\gamma}}. \tag{4}$$

Choi *et al.* [275] presented several general infinite product formulas, which include, as their special cases, the product formula (4) of Wilf [1228] and other product formulas given by Choi and Seo [285]. The function $d(n)$ is the number of divisors of $n\in\mathbb{N}$, including 1 and n. The average order of $d(n)$ was proved by Dirichlet in 1838 that

$$\frac{1}{n}\sum_{k=1}^{n}d(k)=\ln n+2\gamma-1+O\left(\frac{1}{\sqrt{n}}\right)\quad(n\to\infty), \tag{5}$$

where γ is the Euler-Mascheroni constant (see [538, p. 264,Theorem 320]). Very recently, Pillichshammer [893] treated this subject in a more general way. A lot of series representations of γ have been presented (see, e.g., [12, 208, 264, 274, 286, 480, 531, 963, 964, 1094, 1179]), such as (see Eq. 3.4(23))

$$\gamma = \sum_{k=2}^{\infty} (-1)^k \frac{\zeta(k)}{k}, \tag{6}$$

where $\zeta(s)$ is the Riemann Zeta function given in 2.3(1).

Sondow [1051] presented an elegant double integral representation:

$$\gamma = \int_0^1 \int_0^1 \frac{x-1}{(1-xy)\ln(xy)}\, dx\, dy. \tag{7}$$

In fact, many integral representations of γ have been developed. Here, in what follows, several further integral representations of γ with mild generalizations of some known formulas of γ are presented.

A Set of Known Integral Representations for γ

Among a variety of known integral representations for γ, here we choose to recall some of them (see, e.g., [53, 54, 220, 264, 284, 365, 505, 1094, 1225]):

$$\gamma = \int_0^1 \frac{1-e^{-x}}{x}\, dx - \int_1^{\infty} \frac{e^{-x}}{x}\, dx. \tag{8}$$

$$\gamma = \int_0^{\infty} e^{-x} \left(\frac{1}{1-e^{-x}} - \frac{1}{x} \right) dx. \tag{9}$$

$$\gamma = 1 + \int_0^{\infty} \left(\frac{e^{-x}-1}{x} + \frac{1}{1+x} \right) \frac{dx}{x}. \tag{10}$$

$$\gamma = \int_0^1 \left(\frac{1-e^{-x}-e^{-\frac{1}{x}}}{x} \right) dx. \tag{11}$$

$$\gamma = \int_0^{\infty} \left(\frac{1}{1+x} - e^{-x} \right) \frac{dx}{x}. \tag{12}$$

$$\gamma = 2\int_0^{\infty} \left(\frac{1}{1+x^2} - e^{-x^2} \right) \frac{dx}{x}. \tag{13}$$

$$\gamma = \int_0^\infty \left(\frac{1}{1+x^2} - e^{-x} \right) \frac{dx}{x}. \tag{14}$$

$$\gamma = 2 \int_0^\infty \left(e^{-x^2} - e^{-x} \right) \frac{dx}{x}. \tag{15}$$

$$\gamma = \frac{4}{3} \int_0^\infty \left(e^{-x^4} - e^{-x} \right) \frac{dx}{x}. \tag{16}$$

$$\gamma = 4 \int_0^\infty \left(e^{-x^4} - e^{-x^2} \right) \frac{dx}{x}. \tag{17}$$

$$\gamma = \frac{1}{1-2^{-n}} \int_0^\infty \left[\exp\left(-x^{2^n}\right) - e^{-x} \right] \frac{dx}{x} \quad (n \in \mathbb{N}). \tag{18}$$

$$\gamma = 2^n \int_0^\infty \left(\frac{1}{1+x^{2^{n+1}}} - e^{-x^2} \right) \frac{dx}{x} \quad (n \in \mathbb{Z}), \tag{19}$$

where $\mathbb{Z}$ denotes the set of integers.

$$\gamma = 2^n \int_0^\infty \left(\frac{1}{1+x^2} - \exp\left(-x^{2^n}\right) \right) \frac{dx}{x} \quad (n \in \mathbb{N}). \tag{20}$$

$$\gamma = \int_0^\infty \left(\frac{1}{1+x} - \cos x \right) \frac{dx}{x}. \tag{21}$$

$$\gamma = \int_0^\infty \left(\frac{1}{1+x^2} - \cos x \right) \frac{dx}{x}. \tag{22}$$

$$\gamma = 1 + \int_0^\infty \left(\frac{1}{1+x} - \frac{\sin x}{x} \right) \frac{dx}{x}. \tag{23}$$

$$\gamma = 2 \int_0^\infty \left(e^{-x^2} - \cos x \right) \frac{dx}{x}. \tag{24}$$

$$\gamma = -\ln p + \int_0^\infty \left(\frac{2}{\pi} \operatorname{arccot} x - e^{-px} \right) \frac{dx}{x} \quad (p > 0). \tag{25}$$

$$\gamma = \frac{3}{2} + 2\int_0^\infty \left(\frac{\cos x - 1}{x^2} + \frac{1}{2(1+x)} \right) \frac{dx}{x}. \tag{26}$$

$$\gamma = \int_0^{\pi/2} \left[1 - \sec^2 x \cos(\tan x) \right] \frac{dx}{\tan x}. \tag{27}$$

$$\gamma = -\ln p - \frac{2}{\pi} \int_0^\infty \sin(px) \ln x \frac{dx}{x} \quad (p > 0). \tag{28}$$

$$\gamma = 1 - \ln(2p) - \frac{2}{p\pi} \int_0^\infty \left(\frac{\sin^2(px)}{x^2} \right) \ln x \, dx \quad (p > 0). \tag{29}$$

$$\gamma = 1 + \frac{1}{p-q} \ln\left(\frac{q^q}{p^p} \right) + \frac{2}{\pi(p-q)} \int_0^\infty \left(\frac{\cos(px) - \cos(qx)}{x^2} \right) \ln x \, dx \quad (p > 0;\ q > 0;\ p \neq q). \tag{30}$$

$$\gamma = \frac{1}{2} + 2\int_0^\infty \frac{x}{1+x^2} \frac{dx}{e^{2\pi x} - 1}. \tag{31}$$

$$\gamma = \int_0^1 \left(x - \frac{1}{1 - \log x} \right) \frac{dx}{x \log x}. \tag{32}$$

$$\gamma = -\int_0^\infty e^{-x} \log x \, dx = -\int_0^1 \log\left(\log \frac{1}{x} \right) dx. \tag{33}$$

$$\gamma = \log 2 - \pi \int_0^1 \int_0^{\frac{1}{2}} \tan\left(\frac{\pi x}{2} \right) \left(\frac{\sin \pi x u}{\sin \pi u} - x \right) du dx. \tag{34}$$

$$\gamma = \frac{1}{2} + 2\int_0^\infty \frac{\sin(\tan^{-1} x)}{(e^{2\pi x} - 1)\sqrt{1+x^2}} dx. \tag{35}$$

$$\gamma = \log 2 - 2\int_0^\infty \frac{\sin(\tan^{-1} x)}{(e^{2\pi x} + 1)\sqrt{1+x^2}} dx. \tag{36}$$

$$\gamma = 1 + \int_0^\infty \left(\cos x - \frac{\sin x}{x} \right) \frac{\log x}{x} dx. \tag{37}$$

$$\gamma = \int_0^\infty \left(\frac{1}{2} - \cos x\right)\frac{dx}{x}. \tag{38}$$

$$\gamma = \frac{1}{2} + \frac{B_2}{2} + \frac{B_4}{4} + \cdots + \frac{B_{2n}}{2n} - (2n+1)! \int_1^\infty \frac{Q_{2n+1}(x)}{x^{2n+2}} dx, \tag{39}$$

where the functions $Q_n(x)$ are defined by

$$Q_n(x) := \begin{cases} x - \dfrac{1}{2} & (n = 1;\ 0 < x < 1), \\ \dfrac{1}{n!} B_n(x - [x]) & (n \in \mathbb{N} \setminus \{1\};\ 0 \leqq x < \infty), \end{cases}$$

$B_n := B_n(0)$ and $B_n(x)$ being the Bernoulli numbers and polynomials, respectively (see [1094, Section 1.6]). As observed by Knopp [676] by an explicit example with $n = 3$ in (2.61), the approximate value of γ can easily be calculated with much greater accuracy than before (and, theoretically, to any degree of accuracy whatever) by means of the formula (39).

Further Integral Representations for γ

Very recently, Choi and Srivastava [302] presented several further integral representations for γ by making use of some formulas in the previous subsection and other known formulas for $\log \Gamma(z)$, $\psi(z)$ (Section 1.3) and the Hurwitz (or generalized) Zeta function $\zeta(s,a)$ (Section 2.2) or the Riemann Zeta function $\zeta(s)$ (Section 2.3) in conjunction with the residue calculus. Here, we choose to record some of them: We begin by recalling an integral formula for $\log z$ (see [1225, p. 248]).

The following integral formula holds true for $\log z$:

$$\int_0^\infty \left(e^{-t} - e^{-tz}\right)\frac{dt}{t} = \int_0^1 \left(t^{z-1} - 1\right)\frac{dt}{\log t} = \log z \quad \left(\Re(z) > 0\right), \tag{40}$$

where the $\log z$ *is an appropriate branch of the multiple-valued function* $\log z$, *such as*

$$\log z = \ln|z| + i \arg z \quad \left(|z| > 0;\ \alpha < \arg(z) < \alpha + 2\pi\right)$$

for some real $\alpha \in \mathbb{R}$ *with possibly* $-\pi \leqq \alpha \leqq -\frac{\pi}{2}$.

If (40) is used in the formulas (25), (28), (29) and (30) and $\tan x$ is replaced by x in (27), the following integral formulas for γ are obtained:

Each of the following integral representations holds true for γ :

$$\gamma = \int_0^\infty \left(\frac{2}{\pi} \operatorname{arccot} x - e^{-x}\right) \frac{dx}{x}, \tag{41}$$

$$\gamma = -\frac{2}{\pi} \int_0^\infty \left[e^{-x} - e^{-p^{\frac{\pi}{2}} x} + \ln x \sin (px)\right] \frac{dx}{x} \quad (p > 0), \tag{42}$$

$$\gamma = 1 - \frac{2}{p\pi} \int_0^\infty \left(e^{-x} - \exp\left[-(2p)^{\frac{p\pi}{2}} x\right] + \ln x \frac{\sin^2 (px)}{x}\right) \frac{dx}{x} \quad (p > 0), \tag{43}$$

$$\gamma = 1 + \frac{2}{\pi (p - q)} \int_0^\infty \left(e^{-x} - \exp\left[-\left(\frac{q^q}{p^p}\right)^{\frac{\pi}{2}} x\right] + \ln x \frac{\cos (px) - \cos (qx)}{x}\right) \frac{dx}{x}$$

$$(p > 0;\ q > 0) \tag{44}$$

and

$$\gamma = \int_0^\infty \left[\cos^2 (\arctan x) - \cos x\right] \frac{dx}{x}. \tag{45}$$

If x is replaced by x^p in (10), (12), (21), (23) and (26), the following mildly more general formulas for γ are obtained.

Each of the following integral representations holds true for γ:

$$\gamma = p \int_0^\infty \left(\frac{1}{1 + x^p} - \exp(-x^p)\right) \frac{dx}{x} \quad (p > 0), \tag{46}$$

$$\gamma = p \int_0^\infty \left(\frac{1}{1 + x^p} - \cos (x^p)\right) \frac{dx}{x} \quad (p > 0), \tag{47}$$

$$\gamma = 1 + p \int_0^\infty \left(\frac{1}{1 + x^p} + \frac{\exp(-x^p) - 1}{x^p}\right) \frac{dx}{x} \quad (p > 0), \tag{48}$$

$$\gamma = 1 + p \int_0^\infty \left(\frac{1}{1 + x^p} - \frac{\sin (x^p)}{x^p}\right) \frac{dx}{x} \quad (p > 0) \tag{49}$$

and

$$\gamma = \frac{3}{2} + 2p \int_0^\infty \left(\frac{\cos\,(x^p) - 1}{x^{2p}} + \frac{1}{2(1+x^p)} \right) \frac{dx}{x} \quad (p > 0). \tag{50}$$

It is noted that the case $p = 2$ of (46) would obviously reduce to (13).

A class of vanishing integrals is provided just below.

The following vanishing integral formula holds true:

$$\int_0^\infty \left(\frac{1}{1+x^p} - \frac{1}{1+x^q} \right) \frac{dx}{x} = \int_0^\infty \left(\frac{x^{q-1} - x^{p-1}}{(1+x^p)(1+x^q)} \right) dx = 0 \tag{51}$$

$$(p > 0;\ q > 0).$$

Proof. The integral (51) is separated into two parts as follows:

$$\int_0^\infty \left(\frac{x^{q-1} - x^{p-1}}{(1+x^p)(1+x^q)} \right) dx = \left(\int_0^1 + \int_1^\infty \right) \left(\frac{x^{q-1} - x^{p-1}}{(1+x^p)(1+x^q)} \right) dx,$$

which, upon replacing x by $1/x$ in the second integral, is seen to vanish to 0. ■

Applying Eq. (51) to Eqs. (46)–(50), Choi and Srivastava [302] derived much more general integral representations for γ, which are recorded here.

$$\gamma = p \int_0^\infty \left(\frac{1}{1+x^q} - \exp(-x^p) \right) \frac{dx}{x} \quad (p > 0;\ q > 0), \tag{52}$$

$$\gamma = p \int_0^\infty \left(\frac{1}{1+x^q} - \cos\,(x^p) \right) \frac{dx}{x} \quad (p > 0;\ q > 0), \tag{53}$$

$$\gamma = 1 + p \int_0^\infty \left(\frac{1}{1+x^q} + \frac{\exp(-x^p) - 1}{x^p} \right) \frac{dx}{x} \quad (p > 0;\ q > 0), \tag{54}$$

$$\gamma = 1 + p \int_0^\infty \left(\frac{1}{1+x^q} - \frac{\sin\,(x^p)}{x^p} \right) \frac{dx}{x} \quad (p > 0;\ q > 0), \tag{55}$$

$$\gamma = \frac{3}{2} + 2p \int_0^\infty \left(\frac{\cos\,(x^p) - 1}{x^{2p}} + \frac{1}{2(1+x^q)} \right) \frac{dx}{x} \quad (p > 0;\ q > 0), \tag{56}$$

$$\gamma = \frac{pq}{q-p} \int_0^\infty \left[\exp(-x^q) - \exp(-x^p) \right] \frac{dx}{x} \quad (p > 0;\ q > 0;\ p \neq q), \tag{57}$$

$$\gamma = \frac{pq}{q-p} \int_0^\infty \left[\cos\left(x^q\right) - \cos\left(x^p\right)\right] \frac{dx}{x} \quad (p > 0;\ q > 0;\ p \neq q), \tag{58}$$

$$\gamma = 1 + \frac{pq}{q-p} \int_0^\infty \left(\frac{\sin\,(x^q)}{x^q} - \frac{\sin\,(x^p)}{x^p}\right) \frac{dx}{x} \quad (p > 0;\ q > 0;\ p \neq q), \tag{59}$$

$$\gamma = \frac{3}{2} + \frac{2pq}{q-p} \int_0^\infty \left(\frac{\cos\,(x^p) - 1}{x^{2p}} - \frac{\cos\,(x^q) - 1}{x^{2q}}\right) \frac{dx}{x} \quad (p > 0;\ q > 0;\ p \neq q), \tag{60}$$

$$\gamma = \frac{pq}{q-p} \int_0^\infty \left[\cos\,(x^p) - \exp(-x^p)\right] \frac{dx}{x} \quad (p > 0;\ q > 0;\ p \neq q), \tag{61}$$

$$\gamma = \frac{p}{p-q} + \frac{pq}{p-q} \int_0^\infty \left(\exp(-x^p) - \frac{\sin\,(x^p)}{x^p}\right) \frac{dx}{x} \quad (p > 0;\ q > 0;\ p \neq q), \tag{62}$$

$$\gamma = \frac{3p}{2(p-q)} + \frac{pq}{p-q} \int_0^\infty \left(\exp(-x^p) + \frac{2\left[\cos\,(x^p) - 1\right]}{x^{2q}}\right) \frac{dx}{x} \quad (p > 0;\ q > 0;\ p \neq q), \tag{63}$$

$$\gamma = \frac{p}{p-q} + \frac{pq}{p-q} \int_0^\infty \left(\cos\,(x^p) - \frac{\sin\,(x^p)}{x^p}\right) \frac{dx}{x} \quad (p > 0;\ q > 0;\ p \neq q), \tag{64}$$

$$\gamma = \frac{3p}{2(p-q)} + \frac{pq}{p-q} \int_0^\infty \left(\cos\,(x^p) + \frac{2\left[\cos\,(x^p) - 1\right]}{x^{2q}}\right) \frac{dx}{x} \quad (p > 0;\ q > 0;\ p \neq q), \tag{65}$$

$$\gamma = \frac{3p-2q}{2(p-q)} + \frac{pq}{p-q} \int_0^\infty \left(\frac{\sin\,(x^p)}{x^p} + \frac{2\left[\cos\,(x^p) - 1\right]}{x^{2q}}\right) \frac{dx}{x} \quad (p > 0;\ q > 0;\ p \neq q), \tag{66}$$

$$\gamma = \frac{p}{p-q} + \frac{pq}{p-q} \int_0^\infty \left(\frac{\exp\left(-x^p\right) - 1}{x^p} + \exp(-x^p)\right) \frac{dx}{x} \quad (p > 0;\ q > 0;\ p \neq q), \tag{67}$$

$$\gamma = \frac{p}{p-q} + \frac{pq}{p-q} \int_0^\infty \left(\frac{\exp(-x^p) - 1}{x^p} + \cos\left(x^p\right) \right) \frac{dx}{x} \tag{68}$$

$$(p > 0;\ q > 0;\ p \neq q),$$

$$\gamma = 1 + \frac{pq}{p-q} \int_0^\infty \left(\frac{\exp(-x^p) - 1}{x^p} + \frac{\sin\ (x^p)}{x^p} \right) \frac{dx}{x} \tag{69}$$

$$(p > 0;\ q > 0;\ p \neq q),$$

$$\gamma = \frac{2p - 3q}{2(p-q)} + \frac{pq}{p-q} \int_0^\infty \left(\frac{\exp(-x^p) - 1}{x^p} + \frac{2\,[1 - \cos\ (x^p)]}{x^{2p}} \right) \frac{dx}{x} \tag{70}$$

$$(p > 0;\ q > 0;\ p \neq q).$$

It is noted that the integral formula (57) is recorded in [505, p. 364, Entry 3.476-2] and many (if not all) of the integral formulas in the previous subsection can be seen to be special cases of the corresponding integral formulas asserted in this subsection.

From an Application of the Residue Calculus

Consider a function $f(z)$ given by

$$f(z) = \frac{1}{z} \left(\frac{1}{1 + z^n} - e^{iz^n} \right) \quad (n \in \mathbb{N}).$$

Since

$$\lim_{z \to 0} f(z) = \begin{cases} -1 - i & (n = 1) \\ 0 & (n \in \mathbb{N} \setminus \{1\}), \end{cases}$$

the function $f(z)$ has a removable singularity at $z = 0$ and simple poles at

$$z = \exp\left(\frac{(2k+1)\pi i}{n} \right) \quad (k = 0, 1, \ldots, n-1).$$

We now consider a counterclockwise-oriented simple closed contour:

$$C := C_\delta \cup L_1 \cup C_R \cup L_2 \quad (0 < \delta < 1 < R),$$

where

$$C_\delta : z = \delta\, e^{i\theta} \quad \left(\theta \text{ varies from } \frac{\pi}{2n} \text{ to } 0 \right),$$

L_1 a line segment from δ to R on the positive real axis,

$$C_R : z = Re^{i\theta} \quad \left(\theta \text{ varies from 0 to } \frac{\pi}{2n}\right)$$

and

$$L_2 : z = x\exp\left(\frac{i\pi}{2n}\right) \quad (x \text{ varies from } R \text{ to } \delta),$$

that is, a line segment on the half-line beginning at the origin with the argument $\frac{\pi}{2n}$. Since $f(z)$ is analytic throughout the domain interior to and on the closed contour C, it follows from the Cauchy-Goursat theorem that

$$\left(\int_{C_\delta} + \int_{L_1} + \int_{C_R} + \int_{L_2}\right) f(z)\,dz = 0,$$

which, upon taking the limits as

$$\delta \to 0+ \quad \text{and} \quad R \to \infty$$

and equating the real and imaginary parts of the last resulting equation, yields the following two interesting integral identities:

$$\int_0^\infty \left(\frac{1}{1+x^n} - \cos\left(x^n\right)\right)\frac{dx}{x} = \int_0^\infty \left(\frac{1}{1+x^{2n}} - \exp\left(-x^n\right)\right)\frac{dx}{x} \quad (n \in \mathbb{N}) \tag{71}$$

and

$$\int_0^\infty \frac{x^{n-1}}{1+x^{2n}}\,dx = \int_0^\infty \frac{\sin\left(x^n\right)}{x}\,dx = \frac{\pi}{2n} \quad (n \in \mathbb{N}). \tag{72}$$

It is noted that the integral identity (71) is a special case of (52) or (53). Moreover, (72) can be evaluated, as above, by applying the residue calculus to another function

$$\mathfrak{f}(z) = \frac{\exp\left(iz^n\right)}{z} \quad (n \in \mathbb{N})$$

and a counterclockwise-oriented simple closed contour

$$\mathcal{C} := C_\delta \cup L_1 \cup C_R \cup L_2 \quad (0 < \delta < R),$$

where C_δ and L_1 are the same as above,

$$C_R : z = R e^{i\theta} \quad \left(\theta \text{ varies from } 0 \text{ to } \frac{\pi}{n}\right)$$

and

$$L_2 : z = x \exp\left(\frac{i\pi}{n}\right) \quad (x \text{ varies from } R \text{ to } \delta).$$

We conclude this section by remarking that more integral representations for γ can be obtained by applying the same techniques employed here (see [302]) or other methods (if any) to some other known formulas that have not been used (see [572]).

1.3 Polygamma Functions

The Psi (or Digamma) Function

The *Psi* (or *Digamma*) *function* $\psi(z)$ defined by

$$\psi(z) := \frac{d}{dz}\{\log \Gamma(z)\} = \frac{\Gamma'(z)}{\Gamma(z)} \quad \text{or} \quad \log \Gamma(z) = \int_1^z \psi(t)dt \tag{1}$$

possesses the following properties:

$$\psi(z) = \lim_{n\to\infty}\left(\log n - \sum_{k=0}^{n} \frac{1}{z+k}\right);$$

$$\psi(z) = -\gamma - \frac{1}{z} + \sum_{n=1}^{\infty} \frac{z}{n(z+n)} \tag{2}$$

$$= -\gamma + (z-1)\sum_{n=0}^{\infty} \frac{1}{(n+1)(z+n)}, \tag{3}$$

where γ is the Euler-Mascheroni constant defined by 1.1(3) (or 1.2(2)).

These results clearly imply that $\psi(z)$ is meromorphic (that is, analytic everywhere in the bounded complex z–plane, except for poles) with simple poles at $z = -n (n \in \mathbb{N}_0)$ with its residue -1. Also we have

$$\psi(1) = -\gamma, \tag{4}$$

which follows at once from (3). It is noted that, very recently, Bagby [83] proved (4) in another way.

The following additional properties of $\psi(z)$ can be deduced from known results for $\Gamma(z)$:

$$\log \frac{\Gamma(z+1)}{\Gamma(z)} = \log z = -\gamma + \sum_{n=0}^{\infty} \left[\frac{1}{n+1} - \log\left(1 + \frac{1}{n+z}\right)\right]; \tag{5}$$

$$\psi(z) = \log z - \sum_{n=0}^{\infty} \left[\frac{1}{n+z} - \log\left(1 + \frac{1}{n+z}\right)\right], \tag{6}$$

which follows from (3) and (5);

$$\psi(z+n) = \psi(z) + \sum_{k=1}^{n} \frac{1}{z+k-1} \quad (n \in \mathbb{N}); \tag{7}$$

$$\psi(z) - \psi(-z) = -\pi \cot \pi z - \frac{1}{z}; \tag{8}$$

$$\psi(1+z) - \psi(1-z) = \frac{1}{z} - \pi \cot \pi z; \tag{9}$$

$$\psi(z) - \psi(1-z) = -\pi \cot \pi z; \tag{10}$$

$$\psi\left(\frac{1}{2} + z\right) - \psi\left(\frac{1}{2} - z\right) = \pi \tan \pi z; \tag{11}$$

$$\psi(mz) = \log m + \frac{1}{m} \sum_{k=0}^{m-1} \psi\left(z + \frac{k}{m}\right) \quad (m \in \mathbb{N}). \tag{12}$$

Integral Representations for $\psi(z)$

Expanding $(1-t)^{-1}$ into a series, integrating term by term and using (3), we get

$$\psi(z) = -\gamma + \int_0^1 \left(1 - t^{z-1}\right)(1-t)^{-1}\, dt \quad (\Re(z) > 0), \tag{13}$$

which, upon replacing t by e^{-t}, yields

$$\psi(z) = -\gamma + \int_0^{\infty} \left(e^{-t} - e^{-tz}\right)\left(1 - e^{-t}\right)^{-1} dt \quad (\Re(z) > 0). \tag{14}$$

Making use of (10), it follows from (13) and (14) that

$$\psi(z) = -\gamma - \pi \cot \pi z + \int_0^1 \left(1 - t^{-z}\right)(1-t)^{-1}\, dt \quad (\Re(z) < 1) \tag{15}$$

and

$$\psi(z) = -\gamma - \pi \cot \pi z + \int_0^\infty \left(1 - e^{tz}\right)\left(e^t - 1\right)^{-1} dt \quad (\Re(z) < 1). \tag{16}$$

It is not difficult to derive each of the following integral representations for $\psi(z)$.

$$\psi(z) = \int_0^\infty \left[t^{-1} e^{-t} - \left(1 - e^{-t}\right)^{-1} e^{-tz}\right] dt \quad (\Re(z) > 0), \tag{17}$$

which is due to Gauss and reduces, when $z = 1$, to the integral expression 1.2(9) for the Euler-Mascheroni constant γ;

$$\psi(z) = \int_0^\infty \left[e^{-t} - (1+t)^{-z}\right] t^{-1} dt \quad (\Re(z) > 0), \tag{18}$$

which is due to Dirichlet and reduces, when $z = 1$, to 1.2(11);

$$\psi(z) = -\gamma + \int_0^\infty \left[(1+t)^{-1} - (1+t)^{-z}\right] t^{-1} dt \quad (\Re(z) > 0); \tag{19}$$

$$\psi(z) = \log z + \int_0^\infty \left[t^{-1} - \left(1 - e^{-t}\right)^{-1}\right] e^{-tz} dt \quad (\Re(z) > 0), \tag{20}$$

$$\psi(z) = \log z - \frac{1}{2z} - \int_0^\infty \left[\left(1 - e^{-t}\right)^{-1} - t^{-1} - \frac{1}{2}\right] e^{-tz} \quad (\Re(z) > 0), \tag{21}$$

$$\psi(z) = \log z + \int_0^\infty \left[\left(1 - e^{-t}\right)^{-1} + t^{-1} - 1\right] e^{-tz} dt \quad (\Re(z) > 0), \tag{22}$$

and

$$\psi(z) = \log z - \frac{1}{2z} - \int_0^\infty \left[\left(e^t - 1\right)^{-1} - t^{-1} + \frac{1}{2}\right] e^{-tz} dt \quad (\Re(z) > 0), \tag{23}$$

all four being attributed to Binet;

$$\psi(z) = \log z - \frac{1}{2z} - 2\int_0^\infty \left(t^2 + z^2\right)^{-1}\left(e^{2\pi t} - 1\right)^{-1} t\, dt \quad (\Re(z) > 0), \tag{24}$$

which, in the special case when $z = 1$, yields 1.2(31).

Gauss's integral representation (17) also yields Malmstén's formula:

$$\log\Gamma(z) = \int_1^z \psi(t)\,dt = \int_0^\infty \left[(z-1) - \frac{1-e^{-(z-1)t}}{1-e^{-t}}\right]\frac{e^{-t}}{t}\,dt \quad (\Re(z)>0). \tag{25}$$

From (23), we can deduce that

$$\log\Gamma(z) = \left(z-\frac{1}{2}\right)\log z - z + 1 + \int_0^\infty \left(\frac{1}{2}-\frac{1}{t}+\frac{1}{e^t-1}\right)\frac{e^{-tz}}{t}\,dt - \int_0^\infty \left(\frac{1}{2}-\frac{1}{t}+\frac{1}{e^t-1}\right)\frac{e^{-t}}{t}\,dt \quad (\Re(z)>0). \tag{26}$$

The second integral in (26) can readily be evaluated:

$$\int_0^\infty \left(\frac{1}{2}-\frac{1}{t}+\frac{1}{e^t-1}\right)\frac{e^{-t}}{t}\,dt = 1 - \frac{1}{2}\log(2\pi) \tag{27}$$

by using an elementary (yet technical) separation and recalling a formula for $\log z$:

$$\log z = \int_0^\infty \left(e^{-t}-e^{-zt}\right)\frac{dt}{t} \quad (\Re(z)>0). \tag{28}$$

Now, putting (27) into (26) yields *Binet's first expression for* $\log\Gamma(z)$:

$$\log\Gamma(z) = \left(z-\frac{1}{2}\right)\log z - z + \frac{1}{2}\log(2\pi) + \int_0^\infty \left(\frac{1}{2}-\frac{1}{t}+\frac{1}{e^t-1}\right)\frac{e^{-zt}}{t}\,dt \quad (\Re(z)>0), \tag{29}$$

which may also give an approximate expression for $\log\Gamma(z)$, just as the special case of 1.1(34), when $n=0$.

Binet's second expression for $\log\Gamma(z)$ may be written as follows:

$$\log\Gamma(z)=\left(z-\frac{1}{2}\right)\log z-z+\frac{1}{2}\log(2\pi)$$
$$+2\int_0^\infty\frac{\arctan(t/z)}{e^{2\pi t}-1}\,dt\quad(\Re(z)>0).\tag{30}$$

From 1.1(12) and (25), we obtain *Kummer's expression for* $\log\Gamma(z)$:

$$\log\Gamma(z)=\frac{1}{2}\log\pi-\frac{1}{2}\log(\sin\pi z)$$
$$+\frac{1}{2}\int_0^\infty\left(\frac{\sinh\left[\left(\frac{1}{2}-z\right)t\right]}{\sinh\left(\frac{1}{2}t\right)}-(1-2z)\,e^{-t}\right)\frac{dt}{t}\quad(0<\Re(z)<1).\tag{31}$$

By applying (31), Kummer derived the following Fourier series for $\log\Gamma(x)$:

$$\log\Gamma(x)=\frac{1}{2}\log\pi-\frac{1}{2}\log(\sin\pi x)+2\sum_{n=1}^\infty a_n\sin(2n\pi x)\quad(0<x<1),\tag{32}$$

where

$$a_n=\int_0^\infty\left(\frac{2n\pi}{t^2+4n^2\pi^2}-\frac{e^{-t}}{2n\pi}\right)\frac{dt}{t}$$
$$=\frac{1}{2n\pi}\left[\int_0^\infty\left(\frac{1}{1+t^2}-\cos t\right)\frac{dt}{t}+\int_0^\infty\frac{e^{-t}-e^{-2\pi nt}}{t}dt+\int_0^\infty\left(\cos t-e^{-t}\right)\frac{dt}{t}\right]$$
$$=\frac{1}{2n\pi}(\gamma+\log 2\pi+\log n).\tag{33}$$

Since

$$\log(\sin\pi x)=-\log 2-\sum_{n=1}^\infty\frac{1}{n}\cos(2\pi nx)\quad(0<x<1),\tag{34}$$

by rewriting (32) in the form:

$$\log\Gamma(x)=\left(\frac{1}{2}-x\right)(\gamma+\log 2)+(1-x)\log\pi-\frac{1}{2}\log(\sin\pi x)$$
$$+\sum_{n=1}^\infty\frac{\log n}{\pi n}\sin(2\pi nx)\quad(0<x<1),\tag{35}$$

we finally obtain

$$\log\Gamma(x) = \frac{1}{2}\log(2\pi) + \sum_{n=1}^{\infty}\left(\frac{1}{2n}\cos(2\pi nx) + \frac{\gamma+\log(2\pi n)}{\pi n}\sin(2\pi nx)\right) \quad (0 < x < 1). \tag{36}$$

The Fourier series (36) readily implies each of the integral formulas:

$$\int_0^1 \sin(2\pi nx)\log\Gamma(x)\,dx = \frac{\gamma+\log(2\pi n)}{2\pi n} \quad (n \in \mathbb{N}), \tag{37}$$

$$\int_0^1 \cos(2\pi nx)\log\Gamma(x)\,dx = \frac{1}{4n} \quad (n \in \mathbb{N}), \tag{38}$$

and

$$\int_0^1 \log\Gamma(x)\,dx = \frac{1}{2}\log(2\pi). \tag{39}$$

Considering 1.1(31) and 1.1(34), when $n = 0$, we readily obtain a more general integral formula than (39) above:

$$\int_x^{x+1} \log\Gamma(t)\,dt = x\log x - x + \frac{1}{2}\log(2\pi) \quad (x \geqq 0), \tag{40}$$

by observing the following relation:

$$\lim_{m\to\infty}\sum_{k=0}^{m-1}\frac{1}{m}\log\Gamma\left(x+\frac{k}{m}\right) = \int_0^1 \log\Gamma(x+u)\,du = \int_x^{x+1}\log\Gamma(t)\,dt.$$

The formula (40) can be extended, without difficulty, to the case, when $x \in \mathbb{C}\setminus\mathbb{Z}^-$ $\left(\mathbb{Z}^- := \{-1,-2,\ldots\}\right)$ and the principal values of the involved logarithms are taken, and can also be further generalized (by iteration) as follows:

$$\int_x^{x+n}\log\Gamma(t)\,dt = \sum_{k=0}^{n-1}(x+k)\log(x+k) - nx - \frac{1}{2}n(n-1) + \frac{1}{2}n\log(2\pi) \quad (n \in \mathbb{N}). \tag{41}$$

Gauss's Formulas for $\psi\left(\frac{p}{q}\right)$

Taking $z = \frac{p}{q}$ $(0 < p < q;\ p,\ q \in \mathbb{N})$ in (13) and $t = x^p$, we obtain

$$\gamma + \psi\left(\frac{p}{q}\right) = q \int_0^1 \frac{x^{q-1} - x^{p-1}}{1 - x^p} dx, \tag{42}$$

which, upon noting that

$$q \frac{x^{p-1} - x^{q-1}}{x^p - 1} = \sum_{k=1}^{q-1} \frac{\omega_k^p - 1}{x - \omega_k} \quad \left(\omega_k := \exp\left(\frac{2\pi ik}{q}\right)\right), \tag{43}$$

readily yields

$$\gamma + \psi\left(\frac{p}{q}\right) = \frac{\pi}{2q} \sum_{k=1}^{q-1} (2k - q)\ \sin \frac{2kp\pi}{q} - \sum_{k=1}^{q-1} \left(1 - \cos \frac{2kp\pi}{q}\right) \log\left(2 \sin \frac{k\pi}{q}\right). \tag{44}$$

By applying the known trigonometric identities:

$$\sum_{k=0}^{q-1} \cos 2k\alpha = \frac{1}{2} + \frac{\sin\,(2q-1)\alpha}{2 \sin\,\alpha} \tag{45}$$

and

$$\sum_{k=1}^{q-1} \sin\ 2k\alpha = \frac{\cos\alpha - \cos\,(2q-1)\alpha}{2 \sin\alpha} \tag{46}$$

in (44), we obtain *Gauss's formula*:

$$\psi\left(\frac{p}{q}\right) = -\gamma - \frac{\pi}{2} \cot \frac{p\pi}{q} - \log q + \sum_{k=1}^{q-1} \cos \frac{2kp\pi}{q} \log\left(2 \sin \frac{k\pi}{q}\right) \tag{47}$$

$$(0 < p < q;\ p,\ q \in \mathbb{N}),$$

which implies that, for a positive proper fraction z, the value of $\psi(z)$ can be expressed as a finite combination of elementary functions, and, yet, by means of 1.1(10), may be extended to every rational value of z.

By reversing the order of summation, Gauss's formula (47) can be rewritten in the following form:

$$\begin{aligned}\psi\left(\frac{p}{q}\right) &= -\gamma - \frac{\pi}{2}\cot\frac{p\pi}{q} - \log q \\ &\quad + 2\sum_{k=1}^{[(q-1)/2]} \cos\left(\frac{2kp\pi}{q}\right)\log\left(2\sin\frac{k\pi}{q}\right) + r_q(p),\end{aligned} \tag{48}$$

where $[x]$ denotes the greatest integer $\leq x$, and

$$r_q(p) = \begin{cases} (-1)^p \log 2 & \text{if } q \text{ is even,} \\ 0 & \text{if } q \text{ is odd.} \end{cases}$$

If we rewrite (42) as

$$\gamma - q\int_0^1 \frac{x^{q-1}}{1-x^p}dx = -\psi\left(\frac{p}{q}\right) - q\int_0^1 \frac{x^{p-1}}{1-x^p}dx,$$

where the left-hand side is independent of p and only the right-hand side is dependent on p, we can derive *Gauss's second formula*:

$$\sum_{p=1}^{q} \psi\left(\frac{p}{q}\right)\exp\left(\frac{2pk\pi i}{q}\right) = q\log\left[1-\exp\left(\frac{2k\pi i}{q}\right)\right] \quad (q \in \mathbb{N};\ k \in \mathbb{Z}). \tag{49}$$

Special Values of $\psi(z)$

Setting $z = 1$ and $z = \frac{1}{2}$ in (7), we obtain

$$\psi(n) = -\gamma + \sum_{k=1}^{n-1}\frac{1}{k} \quad (n \in \mathbb{N}) \tag{50}$$

and

$$\psi\left(n+\frac{1}{2}\right) = -\gamma - 2\log 2 + 2\sum_{k=0}^{n-1}\frac{1}{2k+1} \quad (n \in \mathbb{N}_0), \tag{51}$$

it being understood (here as well as throughout this book) that an empty sum is nil.

By suitably applying the various formulas for the ψ–function, we [1094, pp. 20–22] could derive the following special values of $\psi(z)$, some of which were corrected and simplified by Tee [1146] as follows:

$$\psi\left(\frac{1}{2}\right) = -\gamma - 2\log 2;$$
$$\psi\left(\frac{1}{3}\right) = -\gamma - \frac{1}{6}\pi\sqrt{3} - \frac{3}{2}\log 3;$$
$$\psi\left(\frac{2}{3}\right) = -\gamma + \frac{1}{6}\pi\sqrt{3} - \frac{3}{2}\log 3;$$
$$\psi\left(\frac{1}{4}\right) = -\gamma - \frac{1}{2}\pi - 3\log 2;$$
$$\psi\left(\frac{3}{4}\right) = -\gamma + \frac{1}{2}\pi - 3\log 2;$$
$$\psi\left(\frac{1}{5}\right) = -\gamma - \frac{\pi}{2}\sqrt{1+\frac{2}{5}\sqrt{5}} - \frac{5}{4}\log 5 - \frac{\sqrt{5}}{2}\log\frac{1+\sqrt{5}}{2};$$
$$\psi\left(\frac{2}{5}\right) = -\gamma - \frac{\pi}{2}\sqrt{1-\frac{2}{5}\sqrt{5}} - \frac{5}{4}\log 5 + \frac{\sqrt{5}}{2}\log\frac{1+\sqrt{5}}{2};$$
$$\psi\left(\frac{3}{5}\right) = -\gamma + \frac{\pi}{2}\sqrt{1-\frac{2}{5}\sqrt{5}} - \frac{5}{4}\log 5 + \frac{\sqrt{5}}{2}\log\frac{1+\sqrt{5}}{2};$$
$$\psi\left(\frac{4}{5}\right) = -\gamma + \frac{\pi}{2}\sqrt{1+\frac{2}{5}\sqrt{5}} - \frac{5}{4}\log 5 - \frac{\sqrt{5}}{2}\log\frac{1+\sqrt{5}}{2};$$
$$\psi\left(\frac{1}{6}\right) = -\gamma - \frac{1}{2}\pi\sqrt{3} - \frac{3}{2}\log 3 - 2\log 2;$$
$$\psi\left(\frac{5}{6}\right) = -\gamma + \frac{1}{2}\pi\sqrt{3} - \frac{3}{2}\log 3 - 2\log 2;$$
$$\psi\left(\frac{1}{8}\right) = -\gamma - \frac{\pi}{2}(\sqrt{2}+1) - 4\log 2 - \sqrt{2}\log(\sqrt{2}+1);$$
$$\psi\left(\frac{3}{8}\right) = -\gamma - \frac{\pi}{2}(\sqrt{2}-1) - 4\log 2 + \sqrt{2}\log(\sqrt{2}+1);$$
$$\psi\left(\frac{5}{8}\right) = -\gamma + \frac{\pi}{2}(\sqrt{2}-1) - 4\log 2 + \sqrt{2}\log(\sqrt{2}+1);$$
$$\psi\left(\frac{7}{8}\right) = -\gamma + \frac{\pi}{2}(\sqrt{2}+1) - 4\log 2 - \sqrt{2}\log(\sqrt{2}+1);$$
$$\psi\left(\frac{1}{10}\right) = -\gamma - \frac{\pi}{2}\sqrt{5+2\sqrt{5}} - 2\log 2 - \frac{5}{4}\log 5 - \frac{\sqrt{5}}{2}\log\left(2+\sqrt{5}\right);$$
$$\psi\left(\frac{3}{10}\right) = -\gamma - \frac{\pi}{2}\sqrt{5-2\sqrt{5}} - 2\log 2 - \frac{5}{4}\log 5 + \frac{\sqrt{5}}{2}\log\left(2+\sqrt{5}\right);$$
$$\psi\left(\frac{7}{10}\right) = -\gamma + \frac{\pi}{2}\sqrt{5-2\sqrt{5}} - 2\log 2 - \frac{5}{4}\log 5 + \frac{\sqrt{5}}{2}\log\left(2+\sqrt{5}\right);$$

$$\psi\left(\frac{9}{10}\right) = -\gamma + \frac{\pi}{2}\sqrt{5+2\sqrt{5}} - 2\log 2 - \frac{5}{4}\log 5 - \frac{\sqrt{5}}{2}\log\left(2+\sqrt{5}\right);$$

$$\psi\left(\frac{1}{12}\right) = -\gamma - \frac{\pi}{2}(2+\sqrt{3}) - \sqrt{3}\log(2+\sqrt{3}) - \frac{3}{2}\log 3 - 3\log 2;$$

$$\psi\left(\frac{5}{12}\right) = -\gamma - \frac{\pi}{2}(2-\sqrt{3}) + \sqrt{3}\log(2+\sqrt{3}) - \frac{3}{2}\log 3 - 3\log 2;$$

$$\psi\left(\frac{7}{12}\right) = -\gamma + \frac{\pi}{2}(2-\sqrt{3}) + \sqrt{3}\log(2+\sqrt{3}) - \frac{3}{2}\log 3 - 3\log 2;$$

$$\psi\left(\frac{11}{12}\right) = -\gamma + \frac{\pi}{2}(2+\sqrt{3}) - \sqrt{3}\log(2+\sqrt{3}) - \frac{3}{2}\log 3 - 3\log 2;$$

$$\psi\left(-\frac{1}{2}\right) = 2 - \gamma - 2\log 2;$$

$$\psi\left(-\frac{1}{3}\right) = 3 - \gamma + \frac{\sqrt{3}}{6}\pi - \frac{3}{2}\log 3;$$

$$\psi\left(-\frac{5}{6}\right) = \frac{6}{5} - \gamma - \frac{\sqrt{3}}{2}\pi - \frac{3}{2}\log 3 - 2\log 2.$$

The Polygamma Functions

The *Polygamma functions* $\psi^{(n)}(z)$ $(n \in \mathbb{N})$ are defined by

$$\psi^{(n)}(z) := \frac{d^{n+1}}{dz^{n+1}}\log\Gamma(z) = \frac{d^n}{dz^n}\psi(z) \quad (n \in \mathbb{N}_0;\ z \notin \mathbb{Z}_0^-). \tag{52}$$

In terms of the generalized (or Hurwitz) Zeta function $\zeta(s,a)$ (see Section 2.2), we can write

$$\psi^{(n)}(z) = (-1)^{n+1}\, n! \sum_{k=0}^{\infty} \frac{1}{(k+z)^{n+1}} = (-1)^{n+1}\, n!\, \zeta(n+1, z) \quad (n \in \mathbb{N};\ z \notin \mathbb{Z}_0^-), \tag{53}$$

which may be used to deduce the properties of $\psi^{(n)}(z)$ $(n \in \mathbb{N})$ from those of $\zeta(s,z)$ $(s = n+1;\ n \in \mathbb{N})$.

It is also easy to have the following expressions:

$$\psi^{(n)}(z+m) - \psi^{(n)}(z) = (-1)^n\, n! \sum_{k=1}^{m} \frac{1}{(z+k-1)^{n+1}} \quad (m,\ n \in \mathbb{N}_0) \tag{54}$$

and

$$\psi^{(n)}(z) - (-1)^n\, \psi^{(n)}(1-z) = -\pi \frac{d^n}{dz^n}\{\cot \pi z\} \quad (n \in \mathbb{N}_0), \tag{55}$$

it being understood (as usual) that

$$\psi^{(0)}(z) = \psi(z).$$

By using Gauss's multiplication formula for $\Gamma(z)$ in 1.1(31), it is easy to get the multiplication formula for $\psi^{(n)}(z)$:

$$\psi^{(n)}(mz) = \sum_{j=1}^{m} \psi^{(n)}\left(z + \frac{j-1}{m}\right) \quad (n,\ m \in \mathbb{N}). \tag{56}$$

Davis [368] extended Gauss's result (47) (or (48)) to the Polygamma functions $\psi^{(n)}(z)$ $(n \in \mathbb{N})$, by using a known series representation of $\psi^{(n)}(z)$ in an elementary (yet technical) way. Kölbig [687, 688], in his CERN technical report, also gave two extensions of Gauss's result to $\psi^{(n)}(z)$, by using the series definition of Polylogarithm function and the above-known series representation of $\psi^{(n)}(z)$. Recently, Choi and Cvijović [271] presented a unified formula of $\psi^{(n)}(p/q)$ $(p, q, n \in \mathbb{N};\ 1 \leqq p < q)$, which can be specialized in those formulas of Davis [368] and Kölbig [687, 688], expressed in terms of the Bernoulli polynomials given in 1.7 and the generalized Zeta functions in 2.2, which was shown in the following two ways:

$$\begin{aligned}\psi^{(n)}\left(\frac{p}{q}\right) =& (-1)^{n+1}\, n!\, q^n \\ &\cdot \sum_{s=0}^{q-1} \left[\mathcal{E}_n(s;p;q)\,(-1)^{1+\left[\frac{1}{2}(n+1)\right]} \frac{(2\pi)^{n+1}}{2\cdot(n+1)!} B_{n+1}\left(\frac{s}{q}\right) \right. \\ &\left. + \frac{1}{q^{n+1}} \mathcal{F}_n(s;p;q) \sum_{k=1}^{q} \zeta\left(n+1, \frac{k}{q}\right) \mathcal{E}_{n+1}(k;s;q) \right] \\ & \qquad (p,\ q,\ n \in \mathbb{N};\ 1 \leqq p < q),\end{aligned} \tag{57}$$

where $[x]$ denotes the greatest integer $\leqq x$, and

$$\mathcal{E}_n(s;p;q) := \frac{1+(-1)^n}{2} \sin\left(\frac{2\pi sp}{q}\right) + \frac{1-(-1)^n}{2} \cos\left(\frac{2\pi sp}{q}\right)$$

and

$$\mathcal{F}_n(s;p;q) := \frac{1+(-1)^n}{2} \cos\left(\frac{2\pi sp}{q}\right) + \frac{1-(-1)^n}{2} \sin\left(\frac{2\pi sp}{q}\right).$$

Special Values of $\psi^{(n)}(z)$

Since $\psi^{(n)}\left(\frac{p}{q}\right)$ in (57) is expressed in terms of $B_n(x)$ and $\zeta(s,a)$, in order to give special cases of (57) (see [688] and [9]), it is natural to know some of their properties. We demonstrate to give the value of the simple case of (57) when $p = 1$ and $q = 2$:

$$\psi^{(n)}\left(\frac{1}{2}\right) = (-1)^{n+1}\, n!\left(2^{n+1} - 1\right) \zeta(n+1) \quad (n \in \mathbb{N}), \tag{58}$$

which is obtained from the aid of the formulas 1.7(19), 2.3(3) and 2.3(18). It is noted that (58) is easily derived from (47) and 2.3(3). Likewise, we have (see [272])

$$\left.\begin{array}{l}\psi^{(2n)}\left(\frac{1}{3}\right)\\ \psi^{(2n)}\left(\frac{2}{3}\right)\end{array}\right\} = \pm(-1)^n\sqrt{3}\,3^{2n}\frac{(2\pi)^{2n+1}}{2(2n+1)}B_{2n+1}\left(\frac{1}{3}\right) + \frac{(2n)!}{2}\left(1-3^{2n+1}\right)\zeta(2n+1) \quad (n\in\mathbb{N}) \tag{59}$$

and

$$\left.\begin{array}{l}\psi^{(2n)}\left(\frac{1}{4}\right)\\ \psi^{(2n)}\left(\frac{3}{4}\right)\end{array}\right\} = \pm(-1)^n 4^{2n}\frac{(2\pi)^{2n+1}}{2n+1}B_{2n+1}\left(\frac{1}{4}\right) + (2n)!\left(1-2^{2n+1}\right)2^{2n}\zeta(2n+1) \quad (n\in\mathbb{N}). \tag{60}$$

We give the relations

$$\psi^{(2n)}\left(\frac{1}{6}\right) = 2\,\psi^{(2n)}\left(\frac{1}{3}\right) \quad (n\in\mathbb{N}) \tag{61}$$

and

$$\psi^{(2n)}\left(\frac{5}{6}\right) = -2\,\psi^{(2n)}\left(\frac{1}{3}\right) - \psi^{(2n)}\left(\frac{1}{2}\right) \quad (n\in\mathbb{N}), \tag{62}$$

which is easily obtained by recalling the multiplication formula (56) for $\psi^{(n)}(z)$.

Finally, several further special values are:

$$\left.\begin{array}{l}\psi^{(2n-1)}\left(\frac{1}{4}\right)\\ \psi^{(2n-1)}\left(\frac{3}{4}\right)\end{array}\right\} = \pm(2n-1)!\,2^{4n-1}\beta(2n) + (-1)^{n-1}2^{2n-2}(2^{2n}-1)B_{2n}\frac{(2\pi)^{2n}}{2n}, \tag{63}$$

where

$$\beta(s) := \frac{\zeta\left(s,\frac{1}{4}\right)-\zeta\left(s,\frac{1}{4}\right)}{4^s}. \tag{64}$$

The Asymptotic Expansion for $\psi(z)$

From 1.1(34), we obtain the following asymptotic expansion for $\psi(z)$:

$$\psi(z) = \log z - \frac{1}{2z} - \sum_{k=1}^{n} \frac{B_{2k}}{2k\, z^{2k}} + O\left(z^{-2n-2}\right) \tag{65}$$
$$(|z| \to \infty;\ |\arg(z)| \leqq \pi - \epsilon\ (0 < \epsilon < \pi);\ n \in \mathbb{N}_0).$$

Now we shall show (see Barnes [94]) that

$$\sum_{k=1}^{n} \psi(k) = n \log n - n + \frac{1}{2} + O\left(n^{-1}\right) \quad (n \to \infty) \tag{66}$$

and

$$\sum_{k=1}^{n} \psi'(k) = \log n + 1 + \gamma + O\left(n^{-1}\right) \quad (n \to \infty), \tag{67}$$

where γ denotes the Euler-Mascheroni constant given in 1.1(3).

Indeed, setting $f(x) = 1/x$ and $f(x) = 1/x^2$ with $a = 1$ in the Euler-Maclaurin summation formula 1.4(68), we obtain

$$1 + \frac{1}{2} + \cdots + \frac{1}{n} = \gamma + \log n + \frac{1}{2n} - \frac{1}{8n^2} + \frac{15}{2n^4} - \cdots \tag{68}$$

and

$$1 + \frac{1}{2^2} + \cdots + \frac{1}{n^2} = \frac{\pi^2}{6} - \frac{1}{n} + \frac{1}{2n^2} - \frac{1}{6n^3} + \frac{1}{30n^5} - \cdots, \tag{69}$$

respectively. We can readily deduce from (69) that

$$\frac{1}{n^2} + \frac{1}{(n+1)^2} + \cdots = \frac{1}{n} + \frac{1}{2n^2} + O\left(n^{-3}\right) \quad (n \to \infty). \tag{70}$$

Since

$$\psi(m) = \frac{1}{1} + \frac{1}{2} + \cdots + \frac{1}{m-1} - \gamma \quad (m \in \mathbb{N} \setminus \{1\}), \tag{71}$$

we can also write (66) in the form:

$$\sum_{k=1}^{n} \psi(k) = -n\gamma + \frac{n-1}{1} + \frac{n-2}{2} + \cdots + \frac{1}{n-1}.$$

We, thus, obtain

$$\begin{aligned}\sum_{k=1}^{n}\psi(k) &= n\left(-\gamma+\frac{1}{1}+\frac{1}{2}+\cdots+\frac{1}{n-1}\right)-\frac{1}{1}-\frac{2}{2}-\cdots-\frac{n-1}{n-1}\\ &= n\left(-\gamma+\sum_{k=1}^{n-1}\frac{1}{k}\right)-n+1,\end{aligned}$$

which, in view of (68), yields

$$\sum_{k=1}^{n}\psi(k)=n\left[-\gamma+\gamma+\log(n-1)+\frac{1}{2(n-1)}+O\left(n^{-2}\right)\right]-n+1 \quad (n\to\infty),$$

which implies the result (66).

In order to prove (67), we consider (54) with $m=n=1$:

$$\psi'(z+1)=\psi'(z)-\frac{1}{z^2}, \tag{72}$$

the repeated application of which gives

$$\psi'(m)=-\left[\frac{1}{1^2}+\frac{1}{2^2}+\cdots+\frac{1}{(m-1)^2}\right]+\psi'(1),$$

which may also follow directly from (54). Hence, we have

$$\sum_{k=1}^{n}\psi'(k)=n\psi'(1)-\frac{n-1}{1^2}-\frac{n-2}{2^2}-\cdots-\frac{1}{(n-1)^2}. \tag{73}$$

Now apply the following special case of (53) when $n=z=1$;

$$\psi'(1)=\zeta(2)=\frac{\pi^2}{6}, \tag{74}$$

in conjunction with (68), and (73) readily yields

$$\begin{aligned}\sum_{k=1}^{n}\psi'(k) &= n\frac{\pi^2}{6}-\left[\frac{1}{1^2}+\frac{1}{2^2}+\cdots+\frac{1}{(n-1)^2}\right]+\frac{1}{1^2}+\frac{2}{2^2}+\cdots+\frac{n-1}{(n-1)^2}\\ &= n\left[\frac{1}{n^2}+\frac{1}{(n+1)^2}+\cdots\right]+\gamma+\log(n-1)+O\left(n^{-1}\right) \quad (n\to\infty),\end{aligned} \tag{75}$$

which, by applying (70), produces the desired result (67).

1.4 The Multiple Gamma Functions

The double Gamma function Γ_2 and the multiple Gamma functions Γ_n were defined and studied systematically by Barns [94–97] in about 1900. Before their investigation by Barnes, these functions had been introduced in a different form by (for example) Hölder [563], Alexeiewsky [18] and Kinkelin [666]. Although these functions did not appear in the tables of the most well-known special functions, the double Gamma function was cited in the exercises by Whittaker and Watson [1225, p. 264] and recorded also by Gradshteyn and Ryzhik [505, p. 661, Entry 6.441(4); p. 937, Entry 8.333]. In about the middle of the 1980s, these functions were revived in the study of the determinants of the Laplacians on the n–dimensional unit sphere $\mathbf{S}^n$ (see [260], [706], [881], [954], [1190], [1201]). Shintani [1026] also used the double Gamma function to prove the classical Kronecker limit formula. Friedman and Ruijsenaars [463] showed that Shintani's work on multiple Zeta and Gamma functions can be simplified and extended by making use of difference equations. Its p-adic analytic extension appeared in a formula of Cassou-Noguès [221] for the p-adic L-functions at the point 0. Choi *et al.* (see [269, 291, 292]) used these functions to evaluate some families of series involving the Riemann Zeta function, as well as to compute the determinants of the Laplacians. Choi *et al.* [269] addressed the converse problem and applied various formulas for series associated with the Zeta and related functions with a view to developing the corresponding theory of multiple Gamma functions. Adamchik [8] discussed some theoretical aspects of the multiple Gamma functions and their applications to summation of series and infinite products. Matsumoto [804] proved several asymptotic expansions of the Barnes double Zeta function and the double Gamma function and presented an application to the Hecke L-functions of real quadratic fields. Ruijsenaars [990] showed how various known results concerning the Barnes multiple Zeta and Gamma functions can be obtained as specializations of the simple features shared by a quite remarkably extensive class of functions.

The Double Gamma Function Γ_2

Here, we summarize some properties of the double Gamma and related functions. We also introduce some mathematical constants associated with the double and triple Gamma functions.

Barnes [94] defined the double Gamma function $\Gamma_2 = 1/G$, satisfying each of the following properties:

(a) $G(z+1) = \Gamma(z)G(z) \quad (z \in \mathbb{C})$;
(b) $G(1) = 1$;
(c) Asymptotically,

$$\begin{aligned}\log G(z+n+2) = \frac{n+1+z}{2}\log(2\pi) + \left[\frac{n^2}{2} + n + \frac{5}{12} + \frac{z^2}{2} + (n+1)z\right]\log n \\ - \frac{3n^2}{4} - n - nz - \log A + \frac{1}{12} + O\left(n^{-1}\right) \quad (n \to \infty),\end{aligned} \tag{1}$$

where Γ is the familiar Gamma function introduced in Section 1.1 and A is called the Glaisher-Kinkelin constant, defined by

$$\log A = \lim_{n\to\infty}\left\{\sum_{k=1}^{n} k\log k - \left(\frac{n^2}{2}+\frac{n}{2}+\frac{1}{12}\right)\log n + \frac{n^2}{4}\right\}, \tag{2}$$

the numerical value of A being given by

$$A \cong 1.282427130\cdots.$$

From this definition, Barnes [94] deduced several explicit Weierstrass canonical product forms of the double Gamma function Γ_2, one of which is recalled here in the form:

$$\begin{aligned}\{\Gamma_2(z+1)\}^{-1} &= G(z+1)\\ &= (2\pi)^{\frac{1}{2}z}\exp\left(-\frac{1}{2}z-\frac{1}{2}(\gamma+1)z^2\right)\prod_{k=1}^{\infty}\left\{\left(1+\frac{z}{k}\right)^k\exp\left(-z+\frac{z^2}{2k}\right)\right\},\end{aligned} \tag{3}$$

where γ denotes the Euler-Mascheroni constant given by 1.1(3).

Barnes [94] also gave the following two more equivalent forms of the double Gamma function Γ_2:

$$\begin{aligned}\{\Gamma_2(z+1)\}^{-1} &= G(z+1)\\ &= (2\pi)^{\frac{1}{2}z}\exp\left(-\frac{1}{2}z(z+1)-\frac{1}{2}\gamma z^2\right)\prod_{k=1}^{\infty}\frac{\Gamma(k)}{\Gamma(z+k)}\exp\left[z\psi(k)+\frac{1}{2}z^2\,\psi'(k)\right];\end{aligned} \tag{4}$$

$$\begin{aligned}\{\Gamma_2(z+1)\}^{-1} = G(z+1) &= (2\pi)^{\frac{1}{2}z}\exp\left[\left(\gamma-\frac{1}{2}\right)z-\left(\frac{\pi^2}{6}+1+\gamma\right)\frac{z^2}{2}\right]\Gamma(z+1)\\ &\cdot\prod_{m=0}^{\infty}{\prod_{n=0}^{\infty}}'\left(1+\frac{z}{m+n}\right)\exp\left(-\frac{z}{m+n}+\frac{z^2}{2(m+n)^2}\right),\end{aligned} \tag{5}$$

where the prime denotes the exclusion of the case $n=m=0$ and the Psi (or Digamma) function ψ is given by 1.2(1). Each form of these products is convergent for all finite values of $|z|$, by the Weierstrass factorization theorem (see Conway [339, p. 170]).

The double Gamma function satisfies the following relations:

$$G(1)=1 \quad \text{and} \quad G(z+1)=\Gamma(z)G(z) \quad (z\in\mathbb{C}). \tag{6}$$

For sufficiently large real x and $a \in \mathbb{C}$, we have the Stirling formula for the G-function:

$$\log G(x+a+1) = \frac{x+a}{2}\log(2\pi) - \log A + \frac{1}{12} - \frac{3x^2}{4} - ax + \left(\frac{x^2}{2} - \frac{1}{12} + \frac{a^2}{2} + ax\right)\log x + O\left(x^{-1}\right) \quad (x \to \infty). \tag{7}$$

The following special values of G (see Barnes [94]) may be recalled here:

$$G\left(\frac{1}{2}\right) = 2^{\frac{1}{24}} \cdot \pi^{-\frac{1}{4}} \cdot e^{\frac{1}{8}} \cdot A^{-\frac{3}{2}}; \tag{8}$$

$$G(n+2) = 1!2!\cdots n! \quad \text{and} \quad G(n+1) = \frac{(n!)^n}{1\cdot 2\cdot 3^2\cdot 4^3\cdots n^{n-1}} \quad (n \in \mathbb{N}). \tag{9}$$

We shall deduce only the expression (4) (see Barnes [94]). Indeed, taking the logarithmic derivative on both sides of the fundamental functional relation in (6), with respect to z, we obtain

$$\frac{G'(z+1)}{G(z+1)} = \frac{\Gamma'(z)}{\Gamma(z)} + \frac{G'(z)}{G(z)},$$

from which

$$\frac{G'(z+n+2)}{G(z+n+2)} = \sum_{k=0}^{n} \frac{\Gamma'(z+k+1)}{\Gamma(z+k+1)} + \frac{G'(z+1)}{G(z+1)}.$$

For sufficiently small values of $|z|$, by Taylor's theorem, we have

$$\frac{\Gamma'(z+1+k)}{\Gamma(z+1+k)} = \frac{\Gamma'(1+k)}{\Gamma(1+k)} + z\frac{d}{dk}\frac{\Gamma'(1+k)}{\Gamma(1+k)} + \frac{z^2}{2!}\frac{d^2}{dk^2}\frac{\Gamma'(1+k)}{\Gamma(1+k)} + \cdots,$$

provided that the coefficients in the expansion are finite.

Conversely, it follows from 1.1(2) or 1.3(3) that

$$-\frac{\Gamma'(1+z)}{\Gamma(1+z)} = \gamma + \sum_{m=1}^{\infty}\left(\frac{1}{z+m} - \frac{1}{m}\right),$$

which, upon differentiating r times with respect to z and setting $z = k$ in the resulting equation, yields

$$\frac{d^r}{dk^r}\frac{\Gamma'(1+k)}{\Gamma(1+k)} = (-1)^{r-1}\, r! \sum_{m=1}^{\infty} \frac{1}{(m+k)^{r+1}} \quad (r \in \mathbb{N}\setminus\{1\}),$$

so that

$$\frac{\Gamma'(z+1+k)}{\Gamma(z+1+k)} - \frac{\Gamma'(1+k)}{\Gamma(1+k)} - z\frac{d}{dk}\frac{\Gamma'(1+k)}{\Gamma(1+k)}$$
$$= \sum_{r=2}^{\infty}\left\{(-1)^{r-1}\sum_{m=1}^{\infty}\frac{1}{(m+k)^{r+1}}z^r\right\},$$

from which we find that

$$\sum_{k=0}^{n}\left\{\frac{\Gamma'(z+1+k)}{\Gamma(z+1+k)} - \frac{\Gamma'(1+k)}{\Gamma(1+k)} - z\frac{d}{dk}\frac{\Gamma'(1+k)}{\Gamma(1+k)}\right\}$$
$$= \sum_{r=2}^{\infty}(-1)^{r-1}\left\{\sum_{k=0}^{n}\sum_{m=1}^{\infty}\frac{1}{(m+k)^{r+1}}z^r\right\},$$

where we are not rearranging a double series, if n is not actually infinite.

Now it is not difficult to see from Eisenstein's theorem (see Forsyth [457, p. 87]) that

$$\lim_{n\to\infty}\sum_{k=0}^{n}\sum_{m=1}^{\infty}\frac{1}{(m+k)^{r+1}}$$

is convergent when $r \geqq 2$. Hence,

$$\lim_{n\to\infty}\sum_{k=0}^{n}\left\{\frac{\Gamma'(z+1+k)}{\Gamma(z+1+k)} - \frac{\Gamma'(1+k)}{\Gamma(1+k)} - z\frac{d}{dk}\frac{\Gamma'(1+k)}{\Gamma(1+k)}\right\}$$

is finite, when $|z| < 1$ $(z \in \mathbb{C})$.

We also have the following relation:

$$-\frac{G'(z+1)}{G(z+1)} = \sum_{k=0}^{n}\left\{\frac{\Gamma'(z+1+k)}{\Gamma(z+1+k)} - \frac{\Gamma'(1+k)}{\Gamma(1+k)} - z\frac{d}{dk}\frac{\Gamma'(1+k)}{\Gamma(1+k)}\right\}$$
$$-\frac{G'(z+n+2)}{G(z+n+2)} + \sum_{k=0}^{n}\left\{\frac{\Gamma'(1+k)}{\Gamma(1+k)} + z\frac{d}{dk}\frac{\Gamma'(1+k)}{\Gamma(1+k)}\right\}.$$

Therefore, for sufficiently small values of $|z|$, we may take

$$-\frac{G'(z+1)}{G(z+1)} = \alpha + 2\beta z$$
$$+\frac{d}{dz}\log\left[\prod_{k=0}^{\infty}\left\{\frac{\Gamma(z+1+k)}{\Gamma(1+k)}\exp\left(-z\frac{\Gamma'(1+k)}{\Gamma(1+k)} - \frac{z^2}{2}\frac{d}{dk}\frac{\Gamma'(1+k)}{\Gamma(1+k)}\right)\right\}\right], \tag{10}$$

provided that

$$-\lim_{n\to\infty}\frac{G'(z+n+2)}{G(z+n+2)}=-\alpha-2\beta z+\lim_{n\to\infty}\sum_{k=0}^{n}\left\{\frac{\Gamma'(1+k)}{\Gamma(1+k)}+z\frac{d}{dk}\frac{\Gamma'(1+k)}{\Gamma(1+k)}\right\},$$

and provided also that the expression on the right-hand side tends to the same value as that obtainable from (1). Thus, we have

$$\frac{G'(z+n+2)}{G(z+n+2)}=(n+1+z)\log n+\frac{1}{2}\log(2\pi)-n+O\left(n^{-1}\right)\quad(n\to\infty).$$

If, then, we can prove that, for suitable values of α and β,

$$\begin{aligned}&-\alpha-2\beta z+\sum_{k=0}^{n}\left(\frac{\Gamma'(1+k)}{\Gamma(1+k)}+z\frac{d}{dk}\frac{\Gamma'(1+k)}{\Gamma(1+k)}\right)\\&=(n+1+z)\log n+\frac{1}{2}\log(2\pi)-n+O\left(n^{-1}\right)\quad(n\to\infty),\end{aligned}\tag{11}$$

we shall have shown that, for suitable values of α and β, (10) holds true from the fact (see Barnes [94, p. 269]) that $G'(z+1)/G(z+1)$ is the only solution of

$$f(z+1)=f(z)+\frac{\Gamma'(z)}{\Gamma(z)},$$

which has such an asymptotic expansion near at infinity.

Now the left-hand side of (11) may be written as

$$-\alpha-2\beta z+\sum_{k=1}^{n+1}\left\{\psi(k)+z\psi'(k)\right\},$$

which, in view of 1.3(57) and 1.3(58), reduces at once to

$$(n+1)\log n-n+z(\log n+1+\gamma-2\beta)+\frac{1}{2}-\alpha+O\left(n^{-1}\right)\quad(n\to\infty),$$

which, upon putting

$$\alpha=\frac{1}{2}-\frac{1}{2}\log(2\pi)\quad\text{and}\quad\beta=\frac{1+\gamma}{2},\tag{12}$$

becomes the right-hand side of (7).

If we now substitute the values of α and β from (12) into (10), we obtain the following expression:

$$-\frac{G'(z+1)}{G(z+1)} = \frac{1}{2} - \frac{1}{2}\log(2\pi) + (1+\gamma)z$$
$$+\frac{d}{dz}\log\left[\prod_{k=0}^{\infty}\left\{\frac{\Gamma(z+1+k)}{\Gamma(1+k)}\exp\left(-z\,\psi(1+k) - \frac{1}{2}z^2\,\psi'(1+k)\right)\right\}\right]$$
$$(|z| < 1;\; z \in \mathbb{C}),$$

which, upon integrating with respect to z and obtaining the value of the constant of integration by making $z = 0$, finally yields the desired expression (4) valid for $|z| < 1$. In fact, (4) holds true for all finite values of $|z|$ by the principle of analytic continuation.

From (4), we may also evaluate a series involving the ψ-function:

$$\sum_{k=1}^{\infty}\left\{\psi(k) + \frac{1}{2}\psi'(k) - \log k\right\} = 1 + \frac{\gamma}{2} - \frac{1}{2}\log(2\pi) \tag{13}$$

by taking logarithms on both sides of (4) and setting $z = 1$ in the resulting equation. In view of 1.2(53), the left-hand side of (13) can be written in its equivalent form:

$$\sum_{k=1}^{\infty}\left\{\psi(k) + \frac{1}{2}\zeta(2,k) - \log k\right\} = 1 + \frac{\gamma}{2} - \frac{1}{2}\log(2\pi). \tag{14}$$

In view of their need in evaluating integrals involving the double Gamma function, it may be important to know other special values of the G-function. Here, we give two special values of the G-function as an illustration. The following special value of the Γ-function is known (see Spiegel [1058, p. 1]):

$$\Gamma\left(\frac{1}{4}\right) \cong 3.62560990822190 8\cdots. \tag{15}$$

The Catalan constant G is defined by

$$G := \frac{1}{2}\int_0^1 \mathbf{K}(k)dk = \sum_{m=0}^{\infty}\frac{(-1)^m}{(2m+1)^2} \cong 0.91596559417721901 5\cdots, \tag{16}$$

where $\mathbf{K}$ is the complete elliptic integral of the first kind, given by (*cf.* Equation 1.5(33))

$$\mathbf{K}(k) := \int_0^{\pi/2}\frac{dt}{\sqrt{1-k^2\sin^2 t}} \qquad (|k| < 1). \tag{17}$$

The following integral is known (see Gradshteyn and Ryzhik [505, p. 526, Entry 4.224]):

$$\int_0^{\pi/4} \log \sin t \, dt = -\frac{\pi}{4} \log 2 - \frac{1}{2} G. \tag{18}$$

Considering (6), (18) and (28), we obtain

$$G\left(\frac{3}{4}\right) = 2^{-\frac{1}{8}} \cdot \pi^{-\frac{1}{4}} \cdot e^{\frac{G}{2\pi}} \cdot \Gamma\left(\frac{1}{4}\right) G\left(\frac{1}{4}\right). \tag{19}$$

We also recall a duplication formula for the G-function (see Choi [263, p. 290]):

$$G(z)\left\{G\left(z+\frac{1}{2}\right)\right\}^2 G(z+1) = e^{\frac{1}{4}} \cdot A^{-3} \cdot 2^{-2z^2+3z-\frac{11}{12}} \cdot \pi^{z-\frac{1}{2}} \cdot G(2z), \tag{20}$$

which is an obvious special case ($n = 2$) of the following multiplication formula given by Barnes [94, p. 291]:

$$\prod_{i=0}^{n-1} \prod_{j=0}^{n-1} G\left(z + \frac{i+j}{n}\right) = \mathcal{K}(n)\,(2\pi)^{\frac{1}{2}n(n-1)z}\, n^{nz-\frac{1}{2}n^2z^2}\, G(nz), \tag{21}$$

where, for convenience,

$$\mathcal{K}(n) := A^{1-n^2}\, e^{\frac{1}{12}(n^2-1)}\, (2\pi)^{-\frac{1}{2}(n-1)}\, n^{-\frac{5}{12}}.$$

Setting $z = \frac{1}{4}$ in (20), and using (6) and (8), we obtain

$$\Gamma\left(\frac{1}{4}\right)\left\{G\left(\frac{1}{4}\right) G\left(\frac{3}{4}\right)\right\}^2 = 2^{-\frac{1}{4}} \cdot \pi^{-\frac{1}{2}} \cdot e^{\frac{3}{8}} \cdot A^{-\frac{9}{2}}. \tag{22}$$

Combining (19) and (22), we obtain

$$G\left(\frac{1}{4}\right) = e^{\frac{3}{32}-\frac{G}{4\pi}} \cdot A^{-\frac{9}{8}} \left\{\Gamma\left(\frac{1}{4}\right)\right\}^{-\frac{3}{4}} \cong 0.293756\cdots \tag{23}$$

or, equivalently,

$$G\left(\frac{3}{4}\right) = 2^{-\frac{1}{8}} \cdot \pi^{-\frac{1}{4}} \cdot e^{\frac{3}{32}+\frac{G}{\pi}} \cdot A^{-\frac{9}{8}} \left\{\Gamma\left(\frac{1}{4}\right)\right\}^{\frac{1}{4}} \cong 0.848718\cdots. \tag{24}$$

It follows from (23), (24) and (6) that

$$\frac{G\left(\frac{3}{4}\right)}{G\left(\frac{5}{4}\right)} = 2^{-\frac{1}{8}} \cdot \pi^{-\frac{1}{4}} \cdot e^{\frac{G}{2\pi}}. \tag{25}$$

Integral Formulas Involving the Double Gamma Function

We begin by recalling the following integral formula:

$$\int_0^z \pi t \cot \pi t \, dt = z \log(2\pi) + \log \frac{G(1-z)}{G(1+z)}, \tag{26}$$

which is due, originally, to Kinkelin [666]. Indeed, in view of (3), if we set

$$\Phi(z) := \frac{G(1+z)}{G(1-z)} = (2\pi)^z e^{-z} \prod_{k=1}^{\infty} \left\{ \left(\frac{1+\frac{z}{k}}{1-\frac{z}{k}} \right)^k e^{-2z} \right\}$$

and differentiate logarithmically with respect to z, and apply the known expansion (see Ahlfors [13, p. 188]):

$$\pi z \cot \pi z = 1 + 2 \sum_{n=1}^{\infty} \frac{z^2}{z^2 - n^2} \quad (z \notin \mathbb{Z}), \tag{27}$$

we shall readily obtain

$$\frac{d}{dz} \log \Phi(z) = \log(2\pi) - \pi z \cot \pi z \quad (z \notin \mathbb{Z}),$$

which, upon integration, yields (26), since $\Phi(0) = 1$.

Some simple consequences of Kinkelin's formula (26) are worthy of note here (see also Barnes [94, p. 279]). First, by using integration by parts in (26), we have

$$\int_0^z \log \sin \pi t \, dt = z \log\left(\frac{\sin \pi z}{2\pi} \right) + \log \frac{G(1+z)}{G(1-z)}, \tag{28}$$

which, upon setting $t = \frac{1}{2} - u$ and replacing z by $\frac{1}{2} - z$, yields

$$\int_0^z \log \cos \pi t \, dt = \left(z - \frac{1}{2} \right) \log\left(\frac{\cos \pi z}{2\pi} \right) - \frac{1}{2} \log 2 - \log \Gamma\left(\frac{1}{2} - z \right) + \log \frac{G\left(\frac{1}{2}+z\right)}{G\left(\frac{1}{2}-z\right)}. \tag{29}$$

Making use of (29), we obtain the following analogue of (26):

$$\int_0^z \pi t \tan \pi t \, dt = -\frac{1}{2} \log\left(\frac{\cos \pi z}{\pi} \right) - z \log(2\pi) - \log \Gamma\left(\frac{1}{2} - z \right) + \log \frac{G\left(\frac{1}{2}+z\right)}{G\left(\frac{1}{2}-z\right)}, \tag{30}$$

which would follow also from (26) by setting $t = \frac{1}{2} - u$ and replacing z by $\frac{1}{2} - z$.

Combining (28) and (29), we readily have the integral formula:

$$\int_0^z \log\tan\pi t\,dt = z\log\tan\pi z + \frac{1}{2}\log\frac{\cos\pi z}{\pi} + \log\Gamma\left(\frac{1}{2} - z\right) + \log\frac{G(1+z)}{G(1-z)} - \log\frac{G\left(\frac{1}{2}+z\right)}{G\left(\frac{1}{2}-z\right)}. \tag{31}$$

Similarly, by using various trigonometric identities, we can obtain the following integral formulas:

$$\int_0^z \left(\frac{\pi t}{\cos\pi t}\right)^2 dt = \pi z^2\tan\pi z + \log\frac{\cos\pi z}{\pi} + 2z\log(2\pi) + 2\log\Gamma\left(\frac{1}{2}-z\right) - 2\log\frac{G\left(\frac{1}{2}+z\right)}{G\left(\frac{1}{2}-z\right)}; \tag{32}$$

$$\int_0^z \left(\frac{\pi t}{\sin\pi t}\right)^2 dt = -\pi z^2\cot\pi z + 2z\log(2\pi) - 2\log\frac{G(1-z)}{G(1+z)}. \tag{33}$$

$$\int_0^z \pi t\tan\pi t\,dt = \log\frac{G(1-z)}{G(1+z)} - \frac{1}{2}\log\frac{G(1-2z)}{G(1+2z)}, \tag{34}$$

which, in view of the known duplication formula (20) for the G-function, is the same as (31);

$$\int_0^z \frac{\pi t}{\sin\pi t}dt = z\log(2\pi) + 4\log\frac{G\left(1-\frac{1}{2}z\right)}{G\left(1+\frac{1}{2}z\right)} - \log\frac{G(1-z)}{G(1+z)}; \tag{35}$$

$$\int_0^z (\pi t\tan\pi t)^2 dt = -\frac{\pi^2 z^3}{3} + \pi z^2\tan\pi z + \log\frac{\cos\pi z}{\pi} + 2z\log(2\pi) + 2\log\frac{G\left(\frac{3}{2}-z\right)}{G\left(\frac{1}{2}+z\right)}; \tag{36}$$

$$\int_0^z (\pi t\cot\pi t)^2 dt = -\frac{\pi^2 z^3}{3} - \pi z^2\cot\pi z + 2z\log(2\pi) + 2\log\frac{G(1-z)}{G(1+z)}; \tag{37}$$

$$\int_0^z \left(\frac{\pi t}{\sin \pi t}\right)^2 \cos \pi t \, dt = 2\int_0^{z/2} \left(\frac{\pi t}{\sin \pi t}\right)^2 dt - 2\int_0^{z/2} \left(\frac{\pi t}{\cos \pi t}\right)^2 dt$$

$$= -\frac{\pi z^2}{\sin \pi z} - 2\log \frac{\cos \frac{\pi z}{2}}{\pi} + 4\log \frac{G\left(1-\frac{z}{2}\right)}{G\left(1+\frac{z}{2}\right)} + 4\log \frac{G\left(\frac{1}{2}+\frac{z}{2}\right)}{G\left(\frac{3}{2}-\frac{z}{2}\right)}; \tag{38}$$

$$\int_0^z \frac{\pi t}{\sin \pi t \cos \pi t} dt = -\frac{1}{2}\log \frac{\cos \pi z}{\pi} + \log \frac{G(1-z)}{G(1+z)} + \log \frac{G\left(\frac{1}{2}+z\right)}{G\left(\frac{3}{2}-z\right)}. \tag{39}$$

Replacing t by at/π in (28), we obtain

$$\int_0^z \log \sin at \, dt = z \log \frac{\sin az}{2\pi} + \frac{\pi}{a} \log \frac{G\left(1+\frac{a}{\pi}z\right)}{G\left(1-\frac{a}{\pi}z\right)}, \tag{40}$$

which, in view of the trigonometric identity:

$$\cos bt - \cos at = 2\sin\left[\frac{1}{2}(a+b)t\right] \sin\left[\frac{1}{2}(a-b)t\right],$$

yields

$$\begin{aligned} \int_0^z \log(\cos bt - \cos at)\, dt &= z\log \frac{\sin\left[\frac{1}{2}(a+b)z\right] \sin\left[\frac{1}{2}(a-b)z\right]}{2\pi^2} \\ &+ \frac{2\pi}{a+b}\log \frac{G\left(1+\frac{a+b}{2\pi}z\right)}{G\left(1-\frac{a+b}{2\pi}z\right)} + \frac{2\pi}{a-b}\log \frac{G\left(1+\frac{a-b}{2\pi}z\right)}{G\left(1-\frac{a-b}{2\pi}z\right)}. \end{aligned} \tag{41}$$

By differentiating Alexeiewsky's theorem (see Barnes [94, p. 281]):

$$\int_0^z \log \Gamma(t+1)\, dt = \frac{1}{2}[\log(2\pi) - 1]z - \frac{z^2}{2} + z\log\Gamma(z+1) - \log G(z+1) \tag{42}$$

with respect to z, we obtain

$$\frac{d}{dz}\log G(z) = \frac{1}{2}[\log(2\pi)+1] - z + (z-1)\psi(z), \tag{43}$$

and so we have

$$\begin{aligned} &\frac{\partial}{\partial a}\log G(h(az)) \\ &= \left\{\frac{1}{2}[\log(2\pi)+1] - h(az) + [h(az)-1]\psi(h(az))\right\} \frac{\partial}{\partial a} h(az), \end{aligned} \tag{44}$$

where $h(z)$ is a function of z and a is a constant.

Differentiating each side of (41) with respect to a and using (44) and 1.3(9), we obtain

$$\begin{aligned}\int_0^z \frac{t\sin at}{\cos bt - \cos at}\,dt &= \frac{2az}{a^2-b^2}\log(2\pi)\\ &\quad - \frac{2\pi}{(a+b)^2}\log\frac{G\left(1-\frac{a+b}{2\pi}z\right)}{G\left(1+\frac{a+b}{2\pi}z\right)} - \frac{2\pi}{(a-b)^2}\log\frac{G\left(1-\frac{a-b}{2\pi}z\right)}{G\left(1+\frac{a-b}{2\pi}z\right)}.\end{aligned} \tag{45}$$

Replacing t by $\frac{1}{\pi}\arctan at$ in (31), we obtain

$$\begin{aligned}\int_0^z \frac{\log t}{1+a^2t^2}\,dt &= \left(\frac{\log z}{a}\right)\arctan az + \frac{\pi}{2a}\log\left(\frac{\cos\,\arctan az}{\pi}\right)\\ &\quad + \frac{\pi}{a}\log\frac{G\left(1+\frac{1}{\pi}\arctan az\right)G\left(\frac{3}{2}-\frac{1}{\pi}\arctan az\right)}{G\left(1-\frac{1}{\pi}\arctan az\right)G\left(\frac{1}{2}+\frac{1}{\pi}\arctan az\right)}.\end{aligned} \tag{46}$$

Setting $a=1$ and $z=1$ in (46) and using 1.3(4), we obtain

$$\int_0^1 \frac{\log t}{1+t^2}\,dt = -G. \tag{47}$$

It is easy to verify that

$$\int_0^z \frac{a}{1+a^2+2a\cos t}\,dt = \frac{2a}{1-a^2}\arctan\left(\frac{1-a}{1+a}\tan\frac{1}{2}z\right)$$

and

$$\int_0^z \frac{\cos t}{1+a^2+2a\cos t}\,dt = \frac{z}{2a} - \frac{1+a^2}{a(1-a^2)}\arctan\left(\frac{1-a}{1+a}\tan\frac{1}{2}z\right).$$

Using these identities and applying the Leibniz rule (see De Lillo [372, p. 665]) to the following integral and again integrating the resulting equation with respect to a

from 1 to a, we obtain

$$\int_0^z \log\left(\frac{1+a^2+2a\cos t}{2+2\cos t}\right)dt = z\log a - 2\int_1^a \arctan\left(\frac{1-t}{1+t}\tan\frac{1}{2}z\right)\frac{dt}{t}$$
$$= (2z-\pi)\log a + 2\int_1^a \arctan\left(\frac{t+\cos z}{\sin z}\right)\frac{dt}{t}.$$

Equivalently, from (29) and the following integral:

$$\int_0^z \log(2+2\cos t)\,dt = (2z-2\pi)\log\left(\frac{\cos\left(\frac{1}{2}z\right)}{\pi}\right) + 4\pi\log\frac{G\left(\frac{1}{2}+\frac{z}{2\pi}\right)}{G\left(\frac{3}{2}-\frac{z}{2\pi}\right)},$$

we have

$$\int_0^z \log(1+a^2+2a\cos t)\,dt = \pi\log a + (2z-2\pi)\log\left(\frac{a\cos\left(\frac{1}{2}z\right)}{\pi}\right)$$
$$+4\pi\log\frac{G\left(\frac{1}{2}+\frac{z}{2\pi}\right)}{G\left(\frac{3}{2}-\frac{z}{2\pi}\right)} + 2\int_1^a \arctan\left(\frac{t+\cos z}{\sin z}\right)\frac{dt}{t}.$$

Applying integration by parts to the last integral gives us the following equivalent form:

$$\int_0^z \log(1+a^2+2a\cos t)\,dt = \pi\log a + (z-\pi)\log\left(\frac{a^2\cot\left(\frac{1}{2}z\right)}{2\pi^2}\right)$$
$$+2\arctan\left(\frac{a+\cos z}{\sin z}\right)\log\left(\frac{a}{\sin z}\right) + 4\pi\log\frac{G\left(\frac{1}{2}+\frac{z}{2\pi}\right)}{G\left(\frac{3}{2}-\frac{z}{2\pi}\right)} \tag{48}$$
$$-2\int_{\cot\frac{1}{2}z}^{\frac{a+\cos z}{\sin z}} \frac{\log(t-\cot z)}{1+t^2}\,dt.$$

Setting $z = \frac{1}{2}\pi$ in (48) and using 1.3(4), (46) and (47), we obtain

$$\int_0^{\pi/2} \log(1 + a^2 + 2a\cos\, t)\, dt = -\pi \log\left(\frac{\cos\,(\arctan a)}{\pi}\right) - 2\pi \log \frac{G\left(1 + \frac{1}{\pi}\arctan a\right) G\left(\frac{3}{2} - \frac{1}{\pi}\arctan a\right)}{G\left(1 - \frac{1}{\pi}\arctan a\right) G\left(\frac{1}{2} + \frac{1}{\pi}\arctan a\right)}. \tag{49}$$

Integrating by parts and using (46), we have

$$\int_0^z \frac{\arctan at}{t} dt = -\frac{\pi}{2}\log\left(\frac{\cos\,(\arctan az)}{\pi}\right) - \pi \log \frac{G\left(1 + \frac{1}{\pi}\arctan az\right) G\left(\frac{3}{2} - \frac{1}{\pi}\arctan az\right)}{G\left(1 - \frac{1}{\pi}\arctan az\right) G\left(\frac{1}{2} + \frac{1}{\pi}\arctan az\right)}. \tag{50}$$

Replacing t by $\frac{1}{\pi}\arcsin at$ in (28), we get

$$\int_0^z \frac{\log t\, dt}{\sqrt{1 - a^2 t^2}} = \left(\frac{\arcsin az}{a}\right) \log\left(\frac{z}{2\pi}\right) + \frac{\pi}{a} \log \frac{G\left(1 + \frac{1}{\pi}\arcsin az\right)}{G\left(1 - \frac{1}{\pi}\arcsin az\right)}. \tag{51}$$

Using (51) and the following relation:

$$\int_0^z \frac{\arcsin at}{t} dt = (\arcsin az)\, \log z - a \int_0^z \frac{\log t\, dt}{\sqrt{1 - a^2 t^2}},$$

we obtain

$$\int_0^z \frac{\arcsin at}{t} dt = \log(2\pi)(\arcsin az) + \pi \log \frac{G\left(1 - \frac{1}{\pi}\arcsin az\right)}{G\left(1 + \frac{1}{\pi}\arcsin az\right)}. \tag{52}$$

Integrating by parts and using (46), we find that

$$\int_0^z \frac{(\arctan at)^2 dt}{t^2}$$
$$= -\frac{(\arctan az)^2}{z} - \pi a \log\left(\frac{\cos\,(\arctan az)}{\pi}\right) - a\log\left(1+a^2z^2\right)\arctan az$$
$$+2\pi a \log \frac{G\left(1-\frac{1}{\pi}\arctan az\right)G\left(\frac{1}{2}+\frac{1}{\pi}\arctan az\right)}{G\left(1+\frac{1}{\pi}\arctan az\right)G\left(\frac{3}{2}-\frac{1}{\pi}\arctan az\right)}$$
$$+a^2\int_0^z \frac{\log(1+a^2t^2)}{1+a^2t^2}dt. \tag{53}$$

Numerous special cases of the integral formulas (considered in this section) include the following results:

$$\int_0^{\pi/2} \left(\frac{x}{\sin\,x}\right)^2 \cos\,x\,dx = -\frac{\pi^2}{4}+4G; \tag{54}$$

$$\int_0^{1/2} \log\Gamma(t+1)dt = -\frac{1}{2}-\frac{7}{24}\log 2+\frac{1}{4}\log\pi+\frac{3}{2}\log A; \tag{55}$$

$$\int_0^{1/4} \log\Gamma(t+1)dt = -\frac{1}{4}-\frac{3}{8}\log 2+\frac{1}{8}\log\pi+\frac{9}{8}\log A+\frac{G}{\pi}; \tag{56}$$

$$\int_0^1 \frac{\arctan x}{x}dx = G; \tag{57}$$

$$\int_0^{1/\sqrt{2}} \frac{\arcsin x}{x}dx = \frac{\pi}{8}\log 2+\frac{G}{2}; \tag{58}$$

$$\int_0^{\pi/4} x\cot\,x\,dx = \frac{\pi}{8}\log 2+\frac{G}{2}; \tag{59}$$

$$\int_0^{\pi/4} \log\cos\,x\,dx = -\frac{\pi}{4}\log 2+\frac{G}{2}; \tag{60}$$

$$\int_0^{\pi/4} x\tan x\,dx = -\frac{\pi}{8}\log 2 + \frac{G}{2}; \tag{61}$$

$$\int_0^{\pi/4} \log\tan x\,dx = -G; \tag{62}$$

$$\int_0^{\pi/4} \left(\frac{x}{\cos x}\right)^2 dx = \frac{\pi^2}{16} + \frac{\pi}{4}\log 2 - G; \tag{63}$$

$$\int_0^{\pi/4} \left(\frac{x}{\sin x}\right)^2 dx = -\frac{\pi^2}{16} + \frac{\pi}{4}\log 2 + G; \tag{64}$$

$$\int_0^{\pi/4} x^2\tan^2 x\,dx = -\frac{\pi^3}{192} + \frac{\pi^2}{16} + \frac{\pi}{4}\log 2 - G; \tag{65}$$

$$\int_0^{\pi/4} x^2\cot^2 x\,dx = -\frac{\pi^3}{192} - \frac{\pi^2}{16} + \frac{\pi}{4}\log 2 + G; \tag{66}$$

$$\int_0^{\pi/2} \frac{x}{\sin x}\,dx = 2G. \tag{67}$$

The Evaluation of an Integral Involving $\log G(z)$

First, we shall introduce two interesting mathematical constants, in addition to the Glaisher-Kinkelin constant A, by recalling the Euler-Maclaurin summation formula (*cf.* Hardy [537, p. 318]):

$$\sum_{k=1}^{n} f(k) \sim C_0 + \int_a^n f(x)\,dx + \frac{1}{2}f(n) + \sum_{r=1}^{\infty} \frac{B_{2r}}{(2r)!} f^{(2r-1)}(n), \tag{68}$$

where C_0 is an arbitrary constant to be determined in each special case and

$$B_0 = 1,\ B_1 = -\frac{1}{2},\ B_2 = \frac{1}{6},\ B_4 = -\frac{1}{30},\ B_6 = \frac{1}{42},\ B_8 = -\frac{1}{30},\ B_{10} = \frac{5}{66}, \cdots,$$

$$\text{and } B_{2n+1} = 0 \quad (n \in \mathbb{N})$$

are the Bernoulli numbers (see Section 1.7). For another useful summation formula, see Edwards [399, p. 117].

Letting $f(x) = x^2 \log x$ and $f(x) = x^3 \log x$ in (68) with $a = 1$, we obtain

$$\log B = \lim_{n\to\infty} \left[\sum_{k=1}^{n} k^2 \log k - \left(\frac{n^3}{3} + \frac{n^2}{2} + \frac{n}{6} \right) \log n + \frac{n^3}{9} - \frac{n}{12} \right] \tag{69}$$

and

$$\log C = \lim_{n\to\infty} \left[\sum_{k=1}^{n} k^3 \log k - \left(\frac{n^4}{4} + \frac{n^3}{2} + \frac{n^2}{4} - \frac{1}{120} \right) \log n + \frac{n^4}{16} - \frac{n^2}{12} \right], \tag{70}$$

respectively; here, B and C are constants, whose approximate numerical values are given by

$$B \cong 1.03091\,675 \cdots \tag{71}$$

and

$$C \cong 0.97955\,746 \cdots . \tag{72}$$

The constants B and C were considered, recently, by Choi and Srivastava [289, 292]. See also Adamchik [6, p. 199]. Bendersky [114] presented a set of constants, including B and C.

It is known that (*cf.* Barnes [94, p. 288])

$$\int_0^1 \log G(t+1)dt = \frac{1}{12} + \frac{1}{4}\log(2\pi) - 2\log A. \tag{73}$$

Setting $z = t$ in the logarithmic form of (3) and integrating both sides of the resulting equation from $t = 0$ to $t = \frac{1}{2}$, we obtain

$$\int_0^{\frac{1}{2}} \log G(t+1)dt = \frac{1}{16}\log(2\pi) - \frac{\gamma}{48} - \frac{1}{12}$$
$$+ \frac{1}{4}\sum_{k=1}^{\infty} \left[\left(4k^2 + 2k\right) \log(2k+1) - \left(4k^2 + 2k\right) \log(2k) - 2k - \frac{1}{2} + \frac{1}{12k} \right]. \tag{74}$$

Consider the sum S_n given by

$$\begin{aligned}
S_n &:= \sum_{k=1}^{n}\left[\left(4k^2+2k\right)\log(2k+1)-\left(4k^2+2k\right)\log(2k)\right]\\
&= \sum_{k=1}^{n}\left[(2k+1)^2\log(2k+1)-(2k+1)\log(2k+1)-(2k)^2\log(2k)-2k\log(2k)\right]\\
&= \left[\sum_{k=1}^{2n+1}k^2\log k-2\sum_{k=1}^{n}(2k)^2\log(2k)-\sum_{k=1}^{2n+1}k\log k\right]\\
&= \left[\sum_{k=1}^{2n+1}k^2\log k-8\sum_{k=1}^{n}k^2\log k-\sum_{k=1}^{2n+1}k\log k-8\log 2\sum_{k=1}^{n}k^2\right],
\end{aligned}$$

which, in view of (2) and (69), yields

$$\begin{aligned}
S_n = \Bigg[&\log B+\left\{\frac{(2n+1)^3}{3}+\frac{(2n+1)^2}{2}+\frac{(2n+1)}{6}\right\}\log(2n+1)\\
&-\frac{(2n+1)^3}{9}+\frac{(2n+1)}{12}-8\log B-8\left(\frac{n^3}{3}+\frac{n^2}{2}+\frac{n}{6}\right)\log n+\frac{8}{9}n^3-\frac{8}{12}n
\end{aligned} \tag{75}$$

$$\begin{aligned}
&-\log A-\left\{\frac{(2n+1)^2}{2}+\frac{(2n+1)}{2}+\frac{1}{12}\right\}\log(2n+1)\\
&+\frac{(2n+1)^2}{4}-8\log 2\frac{n(n+1)(2n+1)}{6}\Bigg]+O\left(n^{-1}\right)\quad (n\to\infty)\\
&= -\log A-7\log B+\frac{2}{9}+\lim_{n\to\infty}\Bigg[\left(\frac{8}{3}n^3+4n^2+\frac{4}{3}n-\frac{1}{12}\right)\log(2n+1)\\
&-\left(\frac{8}{3}n^3+4n^2+\frac{4}{3}n\right)\log n-\frac{1}{3}n^2-\frac{1}{6}n\\
&-\left(\frac{8}{3}\log 2\right)n^3-(4\log 2)n^2-\left(\frac{4}{3}\log 2\right)n\Bigg]+O\left(n^{-1}\right)\quad (n\to\infty).
\end{aligned}$$

Now, substituting from (75) and recalling 1.1(3) and applying the following expansion:

$$\log\left(1+\frac{1}{2n}\right)=\frac{1}{2n}-\frac{1}{8n^2}+\frac{1}{24n^3}-\frac{d_n}{64n^4}\quad (0<d_n<1;\ n\in\mathbb{N}), \tag{76}$$

we evaluate the summation part of (74) as follows:

$$\begin{aligned}
T_n :=& -\log A - 7\log B + \frac{2}{9} \\
&+ \left[\left(\frac{8}{3}n^3 + 4n^2 + \frac{4}{3}n - \frac{1}{12} \right) \log(2n+1) \right. \\
&- \left(\frac{8}{3}n^3 + 4n^2 + \frac{4}{3}n \right) \log n - \frac{1}{3}n^2 - \frac{1}{6}n \\
&- \left(\frac{8}{3}\log 2 \right) n^3 - (4\log 2)n^2 - \left(\frac{4}{3}\log 2 \right) n \\
&\left. -2\frac{n(n+1)}{2} - \frac{n}{2} + \frac{1}{12}(\gamma + \log n) \right] + O\left(n^{-1}\right) \quad (n \to \infty) \\
=& \frac{\gamma}{12} - \log A - 7\log B - \frac{1}{12} \\
&+ \left[\left(\frac{8}{3}n^3 + 4n^2 + \frac{4}{3}n - \frac{1}{12} \right) \log 2 \right. \\
&+ \left(\frac{8}{3}n^3 + 4n^2 + \frac{4}{3}n - \frac{1}{12} \right) \log\left(1 + \frac{1}{2n}\right) \\
&\left. - \left(\frac{8}{3}\log 2 \right) n^3 - \left(4\log 2 + \frac{4}{3} \right) n^2 - \left(\frac{4}{3}\log 2 + \frac{5}{3} \right) n \right] + O\left(n^{-1}\right) \quad (n \to \infty).
\end{aligned}$$

Thus, we have

$$\begin{aligned}
T_n =& \frac{\gamma}{12} - \log A - 7\log B - \frac{1}{12} \\
&+ \left[\left(\frac{8}{3}n^3 + 4n^2 + \frac{4}{3}n - \frac{1}{12} \right) \log 2 + \frac{4}{3}n^2 + \frac{5}{3}n + \frac{5}{18} \right. \\
&\left. - \left(\frac{8}{3}\log 2 \right) n^3 - \left(4\log 2 + \frac{4}{3} \right) n^2 - \left(\frac{4}{3}\log 2 + \frac{5}{3} \right) n \right] + O\left(n^{-1}\right) \quad (n \to \infty),
\end{aligned}$$

which, upon taking the limit as $n \to \infty$ and using (74), yields our desired result:

$$\int_0^{\frac{1}{2}} \log G(t+1)dt = \frac{1}{24}(\log 2 + 1) + \frac{1}{16}\log \pi - \frac{1}{4}\log A - \frac{7}{4}\log B, \tag{77}$$

in terms of the mathematical constant B defined by (69).

Vignéras [1199] gave an integral representation for the double Gamma function Γ_2:

$$\begin{aligned}
\log \Gamma_2(z+1) =& -\int_0^\infty \frac{e^{-t}}{t\left(1-e^{-t}\right)^2} \left(1 - zt - \frac{z^2t^2}{2} - e^{-zt} \right) dt \\
&+ (1+\gamma)\frac{z^2}{2} - \frac{3}{2}\log \pi \quad (\Re(z) > -1).
\end{aligned} \tag{78}$$

The Multiple Gamma Functions

There are two known ways to define an n-ple Gamma functions Γ_n: Barnes [97] (see also Vardi [1190]) defined Γ_n by using the n-ple Hurwitz Zeta functions (see Choi and Quine [278]; also Seo *et al.* [1019]); a recurrence formula of the Weierstrass canonical product forms of the n-ple Gamma functions Γ_n was given by Vignéras [1199], who used the theorem of Dufresnoy and Pisot [395], which provides the existence, uniqueness and expansion of the series of Weierstrass satisfying a certain functional equation.

By making use of the aforementioned Dufresnoy-Pisot theorem and starting with

$$f_1(x) = -\gamma x + \sum_{n=1}^{\infty} \left\{ \frac{x}{n} - \log\left(1+\frac{x}{n}\right) \right\}, \tag{79}$$

Vignéras [1199] obtained a recurrence formula of Γ_n $(n \in \mathbb{N})$, which is given by

Theorem 1.4 *The n-ple Gamma functions* Γ_n *are defined by*

$$\Gamma_n(z) = \{G_n(z)\}^{(-1)^{n-1}} \quad (n \in \mathbb{N}), \tag{80}$$

where

$$G_n(z+1) = \exp(f_n(z)) \tag{81}$$

and the functions $f_n(z)$ *are given by*

$$f_n(z) = -zA_n(1) + \sum_{k=1}^{n-1} \frac{p_k(z)}{k!} \left[f_{n-1}^{(k)}(0) - A_n^{(k)}(1) \right] + A_n(z), \tag{82}$$

with

$$A_n(z) = \sum_{m \in \mathbb{N}_0{}^{n-1} \times \mathbb{N}} \left[\frac{1}{n} \left(\frac{z}{L(m)} \right)^n - \frac{1}{n-1} \left(\frac{z}{L(m)} \right)^{n-1} + \cdots + (-1)^{n-1} \frac{z}{L(m)} + (-1)^n \log\left(1 + \frac{z}{L(m)}\right) \right], \tag{83}$$

where

$$L(m) = m_1 + m_2 + \cdots + m_n,$$

if

$$m = (m_1, m_2, \ldots, m_n) \in \mathbb{N}_0{}^{n-1} \times \mathbb{N}$$

and the polynomials $p_n(z)$ given by

$$p_n(z) := \begin{cases} 1^n + 2^n + 3^n + \cdots + (N-1)^n & (z = N;\ N \in \mathbb{N} \setminus \{1\}) \\ \dfrac{B_{n+1}(z) - B_{n+1}}{n+1} & (z \in \mathbb{C}) \end{cases} \tag{84}$$

satisfy the following relations:

$$p_n'(z) = \frac{B_{n+1}'(z)}{n+1} = B_n(z) \quad \textit{and} \quad p_n(0) = 0,$$

$B_n(z)$ being the Bernoulli polynomial of degree n in z.

Clearly, we find from the definition (84) that

$$p_n(z) = \frac{1}{n+1} \sum_{k=1}^{n+1} \binom{n+1}{k} B_{n+1-k} z^k \quad (n \in \mathbb{N}). \tag{85}$$

By analogy with the Bohr-Mollerup (Theorem 1.1), which guarantees the uniqueness of the Gamma function Γ, one can give for the double Gamma function and (more generally) for the multiple Gamma functions of order n ($n \in \mathbb{N}$) a definition of Artin by means of the following theorem (see Vignéras [1199, p. 239]):

Theorem 1.5 *For all $n \in \mathbb{N}$, there exists a unique meromorphic function $G_n(z)$ satisfying each of the following properties:*

(a) $G_n(z+1) = G_{n-1}(z)\, G_n(z) \quad (z \in \mathbb{C})$;
(b) $G_n(1) = 1$;
(c) *For $x \geqq 1$, $G_n(x)$ are infinitely differentiable and*

$$\frac{d^{n+1}}{dx^{n+1}} \{\log G_n(x)\} \geqq 0;$$

(d) $G_0(x) = x$.

It may be remarked in passing that

$$G_1(z) = \Gamma_1(z) = \Gamma(z)$$

and the case $n = 2$ of (80) can readily be seen to produce the double Gamma function (5).

We observe that the function $\{\Gamma_n(z)\}^{-1}$ defined in (80) is an entire function of order n, all of whose zeros are the nonpositive integers $0, -1, -2, \ldots$, just as the Gamma function $\{\Gamma(z)\}^{-1}$, and the multiplicity of the zeros of $\{\Gamma_n(z)\}^{-1}$ at $z = -k$ ($k \in \mathbb{N}_0$) is equal to the number of solutions of the equation:

$$L(m) = k + 1 \quad \left(m \in {\mathbb{N}_0}^{n-1} \times \mathbb{N}\right).$$

The number of the solutions for $m \in \mathbb{N}_0{}^n$ is the coefficient of x^{k+1} in the Maclaurin series expansion of

$$(1-x)^{-n}, \quad \text{that is,} \quad \binom{n+k}{k+1}.$$

Therefore, the number of the solutions for $m \in \mathbb{N}_0{}^{n-1} \times \mathbb{N}$ is equal to

$$\binom{n+k}{k+1} - \binom{n+k-1}{k+1} = \binom{n+k-1}{n-1}.$$

We, thus, conclude that $\{\Gamma_n(z)\}^{-1}$ is an entire function with zeros at $z=-k$ $(k \in \mathbb{N}_0)$, whose multiplicities are

$$\binom{n+k-1}{n-1} \quad (n \in \mathbb{N};\ k \in \mathbb{N}_0). \tag{86}$$

The Triple Gamma Function Γ_3

When $n=3$ in (80), we readily obtain an explicit form of the *triple Gamma function* Γ_3:

$$\begin{aligned}\Gamma_3(1+z) =& G_3(1+z)\\ =& \exp\left[-\frac{1}{6}\left(\gamma+\frac{\pi^2}{6}+\frac{3}{2}\right)z^3+\frac{1}{4}\left(\gamma+\log(2\pi)+\frac{1}{2}\right)z^2+\Omega z\right]\\ &\cdot \prod_{m\in\mathbb{N}_0{}^2\times\mathbb{N}}\left\{\left(1+\frac{z}{L(m)}\right)^{-1}\exp\left[\frac{z}{L(m)}-\frac{1}{2}\left(\frac{z}{L(m)}\right)^2+\frac{1}{3}\left(\frac{z}{L(m)}\right)^3\right]\right\},\end{aligned} \tag{87}$$

where

$$\Omega = \frac{1}{12}\left(\frac{3}{2}-\gamma-3\log(2\pi)+\frac{\pi^2}{12}\right)+\frac{1}{2}\sum_{n=1}^{\infty}(-1)^n\frac{\zeta(n+2)}{(n+3)(n+4)}$$

and

$$L(m) = m_1+m_2+m_3 \quad \text{with} \quad m=(m_1,m_2,m_3) \in \mathbb{N}_0{}^2 \times \mathbb{N},$$

$\zeta(s)$ being the Riemann zeta function (see Section 2.3). Now the infinite sum in Ω can be evaluated explicitly by using a known formula (see Choi and Srivastava [288, p. 116, Eq. (2.63)]):

$$\sum_{k=3}^{\infty}(-1)^k\frac{\zeta(k)}{(k+1)(k+2)} = \frac{1}{2}+\frac{\gamma}{6}-\frac{\pi^2}{72}-2\log A. \tag{88}$$

We, thus, find that

$$\Omega = \frac{3}{8} - \frac{1}{4}\log(2\pi) - \log A \tag{89}$$

in terms of the Glaisher-Kinkelin constant A defined by (2).

If we set $n = 3$ in (86), we observe that $\{\Gamma_3(z)\}^{-1}$ is an entire function with zeros at $z = -k$ $(k \in \mathbb{N}_0)$, whose multiplicity is

$$\frac{1}{2}\left(k^2 + 3k + 2\right) \quad (k \in \mathbb{N}_0). \tag{90}$$

Furthermore, in view of (90), (87) can be written in the following equivalent form analogous to (3):

$$\begin{aligned}\Gamma_3(1+z) &= G_3(1+z)\\ &= e^{R(z)} \prod_{k=1}^{\infty} \left\{ \left(1+\frac{z}{k}\right)^{-\frac{1}{2}k(k+1)} \right.\\ &\quad \left. \cdot \exp\left[\frac{1}{2}(k+1)z - \frac{1}{4}\left(1+\frac{1}{k}\right)z^2 + \frac{1}{6k}\left(1+\frac{1}{k}\right)z^3\right]\right\},\end{aligned} \tag{91}$$

where, for convenience,

$$R(z) := -\frac{1}{6}\left(\gamma + \frac{\pi^2}{6} + \frac{3}{2}\right)z^3 + \frac{1}{4}\left(\gamma + \log(2\pi) + \frac{1}{2}\right)z^2 + \left(\frac{3}{8} - \frac{1}{4}\log(2\pi) - \log A\right)z.$$

It follows that Γ_3 satisfies several basic properties and characteristics, which are summarized here in

Theorem 1.6 *The triple Gamma function Γ_3 is the unique meromorphic function satisfying each of the following properties:*

(a) $\Gamma_3(1) = 1$;
(b) $\Gamma_3(z+1) = G(z)\Gamma_3(z)$ $(z \in \mathbb{C})$;
(c) *For $x \geqq 1$, $\Gamma_3(x)$ is infinitely differentiable and*

$$\frac{d^4}{dx^4}\{\log \Gamma_3(x)\} \geqq 0.$$

Just as in (8), we now proceed to express the value of $\Gamma_3\left(\frac{1}{2}\right)$ in terms of the mathematical constants π, e, A and B. We begin by recalling the following known results:

$$\frac{1}{2}\log(2\pi) = \lim_{n\to\infty}\left[\sum_{k=1}^{n}\log k - \left(n+\frac{1}{2}\right)\log n + n\right] \tag{92}$$

and

$$\log\left(1+\frac{1}{n}\right)=\frac{1}{n}-\frac{1}{2n^2}+\frac{1}{3n^3}+O\left(n^{-4}\right)\quad(n\to\infty). \tag{93}$$

By taking logarithms on both sides of (91) and setting $z=\frac{1}{2}$ in the resulting equation, if we make use of 1.1(3), we obtain

$$\begin{aligned}\log\Gamma_3\left(1+\frac{1}{2}\right)&=\frac{3}{16}-\frac{1}{16}\log(2\pi)-\frac{1}{2}\log A\\&+\lim_{n\to\infty}\left[-\sum_{k=1}^{n}\frac{k(k+1)}{2}\log\left(1+\frac{1}{2k}\right)+\frac{n^2}{8}+\frac{5}{16}n-\frac{1}{24}\log n\right].\end{aligned} \tag{94}$$

We, first, consider the following sum:

$$\begin{aligned}U_n:&=\sum_{k=1}^{n}\frac{k(k+1)}{2}\log\left(1+\frac{1}{2k}\right)\\&=\frac{1}{2}\sum_{k=1}^{n}k(k+1)\log(2k+1)\\&\quad-\frac{\log 2}{2}\sum_{k=1}^{n}k(k+1)-\frac{1}{2}\sum_{k=1}^{n}k^2\log k-\frac{1}{2}\sum_{k=1}^{n}k\log k\\&=\frac{1}{8}\left(\sum_{k=1}^{n}(2k+1)^2\log(2k+1)-\sum_{k=1}^{n}\log(2k+1)\right)\\&\quad-\frac{\log 2}{2}\sum_{k=1}^{n}k(k+1)-\frac{1}{2}\sum_{k=1}^{n}k^2\log k-\frac{1}{2}\sum_{k=1}^{n}k\log k\end{aligned}$$

and so

$$\begin{aligned}U_n=&\frac{1}{8}\left(\sum_{k=1}^{2n+1}k^2\log k-4\sum_{k=1}^{n}k^2\log k-4\log 2\sum_{k=1}^{n}k^2\right.\\&\left.-\sum_{k=1}^{2n+1}\log k+\sum_{k=1}^{n}\log k+n\log 2\right)\\&-\frac{\log 2}{2}\sum_{k=1}^{n}k(k+1)-\frac{1}{2}\sum_{k=1}^{n}k^2\log k-\frac{1}{2}\sum_{k=1}^{n}k\log k,\end{aligned}$$

which immediately leads us to

$$U_n = \frac{1}{8}\sum_{k=1}^{2n+1} k^2 \log k - \frac{1}{8}\sum_{k=1}^{2n+1} \log k - \sum_{k=1}^{n} k^2 \log k - \frac{1}{2}\sum_{k=1}^{n} k \log k + \frac{1}{8}\sum_{k=1}^{n} \log k - \left(\frac{n^3}{3} + \frac{3}{4}n^2 + \frac{7}{24}n\right)\log 2. \tag{95}$$

Upon substituting from (95) into (94), if we apply (2), (69) and (92), we obtain

$$\begin{aligned}\log\Gamma_3\left(1+\frac{1}{2}\right) &= \frac{3}{16} - \frac{1}{16}\log(2\pi) + \frac{7}{8}\log B \\ &\quad + \lim_{n\to\infty}\left[-\left(\frac{n^3}{3} + \frac{3}{4}n^2 + \frac{7}{24}n - \frac{1}{16}\right)\log\left(1+\frac{1}{2n}\right)\right. \\ &\quad \left. + \frac{n^2}{6} + \frac{n}{3} - \frac{35}{288} + \frac{1}{16}\log 2\right] \\ &= \frac{3}{16} - \frac{1}{16}\log\pi + \frac{7}{8}\log B \\ &\quad + \lim_{n\to\infty}\left[-\frac{n^2}{6} - \frac{n}{3} - \frac{19}{288} + O\left(n^{-1}\right) + \frac{n^2}{6} + \frac{n}{3} - \frac{35}{288}\right] \\ &= -\frac{1}{16}\log\pi + \frac{7}{8}\log B,\end{aligned}$$

where we have also used (93) for the second equality. Thus, we find that

$$\Gamma_3\left(1+\frac{1}{2}\right) = \pi^{-\frac{1}{16}} \cdot B^{\frac{7}{8}}, \tag{96}$$

which, in view of (8) and the assertion (b) of Theorem 1.5, yields

$$\Gamma_3\left(\frac{1}{2}\right) = 2^{-\frac{1}{24}} \cdot \pi^{\frac{3}{16}} \cdot e^{-\frac{1}{8}} \cdot A^{\frac{3}{2}} \cdot B^{\frac{7}{8}}. \tag{97}$$

A Multiplication Formula for the Γ_n

Nishizawa [867] obtained a multiplication formula for the n-ple Gamma function Γ_n, by using his product formula for the multiple Gamma function Γ_n and other asymptotic formulas.

Here, by employing the same method used by Choi and Quine [278], Choi and Srivastava [300] showed how the following multiplication formula for the multiple Gamma function Γ_n can be obtained rather easily and nicely:

$$\Gamma_n(z) = \frac{f_n(z,p,r)}{p^{\zeta_n(0,z)}} \prod_{q_1,\cdots,q_r=0}^{p-1} \Gamma_n\left(\frac{z+q_1+q_2+\cdots+q_r}{p}\right) \quad (p,\, r,\, n \in \mathbb{N}), \tag{98}$$

where

$$f_n(z,p,r) := \prod_{m=1}^{n} R_{n-m+1}^{(-1)^m\left\{\binom{z}{m-1}-\sum_{q_1,\ldots,q_r=0}^{p-1}\binom{\frac{z+q_1+q_2+\cdots+q_r}{p}}{m-1}\right\}}$$

and R_m are given as in 2.1(28). Thus, in view of 2.1(16), $\zeta_n(0,z)$ can be expressed in terms of the generalized Bernoulli polynomials of degree n.

The special case of (98), when $n=2$ and $r=2$, reduces to 1.4(21) by letting $z=pa$. Indeed, starting with

$$\sum_{q_1,\ldots,q_r=0}^{p-1} \zeta_n\left(s,\, a+\frac{q_1+q_2+\cdots+q_r}{p}\right)$$

$$= \sum_{q_1,\ldots,q_r=0}^{p-1} \sum_{k_1,\ldots,k_n}^{\infty} \left(a+\frac{q_1+q_2+\cdots+q_r}{p}+k_1+k_2+\cdots+k_n\right)^{-s}$$

$$= p^s \sum_{k_1,\ldots,k_n}^{\infty} \sum_{q_1,\ldots,q_r=0}^{p-1} (pa+q_1+q_2+\cdots+q_r+pk_1+pk_2+\cdots+pk_n)^{-s}$$

$$= p^s \sum_{k_1,\ldots,k_n}^{\infty} (pa+k_1+k_2+\cdots+k_n)^{-s} = p^s\,\zeta_n(s,\,pa),$$

which yields

$$\sum_{q_1,\ldots,q_r=0}^{p-1} \zeta_n\left(s,\, a+\frac{q_1+q_2+\cdots+q_r}{p}\right) = p^s\,\zeta_n(s,\,pa). \tag{99}$$

Differentiating each side of (99) with respect to s and setting $s=0$ in the resulting equation, and if we take the exponential in the last identity and then consider 2.1(27) with $pa=z$, we obtain our desired multiplication formula (98).

A special case of (98), when $n=1$, yields a multiplication formula for the Gamma function Γ as follows:

$$\prod_{q_1,\ldots,q_r=0}^{p-1} \Gamma\left(\frac{z+q_1+q_2+\cdots+q_r}{p}\right) = (2\pi)^{\frac{1}{2}(p^r-1)}\,p^{\frac{1}{2}-z}\,\Gamma(z) \quad (p,\, r\in\mathbb{N}). \tag{100}$$

1.5 The Gaussian Hypergeometric Function and its Generalization

The Gauss Hypergeometric Equation

The second-order linear ordinary differential equation:

$$z(1-z)\frac{d^2w}{dz^2}+[c-(a+b+1)z]\frac{dw}{dz}-abw=0 \tag{1}$$

or, equivalently,

$$\{\delta(\delta+c-1)-z(\delta+a)(\delta+b)\}w=0 \qquad \left(\delta:=z\frac{d}{dz}\right), \tag{2}$$

in which a, b and c are real or complex parameters, is called the *Gauss hypergeometric equation.* Its only singularities are at

$$z=0,\ 1,\ \infty;$$

each singularity is easily seen to be of *regular* kind.

The hypergeometric equation (1) or (2) is the most celebrated equation of the *Fuchsian class*, which consists of differential equations, whose only singularities (including the point at infinity) are regular singular points. Its importance stems, in part, from a well-known theorem that *every homogeneous linear differential equation of the second order, whose singularities (including the point at infinity) are regular and at most three in number, can be transformed into the hypergeometric equation.*

Power-series solutions of the hypergeometric equation (1) valid in the neighborhoods of the regular singular points $z=0,\ 1$ or ∞ can be developed by direct application of the classical method of Frobenius. Thus, if c is not an integer, the general solution of (1) valid in a neighborhood of the origin ($z=0$) is found to be

$$w=C_1\,{}_2F_1(a,\,b;\,c;\,z)+C_2\,z^{1-c}\,{}_2F_1(a-c+1,\,b-c+1;\,2-c;\,z) \qquad (c\notin\mathbb{Z}), \tag{3}$$

where C_1 and C_2 are arbitrary constants, and (for convenience)

$$\begin{aligned} {}_2F_1(a,\,b;\,c;\,z) &:= 1+\frac{ab}{1\cdot c}z+\frac{a(a+1)b(b+1)}{1\cdot 2\cdot c(c+1)}z^2+\cdots \\ &=\sum_{n=0}^{\infty}\frac{(a)_n\,(b)_n}{(c)_n}\frac{z^n}{n!} \qquad \left(c\notin\mathbb{Z}_0^-\right), \end{aligned} \tag{4}$$

in terms of the Pochhammer symbol $(\lambda)_n$ defined by 1.1(5).

Gauss's Hypergeometric Series

The infinite series in (4) obviously reduces to the elementary geometric series:

$$\sum_{n=0}^{\infty} z^n = 1 + z + z^2 + \cdots + z^n + \cdots \tag{5}$$

in its special cases, when

$$\text{(i)} \quad a = c \quad \text{and} \quad b = 1 \quad \text{or} \quad \text{(ii)} \quad a = 1 \quad \text{and} \quad b = c. \tag{6}$$

Hence, it is called the *hypergeometric series* or, more precisely, *Gauss's hypergeometric series* after the famous German mathematician, Carl Friedrich Gauss (1777–1855), who in the year 1812 introduced this series into analysis and gave the F–notation for it. (See also Equation 1.5(51) in which the general notation ${}_pF_q$ is introduced, p and q being any nonnegative integers.)

By d'Alembert's ratio test, it is easily seen that the hypergeometric series in (4) converges absolutely *within* the unit circle, that is, when $|z| < 1$, provided that the denominator parameter c is neither zero nor a negative integer. Notice, however, that, if either or both of the numerator parameters a and b in (4) is zero or a negative integer, the hypergeometric series terminates, and the question of convergence does not enter into the discussion.

Further tests show that the hypergeometric series in (4), when $|z| = 1$ (that is, on the unit circle), is

(i) absolutely convergent, if $\Re(c - a - b) > 0$;
(ii) conditionally convergent, if $-1 < \Re(c - a - b) \leqq 0 \quad (z \neq 1)$;
(iii) divergent, if $\Re(c - a - b) \leqq -1$.

As a matter of fact, in Case (i), we are led to the well-known *Gauss's summation theorem*:

$$\begin{gathered} {}_2F_1(a, b; c; 1) = \frac{\Gamma(c)\,\Gamma(c-a-b)}{\Gamma(c-a)\,\Gamma(c-b)} \\ \left(\Re(c-a-b) > 0;\ c \notin \mathbb{Z}_0^-\right). \end{gathered} \tag{7}$$

An obvious special case of (7) occurs when the numerator parameter a or b is a nonpositive integer $-n$. We, thus, have the *Chu-Vandermonde summation formula*:

$${}_2F_1(-n, b; c; 1) = \frac{(c-b)_n}{(c)_n} \qquad \left(n \in \mathbb{N}_0;\ c \notin \mathbb{Z}_0^-\right), \tag{8}$$

which is, in fact, equivalent to *Vandermonde's convolution theorem*:

$$\begin{gathered} \sum_{k=0}^{n} \binom{\lambda}{k}\binom{\mu}{n-k} = \binom{\lambda+\mu}{n} = \sum_{k=0}^{n} \binom{\lambda}{n-k}\binom{\mu}{k} \\ (n \in \mathbb{N}_0;\ \lambda, \mu \in \mathbb{C}). \end{gathered} \tag{9}$$

For a number of summation theorems for the hypergeometric series (4), when z takes on other special values, see Bailey [87, pp. 9–11], Erdélyi *et al.* [421, pp. 104–105], Slater [1040, p. 243] and Luke [779, pp. 271–273].

The Hypergeometric Series and Its Analytic Continuation

We have seen that the hypergeometric series in (4) converges absolutely, when $|z| < 1$ and, thus, defines a function:

$$ {}_2F_1\,(a,\,b;\,c;\,z), $$

which is analytic, when $|z| < 1$, provided that c is neither zero nor a negative integer. This function is correspondingly called the *hypergeometric function* or Gauss's hypergeometric function. Indeed, it is the only solution of the Fuchsian equation (1) that is analytic at the point $z = 0$ *and* assumes the value 1 at that point.

The hypergeometric function ${}_2F_1\,(a,\,b;\,c;\,z)$ can, in fact, be continued analytically outside the unit circle in a number of ways. If, for convenience, we use this same notation for the analytically continued function, one way is to employ *Euler's integral representation*:

$$ {}_2F_1\,(a,\,b;\,c;\,z) = \frac{\Gamma(c)}{\Gamma(a)\,\Gamma(c-a)} \int_0^1 t^{a-1}\,(1-t)^{c-a-1}\,(1-zt)^{-b}\,dt \tag{10} $$
$$ (\Re(c) > \Re(a) > 0;\ |\arg(1-z)| \leqq \pi - \epsilon\ (0 < \epsilon < \pi)) $$

or, equivalently,

$$ {}_2F_1\,(a,\,b;\,c;\,z) = \frac{\Gamma(c)}{\Gamma(b)\,\Gamma(c-b)} \int_0^1 t^{b-1}\,(1-t)^{c-b-1}\,(1-zt)^{-a}\,dt \tag{11} $$
$$ (\Re(c) > \Re(b) > 0;\ |\arg(1-z)| \leqq \pi - \epsilon\ (0 < \epsilon < \pi)), $$

since, by the series definition (4),

$$ {}_2F_1\,(a,\,b;\,c;\,z) \equiv {}_2F_1\,(b,\,a;\,c;\,z). \tag{12} $$

An analytic continuation without such parametric constraints as in (10) and (11) can be provided by employing the following *Mellin-Barnes contour integral representation* of the hypergeometric function:

$$ \frac{\Gamma(a)\,\Gamma(b)}{\Gamma(c)}\,{}_2F_1\,(a,\,b;\,c;\,z) $$
$$ = \frac{1}{2\pi i} \int_{-i\infty}^{i\infty} \frac{\Gamma(a+\zeta)\,\Gamma(b+\zeta)\,\Gamma(-\zeta)}{\Gamma(c+\zeta)}\,(-z)^{\zeta}\,d\zeta \tag{13} $$
$$ \left(|\arg(-z)| \leqq \pi - \epsilon\ (0 < \epsilon < \pi);\ a,\, b \notin \mathbb{Z}_0^-\right), $$

where the path of integration is the imaginary axis (in the complex ζ-plane), which is indented, if necessary, to ensure that the poles of $\Gamma(a+\zeta)\Gamma(b+\zeta)$, *viz*

$$\zeta = -a-n \quad \text{and} \quad \zeta = -b-n \quad (n \in \mathbb{N}_0), \tag{14}$$

lie to the left of the path and the poles of $\Gamma(-\zeta)$, *viz*

$$\zeta = 0,\, 1,\, 2, \ldots, \tag{15}$$

lie to the right of the path.

The integrals in (10), (11) and (13) define analytic functions of z, which are single-valued in the domain $|\arg(-z)| < \pi$, that is, in the whole z-plane, with the exception of the points on the negative real axis. Since the Gaussian function ${}_2F_1\,(a,\, b;\, c;\, z)$ is defined by the hypergeometric series (4) everywhere within the unit circle $|z| = 1$ (including, for example, at points on the negative real axis from $z = 0$ to $z = 1-\epsilon$, ϵ being an arbitrarily small positive number), we can use either of these integrals to provide the analytic continuation of the hypergeometric function to the whole complex z-plane cut along the real axis from $z = 1$ to $z = \infty$. This analytic continuation of ${}_2F_1\,(a,\, b;\, c;\, z)$ is conveniently denoted by the same symbol ${}_2F_1\,(a,\, b;\, c;\, z)$, which will henceforth represent a branch (the *principal branch*) of the analytic function generated by the Gaussian hypergeometric series in (4).

By evaluating the contour integral in (13) using the sum of the residues of the integrand at the poles given by (14), it can be shown that

$$\begin{aligned}
&\frac{1}{\Gamma(c)}\,{}_2F_1\,(a,\, b;\, c;\, z) \\
&\quad = \frac{\Gamma(b-a)}{\Gamma(b)\,\Gamma(c-a)}\,(-z)^{-a}\,{}_2F_1\left(a,\, 1-c+a;\, 1-b+a;\, \frac{1}{z}\right) \\
&\qquad + \frac{\Gamma(a-b)}{\Gamma(a)\,\Gamma(c-b)}\,(-z)^{-b}\,{}_2F_1\left(b,\, 1-c+b;\, 1-a+b;\, \frac{1}{z}\right) \\
&\qquad\qquad (|\arg(-z)| \leqq \pi-\epsilon\ (0<\epsilon<\pi);\ a-b \notin \mathbb{Z}).
\end{aligned} \tag{16}$$

This is an important result, for it may be used to provide the analytic continuation of ${}_2F_1\,(a,\, b;\, c;\, z)$ into the domain $|z| > 1$, that is, outside the circle of convergence of the hypergeometric series (4). It also shows that the function so defined possesses a branch point at infinity and that, asymptotically,

$$\begin{aligned}
&{}_2F_1\,(a,\, b;\, c;\, z) \sim \mathcal{A}\,(-z)^{-a} + \mathcal{B}\,(-z)^{-b} \\
&\qquad\qquad (|z| \to \infty;\ |\arg(-z)| \leqq \pi-\epsilon\ (0<\epsilon<\pi)),
\end{aligned} \tag{17}$$

where, for convenience,

$$\mathcal{A} := \frac{\Gamma(c)\,\Gamma(b-a)}{\Gamma(b)\,\Gamma(c-a)} \quad \text{and} \quad \mathcal{B} := \frac{\Gamma(c)\,\Gamma(a-b)}{\Gamma(a)\,\Gamma(c-b)}. \tag{18}$$

Linear, Quadratic and Cubic Transformations

Pfaff-Kummer transformations:

$$ {}_2F_1(a, b; c; z) = (1-z)^{-a}\,{}_2F_1\left(a, c-b; c; \frac{z}{z-1}\right) \tag{19} $$
$$ \left(c \notin \mathbb{Z}_0^-;\ |\arg(1-z)| \leqq \pi - \epsilon\ (0 < \epsilon < \pi)\right); $$

$$ {}_2F_1(a, b; c; z) = (1-z)^{-b}\,{}_2F_1\left(b, c-a; c; \frac{z}{z-1}\right) \tag{20} $$
$$ (c \notin \mathbb{Z}_0^-;\ |\arg(1-z)| \leqq \pi - \epsilon\ (0 < \epsilon < \pi)). $$

Euler's transformation:

$$ \begin{gathered} {}_2F_1(a, b; c; z) = (1-z)^{c-a-b}\,{}_2F_1(c-a, c-b; c; z) \\ (c \notin \mathbb{Z}_0^-;\ |\arg(1-z)| \leqq \pi - \epsilon\ (0 < \epsilon < \pi)). \end{gathered} \tag{21} $$

The (Pfaff-Kummer) transformations (19) and (20) can be proven, for example, by setting $t = 1 - s$ in Euler's integral representations (10) and (11), respectively, and then appealing to the principle of analytic continuation. The Euler transformation (21) follows, if we perform the transformations (19) and (20) successively.

For an extensive list of quadratic and cubic transformations of the Gaussian hypergeometric function, see Erdélyi *et al.* [421, pp. 110–114] and Luke [779, Vol. I, p. 92 *et seq.*].

Hypergeometric Representations of Elementary Functions

$$ (1-z)^{-a} = \sum_{n=0}^{\infty} (a)_n \frac{z^n}{n!} = {}_2F_1(a, b; b; z); \tag{22} $$

$$ \log(1+z) = z\,{}_2F_1(1, 1; 2; -z); \tag{23} $$

$$ \log\left(\frac{1+z}{1-z}\right) = 2z\,{}_2F_1\left(\frac{1}{2}, 1; \frac{3}{2}; z^2\right); \tag{24} $$

$$ \sin^{-1} z = z\,{}_2F_1\left(\frac{1}{2}, \frac{1}{2}; \frac{3}{2}; z^2\right); \tag{25} $$

$$ \tan^{-1} z = z\,{}_2F_1\left(\frac{1}{2}, 1; \frac{3}{2}; -z^2\right); \tag{26} $$

$$ \begin{aligned} \sinh^{-1} z &= \log\left(z + \sqrt{1+z^2}\right) \\ &= z\,{}_2F_1\left(\frac{1}{2}, \frac{1}{2}; \frac{3}{2}; -z^2\right); \end{aligned} \tag{27} $$

$$\left\{\frac{1}{2}\left(1+\sqrt{1-z}\right)\right\}^{1-2a} = {}_2F_1\left(a,\, a-\frac{1}{2};\, 2a;\, z\right) = \sqrt{1-z}\,{}_2F_1\left(a,\, a+\frac{1}{2};\, 2a;\, z\right); \tag{28}$$

$$\left(1+\sqrt{z}\right)^{-2a} + \left(1-\sqrt{z}\right)^{-2a} = 2\,{}_2F_1\left(a,\, a+\frac{1}{2};\, \frac{1}{2};\, z\right). \tag{29}$$

Hypergeometric Representations of Other Functions

Legendre functions of the first and second kind:

$$P_\nu^\mu(z) = \frac{1}{\Gamma(1-\mu)}\left(\frac{z+1}{z-1}\right)^{\frac{1}{2}\mu} {}_2F_1\left(-\nu,\, \nu+1;\, 1-\mu;\, \frac{1-z}{2}\right); \tag{30}$$

$$Q_\nu^\mu(z) = \frac{\sqrt{\pi}\, e^{i\mu\pi}}{2^{\nu+1}}\,\frac{\Gamma(\mu+\nu+1)}{\Gamma\left(\nu+\frac{3}{2}\right)}\,\frac{(z^2-1)^{\frac{1}{2}\mu}}{z^{\mu+\nu+1}} \cdot {}_2F_1\left[\frac{1}{2}(\mu+\nu+1),\, \frac{1}{2}(\mu+\nu)+1;\, \nu+\frac{3}{2};\, z^{-2}\right]. \tag{31}$$

The incomplete Beta function:

$$B_z(\alpha,\beta) = \alpha^{-1} z^\alpha\, {}_2F_1\,(\alpha,\, 1-\beta;\, \alpha+1;\, z)\,. \tag{32}$$

Complete Elliptic integrals of the first and second kind:

$$\mathbf{K}(k) = \int_0^{\pi/2} \frac{d\theta}{\sqrt{1-k^2\sin^2\theta}} = \frac{1}{2}\pi\, {}_2F_1\left(\frac{1}{2},\, \frac{1}{2};\, 1;\, k^2\right); \tag{33}$$

$$\mathbf{E}(k) = \int_0^{\pi/2} \sqrt{1-k^2\sin^2\theta}\, d\theta = \frac{1}{2}\pi\, {}_2F_1\left(-\frac{1}{2},\, \frac{1}{2};\, 1;\, k^2\right). \tag{34}$$

Jacobi polynomials:

$$P_n^{(\alpha,\beta)}(z) = (-1)^n P_n^{(\beta,\alpha)}(-z) = \binom{\alpha+n}{n}\, {}_2F_1\left[\begin{matrix} -n,\, \alpha+\beta+n+1; & \dfrac{1-z}{2} \\ \alpha+1; & \end{matrix}\right]. \tag{35}$$

Gegenbauer (or ultraspherical) polynomials:

$$C_n^\nu(z) = \binom{n+\nu-\frac{1}{2}}{n}^{-1}\binom{n+2\nu-1}{n}P_n^{\left(\nu-\frac{1}{2},\nu-\frac{1}{2}\right)}(z)$$
$$= \binom{n+2\nu-1}{n}{}_2F_1\left[\begin{matrix}-n,\ 2\nu+n; \\ \nu+\frac{1}{2};\end{matrix}\ \frac{1-z}{2}\right]. \tag{36}$$

Legendre (or spherical) polynomials:

$$P_n(z) = P_n^{(0,0)}(z) = C_n^{\frac{1}{2}}(z) = {}_2F_1\left(-n,\ n+1;\ 1;\ \frac{1-z}{2}\right). \tag{37}$$

Tchebycheff polynomials of the first and second kind:

$$T_n(z) = \binom{n-\frac{1}{2}}{n}^{-1}P_n^{\left(-\frac{1}{2},-\frac{1}{2}\right)}(z) = \frac{1}{2}n\,C_n^0(z) = {}_2F_1\left(-n,\ n;\ \frac{1}{2};\ \frac{1-z}{2}\right), \tag{38}$$

where

$$C_n^0(z) := \lim_{\nu\to 0}\left\{\nu^{-1}\,C_n^\nu(z)\right\}; \tag{39}$$

$$U_n(z) = \frac{1}{2}\binom{n+\frac{1}{2}}{n+1}^{-1}P_n^{\left(\frac{1}{2},\frac{1}{2}\right)}(z) = C_n^1(z)$$
$$= (n+1)\,{}_2F_1\left(-n,\ n+2;\ \frac{3}{2};\ \frac{1-z}{2}\right). \tag{40}$$

The Confluent Hypergeometric Function

If, in Gauss's hypergeometric equation (1), we replace z by z/b, the resulting equation will have three regular singularities at

$$z = 0,\ b,\ \infty.$$

By letting $|b| \to \infty$, this transformed equation reduces to the form:

$$z\frac{d^2w}{dz^2} + (c-z)\frac{dw}{dz} - aw = 0 \tag{41}$$

or, equivalently,

$$\{\delta(\delta+c-1) - z(\delta+a)\}\,w = 0 \quad \left(\delta := z\frac{d}{dz}\right). \tag{42}$$

Equation (41) or (42) has a regular singularity at $z = 0$ and an *irregular* singularity at $z = \infty$, which is formed by the *confluence* of two regular singularities at b and ∞ of Gauss's equation (1) with z replaced by z/b. Consequently, (41) or (42) is called the *confluent hypergeometric equation* or Kummer's differential equation after Ernst Eduard Kummer (1810–1893), who presented a detailed study of its solutions in 1836 (*cf.* Kummer [707]).

The simplest solution of (1) or (2) is Kummer's function $F(a, c, z)$, which, in the notations of this section, is

$$\begin{aligned} {}_1F_1(a; c; z) &:= 1 + \frac{a}{1 \cdot c} z + \frac{a(a+1)}{1 \cdot 2 \cdot c(c+1)} z^2 + \cdots \\ &= \sum_{n=0}^{\infty} \frac{(a)_n}{(c)_n} \frac{z^n}{n!} \quad (c \notin \mathbb{Z}_0^-;\ |z| < \infty). \end{aligned} \tag{43}$$

Moreover, since

$$\lim_{|\lambda| \to \infty} \left\{ (\lambda)_n \left(\frac{z}{\lambda} \right)^n \right\} = z^n = \lim_{|\mu| \to \infty} \left\{ \frac{(\mu z)^n}{(\mu)_n} \right\} \quad (n \in \mathbb{N}_0;\ |z| < \infty), \tag{44}$$

we readily have

$${}_1F_1(a; c; z) = \lim_{|b| \to \infty} {}_2F_1\left(a, b; c; \frac{z}{b}\right). \tag{45}$$

In view of the principle of confluence involved in (45), the solution (43) is also called the *confluent hypergeometric function*.

Other popular notations for the series solution (43) are (i) $M(a, c, z)$, used by Airey and Webb [15], and (ii) $\Phi(a; c; z)$ or $\Phi(a, c, z)$, introduced by Humbert [573, 574].

Important Properties of Kummer's Confluent Hypergeometric Function

The principle of confluence exhibited by (45) is useful to deduce properties of the confluent hypergeometric ${}_1F_1$ function from those of Gauss's hypergeometric ${}_2F_1$ function. Thus, from (10) and (20), we obtain

$${}_1F_1(a; c; z) = \frac{\Gamma(c)}{\Gamma(a)\Gamma(c-a)} \int_0^1 t^{a-1} (1-t)^{c-a-1} e^{zt}\, dt \quad (\Re(c) > \Re(a) > 0); \tag{46}$$

$${}_1F_1(a; c; z) = e^z \, {}_1F_1(c-a; c; -z), \tag{47}$$

which is known as *Kummer's first transformation.* We also have *Kummer's second transformation*:

$$e^{-z}\,{}_1F_1(a;\,2a;\,2z) = {}_0F_1\left(-;\,a+\frac{1}{2};\,\frac{1}{4}z^2\right)$$
$$(2a \neq -1,\,-3,\,-5,\,\ldots), \tag{48}$$

where

$$\begin{aligned} {}_0F_1(-;\,c;\,z) &= \lim_{|a|\to\infty} {}_1F_1(a;\,c;\,z/a) \\ &= \sum_{n=0}^{\infty} \frac{z^n}{n!\,(c)_n} \quad (c \notin \mathbb{Z}_0^-;\ |z| < \infty). \end{aligned} \tag{49}$$

It follows immediately from the definition (43) that

$${}_1F_1(a;\,a;\,z) = {}_0F_0(-;\,-;\,z) = e^z. \tag{50}$$

The Generalized (Gauss and Kummer) Hypergeometric Function

A natural generalization of the hypergeometric functions ${}_2F_1$, ${}_1F_1$, *et cetera* (considered in this section) is accomplished by the introduction of an arbitrary number of numerator and denominator parameters. The resulting series:

$$\begin{aligned} {}_pF_q\left[\begin{matrix}\alpha_1,\,\ldots,\,\alpha_p; \\ \beta_1,\,\ldots,\,\beta_q;\end{matrix}\ z\right] &= \sum_{n=0}^{\infty} \frac{(\alpha_1)_n \cdots (\alpha_p)_n}{(\beta_1)_n \cdots (\beta_q)_n} \frac{z^n}{n!} \\ &= {}_pF_q(\alpha_1,\,\ldots,\,\alpha_p;\,\beta_1,\,\ldots,\,\beta_q;\,z), \end{aligned} \tag{51}$$

where $(\lambda)_n$ is the Pochhammer symbol defined by 1.1(5), is known as the *generalized Gauss (and Kummer) series*, or simply, the *generalized hypergeometric series.* Here, p and q are positive integers or zero (interpreting an empty product as 1), and we assume (for simplicity) that the variable z, the numerator parameters $\alpha_1, \cdots, \alpha_p$ and the denominator parameters $\beta_1, \ldots, \beta_q$ take on complex values, provided that no zeros appear in the denominator of (51), that is, that

$$(\beta_j \notin \mathbb{Z}_0^-;\ j = 1, \ldots, q). \tag{52}$$

Thus, if a numerator parameter is a negative integer or zero, the ${}_pF_q$ series terminates in view of the identity 1.1(47), and we are led to a (generalized hypergeometric) polynomial of the type:

$$
{}_{p+1}F_q\left[\begin{matrix}-n,\,\alpha_1,\,\ldots,\,\alpha_p;\\ \beta_1,\,\ldots,\,\beta_q;\end{matrix}\ z\right]=\sum_{k=0}^{n}\frac{(-n)_k(\alpha_1)_k\ldots(\alpha_p)_k}{(\beta_1)_k\ldots(\beta_q)_k}\frac{z^k}{k!}
$$

$$
=\frac{(\alpha_1)_n\ldots(\alpha_p)_n}{(\beta_1)_n\ldots(\beta_q)_n}(-z)^n \tag{53}
$$

$$
\cdot\,{}_{q+1}F_p\left[\begin{matrix}-n,\,1-\beta_1-n,\,\ldots,\,1-\beta_q-n;\\ 1-\alpha_1-n,\,\ldots,\,1-\alpha_p-n;\end{matrix}\ \frac{(-1)^{p+q}}{z}\right]\quad (n\in\mathbb{N}_0),
$$

where we have reversed the order of the terms of the polynomial by using 1.1(25) and 1.1(26).

Assuming that none of the numerator parameters is zero or a negative integer (otherwise the question of convergence will not arise) and with the usual restriction given by (52), the ${}_pF_q$ series in (51)

(i) converges for $|z|<\infty$, if $p\leq q$,
(ii) converges for $|z|<1$, if $p=q+1$, and
(iii) diverges for all z, $z\neq 0$, if $p>q+1$.

Furthermore, if we set

$$
\omega:=\sum_{j=1}^{q}\beta_j-\sum_{j=1}^{p}\alpha_j, \tag{54}
$$

then it is known that the ${}_pF_q$ series in (51), with $p=q+1$, is

(i) absolutely convergent for $|z|=1$, if $\Re(\omega)>0$,
(ii) conditionally convergent for $|z|=1$ $(z\neq 1)$, if $-1<\Re(\omega)\leqq 0$, and
(iii) divergent for $|z|=1$, if $\Re(\omega)\leqq -1$.

Analytic Continuation of the Generalized Hypergeometric Function

The generalized hypergeometric ${}_pF_q$ function is defined by the (absolutely) convergent series (51), whenever

(i) $p\leqq q$ and $|z|<\infty$

or

(ii) $p=q+1$ and $|z|<1$.

As in the above-detailed case of the ${}_2F_1$ function, the generalized hypergeometric function ${}_pF_q$ (with $p=q+1$) can be continued analytically to the domain $|\arg(1-z)|<\pi$, that is, to the whole complex z-plane cut along the real axis from $z=1$ to

$z = \infty$, by using (51) in conjunction with the Mellin–Barnes contour integral representation:

$$\frac{\Gamma(\alpha_1)\ldots\Gamma(\alpha_p)}{\Gamma(\beta_1)\ldots\Gamma(\beta_q)}\,{}_pF_q\left[\begin{matrix}\alpha_1,\ \ldots,\ \alpha_p;\\ \beta_1,\ \ldots,\ \beta_q;\end{matrix}\ z\right]$$

$$= \frac{1}{2\pi i}\int_{-i\infty}^{i\infty}\frac{\Gamma(\alpha_1+\zeta)\ldots\Gamma(\alpha_p+\zeta)\,\Gamma(-\zeta)}{\Gamma(\beta_1+\zeta)\ldots\Gamma(\beta_q+\zeta)}(-z)^{\zeta}\,d\zeta \tag{55}$$

$$(\alpha_j \notin \mathbb{Z}_0^-;\ j = 1,\ \ldots,\ p;\ |\arg(-z)| < \pi),$$

where the path of integration is the imaginary axis (in the complex ζ-plane), which is indented, if necessary, to separate the poles of $\Gamma(\alpha_j+\zeta)$ $(j = 1, \cdots, p)$ from those of $\Gamma(-\zeta)$.

Functions Expressible in Terms of the ${}_pF_q$ Function

The following list is in addition to the important functions (considered above in this section) that can be expressed as a hypergeometric ${}_2F_1$ function.

Elementary functions:

$$e^z = {}_0F_0[-;\ -;\ z]; \tag{56}$$

$$(1-z)^{-\alpha} = {}_1F_0[\alpha;\ -;\ z]; \tag{57}$$

$$\cos z = {}_0F_1\left[-;\ \frac{1}{2};\ -\frac{1}{4}z^2\right]; \tag{58}$$

$$\sin z = z\,{}_0F_1\left[-;\ \frac{3}{2};\ -\frac{1}{4}z^2\right]. \tag{59}$$

Bessel functions:

$$\begin{aligned}J_\nu(z) &= \frac{\left(\frac{1}{2}z\right)^\nu}{\Gamma(\nu+1)}\,{}_0F_1\left[-;\ \nu+1;\ -\frac{1}{4}z^2\right]\\ &= \frac{\left(\frac{1}{2}z\right)^\nu}{\Gamma(\nu+1)}\exp(\pm iz)\,{}_1F_1\left[\nu+\frac{1}{2};\ 2\nu+1;\ \mp 2iz\right];\end{aligned} \tag{60}$$

$$\begin{aligned}I_\nu(z) &= \frac{\left(\frac{1}{2}z\right)^\nu}{\Gamma(\nu+1)}\,{}_0F_1\left[-;\ \nu+1;\ \frac{1}{4}z^2\right]\\ &= \frac{\left(\frac{1}{2}z\right)^\nu}{\Gamma(\nu+1)}\exp(\pm z)\,{}_1F_1\left[\nu+\frac{1}{2};\ 2\nu+1;\ \mp 2z\right].\end{aligned} \tag{61}$$

Lommel functions:

$$s_{\mu,\nu}(z) = \frac{z^{\mu+1}}{(\mu-\nu+1)(\mu+\nu+1)} \cdot {}_1F_2\left[\begin{matrix} 1; \\ \frac{1}{2}(\mu-\nu+3),\ \frac{1}{2}(\mu+\nu+3); \end{matrix}\ -\frac{1}{4}z^2\right]; \tag{62}$$

$$S_{\mu,\nu}(z) = s_{\mu,\nu}(z) + 2^{\mu-1}\,\Gamma\left(\frac{\mu-\nu+1}{2}\right)\Gamma\left(\frac{\mu+\nu+1}{2}\right) \cdot \left\{\sin\left[\frac{1}{2}(\mu-\nu)\pi\right]J_\nu(z) - \cos\left[\frac{1}{2}(\mu-\nu)\pi\right]Y_\nu(z)\right\}, \tag{63}$$

where $J_\nu(z)$ is given by (60), and

$$Y_\nu(z) = \csc(\nu\pi)\,[\cos(\nu\pi)J_\nu(z) - J_{-\nu}(z)]. \tag{64}$$

Struve functions:

$$\begin{aligned} \mathbf{H}_\nu(z) &= \frac{\left(\frac{1}{2}z\right)^{\nu+1}}{\Gamma\left(\frac{3}{2}\right)\Gamma\left(\nu+\frac{3}{2}\right)}\,{}_1F_2\left[\begin{matrix} 1; \\ \frac{3}{2},\ \nu+\frac{3}{2}; \end{matrix}\ -\frac{1}{4}z^2\right] \\ &= \frac{2^{1-\nu}}{\sqrt{\pi}\,\Gamma\left(\nu+\frac{1}{2}\right)}\,s_{\nu,\nu}(z) \\ &= Y_\nu(z) + \frac{2^{1-\nu}}{\sqrt{\pi}\,\Gamma\left(\nu+\frac{1}{2}\right)}\,S_{\nu,\nu}(z); \end{aligned} \tag{65}$$

$$\begin{aligned} \mathbf{L}_\nu(z) &= \frac{\left(\frac{1}{2}z\right)^{\nu+1}}{\Gamma\left(\frac{3}{2}\right)\Gamma\left(\nu+\frac{3}{2}\right)}\,{}_1F_2\left[\begin{matrix} 1; \\ \frac{3}{2},\ \nu+\frac{3}{2}; \end{matrix}\ \frac{1}{4}z^2\right] \\ &= \exp\left[-\frac{1}{2}(\nu+1)i\pi\right]\mathbf{H}_\nu(iz). \end{aligned} \tag{66}$$

Whittaker function of the first kind:

$$\begin{aligned} M_{\kappa,\mu}(z) &= z^{\mu+\frac{1}{2}}\exp\left(-\frac{1}{2}z\right){}_1F_1\left(\mu-\kappa+\frac{1}{2};\ 2\mu+1;\ z\right) \\ &= z^{\mu+\frac{1}{2}}\exp\left(\frac{1}{2}z\right){}_1F_1\left(\mu+\kappa+\frac{1}{2};\ 2\mu+1;\ -z\right). \end{aligned} \tag{67}$$

Incomplete Gamma function:

$$\gamma(\alpha,z) = \alpha^{-1}z^{\alpha}\,{}_1F_1(\alpha;\ \alpha+1;\ -z). \tag{68}$$

Error function:

$$\operatorname{Erf}(z) = \frac{\sqrt{\pi}}{2}\operatorname{erf}(z) = \frac{1}{2}\gamma\left(\frac{1}{2}, z^2\right)$$
$$= z\,{}_1F_1\left(\frac{1}{2}; \frac{3}{2}; -z^2\right). \tag{69}$$

Hermite polynomials:

$$H_n(z) = \sum_{k=0}^{[n/2]} \frac{(-1)^k n!\,(2z)^{n-2k}}{k!\,(n-2k)!}$$
$$= (2z)^n\,{}_2F_0\left(-\frac{1}{2}n, \frac{1}{2} - \frac{1}{2}n; -; -z^{-2}\right) \quad (n \in \mathbb{N}_0). \tag{70}$$

Laguerre functions and polynomials:

$$L_\nu^{(\alpha)}(z) = \frac{\Gamma(\nu+\alpha+1)}{\Gamma(\nu+1)\,\Gamma(\alpha+1)}\,{}_1F_1(-\nu; \alpha+1; z); \tag{71}$$

$$L_n^{(\alpha)}(z) = \sum_{k=0}^{n} \binom{n+\alpha}{n-k} \frac{(-z)^k}{k!}$$
$$= \binom{n+\alpha}{n}\,{}_1F_1(-n; \alpha+1; z)$$
$$= \frac{(-z)^n}{n!}\,{}_2F_0\left(-n, -\alpha-n; -; -\frac{1}{z}\right) \quad (n \in \mathbb{N}_0). \tag{72}$$

Poisson–Charlier polynomials:

$$c_n(x;\alpha) = \sum_{k=0}^{n} (-1)^k \binom{n}{k}\binom{x}{k} k!\,\alpha^{-k} \quad (\alpha > 0;\ x \in \mathbb{N}_0)$$
$$= n!\,(-\alpha)^{-n} L_n^{(x-n)}(\alpha)$$
$$= (-\alpha)^{-n} n! \binom{x}{n}\,{}_1F_1(-n; x-n+1; \alpha). \tag{73}$$

In an alternative notation, we have (see Szegö [1141, p. 35])

$$p_n(x) = \alpha^{-\frac{1}{2}n} \frac{(x-n+1)_n}{\sqrt{n!}}\,{}_1F_1(-n; x-n+1; \alpha)$$
$$= \alpha^{-\frac{1}{2}n}\sqrt{n!}\binom{x}{n}\,{}_1F_1(-n; x-n+1; \alpha). \tag{74}$$

1.6 Stirling Numbers of the First and Second Kind

Stirling Numbers of the First Kind

The *Stirling numbers* $s(n,k)$ *of the first kind* are defined by the generating functions:

$$z(z-1)\cdots(z-n+1)=\sum_{k=0}^{n} s(n,k)\,z^k \tag{1}$$

and

$$\{\log(1+z)\}^k = k!\sum_{n=k}^{\infty} s(n,k)\frac{z^n}{n!} \quad (|z|<1). \tag{2}$$

We have the following recurrence relations satisfied by $s(n,k)$:

$$s(n+1,k)=s(n,k-1)-n\,s(n,k) \quad (n \geqq k \geqq 1); \tag{3}$$

$$\binom{k}{j} s(n,k)=\sum_{l=k-j}^{n-j}\binom{n}{l} s(n-l,j)\,s(l,k-j) \quad (n \geqq k \geqq j). \tag{4}$$

From the definition (1) of $s(n,k)$, the Pochhammer symbol in 1.1(5) can be written in the form:

$$(z)_n = z(z+1)\cdots(z+n-1)=\sum_{k=0}^{n}(-1)^{n+k}s(n,k)\,z^k, \tag{5}$$

where $(-1)^{n+k}s(n,k)$ denotes the number of permutations of n symbols, which has exactly k cycles.

It is not difficult to see also that

$$\begin{gathered} s(n,0)=\begin{cases}1 & (n=0)\\ 0 & (n\in\mathbb{N}),\end{cases} \qquad s(n,n)=1, \\ s(n,1)=(-1)^{n+1}(n-1)!, \quad s(n,n-1)=-\binom{n}{2} \end{gathered} \tag{6}$$

and

$$\begin{gathered} \sum_{k=1}^{n} s(n,k)=0 \quad (n\in\mathbb{N}\setminus\{1\}); \quad \sum_{k=0}^{n}(-1)^{n+k}s(n,k)=n!; \\ \sum_{j=k}^{n} s(n+1,j+1)\,n^{j-k}=s(n,k). \end{gathered} \tag{7}$$

Yet another recursion formula for $s(n,k)$ is given by (see Shen [1024]):

$$(k-1)\,s(n,k) = -\sum_{m=1}^{k-1} s(n,k-m)\,H_{n-1}^{(m)}, \tag{8}$$

where $H_n^{(s)}$ is the generalized harmonic numbers of order s, defined by

$$H_n^{(s)} := \sum_{k=1}^{n} \frac{1}{k^s} \quad (n \in \mathbb{N};\ s \in \mathbb{C}) \tag{9}$$

and $H_n^{(1)} := H_n$ $(n \in \mathbb{N})$ is the harmonic numbers.
Here, for (8), we assume $H_0^{(0)} := 1$ and $H_0^{(m)} := 0$ $(m \in \mathbb{N})$.
It readily follows from the recursion formula (8) that

$$\begin{aligned}
s(n,2) &= (-1)^n\,(n-1)!\,H_{n-1};\\
s(n,3) &= (-1)^{n+1}\,\frac{(n-1)!}{2}\left[(H_{n-1})^2 - H_{n-1}^{(2)}\right];\\
s(n,4) &= (-1)^n\,\frac{(n-1)!}{6}\left[(H_{n-1})^3 - 3H_{n-1}\,H_{n-1}^{(2)} + 2H_{n-1}^{(3)}\right];\\
s(n,5) &= (-1)^{n+1}\,\frac{(n-1)!}{24}\\
&\quad\cdot\left[(H_{n-1})^4 + 8\,H_{n-1}\,H_{n-1}^{(3)} - 6\,(H_{n-1})^2\,H_{n-1}^{(2)} + 3\left(H_{n-1}^{(2)}\right)^2 - 6\,H_{n-1}^{(4)}\right].
\end{aligned} \tag{10}$$

In view of (5), by logarithmically differentiating 1.1(19) and then using 1.2(7), we obtain

$$\begin{aligned}
\frac{d}{dz}\{(z)_n\} &= \sum_{k=1}^{n} (-1)^{n+k}\,k\,s(n,k)\,z^{k-1}\\
&= (z)_n\left(\sum_{k=1}^{n}\frac{1}{z+k-1}\right),
\end{aligned} \tag{11}$$

which, upon employing Leibniz's rule for differentiation, yields a more general formula:

$$\begin{aligned}
\frac{d^{j+1}}{dz^{j+1}}\{(z)_n\} &= \sum_{k=j+1}^{n} (-1)^{n+k+j+1}\,(-k)_{j+1}\,s(n,k)\,z^{k-j-1}\\
&= \sum_{l=0}^{j} (-1)^l\,\frac{j!}{(j-l)!}\left(\sum_{k=1}^{n}\frac{1}{(z+k-1)^{l+1}}\right)\left(\frac{d^{j-l}}{dz^{j-l}}\{(z)_n\}\right)\\
&\quad (j \in \mathbb{N}_0;\ n \in \mathbb{N}).
\end{aligned} \tag{12}$$

For $j = 1$ and $j = 2$, (12) immediately yields

$$\sum_{k=2}^{n}(-1)^{n+k}\,k(k-1)\,s(n,k)\,z^{k-2} = (z)_n\left[\left(\sum_{k=1}^{n}\frac{1}{z+k-1}\right)^2-\sum_{k=1}^{n}\frac{1}{(z+k-1)^2}\right] \tag{13}$$

and

$$\sum_{k=3}^{n}(-1)^{n+k}\,k(k-1)(k-2)\,s(n,k)\,z^{k-3} = (z)_n\left[\left(\sum_{k=1}^{n}\frac{1}{z+k-1}\right)^3-3\left(\sum_{k=1}^{n}\frac{1}{z+k-1}\right)\left(\sum_{k=1}^{n}\frac{1}{(z+k-1)^2}\right)+\sum_{k=1}^{n}\frac{1}{(z+k-1)^3}\right]. \tag{14}$$

Stirling Numbers of the Second Kind

The *Stirling numbers* $S(n,k)$ *of the second kind* are defined by the generating functions:

$$z^n = \sum_{k=0}^{n} S(n,k)\,z(z-1)\cdots(z-k+1), \tag{15}$$

$$(e^z-1)^k = k!\sum_{n=k}^{\infty} S(n,k)\,\frac{z^n}{n!}, \tag{16}$$

and

$$(1-z)^{-1}(1-2z)^{-1}\cdots(1-kz)^{-1} = \sum_{n=k}^{\infty} S(n,k)\,z^{n-k} \quad (|z|<k^{-1}), \tag{17}$$

where $S(n,k)$ denotes the number of ways of partitioning a set of n elements into k nonempty subsets.

It is not difficult to see also that

$$S(n,0) = \begin{cases} 1 & (n=0) \\ 0 & (n\in\mathbb{N}), \end{cases}$$

$$S(n,1) = S(n,n) = 1, \quad \text{and} \quad S(n,n-1) = \binom{n}{2}. \tag{18}$$

The recurrence relations for $S(n,k)$ are given by

$$S(n+1,k) = k\,S(n,k) + S(n,k-1) \quad (n \geqq k \geqq 1) \tag{19}$$

and

$$\binom{k}{j} S(n,k) = \sum_{i=k-j}^{n-j} \binom{n}{i} S(n-i,j)\,S(i,k-j) \quad (n \geqq k \geqq j). \tag{20}$$

The numbers $S(n,k)$ can be expressed in an explicit form:

$$S(n,k) = \frac{1}{k!} \sum_{j=0}^{k} (-1)^{k-j} \binom{k}{j} j^n. \tag{21}$$

Some additional properties of $S(n,k)$ are recalled here as follows:

$$\sum_{k=0}^{n} (-1)^{n-k}\, k!\, S(n,k) = 1; \tag{22}$$

$$\sum_{j=k}^{n} S(j-1,k-1)\, k^{n-j} = S(n,k); \tag{23}$$

$$S(n,k) = \sum_{j=0}^{n-k} (-1)^j \binom{n-1+j}{n-k+j} \binom{2n-k}{n-k-j} S(n-k+j,j); \tag{24}$$

$$\sum_{j=k}^{n} S(j,k)\, S(n,j) = \delta_{kn}, \tag{25}$$

where δ_{mn} denotes the Kronecker delta defined by

$$\delta_{mn} = \begin{cases} 0 & (m \neq n), \\ 1 & (m = n). \end{cases} \quad \text{and} \quad \delta_m = \begin{cases} 0 & (m \neq 0), \\ 1 & (m = 0). \end{cases} \tag{26}$$

Relationships Among Stirling Numbers of the First and Second Kind and Bernoulli Numbers

Akiyama and Tanigawa [17] presented, to evaluate multiple zeta values at nonpositive integers, the following identities:

$$\sum_{\ell=k}^{n} \frac{1}{\ell} S(n,\ell)\, s(\ell,k) = \frac{1}{n}\binom{n}{k} B_{n-k} + \delta_{n-k-1} \quad (n,\, k \in \mathbb{N};\; n \geqq k), \tag{27}$$

where B_k is the Bernoulli number given in 1.6.

$$\sum_{\ell=0}^{n} \binom{n}{\ell} B_{n-\ell}\, S(\ell,k) = \frac{n}{k}\, S(n-1,k-1) \quad (n,\, k \in \mathbb{N}). \tag{28}$$

$$\sum_{\ell=k-1}^{n} \frac{S(n,\ell)\, s(\ell+1,k)}{\ell+1} = \frac{B_{n+1-k}}{n+1} \binom{n+1}{k} \quad (n,\, k \in \mathbb{N};\ n \geqq k-1), \tag{29}$$

which, upon setting $k=1$ and using (21), yields

$$\begin{aligned} B_n &= \sum_{\ell=0}^{n} \frac{(-1)^{\ell}\, \ell!}{\ell+1}\, S(n,\ell) \\ &= \sum_{\ell=0}^{n} \frac{1}{\ell+1} \sum_{j=0}^{\ell} (-1)^j \binom{\ell}{j} j^n \quad (n \in \mathbb{N}_0), \end{aligned} \tag{30}$$

or, equivalently,

$$\sum_{\ell=0}^{n} s(n,\ell)\, B_{\ell} = \frac{(-1)^n\, n!}{n+1} \quad (n \in \mathbb{N}_0). \tag{31}$$

We also recall some known formulas (see, e.g., [982]):

$$\sum_{\ell=0}^{\max\{k,j\}+1} s(\ell,j)\, S(k,\ell) = \delta_{jk}. \tag{32}$$

$$\sum_{\ell=0}^{\max\{k,j\}+1} s(k,\ell)\, S(\ell,j) = \delta_{jk}. \tag{33}$$

$$s(n,i) = \sum_{k=i}^{n} \sum_{j=0}^{k} s(n,k)\, s(k,j)\, S(j,i). \tag{34}$$

$$S(n,i) = \sum_{k=i}^{n} \sum_{j=0}^{k} S(n,k)\, S(k,j)\, s(j,i). \tag{35}$$

$$S(n,m) = \sum_{k=0}^{n-m} (-1)^k \binom{k+n-1}{k+n-m} \binom{2n-m}{n-k-m} s(k-m+n,k). \tag{36}$$

$$s(n,m) = \sum_{k=0}^{n-m} (-1)^k \binom{k+n-1}{k+n-m} \binom{2n-m}{n-k-m} S(k-m+n,k). \tag{37}$$

1.7 Bernoulli, Euler and Genocchi Polynomials and Numbers

Bernoulli Polynomials and Numbers

The *Bernoulli polynomials* $B_n(x)$ are defined by the generating function:

$$\frac{z\,e^{xz}}{e^z-1}=\sum_{n=0}^{\infty} B_n(x)\,\frac{z^n}{n!} \quad (|z|<2\pi). \tag{1}$$

The numbers $B_n := B_n(0)$ are called the *Bernoulli numbers* generated by

$$\frac{z}{e^z-1}=\sum_{n=0}^{\infty} B_n\,\frac{z^n}{n!} \quad (|z|<2\pi). \tag{2}$$

It easily follows from (1) and (2) that

$$B_n(x)=\sum_{k=0}^{n}\binom{n}{k} B_k\,x^{n-k}. \tag{3}$$

The Bernoulli polynomials $B_n(x)$ satisfy the difference equation:

$$B_n(x+1)-B_n(x)=n\,x^{n-1} \quad (n\in\mathbb{N}_0), \tag{4}$$

which yields

$$B_n(0)=B_n(1) \quad (n\in\mathbb{N}\setminus\{1\}). \tag{5}$$

Setting $x=1$ in (3), in view of (5), we have

$$B_n=\sum_{k=0}^{n}\binom{n}{k} B_k, \tag{6}$$

which gives a recursion formula for computing Bernoulli numbers. The first few of the Bernoulli numbers are already listed with the Euler-Maclaurin summation formula 1.4(68), and (for the sake of completeness) we have the following list:

$$\begin{gathered}
B_0=1,\quad B_1=-\frac{1}{2},\quad B_2=\frac{1}{6},\quad B_4=-\frac{1}{30},\quad B_6=\frac{1}{42},\quad B_8=-\frac{1}{30},\quad B_{10}=\frac{5}{66},\\
B_{12}=-\frac{691}{2730},\quad B_{14}=\frac{7}{6},\quad B_{16}=-\frac{3617}{510},\quad B_{18}=\frac{43867}{798},\quad B_{20}=-\frac{174611}{330},\\
B_{22}=\frac{854513}{138},\quad B_{24}=-\frac{236364091}{2730},\quad B_{26}=\frac{8553103}{6},\ldots,B_{2n+1}=0 \quad (n\in\mathbb{N}).
\end{gathered} \tag{7}$$

The first few of the Bernoulli polynomials are given below:

$$\begin{aligned}
&B_0(x)=1, \quad B_1(x)=x-\frac{1}{2}, \quad B_2(x)=x^2-x+\frac{1}{6},\\
&B_3(x)=x^3-\frac{3}{2}x^2+\frac{1}{2}x, \quad B_4(x)=x^4-2x^3+x^2-\frac{1}{30},\\
&B_5(x)=x^5-\frac{5}{2}x^4+\frac{5}{3}x^3-\frac{1}{6}x,\\
&B_6(x)=x^6-3x^5+\frac{5}{2}x^4-\frac{1}{2}x^2+\frac{1}{42},\\
&B_7(x)=x^7-\frac{7}{2}x^6+\frac{7}{2}x^5-\frac{7}{6}x^3+\frac{1}{6}x, \ldots .
\end{aligned} \tag{8}$$

It is not difficult to derive the following identities for the Bernoulli polynomials:

$$B_n'(x)=nB_{n-1}(x) \quad (n\in\mathbb{N}); \tag{9}$$

$$B_n(1-x)=(-1)^nB_n(x) \quad (n\in\mathbb{N}_0); \tag{10}$$

$$(-1)^nB_n(-x)=B_n(x)+nx^{n-1} \quad (n\in\mathbb{N}_0). \tag{11}$$

Multiplication formula:

$$B_n(mx)=m^{n-1}\sum_{k=0}^{m-1}B_n\left(x+\frac{k}{m}\right) \quad (n\in\mathbb{N}_0,\ m\in\mathbb{N}). \tag{12}$$

Addition formula:

$$B_n(x+y)=\sum_{k=0}^{n}\binom{n}{k}B_k(x)\,y^{n-k} \quad (n\in\mathbb{N}_0). \tag{13}$$

Integral formulas:

$$\int_x^y B_n(t)\,dt=\frac{B_{n+1}(y)-B_{n+1}(x)}{n+1}; \tag{14}$$

$$\int_x^{x+1} B_n(t)\,dt=x^n; \tag{15}$$

$$\int_0^1 B_n(t)\,B_m(t)\,dt=(-1)^{n-1}\frac{m!\,n!}{(m+n)!}B_{m+n} \quad (m,\,n\in\mathbb{N}). \tag{16}$$

It follows from (14) and (15) that the finite sum of powers is expressed as Bernoulli polynomials and numbers:

$$\sum_{k=1}^{m} k^n = \frac{B_{n+1}(m+1) - B_{n+1}}{n+1} \quad (m,\, n \in \mathbb{N}). \tag{17}$$

By writing $2n$ for n in (3), we can deduce that

$$B_{2n}(x) + n\,x^{2n-1} = \sum_{k=0}^{n} \binom{2n}{2k} B_{2k}\, x^{2n-2k},$$

which, upon integrating from 0 to $\frac{1}{2}$, yields

$$\sum_{k=0}^{n} \frac{2^{2k} B_{2k}}{(2k)!(2n-2k+1)!} = \frac{1}{(2n)!} \quad (n \in \mathbb{N}_0), \tag{18}$$

where we have applied (14).

It is readily shown that

$$B_{2n}\left(\frac{1}{2}\right) = \left(2^{1-2n} - 1\right) B_{2n} \quad \text{and} \quad B_{2n+1}\left(\frac{1}{2}\right) = 0 \quad (n \in \mathbb{N}). \tag{19}$$

The Generalized Bernoulli Polynomials and Numbers

The *generalized Bernoulli polynomials* $B_n^{(\alpha)}(x)$ of degree n in x are defined by the generating function:

$$\left(\frac{z}{e^z - 1}\right)^{\alpha} e^{xz} = \sum_{n=0}^{\infty} B_n^{(\alpha)}(x) \frac{z^n}{n!} \quad \left(|z| < 2\pi;\ 1^{\alpha} := 1\right) \tag{20}$$

for arbitrary (real or complex) parameter α. Clearly, we have

$$B_n^{(\alpha)}(x) = (-1)^n B_n^{(\alpha)}(\alpha - x), \tag{21}$$

so that

$$B_n^{(\alpha)}(\alpha) = (-1)^n B_n^{(\alpha)}(0) =: (-1)^n B_n^{(\alpha)}, \tag{22}$$

in terms of the *generalized Bernoulli numbers* $B_n^{(\alpha)}$ defined by the generating function:

$$\left(\frac{z}{e^z - 1}\right)^{\alpha} = \sum_{n=0}^{\infty} B_n^{(\alpha)} \frac{z^n}{n!} \quad (|z| < 2\pi;\ 1^{\alpha} := 1). \tag{23}$$

It is easily observed that

$$B_n^{(1)}(x) = B_n(x) \quad \text{and} \quad B_n^{(1)} = B_n \quad (n \in \mathbb{N}_0). \tag{24}$$

From the generating function (20), it is fairly straightforward to deduce the addition theorem:

$$B_n^{(\alpha+\beta)}(x+y) = \sum_{k=0}^{n} \binom{n}{k} B_k^{(\alpha)}(x)\, B_{n-k}^{(\beta)}(y), \tag{25}$$

which, for $x = \beta = 0$, corresponds to the elegant representation:

$$B_n^{(\alpha)}(x) = \sum_{k=0}^{n} \binom{n}{k} B_k^{(\alpha)}\, x^{n-k} \tag{26}$$

for the generalized Bernoulli polynomials as a finite sum of the generalized Bernoulli numbers.

Srivastava *et al.* [1101, p. 442, Eqs. (4.4) and (4.5)] gave two new classes of addition theorems for the generalized Bernoulli polynomials:

$$B_n^{(\alpha+\lambda\gamma)}(x+\gamma y) = \sum_{k=0}^{n} \frac{\gamma+n}{\gamma+k} \binom{n}{k} B_k^{(\alpha-\lambda k)}(x-ky)\, B_{n-k}^{(\lambda k+\lambda\gamma)}(ky+\gamma y) \quad (\Re(\gamma) > 0); \tag{27}$$

$$B_n^{(\alpha+\beta+n+1)}(x+y+n) = \sum_{k=0}^{n} \binom{n}{k} B_k^{(\alpha+k+1)}(x+k)\, B_{n-k}^{(\beta+n-k+1)}(y+n-k). \tag{28}$$

Srivastava and Todorov [1110, p. 510, Eq. (3)] proved the following explicit formula for the generalized Bernoulli polynomials:

$$\begin{aligned} B_n^{(\alpha)}(x) = {} & \sum_{k=0}^{n} \binom{n}{k} \binom{\alpha+k-1}{k} \frac{k!}{(2k)!} \sum_{j=0}^{k} (-1)^j \binom{k}{j} j^{2k} (x+j)^{n-k} \\ & \cdot {}_2F_1[k-n,\, k-\alpha;\, 2k+1;\, j/(x+j)], \end{aligned} \tag{29}$$

in terms of the Gaussian hypergeometric function (see Section 1.5). They also applied the representation (29) to derive certain interesting special cases considered earlier by Gould [499] and Todorov [1153]. Indeed, by the Chu-Vandermonde theorem 1.5(9), we have

$${}_2F_1(-N, b; c; 1) = \binom{c-b+N-1}{N} \binom{c+N-1}{N}^{-1} \quad (N \in \mathbb{N}_0),$$

which, for $N = n-k$, $b = k-\alpha$ and $c = 2k+1$, readily yields

$$ {}_2F_1(k-n,\, k-\alpha;\, 2k+1;\, 1) = \binom{\alpha+n}{n-k} \frac{(n-k)!\,(2k)!}{(n+k)!} \qquad (0 \leq k \leq n). \tag{30}$$

In view of (30), the special case of the Srivastava-Todorov formula (29), when $x = 0$, gives the following representation for the generalized Bernoulli numbers:

$$B_n^{(\alpha)} = \sum_{k=0}^{n} \binom{\alpha+n}{n-k} \binom{\alpha+k-1}{k} \frac{n!}{(n+k)!} \sum_{j=0}^{k} (-1)^j \binom{k}{j} j^{n+k} \tag{31}$$

or, equivalently,

$$B_n^{(\alpha)} = \sum_{k=0}^{n} (-1)^k \binom{\alpha+n}{n-k} \binom{\alpha+k-1}{k} \frac{n!}{(n+k)!} \Delta^k 0^{n+k}, \tag{32}$$

where, for convenience,

$$\Delta^k a^r = \Delta^k x^r \Big|_{x=a} = \sum_{j=0}^{k} (-1)^{k-j} \binom{k}{j} (a+j)^r, \tag{33}$$

Δ being the difference operator defined by (*cf.* Comtet [337, p. 13 *et seq.*])

$$\Delta f(x) = f(x+1) - f(x), \tag{34}$$

so that, in general,

$$\Delta^k f(x) = \sum_{j=0}^{k} (-1)^{k-j} \binom{k}{j} f(x+j). \tag{35}$$

Alternatively, since (Comtet [337, p. 204, Theorem A]; see also Eq. 1.5(20))

$$S(n,k) = \frac{1}{k!} \Delta^k 0^n, \tag{36}$$

where $S(n,k)$ denotes the Stirling number of the second kind defined by 1.6(14), that is, by

$$z^n = \sum_{k=0}^{n} \binom{z}{k} k!\, S(n,k), \tag{37}$$

the representation (31) or (32) can be written also as (Todorov [1153, p. 665, Eq. (3)])

$$B_n^{(\alpha)} = \sum_{k=0}^{n} (-1)^k \binom{\alpha+n}{n-k} \binom{\alpha+k-1}{k} \binom{n+k}{k}^{-1} S(n+k,k). \tag{38}$$

Formula (38) provides an interesting generalization of the following known result for the Bernoulli numbers B_n:

$$B_n = \sum_{k=0}^{n} (-1)^k \binom{n+1}{k+1} \binom{n+k}{k}^{-1} \frac{\Delta^k 0^{n+k}}{k!}, \tag{39}$$

which was considered, for example, by Gould [499, p. 49, Eq. (17)].

Euler Polynomials and Numbers

The *Euler polynomials* $E_n(x)$ and the *Euler numbers* E_n are defined by the following generating functions:

$$\frac{2e^{xz}}{e^z+1} = \sum_{n=0}^{\infty} E_n(x) \frac{z^n}{n!} \quad (|z| < \pi) \tag{40}$$

and

$$\frac{2e^z}{e^{2z}+1} = \operatorname{sech} z = \sum_{n=0}^{\infty} E_n \frac{z^n}{n!} \quad \left(|z| < \frac{\pi}{2}\right), \tag{41}$$

respectively.

The following formulas are readily derivable from (40) and (41):

$$E_n(x+1) + E_n(x) = 2x^n \quad (n \in \mathbb{N}_0); \tag{42}$$

$$E_n'(x) = n E_{n-1}(x) \quad (n \in \mathbb{N}); \tag{43}$$

$$E_n(1-x) = (-1)^n E_n(x) \quad (n \in \mathbb{N}_0); \tag{44}$$

$$(-1)^{n+1} E_n(-x) = E_n(x) - 2x^n \quad (n \in \mathbb{N}_0); \tag{45}$$

$$E_n(x+y) = \sum_{k=0}^{n} \binom{n}{k} E_k(x) y^{n-k} \quad (n \in \mathbb{N}_0); \tag{46}$$

$$E_n(x) = \sum_{k=0}^{n} \binom{n}{k} \frac{E_k}{2^k} \left(x - \frac{1}{2}\right)^{n-k} \quad (n \in \mathbb{N}_0), \tag{47}$$

which, upon taking $x = \frac{1}{2}$, yields

$$E_n = 2^n E_n\left(\frac{1}{2}\right) \quad (n \in \mathbb{N}_0); \tag{48}$$

$$\sum_{k=0}^{n} \binom{2n}{2k} E_{2k} = 0 \quad (n \in \mathbb{N}). \tag{49}$$

Multiplication formulas:

$$E_n(mx) = m^n \sum_{k=0}^{m-1} (-1)^k E_n\left(x + \frac{k}{m}\right) \quad (n \in \mathbb{N}_0;\ m = 1, 3, 5, \ldots); \tag{50}$$

$$E_n(mx) = -\frac{2}{n+1} m^n \sum_{k=0}^{m-1} (-1)^k B_{n+1}\left(x + \frac{k}{m}\right) \quad (n \in \mathbb{N}_0;\ m = 2, 4, 6, \ldots). \tag{51}$$

Integral formulas:

$$\int_x^y E_n(t)\, dt = \frac{E_{n+1}(y) - E_{n+1}(x)}{n+1} \quad (m,\ n \in \mathbb{N}_0); \tag{52}$$

$$\int_0^1 E_n(t)\, E_m(t)\, dt = (-1)^n 4 \left(2^{m+n+2} - 1\right) \frac{m!\, n!}{(m+n+2)!} B_{m+n+2} \quad (m,\ n \in \mathbb{N}_0). \tag{53}$$

An alternating finite sum of powers can be expressed as the Euler polynomials:

$$\sum_{k=1}^{m} (-1)^{m-k} k^n = \frac{1}{2}\left[E_n(m+1) + (-1)^m E_n(0)\right] \quad (m,\ n \in \mathbb{N}). \tag{54}$$

Fourier Series Expansions of Bernoulli and Euler Polynomials

By employing suitable contour integrations in complex function theory, we can obtain the following Fourier series expansions of Bernoulli and Euler polynomials:

$$B_{2n}(x) = \frac{(-1)^{n-1}\, 2\, (2n)!}{(2\pi)^{2n}} \sum_{k=1}^{\infty} \frac{\cos\, 2k\pi x}{k^{2n}} \quad (n \in \mathbb{N};\ 0 \leqq x \leqq 1); \tag{55}$$

$$B_{2n-1}(x) = \frac{(-1)^n\, 2\, (2n-1)!}{(2\pi)^{2n-1}} \sum_{k=1}^{\infty} \frac{\sin\, 2k\pi x}{k^{2n-1}}$$
$$(n = 1 \text{ and } 0 < x < 1;\ n \in \mathbb{N} \setminus \{1\} \text{ and } 0 \leqq x \leqq 1); \tag{56}$$

$$E_{2n}(x) = \frac{(-1)^n\, 4\, (2n)!}{\pi^{2n+1}} \sum_{k=0}^{\infty} \frac{\sin\, (2k+1)\pi x}{(2k+1)^{2n+1}}$$
$$(n = 0 \text{ and } 0 < x < 1;\ n \in \mathbb{N} \text{ and } 0 \leqq x \leqq 1); \tag{57}$$

$$E_{2n-1}(x) = \frac{(-1)^n 4\,(2n-1)!}{\pi^{2n}} \sum_{k=0}^{\infty} \frac{\cos\,(2k+1)\pi x}{(2k+1)^{2n}} \quad (n \in \mathbb{N};\ 0 \leqq x \leqq 1), \tag{58}$$

which, for $x = \frac{1}{2}$, yields

$$E_{2n+1} = 0 \quad (n \in \mathbb{N}_0), \tag{59}$$

where use is made of the relationship (48).

The first few of the Euler numbers E_n are given below:

$$E_0 = 1,\ E_2 = -1,\ E_4 = 5,\ E_6 = -61,\ E_8 = 1385,\ E_{10} = -50521,\ \ldots. \tag{60}$$

Relations Between Bernoulli and Euler Polynomials

The following relationships between the Bernoulli and Euler polynomials follow easily from the definitions (1) and (40):

$$\begin{aligned} E_n(x) &= \frac{2^{n+1}}{n+1}\left\{B_{n+1}\left(\frac{x+1}{2}\right) - B_{n+1}\left(\frac{x}{2}\right)\right\} \\ &= \frac{2}{n+1}\left\{B_{n+1}(x) - 2^{n+1} B_{n+1}\left(\frac{x}{2}\right)\right\} \quad (n \in \mathbb{N}_0), \end{aligned} \tag{61}$$

which, in view of (10), can also be written in the form:

$$E_n(1-x) = (-1)^n \frac{2}{n+1}\left[2^{n+1} B_{n+1}\left(\frac{x+1}{2}\right) - B_{n+1}(x)\right] \quad (n \in \mathbb{N}_0). \tag{62}$$

Two additional formulas involving these polynomials are given below:

$$E_{n-2}(x) = 2\binom{n}{2}^{-1} \sum_{k=0}^{n-2} \binom{n}{k} \left(2^{n-k} - 1\right) B_{n-k} B_k(x) \quad (n \in \mathbb{N} \setminus \{1\}); \tag{63}$$

$$B_n(x) = 2^{-n} \sum_{k=0}^{n} \binom{n}{k} B_{n-k} E_k(2x) \quad (n \in \mathbb{N}_0). \tag{64}$$

The Generalized Euler Polynomials and Numbers

The *generalized Euler polynomials* $E_n^{(\alpha)}(x)$ and the *generalized Euler numbers* $E_n^{(\alpha)}$ are defined by the generating functions:

$$\left(\frac{2}{e^z+1}\right)^{\alpha} \exp(xz) = \sum_{n=0}^{\infty} E_n^{(\alpha)}(x) \frac{z^n}{n!} \quad (|z| < \pi;\ 1^{\alpha} := 1) \tag{65}$$

and

$$\left(\frac{2\,e^{z}}{e^{2z}+1}\right)^{\alpha}=\sum_{n=0}^{\infty}E_n^{(\alpha)}\,\frac{z^n}{n!}\qquad\left(|z|<\frac{\pi}{2};\;1^{\alpha}:=1\right)\tag{66}$$

for arbitrary (real or complex) parameter α. Clearly, we have

$$E_n^{(1)}(x)=E_n(x)\quad\text{and}\quad E_n^{(1)}=E_n\quad(n\in\mathbb{N}_0).\tag{67}$$

It is easy to find from (65) and (66) that

$$2^n\,E_n^{(\alpha)}\left(\frac{\alpha}{2}\right)=E_n^{(\alpha)}\quad(n\in\mathbb{N}_0),\tag{68}$$

which, for $\alpha=1$, reduces to (48).

Srivastava *et al.* [1101, p. 443, Eq. (4.12)] proved the following interesting addition theorem for the generalized Euler polynomials:

$$E_n^{(\alpha+\lambda\gamma)}(x+\gamma y)=\sum_{k=0}^{n}\frac{\gamma+n}{\gamma+k}\binom{n}{k}E_k^{(\alpha-\lambda k)}(x-ky)\,E_{n-k}^{(\lambda k+\lambda\gamma)}(ky+\gamma y)\quad(\Re(\gamma)>0).\tag{69}$$

From the generating functions (20) and (65), it is easily seen that

$$B_n^{(0)}(x)=E_n^{(0)}(x)=x^n\quad(n\in\mathbb{N}_0).\tag{70}$$

Recently, by making use of some fairly standard techniques based on series rearrangement, Srivastava and Pintér [1105] derived each of the following elegant theorems (*cf.* [1105, p. 379, Theorem 1; p. 380, Theorem 2]). The following relationship (*cf.* [1105, p. 379, Theorem 1]):

$$B_n^{(\alpha)}(x+y)=\sum_{k=0}^{n}\binom{n}{k}\left[B_k^{(\alpha)}(y)+\frac{k}{2}B_{k-1}^{(\alpha-1)}(y)\right]E_{n-k}(x)(\alpha\in\mathbb{C};\;n\in\mathbb{N}_0)\tag{71}$$

holds true between the generalized Bernoulli polynomials and the classical Euler polynomials.

The following relationship (*cf.* [1105, p. 380, Theorem 2]):

$$E_n^{(\alpha)}(x+y)=\sum_{k=0}^{n}\frac{2}{k+1}\binom{n}{k}\left[E_{k+1}^{(\alpha-1)}(y)-E_{k+1}^{(\alpha)}(y)\right]B_{n-k}(x)(\alpha\in\mathbb{C};\;n\in\mathbb{N}_0),\tag{72}$$

holds true between the generalized Euler polynomials and the classical Bernoulli polynomials.

Upon setting $\alpha = 1$ in (71), if we let $y \to 0$ and make use of (70), we can deduce the aforementioned *main* relationship in Cheon's work (*cf.* [254, p. 368, Theorem 3]):

$$B_n(x) = \sum_{\substack{k=0 \\ (k \neq 1)}}^{n} \binom{n}{k} B_k E_{n-k}(x) \quad (n \in \mathbb{N}_0), \tag{73}$$

just as it was accomplished by Srivastava and Pintér [1105, p. 379].

Genocchi Polynomials and Numbers

The Genocchi polynomials $G_n^{(\alpha)}(x)$ of (real or complex) order α are usually defined by means of the following generating function:

$$\left(\frac{2z}{e^z+1}\right)^{\alpha} \cdot e^{xz} = \sum_{n=0}^{\infty} G_n^{(\alpha)}(x) \frac{z^n}{n!} \qquad \left(|z| < \pi;\ 1^{\alpha} := 1\right), \tag{74}$$

so that, obviously, the classical Genocchi polynomials $G_n(x)$, given by

$$G_n(x) := G_n^{(1)}(x) \qquad (n \in \mathbb{N}_0), \tag{75}$$

are defined by the following generating function:

$$\frac{2ze^{xz}}{e^z+1} = \sum_{n=0}^{\infty} G_n(x) \frac{z^n}{n!} \qquad (|z| < \pi). \tag{76}$$

For the classical Genocchi numbers G_n, we have (see also Problem 54)

$$G_n := G_n(0) = G_n^{(1)}(0). \tag{77}$$

Finally, in light of the definition (74), we find for the Genocchi numbers $G_n^{(\alpha)}$ of (real or complex) order α that

$$\left(\frac{2z}{e^z+1}\right)^{\alpha} = \sum_{n=0}^{\infty} G_n^{(\alpha)} \frac{z^n}{n!} \qquad \left(|z| < \pi;\ 1^{\alpha} := 1\right), \tag{78}$$

so that, just as we observed above in Equation (77),

$$G_n = G_n^{(1)} \quad (n \in \mathbb{N}_0).$$

Various properties and characteristics of the above-defined Genocchi polynomials and numbers, analogous to those that hold true for the Bernoulli and Euler polynomials and numbers, can be found in the widely-scattered literature, which is cited by (for example) Luo and Srivastava [791] (see also Section 1.8 below).

1.8 Apostol-Bernoulli, Apostol-Euler and Apostol-Genocchi Polynomials and Numbers

Apostol-Bernoulli Polynomials and Numbers

Some interesting analogues of the classical Bernoulli polynomials and numbers were investigated by Apostol [58, p. 165, Eq. (3.1)] and (more recently) by Srivastava [1087, pp. 83–84]. We begin by recalling, here, Apostol's definitions as follows:

The Apostol-Bernoulli polynomials $\mathcal{B}_n(x;\lambda)$ are defined by means of the generating function (Apostol [58]; see also Srivastava [1087]):

$$\frac{ze^{xz}}{\lambda e^z-1}=\sum_{n=0}^{\infty}\mathcal{B}_n(x;\lambda)\,\frac{z^n}{n!} \tag{1}$$

$$(|z|<2\pi, \quad \text{when} \quad \lambda=1;\ |z|<|\log\lambda|, \quad \text{when} \quad \lambda\neq 1)$$

with, of course,

$$B_n(x)=\mathcal{B}_n(x;1) \qquad \text{and} \qquad \mathcal{B}_n(\lambda):=\mathcal{B}_n(0;\lambda) \tag{2}$$

where $\mathcal{B}_n(\lambda)$ denotes the so-called Apostol-Bernoulli numbers.

Under the assumption $\lambda\neq 1$, Apostol [58] gave the main properties of $\mathcal{B}_n(a,\lambda)$, including, for example, the summation formula:

$$\mathcal{B}_n(a,\lambda)=\sum_{k=0}^{n}\binom{n}{k}\mathcal{B}_k(\lambda)\,a^{n-k} \quad (n\in\mathbb{N}_0), \tag{3}$$

from which it is seen that $\mathcal{B}_n(a,\lambda)$ are polynomials in a, satisfying the difference equation:

$$\lambda\,\mathcal{B}_n(a+1,\lambda)-\mathcal{B}_n(a,\lambda)=n\,a^{n-1} \quad (n\in\mathbb{N}), \tag{4}$$

which yields the following special cases:

$$\lambda\,\mathcal{B}_1(1,\lambda)=1+\mathcal{B}_1(\lambda) \tag{5}$$

and

$$\lambda\,\mathcal{B}_n(1,\lambda)=\mathcal{B}_n(\lambda) \quad (n\in\mathbb{N}\backslash\{1\}). \tag{6}$$

Equations (5), (6) and (3), with $a = 1$, together, aid us in computing $\mathcal{B}_n(\lambda)$ recursively. Thus, we have

$$\begin{aligned}
&\mathcal{B}_0(\lambda) = 0, \quad \mathcal{B}_1(\lambda) = \frac{1}{\lambda - 1}, \quad \mathcal{B}_2(\lambda) = -\frac{2\lambda}{(\lambda-1)^2}, \quad \mathcal{B}_3(\lambda) = \frac{3\lambda(\lambda+1)}{(\lambda-1)^3},\\
&\mathcal{B}_4(\lambda) = -\frac{4\lambda\left(\lambda^2+4\lambda+1\right)}{(\lambda-1)^4}, \quad \mathcal{B}_5(\lambda) = \frac{5\lambda\left(\lambda^3+11\lambda^2+11\lambda+1\right)}{(\lambda-1)^5},\\
&\mathcal{B}_6(\lambda) = -\frac{6\lambda\left(\lambda^4+26\lambda^3+66\lambda^2+26\lambda+1\right)}{(\lambda-1)^6}
\end{aligned} \tag{7}$$

and (in general)

$$\mathcal{B}_n(\lambda) = \frac{n\lambda}{(\lambda-1)^n} \sum_{k=1}^{n-1} (-1)^k k!\, \lambda^{k-1} (\lambda-1)^{n-1-k} S(n-1,k), \tag{8}$$

in terms of the Stirling numbers of the second kind (see Section 1.6).

We recall here several further properties of the functions $\mathcal{B}_n(a,\lambda)$ as follows:

$$\frac{\partial^p}{\partial a^p}\{\mathcal{B}_n(a,\lambda)\} = \frac{n!}{(n-p)!}\,\mathcal{B}_{n-p}(a,\lambda) \quad (p = 0,\, 1,\, \ldots,\, n); \tag{9}$$

$$\mathcal{B}_n(a+b,\lambda) = \sum_{k=0}^{n} \binom{n}{k} \mathcal{B}_k(a,\lambda)\, b^{n-k}; \tag{10}$$

$$\int_a^b \mathcal{B}_n(t,\lambda)\, dt = \frac{\mathcal{B}_{n+1}(b,\lambda) - \mathcal{B}_{n+1}(a,\lambda)}{n+1}; \tag{11}$$

$$\sum_{k=0}^{m-1} k^n = \frac{\lambda-1}{n+1} \sum_{k=1}^{m} \mathcal{B}_{n+1}(k,\lambda) + \frac{\mathcal{B}_{n+1}(m,\lambda) - \mathcal{B}_{n+1}(\lambda)}{n+1}, \tag{12}$$

which obviously generalizes the familiar result 1.7(17).

Motivated by the success of the generalizations in 1.7(20) and 1.7(66) of the classical Bernoulli polynomials and the classical Euler polynomials involving a real or complex parameter α, Luo and Srivastava [788, 789] introduced and investigated the so-called Apostol-Bernoulli polynomials $\mathcal{B}_n^{(\alpha)}(x;\lambda)$ of order α and the Apostol-Euler polynomials $\mathcal{E}_n^{(\alpha)}(x;\lambda)$ of order α, which are defined as follows:

The Apostol-Bernoulli polynomials $\mathcal{B}_n^{(\alpha)}(x;\lambda)$ of (real or complex) order α are defined by means of the following generating function:

$$\left(\frac{z}{\lambda e^z - 1}\right)^{\alpha} \cdot e^{xz} = \sum_{n=0}^{\infty} \mathcal{B}_n^{(\alpha)}(x;\lambda)\, \frac{z^n}{n!} \tag{13}$$

$$\left(|z| < 2\pi, \quad \text{when} \quad \lambda = 1; \; |z| < |\log\lambda|, \quad \text{when} \quad \lambda \neq 1; \; 1^{\alpha} := 1\right),$$

with, of course,

$$B_n^{(\alpha)}(x) = \mathcal{B}_n^{(\alpha)}(x;1) \quad \text{and} \quad \mathcal{B}_n^{(\alpha)}(\lambda) := \mathcal{B}_n^{(\alpha)}(0;\lambda), \tag{14}$$

where $\mathcal{B}_n^{(\alpha)}(\lambda)$ denotes the so-called Apostol-Bernoulli numbers of order α.

The Apostol-Euler polynomials $\mathcal{E}_n^{(\alpha)}(x;\lambda)$ of (real or complex) order α are defined by means of the following generating function (*cf.* Luo [781]):

$$\left(\frac{2}{\lambda e^z + 1}\right)^{\alpha} \cdot e^{xz} = \sum_{n=0}^{\infty} \mathcal{E}_n^{(\alpha)}(x;\lambda)\,\frac{z^n}{n!} \qquad \big(|z| < |\log(-\lambda)|;\ 1^{\alpha} := 1\big), \tag{15}$$

with, of course,

$$E_n^{(\alpha)}(x) = \mathcal{E}_n^{(\alpha)}(x;1) \quad \text{and} \quad \mathcal{E}_n^{(\alpha)}(\lambda) := \mathcal{E}_n^{(\alpha)}(0;\lambda), \tag{16}$$

where $\mathcal{E}_n^{(\alpha)}(\lambda)$ denotes the so-called Apostol-Euler numbers of order α.

Luo and Srivastava [788, 789] presented a variety of properties and relations between other mathematical functions. Among them, we choose to record, here, the following results:

The Apostol-Euler polynomials $\mathcal{E}_n^{(\alpha)}(x;\lambda)$ of order α is represented by

$$\mathcal{E}_n^{(\alpha)}(x;\lambda) = e^{-x\log\lambda} \sum_{k=0}^{\infty} E_{n+k}^{(\alpha)}(x)\,\frac{(\log\lambda)^k}{k!} \qquad (n \in \mathbb{N}_0) \tag{17}$$

in series of the familiar Euler polynomials $E_n^{(\alpha)}(x)$ of order α.

The Aposto-Bernoulli polynomials $\mathcal{B}_n^{(\ell)}(x;\lambda)$ of order ℓ is represented by

$$\mathcal{B}_n^{(\ell)}(x;\lambda) = e^{-x\log\lambda} \sum_{k=0}^{\infty} \binom{n+k-\ell}{k} \binom{n+k}{k}^{-1} B_{n+k}^{(\ell)}(x)\,\frac{(\log\lambda)^k}{k!} \qquad (n,\ \ell \in \mathbb{N}_0) \tag{18}$$

in series of the familiar Bernoulli polynomials $B_n^{(\ell)}(x)$ of order ℓ.

An explicit representation of $\mathcal{B}_n^{(\ell)}(\lambda)$ is given:

$$\mathcal{B}_n^{(\ell)}(\lambda) = \ell!\binom{n}{\ell} \sum_{k=0}^{n-\ell} \binom{\ell+k-1}{k} \frac{k!\,(-\lambda)^k}{(\lambda-1)^{k+\ell}}\, S(n-\ell,k), \qquad \big(n,\ \ell \in \mathbb{N}_0;\ \lambda \in \mathbb{C}\setminus\{1\}\big). \tag{19}$$

An explicit formula for the Apostol-Bernoulli polynomials $\mathcal{B}_n^{(\alpha)}(x;\lambda)$ involving the Stirling numbers of the second kind is also given as follows:

$$\mathcal{B}_n^{(\ell)}(x;\lambda)=\ell!\sum_{k=\ell}^{n}\binom{n}{k}\binom{k}{\ell}x^{n-k}\sum_{j=0}^{k-\ell}\binom{\ell+j-1}{j}\frac{j!\,(-\lambda)^j}{(\lambda-1)^{j+\ell}}\,S(k-\ell,j) \tag{20}$$

$$\left(n,\,\ell\in\mathbb{N}_0;\ \lambda\in\mathbb{C}\setminus\{1\}\right).$$

The following relationship:

$$\mathcal{B}_n^{(\alpha)}(x;\lambda)=\sum_{k=0}^{n}\binom{n}{k}\mathcal{B}_{n-k}^{(\alpha-1)}(\lambda)\,\mathcal{B}_k(x;\lambda)\qquad(n\in\mathbb{N}_0) \tag{21}$$

holds true between the Apostol-Bernoulli polynomials $\mathcal{B}_n^{(\alpha)}(x;\lambda)$ of order α and the Apostol-Bernoulli numbers $\mathcal{B}_n^{(\alpha-1)}(\lambda)$ of order $\alpha-1$.

Let $n\in\mathbb{N}_0$. Suppose also that α and λ are suitable (real or complex) parameters. Then,

$$\mathcal{B}_n^{(\alpha)}(x;\lambda)=\sum_{k=0}^{n}\binom{n}{k}\mathcal{B}_k^{(\alpha)}(\lambda)\,x^{n-k}\quad\textit{and}\quad\mathcal{B}_n^{(0)}(x;\lambda)=x^n, \tag{22}$$

$$\mathcal{E}_n^{(\alpha)}(x;\lambda)=\sum_{k=0}^{n}\binom{n}{k}\frac{\mathcal{E}_k^{(\alpha)}(\lambda)}{2^k}\left(x-\frac{\alpha}{2}\right)^{n-k}, \tag{23}$$

$$\lambda\mathcal{B}_n^{(\alpha)}(x+1;\lambda)-\mathcal{B}_n^{(\alpha)}(x;\lambda)=n\mathcal{B}_{n-1}^{(\alpha-1)}(x;\lambda), \tag{24}$$

$$\frac{\partial}{\partial x}\left\{\mathcal{B}_n^{(\alpha)}(x;\lambda)\right\}=n\mathcal{B}_{n-1}^{(\alpha)}(x;\lambda), \tag{25}$$

$$\int_a^b\mathcal{B}_n^{(\alpha)}(x;\lambda)\,,dx=\frac{\mathcal{B}_{n+1}^{(\alpha)}(b;\lambda)-\mathcal{B}_{n+1}^{(\alpha)}(a;\lambda)}{n+1}, \tag{26}$$

$$\mathcal{B}_n^{(\alpha+\beta)}(x+y;\lambda)=\sum_{k=0}^{n}\binom{n}{k}\mathcal{B}_k^{(\alpha)}(x;\lambda)\,\mathcal{B}_{n-k}^{(\beta)}(y;\lambda), \tag{27}$$

$$\mathcal{E}_n^{(\alpha+\beta)}(x+y;\lambda)=\sum_{k=0}^{n}\binom{n}{k}\mathcal{E}_k^{(\alpha)}(x;\lambda)\mathcal{E}_{n-k}^{(\beta)}(y;\lambda), \tag{28}$$

$$\mathcal{B}_n^{(\alpha)}(\alpha-x;\lambda)=\frac{(-1)^n}{\lambda^\alpha}\mathcal{B}_n^{(\alpha)}\left(x;\lambda^{-1}\right), \tag{29}$$

$$\mathcal{E}_n^{(\alpha)}(\alpha-x;\lambda)=\frac{(-1)^n}{\lambda^\alpha}\mathcal{E}_n^{(\alpha)}(x;\lambda^{-1}), \tag{30}$$

$$\mathcal{B}_n^{(\alpha)}(\alpha+x;\lambda)=\frac{(-1)^n}{\lambda^\alpha}\mathcal{B}_n^{(\alpha)}\left(-x;\lambda^{-1}\right), \tag{31}$$

$$nx\mathcal{B}_{n-1}^{(\alpha)}(x;\lambda)=(n-\alpha)\mathcal{B}_n^{(\alpha)}(x;\lambda)+\alpha\lambda\mathcal{B}_n^{(\alpha+1)}(x+1;\lambda) \tag{32}$$

and

$$\mathcal{B}_n^{(\alpha+1)}(x;\lambda) = \left(1-\frac{n}{\alpha}\right)\mathcal{B}_n^{(\alpha)}(x;\lambda) + n\left(\frac{x}{\alpha}-1\right)\mathcal{B}_{n-1}^{(\alpha)}(x;\lambda). \tag{33}$$

From the generating functions (13) and (15), it follows also that (see [788] and [781])

$$\lambda\mathcal{B}_n^{(\alpha)}(x+1;\lambda) - \mathcal{B}_n^{(\alpha)}(x;\lambda) = n\mathcal{B}_{n-1}^{(\alpha-1)}(x;\lambda) \tag{34}$$

and

$$\lambda\mathcal{E}_n^{(\alpha)}(x+1;\lambda) + \mathcal{E}_n^{(\alpha)}(x;\lambda) = 2\mathcal{E}_n^{(\alpha-1)}(x;\lambda), \tag{35}$$

respectively. Now, since

$$\mathcal{B}_n^{(0)}(x;\lambda) = \mathcal{E}_n^{(0)}(x;\lambda) = x^n \quad (n \in \mathbb{N}_0), \tag{36}$$

upon setting $\beta = 0$ in the addition theorems (27) *and* (28), if we interchange x and y, we obtain

$$\mathcal{B}_n^{(\alpha)}(x+y;\lambda) = \sum_{k=0}^{n}\binom{n}{k}\mathcal{B}_k^{(\alpha)}(y;\lambda)x^{n-k} \tag{37}$$

and

$$\mathcal{E}_n^{(\alpha)}(x+y;\lambda) = \sum_{k=0}^{n}\binom{n}{k}\mathcal{E}_k^{(\alpha)}(y;\lambda)x^{n-k}, \tag{38}$$

respectively.

Next, by combining (34) and (37) (with $x = 1$ *and* $y \longmapsto x$), we find that

$$\mathcal{B}_n^{(\alpha-1)}(x;\lambda) = \frac{1}{n+1}\left[\lambda\sum_{k=0}^{n+1}\binom{n+1}{k}\mathcal{B}_k^{(\alpha)}(x;\lambda) - \mathcal{B}_{n+1}^{(\alpha)}(x;\lambda)\right] \quad (n \in \mathbb{N}_0), \tag{39}$$

which, in the special case when $\alpha = 1$, yields the following expansion:

$$x^n = \frac{1}{n+1}\left[\lambda\sum_{k=0}^{n+1}\binom{n+1}{k}\mathcal{B}_k(x;\lambda) - \mathcal{B}_{n+1}(x;\lambda)\right] \quad (n \in \mathbb{N}_0), \tag{40}$$

in series of the Apostol-Bernoulli polynomials $\{\mathcal{B}_n(x;\lambda)\}_{n=0}^{\infty}$.

In the special case of (40), when $\lambda = 1$, we obtain the following familiar expansion (*cf.*, e.g., [795, p. 26]):

$$x^n = \frac{1}{n+1}\sum_{k=0}^{n}\binom{n+1}{k}B_k(x) \quad (n \in \mathbb{N}_0) \tag{41}$$

in series of the classical Bernoulli polynomials $\{B_n(x)\}_{n=0}^{\infty}$.

In precisely the same manner, the addition theorem (38) in conjunction with (35) would lead us to the following companions of (39) and (40):

$$\mathcal{E}_n^{(\alpha-1)}(x;\lambda) = \frac{1}{2}\left[\lambda \sum_{k=0}^{n} \binom{n}{k} \mathcal{E}_k^{(\alpha)}(x;\lambda) + \mathcal{E}_n^{(\alpha)}(x;\lambda)\right] \quad (n \in \mathbb{N}_0) \tag{42}$$

and

$$x^n = \frac{1}{2}\left[\lambda \sum_{k=0}^{n} \binom{n}{k} \mathcal{E}_k(x;\lambda) + \mathcal{E}_n(x;\lambda)\right] \quad (n \in \mathbb{N}_0). \tag{43}$$

In view of (36), this last expansion (43) in series of the Apostol-Euler polynomials $\{\mathcal{E}_n(x;\lambda)\}_{n=0}^{\infty}$ is indeed an immediate consequence of (42), when $\alpha = 1$.

By using (13) (with $\alpha = 1$) and (15) (with $\alpha = 1$), we have

$$\begin{aligned}
\sum_{n=0}^{\infty} \mathcal{B}_n(x;\lambda^2)\frac{t^n}{n!} &= \frac{te^{xt}}{\lambda^2 e^t - 1} = \frac{t/2}{\lambda e^{t/2} - 1} \cdot \frac{2e^{xt}}{\lambda e^{t/2} + 1} \\
&= \sum_{n=0}^{\infty} 2^{-n} \mathcal{B}_n(\lambda) \frac{t^n}{n!} \cdot \sum_{n=0}^{\infty} 2^{-n} \mathcal{E}_n(2x;\lambda) \frac{t^n}{n!} \\
&= \sum_{n=0}^{\infty} \left[2^{-n} \sum_{k=0}^{n} \binom{n}{k} \mathcal{B}_{n-k}(\lambda) \mathcal{E}_k(2x;\lambda)\right] \frac{t^n}{n!},
\end{aligned}$$

which yields the following relationship between the Apostol-Bernoulli and Apostol-Euler polynomials:

$$\mathcal{B}_n(x;\lambda^2) = 2^{-n} \sum_{k=0}^{n} \binom{n}{k} \mathcal{B}_{n-k}(\lambda) \mathcal{E}_k(2x;\lambda) \tag{44}$$

or, equivalently,

$$2^n \mathcal{B}_n\left(\frac{x}{2};\lambda^2\right) = \sum_{k=0}^{n} \binom{n}{k} \mathcal{B}_k(\lambda) \mathcal{E}_{n-k}(x;\lambda). \tag{45}$$

By applying similar arguments, it is not difficult to get the following explicit representation for the Apostol-Euler polynomials $\mathcal{E}_n(x;\lambda)$ in terms of the Apostol-Bernoulli polynomials $\mathcal{B}_n(x;\lambda)$:

$$\mathcal{E}_{n-1}(x;\lambda) = \frac{2^n}{n}\left[\lambda \mathcal{B}_n\left(\frac{x+1}{2};\lambda^2\right) - \mathcal{B}_n\left(\frac{x}{2};\lambda^2\right)\right] \tag{46}$$

or, equivalently,

$$\mathcal{E}_{n-1}(x;\lambda) = \frac{2}{n}\left[\mathcal{B}_n(x;\lambda) - 2^n\mathcal{B}_n\left(\frac{x}{2};\lambda^2\right)\right]. \tag{47}$$

In addition, from the relationships (46) (with $x = 0$) and (47) (with $x = 0$), we find that

$$\lambda\mathcal{B}_n\left(\frac{1}{2};\lambda^2\right) = 2^{-n}\mathcal{B}_n(\lambda) + n\cdot 2^{-n-1}\mathcal{E}_{n-1}(0;\lambda). \tag{48}$$

Thus, by substituting for $\mathcal{E}_{n-1}(0;\lambda)$ from (47) (with $x = 0$) into (48), we obtain the above-asserted relationship:

$$\begin{aligned}\mathcal{B}_n(\lambda) := \mathcal{B}_n(0;\lambda) &= \frac{(-1)^n}{\lambda}\mathcal{B}_n\left(1;\frac{1}{\lambda}\right)\\ &= 2^n\left[\frac{1}{2}\mathcal{B}_n(\lambda^2) + \frac{\lambda}{2}\mathcal{B}_n\left(\frac{1}{2};\lambda^2\right)\right] \quad (n \in \mathbb{N}_0),\end{aligned} \tag{49}$$

that is,

$$\mathcal{B}_n(\lambda) = 2^{n-1}\left[\mathcal{B}_n(\lambda^2) + \lambda\mathcal{B}_n\left(\frac{1}{2};\lambda^2\right)\right] \quad (n \in \mathbb{N}_0). \tag{50}$$

The following relationship:

$$\mathcal{B}_n^{(\alpha)}(x+y;\lambda) = \sum_{k=0}^{n}\binom{n}{k}\left[\mathcal{B}_k^{(\alpha)}(y;\lambda) + \frac{k}{2}\mathcal{B}_{k-1}^{(\alpha-1)}(y;\lambda)\right]\mathcal{E}_{n-k}(x;\lambda) \tag{51}$$

$$(\alpha,\lambda \in \mathbb{C};\ n \in \mathbb{N}_0)$$

holds true between the generalized Apostol-Bernoulli polynomials and the Apostol-Euler polynomials.

Luo [781] obtained the following general recursion formulas for the generalized Apostol-Euler polynomials $\mathcal{E}_n^{(\alpha)}(x;\lambda)$ and the generalized Apostol-Euler numbers $\mathcal{E}_n^{(\alpha)}(\lambda)$ (see [781, Equations (20) and (29)]):

$$\mathcal{E}_n^{(\alpha)}(x;\lambda) = 2^\alpha\sum_{k=0}^{n}\binom{n}{k}x^{n-k}\sum_{j=0}^{k}\binom{\alpha+j-1}{j}\frac{j!(-\lambda)^j}{(\lambda+1)^{j+\alpha}}S(k,j) \qquad (\alpha,\lambda \in \mathbb{C};\ n \in \mathbb{N}_0) \tag{52}$$

and

$$\mathcal{E}_n^{(\alpha)}(\lambda) = (-1)^n\sum_{k=0}^{n}\binom{n}{k}2^{k+\alpha}\alpha^{n-k}\sum_{j=0}^{k}\binom{\alpha+j-1}{j}\frac{j!(-\lambda)^j}{(\lambda+1)^{j+\alpha}}S(k,j) \qquad (\alpha,\lambda \in \mathbb{C};\ n \in \mathbb{N}_0). \tag{53}$$

Luo and Srivastava [789] gave an addition formula for each of the generalized Apostol-Bernoulli and the generalized Apostol-Euler polynomials:

$$\mathcal{B}_n^{(\alpha)}(x+y;\lambda) = \sum_{k=0}^{n} \binom{x}{k} k! \sum_{j=0}^{n-k} \binom{n}{j} \mathcal{B}_j^{(\alpha)}(y;\lambda) S(n-j,k) \tag{54}$$

and

$$\mathcal{E}_n^{(\alpha)}(x+y;\lambda) = \sum_{k=0}^{n} \binom{x}{k} k! \sum_{j=0}^{n-k} \binom{n}{j} \mathcal{E}_j^{(\alpha)}(y;\lambda) S(n-j,k) \tag{55}$$

$$(\alpha, \lambda \in \mathbb{C};\ n \in \mathbb{N}_0)$$

hold true between the generalized Apostol-Bernoulli polynomials, the generalized Apostol-Euler polynomials and the Stirling numbers of the second kind. By setting $\lambda = 1$ in (54) and (55), it is easy to deduce the following interesting identities:

$$B_n^{(\alpha)}(x+y) = \sum_{k=0}^{n} \binom{x}{k} k! \sum_{j=0}^{n-k} \binom{n}{j} B_j^{(\alpha)}(y) S(n-j,k) \quad (\alpha \in \mathbb{C};\ n \in \mathbb{N}_0) \tag{56}$$

and

$$E_n^{(\alpha)}(x+y) = \sum_{k=0}^{n} \binom{x}{k} k! \sum_{j=0}^{n-k} \binom{n}{j} E_j^{(\alpha)}(y) S(n-j,k) \quad (\alpha \in \mathbb{C};\ n \in \mathbb{N}_0) \tag{57}$$

for the generalized Bernoulli polynomials and the generalized Euler polynomials of order α.

Apostol-Genocchi Polynomials and Numbers

Since the publication of the works by Luo and Srivastava (see [780], [781], [788] and [789]), many further investigations of the above-mentioned Apostol type polynomials have appeared in the literature. Boyadzhiev [162] gave some properties and representations of the Apostol-Bernoulli polynomials and the Eulerian polynomials. Garg *et al.* [467] studied the Apostol-Bernoulli polynomials of order α and obtained some new relations and formulas involving the Apostol type polynomials and the Hurwitz (or generalized) zeta function $\zeta(s,a)$ defined by 2.2(1) below. Luo (see [782] and [783]) obtained the Fourier expansions and integral representations for the Apostol-Bernoulli and the Apostol-Euler polynomials and gave the multiplication formulas for the Apostol-Bernoulli and the Apostol-Euler polynomials of order α. Prévost [915] investigated the Apostol-Bernoulli and the Apostol-Euler polynomials by using the

Padé approximation methods. Wang *et al.* (see [1207] and [1208]) further developed some results of Luo and Srivastava [789] and obtained some formulas involving power sums of the Apostol type polynomials. Zhang and Yang [1258] gave several identities for the generalized Apostol-Bernoulli polynomials. Conversely, Cenkci and Can [227] gave a q-analogue of the Apostol-Bernoulli polynomials $\mathcal{B}_n(x;\lambda)$. Choi *et al.* [267] gave the q-extensions of the Apostol-Bernoulli polynomials of order α and the Apostol-Euler polynomials of order α (see also [268]). Hwang *et al.* [579] and Kim *et al.* [663] also gave q-extensions of Apostol's type Euler polynomials.

On the subject of the Genocchi polynomials $G_n(x)$ and their various extensions, a remarkably large number of investigations have appeared in the literature (see, e.g., [229], [268], [566], [567], [568], [604], [655], [656], [657], [720], [721], [722], [773], [784], [785], [786], [886] and [1257]; see also the references cited in each of these works). Moreover, Luo (see [784] and [786]) introduced and investigated the Apostol-Genocchi polynomials of (real or complex) order α, which are defined as follows.

The Apostol-Genocchi polynomials

$$\mathcal{G}_n^{(\alpha)}(x;\lambda) \quad (\lambda \in \mathbb{C})$$

of (real or complex) order α are defined by means of the following generating function:

$$\left(\frac{2z}{\lambda e^z+1}\right)^{\alpha} \cdot e^{xz} = \sum_{n=0}^{\infty} \mathcal{G}_n^{(\alpha)}(x;\lambda)\frac{z^n}{n!} \quad \big(|z| < |\log(-\lambda)|;\ 1^{\alpha} := 1\big) \tag{58}$$

with, of course,

$$G_n^{(\alpha)}(x) = \mathcal{G}_n^{(\alpha)}(x;1), \quad \mathcal{G}_n^{(\alpha)}(\lambda) := \mathcal{G}_n^{(\alpha)}(0;\lambda), \tag{59}$$

$$\mathcal{G}_n(x;\lambda) := \mathcal{G}_n^{(1)}(x;\lambda) \quad \text{and} \quad \mathcal{G}_n(\lambda) := \mathcal{G}_n^{(1)}(\lambda), \tag{60}$$

where $\mathcal{G}_n(\lambda)$, $\mathcal{G}_n^{(\alpha)}(\lambda)$ and $\mathcal{G}_n(x;\lambda)$ denote the so-called Apostol-Genocchi numbers, the Apostol-Genocchi numbers of order α and the Apostol-Genocchi polynomials, respectively.

Important Remarks and Observations

The constraints on $|z|$, which we have used in the definitions (1), (13), (15) and (58) above, are meant to ensure that the generating functions in (1), (13), (15) and (58) are *analytic* throughout the prescribed *open* disks in the complex z-plane (centred at the origin $z = 0$) to have the corresponding convergent Taylor-Maclaurin series expansions (about the origin $z = 0$) occurring on their right-hand sides (each with a *positive* radius of convergence). Moreover, throughout this investigation, $\log z$ is tacitly assumed to denote the *principal* branch of the many-valued function $\log z$

with the imaginary part $\Im(\log z)$ constrained by $-\pi < \Im(\log z) \leqq \pi$. More importantly, throughout this presentation, wherever $|\log \lambda|$ and $|\log(-\lambda)|$ appear as the radii of the open disks in the complex z-plane (centred at the origin $z = 0$) in which the defining generating functions are analytic, it is *tacitly* assumed that the obviously exceptional cases when $\lambda = 1$ and $\lambda = -1$, respectively, are to be treated *separately*. Naturally, therefore, the corresponding constraints on $|z|$ in the earlier investigations (see, e.g., [781], [788], [789] and [1087]) and elsewhere in the literature dealing with one or the other of these Apostol type polynomials (see also [58]) should also be modified accordingly.

Generalizations and Unified Presentations of the Apostol Type Polynomials

The mutual relationships among the families of the generalized Apostol-Bernoulli polynomials, the generalized Apostol-Euler polynomials and the generalized Apostol-Genocchi polynomials, which are asserted by Problems 70, 71 and 72 below, can be appropriately applied with a view to translating various formulas involving one family of these generalized polynomials into the corresponding results involving each of the other two families of these generalized polynomials. Nevertheless, we find it useful to investigate properties and results involving these three families of generalized Apostol type polynomials in a *unified* manner. In fact, the following interesting unification (and generalization) of the generating functions of the three families of Apostol type polynomials was recently investigated, rather systematically, by Ozden *et al.* (*cf.* [886, p. 2779, Equation (1.1)]):

$$\frac{2^{1-\kappa}\, z^{\kappa}\, e^{xz}}{\beta^{b} e^{z} - a^{b}} = \sum_{n=0}^{\infty} \mathcal{Y}_{n,\beta}(x;\kappa,a,b)\, \frac{z^n}{n!} \tag{61}$$

$$\left(|z| < 2\pi \quad \text{when} \quad \beta = a;\ |z| < \left| b \log\left(\frac{\beta}{a}\right)\right| \quad \text{when} \quad \beta \neq a;\right.$$

$$\left. 1^{\alpha} := 1;\ \kappa, \beta \in \mathbb{C};\ a, b \in \mathbb{C} \setminus \{0\}\right),$$

where we have not only suitably relaxed the constraints on the parameters κ, a and b, but we have also strictly followed the above remarks and observations regarding the open disk in the complex z-plane (centred at the origin $z = 0$) within which the generating function in (61) is analytic to have the corresponding convergent Taylor-Maclaurin series expansion (about the origin $z = 0$) occurring on the right-hand side (with a *positive* radius of convergence).

Here, in conclusion of our present section, we first define the following unification (and generalization) of the generating functions of the above-mentioned three families of the generalized Apostol type polynomials.

Definition 1.1 The *generalized Apostol type polynomials*

$$\mathcal{F}_n^{(\alpha)}(x; \lambda; \mu; \nu) \quad (\alpha, \lambda, \mu, \nu \in \mathbb{C})$$

of (real or complex) order α are defined by means of the following generating function:

$$\left(\frac{2^{\mu} z^{\nu}}{\lambda e^z + 1}\right)^{\alpha} \cdot e^{xz} = \sum_{n=0}^{\infty} \mathcal{F}_n^{(\alpha)}(x; \lambda; \mu; \nu) \frac{z^n}{n!} \quad \left(|z| < |\log(-\lambda)|;\ 1^{\alpha} := 1\right), \tag{62}$$

so that, by comparing Definition 1 with the corresponding definitions given above, we have

$$\mathcal{B}_n^{(\alpha)}(x; \lambda) = (-1)^{\alpha}\, \mathcal{F}_n^{(\alpha)}(x; -\lambda; 0; 1), \tag{63}$$

$$\mathcal{E}_n^{(\alpha)}(x; \lambda) = \mathcal{F}_n^{(\alpha)}(x; \lambda; 1; 0) \tag{64}$$

and

$$\mathcal{G}_n^{(\alpha)}(x; \lambda) = \mathcal{F}_n^{(\alpha)}(x; \lambda; 1; 1). \tag{65}$$

Furthermore, if we compare the generating functions (61) and (62), we have

$$\mathcal{Y}_{n,\beta}(x; \kappa, a, b) = -\frac{1}{a^b}\, \mathcal{F}_n^{(1)}\left(x; -\left(\frac{\beta}{a}\right)^b; 1-\kappa; \kappa\right). \tag{66}$$

We, thus, see from the relationships (63), (64), (65) and (66), that the generating function of $\mathcal{F}_n^{(\alpha)}(x; \lambda; \mu; \nu)$ in (62) includes, as its special cases, not only the generating function of the polynomials $\mathcal{Y}_{n,\beta}(x; \kappa, a, b)$ in (61) and the generating functions of *all* three of the generalized Apostol type polynomials $\mathcal{B}_n^{(\alpha)}(x; \lambda)$, $\mathcal{E}_n^{(\alpha)}(x; \lambda)$ and $\mathcal{G}_n^{(\alpha)}(x; \lambda)$, but also the generating functions of their special cases $B_n^{(\alpha)}(x)$, $E_n^{(\alpha)}(x)$ and $G_n^{(\alpha)}(x)$.

The various interesting properties and results involving the *new* unified family of generalized Apostol type polynomials $\mathcal{F}_n^{(\alpha)}(x; \lambda; \mu; \nu)$, given by Definition 1 above, can also be derived in a manner analogous to that of our investigation in this presentation.

The following natural generalization and unification of the Apostol-Bernoulli polynomials $\mathcal{B}_n^{(\alpha)}(x; \lambda)$ of order α, as well as the generalized Bernoulli numbers $B_n(a, b)$ studied by Guo and Qi [519] and the generalized Bernoulli polynomials $B_n(x; a, b)$ studied by Luo *et al.* [787], was introduced and investigated recently by Srivastava *et al.* [1095].

Definition 1.2 (*cf.* [1095, p. 254, Equation (20)]). The *generalized Apostol-Bernoulli type polynomials* $\mathfrak{B}_n^{(\alpha)}(x;\lambda;a,b,c)$ *of order* $\alpha \in \mathbb{C}$ are defined by the following generating function:

$$\left(\frac{z}{\lambda b^z - a^z}\right)^{\alpha} \cdot c^{xz} = \sum_{n=0}^{\infty} \mathfrak{B}_n^{(\alpha)}(x;\lambda;a,b,c)\,\frac{z^n}{n!} \tag{67}$$

$$\left(|z| < \left|\frac{\log \lambda}{\log\left(\frac{b}{a}\right)}\right|;\ a \in \mathbb{C}\setminus\{0\};\ b,c \in \mathbb{R}^+;\ a \neq b;\ 1^{\alpha} := 1\right).$$

In a sequel to the work by Srivastava *et al.* [1095], a similar generalization of each of the families of Euler and Genocchi polynomials were introduced and investigated (see, for details, [1096, Section 4]).

Definition 1.3 (*cf.* [1096, Section 2]). The *generalized Apostol-Euler type polynomials* $\mathfrak{E}_n^{(\alpha)}(x;\lambda;a,b,c)$ *of order* $\alpha \in \mathbb{C}$ are defined by the following generating function:

$$\left(\frac{2}{\lambda b^z + a^z}\right)^{\alpha} \cdot c^{xz} = \sum_{n=0}^{\infty} \mathfrak{E}_n^{(\alpha)}(x;\lambda;a,b,c)\,\frac{z^n}{n!} \tag{68}$$

$$\left(|z| < \left|\frac{\log(-\lambda)}{\log\left(\frac{b}{a}\right)}\right|;\ a \in \mathbb{C}\setminus\{0\};\ b,c \in \mathbb{R}^+;\ a \neq b;\ 1^{\alpha} := 1\right).$$

Definition 1.4 (*cf.* [1096, Section 4]). The *generalized Apostol-Genocchi type polynomials* $\mathfrak{G}_n^{(\alpha)}(x;\lambda;a,b,c)$ *of order* $\alpha \in \mathbb{C}$ are defined by the following generating function:

$$\left(\frac{2z}{\lambda b^z + a^z}\right)^{\alpha} \cdot c^{xz} = \sum_{n=0}^{\infty} \mathfrak{G}_n^{(\alpha)}(x;\lambda;a,b,c)\,\frac{z^n}{n!} \tag{69}$$

$$\left(|z| < \left|\frac{\log(-\lambda)}{\log\left(\frac{b}{a}\right)}\right|;\ a \in \mathbb{C}\setminus\{0\};\ b,c \in \mathbb{R}^+;\ a \neq b;\ 1^{\alpha} := 1\right).$$

Remark 1 In their special case when

$$a = 1 \quad \text{and} \quad b = c = e,$$

the generalized Apostol-Bernoulli type polynomials $\mathfrak{B}_n^{(\alpha)}(x;\lambda;a,b,c)$ defined by (67), the generalized Apostol-Euler type polynomials $\mathfrak{E}_n^{(\alpha)}(x;\lambda;a,b,c)$ defined by (68) and the generalized Apostol-Genocchi type polynomials $\mathfrak{G}_n^{(\alpha)}(x;\lambda;a,b,c)$ defined by (69) would reduce at once to the Apostol-Bernoulli polynomials $\mathcal{B}_n^{(\alpha)}(x;\lambda)$, the Apostol-Euler polynomials $\mathcal{E}_n^{(\alpha)}(x;\lambda)$ and the Apostol-Genocchi polynomials $\mathcal{G}_n^{(\alpha)}(x;\lambda)$, respectively.

Since the parameter $\lambda \in \mathbb{C}$, by comparing Definitions 2, 3 and 4 above, we can easily deduce the following potentially useful lemma (see also Lemmas 1, 2 and 3).

Lemma 1.7 *The families of the generalized Apostol-Bernoulli type polynomials*

$$\mathfrak{B}_n^{(l)}(x;\lambda;a,b,c) \quad (l \in \mathbb{N}_0)$$

and the generalized Apostol-Euler type polynomials

$$\mathfrak{E}_n^{(l)}(x;\lambda;a,b,c) \quad (l \in \mathbb{N}_0)$$

are related by

$$\mathfrak{B}_n^{(l)}(x;\lambda;a,b,c) = \left(-\frac{1}{2}\right)^l \frac{n!}{(n-l)!}\,\mathfrak{E}_{n-l}^{(l)}(x;-\lambda;a,b,c) \quad (n,l \in \mathbb{N}_0;\ n \geqq l) \tag{70}$$

or, equivalently, by

$$\mathfrak{E}_n^{(l)}(x;\lambda;a,b,c) = (-2)^l \frac{n!}{(n+l)!}\,\mathfrak{B}_{n+l}^{(l)}(x;-\lambda;a,b,c) \quad (n,l \in \mathbb{N}_0). \tag{71}$$

Furthermore, the families of the generalized Apostol-Bernoulli type polynomials

$$\mathfrak{B}_n^{(l)}(x;\lambda;a,b,c) \quad (l \in \mathbb{N}_0)$$

and the generalized Apostol-Euler type polynomials

$$\mathfrak{E}_n^{(l)}(x;\lambda;a,b,c) \quad (l \in \mathbb{N}_0)$$

are related to the generalized Apostol-Genocchi type polynomials

$$\mathfrak{G}_n^{(l)}(x;\lambda;a,b,c) \quad (l \in \mathbb{N}_0)$$

as follows:

$$\mathfrak{G}_n^{(\alpha)}(x;\lambda;a,b,c) = (-2)^\alpha\ \mathfrak{B}_n^{(\alpha)}(x;-\lambda;a,b,c) \quad \left(\alpha \in \mathbb{C};\ 1^\alpha := 1\right) \tag{72}$$

and

$$\mathfrak{G}_n^{(l)}(x;\lambda;a,b,c) = (-1)^l\,(-n)_l\,\mathfrak{E}_{n-l}^{(l)}(x;\lambda;a,b,c) = \frac{n!}{(n-l)!}\,\mathfrak{E}_{n-l}^{(l)}(x;\lambda;a,b,c)$$

$$\left(n,l \in \mathbb{N}_0;\ n \geqq l;\ \lambda \in \mathbb{C}\right).$$

The inter-relationships asserted by the above Lemma do aid in translating the various properties and results involving any one of these three families of generalized Apostol type polynomials in terms of the corresponding properties and results involving the other two families. Nonetheless, it would occasionally seem more appropriately convenient to investigate these three families in a *unified* manner by means of Definition 5 below.

Definition 1.5 A unification of the *generalized Apostol-Bernoulli type polynomials*

$$\mathfrak{B}_n^{(\alpha)}(x;\lambda;a,b,c),$$

the *generalized Apostol-Euler type polynomials*

$$\mathfrak{E}_n^{(\alpha)}(x;\lambda;a,b,c)$$

and the *generalized Apostol-Genocchi type polynomials*

$$\mathfrak{G}_n^{(\alpha)}(x;\lambda;a,b,c)$$

of order $\alpha \in \mathbb{C}$ is defined by the following generating function:

$$\left(\frac{2^{\mu}\, z^{\nu}}{\lambda b^z + a^z}\right)^{\alpha} \cdot c^{xz} = \sum_{n=0}^{\infty} \mathcal{Z}_n^{(\alpha)}(x;\lambda;a,b,c;\mu;\nu)\,\frac{z^n}{n!} \tag{73}$$

$$\left(|z| < \left|\frac{\log(-\lambda)}{\log\left(\frac{b}{a}\right)}\right|;\ a \in \mathbb{C}\setminus\{0\};\ b,c \in \mathbb{R}^{+};\ a \neq b;\ \alpha,\lambda,\mu,\nu \in \mathbb{C};\ 1^{\alpha} := 1\right),$$

so that, by comparing Definition 5 with Definitions 1 through 4, we have

$$\mathcal{F}_n^{(\alpha)}(x;\lambda;\mu;\nu) = \mathcal{Z}_n^{(\alpha)}(x;\lambda;1,e,e;\mu;\nu), \tag{74}$$

$$\mathfrak{B}_n^{(\alpha)}(x;\lambda;a,b,c) = (-1)^{\alpha}\, \mathcal{Z}_n^{(\alpha)}(x;-\lambda;a,b,c;0;1), \tag{75}$$

$$\mathfrak{E}_n^{(\alpha)}(x;\lambda;a,b,c) = \mathcal{Z}_n^{(\alpha)}(x;\lambda;a,b,c;1;0) \tag{76}$$

and

$$\mathfrak{G}_n^{(\alpha)}(x;\lambda;a,b,c) = \mathcal{Z}_n^{(\alpha)}(x;-\lambda;a,b,c;1;1). \tag{77}$$

Thus, clearly, Definitions 1 and 5 above provide us with remarkably powerful and extensive generalizations of the various families of the Apostol type polynomials and Apostol type numbers. Properties and results involving these generalizations deserve to be investigated further (see also [791], [886], [1095] and [1096]; see also [1091] and Problem 73 onwards of this chapter).

1.9 Inequalities for the Gamma Function and the Double Gamma Function

The Gamma Function and Its Relatives

Recently, many research articles were published providing inequalities for the Gamma function and its relatives. We refer to Gautschi's survey paper [476] and the comprehensive bibliography complied by Sándor [1003].

We begin by proving that $\log\Gamma$ is convex on $(0, \infty)$ (*cf.* Theorem 1.1):

$$\Gamma\left(\frac{x}{p}+\frac{y}{q}\right) \leqq [\Gamma(x)]^{1/p}\,[\Gamma(y)]^{1/q} \quad \left(1<p<\infty;\ \frac{1}{p}+\frac{1}{q}=1\right). \tag{1}$$

Indeed, if $1<p<\infty$ and $1/p+1/q=1$, then we have

$$\begin{aligned}
\Gamma\left(\frac{x}{p}+\frac{y}{q}\right) &= \int_0^\infty t^{\frac{x}{p}+\frac{y}{q}-1}\,e^{-t}\,dt \\
&= \int_0^\infty t^{\frac{x-1}{p}}\,t^{\frac{y-1}{q}}\,e^{-(1/p+1/q)t}\,dt \\
&\leqq \left(\int_0^\infty \left(t^{\frac{x-1}{p}}\,e^{-t/p}\right)^p dt\right)^{1/p} \left(\int_0^\infty \left(t^{\frac{y-1}{q}}\,e^{-t/q}\right)^q dt\right)^{1/q} \\
&= \left(\int_0^\infty t^{x-1}\,e^{-t}\,dt\right)^{1/p} \left(\int_0^\infty t^{y-1}\,e^{-t}\,dt\right)^{1/q} \\
&= [\Gamma(x)]^{1/p}\,[\Gamma(y)]^{1/q},
\end{aligned}$$

where we made use of the well-known Hölder's inequality:

For $1 \leqq p \leqq \infty$, the spaces $L^p(\mu)$ are Banach spaces and, if $f \in L^p(\mu)$ and $g \in L^q(\mu)$ (with $1/p+1/q=1$), then $fg \in L^1(\mu)$ and

$$\int |f(\mu)g(\mu)|\,d\mu \leqq \left(\int |f(\mu)|^p\,d\mu\right)^{1/p} \left(\int |g(\mu)|^q\,d\mu\right)^{1/q}. \tag{2}$$

We recall, here (see, e.g., [115, 406, 932, 939, 1184, 1226]), that a function f is said to be *strictly completely monotonic* on an interval $I \subset \mathbb{R}$, if

$$(-1)^n f^{(n)}(x) > 0 \quad (x \in I;\ n \in \mathbb{N}_0). \tag{3}$$

If $(-1)^n f^{(n)}(x) \geqq 0$ for all $x \in I$ and $n \in \mathbb{N}_0$, then f is called *completely monotonic* on I. We recall also (see, e.g., [76, 935, 939, 943, 944, 948, 950]) that a positive function

f is said to be *logarithmically completely monotonic* on an interval $I \subset \mathbb{R}$, if

$$(-1)^n \{\ln f(x)\}^{(n)} \geqq 0 \quad (x \in I;\ n \in \mathbb{N}). \tag{4}$$

For convenience, $\mathcal{C}[I]$ and $\mathcal{L}[I]$ denote, respectively, the sets of completely monotonic functions and the logarithmically completely monotonic functions on an interval $I \subset \mathbb{R}$. It is known (see [115, 926, 935, 939, 943, 944]) that $\mathcal{L}[I] \subset \mathcal{C}[I]$.

Alzer [22] remarked that completely monotonic functions play a dominant rôle in areas, such as numerical analysis, probability theory and physics. The concept of complete and logarithmically complete monotonicities has also played an important rôle to prove some inequalities involving the gamma function. Here, we investigate some known inequalities involving the gamma function, by, mainly, focusing on $\mathcal{C}[I]$ and $\mathcal{L}[I]$.

Wendel [1223] proved the following double inequality:

$$\left(\frac{x}{x+a}\right)^{1-a} \leqq \frac{\Gamma(x+a)}{x^a\,\Gamma(x)} \leqq 1 \quad (0 < a < 1;\ x > 0), \tag{5}$$

which can be rewritten as follows:

$$x^{1-a} \leqq \frac{\Gamma(x+1)}{\Gamma(x+a)} \leqq (x+a)^{1-a} \quad (0 < a < 1;\ x > 0), \tag{6}$$

to establish the well-known asymptotic relation (see 1.1(57)):

$$\lim_{x\to\infty} \frac{\Gamma(x+a)}{x^a\,\Gamma(x)} = 1 \quad (a,\ x \in \mathbb{R}), \tag{7}$$

by using the Hölder's inequality (2).

Komatu [689] proved the inequality

$$\frac{1}{x+\sqrt{x^2+2}} < e^{x^2} \int_x^\infty e^{-t^2}\,dt \leqq \frac{1}{x+\sqrt{x^2+1}} \quad (0 \leqq x < \infty). \tag{8}$$

Pollak [905] has improved the upper bound in (8), by showing that

$$e^{x^2} \int_x^\infty e^{-t^2}\,dt \leqq \frac{1}{x+\sqrt{x^2+4/\pi}}. \tag{9}$$

Gautschi [470] proved more general inequalities than those in (8) and (9):

$$\frac{1}{2}\left[\left(x^p+2\right)^{1/p} - x\right] < e^{x^p} \int_x^\infty e^{-t^p}\,dt \leqq c_p \left[\left(x^p + \frac{1}{c_p}\right) - x\right] \quad (0 \leqq x < \infty), \tag{10}$$

where

$$c_p := \left\{\Gamma\left(1+\frac{1}{p}\right)\right\}^{p/(p-1)} \qquad (p \in \mathbb{N} \setminus \{1\}).$$

For $p = 2$, the right-hand inequality of (10) reduces to (9), whereas the left-hand inequality reduces to the corresponding inequality in (8). The integral in (10) for $p = 3$ occurs in heat transfer problems [1242], for $p = 4$ in the study of electrical discharge through gases [1009]. An application of (10) for general p is given in [840].

Gautschi [470] also derived inequalities for the following Gamma-function ratio:

$$\frac{\Gamma(n+1)}{\Gamma(n+s)},$$

which are given by

$$n^{1-s} \leqq \frac{\Gamma(n+1)}{\Gamma(n+s)} \leqq \exp[(1-s)\,\psi(n+1)] \quad (0 \leqq s \leqq 1;\ n \in \mathbb{N}) \tag{11}$$

and

$$n^{1-s} \leqq \frac{\Gamma(n+1)}{\Gamma(n+s)} \leqq (n+1)^{1-s} \quad (0 \leqq s \leqq 1;\ n \in \mathbb{N}), \tag{12}$$

where ψ denotes the Psi-function defined in 1.3(1). The inequalities (11) have attracted remarkable interest, and several intriguing papers on the subject were subsequently published by, for example, Erber [420], Kečkić and Vasić [642], Laforgia [724] and Zimering [1266], providing new bounds for the following Gamma-function ratio:

$$\frac{\Gamma(n+1)}{\Gamma(n+s)}.$$

However, it is observed that the upper bound in (12) is not better and the range in (12) is not broader than the corresponding ones in (6).

Kershaw [645] gave proofs of the following closer bounds than (11):

$$\exp\left[(1-s)\,\psi\left(x+s^{1/2}\right)\right] < \frac{\Gamma(x+1)}{\Gamma(x+s)} < \exp\left[(1-s)\,\psi\left(x+\frac{s+1}{2}\right)\right] \tag{13}$$

and

$$\left(x+\frac{s}{2}\right)^{1-s} < \frac{\Gamma(x+1)}{\Gamma(x+s)} < \left[x-\frac{1}{2}+\left(s+\frac{1}{4}\right)^{1/2}\right]^{1-s}, \tag{14}$$

each being valid for $x > 0$ and $0 < s < 1$.

Bustoz and Ismail [196] established a remarkably more general result. They showed that the two functions

$$f_1(x) = \frac{\Gamma(x+s)}{\Gamma(x+1)} \exp\left[(1-s)\,\psi\left(x+\frac{s+1}{2}\right)\right] \quad (0 < s < 1) \tag{15}$$

and

$$f_2(x) = \frac{\Gamma(x+1)}{\Gamma(x+s)} \left(x+\frac{s}{2}\right)^{s-1} \quad (0 < s < 1) \tag{16}$$

are strictly completely monotonic on $(0,\, \infty)$. Since

$$\lim_{x\to\infty} f_1(x) = \lim_{x\to\infty} f_2(x) = 1,$$

the inequalities (13) and (14) are immediate consequences of the fact that f_1 and f_2 are strictly decreasing on $(0,\, \infty)$. Alzer [22] refined one of the Kečkić-Vasić's inequalities in [642] and also gave the following interesting result:

For every $s \in (0,\, 1)$, *the function*

$$\begin{aligned} x \longmapsto f_\alpha(x,s) =& \frac{\Gamma(x+s)}{\Gamma(x+1)} \frac{(x+1)^{x+1/2}}{(x+s)^{x+s-1/2}} \\ &\cdot \exp\left(s-1+\frac{1}{12}\left[\psi'(x+1+\alpha)-\psi'(x+s+\alpha)\right]\right) \quad (\alpha > 0) \end{aligned} \tag{17}$$

is strictly completely monotonic on $(0,\, \infty)$, *if and only if* $\alpha \geqq \frac{1}{2}$. *Furthermore, for every* $s \in (0,\, 1)$, *the function*

$$x \longmapsto \frac{1}{f_\beta(x,s)} \quad (\beta \geqq 0)$$

is strictly completely monotonic on $(0,\, \infty)$, *if and only if* $\beta = 0$.

Gurland [527] obtained the following inequality

$$\frac{[\Gamma(\lambda+\alpha)]^2}{\Gamma(\lambda)\,\Gamma(\lambda+2\alpha)} < \frac{\lambda}{\alpha^2+\lambda} \quad (\alpha,\, \lambda \in \mathbb{R};\ \alpha+\lambda > 0;\ \lambda > 0;\ \alpha \neq 0;\ \alpha \neq 1), \tag{18}$$

by making a novel use of the Cramér-Rao lower bound for the variance of an unbiased estimator (see Cramér [343]). Indeed, we consider the density function given by

$$f(x) = \frac{1}{\theta^\lambda\,\Gamma(\lambda)} \exp\left(-\frac{x}{\theta}\right) x^{\lambda-1} \quad (x > 0;\ \lambda > 0;\ \theta > 0). \tag{19}$$

From the fact that the expression:

$$\frac{\Gamma(\lambda)}{\Gamma(\alpha+\lambda)}\, x^\alpha$$

is an unbiased estimator of θ^{α}, we have the following Cramér-Rao bound for the variance:

$$V\left[\frac{\Gamma(\lambda)}{\Gamma(\alpha+\lambda)}x^{\alpha}\right] \geqq \frac{1}{E\left[\frac{\partial}{\partial\theta^{\alpha}}\{\log f(x)\}\right]^2},$$

which yields the inequality (18).

A special case of (18) with $\lambda = n/2$ and $\alpha = 1/2$ is seen to be reduced to Gurland's formula [526]:

$$\frac{4n+3}{(2n+1)^2}\left[\frac{(2n)!!}{(2n-1)!!}\right]^2 < \pi < \frac{4}{4n+1}\left[\frac{(2n)!!}{(2n-1)!!}\right]^2 \quad (n \in \mathbb{N}), \tag{20}$$

where

$$(2n)!! := 2\cdot 4\cdots(2n-2)(2n) \quad \text{and} \quad (2n-1)!! := 1\cdot 3\cdot 5\cdots(2n-3)(2n-1).$$

The inequality (20) is a sharper inequality for approximating π than that given by Wallis:

$$\frac{2}{2n+1}\left[\frac{(2n)!!}{(2n-1)!!}\right]^2 < \pi < \frac{1}{n}\left[\frac{(2n)!!}{(2n-1)!!}\right]^2 \quad (n \in \mathbb{N}). \tag{21}$$

Gokhale [492] obtained a similar inequality to (18) as follows:

$$\frac{\Gamma(\lambda)\,\Gamma(\lambda+2\alpha)}{[\Gamma(\lambda+\alpha)]^2} > 1 + \frac{\alpha^2(\lambda-2)}{(\lambda+\alpha-1)^2} \quad (\alpha,\,\lambda \in \mathbb{R};\; \lambda+2\alpha > 0,\; \lambda > 2,\; \alpha \neq 0, \alpha \neq -1). \tag{22}$$

Gurland's inequality (18) can be written in the form:

$$\frac{\Gamma(\lambda)\,\Gamma(\lambda+2\alpha)}{[\Gamma(\lambda+\alpha)]^2} > 1 + \frac{\alpha^2}{\lambda}. \tag{23}$$

We note that the inequality (22) is seen to be stronger than that in (23) for a certain range of α. Indeed, when

$$\frac{\alpha^2(\lambda-2)}{(\lambda+\alpha-1)^2} > \frac{\alpha^2}{\lambda}, \quad \text{that is,}$$

$$-(\lambda-1) - \sqrt{\lambda(\lambda-2)} < \alpha < -(\lambda-1) + \sqrt{\lambda(\lambda-2)}.$$

By employing the multivariate generalization of (19), that is, the Wishart distribution, Olkin [878] obtained

$$\prod_{j=0}^{p-1} \frac{[\Gamma(\lambda+\alpha-j/2)]^2}{\Gamma(\lambda-j/2)\,\Gamma(\lambda+2\alpha-j/2)} \leqq \frac{\lambda^2\left(p^2-1\right)^2+\lambda p^4}{\lambda^2\left(p^2-1\right)^2+\lambda p^4+\alpha^2} \tag{24}$$
$$\left(\alpha,\,\lambda\in\mathbb{R};\ \alpha+\lambda>\frac{p-1}{2};\ \lambda>0;\ p>0\right).$$

The special case of (24) with $p=1$ reduces to (18). Selliah [1018], by employing the multiparameter version of the Cramér-Rao lower bound, using the information matrix for the same problem, obtained the following inequality:

$$\prod_{j=0}^{p-1} \frac{[\Gamma(\lambda+\alpha-j/2)]^2}{\Gamma(\lambda-j/2)\,\Gamma(\lambda+2\alpha-j/2)} \leqq \frac{\lambda}{\lambda+p\alpha^2} \tag{25}$$
$$\left(\alpha,\,\lambda\in\mathbb{R};\ \alpha+\lambda>\frac{p-1}{2};\ \lambda>0;\ p>0\right),$$

which is seen to be sharper than that given in (24).

For $x\in(-\alpha,\,\infty)$, define the function $z_{s,t}(x)$ by

$$z_{s,t}(x):=\begin{cases}\left[\dfrac{\Gamma(x+t)}{\Gamma(x+s)}\right]^{1/(t-s)}-x & (s\neq t)\\[2ex] e^{\psi(x+s)}-x & (s=t),\end{cases} \tag{26}$$

where $s,\,t\in\mathbb{R}_0^+$, $\mathbb{R}_0^+$ being the set of nonnegative real numbers, and $\alpha=\min\{s,\,t\}$. A monotonicity and convexity of $z_{s,t}(x)$ was proved (see [241, 406, 929, 945]) so that the function $z_{s,t}(x)$ is either convex and decreasing for $|t-s|<1$ or concave and increasing for $|t-s|>1$. From this fact, the best bounds in the Kershaw's double inequality (14) could be deduced. Qi [930, 931] further generalized this result.

For $x\in(-\rho,\,\infty)$, define the function $H_{a,b,c}(x)$ by

$$H_{a,b,c}(x):=(x+c)^{b-a}\,\frac{\Gamma(x+a)}{\Gamma(x+b)}\quad (a,\,b,\,c\in\mathbb{R};\ \rho=\min\{a,\,b,\,c\}). \tag{27}$$

Very recently, Qi [932, Theorem 1] proved the following results:

$$H_{a,b,c}(x)\in\mathcal{L}[(-\rho,\,\infty)]\quad \big((a,\,b,\,c)\in D_1(a,\,b,\,c)\big) \tag{28}$$

and

$$\big[H_{a,b,c}(x)\big]^{-1}\in\mathcal{L}[(-\rho,\,\infty)]\quad \big((a,\,b,\,c)\in D_2(a,\,b,\,c)\big), \tag{29}$$

where, for convenience,

$$\begin{aligned}D_1(a,\, b,\, c) :=& \left\{(a,\, b,\, c) \mid a+b \geqq 1,\ c \leqq b < c+\frac{1}{2}\right\} \cup \left\{(a,\, b,\, c) \mid a > b \geqq c+\frac{1}{2}\right\}\\ &\cup \left\{(a,\, b,\, c) \mid 2a+1 \leqq a+b \leqq 1,\ a < c\right\} \cup \left\{(a,\, b,\, c) \mid b-1 \leqq a < b \leqq c\right\}\\ &\setminus \{(a,\, b,\, c) \mid a = c+1,\ b = c\}\end{aligned}$$

and

$$\begin{aligned}D_2(a,\, b,\, c) :=& \left\{(a,\, b,\, c) \mid a+b \geqq 1,\ c \leqq a < c+\frac{1}{2}\right\} \cup \left\{(a,\, b,\, c) \mid b > a \geqq c+\frac{1}{2}\right\}\\ &\cup \left\{(a,\, b,\, c) \mid b < a \leqq c\right\} \cup \left\{(a,\, b,\, c) \mid b+1 \leqq a,\ c \leqq a \leqq c+1\right\}\\ &\cup \left\{(a,\, b,\, c) \mid b+c+1 \leqq a+b \leqq 1\right\} \setminus \{(a,\, b,\, c) \mid a = c+1,\ b = c\}\\ &\setminus \{(a,\, b,\, c) \mid b = c+1,\ a = c\}.\end{aligned}$$

Qi [932, Theorem 2] made use of (28), (29) and 1.1(37) to prove the following inequalities:

$$(x+c)^{a-b} < \frac{\Gamma(x+a)}{\Gamma(x+b)} \quad (x \in (-\rho,\, \infty);\ (a,\, b,\, c) \in D_1(a,\, b,\, c)) \tag{30}$$

and

$$\frac{\Gamma(x+a)}{\Gamma(x+b)} \leqq \frac{\Gamma(\delta+a)}{\Gamma(\delta+b)} \left(\frac{x+c}{\delta+c}\right)^{a-b} \quad (x \in [\delta,\, \infty);\ (a,\, b,\, c) \in D_1(a,\, b,\, c)), \tag{31}$$

where $a,\, b,\, c \in \mathbb{R}$, $\rho = \min\{a,\, b,\, c\}$, and δ is a constant greater than $-\rho$. If $(a,\, b,\, c) \in D_2(a,\, b,\, c)$, then inequalities in (30) and (31) are reversed, respectively, in $(-\rho,\, \infty)$ and $[\delta,\, \infty)$.

Qi [932] then observed the following facts: setting $a = 1$ and $0 < b < 1$ in (30) reveals that

$$(x+b)^{1-b} < \frac{\Gamma(x+1)}{\Gamma(x+b)} \quad (0 < b < 1;\ x \in (-b,\, \infty)) \tag{32}$$

holds true. It is obvious that the inequality in (32) not only refines the lower bound, but also extends the range of the left-hand side of the inequality in (14).

Taking $a = 1,\, 0 < b < 1$ and $\delta = 1$ in (31) shows that

$$\frac{\Gamma(x+1)}{\Gamma(x+b)} \leqq \frac{1}{\Gamma(1+b)} \left(\frac{x+b}{1+b}\right)^{1-b} \quad (0 < b < 1;\ x \in [1,\, \infty)) \tag{33}$$

holds true. A usual argument shows that, if

$$x \geqq \frac{\left(\frac{1}{2} - \sqrt{b + \frac{1}{4}}\right)(1+b)\ \sqrt[1-b]{\Gamma(1+b)} + 1}{(1+b)\ \sqrt[1-b]{\Gamma(1+b)} - 1} := \lambda(b),$$

then the inequality (33) would be better than the right-hand side of (14). It is easy to find that

$$\lim_{b\to 0+} \lambda(b) \quad \text{and} \quad \lim_{b\to 1-} \lambda(b) = \frac{e + \left(1 - \sqrt{5}\right) e^{\gamma}}{2\,e^{\gamma} - e} \cong 0.6123686\cdots < 1,$$

where γ denotes the Euler-Mascheroni constant defined by 1.1(3). This implies that the inequality in (33) refines the right-hand side of (14), if b is close enough to 1 and that the upper bound in (33) is better than the one in (14), if x is sufficiently large.

The Double Gamma Function

In a striking contrast to an abundant literature on the inequalities for the Gamma function and its relatives, there has been a few known results for the inequalities for the double Gamma function (see, e.g., [446], [105] and [244]). Presumably, Batir [105] initiated the study of the following class of inequalities for the double Gamma function $\Gamma_2(x+1) = 1/G(x+1)$:

$$(\Gamma(x))^{\frac{x}{2}}\, x^x\, (2\pi)^{\frac{x}{2}}\, e^{-\frac{x}{2}-\frac{x^2}{2}} < G(x+1) < \left(\frac{\Gamma(x)}{\Gamma(x/2)}\right)^x (8\pi)^{\frac{x}{2}}\, e^{-\frac{x}{2}-\frac{x^2}{2}} \quad \left(x \in \mathbb{R}^+\right); \tag{34}$$

$$(2\pi)^{\frac{x}{2}}\, e^{\frac{x}{2}-\frac{x^2}{2}+\frac{x^2}{2}\psi(x/2)} < G(x+1) < (2\pi)^{\frac{x}{2}}\, e^{-\frac{x^2}{2}+\frac{x^2}{2}\psi(x)} \quad \left(x \in \mathbb{R}^+\right); \tag{35}$$

$$\begin{aligned}(2\pi)^{\frac{x}{2}}\, (\Gamma(x+1))^x \exp\left(-\frac{x}{2} - \frac{x^2}{2} - \frac{x^2}{2}\,\psi(\alpha(x))\right) \\ < G(x+1) < (2\pi)^{\frac{x}{2}}\, (\Gamma(x+1))^x \exp\left(-\frac{x}{2} - \frac{x^2}{2} - \frac{x^2}{2}\,\psi(\beta(x))\right) \quad \left(x \in \mathbb{R}^+\right),\end{aligned} \tag{36}$$

where $\mathbb{R}^+$ denotes the set of positive real numbers,

$$\alpha(x) = \frac{x}{3} \quad \text{and} \quad \beta(x) = \frac{x^2}{2}\,\frac{1}{(x+1)\log(x+1) - x}.$$

Problems

1. Show that each form of the Gamma function, defined by 1.1(1), 1.1(2) and 1.1(7), is analytic in its given domain.

(Whittaker and Watson [1225, Chapter 12]; Rainville [959, pp. 15–18])

2. Use the Wielandt's theorem [Theorem 1.3] to derive each of the following results:

- the Gauss product 1.1(4) from the Euler integral 1.1(1),
- Gauss's multiplication formula 1.1(31),
- the representation of the Beta function by Gamma functions 1.1(42),
- Stirling's formula 1.1(33).

(Remmert [973])

3. Prove the following series representations for the Euler-Mascheroni constant γ:

$$\gamma = 1 + \sum_{n=3}^{\infty} \frac{(-1)^n}{n} \left(\frac{\log(n-1)}{\log 2} \right)$$

and

$$\gamma = \sum_{n=1}^{\infty} \frac{(-1)^n}{n} \left(\frac{\log n}{\log 2} \right),$$

where the bracket denotes the greatest integer function.

(Gerst [480]; Sandham [1001])

4. Deduce the explicit Weierstrass canonical product form 1.3(3) for the double Gamma function Γ_2, by using the three properties associated with Barnes's definition of Γ_2.

(Barnes [94, pp. 265–269]; Whittaker and Watson [1225, p. 264])

5. Prove the following generalization of Gauss's multiplication formula 1.1(51) for the Gamma function:

$$\prod_{\ell=0}^{m-1} \Gamma\left(kz + \frac{k\ell}{m}\right) = (2\pi)^{\frac{1}{2}(m-k)} \left(\frac{k}{m}\right)^{mkz+\frac{1}{2}(mk-m-k)} \cdot \prod_{n=0}^{k-1} \Gamma\left(mz + \frac{mn}{k}\right) \quad (k,\, m \in \mathbb{N}).$$

(Choi and Quine [278, p. 131]; Magnus *et al.* [795, p. 3])

6. For the Euler-Mascheroni constant γ, the sequence $\{s_n\}_{n=0}^{\infty}$, defined by

$$s_n := \left(1 + \frac{1}{2} + \cdots + \frac{1}{n-1}\right) - \log n \quad (n \in \mathbb{N} \setminus \{1\}),$$

satisfies the asymptotic property:

$$s_n = \gamma + O\left(n^{-1}\right) \quad (n \to \infty).$$

Show that $\gamma - s_n$ $(n \in \mathbb{N} \setminus \{1\})$ can be expressed, as follows, as an infinite sum with rational terms:

$$\gamma - s_n = \frac{1}{n} \sum_{m=0}^{\infty} \frac{t_{m+2}}{\binom{m+n}{m}} \quad (n \in \mathbb{N} \setminus \{1\}),$$

where

$$t_{m+2} = -\frac{1}{(m+1)!}\int_0^1 (0-x)(1-x)\cdots(m-x)\,dx \quad (m \in \mathbb{N}_0).$$

(Elsner [416, p. 1537]; Jolley [613, pp. 14–15]).

7. Prove the von Staudt-Clausen theorem (Theorem 6.1):

$$B_{2n} = I_n - \sum_{p-1|2n} \frac{1}{p} \quad (n \in \mathbb{N}),$$

where B_n is the Bernoulli number, I_n is an integer and the sum is taken over all primes p such that $p-1$ divides $2n$.

(Carathéodory [210, pp. 281–284]; Apostol [65, pp. 274–275])

8. Prove the following rapidly converging series expansion for $\psi(z)$:

$$\psi(z) = \frac{1}{2}\log\left((z+n)^2 + (z+n) + \frac{1}{3}\right) - \sum_{k=0}^{n} \frac{1}{z+k} - \frac{1}{9}\sum_{k=n+1}^{\infty} (k-n)A_k,$$

where

$$A_k := \int_0^1 (z+k+t)^{-2}\left((z+k+t)^4 - \frac{1}{3}(z+k+t)^2 + \frac{1}{9}\right)^{-1} dt.$$

(Shafer and Lossers [1020])

9. Let S_n denote the area of the sphere of radius 1 in $\mathbb{R}^n$. Prove that

$$S_n = \frac{\left(\int_{-\infty}^{\infty} e^{-t^2}\,dt\right)^n}{\int_0^{\infty} e^{-r^2} r^{n-1}\,dr} = \frac{2\left(\sqrt{\pi}\right)^n}{\Gamma\left(\frac{n}{2}\right)} \quad (n \in \mathbb{N}).$$

(Campbell [208, pp. 126–128])

10. Let

$$I_n := \int \cdots \int f\left(\sum_{k=1}^{n} t_k\right) t_1^{\alpha_1 - 1} \cdots t_n^{\alpha_n - 1}\,dt_1 \cdots dt_n$$

$$\left(\alpha_k > 0;\ t_k \geqq 0;\ k = 1, \cdots, n;\ \sum_{k=1}^{n} t_k \leqq 1\right),$$

where f is a continuous function. Prove that

$$I_n = \frac{\Gamma(\alpha_1)\cdots\Gamma(\alpha_n)}{\Gamma(\alpha_1 + \cdots + \alpha_n)} \int_0^1 f(\tau)\,\tau^{\alpha_1 + \cdots + \alpha_n - 1}\,d\tau.$$

(Whittaker and Watson [1225, p. 258]; Campbell [208, pp. 128–133])

11. Prove that

$$\int_0^{\pi} (\sin\,\theta)^{\lambda-1}\, e^{-\mu\theta}\, d\theta = \frac{\pi\,\Gamma(\lambda)\, e^{-\frac{1}{2}\pi\mu}}{2^{\lambda-1}\,\Gamma\left(\frac{\lambda+i\mu+1}{2}\right)\Gamma\left(\frac{\lambda-i\mu+1}{2}\right)} \qquad (\Re(\lambda) > 0;\ \mu \in \mathbb{C}).$$

(Nielsen [862, p. 159])

12. Show that

$$\int_{-\infty}^{\infty} \frac{e^{-2\pi izt}}{\cosh\,(\pi t)}\,dt = \frac{1}{\cosh\,(\pi z)} \qquad \left(-\frac{1}{2} < \Im(z) < \frac{1}{2}\right)$$

(*cf.* Equation 1.1(46))

13. Making use of the Pochhammer symbol defined by 1.1(5), show that the Vandermonde convolution theorem 1.4(9) can be rewritten in the form:

$$\sum_{k=0}^{n} \frac{(\alpha)_k\,(\beta)_{n-k}}{k!\,(n-k)!} = \frac{(\alpha+\beta)_n}{n!} = \sum_{k=0}^{n} \frac{(\alpha)_{n-k}\,(\beta)_k}{(n-k)!\,k!} \qquad (\alpha,\,\beta \in \mathbb{C};\ n \in \mathbb{N}_0)\,.$$

14. Show that Euler's transformation 1.5(21) is equivalent to the Pfaff-Saalschütz theorem:

$${}_3F_2\left[\begin{matrix} a,\,b,\,-n; \\ c,\,a+b-c-n+1; \end{matrix}\ 1\right] = \frac{(c-a)_n\,(c-b)_n}{(c)_n\,(c-a-b)_n} \qquad \left(c \notin \mathbb{Z}_0^-;\ n \in \mathbb{N}_0\right).$$

15. Deduce Gauss's summation theorem 1.5(7) as a limit case of the Pfaff-Saalschütz theorem (see Problem 13 above).
[**Hint**: Let $n \to \infty$, and apply the asymptotic expansion 1.1(37).]

16. For the Bernoulli polynomials $B_n(x)$ generated by 1.7(1), show that

$$x^n = \sum_{k=0}^{n} \frac{1}{n+1}\binom{n+1}{k} B_k(x) \qquad (n \in \mathbb{N}_0)\,.$$

Hence (or otherwise), deduce the following representation for the Laguerre polynomials $L_n^{(\alpha)}(x)$ defined by 1.5(72):

$$L_n^{(\alpha)}(x) = \sum_{k=0}^{n} \frac{(-1)^k}{k!}\binom{n+\alpha}{n-k} B_k(x)\cdot {}_2F_2(-n+k,\,1;\,\alpha+k+1,\,2;\,1).$$

(*cf.* Magnus *et al.* [795, p. 26]; see also Popov [907])

17. Derive the following asymptotic expansion for the Psi (or Digamma) function:

$$\psi(z) = \log z - \frac{1}{2z} - \frac{1}{12z^2} + \frac{1}{120z^4} - \frac{1}{252z^6} + O\left(z^{-8}\right)$$

$$(|z| \to \infty;\ |\arg(z)| \leqq \pi - \epsilon\ (0 < \epsilon < \pi)).$$

18. For $a, b \in \mathbb{N}$, let (a, b) and $[a, b]$ denote the greatest common divisor and least common multiple, respectively, of a and b, and $\{x\} = x - [x]$ denote the fractional part of x. Then, show that, if $\Re(s) > \frac{1}{2}$,

$$\int_0^1 \zeta(1-s, \{ax\})\, \zeta(1-s, \{bx\})\, dx = \frac{2\{\Gamma(s)\}^2 \zeta(2s)}{(2\pi)^{2s}} \left(\frac{(a,b)}{[a,b]}\right)^s,$$

where $\zeta(s, a)$ is the Hurwitz (or generalized) Zeta function defined by 2.2(1).

Show also that

$$\int_0^1 B_n(\{ax\})\, B_n(\{bx\})\, dx = (-1)^{n-1} \frac{B_{2n}}{(2n)!} \left(\frac{(a,b)}{[a,b]}\right)^n \qquad (n \in \mathbb{N}).$$

(*cf.* Mordell [844, p. 372]; see also Mikolás [827])

19. Prove that

$$\int_0^{1/\sqrt{8}} \frac{t^2 \log t}{(t^2+1)^{\frac{1}{2}}} dt = \frac{1}{96} \left[2\pi^2 + 30(\log 2)^2 - 39 \log 2 - 9\right].$$

(Knuth [679, p. 138])

20. Prove that

$$\int_0^\infty \frac{\cosh\, x \cdot \log x}{\cosh\,(2x) - \cos\,(2\pi a)} dx$$

$$= \frac{\pi}{2 \sin\,(\pi a)} \left[\log\left(\frac{\Gamma\left(\frac{1}{2}a + \frac{1}{2}\right)}{\Gamma\left(\frac{1}{2}a\right)}\right) + \frac{1}{2} \log\left(2\pi \cot\, \frac{a\pi}{2}\right)\right] \qquad (0 < a < 1).$$

(Williams and Zhang [1232, p. 44])

21. Prove the following inequality:

$$\operatorname{cosec}^2 x - \frac{1}{2m+1} < \sum_{k=-m}^{m} \frac{1}{(x-k\pi)^2} < \operatorname{cosec}^2 x \quad \left(m \in \mathbb{N};\ 0 < |x| \leqq \frac{1}{2}\pi;\ x \in \mathbb{R}\right).$$

Also, apply this inequality to show that

$$\int_0^\infty \left(\frac{\sin\, x}{x}\right)^2 dx = \frac{\pi}{2}.$$

(Neville [858, pp. 629–630])

22. Prove that

$$\int_0^\infty \frac{e^x \log x}{e^{2x} - 2e^x \cos\,(2\pi a) + 1} dx = \frac{\pi}{2} \frac{1}{\sin\,(2\pi a)} \log\left((2\pi)^{1-2a} \frac{\Gamma(1-a)}{\Gamma(a)}\right) \qquad (0 < a < 1).$$

(Zhang and Williams [1252, p. 377])

23. Prove that

$$\int_{\frac{\pi}{4}}^{\frac{\pi}{2}} \log(\log(\tan x))\, dx = \frac{\pi}{2} \log\left(\frac{\Gamma\left(\frac{3}{4}\right)}{\Gamma\left(\frac{1}{4}\right)} \sqrt{2\pi} \right).$$

(*cf.* Vardi [1189, p. 308]; see also Gradshteyn and Ryzhik [505, p. 532])

24. Let

$$P = \prod_{n=1}^{\infty} \frac{(n-a_1)(n-a_2)\cdots(n-a_k)}{(n-b_1)(n-b_2)\cdots(n-b_k)} \qquad \left(\sum_{j=1}^{k} a_j = \sum_{j=1}^{k} b_j \right),$$

where no a_j or b_j is a positive integer. Show that

$$P = \prod_{j=1}^{k} \frac{\Gamma(1-b_j)}{\Gamma(1-a_j)}.$$

(*cf.* Rainville [959, pp. 13–15]; see also Melzak [821, p. 101])

25. Let M_n and m_n denote the maximum and minimum of the Bernoulli polynomial $B_n(x)$ for $0 \leqq x \leqq 1$. Show that

$$M_{4n} = B_{4n}\left(\tfrac{1}{2}\right) = \left(1 - 2^{1-4n}\right) |B_{4n}| \quad (n \in \mathbb{N});$$
$$M_{4n+2} = B_{4n+2}(0) = B_{4n+2} \quad (n \in \mathbb{N}_0);$$
$$m_{4n} = B_{4n}(0) = -|B_{4n}| \quad (n \in \mathbb{N});$$
$$m_{4n+2} = B_{4n+2}\left(\tfrac{1}{2}\right) = -\left(1 - 2^{-1-4n}\right) B_{4n+2} \quad (n \in \mathbb{N}_0).$$

(Lehmer [740, pp. 534])

26. Prove the following integrals:

$$\int_0^1 \frac{\log u}{u} [\log(1+u)]^2\, du = \mathcal{A} - \frac{\pi^4}{288}$$

and

$$\int_0^{\pi} u \left[\log\left(2\cos\frac{u}{2}\right)\right]^2 du = \mathcal{A} + \frac{31\pi^4}{480},$$

where

$$\mathcal{A} = \sum_{n=1}^{\infty} \frac{(-1)^{n-1}}{(n+1)^2} \left(\sum_{k=1}^{n} \frac{1}{k^2} \right).$$

(Rutledge and Douglass [991, p. 30])

27. Suppose that
(a) $f(z)$ is a meromorphic function of a complex variable z in $\mathbb{C}$;
(b) $f(z)$ is analytic and vanishes nowhere on $\mathbb{C}\setminus\mathbb{Z}_0^-$;
(c) $f(z)$ satisfies the functional equation

$$n^{z-\frac{1}{2}}\prod_{k=0}^{n-1} f\left(\frac{z+k}{n}\right) = (2\pi)^{\frac{n-1}{2}} f(z) \quad \left(z\in\mathbb{C}\setminus\mathbb{Z}_0^-\right),$$

where n is an arbitrarily fixed positive integer greater than 1 and $n^{z-\frac{1}{2}}$ denotes the principal value $\exp\left[\left(z-\frac{1}{2}\right)\operatorname{Log} n\right]$. Then, show that the function f must be of the form:

$$f(z) = \exp\left[a\left(z-\frac{1}{2}\right)+\frac{2m\pi i}{n-1}\right]\Gamma(z),$$

where a is an arbitrary complex constant and m is an arbitrary integer.

(*cf.* Haruki [540, p. 174]; see also Theorem 1.1 (Bohr-Mollerup))

28. Suppose that a function $f:\mathbb{R}^+\to\mathbb{R}$ satisfies the functional equation:

$$f(x+1) = xf(x) \quad \left(x\in\mathbb{R}^+\right)$$

and, moreover, that

$$\lim_{x\to\infty}\left(\frac{e}{x}\right)^x\sqrt{\frac{x}{2\pi}}f(x) = 1.$$

Show that $f(x) = \Gamma(x)$ for all $x\in\mathbb{R}^+$.

(Kuczma [705, p. 129])

29. Suppose that a function $f:\mathbb{R}^+\to\mathbb{R}$ satisfies the following properties:
(a) $f(x+1) = xf(x)$;
(b) $\left(\frac{e}{x}\right)^x f(x)$ is decreasing for $x\in\mathbb{R}^+$;
(c) $f(1) = 1$.
Show that $f(x) = \Gamma(x)$ for $x\in\mathbb{R}^+$.

(Anastassiadis [37, p. 117])

30. Suppose that a function $f:\mathbb{R}^+\to\mathbb{R}^+$ satisfies the following properties:
(a) $f(x+1) = \frac{x}{x+y}f(x) \quad (y>0)$;
(b) $f(x)$ is decreasing for $x\in\mathbb{R}^+$;
(c) $f(1) = \frac{1}{y}$.
Show that $f(x) = B(x,y)$ for $x\in\mathbb{R}^+$.

(Anastassiadis [38, pp. 25–26])

31. Prove that, for $p\geq 5$ and $p\in\mathbb{N}$,

$$B_p^{(p)} \equiv -\frac{1}{2}p^2(p-1)!\left(\operatorname{mod} p^5\right),$$

where $B_p^{(p)}$ denotes the generalized Bernoulli numbers defined by 1.6(22).

(Carlitz [216, p. 112])

32. Prove that, for $m \in \mathbb{N}$ and $i = \sqrt{-1}$,

$$
\begin{aligned}
&{}_2F_1(1,\,1;\,2m;\,i) \\
&\quad = (2m-1)\, i^m\, 2^{m-1} \left[\frac{\log 2}{2} - \frac{i\pi}{4} - \sum_{j=1}^{m-1} (-1)^j \, \frac{i^j}{2^j} \left(\frac{1}{2j} + \frac{1+i}{2j-1} \right) \right]
\end{aligned}
$$

and

$$
\begin{aligned}
&{}_2F_1(1,\,1;\,2m+1;\,i) \\
&\quad = m(1+i)\, i^m\, 2^m \left[\frac{\log 2 - 1}{2} - i\frac{2-\pi}{4} - \sum_{j=1}^{m-1} (-1)^j \, \frac{i^j}{2^j} \left(\frac{1}{2j} + \frac{1-i}{4j+2} \right) \right].
\end{aligned}
$$

(Butzer and Hauss [198, p. 355])

33. Prove that

$$
\sum_{n=0}^{\infty} \frac{1}{(2n+1)^2 \binom{2n}{n}} = \frac{8}{3} G - \frac{\pi}{3} \log\left(2 + \sqrt{3}\right).
$$

(Borwein and Borwein [148, p. 386])

34. Prove that

$$
\int_{-1}^{1} (2-2x)^{\alpha}\, (2+2x)^{\beta} \left(1 - x^2\right)^{-\frac{1}{2}} dx = \frac{\Gamma\left(\alpha + \frac{1}{2}\right) \Gamma\left(\beta + \frac{1}{2}\right) \pi}{\Gamma(\alpha+\beta+1) \left[\Gamma\left(\frac{1}{2}\right)\right]^2}\, 2^{2\alpha+2\beta}.
$$

(Askey [74, p. 357])

35. Prove that

$$
\begin{aligned}
\log \Gamma(z+1) = {} & \frac{1}{2}\left(z + \frac{1}{2}\right) \left[\log\left(z + \frac{1}{2} + \frac{1}{2\sqrt{3}} \right) + \log\left(z + \frac{1}{2} - \frac{1}{2\sqrt{3}} \right) \right] \\
& - z - \frac{1}{2} + \frac{1}{2}\log(2\pi) + \int_0^{\infty} e^{-zt} \phi(t)\, dt,
\end{aligned}
$$

where, for convenience,

$$
\begin{aligned}
\phi(t) := {} & \frac{1}{t}\left(\frac{1}{2} - \frac{1}{t} + \frac{1}{e^t - 1} \right) - \frac{1}{2t}\left[1 - e^{-\frac{1}{2}t} \cosh\left(\frac{t}{2\sqrt{3}} \right) \right] \\
& - \frac{d}{dt}\left\{ \frac{1 - e^{-\frac{1}{2}t} \cosh\left(\frac{t}{2\sqrt{3}} \right)}{t} \right\};
\end{aligned}
$$

moreover,

$$
\phi(t) = \frac{e^{-\frac{1}{2}t}}{t} \left[\frac{1}{2} \operatorname{cosech}\left(\frac{1}{2} t \right) + \frac{1}{2\sqrt{3}} \sinh\left(\frac{t}{2\sqrt{3}} \right) - \frac{1}{t} \cosh\left(\frac{t}{2\sqrt{3}} \right) \right].
$$

(*cf.* Watson [1212, p. 5]; see also Eq. (29))

36. Prove that, for fixed $p \in \mathbb{R}^+$,

$$\gamma = \lim_{n\to\infty} \frac{\sum_{k=0}^{\infty} \left(\frac{n^k}{k!}\right)^p (H_k - \log n)}{\sum_{k=0}^{\infty} \left(\frac{n^k}{k!}\right)^p},$$

where γ is the Euler-Mascheroni constant and H_k denotes the harmonic numbers defined by 3.2(36).

(Brent and McMillan [174, pp. 310])

37. Let $\mathbf{u}$ be the row vector $\{u_k = 1/k : k \in \mathbb{N}\}$ and let $\mathbf{M}$ be the matrix with entries $\{m_{ij} = 1/(i-j+1)$ if $j \leqq i$, $m_{ij} = 0$ if $j > i :\ i, j \in \mathbb{N}\}$. Let $\mathbf{v}$ be the column vector $\{v_n = 1/(n+1) : n \in \mathbb{N}\}$. Show that the product $\mathbf{u}\left(\mathbf{M}^{-1}\mathbf{v}\right)$ exists (as a convergent series), and is equal to Euler-Mascheroni constant γ defined by 1.1(3).

(Kenter [644, p. 452])

38. Let $\Omega(z)$ be the function of the complex variable z defined by

$$\Omega(z) := \frac{\Gamma(z+a_1)\,\Gamma(z+a_2)}{\Gamma(z+b_1)\,\Gamma(z+b_2)}$$

in which a_1, a_2, b_1, b_2 are to be regarded as any constants (real or complex). Show that, for $|\arg z| < \pi$ and large $|z|$,

$$\Omega(z) \sim \left(\frac{1}{z}\right)^p \left(1 + \frac{c_1}{z+1} + \frac{c_2}{(z+1)(z+2)} + \frac{c_3}{(z+1)(z+2)(z+3)} + \cdots\right),$$

where the c_n are constants and $p = b_1 + b_2 - a_1 - a_2$.

(Engen [1183, pp. 124–125])

39. Prove the following Lipschitz summation formula:

$$\sum_{m=-\infty}^{\infty} (2m+\tau)^{-\lambda} = \frac{e^{-\pi i\lambda/2}\,\pi^{\lambda}}{\Gamma(\lambda)} \sum_{n=1}^{\infty} n^{\lambda-1}\, e^{\pi i \tau n} \qquad (\Re(\tau) > 0;\ \lambda > 1).$$

(Knopp [677, p. 65])

40. Prove the following series-integral representation of the Catalan constant G:

$$G = \frac{1}{2} \sum_{n=0}^{\infty} (-1)^n \left(\int_0^{\frac{\pi}{2}} \cos^n t\, dt \right)^2.$$

(Catalan [224, p. 51])

41. Prove that

$$\frac{1}{2}\,\psi'(x)\,\psi'''(x) < \left(\psi''(x)\right)^2 \qquad (x > 0),$$

where $\psi(x)$ denotes the Psi (or Digamma) function.

(*cf.* English and Rousseau [419, p. 432]; see also Alzer and Wells [33, p. 1459])

42. A function f is said to be completely monotonic on an interval I, if $f \in C^{\infty}(I)$ and

$$(-1)^k f^{(k)}(x) \geqq 0 \quad (*)$$

for all $x \in I$ and for all $k \in \mathbb{N}_0$. If the inequality $(*)$ is strict for all $x \in I$ and for all $k \in \mathbb{N}_0$, then f is said to be strictly completely monotonic on I. Consider the function

$$F_n(x;c) := \left(\psi^{(n)}(x)\right)^2 - c\,\psi^{(n-1)}(x)\,\psi^{(n+1)}(x)$$
$$\left(x \in \mathbb{R}^+;\ c \in \mathbb{R};\ n \in \mathbb{N} \setminus \{1\}\right).$$

Let $n \in \mathbb{N} \setminus \{1\}$, and let $\alpha,\ \beta \in \mathbb{R}$. Show that the function.

$$x \mapsto F_n(x;\alpha)$$

is strictly completely monotonic on $(0,\infty)$, if and only if

$$\alpha \leqq \frac{n-1}{n},$$

and that the function:

$$x \mapsto -F_n(x;\beta)$$

is strictly completely monotonic on $(0,\infty)$, if and only if

$$\beta \geqq \frac{n}{n+1}.$$

(Alzer and Wells [33, p. 1460])

43. Let the function $g(x)$ be defined by

$$g(x) := x^2\,\psi'(1+x) - x\,\psi(1+x) + \log\Gamma(x+1) \quad (x > -1).$$

Show that $g(x)$ strictly decreases from ∞ to 0 on $(-1,0]$ and strictly increases from 0 to ∞ on $[0,\infty)$.

(Elbert and Laforgia [404, Theorem 2])

44. Prove that

$$\frac{x^{x-1}\,e^{y}}{y^{y-1}\,e^{x}} \leqq \frac{\Gamma(x)}{\Gamma(y)} \leqq \frac{x^{x-\frac{1}{2}}\,e^{y}}{y^{y-\frac{1}{2}}\,e^{x}} \quad \left(x \geqq y > 1\right).$$

(Kečkić and Vasić [642, p. 107])

45. Prove that

(a) For n, an odd positive integer,

$$G = -T\left(\frac{1}{4}\right) = \frac{2n+1}{n+1}\sum_{j=1}^{n}(-1)^j\,T\left(\frac{2j-1}{8n+4}\right);$$

(b) For n, an even positive integer,

$$G = -T\left(\frac{1}{4}\right) = \frac{2n+1}{n}\sum_{j=1}^{n}(-1)^{j+1}\,T\left(\frac{2j-1}{8n+4}\right),$$

where, for convenience, T is defined by

$$T(r) := \int_0^{r\pi} \log(\tan\theta)\, d\theta \qquad \left(0 \leqq r \leqq \frac{1}{2}\right).$$

(Bradley [165, pp. 164–165])

46. For the Stirling numbers $s(n,k)$ defined by 1.6(1), show that

(a) $\displaystyle\sum_{n=0}^{\infty} \frac{(-1)^n}{(n+k)(a)_n} s(n+k,k) = -\frac{(1-a)_k}{k!}\psi^{(k)}(a-k);$

(b) $\displaystyle\sum_{n=0}^{\infty} \frac{(-1)^n}{(n+k-1)(a)_n} s(n+k,k) = \frac{(-a)_k}{a(k-1)!}\psi^{(k-1)}(a-k).$

Also deduce the special cases of *each* of these summation formulas, when

$$a = k+1 \quad \text{and} \quad a = k+2.$$

(*Cf.* Jordan [614, p. 343]; see also Hansen [531, p. 348])

47. Verify that the special case of Problem 46(a) above, when $a = k+1$ is *precisely* the relationship 3.5(16).

(*Cf.* Problem 5 (Chapter 3))

48. Show that

$$\int_0^{\infty} \frac{\cos x}{x^p}\, dx = \frac{\pi}{2\,\Gamma(p)\cos(p\pi/2)} \qquad (0 < p < 1)$$

and

$$\int_0^{\infty} \frac{\sin x}{x^p}\, dx = \frac{\pi}{2\,\Gamma(p)\sin(p\pi/2)} \qquad (0 < p < 1).$$

(Andrews [49, p. 83 and p. 85])

49. For $p \in \mathbb{N}$, prove the following integral:

$$\int_0^1 \cdots \int_0^1 \left(\prod_{j=1}^{p} u_j\right)^{x-1} \left\{\prod_{j=1}^{p}(1-u_j)\right\}^{y-1} |\Delta(u)|^{2z}\, du_1 \cdots du_p$$

$$= \prod_{\nu=1}^{p} \frac{\Gamma(1+\nu z)\Gamma(x+(\nu-1)z)\,\Gamma(y+(\nu-1)z)}{\Gamma(1+z)\,\Gamma(x+y+(p+\nu-2)z)}$$

$$\left(\Re(x) > 0;\ \Re(y) > 0;\ \Re(z) > -\min\left\{\frac{1}{p}, \frac{\Re(x)}{p-1}, \frac{\Re(y)}{p-1}\right\}\right),$$

where

$$\Delta(u) = \Delta(u_1, u_2, \ldots, u_p) := \prod_{i<j}(u_j - u_i).$$

(Selberg [1013]; see aso [1015, pp. 204–213])

50. Prove the result asserted by Equation 1.3(57) in two ways, as noted there.

(Choi and Cvijović [271, Theorem 1])

51. Show that the generalized Barnes G-function G_m defined by

$$G_m(1+z) = \exp\left\{-\sum_{j=1}^{m} \frac{\zeta(j-m)}{j}(-z)^j - \frac{\gamma+H_m}{m+1}(-z)^{1+m} + \sum_{j=1}^{m}\binom{m}{j}\zeta'(j-m)\,(-z)^j\right\}$$
$$\cdot \prod_{k=1}^{\infty}\left\{\left(1+\frac{z}{k}\right)^{k^m} \exp\left(\left(\sum_{r=1}^{m+1}\frac{1}{r k^r}(-z)^r\right)k^m\right)\right\} \qquad (m \in \mathbb{N}_0)$$

is an entire function of order m with the zeros-k with multiplicity k^m ($k \in \mathbb{N}$). Moreover, there is the zero 0 with multiplicity 1 for $m = 0$. The used infinite product is absolutely convergent. One has $G_m(1) = 1$, and G_m satisfies the functional equation

$$G_m(z+1) = \prod_{r=0}^{m}\{G_{m-r}(z)\}^{(-1)^r\binom{m}{r}} \qquad (m \in \mathbb{N}).$$

Here, as usual, $\zeta(s)$ is the Riemann zeta function, γ the Euler-Mascheroni constant, and H_m the harmonic numbers.

(Schuster [1010, Theorem A])

52. Show that the following integrals:

$$I_k(q) := \int_0^\infty \frac{t}{\left(1+t^2\right)^{k+1}\left(e^{2\pi q t}-1\right)}\,dt \qquad (k \in \mathbb{N})$$

are given by

$$I_k(q) = -\frac{1}{4k} - \frac{\binom{2k}{k}}{2^{2k+2}q} + \frac{1}{k\,2^{2k}}\sum_{j=1}^{k}\frac{(-1)^{j+1}}{(j-1)!}\binom{2k-j-1}{k-j}2^{j-1}q^j\,\psi^{(j)}(q).$$

(See Boros *et al.* [141, Theorem 3.2])

53. For arbitrary $n \in \mathbb{N}$, show that

$$\frac{1}{2n+\frac{1}{1-\gamma}-2} \leqq H_n - \log n - \gamma < \frac{1}{2n+\frac{1}{3}},$$

where γ is the Euler-Mascheroni constant and H_n the harmonic numbers. Here, the constants $\frac{1}{1-\gamma}-2$ and $\frac{1}{3}$ are the best possible.

(See Qi *et al.* [938, Theorem 2])

54. The *Genocchi numbers* G_n are defined by means of the following generating function:

$$\frac{2t}{e^t+1} =: \sum_{n=0}^{\infty} G_n \frac{t^n}{n!} = t + \sum_{n=1}^{\infty} G_{2n}\frac{t^{2n}}{(2n)!},$$

which are directly related to the Bernoulli numbers B_n by means of

$$G_n = 2\left(1 - 2^{2n}\right) B_n.$$

Show that

$$G_{2n} = -n - \frac{1}{2}\sum_{k=1}^{n-1}\binom{2n}{2k} G_{2k} \quad \text{and} \quad G_{2n} = -1 - \sum_{k=1}^{n-1}\binom{2n}{2k-1}\frac{G_{2k}}{2k}.$$

(See Herrmann [555, Proposition 2.1])

55. Let $\alpha > 0$ and $0 < c \neq 1$ be real numbers. Show that the following inequality:

$$\left(\frac{\Gamma(x+y+c)}{\Gamma(x+y)}\right)^{1/\alpha} < \left(\frac{\Gamma(x+c)}{\Gamma(x)}\right)^{1/\alpha} + \left(\frac{\Gamma(y+c)}{\Gamma(y)}\right)^{1/\alpha}$$

holds true for all positive real numbers x and y, if and only if $\alpha \geqq \max\{1,\, c\}$. The reverse inequality is valid for all $x,\, y > 0$, if and only if $\alpha \leqq \min\{1,\, c\}$.

(See Alzer [27])

56. Hadamard's Gamma function $H(x)$ is defined by

$$H(x) = \frac{1}{\Gamma(1-x)}\,\frac{d}{dx}\left\{\log\left(\frac{\Gamma(1/2 - x/2)}{\Gamma(1-x/2)}\right)\right\},$$

where $\Gamma(x)$ denotes the classical Euler's Gamma function given in 1.1(1). $H(x)$ is an entire function and satisfies the following relationship:

$$H(n) = (n-1)! \quad (n \in \mathbb{N}).$$

Show that H can be expressed as follows:

$$H(x) = \Gamma(x)\left[1 + \frac{\sin(\pi x)}{2\pi}\left\{\psi\left(\frac{x}{2}\right) - \psi\left(\frac{x+1}{2}\right)\right\}\right]$$

in terms of the sine, Gamma and Psi (or Digamma) functions. Furthermore, Hadamard's Gamma function $H(x)$ satisfies the following remarkable functional equation, which is a counterpart of 1.1(9):

$$H(x+1) = xH(x) + \frac{1}{\Gamma(1-x)},$$

which is a counterpart of 1.1(9).

(See Alzer [28]; Luschny [792]; Newton [859])

57. Show that the following inequalities hold true:

$$\psi_m(x)\,\psi_n(x) \geqq \left[\psi_{\frac{m+n}{2}}(x)\right]^2 \quad \left(x > 0;\; m,\, n,\, \frac{m+n}{2} \in \mathbb{N}\right)$$

and

$$(s+1)\,\frac{\zeta(s)}{\zeta(s+1)} \geqq s\,\frac{\zeta(s+1)}{\zeta(s+2)} \quad (s > 1),$$

where $\psi_n(x) := \psi^{(n)}(x)$ and $\zeta(s)$ is the Riemann Zeta function given in 2.3(1).

(See Laforgia and Natalini [725])

58. Show that

$$\gamma = \frac{2^n}{e^{2^n}} \sum_{r=0}^{\infty} \frac{2^{rn}}{(r+1)!} \sum_{s=0}^{r} \frac{1}{s+1} - n\ln 2 + O\left(\frac{1}{2^n\, e^{2^n}}\right),$$

where γ denotes the Euler-Mascheroni constant (see Section 1.2).

(See Bailey [84]; Mortici [846])

59. Let $m \in \mathbb{N}$ and $z \in \mathbb{C}$. Suppose also that $\Delta(m;\lambda)$ abbreviates the following array of m parameters:

$$\frac{\lambda}{m}, \frac{\lambda+1}{m}, \ldots, \frac{\lambda+m-1}{m} \qquad (m \in \mathbb{N}).$$

Show that

$$\begin{aligned}\sum_{k=0}^{n} \frac{(2mk+1)!!}{(n-k)!(mk+1)!}\,\frac{z^k}{k!} &= \frac{1}{n!}\sum_{k=0}^{n} \binom{2mk+1}{mk}\binom{n}{k}\left(\frac{z}{2^m}\right)^k \\ &= \frac{1}{n!}\,{}_{m+1}F_m\left[\begin{array}{c} -n,\, \Delta\left(m;\, \frac{3}{2}\right); \\ \\ \Delta\,(m;\, 2)\,; \end{array} \; -2^m \cdot z\right]. \end{aligned} \tag{37}$$

(Sofo and Srivastava [1044]; see, for special cases, Srivastava [1089] and Samoletov [1000])

60. Show that each of the following limit formulas holds true:

$$\lim_{z\to -k} \frac{\Gamma(nz)}{\Gamma(qz)} = \frac{(-1)^{(n-q)k}\cdot q}{n} \cdot \frac{(qk)!}{(nk)!} \qquad (n, q \in \mathbb{N};\; k \in \mathbb{N}_0)$$

and

$$\lim_{z\to -k} \frac{\psi(nz)}{\psi(qz)} = \frac{q}{n} \qquad (n, q \in \mathbb{N};\; k \in \mathbb{N}_0).$$

(Prabhu and Srivastava [912])

61. For a suitably bounded sequence $\{C_n\}_{n\in\mathbb{N}_0}$ of essentially arbitrary complex numbers, show that

$$\sum_{n=0}^{\infty} \frac{C_n}{\alpha+Nn}\,\frac{\omega^n\, z^{\alpha+Nn}}{n!} = \sum_{n=0}^{\infty} \frac{\gamma\,(\alpha+n, z)}{n!} \sum_{k=0}^{[n/N]} \binom{n}{Nk} \frac{(Nk)\,!}{k!}\, C_k\, \omega^k \qquad (N \in \mathbb{N}),$$

provided that each member exists. Also, deduce the following hypergeometric form of this expansion formula:

$$z^{\alpha}\,{}_{p+1}F_{q+1}\left[\begin{matrix}\alpha/N,(\alpha_p);\\(\alpha+N)/N,(\beta_q);\end{matrix}\ \omega\left(-\frac{z}{N}\right)^N\right]$$

$$=\alpha\sum_{n=0}^{\infty}\frac{\gamma(\alpha+n,z)}{n!}\,{}_{p+N}F_q\left[\begin{matrix}\Delta(N;-n),(\alpha_p);\\(\beta_q);\end{matrix}\ \omega\right]\qquad(N\in\mathbb{N}).$$

(Lin *et al.* [764, p. 518]; see, for special cases, Gautschi *et al.* [477])

62. For a suitably bounded sequence $\{\Omega(n)\}_{n\in\mathbb{N}_0}$ of essentially arbitrary complex numbers, show that

$$\sum_{k=0}^{\infty}\sum_{k_1,\cdots,k_r=0}^{n}\Omega(k)\frac{(-n)_{k_1}\ldots(-n)_{k_r}(\gamma+n)_{k+k_1+\cdots+k_r}}{k_1!\cdots k_r!(\gamma)_{k+k_1+\cdots+k_r}}$$

$$\cdot\frac{(\beta_1)_{k_1}\ldots(\beta_r)_{k_r}}{(\beta_1+1)_{k_1}\cdots(\beta_r+1)_{k_r}}$$

$$=\frac{(n!)^r(\gamma-\beta_1-\cdots-\beta_r)_n}{(\gamma)_n(\beta_1+1)_n\ldots(\beta_r+1)_n}\sum_{k=0}^{\infty}\Omega(k)\frac{(\gamma-\beta_1-\cdots-\beta_r+n)_k}{(\gamma-\beta_1-\cdots-\beta_r)_k},$$

provided that the series involved are absolutely convergent.

(Carlitz [219, p. 169]; see also Lin and Srivastava [766, p. 310] and Problem 26 of Chapter 6)

63. Let the harmonic numbers H_n and the generalized harmonic numbers $H_n^{(s)}$ be defined, as usual, by

$$H_n:=H_n^{(1)}\quad\text{and}\quad H_n^{(s)}:=\sum_{k=1}^{n}\frac{1}{k^s}\qquad(s\in\mathbb{C};\ n\in\mathbb{N}:=\{1,2,3,\ldots\}),$$

(a) For $\alpha_j,\beta_j\geqq 0$ $(j=1,2,3,4)$ and $m\in\mathbb{N}_0$, show that

$$\sum_{n=1}^{\infty}\frac{t^n\binom{n+m-1}{n-1}[\psi(\beta_1+1+\alpha_1 n)-\psi(\beta_1+1)]}{n^4\prod_{i=1}^{4}\binom{\alpha_i n+\beta_i}{\beta_i}}$$

$$=-\alpha_1\alpha_2\alpha_3\alpha_4\int_0^1\int_0^1\int_0^1\int_0^1\frac{(1-x)^{\beta_1}\ln(1-x)\cdot(1-y)^{\beta_2}(1-z)^{\beta_3}(1-w)^{\beta_4}}{xyzw}$$

$$\cdot\frac{x^{\alpha_1}y^{\alpha_2}z^{\alpha_3}w^{\alpha_4}}{(1-tx^{\alpha_1}y^{\alpha_2}z^{\alpha_3}w^{\alpha_4})^{m+1}}\,dx\,dy\,dz\,dw.$$

(b) Let p and q be positive integers. In terms of the Bernoulli polynomials $B_n(x)$ and the generalized Clausen functions $\mathrm{Cl}_n(z)$, show that

$$\left.\begin{matrix}\psi^{(n)}\left(\dfrac{p}{q}\right)\\[2ex] \psi^{(n)}\left(\dfrac{p}{q}\right)\end{matrix}\right\} = n!q^n \left[\begin{matrix}\dfrac{(-1)^{\left\lfloor \frac{n}{2}\right\rfloor}(2\pi)^{n+1}}{2(n+1)!}\displaystyle\sum_{s=1}^{q} B_{n+1}\left(\frac{s}{q}\right)\left\{\begin{matrix}\cos\left(\dfrac{2\pi ps}{q}\right)\\[2ex] \sin\left(\dfrac{2\pi ps}{q}\right)\end{matrix}\right\}\\[6ex] \pm\displaystyle\sum_{s=1}^{q}\mathrm{Cl}_{n+1}\left(\frac{2\pi s}{q}\right)\left\{\begin{matrix}\sin\left(\dfrac{2\pi ps}{q}\right)\\[2ex] \cos\left(\dfrac{2\pi ps}{q}\right)\end{matrix}\right\}\end{matrix}\right]$$

$$\left(\left\{\begin{matrix} n = 2m-1 \\ n = 2m\end{matrix}\right\};\ m \in \mathbb{N};\ 1 \leqq p \leqq q\right),$$

where the Bernoulli polynomials $B_n(x)$ are generated, as usual, by

$$\frac{te^{xt}}{e^t - 1} = \sum_{n=0}^{\infty} B_n(x)\,\frac{t^n}{n!} \qquad (|t| < 2\pi)$$

and the generalized Clausen functions $\mathrm{Cl}_n(\theta)$ are given by

$$\mathrm{Cl}_n(\theta) = \begin{cases} \displaystyle\sum_{k=1}^{\infty} \frac{\sin(k\theta)}{k^n} & (n \quad \text{even}) \\[3ex] \displaystyle\sum_{k=1}^{\infty} \frac{\cos(k\theta)}{k^n} & (n \quad \text{odd}). \end{cases}$$

(c) Let

$$\alpha_1 \geqq 0,\ \alpha_2 \geqq 0,\ \alpha_3 \geqq 0 \quad \text{and} \quad \alpha_4 \geqq 0$$

be positive real numbers. Also, let $p \in \mathbb{N}_0$ and

$$j, k, l, m \in \mathbb{N}_0,\ 0 \leqq p \leqq 3k - 4 \quad \text{and} \quad |t| \leqq 1.$$

Then show that

$$\sum_{n=1}^{\infty} \frac{n^p t^n\,[\psi(j+1+\alpha_1 n) - \psi(j+1)]}{\binom{\alpha_1 n + j}{j}\binom{\alpha_2 n + k}{k}\binom{\alpha_3 n + l}{l}\binom{\alpha_4 n + m}{m}}$$

$$= -\alpha_1 klm \int_0^1\!\int_0^1\!\int_0^1\!\int_0^1 \frac{(1-x)^j}{x}(1-y)^{k-1}(1-z)^{l-1}(1-w)^{m-1}$$

$$\cdot \ln(1-x)\,\mathrm{Li}_{-p-1}\left(tx^{\alpha_1}y^{\alpha_2}z^{\alpha_3}w^{\alpha_4}\right)\,dx\,dy\,dz\,dw,$$

where the Polylogarithmic function $\mathrm{Li}_q(\beta)$ is given by

$$\mathrm{Li}_q(z) = \sum_{r=1}^{\infty} \frac{z^r}{r^q}.$$

(See Sofo and Srivastava [1045])

64. Let $[\tau]$ denote the greatest integer in $\tau \in \mathbb{R}$. Then, for an essentially arbitrary sequence $\{\Omega_n\}_{n\in\mathbb{N}_0}$ of complex numbers, show that the following general combinatorial series relationship holds true:

$$\sum_{k=0}^{[n/m]} \binom{\lambda+k}{k} \sum_{j=0}^{[k/m]} \binom{k}{mj} \Omega_j\, z^j = \sum_{k=0}^{[n/m]} \binom{\lambda+n+1}{n-mk} \binom{\lambda+mk}{mk} \Omega_k\, z^k$$
$$(\lambda \in \mathbb{C};\ m \in \mathbb{N};\ n \in \mathbb{N}_0)$$

or, equivalently,

$$\sum_{k=0}^{[n/m]} \binom{\lambda+k}{k} \sum_{j=0}^{[k/m]} \binom{k}{mj} \Omega_j\, z^j = \binom{\lambda+n+1}{n} \sum_{k=0}^{[n/m]} \frac{\lambda+1}{\lambda+mk+1} \binom{n}{mk} \Omega_k\, z^k$$
$$(\lambda \in \mathbb{C};\ m \in \mathbb{N};\ n \in \mathbb{N}_0),$$

provided that both members of each of these assertions exist.

Also, deduce the following general Fox-Wright hypergeometric series relationship:

$$_{p+3}\Psi_{q+1}^{*}\left[\begin{array}{c}(-n,m),(\alpha,m),(a_1,A_1),\ldots,(a_p,A_p);\\ \\ (\alpha+1,m),(b_1,B_1),\ldots,(b_q,B_q);\end{array} z\right]$$

$$= \binom{\alpha+n}{n}^{-1} \sum_{k=0}^{[n/m]} \binom{\alpha+k-1}{k}\, {}_{p+1}\Psi_q^{*}\left[\begin{array}{c}(-k,m),(a_1,A_1),\ldots,(a_p,A_p);\\ \\ (b_1,B_1),\ldots,(b_q,B_q);\end{array} z\right],$$

provided that each member of this assertion exists. Here, the Fox-Wright function ${}_p\Psi_q$ $(p,q \in \mathbb{N}_0)$ or ${}_p\Psi_q^{*}$ $(p,q \in \mathbb{N}_0)$, with p numerator parameters $a_1,\cdots,a_p$ and q denominator parameters $b_1,\cdots,b_q$, such that

$$a_j \in \mathbb{C}\ (j=1,\ \ldots,p) \quad \text{and} \quad b_j \in \mathbb{C}\setminus\mathbb{Z}_0^-\ \ (j=1,\ \ldots,q),$$

is defined by

$${}_p\Psi_q^{*}\left[\begin{array}{c}(a_1,A_1),\ldots,(a_p,A_p);\\ \\ (b_1,B_1),\ldots,(b_q,B_q);\end{array} z\right] := \sum_{k=0}^{\infty} \frac{(a_1)_{A_1k}\cdots(a_p)_{A_pk}}{(b_1)_{B_1k}\cdots(b_q)_{B_qk}} \frac{z^k}{k!}$$

$$= \frac{\Gamma(b_1)\cdots\Gamma(b_q)}{\Gamma(a_1)\cdots\Gamma(a_p)}\, {}_p\Psi_q\left[\begin{array}{c}(a_1,A_1),\ldots,(a_p,A_p);\\ \\ (b_1,B_1),\ldots,(b_q,B_q);\end{array} z\right]$$

$$\left(A_j>0 \quad (j=1,\ldots,p);\ B_j>0 \quad (j=1,\ldots,q);\ 1+\sum_{j=1}^{q}B_j-\sum_{j=1}^{p}A_j \geqq 0\right),$$

where the equality in the convergence condition holds true for suitably bounded values of $|z|$, given by

$$|z| < \left(\prod_{j=1}^{p} A_j^{-A_j}\right) \cdot \left(\prod_{j=1}^{q} B_j^{B_j}\right).$$

Clearly, in terms of the generalized hypergeometric function ${}_pF_q$ $(p,q \in \mathbb{N}_0)$, we have the following relationship (see, for details, Section 1.5):

$$
\begin{aligned}
{}_p\Psi_q^* \left[\begin{array}{l}(a_1,1),\ldots,(a_p,1);\\ \\ (b_1,1),\ldots,(b_q,1);\end{array} z\right] &= {}_pF_q\left[\begin{array}{l}a_1,\ldots,a_p;\\ \\ b_1,\ldots,b_q;\end{array} z\right]\\
&= \frac{\Gamma(b_1)\cdots\Gamma(b_q)}{\Gamma(a_1)\cdots\Gamma(a_p)}\, {}_p\Psi_q\left[\begin{array}{l}(a_1,1),\ldots,(a_p,1);\\ \\ (b_1,1),\ldots,(b_q,1);\end{array} z\right].
\end{aligned}
$$

(See R. Srivastava [1118])

65. For the Bernoulli numbers B_n, which are usually given by the recurrence relation 1.7(6), that is, by

$$B_n = (-1)^n \sum_{k=0}^{n} \binom{n}{k} B_k \quad (n \in \mathbb{N}_0)$$

and

$$B_n = -\frac{1}{n+1} \sum_{k=0}^{n-1} \binom{n+1}{k} B_k \quad (n \in \mathbb{N}),$$

derive each of the following computationally more advantageous recursion formulas:

$$B_{2n} = -\frac{1}{(n+1)(2n+1)} \sum_{k=0}^{n-1} \binom{n+1}{k} (n+k+1) B_{n+k} \quad (n \in \mathbb{N})$$

or, equivalently,

$$\mathfrak{B}_n = -\frac{1}{n+1} \sum_{k=0}^{n-1} \binom{n+1}{k} \mathfrak{B}_{n+k} \quad \big(\mathfrak{B}_n := (n+1)B_n;\ n \in \mathbb{N}\big),$$

$$B_{2n} = \frac{1}{2n+1} - \frac{1}{(n+1)(2n+1)} \sum_{k=0}^{n-1} \binom{2n+2}{2k} B_{2k} \quad (n \in \mathbb{N})$$

and

$$B_{2n} = \frac{1}{2} - \frac{1}{2n+1} \sum_{k=0}^{n-1} \binom{2n+1}{2k} B_{2k} \quad (n \in \mathbb{N}).$$

(*Cf.* Srivastava and Miller [1104]; see also [625])

66. In terms of the sequences $\left\{a_n^{(j)}\right\}_{n\in\mathbb{N}_0}$ $(j=1,\ldots,r)$, let

$$f_j(z)=\sum_{n=0}^{\infty} a_n^{(j)}\, z^n \quad (j=1,\ldots,r).$$

Denote also the familiar *multinomial coefficient* by

$$\binom{n}{n_1,\ldots,n_r}:=\frac{n!}{n_1!\ldots n_r!} \quad (n,n_j\in\mathbb{N}_0;\ j=1,\ldots,r;\ r\in\mathbb{N}).$$

By applying the following known series relationship involving product of power series:

$$\prod_{j=1}^{r}\{f_j(z)\}=\sum_{n=0}^{\infty} b_n\, z^n$$
$$\left(b_n:=\sum_{n_1+\cdots+n_r=n}\prod_{j=1}^{r}\left\{\frac{a_{n_j}^{(j)}}{n_j!}\right\}\right),$$

or otherwise, derive several properties of the generalized Bernoulli polynomials $B_n^{(\alpha)}(x)$ and the generalized Euler polynomials $E_n^{(\alpha)}(x)$ of order α as follows:

$$B_n^{(\alpha_1+\cdots+\alpha_r)}(x_1+\cdots+x_r)=\sum_{n_1+\cdots+n_r=n}\binom{n}{n_1,\ldots,n_r}\prod_{j=1}^{r}\left\{B_{n_j}^{(\alpha_j)}(x_j)\right\},$$

$$E_n^{(\alpha_1+\cdots+\alpha_r)}(x_1+\cdots+x_r)=\sum_{n_1+\cdots+n_r=n}\binom{n}{n_1,\ldots,n_r}\prod_{j=1}^{r}\left\{E_{n_j}^{(\alpha_j)}(x_j)\right\}$$

and

$$\sum_{k=0}^{n}\binom{n}{k}B_{n-k}^{(\alpha)}(x)E_k^{(\alpha)}(y)=2^n\, B_n^{(\alpha)}\left(\frac{x+y}{2}\right).$$

(*Cf.* Brychkov [193])

67. In terms of the sequence $\{\Omega_n\}_{n\in\mathbb{N}_0}$, let the function $\Phi(z)$ be defined by

$$\Phi(z)=\sum_{k=0}^{\infty}\Omega_k\, z^k \quad (|z|<R;\ R\in\mathbb{R}^+).$$

Suppose also that

$$\omega=\exp\left(\frac{2\pi i}{n}\right) \quad \text{and} \quad m\in\{0,1,\ldots,n-1\}\ (n\in\mathbb{N}).$$

Then show that

$$\sum_{k=0}^{\infty} \Omega_{nk+m}\, z^{nk+m} = \frac{1}{n}\sum_{k=0}^{n-1} \omega^{(n-m)k}\, \Phi\left(\omega^k z\right) = \frac{1}{n}\sum_{k=1}^{n} \omega^{(n-m)k}\, \Phi\left(\omega^k z\right) \quad (|z| < R)$$

and apply these identities to derive the corresponding results involving the Fox-Wright function ${}_p\Psi_q$ $(p, q \in \mathbb{N}_0)$ or ${}_p\Psi_q^*$ $(p, q \in \mathbb{N}_0)$ (see Problem 64 above).

(See, for details, Srivastava [1081])

68. In connection with the so-called *Littlewood's teaser* about the hitherto nonexistent formula for the sum:

$$\sum_{r=0}^{n} \binom{n}{r}^3,$$

show that

$$\sum_{r=0}^{n} \binom{n}{r}^k \sim \frac{2^{kn}}{\sqrt{k}} \left(\frac{2}{\pi n}\right)^{\frac{1}{2}(k-1)} \qquad (n \to \infty;\ k \in \mathbb{N})$$

and

$$\sum_{r=0}^{n} \binom{n}{r}^k x^{n-r} y^r = \sum_{r=0}^{[n/2]} \binom{n}{2r}\binom{2r}{r}\binom{n+r}{r} x^r y^r (x+y)^{n-2r}.$$

(See Pólya and Szegö [906, p. 65, Theorem 40; p. 239, Entry 40] and MacMahon [794, p. 122]; see also Nanjundiah [854] and Srivastava [1075])

69. For the Bernoulli polynomials $B_n(x)$, show that

$$B_n(x) = \sum_{k=0}^{n} \frac{1}{k+1} \binom{(\mu+1)k - x}{k} \sum_{j=0}^{k} (-1)^j \binom{k}{j} (\mu k + j)^n$$
$$+ (\mu + x) \sum_{k=0}^{n} \frac{1}{k+1} \binom{(\mu+1)k - x}{k} G_k(\mu; x) \sum_{j=0}^{k} (-1)^j \binom{k}{j} (\mu k + j)^n,$$

where, for convenience,

$$G_0(\mu; x) = 0 \quad \text{and} \quad G_k(\mu; x) = \sum_{\ell=1}^{k} \frac{1}{x - \mu k - \ell} \qquad (k = 1, \ldots, n;\ \mu \in \mathbb{C}).$$

(See Srivastava [1074, p. 81, Eq. (27)])

70. Show that the following relationship holds true:

$$\mathcal{G}_n^{(l)}(x; \lambda) = \{n\}_l\, \mathcal{E}_{n-l}^{(l)}(x; \lambda) = \frac{n!}{(n-l)!}\, \mathcal{E}_{n-l}^{(l)}(x; \lambda) \quad \left(n, l \in \mathbb{N}_0;\ 0 \leqq l \leqq n;\ \lambda \in \mathbb{C}\right)$$

or, equivalently,

$$\mathcal{E}_n^{(l)}(x;\lambda) = \frac{1}{\{n+l\}_l}\,\mathcal{G}_{n+l}^{(l)}(x;\lambda) = \frac{n!}{(n+l)!}\,\mathcal{G}_{n+l}^{(l)}(x;\lambda) \quad (n,l \in \mathbb{N}_0;\ \lambda \in \mathbb{C})$$

between the Apostol-Genocchi polynomial of order l and the Apostol-Euler polynomial of order $n-l$, which are defined, respectively, by 1.8(58) and 1.8(15) with $\alpha = l \ (l \in \mathbb{N}_0)$.

(See Luo and Srivastava [791, Lemma 2])

71. Derive the following relationship:

$$\mathcal{G}_n^{(\alpha)}(x;\lambda) = (-2)^{\alpha}\ \mathcal{B}_n^{(\alpha)}(x;-\lambda) \quad \left(\alpha,\lambda \in \mathbb{C};\ 1^{\alpha} := 1\right)$$

or, equivalently,

$$\mathcal{B}_n^{(\alpha)}(x;\lambda) = \frac{1}{(-2)^{\alpha}}\ \mathcal{G}_n^{(\alpha)}(x;-\lambda) \quad \left(\alpha \in \mathbb{C};\ 1^{\alpha} := 1\right)$$

between the Apostol-Genocchi polynomials $\mathcal{G}_n^{(\alpha)}(x;\lambda)$ and the Apostol-Bernoulli polynomials $\mathcal{B}_n^{(\alpha)}(x;\lambda)$, which are defined by 1.8(58) and 1.8(13), respectively.

(See Luo and Srivastava [791, Lemma 3])

72. Verify that the following relationship holds true:

$$\mathcal{B}_n^{(l)}(x;\lambda) = \frac{n!}{(n-l)!(-2)^l}\ \mathcal{E}_{n-l}^{(l)}(x;-\lambda) \quad \left(n,l \in \mathbb{N}_0;\ 0 \leqq l \leqq n;\ \lambda \in \mathbb{C}\right)$$

or, equivalently,

$$\mathcal{E}_n^{(l)}(x;\lambda) = \frac{n!(-2)^l}{(n+l)!}\ \mathcal{B}_{n+l}^{(l)}(x;-\lambda) \quad (n,l \in \mathbb{N}_0;\ \lambda \in \mathbb{C})$$

between the Apostol-Bernoulli polynomial of order l and Apostol-Euler polynomial of order l, which are defined, respectively, by 1.8(13) and 1.8(15), with $\alpha = l \ (l \in \mathbb{N}_0)$.

(See Luo and Srivastava [791, Lemma 4])

73. For the generalized Bernoulli polynomials $\mathfrak{B}_n^{(\alpha)}(x;\lambda;a,b,c)$ of order $\alpha \in \mathbb{C}$, defined by 1.8(66), derive each of the following identities:

$$\mathfrak{B}_n^{(\alpha)}(x+1;\lambda;a,b,c) = \sum_{k=0}^{n} \binom{n}{k} (\ln c)^{n-k}\ \mathfrak{B}_k^{(\alpha)}(x;\lambda;a,b,c);$$

$$\mathfrak{B}_n^{(\alpha)}(x+\alpha;\lambda;a,b,c) = \mathfrak{B}_n^{(\alpha)}\left(x;\lambda;\frac{a}{c},\frac{b}{c},c\right);$$

$$\mathfrak{B}_n^{(\alpha)}(\alpha - x; \lambda; a, b, c) = \mathfrak{B}_n^{(\alpha)}\left(-x; \lambda; \frac{a}{c}, \frac{b}{c}, c\right);$$

$$\alpha\lambda \ln\left(\frac{b}{a}\right) \sum_{k=0}^{n} \binom{n}{k} (\ln b)^k \, \mathfrak{B}_{n-k}^{(\alpha+1)}(x; \lambda; a, b, c) = (\alpha - n)\, \mathfrak{B}_n^{(\alpha)}(x; \lambda; a, b, c)$$
$$+ n(x \ln c - \alpha \ln a)\, \mathfrak{B}_{n-1}^{(\alpha)}(x; \lambda; a, b, c)$$

or, equivalently,

$$\alpha \ln\left(\frac{b}{a}\right) \sum_{k=0}^{n} \binom{n}{k} (\ln a)^k \, \mathfrak{B}_{n-k}^{(\alpha+1)}(x; \lambda; a, b, c) = (\alpha - n)\, \mathfrak{B}_n^{(\alpha)}(x; \lambda; a, b, c)$$
$$+ n(x \ln c - \alpha \ln b)\, \mathfrak{B}_{n-1}^{(\alpha)}(x; \lambda; a, b, c);$$

$$\mathfrak{B}_k^{(\alpha+\beta)}(x + y; \lambda; a, b, c) = \sum_{r=0}^{k} \binom{k}{r} \mathfrak{B}_{k-r}^{(\alpha)}(x; \lambda; a, b, c)\, \mathfrak{B}_r^{(\beta)}(y; \lambda; a, b, c);$$

$$\mathfrak{B}_k^{(\alpha)}(x + y; \lambda; a, b, c) = \sum_{r=0}^{k} \binom{k}{r} (y \ln c)^r \, \mathfrak{B}_{k-r}^{(\alpha)}(x; \lambda; a, b, c);$$

$$\frac{\partial^l}{\partial x^l}\left\{\mathfrak{B}_n^{(\alpha)}(x; \lambda; a, b, c)\right\} = \frac{n!}{(n-l)!} (\ln c)^l \, \mathfrak{B}_{n-l}^{(\alpha)}(x; \lambda; a, b, c) \quad (l \in \mathbb{N}_0)$$

and

$$\int_{\xi}^{\eta} \mathfrak{B}_n^{(\alpha)}(x; \lambda; a, b, c)\, dx$$
$$= \frac{1}{(n+1)\ln c}\left[\mathfrak{B}_{n+1}^{(\alpha)}(\eta; \lambda; a, b, c) - \mathfrak{B}_{n+1}^{(\alpha)}(\xi; \lambda; a, b, c)\right] \quad (\eta > \xi).$$

(See, for details, Srivastava *et al.* [1095])

74. For the generalized Bernoulli polynomials $\mathfrak{B}_n^{(\alpha)}(x; \lambda; a, b, c)$ of order $\alpha \in \mathbb{C}$, defined by 1.8(66), show that each of the following explicit series representations holds true:

$$\mathfrak{B}_n^{(l)}(x; \lambda; a, b, c) = l! \binom{n}{l} \sum_{r=0}^{n-l} \binom{n-l}{r} \binom{l+r-1}{r} \frac{\lambda^r}{(\lambda - 1)^{r+l}} (x \ln c - l \ln a)^{n-r-l}$$
$$\cdot \sum_{j=0}^{r} (-1)^j \binom{r}{j} \left[j \ln\left(\frac{b}{a}\right)\right]^r {}_2F_1\left(l + r - n, 1; r + 1; -\frac{j \ln\left(\frac{b}{a}\right)}{x \ln c - l \ln a}\right)$$
$$\left(a, b, c \in \mathbb{R}^+ \;\; (a \neq b);\; l \in \mathbb{N}_0;\; \lambda \in \mathbb{C} \setminus \{1\}\right)$$

and

$$\mathfrak{B}_n^{(l)}(x;\lambda;a,b,c) = e^{-x\left(\frac{\ln c \ln \lambda}{\ln b - \ln a}\right)} (\ln b - \ln a)^{n-l}$$
$$\cdot \sum_{k=0}^{\infty} \sum_{r=0}^{n+k} (-l)^r \left(\frac{\ln a}{\ln b - \ln a}\right)^r \binom{k}{r}$$
$$\cdot \binom{n+k-r-l}{k} \binom{n+k-r}{k-r}^{-1} \frac{(\ln\lambda)^k}{k!} \sum_{p=0}^{n+k-r} \binom{n+k-r}{p} \binom{l+p-1}{p} \frac{p!}{(2p)!}$$
$$\cdot \sum_{j=0}^{p} (-1)^j \binom{p}{j} j^{2p} \left(x \frac{\ln c}{\ln b - \ln a} + j\right)^{n+k-r-p}$$
$$\cdot {}_2F_1\left(p-n-k+r, p-l; 2p+1; \frac{j \ln\left(\frac{b}{a}\right)}{x \ln c + j \ln\left(\frac{b}{a}\right)}\right)$$
$$\left(a,b,c \in \mathbb{R}^+ \ (a \neq b);\ l \in \mathbb{N}_0\right),$$

where ${}_2F_1(a,b;c;z)$ denotes the Gauss hypergeometric function, defined by 1.5(4).

(Srivastava *et al.* [1095, p. 258, Theorem 6; p. 260, Theorem 7])

75. For the generalized Apostol type polynomials $\mathcal{Y}_{n,\beta}(x;k,a,b)$, defined by 1.8(60), derive each of the following properties:

$$\frac{d}{dx}\left\{\mathcal{Y}_{n,\beta}(x;k,a,b)\right\} = n\mathcal{Y}_{n-1,\beta}(x;k,a,b),$$

$$a^{nb(m-1)} m^{\nu-k} \sum_{j=0}^{m-1} \left(\frac{\beta}{a}\right)^{bjn} \mathcal{Y}_{\nu,\beta^m}\left(\frac{x}{m} + \frac{nj}{m}; k, a^m, b\right)$$
$$= a^{mb(n-1)} n^{\nu-k} \sum_{l=0}^{n-1} \left(\frac{\beta}{a}\right)^{blm} \mathcal{Y}_{\nu,\beta^n}\left(\frac{x}{n} + \frac{ml}{n}; k, a^n, b\right),$$

$$\int_0^y \mathcal{Y}_{n,\beta}(x;k,a,b)dx = \begin{cases} \dfrac{\mathcal{Y}_{n+1,\beta}(y;k,a,b) - \mathcal{Y}_{n+1,\beta}(k,a,b)}{n+1} & (n \in \mathbb{N}) \\ 0 & (n = 0) \end{cases}$$

and

$$\mathcal{Y}_{n+k-1,\beta}(x;k,a,b) = \left(-\frac{1}{2}\right)^{k-1} N^{n-1} \sum_{j=1}^{N} \left(\frac{\beta^{j-1}}{a^{j-N}}\right)^b$$
$$\cdot \mathcal{Y}_{n+k-1,\beta^N}\left(\frac{x+j-1}{N}; k, a^N, b\right).$$

(See, for details, Ozden *et al.* [886])

76. For the generalized Euler polynomials $\mathfrak{E}_n^{(\alpha)}(x;\lambda;a,b,c)$ of order $\alpha \in \mathbb{C}$, defined by 1.8(67), derive each of the following identities:

$$\mathfrak{E}_n^{(\alpha)}(x+1;\lambda;a,b,c) = \sum_{k=0}^{n} \binom{n}{k} (\ln c)^{n-k}\, \mathfrak{E}_k^{(\alpha)}(x;\lambda;a,b,c),$$

$$\mathfrak{E}_n^{(\alpha)}(x+\alpha;\lambda;a,b,c) = \mathfrak{E}_n^{(\alpha)}\left(x;\lambda;\frac{a}{c},\frac{b}{c},c\right),$$

$$\mathfrak{E}_n^{(\alpha)}(\alpha-x;\lambda;a,b,c) = \mathfrak{E}_n^{(\alpha)}\left(-x;\lambda;\frac{a}{c},\frac{b}{c},c\right),$$

$$\mathfrak{E}_n^{(\alpha)}(\alpha-x;\lambda;a,b,c) = (-1)^n\, \mathfrak{E}_n^{(\alpha)}\left(x;\lambda;\frac{c}{a},\frac{c}{b},c\right),$$

$$\frac{\alpha}{2}\ln\left(\frac{b}{a}\right)\sum_{k=0}^{n}\binom{n}{k}(\ln a)^k\, \mathfrak{E}_{n-k}^{(\alpha+1)}(x;\lambda;a,b,c) = \mathfrak{E}_{n+1}^{(\alpha)}(x;\lambda;a,b,c) - (x\ln c - \alpha\ln b)\,\mathfrak{E}_n^{(\alpha)}(x;\lambda;a,b,c),$$

$$\frac{\alpha\lambda}{2}\ln\left(\frac{b}{a}\right)\sum_{k=0}^{n}\binom{n}{k}\left(\ln\frac{b}{c}\right)^k \mathfrak{E}_{n-k}^{(\alpha+1)}(x+1;\lambda;a,b,c) = \mathfrak{E}_{n+1}^{(\alpha)}(x;\lambda;a,b,c) + (x\ln c - \alpha\ln a)\,\mathfrak{E}_n^{(\alpha)}(x;\lambda;a,b,c),$$

$$\frac{\alpha\lambda}{2}\ln\left(\frac{b}{a}\right)\sum_{k=0}^{n}\binom{n}{k}(\ln b)^k\, \mathfrak{E}_{n-k}^{(\alpha+1)}(x;\lambda;a,b,c) = \mathfrak{E}_{n+1}^{(\alpha)}(x;\lambda;a,b,c) - (x\ln c - \alpha\ln a)\,\mathfrak{E}_n^{(\alpha)}(x;\lambda;a,b,c),$$

$$\mathfrak{E}_n^{(\alpha+\beta)}(x+y;\lambda;a,b,c) = \sum_{k=0}^{n}\binom{n}{k}\mathfrak{E}_{n-k}^{(\alpha)}(x;\lambda;a,b,c)\,\mathfrak{E}_k^{(\beta)}(y;\lambda;a,b,c),$$

$$\mathfrak{E}_n^{(\alpha)}(x+y;\lambda;a,b,c) = \sum_{k=0}^{n}\binom{n}{k}(y\ln c)^{n-k}\,\mathfrak{E}_k^{(\alpha)}(x;\lambda;a,b,c),$$

$$\frac{\partial^l}{\partial x^l}\left\{\mathfrak{E}_n^{(\alpha)}(x;\lambda;a,b,c)\right\} = \frac{n!}{(n-l)!}(\ln c)^l\, \mathfrak{E}_{n-l}^{(\alpha)}(x;\lambda;a,b,c)$$

and

$$\int_{\xi}^{\eta}\mathfrak{E}_n^{(\alpha)}(x;\lambda;a,b,c)\,dx = \frac{1}{(n+1)\ln c}\left[\mathfrak{E}_{n+1}^{(\alpha)}(\eta;\lambda;a,b,c) - \mathfrak{E}_{n+1}^{(\alpha)}(\xi;\lambda;a,b,c)\right] \quad (\eta > \xi),$$

it being understood (*wherever needed*) that

$$a,b,c \in \mathbb{R}^+ \ (a \neq b), \quad x \in \mathbb{R} \quad \text{and} \quad l \in \mathbb{N}_0.$$

(See, for details, Srivastava *et al.* [1096])

77. For the generalized Euler polynomials $\mathfrak{E}_n^{(\alpha)}(x;\lambda;a,b,c)$ of order $\alpha \in \mathbb{C}$, defined by 1.8(67), show that each of the following explicit series representations holds true:

$$\mathfrak{E}_n^{(\alpha)}(x;\lambda;a,b,c) = 2^\alpha \sum_{k=0}^{n} \binom{n}{k}\binom{\alpha+k-1}{k} \frac{\lambda^k}{(\lambda+1)^{\alpha+k}} \left[\ln\left(\frac{b}{a}\right)\right]^k (x\ln c - \alpha \ln a)^{n-k}$$

$$\cdot \sum_{j=0}^{k} (-1)^j \binom{k}{j} j^k \, {}_2F_1\left(k-n, 1; k+1; -\frac{j\ln\left(\frac{b}{a}\right)}{x\ln c - \alpha\ln a}\right)$$

$$\left(a,b,c \in \mathbb{R}^+ \ (a \neq b);\ \alpha \in \mathbb{C};\ \lambda \in \mathbb{C}\setminus\{-1\}\right),$$

$$\mathfrak{E}_n^{(\alpha)}(x;\lambda;a,b,c) = 2^\alpha \sum_{k=0}^{n} \binom{n}{k}\binom{\alpha+k-1}{k} \frac{\lambda^k}{(\lambda+1)^{\alpha+k}} \left[\ln\left(\frac{b}{a}\right)\right]^k \sum_{j=0}^{k} (-1)^j \binom{k}{j} j^k$$

$$\cdot \left[x\ln c - \alpha\ln a + j\ln\left(\frac{b}{a}\right)\right]^{n-k} {}_2F_1\left(k-n, k; k+1; -\frac{j\ln\left(\frac{b}{a}\right)}{x\ln c - \alpha\ln a + j\ln\left(\frac{b}{a}\right)}\right)$$

$$\left(a,b,c \in \mathbb{R}^+ \ (a \neq b);\ \alpha \in \mathbb{C};\ \lambda \in \mathbb{C}\setminus\{-1\}\right)$$

and

$$\mathfrak{E}_n^{(\alpha)}(x;\lambda;a,b,c) = e^{-x\left(\frac{\ln c \ln \lambda}{\ln b - \ln a}\right)} \left[\ln\left(\frac{b}{a}\right)\right]^k \sum_{k=0}^{\infty}\sum_{j=0}^{n} (-\alpha)^j \binom{n}{j} \left(\frac{\ln a}{\ln b - \ln a}\right)^j$$

$$\cdot \frac{(\ln\lambda)^k}{k!} \sum_{m=0}^{n+k-j} \binom{n+k-j}{m}\binom{\alpha+m-1}{m} \sum_{\ell=0}^{m} (-1)^\ell \binom{m}{\ell}\left(\frac{\ell}{2}\right)^m$$

$$\cdot \left(\frac{x\ln c}{\ln b - \ln a} + \ell\right)^{n+k-j-m} {}_2F_1\left(m-n-k+j, m; m+1; \frac{\ell\ln\left(\frac{b}{a}\right)}{x\ln c + \ell\ln\left(\frac{b}{a}\right)}\right),$$

$$\left(a,b,c \in \mathbb{R}^+ \ (a \neq b);\ \alpha \in \mathbb{C}\right),$$

where ${}_2F_1(a,b;c;z)$ denotes the Gauss hypergeometric function, defined by 1.5(4).
(Srivastava *et al.* [1096, pp. 294–295, Theorem 6; p. 298, Theorem 7])

78. Show that the following relationships hold true:

$$\mathfrak{G}_n^{(\alpha)}(x;\lambda;a,b,c) = (-2)^\alpha\, \mathfrak{B}_n^{(\alpha)}(x;-\lambda;a,b,c) \quad \left(\alpha \in \mathbb{C};\ 1^\alpha := 1\right)$$

and

$$\mathfrak{G}_n^{(l)}(x;\lambda;a,b,c) = (-1)^l(-n)_l\, \mathfrak{E}_{n-l}^{(l)}(x;\lambda;a,b,c) = \frac{n!}{(n-l)!}\, \mathfrak{E}_{n-l}^{(l)}(x;\lambda;a,b,c)$$

$$\left(n,l \in \mathbb{N}_0;\ n \geqq l;\ \lambda \in \mathbb{C}\right)$$

between the generalized Bernoulli polynomials $\mathfrak{B}_n^{(\alpha)}(x;\lambda;a,b,c)$, the generalized Euler polynomials $\mathfrak{E}_n^{(\alpha)}(x;\lambda;a,b,c)$ and the generalized Genocchi polynomials $\mathfrak{G}_n^{(\alpha)}(x;\lambda;a,b,c)$ of order $\alpha \in \mathbb{C}$, defined by 1.8(66), 1.8(67) and 1.8(68), respectively.
(Srivastava *et al.* [1096, p. 300, Lemma 3])

79. For the generalized Genocchi polynomials $\mathfrak{G}_n^{(\alpha)}(x;\lambda;a,b,c)$ of order $\alpha\in\mathbb{C}$, defined by 1.8(68), show that each of the following explicit series representations holds true:

$$\mathfrak{G}_n^{(l)}(x;\lambda;a,b,c)=\frac{2^l\cdot n!}{(n-l)!}\sum_{k=0}^{n-l}\binom{n-l}{k}\binom{k+l-1}{k}\frac{\lambda^k}{(\lambda+1)^{k+l}}\left[\ln\left(\frac{b}{a}\right)\right]^k$$

$$\cdot(x\ln c-l\ln a)^{n-k-l}\sum_{j=0}^{k}(-1)^j\binom{k}{j}j^k\,{}_2F_1\left(k+l-n,1;k+1;-\frac{j\ln\left(\frac{b}{a}\right)}{x\ln c-l\ln a}\right)$$

$$\left(a,b,c\in\mathbb{R}^+\ (a\neq b);\ l\in\mathbb{N}_0;\ \lambda\in\mathbb{C}\setminus\{1\}\right)$$

and

$$\mathfrak{G}_n^{(l)}(x;\lambda;a,b,c)=\frac{2^l\cdot n!}{(n-l)!}\sum_{k=0}^{n-l}\binom{n-l}{k}\binom{k+l-1}{k}\frac{\lambda^k}{(\lambda+1)^{k+l}}\left[\ln\left(\frac{b}{a}\right)\right]^k$$

$$\cdot\sum_{j=0}^{k}(-1)^j\binom{k}{j}j^k\left[x\ln c-l\ln a+j\ln\left(\frac{b}{a}\right)\right]^{n-k-l}$$

$$\cdot\,{}_2F_1\left(k+l-n,k;k+1;-\frac{j\ln\left(\frac{b}{a}\right)}{x\ln c-l\ln a+j\ln\left(\frac{b}{a}\right)}\right)$$

$$\left(a,b,c\in\mathbb{R}^+\ (a\neq b);\ l\in\mathbb{N}_0;\ \lambda\in\mathbb{C}\setminus\{1\}\right),$$

where ${}_2F_1(a,b;c;z)$ denotes the Gauss hypergeometric function, defined by 1.5(4).

(Srivastava *et al.* [1096, p. 301, Theorem 9])

80. For the Apostol-Genocchi polynomials $\mathcal{G}_n(x;\lambda)$ $(\lambda\in\mathbb{C})$, defined by 1.8(58) (*with* $\alpha=1$), derive the following exponential Fourier series representations:

$$\mathcal{G}_n(x;\lambda)=\frac{2\cdot n!}{\lambda^x}\sum_{k=-\infty}^{\infty}\frac{e^{(2k-1)\pi ix}}{[(2k-1)\pi i-\log\lambda]^n}$$

$$=\frac{(2\cdot n!)i^n}{\lambda^x}\left(\sum_{k=0}^{\infty}\frac{\exp\left[\left(\frac{n\pi}{2}-(2k+1)\pi x\right)i\right]}{[(2k+1)\pi i+\log\lambda]^n}+\sum_{k=0}^{\infty}\frac{\exp\left[\left(-\frac{n\pi}{2}+(2k+1)\pi x\right)i\right]}{[(2k+1)\pi i-\log\lambda]^n}\right)$$

$$(n\in\mathbb{N};\ 0\leqq x\leqq 1;\ \lambda\in\mathbb{C}\setminus\{0,-1\}).$$

(See Luo and Srivastava [791, p. 5726, Theorem 20]; see also [786])

81. Derive each of the following properties of the Apostol-Genocchi polynomials $\mathcal{G}_n^{(\alpha)}(x;\lambda)$ of order $\alpha\in\mathbb{C}$, defined by 1.8(58):

$$\mathcal{G}_n^{(\alpha)}(\lambda)=\mathcal{G}_n^{(\alpha)}(0;\lambda),\quad \mathcal{G}_n^{(0)}(x;\lambda)=x^n,$$

$$\mathcal{G}_n^{(0)}(\lambda)=\delta_{n,0}\quad \textit{and}\quad \mathcal{G}_0^{(\alpha)}(x;\lambda)=\mathcal{G}_0^{(\alpha)}(\lambda)=\delta_{\alpha,0}\quad (n\in\mathbb{N}_0;\ \alpha\in\mathbb{C}),$$

where $\delta_{n,k}$ denotes the Kronecker symbol;

$$\mathcal{G}_n^{(\alpha)}(x;\lambda)=\sum_{k=0}^{n}\binom{n}{k}\mathcal{G}_k^{(\alpha)}(\lambda)\,x^{n-k}$$

and

$$\mathcal{G}_n^{(\alpha)}(x;\lambda)=\sum_{k=0}^{n}\binom{n}{k}\mathcal{G}_{n-k}^{(\alpha-1)}(\lambda)\mathcal{G}_k(x;\lambda);$$

$$\lambda\mathcal{G}_n^{(\alpha)}(x+1;\lambda)+\mathcal{G}_n^{(\alpha)}(x;\lambda)=2n\,\mathcal{G}_{n-1}^{(\alpha-1)}(x;\lambda)\quad(n\in\mathbb{N});$$

$$\frac{\partial}{\partial x}\left\{\mathcal{G}_n^{(\alpha)}(x;\lambda)\right\}=n\mathcal{G}_{n-1}^{(\alpha)}(x;\lambda)\quad(n\in\mathbb{N})$$

and

$$\frac{\partial^p}{\partial x^p}\left\{\mathcal{G}_n^{(\alpha)}(x;\lambda)\right\}=\frac{n!}{(n-p)!}\,\mathcal{G}_{n-p}^{(\alpha)}(x;\lambda)\quad(n,p\in\mathbb{N}_0;\ 0\leqq p\leqq n);$$

$$\int_a^b\mathcal{G}_n^{(\alpha)}(x;\lambda)dx=\frac{\mathcal{G}_{n+1}^{(\alpha)}(b;\lambda)-\mathcal{G}_{n+1}^{(\alpha)}(a;\lambda)}{n+1}$$

and

$$\int_a^b\mathcal{G}_n^{(\alpha)}(x;\lambda)dx=\sum_{k=0}^{n}\frac{1}{n-k+1}\binom{n}{k}\mathcal{G}_k^{(\alpha)}(\lambda)\,(b^{n-k+1}-a^{n-k+1});$$

$$\mathcal{G}_n^{(\alpha+\beta)}(x+y;\lambda)=\sum_{k=0}^{n}\binom{n}{k}\mathcal{G}_k^{(\alpha)}(x;\lambda)\mathcal{G}_{n-k}^{(\beta)}(y;\lambda);$$

$$\mathcal{G}_n^{(\alpha)}(\alpha-x;\lambda)=\frac{(-1)^{n+\alpha}}{\lambda^\alpha}\mathcal{G}_n^{(\alpha)}(x;\lambda^{-1})$$

and

$$\mathcal{G}_n^{(\alpha)}(\alpha+x;\lambda)=\frac{(-1)^{n+\alpha}}{\lambda^\alpha}\mathcal{G}_n^{(\alpha)}(-x;\lambda^{-1});$$

$$(n-\alpha)\,\mathcal{G}_n^{(\alpha)}(x;\lambda)=nx\,\mathcal{G}_{n-1}^{(\alpha)}(x;\lambda)-\frac{\alpha\lambda}{2}\,\mathcal{G}_n^{(\alpha+1)}(x+1;\lambda)$$

and

$$\frac{\alpha}{2}\,\mathcal{G}_n^{(\alpha+1)}(x;\lambda)=n(\alpha-x)\,\mathcal{G}_{n-1}^{(\alpha)}(x;\lambda)+(n-\alpha)\,\mathcal{G}_n^{(\alpha)}(x;\lambda).$$

(See Luo and Srivastava [791, pp. 5706–5607])

82. For the Apostol-Genocchi polynomials $\mathcal{G}_n^{(\alpha)}(x;\lambda)$ of order l $(l\in\mathbb{N}_0)$, defined by 1.8(58) with $\alpha=l$ $(l\in\mathbb{N}_0)$, show that the following explicit series representations hold true:

$$\mathcal{G}_n^{(l)}(x;\lambda)=2^l\,l!\binom{n}{l}\sum_{k=0}^{n-l}\binom{n-l}{k}\binom{l+k-1}{k}\frac{\lambda^k}{(\lambda+1)^{l+k}}$$

$$\cdot\sum_{j=0}^{k}(-1)^j\binom{k}{j}j^k\,(x+j)^{n-k-l}\;{}_2F_1\left(l+k-n,k;k+1;\frac{j}{x+j}\right)$$

$$(n,l\in\mathbb{N}_0;\ \lambda\in\mathbb{C}\setminus\{-1\})$$

and

$$\mathcal{G}_n^{(l)}(x;\lambda) = e^{-x\log\lambda}\sum_{k=0}^{\infty}\binom{n+k-l}{k}\binom{n+k}{k}^{-1}\binom{n+k}{l}\frac{l!\,(\log\lambda)^k}{k!}$$

$$\cdot\sum_{r=0}^{n+k-l}\frac{1}{2^r}\binom{n+k-l}{r}\binom{l+r-1}{r}$$

$$\cdot\sum_{j=0}^{r}(-1)^j\binom{r}{j}j^r\,(x+j)^{n+k-r-l}\;{}_2F_1\left(r+l-n-k,r;r+1;\frac{j}{x+j}\right)$$

$$(n,l\in\mathbb{N}_0;\ \lambda\in\mathbb{C}),$$

where ${}_2F_1(a,b;c;z)$ denotes the Gauss hypergeometric function defined by 1.5(4).

(See Luo and Srivastava [791, p. 5708, Theorem 1])

83. For the Apostol-Genocchi polynomials $\mathcal{G}_n^{(\alpha)}(x;\lambda)$ of order l $(l\in\mathbb{N}_0)$ defined by 1.8(58), derive the following relationship:

$$\mathcal{G}_n^{(\alpha)}(x+y;\lambda) = \sum_{k=0}^{n}\frac{2}{k+1}\binom{n}{k}\left[(k+1)\mathcal{G}_k^{(\alpha-1)}(y;\lambda)-\mathcal{G}_{k+1}^{(\alpha)}(y;\lambda)\right]\mathcal{B}_{n-k}(x;\lambda)$$

$$\left(\alpha,\lambda\in\mathbb{C};\ n\in\mathbb{N}_0\right)$$

with the Apostol-Bernoulli polynomials $\mathcal{B}_{n-k}(x;\lambda)$ defined by 1.8(1).

(See Luo and Srivastava [791, p. 5710, Theorem 2])

84. For the Apostol-Genocchi polynomials $\mathcal{G}_n(x;\lambda)$, defined by 1.8(58) with $\alpha=1$, show that the following integral representation holds true:

$$\mathcal{G}_n(z;e^{2\pi i\xi}) = 2ne^{-2\pi iz\xi}\int_0^{\infty}\left(\frac{M(n;z,t)\cosh(2\pi\xi t)+iN(n;x,t)\sinh(2\pi\xi t)}{\cosh(2\pi t)-\cos(2\pi z)}\right)t^{n-1}\,dt$$

$$\left(n\in\mathbb{N};\ 0\leqq\Re(z)\leqq 1;\ |\xi|<\frac{1}{2}\ (\xi\in\mathbb{R})\right),$$

where

$$M(n;z,t) = \left[e^{\pi t}\cos\left(\pi z-\frac{n\pi}{2}\right)-e^{-\pi t}\cos\left(\pi z+\frac{n\pi}{2}\right)\right]$$

and

$$N(n;z,t) = \left[e^{\pi t}\sin\left(\pi z-\frac{n\pi}{2}\right)+e^{-\pi t}\sin\left(\pi z+\frac{n\pi}{2}\right)\right].$$

(See Luo and Srivastava [791, p. 5724, Theorem 19])

85. Let the so-called λ-Stirling numbers $\mathcal{S}(n,k;\lambda)$ of the second kind be defined by means of the following generating function:

$$\frac{(\lambda e^z-1)^k}{k!} = \sum_{n=0}^{\infty}\mathcal{S}(n,k;\lambda)\,\frac{z^n}{n!}\quad (k\in\mathbb{N}_0;\ \lambda\in\mathbb{C}),$$

so that, obviously,

$$S(n,k) := \mathcal{S}(n,k;1)$$

for the Stirling numbers $S(n,k)$ of the second kind, defined by 1.6(15). Show that each of the following results holds true:

$$\lambda^x x^n = \sum_{k=0}^{\infty} \binom{x}{k} k!\, \mathcal{S}(n,k;\lambda) \quad (k \in \mathbb{N}_0;\ \lambda \in \mathbb{C}),$$

$$\mathcal{S}(n,k;\lambda) = \frac{1}{k!} \sum_{j=0}^{k} (-1)^{k-j} \binom{k}{j} \lambda^j j^n \quad (n,k \in \mathbb{N}_0;\ \lambda \in \mathbb{C}),$$

$$\mathcal{S}(n,k;\lambda) = \frac{1}{k!} \sum_{j=0}^{k} (-1)^{j} \binom{k}{j} \lambda^{k-j} (k-j)^n \quad (n,k \in \mathbb{N}_0;\ \lambda \in \mathbb{C}),$$

$$\mathcal{S}(n,k;\lambda) = \mathcal{S}(n-1,k-1;\lambda) + k\, \mathcal{S}(n-1,k;\lambda) \quad (n,k \in \mathbb{N})$$

and

$$\mathcal{S}(n,k;\lambda) = \sum_{j=0}^{n-1} \binom{n-1}{j} \lambda^{n-j-1}\, \mathcal{S}(j,k-1;\lambda) \quad (n,k \in \mathbb{N}).$$

(See Luo and Srivastava [791, p. 5716, Theorems 9 to 11])

86. For the λ-Stirling numbers $\mathcal{S}(n,k;\lambda)$ of the second kind, defined by means of the generating function in Problem 84 above, show that each of the following explicit relationships holds true:

$$\mathcal{S}(n,k;\lambda) = n! \sum_{j=n}^{\infty} \binom{j}{n} \frac{(\log \lambda)^{j-n}}{j!}\, S(j,k) \quad (n,k \in \mathbb{N}_0;\ \lambda \in \mathbb{C})$$

and

$$\mathcal{S}(n,k;\lambda) = \sum_{j=0}^{k} \frac{\lambda^j (\lambda-1)^{k-j}}{(k-j)!}\, S(n,j) \quad (n,k \in \mathbb{N}_0;\ \lambda \in \mathbb{C}),$$

with the Stirling numbers $S(n,k)$ of the second kind, defined by 1.6(15).

(See Luo and Srivastava [791, pp. 5716–5717, Theorem 12])

2 The Zeta and Related Functions

This chapter aims at providing a self-contained theory of the Zeta and related functions, which will be required in each of the next chapters. We first introduce (and investigate the various properties and relationships satisfied by) the multiple Hurwitz Zeta function $\zeta_n(s,a)$ $(n \in \mathbb{N})$ and consider its relatively more familiar special case when $n = 1$, that is, the Hurwitz (or generalized) Zeta function $\zeta(s,a)$. We then deal rather systematically with the Riemann Zeta function, which (for the *main* purpose of this book) happens to be the most important member of the significantly large family of Zeta functions considered in this chapter. Other functions (introduced in this chapter) include the Polylogarithm functions, Legendre's Chi function, Clausen's integral (or Clausen's function), the Hurwitz–Lerch Zeta function and so on.

2.1 Multiple Hurwitz Zeta Functions

Barnes [97] introduced and studied the generalized multiple Hurwitz Zeta function $\zeta_n(s, a \,|\, w_1, \ldots, w_n)$ defined for $\Re(s) > n$ by the n-ple series

$$\zeta_n(s, a \,|\, w_1, \ldots, w_n) := \sum_{m_1, \ldots, m_n = 0}^{\infty} \frac{1}{(a+\Omega)^s} \quad (\Re(s) > n;\ n \in \mathbb{N}), \tag{1}$$

where $\Omega = m_1 w_1 + \cdots + m_n w_n$ and the general conditions for a and the parameters $w_1, \ldots, w_n$ are given in Barnes [97], who used it in the study of the multiple Gamma functions (see Section 1.4). In this section, we consider only the simple case of (1), when

$$w_j = 1\,(j = 1, \ldots, n;\ j, n \in \mathbb{N})$$

and

$$\zeta_n(s,a) := \sum_{k_1, \ldots, k_n = 0}^{\infty} (a + k_1 + \cdots + k_n)^{-s} \quad (\Re(s) > n;\ a \notin \mathbb{Z}_0^-), \tag{2}$$

which is also referred to as *n-ple* (or, simply, *multiple*) *Hurwitz Zeta function*. We shall give a rather detailed investigation of the properties and characteristics of the function $\zeta_n(s,a)$ in (2), including its analytic continuation.

Zeta and q-Zeta Functions and Associated Series and Integrals. DOI: 10.1016/B978-0-12-385218-2.00002-5

The Analytic Continuation of $\zeta_n(s,a)$

In order to give the analytic continuation of $\zeta_n(s,a)$, we first recall a criterion for the convergence of an n-ple series attributed to Eisenstein in the work of Forsyth [457], who gave the proof for the double series. Here, we shall give a proof of the general case:

Theorem 2.1 *The n-ple series:*

$$\sum_{m_1,\ldots,m_n=-\infty}^{\infty}{}' \left(m_1^2+\cdots+m_n^2\right)^{-\mu} \tag{3}$$

converges, if $\mu > \frac{1}{2}n$, *where the prime denotes the exclusion of the case when* $m_j = 0$ $(j = 1, \ldots, n)$.

Proof. Let the series be arranged in the partial series: for this purpose, choose integers k_j, such that

$$2^{k_j} \leqq m_j < 2^{k_j+1} \quad (1 \leqq j \leqq n_j;\ k_j \in \mathbb{N}_0).$$

Let

$$D := \{(m_1,\ldots,m_n)\,|\,2^{k_j} \leqq m_j < 2^{k_j+1};\ 1 \leqq j \leqq n\}$$

and let $I_j := [2^{k_j}, 2^{k_j+1})$ $(1 \leqq j \leqq n)$ be half-open intervals. Since $2^{k+1} - 2^k = 2^k$, the number of integers m_j in I_j is 2^{k_j} $(1 \leqq j \leqq n)$. So the number of n-tuples in D is

$$2^{k_1}\cdots 2^{k_n} = 2^{k_1+\cdots+k_n} = 2^{nx},$$

where

$$x := \frac{k_1+\cdots+k_n}{n}.$$

Note that, for $(m_1,\ldots,m_n) \in D$,

$$\sum_{j=1}^{n} 2^{2k_j} \leqq m_1^2+\cdots+m_n^2 < \sum_{j=1}^{n} 2^{2k_j+2}$$

and, by comparing the arithmetic and geometric means of 2^{2k_j} $(1 \leqq j \leqq n)$, we also have

$$2^{2x} \leqq \frac{1}{n}\sum_{j=1}^{n} 2^{2k_j} \leqq m_1^2+\cdots+m_n^2.$$

Now, for any $(m_1,\ldots,m_n)$ in D and any positive real μ, we have

$$\frac{1}{(m_1^2+\cdots+m_n^2)^{\mu}} \leqq \frac{1}{2^{2x\mu}}.$$

Then, the sum of all terms as $1/(m_1^2+\cdots+m_n^2)^{\mu}$ in D is less than or equal to

$$\frac{2^{nx}}{2^{2x\mu}} = \frac{1}{2^{x(2\mu-n)}} = \frac{1}{2^{\frac{2\mu-n}{n}k_1}} \cdots \frac{1}{2^{\frac{2\mu-n}{n}k_n}}.$$

Let $k=\max\{k_j \mid 1 \leqq j \leqq n\}$. We find that the sum of all the partial series is less than or equal to

$$\begin{aligned}\sum_{k_1=0}^{k}\cdots\sum_{k_n=0}^{k}\frac{1}{2^{\frac{2\mu-n}{n}k_1}}\cdots\frac{1}{2^{\frac{2\mu-n}{n}k_n}} &= \left(\sum_{k_1=0}^{k}\frac{1}{2^{\frac{2\mu-n}{n}k_1}}\right)^n \\ &= \left[\frac{1-2^{-(k+1)\frac{2\mu-n}{n}}}{1-\frac{1}{2^{\frac{2\mu-n}{n}}}}\right]^n.\end{aligned} \tag{4}$$

Taking the limit in (4) as $k \to \infty$, we observe that the n-ple series in (4) converges, if $2\mu-n>0$, that is, if $\mu > \frac{1}{2}n$ and the sum is

$$\left(\frac{1}{1-\frac{1}{2^{\frac{2\mu-n}{n}}}}\right)^n = \frac{2^{2\mu-n}}{\left(2^{\frac{2\mu-n}{n}}-1\right)^n}.$$

This completes the proof of Theorem 2.1. ■

Let $s=\sigma+it$ ($\sigma,\, t \in \mathbb{R}$). First, for convergence, we consider $\zeta_n(s,a)$ in (2) for the case when $a>0$:

$$\zeta_n(s,a) = \sum_{k_1,\ldots,k_n=0}^{\infty} (a+k_1+\cdots+k_n)^{-s} \quad (\Re(s)=\sigma>n;\ a>0). \tag{5}$$

Theorem 2.2 *The series for $\zeta_n(s,a)$ in (5) converges absolutely for $\sigma>n$. The convergence is uniform in every half-plane $\sigma \geq n+\delta$ ($\delta>0$), so $\zeta_n(s,a)$ is an analytic function of s in the half-plane $\sigma>n$.*

Proof. Observe that, for $\sigma>0$,

$$\begin{aligned}\sum_{k_1,\ldots,k_n=0}^{\infty}{}'(k_1+\cdots+k_n)^{-\sigma} &= \sum_{k_1,\ldots,k_n=0}^{\infty}{}'[(k_1+\cdots+k_n)^2]^{-\frac{1}{2}\sigma} \\ &\leqq \sum_{k_1,\ldots,k_n=0}^{\infty}{}'(k_1^2+\cdots+k_n^2)^{-\frac{1}{2}\sigma},\end{aligned}$$

in which the prime denotes the exclusion of the case when $k_j = 0$ $(1 \leqq j \leqq n)$ and that the last series converges for $\sigma > n$ by Eisentein's theorem (Theorem 2.1). Thus, all statements in Theorem 2.2 follow from the inequality:

$$\sum_{k_1,\ldots,k_n=0}^{\infty} |(a+k_1+\cdots+k_n)^{-s}| = \sum_{k_1,\ldots,k_n=0}^{\infty} (a+k_1+\cdots+k_n)^{-\sigma} \leqq \sum_{k_1,\ldots,k_n=0}^{\infty} (a+k_1+\cdots+k_n)^{-n-\delta}.$$

We now choose to recall a convergence theorem concerning term-by-term integration of monotonic sequences of functions, which is due to Lévi (see Apostol [64, p. 268, Theorem 10.25]). ∎

Theorem 2.3 *Let $L(I)$ denote the set of all Lebesgue-integrable functions on an interval I. Also, let $\{g_n\}$ be a sequence of functions in $L(I)$, such that*

(a) *each g_n is nonnegative almost everywhere on I*
(b) *the series $\sum_{n=1}^{\infty} \int_I g_n$ converges.*

Then, the series $\sum_{n=1}^{\infty} g_n$ converges almost everywhere on I to a sum function g in $L(I)$. Moreover,

$$\int_I g = \int_I \sum_{n=1}^{\infty} g_n = \sum_{n=1}^{\infty} \int_I g_n.$$

Next, we present an integral representation of $\zeta_n(s,a)$, which is given by

Theorem 2.4 *If $\Re(s) = \sigma > n$, then*

$$\Gamma(s)\,\zeta_n(s,a) = \int_0^{\infty} \frac{x^{s-1}e^{-ax}}{(1-e^{-x})^n}dx \quad (\Re(s) > n;\ n \in \mathbb{N}). \tag{6}$$

Proof. It follows from 1.1(1) that, for $\Re(s) = \sigma > 0$,

$$\Gamma(s) = \int_0^{\infty} x^{s-1}e^{-x}dx.$$

We first keep s real, $s > n$ $(n \in \mathbb{N})$, and then extend the result to complex s by analytic continuation. In this Eulerian integral for $\Gamma(s)$, we set

$$x = (a+k_1+\cdots+k_n)t \quad (k_j \in \mathbb{N}_0;\ j = 1,\ \ldots,\ n),$$

and we find that

$$\Gamma(s) = (a + k_1 + \cdots + k_n)^s \int_0^\infty e^{-(a+k_1+\cdots+k_n)t} \; t^{s-1} dt,$$

so that

$$(a + k_1 + \cdots + k_n)^{-s} \, \Gamma(s) = \int_0^\infty e^{-(k_1+\cdots+k_n)t} e^{-at} t^{s-1} dt.$$

Summing over all $k_j \in \mathbb{N}_0$ $(1 \leqq j \leqq n)$, we obtain

$$\zeta_n(s, a) \, \Gamma(s) = \sum_{k_1, \ldots, k_n = 0}^{\infty} \int_0^\infty e^{-(k_1+\cdots+k_n)t} e^{-at} t^{s-1} \, dt,$$

the series on the right being convergent, if $s > n$.

Now, we wish to interchange the order of summation and integration. The simplest way to justify this process is to regard the integral as a Lebesgue integral. Since the integrand is nonnegative, Lévi's convergence theorem (Theorem 2.3) implies that the series:

$$\sum_{k_1, \ldots, k_n = 0}^{\infty} \int_0^\infty e^{-(a+k_1+\cdots+k_n)t} e^{-at} t^{s-1} dt$$

converges almost everywhere to a sum function, which itself is Lebesgue-integrable on $[0, \infty)$ and that

$$\zeta_n(s, a) \, \Gamma(s) = \int_0^\infty \sum_{k_1, \ldots, k_n = 0}^{\infty} e^{-(k_1+\cdots+k_n)t} e^{-at} t^{s-1} \, dt.$$

But, if $t > 0$, we have $0 < e^{-t} < 1$, and, hence,

$$\sum_{k=0}^{\infty} e^{-kt} = \frac{1}{1 - e^{-t}},$$

the series being a geometric series. We, therefore, have

$$\sum_{k_1, \ldots, k_n = 0}^{\infty} e^{-(k_1+\cdots+k_n)t} e^{-at} t^{s-1} = \frac{e^{-at} t^{s-1}}{(1 - e^{-t})^n}$$

almost everywhere on $[0, \infty)$; in fact, everywhere except at 0, so we get

$$\zeta_n(s,a)\,\Gamma(s) = \int_0^\infty \frac{e^{-at}t^{s-1}}{(1-e^{-t})^n}\,dt.$$

This proves (6) for real s and $s > n$ $(n \in \mathbb{N})$.

To extend this integral representation to all complex $s = \sigma + it$ with $\sigma > n$, we note that both functions on the left-hand side of (6) are analytic for $\sigma > n$. To show that the right member is also analytic, we assume that $n + \delta \leqq \sigma \leqq c$, where $c > n$ and $\delta > 0$, and write

$$\int_0^\infty \left|\frac{e^{-at}t^{s-1}}{(1-e^{-t})^n}\right| dt = \int_0^\infty \frac{e^{-at}t^{\sigma-1}}{(1-e^{-t})^n}\,dt$$
$$= \left(\int_0^1 + \int_1^\infty\right) \frac{e^{-at}t^{\sigma-1}}{(1-e^{-t})^n}\,dt.$$

If $0 \leqq t \leqq 1$, we have $t^{\sigma-n} \leqq t^{\delta}$, and, if $t \geqq 1$, we have $t^{\sigma-n} \leqq t^{c-n}$. Also, since $e^t - 1 \geqq t$ for $t \geqq 0$, we have

$$\int_0^1 \frac{e^{-at}t^{\sigma-1}}{(1-e^{-t})^n}\,dt \leqq \int_0^1 \frac{e^{(n-a)t}t^{\delta+n-1}}{(e^t-1)^n}\,dt$$
$$\leqq \begin{cases} e^{n-a}\int_0^1 t^{\delta-1}\,dt = \frac{e^{n-a}}{\delta} & (0 < a \leqq n) \\ \int_0^1 t^{\delta-1}dt = \frac{1}{\delta} & (a > n) \end{cases}$$

and

$$\int_1^\infty \frac{e^{-at}t^{\sigma-1}}{(1-e^{-t})^n}\,dt \leqq \int_0^\infty \frac{e^{-at}t^{c-1}}{(1-e^{-t})^n}\,dt = \Gamma(c)\,\zeta_n(c,a).$$

This shows that the integral in (6) converges uniformly in every strip $n + \delta \leqq \sigma \leqq c$, where $\delta > 0$ and, therefore, represents an analytic function in every such strip; hence, also in the half-plane $\sigma > n$. Thus, by the principle of analytic continuation, (6) holds true for all s with $\Re(s) = \sigma > n$ $(n \in \mathbb{N})$. ∎

In order to extend $\zeta_n(s,a)$ to the half-plane on the left of the line $\sigma = n$, we derive another representation in terms of a contour integral. The contour C is essentially a Hankel's loop (*cf.*, e.g., Whittaker and Watson [1225, p. 245]), which starts from ∞

along the upper side of the positive real axis, encircles the origin once in the positive (counter-clockwise) direction and then returns to ∞ along the lower side of the positive real axis.

The loop C is composed of three parts C_1, C_2 and C_3, where C_2 is a positively-oriented circle of radius $c < 2\pi$ about the origin and C_1 and C_3 are the upper and lower edges of a cut in the complex z-plane along the positive real axis, traversed as described above. Thus, we can use the parameterizations $-z = re^{-\pi i}$ on C_1 and $-z = re^{\pi i}$ on C_3, where r varies from c to ∞.

Theorem 2.5 *If $a > 0$, then the function defined by the following contour integral:*

$$I_n(s,a) = -\frac{1}{2\pi i}\int_{\infty}^{(0+)} \frac{(-z)^{s-1}e^{-az}}{(1-e^{-z})^n}\,dz \tag{7}$$

is an entire function of s. Moreover,

$$\zeta_n(s,a) = \Gamma(1-s)\,I_n(s,a) \quad (\Re(s) = \sigma > n). \tag{8}$$

Proof. Here $(-z)^s$ means $r^s e^{-\pi i s}$ on C_1 and $r^s e^{\pi i s}$ on C_3. We consider an arbitrary compact disk $|s| \leqq M$ and prove that the integrals along C_1 and C_3 converge uniformly on every such disk. Since the integrand in (7) is an entire function of s, this will prove that $I_n(s,a)$ is also an entire function of s for $a > 0$. Along C_1, we have, for $r \geqq 1$,

$$|(-z)^{s-1}| = r^{\sigma-1}\left|e^{-\pi i(\sigma-1+it)}\right| = r^{\sigma-1}\,e^{\pi t} \leqq r^{M-1}e^{\pi M},$$

since $|s| \leqq M$. Similarly, along C_3, we have, for $r \geqq 1$,

$$\left|(-z)^{s-1}\right| = r^{\sigma-1}|e^{\pi i(\sigma-1+it)}| = r^{\sigma-1}e^{-\pi t} \leqq r^{M-1}e^{\pi M}.$$

Hence, on either C_1 or C_3, we have, for $r \geqq 1$,

$$\left|\frac{(-z)^{s-1}e^{-az}}{(1-e^{-z})^n}\right| \leqq \frac{r^{M-1}e^{\pi M}e^{-ar}}{(1-e^{-r})^n} \leqq \frac{e^{\pi M}}{\left(1-e^{-1}\right)^n}\cdot r^{M-1}\cdot e^{-ar}.$$

But the integral:

$$\int_c^\infty r^{M-1}\,e^{-ar}\,dr$$

converges, if $c > 0$; this shows that the integrals along C_1 and C_3 converge uniformly on every compact disk $|s| \leqq M$, and, hence, $I_n(s,a)$ is an entire function of s. To prove (8), we write

$$-2\pi i I_n(s,a) = \left(\int_{C_1} + \int_{C_2} + \int_{C_3}\right)(-z)^{s-1}g(-z)\,dz,$$

where

$$g(-z) = \frac{e^{-az}}{(1-e^{-z})^n}.$$

On C_1 and C_3, we have $g(-z) = g(-r)$, and, on C_2, we write $-z = ce^{i\theta}$, where θ varies from 2π to 0. This gives us

$$\begin{aligned} -2\pi i I_n(s,a) &= \int_\infty^c r^{s-1} e^{-\pi i(s-1)} g(-r)\,dr - i\int_{2\pi}^0 c^{s-1} e^{(s-1)i\theta} ce^{i\theta} g(ce^{i\theta})\,d\theta \\ &\quad + \int_c^\infty r^{s-1} e^{\pi i(s-1)} g(-r)\,dr \qquad (9) \\ &= -2i\sin(\pi s)\int_c^\infty r^{s-1} g(-r)\,dr - ic^s \int_{2\pi}^0 e^{is\theta} g(ce^{i\theta})\,d\theta. \end{aligned}$$

Dividing both sides of (9) by $-2i$, we obtain

$$\pi I_n(s,a) = \sin(\pi s) I_1(s,c) + I_2(s,c),$$

where

$$I_1(s,c) = \int_c^\infty r^{s-1} g(-r)dr \quad \text{and} \quad I_2(s,c) = \frac{c^s}{2}\int_{2\pi}^0 e^{is\theta} g(ce^{i\theta})d\theta.$$

Now, let $c \to 0$. We find that, in view of (6),

$$\lim_{c\to 0} I_1(s,c) = \int_0^\infty \frac{r^{s-1}e^{-ar}}{(1-e^{-r})^n} dr = \Gamma(s)\,\zeta_n(s,a) \quad (\sigma = \Re(s) > n).$$

We next show that

$$\lim_{c\to 0} I_2(s,c) = 0.$$

To do this, we note that $g(-z)$ is analytic in $|z| < 2\pi$, except for a pole of order n at $z=0$. Therefore, $z^n g(-z)$ is analytic everywhere inside $|z| < 2\pi$ and, hence, is bounded there, that is,

$$|g(-z)| \leqq \frac{A}{|z|^n} \qquad (|z| = c < 2\pi),$$

and A is a positive constant. We, thus, have

$$|I_2(s,c)| \leqq \frac{c^\sigma}{2} \int_0^{2\pi} e^{-t\theta} \cdot \frac{A}{c^n} \, d\theta \leqq \pi A \, e^{2\pi |t|} \, c^{\sigma - n}.$$

If $\sigma > n$ and $c \to 0$, we find that $I_2(s,c) \to 0$; hence

$$\pi I_n(s,a) = \sin(\pi s) \, \Gamma(s) \, \zeta_n(s,a), \tag{10}$$

which is, in terms of 1.1(12), seen to be equivalent to (8). ■

In the equation (8), valid for $\sigma > n$, the function $I_n(s,a)$ is an entire function of s, and $\Gamma(1-s)$ is analytic for every complex s for $s \in \mathbb{C} \setminus \mathbb{N}$. We, therefore, can use this equation to define $\zeta_n(s,a)$ for $\sigma \leqq n$, that is, outside $\sigma > n$ as desired.

Definition 2.1 If $\Re(s) = \sigma \leqq n$, we define $\zeta_n(s,a)$ by

$$\zeta_n(s,a) := \Gamma(1-s) \, I_n(s,a), \tag{11}$$

where $I_n(s,a)$ is given in (7).

This equation (11) provides the analytic continuation of $\zeta_n(s,a)$ to the whole complex s-plane.

Theorem 2.6 *The function $\zeta_n(s,a)$, defined by (11), is analytic for all s except for simple poles at $s = k$ $(1 \leqq k \leqq n)$, with their respective residues given by*

$$\operatorname*{Res}_{s=k} \zeta_n(s,a) = \frac{1}{(n-k)!(k-1)!} \lim_{z \to 0} \frac{d^{n-k}}{dz^{n-k}} \frac{z^n e^{-az}}{(1-e^{-z})^n} \qquad (k = 1, \ldots, n; \; n \in \mathbb{N}). \tag{12}$$

In particular, when $s = n$, its residue is $1/(n-1)!$.

Proof. Since $I_n(s,a)$ is an entire function of s for $a > 0$, the only possible singularities of $\zeta_n(s,a)$ are the poles of $\Gamma(1-s)$. Since $1/\Gamma(1-s)$ has simple zeros at $s \in \mathbb{N}$, $\Gamma(1-s)$ has simple poles at $s \in \mathbb{N}$. But Theorem 2.2 shows that $\zeta_n(s,a)$ is analytic in $\Re(s) = \sigma > n$, and so $s = 1, \ldots, n$ are the only poles of $\zeta_n(s,a)$.

We next prove the assertion (12). If s is any integer, say $s = k$, the integrand in the contour integral for $I_n(s,a)$ in (7) takes the same values on both C_1 and C_3, and, hence, the integrals along C_1 and C_3 cancel, giving us

$$\begin{aligned} I_n(k,a) &= -\frac{1}{2\pi i} \int_{C_2} \frac{(-z)^{k-1} e^{-az}}{\left(1-e^{-z}\right)^n} dz \\ &= -\operatorname*{Res}_{z=0} f(z), \end{aligned}$$

where, for convenience,

$$f(z) := \frac{(-z)^{k-1} e^{-az}}{(1-e^{-z})^n}.$$

We observe that, for $1 \leqq k \leqq n$, the function $f(z)$ has a pole of order $n+1-k$ at $z=0$. We, therefore, have

$$I_n(k,a) = \frac{(-1)^k}{(n-k)!} \lim_{z\to 0} \frac{d^{n-k}}{dz^{n-k}} \frac{z^n e^{-az}}{(1-e^{-z})^n}. \tag{13}$$

To find the residue of $\zeta_n(s,a)$ at $s=k$ $(1 \leqq k \leqq n)$, by using 1.1(12), we compute the limit:

$$\begin{aligned}\lim_{s\to k}(s-k)\zeta_n(s,a) &= \lim_{s\to k}(s-k)\Gamma(1-s)I_n(s,a)\\ &= \frac{\pi\, I_n(k,a)}{\Gamma(k)} \cdot \lim_{s\to k} \frac{s-k}{\sin(\pi s)}\\ &= \frac{I_n(k,a)}{(-1)^k (k-1)!},\end{aligned}$$

which, by virtue of (13), immediately yields (12). ∎

Relationship between $\zeta_n(s,x)$ and $B_n^{(\alpha)}(x)$

By using the multiple Hurwitz Zeta function, Choi [262] derived the following explicit formula for $B_n^{(\alpha)}(x)$ different from 1.7(28):

$$B_{n+k}^{(n)}(x) = n\binom{n+k}{n} \sum_{j=0}^{n-1} (-1)^j \frac{B_{k+j+1}(x)}{k+j+1} \sum_{\ell=j}^{n-1} \binom{\ell}{j} s(n,\ell+1) x^{\ell-j} \tag{14}$$

in terms of the Stirling numbers $s(n,k)$ of the first kind (see Section 1.6).

The value of $\zeta_n(-\ell,x)$ can be calculated explicitly for $\ell \in \mathbb{N}_0$. Taking $s=-\ell$ in the relation (8), with a replaced by x, we find that

$$\zeta_n(-\ell,x) = \Gamma(1+\ell) I_n(-\ell,x) = \ell!\, I_n(-\ell,x). \tag{15}$$

Now, from the definition (19) of the generalized Bernoulli polynomials $B_n^{(\alpha)}(x)$, we have

$$
\begin{aligned}
I_n(-\ell,x) &= -\frac{1}{2\pi i}\int_C \frac{(-z)^{-\ell-1}e^{-xz}}{\left(1-e^{-z}\right)^n}dz \\
&= -\operatorname*{Res}_{z=0}\frac{(-z)^{-\ell-1}e^{-xz}}{\left(1-e^{-z}\right)^n} \\
&= (-1)^{\ell}\operatorname*{Res}_{z=0} z^{-n-\ell-1}\frac{z^n e^{(n-x)z}}{(e^z-1)^n} \\
&= (-1)^{\ell}\operatorname*{Res}_{z=0} z^{-n-\ell-1}\sum_{k=0}^{\infty} B_k^{(n)}(n-x)\frac{z^k}{k!} \\
&= (-1)^{\ell}\frac{B_{n+\ell}^{(n)}(n-x)}{(n+\ell)!},
\end{aligned}
$$

which, in view of 1.7(20) and (15), yields the desired relationship:

$$\zeta_n(-\ell,x) = (-1)^n \frac{\ell!}{(n+\ell)!}B_{n+\ell}^{(n)}(x) \quad (\ell \in \mathbb{N}_0). \tag{16}$$

Setting $n=1$ in (16), we have the well-known result:

$$\zeta(-\ell,x) = -\frac{B_{\ell+1}(x)}{\ell+1} \quad (\ell \in \mathbb{N}_0), \tag{17}$$

where $\zeta(s,x) := \zeta_1(s,x)$ is the Hurwitz (or generalized) Zeta function (see Section 2.2).

The number of solutions of

$$k_1+\cdots+k_n = k \quad (k \in \mathbb{N}_0;\ (k_1,\ldots,k_n) \in \mathbb{N}_0{}^n)$$

is equal to the coefficient of x^k in the Maclaurin series expansion of $(1-x)^{-n}$:

$$(1-x)^{-n} = \sum_{k=0}^{\infty}\binom{-n}{k}(-x)^k = \sum_{k=0}^{\infty}\binom{k+n-1}{n-1}x^k. \tag{18}$$

The multiple Hurwitz Zeta function in (5) can, thus, be expressed as a simple series:

$$\zeta_n(s,x) = \sum_{k=0}^{\infty}\binom{k+n-1}{n-1}(x+k)^{-s}. \tag{19}$$

From 1.6(1) or 1.6(5), we find that

$$\binom{k+n-1}{n-1} = \frac{1}{(n-1)!}\sum_{j=0}^{n-1} |s|(n,j+1)k^j, \tag{20}$$

where $|s|(n,k) := (-1)^{n+k}s(n,k)$ are often called the *unsigned* or *absolute* Stirling numbers of the first kind. Combining (19) and (20), we have

$$\zeta_n(s,x) = \frac{1}{(n-1)!}\sum_{k=0}^{\infty}\left(\sum_{j=0}^{n-1}|s|(n,j+1)k^j\right)(x+k)^{-s},$$

which, by virtue of the identity:

$$k^\ell = \{(-x)+(x+k)\}^\ell = \sum_{j=0}^{\ell}\binom{\ell}{j}(-x)^{\ell-j}(x+k)^j,$$

yields

$$\begin{aligned}\zeta_n(s,x) &= \frac{1}{(n-1)!}\left[\sum_{\ell=0}^{n-1}|s|(n,\ell+1)\left\{\sum_{j=0}^{\ell}\binom{\ell}{j}(-x)^{\ell-j}\right\}\right]\left(\sum_{k=0}^{\infty}\frac{1}{(x+k)^{s-j}}\right)\\ &= \frac{1}{(n-1)!}\sum_{\ell=0}^{n-1}|s|(n,\ell+1)\sum_{j=0}^{\ell}\binom{\ell}{j}(-x)^{\ell-j}\zeta(s-j,x).\end{aligned}$$

Next, it is easy to show that $\zeta_n(s,x)$ is expressible as a finite combination of the generalized Zeta function $\zeta(s,x)$ with polynomial coefficients in x:

$$\zeta_n(s,x) = \sum_{j=0}^{n-1} p_{n,j}(x)\zeta(s-j,x), \tag{21}$$

where

$$p_{n,j}(x) = \frac{1}{(n-1)!}\sum_{\ell=j}^{n-1}(-1)^{n+1-j}\binom{\ell}{j}s(n,\ell+1)x^{\ell-j}. \tag{22}$$

We shall now find $p_{n,j}(x)$ in (22) as a polynomial in x of degree $n-1-j$ with rational coefficients.

Since $\zeta(s,x)$ can be continued analytically to a meromorphic function, having a simple pole at $s=1$ with its residue 1, the representation (21) shows that $\zeta_n(s,x)$ is analytic for all s, except for simple poles only at $s=k$ $(k=1,\ldots,n;\ n\in\mathbb{N})$ with their respective residues given by

$$\operatorname*{Res}_{s=k}\zeta_n(s,x)=p_{n,k-1}(x)\quad (k=1,\ldots,n;\ n\in\mathbb{N}). \tag{23}$$

In view of 1.6(3) and 1.6(6), $\zeta_n(s,x)$ can be expressed explicitly for the first few values of n:

$$\begin{aligned}\zeta_2(s,x)&=(1-x)\zeta(s,x)+\zeta(s-1,x),\\ \zeta_3(s,x)&=\frac{1}{2}\left(x^2-3x+2\right)\zeta(s,x)+\left(\frac{3}{2}-x\right)\zeta(s-1,x)+\frac{1}{2}\zeta(s-2,x),\\ \zeta_4(s,x)&=\frac{1}{6}\Big\{\left(-x^3+6x^2-11x+6\right)\zeta(s,x)+\left(3x^2-12x+11\right)\zeta(s-1,x)\\ &\quad-(3x-6)\zeta(s-2,x)+\zeta(s-3,x)\Big\}.\end{aligned} \tag{24}$$

Letting $s=-\ell$ in (21) and applying (16), we obtain the desired formula (14). Also, upon setting $x=n$ $(n\in\mathbb{N})$ in (14) and making use of 1.7(21), we find that

$$B_{n+k}^{(n)}=(-1)^{n+k}n\binom{n+k}{n}\sum_{j=0}^{n-1}(-1)^j\frac{B_{k+j+1}(n)}{k+j+1}\sum_{\ell=j}^{n-1}\binom{\ell}{j}s(n,\ell+1)n^{\ell-j}. \tag{25}$$

Now, we express $\operatorname{Res}_{s=k}\zeta_n(s,x)$ in (12) in a more recognizable form:

$$\operatorname*{Res}_{s=k}\zeta_n(s,x)=\frac{B_{n-k}^{(n)}(n-x)}{(n-k)!(k-1)!}. \tag{26}$$

Indeed, it follows from 1.7(19) and (12) that

$$\begin{aligned}\lim_{s\to k}(s-k)\zeta_n(s,x)&=\frac{1}{(n-k)!(k-1)!}\lim_{z\to 0}\frac{d^{n-k}}{dz^{n-k}}\sum_{j=0}^{\infty}B_j^{(n)}(n-x)\frac{z^j}{j!}\\ &=\frac{B_{n-k}^{(n)}(n-x)}{(n-k)!(k-1)!}.\end{aligned}$$

The Vardi-Barnes Multiple Gamma Functions

Vardi [1190, p. 498] gave another expression for the multiple Gamma functions $\Gamma_n(a)$ (see Section 1.4), whose general form was also studied by Barnes [97]:

$$\Gamma_n(a)=\left[\prod_{m=1}^{n}R_{n-m+1}^{(-1)^m\binom{a}{m-1}}\right]G_n(a)\quad (n\in\mathbb{N}), \tag{27}$$

where

$$G_n(a) := \exp\left[\zeta_n'(0,a)\right] \quad \text{with} \quad \zeta_n'(s,a) = \frac{\partial}{\partial s}\zeta_n(s,a)$$

and

$$R_m := \exp\left(\sum_{k=1}^{m} \zeta_k'(0,1)\right) \text{ with } R_0 = 1. \tag{28}$$

In particular, the special cases of (27), when $n = 1$ and $n = 2$, give other forms of the simple and double Gamma functions $\Gamma_1 = \Gamma$ and Γ_2:

$$\begin{aligned}\Gamma(a) &= \exp\left[-\zeta'(0) + \zeta'(0,a)\right] \\ &= \sqrt{2\pi}\exp\left[\zeta'(0,a)\right],\end{aligned} \tag{29}$$

where $\zeta(s) := \zeta(s,1)$ is the Riemann Zeta function (see Section 2.3);

$$\Gamma_2(a) = A(2\pi)^{\frac{1}{2}-\frac{1}{2}a}\exp\left(-\frac{1}{12} + \zeta_2'(0,a)\right), \tag{30}$$

where we have used (24) and the known identity (see Voros [1201, p. 462, Eq. (A.11)]):

$$\log A = \frac{1}{12} - \zeta'(-1). \tag{31}$$

Here we can give another proof of the multiplication formula for Γ_2 (see 1.4(21)) different from that of Barnes [94], by using (30) (see Choi and Quine [278]). We consider

$$\begin{aligned}\sum_{\ell=0}^{n-1}\sum_{j=0}^{n-1}\zeta_2\left(s, a + \frac{\ell+j}{n}\right) &= \sum_{\ell=0}^{n-1}\sum_{j=0}^{n-1}\sum_{k_1,k_2=0}^{\infty}\left(a + \frac{\ell+j}{n} + k_1 + k_2\right)^{-s} \\ &= n^s \sum_{k_1,k_2=0}^{\infty}\sum_{\ell=0}^{n-1}\sum_{j=0}^{n-1}(na + \ell + j + nk_1 + nk_2)^{-s} \\ &= n^s \sum_{k_1,k_2=0}^{\infty}(na + k_1 + k_2)^{-s} = n^s\,\zeta_2(s, na),\end{aligned}$$

which, upon differentiating with respect to s, yields

$$\sum_{\ell=0}^{n-1}\sum_{j=0}^{n-1}\zeta_2'\left(s, a + \frac{\ell+j}{n}\right) = (\log n)\,n^s\,\zeta_2(s,na) + n^s\,\zeta_2'(s,na).$$

By virtue of (30), we readily obtain the following multiplication formula for Γ_2:

$$\prod_{\ell=0}^{n-1}\prod_{j=0}^{n-1}\Gamma_2\left(a+\frac{\ell+j}{n}\right)=\mathcal{C}(n)\,(2\pi)^{-\frac{1}{2}n(n-1)a}\,n^{-\frac{n^2a^2}{2}+na}\,\Gamma_2(na), \tag{32}$$

where

$$\mathcal{C}(n):=A^{n^2-1}\cdot e^{\frac{1}{12}(1-n^2)}\cdot(2\pi)^{\frac{1}{2}(n-1)}\cdot n^{\frac{5}{12}}.$$

An interesting identity is also obtained from (32):

$$\prod_{\ell=0}^{n-1}\prod_{j=0}^{n-1}{}'\,\Gamma_2\left(\frac{\ell+j}{n}\right)=\frac{\mathcal{C}(n)}{n}, \tag{33}$$

where the prime denotes the exclusion of the case when $\ell=0=j$.

2.2 The Hurwitz (or Generalized) Zeta Function

The *Hurwitz* (or *generalized*) *Zeta function* $\zeta(s,a)$ is defined by

$$\zeta(s,a):=\sum_{k=0}^{\infty}(k+a)^{-s}\quad(\Re(s)>1;\ a\notin\mathbb{Z}_0^-). \tag{1}$$

It is easy to see that $\zeta(s,a)=\zeta_1(s,a)$ for the case when $n=1$ in 2.1(2). Thus, we can deduce many properties of $\zeta(s,a)$ from those of $\zeta_n(s,a)$ in Section 2.1. Indeed, the series for $\zeta(s,a)$ in (1) converges absolutely for $\Re(s)=\sigma>1$. The convergence is uniform in every half-plane $\sigma\geq 1+\delta$ $(\delta>0)$, so $\zeta(s,a)$ is an analytic function of s in the half-plane $\Re(s)=\sigma>1$. Setting $n=1$ in 2.1(6), we have the integral representation:

$$\begin{aligned}\Gamma(s)\,\zeta(s,a)&=\int_0^{\infty}\frac{x^{s-1}\,e^{-ax}}{1-e^{-x}}dx=\int_0^{\infty}\frac{x^{s-1}\,e^{-(a-1)x}}{e^x-1}dx\\&=\int_0^{1}\frac{x^{a-1}}{1-x}\left(\log\frac{1}{x}\right)^{s-1}dx\quad(\Re(s)>1;\ \Re(a)>0).\end{aligned} \tag{2}$$

Moreover, $\zeta(s,a)$ can be continued meromorphically to the whole complex s-plane (except for a simple pole at $s=1$ with its residue 1) by means of the contour integral representation (see Theorem 2.5):

$$\zeta(s,a) = -\frac{\Gamma(1-s)}{2\pi i}\int_C \frac{(-z)^{s-1}\,e^{-az}}{1-e^{-z}}\,dz, \tag{3}$$

where the contour C is the Hankel loop of Theorem 2.5. The connection between $\zeta(s,a)$ and the Bernoulli polynomials $B_n(x)$ is also given in 2.1(17).

From the definition (1) of $\zeta(s,a)$, it easily follows that

$$\zeta(s,a) = \zeta(s,n+a) + \sum_{k=0}^{n-1}(k+a)^{-s} \quad (n\in\mathbb{N}); \tag{4}$$

$$\zeta\left(s,\frac{1}{2}a\right) - \zeta\left(s,\frac{1}{2}a+\frac{1}{2}\right) = 2^s\sum_{n=0}^{\infty}(-1)^n\,(a+n)^{-s}. \tag{5}$$

Hurwitz's Formula for $\zeta(s,a)$

The series expression $\zeta(s,a)$ was originally meaningful for $\sigma > 1$ $(s=\sigma+it)$. Hurwitz obtained another series representation for $\zeta(s,a)$ valid in the half-plane $\sigma < 0$:

$$\begin{gathered}\zeta(1-s,a) = \frac{\Gamma(s)}{(2\pi)^s}\left\{e^{-\frac{1}{2}\pi is}\,L(a,s) + e^{\frac{1}{2}\pi is}\,L(-a,s)\right\} \\ (0 < a \leqq 1,\ \sigma = \Re(s) > 1;\ 0 < a < 1,\ \sigma > 0),\end{gathered} \tag{6}$$

where the function $L(x,s)$ is defined by

$$L(x,s) := \sum_{n=1}^{\infty}\frac{e^{2\pi inx}}{n^s} \quad (x\in\mathbb{R};\ \sigma = \Re(s) > 1), \tag{7}$$

which is often referred to as the *periodic* (or *Lerch*) *Zeta function*.

We note that the Dirichlet series in (7) is a periodic function of x with period 1 and that $L(1,s) = \zeta(s)$, the Riemann Zeta function (see Section 2.3). The series in (7) converges absolutely for $\sigma > 1$. Yet, if $x \notin \mathbb{Z}$, the series can also be seen to converge conditionally for $\sigma > 0$. So, the formula (6) is also valid for $\sigma > 0$, if $a \neq 1$.

We observe that the function $L(x,s)$ in (7) is a linear combination of the Hurwitz Zeta functions, when x is a rational number. Indeed, setting $x = p/q$ $(1 \leqq p \leqq q;\ p,\, q\in\mathbb{N})$ in (7), the terms in (7) can be rearranged according to the residue classes $\bmod q$, by letting

$$n = kq + r \ (1 \leqq r \leqq q;\ k\in\mathbb{N}_0),$$

which gives us, for $\sigma > 1$,

$$L\left(\frac{p}{q}, s\right) = \sum_{n=1}^{\infty} \frac{1}{n^s} \exp\left(\frac{2\pi inp}{q}\right) = \sum_{r=1}^{q} \sum_{k=0}^{\infty} \frac{\exp\left(\frac{2\pi irp}{q}\right)}{(kq+r)^s}$$
$$= \frac{1}{q^s} \sum_{r=1}^{q} \exp\left(\frac{2\pi irp}{q}\right) \sum_{k=0}^{\infty} \frac{1}{\left(k + \frac{r}{q}\right)^s}$$
$$= \frac{1}{q^s} \sum_{r=1}^{q} \exp\left(\frac{2\pi irp}{q}\right) \zeta\left(s, \frac{r}{q}\right).$$

Therefore, if we take $a = p/q$ in the Hurwitz formula (6), we obtain:

$$\zeta\left(1 - s, \frac{p}{q}\right) = \frac{2\Gamma(s)}{(2\pi q)^s} \sum_{r=1}^{q} \cos\left(\frac{\pi s}{2} - \frac{2\pi rp}{q}\right) \zeta\left(s, \frac{r}{q}\right) \qquad (1 \leqq p \leqq q;\ p, q \in \mathbb{N}), \tag{8}$$

which holds true, by the principle of analytic continuation, for all admissible values of $s \in \mathbb{C}$.

Hermite's Formula for $\zeta(s, a)$

We, first, recall *Plana's summation formula*:

$$\sum_{k=m}^{n} f(k) = \frac{1}{2}[f(m) + f(n)] + \int_{m}^{n} f(\tau)\, d\tau - 2 \int_{0}^{\infty} \frac{q(m,t) - q(n,t)}{e^{2\pi t} - 1} dt, \tag{9}$$

where $f(z)$ is a bounded analytic function in $m \leqq \Re(z) \leqq n$ and

$$q(\lambda, t) = \frac{1}{2i}[f(\lambda + it) - f(\lambda - it)] \ (i = \sqrt{-1}).$$

Choose the function $f(z)$ in (9) as follows:

$$f(z) = (a + z)^{-s} \quad (s = \sigma + it;\ |\arg(a + z)| < \pi). \tag{10}$$

Then, it is easy to see that

$$q(x, y) = -\left[(a + x)^2 + y^2\right]^{-\frac{1}{2}s} \sin\left[s \arctan\left(\frac{y}{x + a}\right)\right].$$

In view of the elementary inequality:

$$\left|\arctan\left(\frac{y}{x + a}\right)\right| \leqq \min\left\{\frac{\pi}{2}, \frac{|y|}{x + a}\right\},$$

we obtain

$$
\begin{aligned}
|q(x,y)| &\leqq \left[(a+x)^2+y^2\right]^{\frac{1}{2}-\frac{1}{2}\sigma} \left|y^{-1}\right| \sinh\left(\frac{1}{2}\pi|s|\right) \quad (|y|>a); \\
|q(x,y)| &\leqq \left[(a+x)^2+y^2\right]^{-\frac{1}{2}\sigma} \left|\sinh\left(\frac{y|s|}{x+a}\right)\right| \quad (|y|<a).
\end{aligned}
\tag{11}
$$

Making use of (11), it is easily seen that the integral:

$$\int_0^\infty q(x,y)\left(e^{2\pi y}-1\right)^{-1} dy \quad (\sigma>0)$$

converges when $x \geqq 0$ and tends to 0 as $x \to \infty$. Also, the improper integral:

$$\int_0^\infty (a+x)^{-s} dx \quad (\sigma = \Re(s) > 1)$$

converges. Therefore, if $\sigma > 1$, it is valid to make $n \to \infty$ $(m=0)$ in (9) with the function $f(z)$ in (10). Thus, we readily obtain *Hermite's formula* for $\zeta(s,a)$:

$$\zeta(s,a) = \frac{1}{2}a^{-s} + \frac{a^{1-s}}{s-1} + 2\int_0^\infty \left(a^2+y^2\right)^{-\frac{1}{2}s} \left\{\sin\left(s\arctan\frac{y}{a}\right)\right\} \frac{dy}{e^{2\pi y}-1}. \tag{12}$$

We note that the integral involved in (12) converges for all admissible values of $s \in \mathbb{C}$. Moreover, the integral is an entire function of s. A special case of the formula (12) when $a=1$ is attributed to Jensen.

Setting $s=0$ in (12), we have

$$\zeta(0,a) = \frac{1}{2} - a, \tag{13}$$

which is also obtained from 2.1(17) in view of 1.7(8). If we set $z=s$ in 1.3(30) and differentiate the resulting equation with respect to s, we find that

$$\psi(s) = \log s - \frac{1}{2s} - 2\int_0^\infty \frac{t\,dt}{(t^2+s^2)\left(e^{2\pi t}-1\right)} \quad (\Re(s)>0). \tag{14}$$

Taking the limit in (12) as $s \to 1$, by virtue of the uniform convergence of the integral in (12), we get

$$\lim_{s\to 1}\left\{\zeta(s,a) - \frac{1}{s-1}\right\} = \lim_{s\to 1}\frac{a^{1-s}-1}{s-1} + \frac{1}{2a} + 2\int_0^\infty \frac{y\,dy}{(a^2+y^2)\left(e^{2\pi y}-1\right)},$$

which, in view of (14), yields

$$\lim_{s\to 1}\left\{\zeta(s,a)-\frac{1}{s-1}\right\}=-\frac{\Gamma'(a)}{\Gamma(a)}=-\psi(a). \tag{15}$$

Differentiating (12) with respect to s and setting $s=0$ in the resulting equation, we have

$$\left\{\frac{d}{ds}\zeta(s,a)\right\}_{s=0}=\left(a-\frac{1}{2}\right)\log a-a+2\int_0^\infty \frac{\arctan\left(\frac{y}{a}\right)}{e^{2\pi y}-1}dy, \tag{16}$$

which, by virtue of 1.2(30), yields

$$\left.\frac{d}{ds}\zeta(s,a)\right|_{s=0}=\log\Gamma(a)-\frac{1}{2}\log(2\pi), \tag{17}$$

which is equivalent to the identity 2.1(29). In addition to (17), it is easy to find from the definition (1) of $\zeta(s,a)$ that

$$\frac{\partial}{\partial a}\zeta(s,a)=-s\,\zeta(s+1,a). \tag{18}$$

The respective special cases of (15) and (17) when $a=1$, by means of 1.2(4) and 1.1(13), become

$$\lim_{s\to 1}\left\{\zeta(s,a)-\frac{1}{s-1}\right\}=\lim_{\epsilon\to 0}\left\{\zeta(1+\epsilon,a)-\frac{1}{\epsilon}\right\}=\gamma \tag{19}$$

and

$$\zeta'(0)=-\frac{1}{2}\log(2\pi), \tag{20}$$

where $\zeta(s)$ is the Riemann Zeta function (see Section 2.3).

Further Integral Representations for $\zeta(s,a)$

In addition to (12), some known integral representations of $\zeta(s,a)$ are recalled here:

$$\begin{aligned}\Gamma(s)\,\zeta(s,a)&=\int_0^\infty \frac{t^{s-1}e^{-at}}{1-e^{-t}}dt=\int_0^\infty \frac{t^{s-1}e^{-(a-1)t}}{e^t-1}dt\\&=\int_0^1 \frac{t^{a-1}}{1-t}\left(\log\frac{1}{t}\right)^{s-1}dt \qquad (\Re(s)>1;\ \Re(a)>0),\end{aligned} \tag{21}$$

which is the same as (2);

$$\zeta(s,a) = \frac{1}{2}a^{-s} - \frac{a^{1-s}}{1-s} + \frac{1}{\Gamma(s)} \int_0^\infty \left(\frac{1}{e^t - 1} - \frac{1}{t} + \frac{1}{2} \right) e^{-at} t^{s-1} \, dt \tag{22}$$

$$(\Re(s) > -1;\ \Re(a) > 0);$$

$$\zeta(s,a) = \frac{\pi\, 2^{s-2}}{s-1} \int_0^\infty \left[t^2 + (2a-1)^2 \right]^{\frac{1}{2}(1-s)} \frac{\cos\left[(s-1) \arctan\left(\frac{t}{2a-1} \right) \right]}{\cosh^2\left(\frac{1}{2}\pi t \right)} \, dt \tag{23}$$

$$\left(\Re(a) > \frac{1}{2} \right);$$

$$\begin{aligned} \zeta(s,a) = {} & \cos\left(\frac{1}{2}\pi s \right) \sin(2\pi a) \int_0^\infty \frac{t^{-s}}{\cosh(2\pi t) - \cosh(2\pi a)} \, dt \\ & + \sin\left(\frac{1}{2}\pi s \right) \int_0^\infty \frac{t^{-s} \left[\cosh(2\pi a) - e^{-2\pi t} \right]}{\cosh(2\pi t) - \cosh(2\pi a)} \, dt \end{aligned} \tag{24}$$

$$(\Re(s) < 1 \text{ when } 0 < \Re(a) < 1;\ \Re(s) < 0 \text{ when } a = 1);$$

$$\begin{aligned} \zeta(s,a) = {} & a^{-s} + \sum_{k=0}^{n} \frac{\Gamma(k+s-1)}{\Gamma(s)} \frac{B_k}{k!} a^{-k-s+1} \\ & + \frac{1}{\Gamma(s)} \int_0^\infty \left(\frac{1}{e^t - 1} - \sum_{k=0}^{n} \frac{B_k}{k!} t^{k-1} \right) e^{-at} t^{s-1} \, dt \end{aligned} \tag{25}$$

$$(\Re(s) > -(2n-1);\ \Re(a) > 0;\ n \in \mathbb{N}_0).$$

Some Applications of the Derivative Formula (17)

We begin by giving another proof of the derivative formula (17). Indeed, by the analytic continuation of $\zeta(s,a)$ and the special case of (4) when $n = 1$, we observe that

$$\zeta(s, a+1) = \zeta(s,a) - a^{-s}, \tag{26}$$

which, upon differentiating with respect to s and setting $s = 0$ in the resulting equation, yields

$$\zeta'(0, a+1) = \zeta'(0,a) + \log a.$$

Let $f(a) := \exp[\zeta'(0,a)]$. We then have $f(a+1) = af(a)$ $(a > 0)$ and

$$\frac{d^2}{da^2} \log f(a) = \frac{d^2}{da^2} \frac{d}{ds} \zeta(s,a) \bigg|_{s=0} = \sum_{k=0}^\infty \frac{1}{(k+a)^2} > 0 \quad (a > 0),$$

which implies that $f(a)$ is logarithmically convex on $(0, \infty)$. Thus, by appealing to the Bohr-Mollerup theorem (Theorem 1.1), we obtain, for some constant C,

$$f(a) = C\,\Gamma(a),$$

which, for $a = 1$, yields

$$C = \exp[\zeta'(0)] = (2\pi)^{-\frac{1}{2}}.$$

This completes our second proof of the derivative formula (17).

Many authors gave seemingly different proofs of Stirling's formula 1.1(52) (see e.g., Blyth and Pathak [135], Choi [261], Diaconis and Freeman [380] and Patin [889]). Here, by taking the limit in (16) as $a \to \infty$, we have

$$\lim_{a\to\infty}\left\{\zeta'(0,a) + a + \frac{1}{2}\log a - a\log a\right\} = 0, \tag{27}$$

which, upon taking the exponential and using (17), immediately yields Stirling's formula 1.1(33).

Combining formulas (17) and 1.1(42), we obtain a formula for the Beta function $B(\alpha, \beta)$:

$$B(\alpha,\beta) = (2\pi)^{\frac{1}{2}}\exp\left[\zeta'(0,\alpha) + \zeta'(0,\beta) - \zeta'(0,\alpha+\beta)\right], \tag{28}$$

where $\zeta'(s,a) = \frac{\partial}{\partial s}\zeta(s,a)$. Applying the formula (16) and the following trigonometric identities:

$$\begin{aligned} \arctan a + \arctan b &= \arctan\frac{a+b}{1-ab} \quad (ab < 1), \\ \arctan a - \arctan b &= \arctan\frac{a-b}{1+ab} \quad (ab > -1) \end{aligned} \tag{29}$$

to (28), we can readily deduce an integral representation of $B(\alpha, \beta)$ (*cf.* Choi and Nam [276]):

$$B(\alpha,\beta) = \frac{\alpha^{\alpha-\frac{1}{2}}\,\beta^{\beta-\frac{1}{2}}}{(\alpha+\beta)^{\alpha+\beta-\frac{1}{2}}}(2\pi)^{\frac{1}{2}}e^{I(\alpha,\beta)} \quad (\alpha > 0,\ \beta > 0), \tag{30}$$

where, for convenience,

$$I(\alpha,\beta) := 2\rho\int_0^\infty \arctan\left\{\frac{(t^3+t)\rho^3}{\alpha\beta(\alpha+\beta)}\right\}\frac{dt}{e^{2\pi t\rho}-1} \quad \left(\rho^2 := \alpha^2 + \alpha\beta + \beta^2\right).$$

Differentiating both sides of (22) with respect to s and letting $s = 0$ in the resulting equation, we obtain

$$\zeta'(0,a) = \left(a - \frac{1}{2}\right)\log a - a + \int_0^\infty \left(\frac{1}{e^t - 1} - \frac{1}{t} + \frac{1}{2}\right)\frac{e^{-at}}{t}\,dt \qquad (\Re(a) > 0). \tag{31}$$

In a similar manner, (31) leads us to another integral representation of $B(\alpha, \beta)$ (*cf.* Choi and Nam [276]):

$$B(\alpha,\beta) = \frac{\alpha^{\alpha-\frac{1}{2}}\beta^{\beta-\frac{1}{2}}}{(\alpha+\beta)^{\alpha+\beta-\frac{1}{2}}}(2\pi)^{\frac{1}{2}}e^{J(\alpha,\beta)} \quad (\alpha > 0;\ \beta > 0), \tag{32}$$

where, for convenience,

$$J(\alpha,\beta) := \int_0^\infty \left(\frac{1}{e^t - 1} - \frac{1}{t} + \frac{1}{2}\right)\left(e^{-\alpha t} + e^{-\beta t} - e^{-(\alpha+\beta)t}\right)\frac{dt}{t}.$$

Another Form for $\Gamma_2(a)$

From 2.1(24) and 2.1(30), by virtue of (17), we obtain another form for the double Gamma function $\Gamma_2(a)$:

$$\Gamma_2(a) = A\{\Gamma(a)\}^{1-a}\exp\left[-\frac{1}{12} + \zeta'(-1,a)\right] \quad (a > 0), \tag{33}$$

where $\zeta'(s,a) = \frac{\partial}{\partial s}\zeta(s,a)$.

In addition to the integral representation 1.4(78), we can express $\log\Gamma_2(a)$ as improper integrals in many ways. For example, we give two integral representations for $\log\Gamma_2(a)$:

$$\begin{aligned}\log\Gamma_2(a) = &-\frac{1}{12} + \log A - \frac{a^2}{4} + \left(\frac{1}{2}a^2 - \frac{1}{2}a\right)\log a + (1-a)\log\Gamma(a)\\ &+ 2\int_0^\infty \left\{\frac{1}{2}(a^2+t^2)^{\frac{1}{2}}\sin\left(\arctan\frac{t}{a}\right)\log(a^2+t^2)\right.\\ &\left. + (a^2+t^2)^{\frac{1}{2}}\cos\left(\arctan\frac{t}{a}\right)\arctan\frac{t}{a}\right\}\frac{dt}{e^{2\pi t}-1} \quad (\Re(a) > 0);\end{aligned} \tag{34}$$

$$\log\Gamma_2(a) = \log A - \frac{a^2}{4} + \left(\frac{a^2}{2} - \frac{a}{2} + \frac{1}{12}\right)\log a + (1-a)\log\Gamma(a)$$
$$- \int_0^\infty \left(\frac{1}{e^t - 1} - \frac{1}{t} + \frac{1}{2} - \frac{t}{12}\right)\frac{e^{-at}}{t^2}\,dt \quad (\Re(a) > 0). \tag{35}$$

Indeed, by differentiating Hermite's formula (12) for $\zeta(s,a)$, with respect to s, letting $s \to -1$ in the resulting equation and applying (17) and the identity for $\zeta'(-1,a)$, we readily obtain (34). Conversely, setting $n = 2$ in (25), we have

$$\zeta(s,a) = \frac{sa^{-s-1}}{12} + \frac{a^{-s}}{2} + \frac{a^{1-s}}{s-1} + \frac{1}{\Gamma(s)}\int_0^\infty \left(\frac{1}{e^t - 1} - \frac{1}{t} + \frac{1}{2} - \frac{t}{12}\right)t^{s-1}e^{-at}\,dt$$
$$(\Re(s) > -3;\ \Re(a) > 0). \tag{36}$$

Employing the same technique as in getting (34), by making use of (36) and considering the following identities:

$$\frac{d}{ds}\frac{1}{\Gamma(s)}\bigg|_{s=-1} = -1 \qquad \text{and} \qquad \frac{1}{\Gamma(s)}\bigg|_{s=-1} = 0,$$

we obtain (35).

Glaisher [484, p. 47] expressed the Glaisher-Kinkelin constant A given in 1.3(2) as an integral:

$$A = 2^{\frac{7}{36}}\pi^{-\frac{1}{6}}\exp\left(\frac{1}{3} + \frac{2}{3}\int_0^{\frac{1}{2}} \log\Gamma(t+1)\,dt\right). \tag{37}$$

By setting $a = 1$ in (34) and (35), we can also obtain integral representations of $\log A$:

$$\log A = \frac{1}{3} - 2\int_0^\infty \left\{\frac{1}{2}(1+t^2)^{\frac{1}{2}}\sin(\arctan t)\log(1+t^2)\right.$$
$$\left. + (1+t^2)^{\frac{1}{2}}\cos(\arctan t)\arctan t\right\}\frac{dt}{e^{2\pi t} - 1} \tag{38}$$

and

$$\log A = \frac{1}{4} + \int_0^\infty \left(\frac{1}{e^t - 1} - \frac{1}{t} + \frac{1}{2} - \frac{t}{12}\right)\frac{e^{-t}}{t^2}\,dt. \tag{39}$$

The formula (35) can be used to obtain an asymptotic formula for $\log\Gamma_2(a)$ by first observing that, for some $M > 0$ and for all $a > 0$,

$$\left|\int_0^\infty \left(\frac{1}{e^t - 1} - \frac{1}{t} + \frac{1}{2} - \frac{t}{12}\right)\frac{e^{-at}}{t^2}\,dt\right| < \frac{M}{a}.$$

Thus, by employing 1.1(34), we have

$$\begin{aligned}\log\Gamma_2(a) = \log A + \frac{3a^2}{4} - a - \frac{1}{12} - \left(\frac{a^2}{2} - a + \frac{5}{12}\right)\log a \\ + \frac{1}{2}(1-a)\log(2\pi) + O\left(a^{-1}\right) \quad (a \to \infty;\ a > 0),\end{aligned} \tag{40}$$

which may be compared with 1.3(7).

2.3 The Riemann Zeta Function

The *Riemann Zeta function* $\zeta(s)$ is defined by

$$\zeta(s) := \begin{cases} \sum\limits_{n=1}^{\infty} \frac{1}{n^s} = \frac{1}{1-2^{-s}} \sum\limits_{n=1}^{\infty} \frac{1}{(2n-1)^s} & (\Re(s) > 1) \\ \frac{1}{1-2^{1-s}} \sum\limits_{n=1}^{\infty} \frac{(-1)^{n-1}}{n^s} & (\Re(s) > 0;\ s \neq 1). \end{cases} \tag{1}$$

It is easy to see from the definitions (1) and 2.2(1) that

$$\zeta(s) = \zeta(s,1) = \left(2^s - 1\right)^{-1} \zeta\left(s, \frac{1}{2}\right) = 1 + \zeta(s,2) \tag{2}$$

and

$$\zeta(s) = \frac{1}{m^s - 1} \sum_{j=1}^{m-1} \zeta\left(s, \frac{j}{m}\right) \quad (m \in \mathbb{N} \setminus \{1\}). \tag{3}$$

In view of (2), we can deduce many properties of $\zeta(s)$ from those of $\zeta(s,a)$ given in Section 2.2. In fact, the series $\zeta(s) = \sum_{n=1}^{\infty} n^{-s}$ in (1) represents an analytic function of s in the half-plane $\Re(s) = \sigma > 1$. Setting $a = 1$ in 2.2(2), we have an integral representation of $\zeta(s)$ in the form:

$$\begin{aligned}\Gamma(s)\zeta(s) &= \int_0^\infty \frac{x^{s-1}e^{-x}}{1-e^{-x}}dx = \int_0^\infty \frac{x^{s-1}}{e^x - 1}dx \\ &= \int_0^1 \frac{1}{1-x}\left(\log\frac{1}{x}\right)^{s-1} dx \quad (\Re(s) > 1).\end{aligned} \tag{4}$$

Furthermore, just as $\zeta(s,a)$, $\zeta(s)$ can be continued meromorphically to the whole complex s-plane (except for a simple pole at $s=1$ with its residue 1) by means of the contour integral representation:

$$\zeta(s) = -\frac{\Gamma(1-s)}{2\pi i}\int_C \frac{(-z)^{s-1}e^{-z}}{1-e^{-z}}\,dz, \tag{5}$$

where the contour C is the Hankel loop of Theorem 2.5.

Now 2.2(19) and (5), together, imply that the Laurent series of $\zeta(s)$ in a neighborhood of its pole $s=1$ has the form:

$$\zeta(s) = \frac{1}{s-1} + \gamma + \sum_{n=1}^{\infty} a_n (s-1)^n, \tag{6}$$

where γ is the Euler-Mascheroni constant given in 1.1(3) and a_n is also expressed as (see Ivić [586, pp. 4–6]):

$$a_n = \lim_{m\to\infty}\left\{\sum_{k=1}^{m}\frac{(\log k)^n}{k} - \frac{(\log m)^{n+1}}{n+1}\right\} \quad (n\in\mathbb{N}). \tag{7}$$

The Riemann Zeta function $\zeta(s)$ in (1) plays a central rôle in the applications of complex analysis to number theory. The number-theoretic properties of $\zeta(s)$ are exhibited by the following result, known as *Euler's formula*, which gives a relationship between the set of primes and the set of positive integers:

$$\zeta(s) = \prod_p \left(1-p^{-s}\right)^{-1} \quad (\Re(s)>1), \tag{8}$$

where the product is taken over all primes.

From 2.2(4), we have [*cf.* Equation (2) for the *special* cases when $n=0$ and $n=1$]

$$\zeta(s) = \zeta(s,n+1) + \sum_{k=1}^{n} k^{-s} \quad (n\in\mathbb{N}_0). \tag{9}$$

The connection between $\zeta(s)$ and the Bernoulli numbers is given as follows:

$$\zeta(-n) = \begin{cases} -\frac{1}{2} & (n=0) \\ \\ -\frac{B_{n+1}}{n+1} & (n\in\mathbb{N}), \end{cases} \tag{10}$$

which is deduced by setting $x=1$ in 2.1(17) and using 1.6(5).

Riemann's Functional Equation for $\zeta(s)$

The special case of 2.2(8), when $p = 1 = q$, yields *Riemann's functional equation* for $\zeta(s)$:

$$\zeta(1-s) = 2(2\pi)^{-s}\,\Gamma(s)\cos\left(\frac{1}{2}\pi s\right)\zeta(s) \tag{11}$$

or, equivalently,

$$\zeta(s) = 2(2\pi)^{s-1}\,\Gamma(1-s)\sin\left(\frac{1}{2}\pi s\right)\zeta(1-s). \tag{12}$$

Taking $s = 2n+1$ $(n \in \mathbb{N})$ in (11), the factor $\cos\left(\frac{1}{2}\pi s\right)$ vanishes, and we find that

$$\zeta(-2n) = 0 \quad (n \in \mathbb{N}), \tag{13}$$

which are often referred to as the *trivial zeros* of $\zeta(s)$. The equation (13) can also be proven by combining (10) and 1.7(7).

By using the Legendre duplication formulas 1.1(29) and 1.1(12), it is not difficult to see that the functional equations (11) or (12) can be written in a simpler form:

$$\Phi(s) = \Phi(1-s), \tag{14}$$

where the function $\Phi(s)$ is defined by

$$\Phi(s) := \pi^{-\frac{1}{2}s}\,\Gamma\left(\frac{1}{2}s\right)\zeta(s). \tag{15}$$

The function $\Phi(s)$ has simple poles at $s = 0$ and $s = 1$. According to Riemann, to remove these poles, we multiply $\Phi(s)$ by $\frac{1}{2}s(1-s)$ and define

$$\xi(s) := \frac{1}{2}s(1-s)\,\Phi(s), \tag{16}$$

which is an entire function of s and satisfies the functional equation:

$$\xi(s) = \xi(1-s). \tag{17}$$

Setting $s = 2n$ $(n \in \mathbb{N})$ in (11) and applying (10), we have the well-known identity:

$$\zeta(2n) = (-1)^{n+1}\frac{(2\pi)^{2n}}{2\,(2n)!}B_{2n} \quad (n \in \mathbb{N}_0), \tag{18}$$

which, in view of 1.7(7), enables us to list the following special values:

$$\begin{aligned}\zeta(2) &= \frac{\pi^2}{6}, \quad \zeta(4) = \frac{\pi^4}{90}, \quad \zeta(6) = \frac{\pi^6}{945}, \\ \zeta(8) &= \frac{\pi^8}{9450}, \quad \zeta(10) = \frac{\pi^{10}}{93555}, \ldots .\end{aligned} \tag{19}$$

We may recall here a known recursion formula for $\zeta(2n)$ (see also Section 4.1):

$$\zeta(2n) = \frac{2}{2n+1} \sum_{k=1}^{n-1} \zeta(2k)\zeta(2n-2k) \quad (n \in \mathbb{N} \setminus \{1\}), \tag{20}$$

which can also be used to evaluate $\zeta(2n)$ $(n \in \mathbb{N} \setminus \{1\})$.

We get no information about $\zeta(2n+1)$ $(n \in \mathbb{N})$ from Riemann's functional equation, since both members of (11) vanish upon setting $s = 2n+1$ $(n \in \mathbb{N})$. In fact, until now, no simple formula analogous to (18) is known for $\zeta(2n+1)$ or even for any special case, such as $\zeta(3)$. It is not even known whether $\zeta(2n+1)$ is rational or irrational, except that the irrationality of $\zeta(3)$ was proven recently by Apéry [56]. Instead, a known integral formula for $\zeta(2n+1)$ is recalled here:

$$\zeta(2n+1) = \frac{(-1)^{n+1}(2\pi)^{2n+1}}{2(2n+1)!} \int_0^1 B_{2n+1}(t) \cot(\pi t)\, dt \quad (n \in \mathbb{N}). \tag{21}$$

It is readily seen that $\zeta(s) \neq 0$ $(\Re(s) = \sigma \geqq 1)$, and (12) shows that $\zeta(s) \neq 0$ $(\sigma \leqq 0)$, except for the trivial zeros in (13). Furthermore, in view of the second series definition of $\zeta(s)$ in (1), we find that $\zeta(s) < 0$ $(s \in \mathbb{R};\ 0 < s < 1)$. The assertion that all the *non-trivial* zeros of $\zeta(s)$ have real part $\frac{1}{2}$ is popularly known as the *Riemann hypothesis*, which was conjectured (but not proven) in the memoir of Riemann [977]. This hypothesis is still one of the most challenging mathematical problems today (see Edwards [398]).

It is easy to derive from (12) and (13) that (*cf.*, e.g., Srivastava [1084, p. 387, Eq. (1.15)])

$$\zeta'(-2n) = \lim_{\epsilon \to 0} \frac{\zeta(-2n+\epsilon)}{\epsilon} = (-1)^n \frac{(2n)!}{2(2\pi)^{2n}} \zeta(2n+1) \quad (n \in \mathbb{N}). \tag{22}$$

Relationship between $\zeta(s)$ and the Mathematical Constants B and C

Just as $\log A$ in 2.1(31), we can also express $\log B$ and $\log C$ introduced in Section 1.4 as special cases of $\zeta'(s)$. Indeed, using the Euler-Maclaurin summation formula 1.4(68), we can obtain a number of analytical representations of $\zeta(s)$, such as (*cf.* Hardy [537, p. 333])

$$\zeta(s) = \lim_{n\to\infty}\left\{\sum_{k=1}^{n} k^{-s} - \frac{n^{1-s}}{1-s} - \frac{1}{2}n^{-s}\right\} \quad (\Re(s) > -1), \tag{23}$$

$$\zeta(s) = \lim_{n\to\infty}\left\{\sum_{k=1}^{n} k^{-s} - \frac{n^{1-s}}{1-s} - \frac{1}{2}n^{-s} + \frac{1}{12}sn^{-s-1}\right\} \quad (\Re(s) > -3), \tag{24}$$

and

$$\begin{aligned}\zeta(s) = \lim_{n\to\infty}\Bigg\{&\sum_{k=1}^{n} k^{-s} - \frac{n^{1-s}}{1-s} - \frac{1}{2}n^{-s} + \frac{1}{12}sn^{-s-1}\\ &-\frac{1}{720}s(s+1)(s+2)n^{-s-3}\Bigg\} \quad (\Re(s) > -5).\end{aligned} \tag{25}$$

Now, it is not difficult to express the mathematical constants B and C as follows:

$$\log B = -\zeta'(-2) \tag{26}$$

and

$$\log C = -\zeta'(-3) - \frac{11}{720}, \tag{27}$$

respectively.

It is clear from 1.4(70) that $\log C$ must be the finite part of the divergent sum $\sum k^3 \log k$, according to some regularization; hence, $\log C$ must be related to $\zeta'(s)$ for some special value of s. By differentiating both sides of (25) with respect to s and letting $s = -3$, we obtain

$$\begin{aligned}-\zeta'(-3) = \lim_{n\to\infty}&\left\{\sum_{k=1}^{n} k^3 \log k - \left(\frac{n^4}{4} + \frac{n^3}{2} + \frac{n^2}{4} - \frac{1}{120}\right)\log n + \frac{n^4}{16} - \frac{n^2}{12}\right\}\\ &+\frac{11}{720},\end{aligned} \tag{28}$$

which, when compared with 1.4(70), yields the desired expression (27). The special case of (22) when $n = 1$ also shows that, in view of (26),

$$\log B = \frac{\zeta(3)}{4\pi^2}. \tag{29}$$

Integral Representations for $\zeta(s)$

In addition to (4) and (5), in terms of (2), we find from Section 2.2 that

$$\zeta(s) = \frac{1}{\Gamma(s)} \int_0^\infty \frac{t^{s-1}}{e^t - 1} dt \quad (\Re(s) > 1); \tag{30}$$

$$\zeta(s) = \frac{1}{2} + \frac{1}{s-1} + \frac{1}{\Gamma(s)} \int_0^\infty \left(\frac{1}{e^t - 1} - \frac{1}{t} + \frac{1}{2} \right) e^{-t} t^{s-1} \, dt \quad (\Re(s) > -1); \tag{31}$$

$$\zeta(s) = \sin\left(\frac{\pi}{2} s\right) \int_0^\infty \frac{\cosh(2\pi) - e^{-2\pi t}}{\cosh(2\pi t) - \cosh(2\pi)} \frac{dt}{t^s} \quad (\Re(s) < 0); \tag{32}$$

$$\zeta(s) = 1 + \frac{1}{s-1} + \sum_{k=1}^{n} \frac{\Gamma(k+s-1)}{\Gamma(s)} \frac{B_k}{k!} + \frac{1}{\Gamma(s)} \int_0^\infty \left(\frac{1}{e^t - 1} - \sum_{k=0}^{n} \frac{B_k}{k!} t^{k-1} \right) e^{-t} t^{s-1} \, dt \quad (\Re(s) > -(2n-1);\ n \in \mathbb{N}_0); \tag{33}$$

$$\zeta(s) = \frac{1}{2} + \frac{1}{s-1} + 2 \int_0^\infty \frac{\sin(s \arctan t)}{\left(1+t^2\right)^{\frac{1}{2}s}} \frac{dt}{e^{2\pi t} - 1}. \tag{34}$$

Now, we recall some summation formulas analogous to the infinite series version of Plana's summation formula 2.2(9). To do this, let $z = \tau + it$,

$$\frac{1}{e^{-2\pi iz} - 1} = X(\tau, t) + iY(\tau, t)$$

and

$$\frac{1}{e^{-\pi iz} - e^{\pi iz}} = X_1(\tau, t) + iY_1(\tau, t),$$

from which

$$X(\tau, t) = \frac{\cos(2\pi\tau) - e^{-2\pi t}}{e^{2\pi t} - 2\cos(2\pi\tau) + e^{-2\pi t}},$$

$$Y(\tau, t) = \frac{\sin(2\pi\tau)}{e^{2\pi t} - 2\cos(2\pi\tau) + e^{-2\pi t}},$$

$$X_1(\tau, t) = \frac{\cos(\pi\tau)\left(e^{\pi t} - e^{-\pi t}\right)}{e^{2\pi t} - 2\cos(2\pi\tau) + e^{-2\pi t}},$$

and

$$Y_1(\tau,t)=\frac{\sin(\pi\tau)\left(e^{\pi t}+e^{-\pi t}\right)}{e^{2\pi t}-2\cos(2\pi\tau)+e^{-2\pi t}}.$$

Conversely, we let

$$p(\tau,t):=\frac{1}{2}[f(\tau+it)+f(\tau-it)]$$

and

$$q(\tau,t):=\frac{1}{2i}[f(\tau+it)-f(\tau-it)],$$

so that, by appealing to the Reflection Principle, $p(\tau,t)$ and $q(\tau,t)$ represent, respectively, the real and imaginary parts of the analytic function $f(\tau+it)$, whose domain is symmetric with respect to the τ-axis, in case the function $f(\tau+i0)$ is real. Then, we have the summation formulas (see Lindelöf [769, Chapter III]):

$$\sum_{n=m}^{\infty} f(n)=\int_{\alpha}^{\infty} f(\tau)\,d\tau-2\int_{0}^{\infty} Q(\alpha,t)\,dt, \tag{35}$$

where, for convenience,

$$Q(\tau,t):=p(\tau,t)\,Y(\tau,t)+q(\tau,t)\,X(\tau,t); \tag{36}$$

$$\sum_{n=m}^{\infty} f(n)=-\frac{1}{2\pi i}\int_{\alpha-i\infty}^{\alpha+i\infty}\left(\frac{\pi}{\sin(\pi z)}\right)^{2} F(z)\,dz, \tag{37}$$

F being a primitive function of f;

$$\sum_{n=m}^{\infty}(-1)^n f(n)=-\int_{\alpha-i\infty}^{\alpha+i\infty}\frac{f(z)\,dz}{e^{\pi iz}-e^{-\pi iz}}=-2\int_{0}^{\infty} Q_1(\alpha,t)\,dt, \tag{38}$$

where, for convenience,

$$Q_1(\tau,t):=p(\tau,t)\,Y_1(\tau,t)+q(\tau,t)\,X_1(\tau,t), \tag{39}$$

α being any number between $m-1$ and m in both (37) and (39).

Moreover, the formulas (35), (37) and (38) are valid, if the function f satisfies each of the following conditions:

(a) $f(z)$ is analytic for $\Re(z) > 0$ $(z = \tau + it)$;
(b) $\lim_{t\to\infty} \exp(-2\pi|t|) f(\tau + it) = 0$ uniformly for $0 \leq \tau < \infty$;
(c)

$$\lim_{\tau\to\infty} \int_{-\infty}^{\infty} \exp(-2\pi|t|)\, |f(\tau + it)|\, dt = 0.$$

Now, we set $f(z) = z^{-s}$, with $m = 1$ and $\alpha = \frac{1}{2}$, in the three summation formulas just introduced. We, thus, obtain

$$\begin{aligned} p(\tau, t) &= (\tau^2 + t^2)^{-\frac{1}{2}s} \cos\left(s \arctan \frac{t}{\tau}\right), \\ q(\tau, t) &= -(\tau^2 + t^2)^{-\frac{1}{2}s} \sin\left(s \arctan \frac{t}{\tau}\right). \end{aligned} \tag{40}$$

We also find, from (35), that

$$\zeta(s) = \frac{2^{s-1}}{s-1} - 2^s \int_0^\infty \left(1 + t^2\right)^{-\frac{1}{2}s} \sin(s \arctan t) \frac{dt}{e^{\pi t} + 1}. \tag{41}$$

Setting $z = \frac{1}{2} + it$ in (37), we have

$$\zeta(s) = \frac{\pi\, 2^{s-2}}{s-1} \int_0^\infty \left(1 + t^2\right)^{\frac{1-s}{2}} \frac{\cos[(s-1)\arctan t]}{\left[\cosh\left(\frac{1}{2}\pi t\right)\right]^2} dt. \tag{42}$$

Starting with the second part of the definition (1) of $\zeta(s)$ and applying (38), we also get

$$\zeta(s) = \frac{2^{s-1}}{1 - 2^{1-s}} \int_0^\infty \left(1 + t^2\right)^{-\frac{1}{2}s} \frac{\cos(s \arctan t)}{\cosh\left(\frac{1}{2}\pi t\right)} dt. \tag{43}$$

Formulas (34), (41), (42) and (43) are due to Jensen. It is observed that the definite integrals appearing in these four expressions represent analytic functions for all bounded values of s. In fact, any one of these expressions gives the analytic continuation of $\zeta(s)$ in the whole complex s-plane, except for a simple pole at $s = 1$ with its residue 1, just as it is with the contour integral representation (5).

Some other integral representations of $\zeta(s)$ are recalled here as follows:

$$\zeta(s) = \frac{1}{s-1} + \frac{\sin(\pi s)}{\pi} \int_0^\infty [\log(1+t) - \psi(1+t)] \frac{dt}{t^s}, \tag{44}$$

$$\begin{aligned}\zeta(s+1) &= \frac{\sin(\pi s)}{\pi s} \int_0^\infty \psi'(1+t)]\, \frac{dt}{t^s} \\ &= \frac{\sin(\pi s)}{\pi s} \int_0^\infty [\psi(1+t)+\gamma]\, \frac{dt}{t^{1+s}},\end{aligned} \tag{45}$$

and

$$\zeta(s+m) = (-1)^{m-1} \frac{\Gamma(s)\sin(\pi s)}{\pi\,\Gamma(s+m)} \int_0^\infty \psi^{(m)}(1+t)\, \frac{dt}{t^s} \quad (m \in \mathbb{N}), \tag{46}$$

all three of which are due to de Bruijn [371] and are valid for $0 < \Re(s) < 1$. Moreover,

$$\zeta(s) = \frac{1}{s-1} + \frac{\sin(\pi s)}{\pi(s-1)} \int_0^\infty \left(\psi'(1+t) - \frac{1}{1+t}\right) \frac{dt}{t^{1-s}} \tag{47}$$

$$(0 < \Re(s) < 2;\ s \neq 1);$$

$$\zeta(s) = \frac{1}{\Gamma(s+1)} \int_0^\infty \frac{e^t\, t^s}{(e^t-1)^2} dt \quad (\Re(s) > 1); \tag{48}$$

$$\zeta(s) = \frac{\left(1-2^{1-s}\right)^{-1}}{\Gamma(s)} \int_0^\infty \frac{t^{s-1}}{e^t+1} dt \quad (\Re(s) > 0); \tag{49}$$

$$\zeta(s) = \frac{\left(1-2^{-s}\right)^{-1}}{2\,\Gamma(s)} \int_0^\infty \frac{t^{s-1}}{\sinh t} dt \quad (\Re(s) > 1); \tag{50}$$

$$\zeta(s) = \frac{\left(1-2^{1-s}\right)^{-1}}{2\,\Gamma(s+1)} \int_0^\infty \frac{t^s\, e^t}{(e^t+1)^2} dt \quad (\Re(s) > 0); \tag{51}$$

$$\zeta(s) = \frac{\pi^{\frac{1}{2}s}}{s(s-1)\,\Gamma\left(\frac{1}{2}s\right)} + \frac{\pi^{\frac{1}{2}s}}{\Gamma\left(\frac{1}{2}s\right)} \int_1^\infty \left(t^{\frac{1}{2}(1-s)} + t^{\frac{1}{2}s}\right) \frac{\omega(t)}{t}\, dt, \tag{52}$$

which is due to Riemann and where

$$\omega(t) := \sum_{n=1}^\infty \exp\left(-n^2 \pi\, t\right). \tag{53}$$

A Summation Identity for $\zeta(n)$

Sitaramachandrarao and Sivaramsarma [1035] proved the following summation identity for $\zeta(n)$, by using a known transformation formula and some reciprocity relations:

$$2\sum_{q=1}^{\infty}\frac{1}{q^n}\left(\sum_{k=1}^{q-1}\frac{1}{k}\right)=n\,\zeta(n+1)-\sum_{k=1}^{n-2}\zeta(n-k)\,\zeta(k+1)\quad(n\in\mathbb{N}\setminus\{1\}),\tag{54}$$

where an empty sum is understood (as usual) to be nil. The formula (54) contains many (known or new) special cases. Here, we give an elementary proof of (54) (*cf.* Williams [1229]). Let us, first, consider the sum:

$$\begin{aligned}\sum_{k=1}^{n-2}\zeta(n-k)\zeta(k+1)&=\zeta(2)\zeta(n-1)+\zeta(3)\zeta(n-2)+\cdots+\zeta(n-1)\zeta(2)\\&=\sum_{p=1}^{\infty}\sum_{q=1}^{\infty}\left\{\frac{1}{p^2}\cdot\frac{1}{q^{n-1}}+\frac{1}{p^3}\cdot\frac{1}{q^{n-2}}+\cdots+\frac{1}{p^{n-1}}\cdot\frac{1}{q^2}\right\}\\&=\lim_{N\to\infty}S_N,\end{aligned}$$

where, for convenience,

$$S_N:=\sum_{p=1}^{N}\sum_{q=1}^{N}\left\{\frac{1}{p^2}\cdot\frac{1}{q^{n-1}}+\frac{1}{p^3}\cdot\frac{1}{q^{n-2}}+\cdots+\frac{1}{p^{n-1}}\cdot\frac{1}{q^2}\right\}\quad(N\in\mathbb{N}).\tag{55}$$

Observing that the sum within braces in (55) is a geometric series and taking care of the exceptional case when $p=q$, the series (55) becomes

$$\sum_{q=1}^{N}\left\{\sum_{\substack{p=1\\p\neq q}}^{N}\frac{1}{p-q}\left(\frac{q^{1-n}}{p}-\frac{p^{1-n}}{q}\right)+(n-2)\frac{1}{q^{n+1}}\right\}.\tag{56}$$

Now, consider the first double sum in (56):

$$\begin{aligned}&\sum_{q=1}^{N}\sum_{\substack{p=1\\p\neq q}}^{N}\frac{1}{p-q}\left(\frac{q^{1-n}}{p}-\frac{p^{1-n}}{q}\right)\\&\quad=\sum_{q=1}^{N}\sum_{\substack{p=1\\p\neq q}}^{N}\frac{1}{p-q}\cdot\frac{q^{1-n}}{p}+\sum_{p=1}^{N}\sum_{\substack{q=1\\q\neq p}}^{N}\frac{1}{q-p}\cdot\frac{p^{1-n}}{q}.\end{aligned}\tag{57}$$

Inverting the order of summation in the second term on the right-hand side of (57), it is fairly straightforward to write (57) in the form:

$$\sum_{q=1}^{N}\sum_{\substack{p=1\\p\neq q}}^{N}\frac{q^{1-n}}{p(p-q)}+\sum_{p=1}^{N}\sum_{\substack{q=1\\q\neq p}}^{N}\frac{p^{1-n}}{q(q-p)}=2\sum_{q=1}^{N}\frac{1}{q^{n-1}}\sum_{\substack{p=1\\p\neq q}}^{N}\frac{1}{p(p-q)}. \tag{58}$$

Combining (55) through (58), we find that

$$\begin{aligned}&\sum_{q=1}^{N}\sum_{p=1}^{N}\left\{\frac{1}{p^2}\cdot\frac{1}{q^{n-1}}+\frac{1}{p^3}\cdot\frac{1}{q^{n-2}}+\cdots+\frac{1}{p^{n-1}}\cdot\frac{1}{q^2}\right\}\\&\quad=(n-2)\sum_{q=1}^{N}\frac{1}{q^{n+1}}+2\sum_{q=1}^{N}\frac{1}{q^{n-1}}\sum_{\substack{p=1\\p\neq q}}^{N}\frac{1}{p(p-q)}.\end{aligned} \tag{59}$$

Now, consider

$$\begin{aligned}-q\sum_{\substack{p=1\\p\neq q}}^{N}\frac{1}{p(p-q)}&=\sum_{\substack{p=1\\p\neq q}}^{N}\left(\frac{1}{p}-\frac{1}{p-q}\right)\\&=\sum_{p=1}^{q-1}\frac{1}{q-p}-\sum_{p=q+1}^{N}\frac{1}{p-q}+\sum_{p=1}^{N}\frac{1}{p}-\frac{1}{q}\\&=\sum_{p=1}^{q-1}\frac{1}{p}-\sum_{p=1}^{N-q}\frac{1}{p}+\sum_{p=1}^{N}\frac{1}{p}-\frac{1}{q}\\&=-\frac{1}{q}+\sum_{p=1}^{q-1}\frac{1}{p}+\sum_{p=N-q+1}^{N}\frac{1}{p},\end{aligned} \tag{60}$$

which, when substituted into (59), yields

$$\begin{aligned}&\sum_{q=1}^{N}\sum_{p=1}^{N}\left\{\frac{1}{p^2}\cdot\frac{1}{q^{n-1}}+\frac{1}{p^3}\cdot\frac{1}{q^{n-2}}+\cdots+\frac{1}{p^{n-1}}\cdot\frac{1}{q^2}\right\}\\&\quad=n\sum_{q=1}^{N}\frac{1}{q^{n+1}}-2\sum_{q=1}^{N}\frac{1}{q^n}\left(\sum_{k=1}^{q-1}\frac{1}{k}\right)-2\sum_{q=1}^{N}\frac{1}{q^n}\left(\sum_{k=N-q+1}^{N}\frac{1}{k}\right).\end{aligned} \tag{61}$$

Finally, we consider the inequality:

$$0\leqq\sum_{k=N-q+1}^{N}\frac{1}{k}=\frac{1}{N-q+1}+\frac{1}{N-q+2}+\cdots+\frac{1}{N}\leqq\frac{q}{N-q+1},$$

so that

$$
\begin{aligned}
0 &< \sum_{q=1}^{N} \frac{1}{q^n} \left(\sum_{k=N-q+1}^{N} \frac{1}{k} \right) < \sum_{q=1}^{N} \frac{1}{q^{n-1}} \cdot \frac{1}{N-q+1} \\
&\leqq \sum_{q=1}^{N} \frac{1}{q(N-q+1)} \qquad (n \in \mathbb{N} \setminus \{1\}) \\
&= \frac{1}{N+1} \sum_{q=1}^{N} \left(\frac{1}{q} + \frac{1}{N-q+1} \right) = \frac{2}{N+1} \sum_{q=1}^{N} \frac{1}{q} \\
&< \frac{2}{N+1} (1 + \log N) \to 0 \ \text{ as } N \to \infty.
\end{aligned} \tag{62}
$$

By taking the limit in (61) as $N \to \infty$ and applying (62), we complete the proof of the summation identity (54).

2.4 Polylogarithm Functions

The so-called *Polylogarithm functions* have been studied rather extensively in the works of (for example) Lewin [751, 752], who was exceedingly fascinated by the following integral formula, even during his school days:

$$
\int_0^{\frac{1}{2}(\sqrt{5}-1)} \frac{\log(1-t)}{t}\, dt = \left\{ \log \left(\frac{\sqrt{5}-1}{2} \right) \right\}^2 - \frac{\pi^2}{10}. \tag{1}
$$

Many of the earlier authors who studied these functions include Euler, Abel, Legendre, Kummer, Spence and so on. These functions have been found useful in physics, electronics, quantum electrodynamics, etc., and are closely related to many other mathematical functions, such as the Clausen integral (or function) $\mathrm{Cl}_2(x)$ given by

$$
\mathrm{Cl}_2(x) = x \log \pi - x \log \sin \left(\frac{x}{2} \right) + 2\pi \log \frac{G\left(1 - \frac{x}{2\pi}\right)}{G\left(1 + \frac{x}{2\pi}\right)}, \tag{2}
$$

where G is the Barnes G-function studied in Section 1.4.

Throughout this work, we choose to follow the notations and conventions used by Lewin [751, 752].

The Dilogarithm Function

The *Dilogarithm function* $\mathrm{Li}_2(z)$ is defined by

$$\begin{aligned}\mathrm{Li}_2(z) &:= \sum_{n=1}^{\infty} \frac{z^n}{n^2} \quad (|z| \leq 1) \\ &= -\int_0^z \frac{\log(1-t)}{t}\, dt.\end{aligned} \tag{3}$$

It follows from (3) that

$$\frac{d}{dx}\mathrm{Li}_2\left(-\frac{1}{x}\right) = \frac{\log(1+x) - \log x}{x},$$

the integration of which yields

$$\mathrm{Li}_2\left(-\frac{1}{x}\right) + \mathrm{Li}_2(-x) = 2\,\mathrm{Li}_2(-1) - \frac{1}{2}(\log x)^2, \tag{4}$$

where the constant of integration is determined by taking $x = 1$.

The special case of (4) when $x = -1 = e^{i\pi}$ gives

$$2\,\mathrm{Li}_2(1) = 2\,\mathrm{Li}_2(-1) + \frac{\pi^2}{2}. \tag{5}$$

It is easy to see also that

$$\mathrm{Li}_2(-1) = -\frac{1}{2}\mathrm{Li}_2(1), \tag{6}$$

which, in view of (5), yields

$$\mathrm{Li}_2(1) = \frac{1}{1^2} + \frac{1}{2^2} + \frac{1}{3^2} + \cdots = \frac{\pi^2}{6} = \zeta(2), \tag{7}$$

where $\zeta(s)$ is the Riemann Zeta function (see Section 2.3). It follows from (6) and (7) that

$$\mathrm{Li}_2(-1) = -\frac{\pi^2}{12}. \tag{8}$$

By taking $x = y e^{i\pi}$ in (4), we obtain

$$\mathrm{Li}_2(y) + \mathrm{Li}_2\left(\frac{1}{y}\right) = \frac{\pi^2}{3} - \frac{1}{2}(\log y)^2 - i\pi \log y \quad (y > 1). \tag{9}$$

From (3), we also find that

$$\begin{aligned}\mathrm{Li}_2(z) &= -\log z \log(1-z) - \int_0^z \frac{\log t}{1-t}\,dt \\ &= -\log z \log(1-z) - \mathrm{Li}_2(1-z) + \mathrm{Li}_2(1),\end{aligned}$$

from which

$$\mathrm{Li}_2(z) + \mathrm{Li}_2(1-z) = \frac{\pi^2}{6} - \log z \log(1-z). \tag{10}$$

By choosing a suitable argument and then employing differentiation and integration, we have Landen's formulas (see Lewin [751, p. 5]):

$$\mathrm{Li}_2(x) + \mathrm{Li}_2\left(\frac{x}{x-1}\right) = -\frac{1}{2}\{\log(1-x)\}^2 \quad (x<1) \tag{11}$$

and

$$\begin{aligned}\mathrm{Li}_2(x) + \mathrm{Li}_2\left(\frac{x}{x-1}\right) &= \frac{\pi^2}{2} - 2i\pi \log x + i\pi \log(x-1) \\ &\quad - \frac{1}{2}\{\log(x-1)\}^2 \quad (x>1).\end{aligned} \tag{12}$$

A factorization formula for $\mathrm{Li}_2(x)$ is given by

$$\frac{1}{n}\mathrm{Li}_2(x^n) = \sum_{k=0}^{n-1} \mathrm{Li}_2\left(\omega^k x\right) \quad \left(\omega := \exp\left(\frac{2\pi i}{n}\right)\right), \tag{13}$$

which follows from the elementary identity:

$$1 - x^n = \prod_{k=0}^{n-1}\left(1 - \omega^k x\right) \quad \left(\omega := \exp\left(\frac{2\pi i}{n}\right)\right).$$

The special case of (13) when $n = 2$ gives us

$$\frac{1}{2}\mathrm{Li}_2\left(x^2\right) = \mathrm{Li}_2(x) + \mathrm{Li}_2(-x). \tag{14}$$

By eliminating $\mathrm{Li}_2(x)$ between (11) and (14), we get

$$\mathrm{Li}_2\left(\frac{x}{x-1}\right) + \frac{1}{2}\mathrm{Li}_2\left(x^2\right) - \mathrm{Li}_2(-x) = -\frac{1}{2}\{\log(1-x)\}^2. \tag{15}$$

Setting $x = \frac{3-\sqrt{5}}{2}$ and $x = \frac{\sqrt{5}-1}{2}$ in (11) yields the following results:

$$\mathrm{Li}_2\left(\frac{1-\sqrt{5}}{2}\right) = -\frac{\pi^2}{15} + \frac{1}{2}\left\{\log\left(\frac{\sqrt{5}-1}{2}\right)\right\}^2 \tag{16}$$

and

$$\mathrm{Li}_2\left(-\frac{1+\sqrt{5}}{2}\right) = -\frac{\pi^2}{10} + \frac{1}{2}\left\{\log\left(\frac{\sqrt{5}+1}{2}\right)\right\}^2, \tag{17}$$

respectively.

The *Legendre's Chi function* $\chi_2(x)$ is defined by

$$\begin{aligned}\chi_2(x) :=& \sum_{n=1}^{\infty} \frac{x^{2n-1}}{(2n-1)^2} \qquad (-1 \leqq x \leqq 1)\\ =& \frac{1}{2}\{\mathrm{Li}_2(x) - \mathrm{Li}_2(-x)\},\end{aligned} \tag{18}$$

which, in view of (3), immediately becomes

$$\chi_2(x) = \frac{1}{2}\int_0^x \log\left(\frac{1+t}{1-t}\right)\frac{dt}{t}. \tag{19}$$

From (19), one readily finds the *Euler-Legendre-Landen* identity:

$$\chi_2\left(\frac{1-x}{1+x}\right) + \chi_2(x) = \frac{\pi^2}{8} + \frac{1}{2}\log x \log\left(\frac{1+x}{1-x}\right). \tag{20}$$

Putting $(1-x)/(1+x) = x$ gives $x^2 + 2x - 1 = 0$ or $x = \sqrt{2} - 1 = \tan\left(\frac{1}{8}\pi\right)$. We, thus, find from (20) that

$$\chi_2\left(\tan\frac{\pi}{8}\right) = \frac{\pi^2}{16} - \frac{1}{4}\left[\log\left(\tan\frac{\pi}{8}\right)\right]^2. \tag{21}$$

Upon integrating by parts twice, it is easily observed that

$$\int_0^x t^{\alpha}\,\mathrm{Li}_2(t)\,dt = \frac{x^{\alpha+1}}{\alpha+1}\mathrm{Li}_2(x) + \frac{x^{\alpha+1}-1}{(\alpha+1)^2}\log(1-x) - \frac{1}{(\alpha+1)^2}\int_0^x \frac{1-t^{\alpha+1}}{1-t}\,dt. \tag{22}$$

In view of 1.2(3), the integral formula (22) with $x = 1$ assumes the form:

$$\int_0^1 t^{\alpha-1} \operatorname{Li}_2(t)\, dt = \frac{\pi^2}{6\alpha} - \frac{\psi(\alpha+1)+\gamma}{\alpha^2}, \tag{23}$$

which, for $\alpha = n \in \mathbb{N}_0$, yields

$$\int_0^1 t^n \operatorname{Li}_2(t)\, dt = \frac{\pi^2}{6(n+1)} - \frac{1}{(n+1)^2} \sum_{k=1}^{n+1} \frac{1}{k}. \tag{24}$$

Nielsen (see Lewin [751, pp. 20–21]) applied (24) to show that

$$\int_0^1 f(tx) \operatorname{Li}_2(t)\, dt = \frac{\pi^2}{6x} \int_0^x f(t)\, dt - \frac{1}{x} \sum_{n=0}^{\infty} \frac{a_n x^{n+1}}{(n+1)^2} \left(\sum_{k=1}^{n+1} \frac{1}{k} \right), \tag{25}$$

where $f(x) := \sum_{n=0}^{\infty} a_n x^n$. Take $a_n = n+1$, so that $f(t) = (1-t)^{-2}$. By making use of the well-known expansion:

$$\frac{1}{2} \{\log(1-x)\}^2 = \sum_{n=1}^{\infty} \frac{x^{n+1}}{n+1} \left(\sum_{k=1}^{n} \frac{1}{k} \right),$$

we find the following integral equation involving the Dilogarithm function:

$$\int_0^1 \frac{\operatorname{Li}_2(t)}{(1-xt)^2}\, dt = \frac{\pi^2}{6(1-x)} - \frac{\operatorname{Li}_2(x)}{x} - \frac{\{\log(1-x)\}^2}{2x} \quad (-1 \leqq x \leqq 1), \tag{26}$$

whose special cases when $x = -1$ and $x = \frac{1}{2}$ become

$$\int_0^1 \frac{\operatorname{Li}_2(t)}{(1+t)^2}\, dt = \frac{1}{2} (\log 2)^2 \tag{27}$$

and

$$\int_0^1 \frac{\operatorname{Li}_2(t)}{(2-t)^2}\, dt = \frac{\pi^2}{24}, \tag{28}$$

respectively.

If z is a pure imaginary number in (3), say $z = iy$ $(y \in \mathbb{R})$, then

$$\mathrm{Li}_2(iy) = \frac{1}{4}\mathrm{Li}_2\left(-y^2\right) + i\,\mathrm{Ti}_2(y) \quad (y \in \mathbb{R}), \tag{29}$$

where the new function $\mathrm{Ti}_2(x)$ is defined by

$$\mathrm{Ti}_2(x) := \sum_{n=1}^{\infty} \frac{(-1)^{n+1}\, x^{2n-1}}{(2n-1)^2} \quad (x \leq 1). \tag{30}$$

From (29), we also see that

$$\mathrm{Ti}_2(y) = \frac{1}{2i}\left[\mathrm{Li}_2(iy) - \mathrm{Li}_2(-iy)\right]. \tag{31}$$

It follows from (18) that

$$\mathrm{Ti}_2(y) = -i\,\chi_2(iy). \tag{32}$$

In view of the following series expansion of $\tan^{-1} x$:

$$\tan^{-1} x = \sum_{n=1}^{\infty} (-1)^{n+1} \frac{x^{2n-1}}{2n-1} \quad (x \leq 1), \tag{33}$$

the definition (30) can be written in the form:

$$\mathrm{Ti}_2(x) = \int_0^x \frac{\tan^{-1} t}{t}\, dt \quad \text{or} \quad \mathrm{Ti}_2(x) = \int_0^x \left(\tan^{-1} t\right) \frac{d(\log t)}{dt}\, dt. \tag{34}$$

If one takes (34) as a definition of $\mathrm{Ti}_2(x)$, the domain of $\mathrm{Ti}_2(x)$ may be extended to ∞.

Some known formulas and special values for $\mathrm{Ti}_2(x)$ are recalled here:

$$\mathrm{Ti}_2(x) - \mathrm{Ti}_2\left(\frac{1}{x}\right) = \mathrm{sgn}(x)\,\frac{\pi}{2}\log|x| \quad (x \in \mathbb{R}), \tag{35}$$

where $\mathrm{sgn}(x)$ is defined by

$$\mathrm{sgn}(x) := \begin{cases} 1 & (x > 0), \\ -1 & (x < 0); \end{cases} \tag{36}$$

$$\mathrm{Ti}_2(1) = G, \tag{37}$$

where G is the Catalan constant given in 1.3(16);

$$\frac{1}{2}\mathrm{Ti}_2\left(\frac{2x}{1-x^2}\right) = \mathrm{Ti}_2(x) + \mathrm{Ti}_2(-x, 1) - \mathrm{Ti}_2(x, 1), \tag{38}$$

where $\mathrm{Ti}_2(x, a)$ is defined by

$$\mathrm{Ti}_2(x, a) := \int_0^x \frac{\tan^{-1} t}{a+t}\, dt; \tag{39}$$

$$\begin{aligned}\frac{1}{3}\mathrm{Ti}_2\left(\frac{3x-x^3}{1-3x^2}\right) &= \mathrm{Ti}_2(x) + \mathrm{Ti}_2\left(\frac{1-\sqrt{3}x}{\sqrt{3}+x}\right) - \mathrm{Ti}_2\left(\frac{1+\sqrt{3}x}{\sqrt{3}-x}\right) \\ &+ \frac{\pi}{6}\log\left(\frac{\sqrt{3}+x}{1-\sqrt{3}x}\cdot\frac{1+\sqrt{3}x}{\sqrt{3}-x}\right) \quad \left(-\frac{1}{\sqrt{3}} < x < \frac{1}{\sqrt{3}}\right),\end{aligned} \tag{40}$$

which, upon setting $x = \tan\theta$, yields a trigonometric form:

$$\begin{aligned}\frac{1}{3}\mathrm{Ti}_2(\tan 3\theta) &= \mathrm{Ti}_2(\tan\theta) + \mathrm{Ti}_2\left[\tan\left(\frac{\pi}{6}-\theta\right)\right] - \mathrm{Ti}_2\left[\tan\left(\frac{\pi}{6}+\theta\right)\right] \\ &+ \frac{\pi}{6}\log\left[\frac{\tan\left(\frac{\pi}{6}+\theta\right)}{\tan\left(\frac{\pi}{6}-\theta\right)}\right].\end{aligned} \tag{41}$$

Clausen's Integral (or Function)

From the series definition in (3) of the Dilogarithm function $\mathrm{Li}_2(z)$, we have

$$\mathrm{Li}_2\left(e^{i\theta}\right) = \sum_{n=1}^{\infty} \frac{\cos n\theta}{n^2} + i\sum_{n=1}^{\infty} \frac{\sin n\theta}{n^2}. \tag{42}$$

From the integral definition in (3) of $\mathrm{Li}_2(z)$, we also have

$$\mathrm{Li}_2\left(e^{i\theta}\right) - \mathrm{Li}_2(1) = -\int_1^{\exp(i\theta)} \frac{\log(1-t)}{t}\, dt,$$

which, upon putting $t = e^{i\eta}$, yields

$$\begin{aligned}\mathrm{Li}_2\left(e^{i\theta}\right) - \mathrm{Li}_2(1) &= -i\int_0^\theta \log\left(1-e^{i\eta}\right) d\eta \\ &= -i\int_0^\theta \log\left[2\sin\left(\frac{1}{2}\eta\right)\exp\left(-\frac{1}{2}(\pi-\eta)\right)i\right] d\eta \\ &= -i\int_0^\theta \log\left[2\sin\left(\frac{1}{2}\eta\right)\right] d\eta \\ &+ \frac{1}{4}(\pi-\theta)^2 - \frac{1}{4}\pi^2 \quad (0 \leqq \theta \leqq 2\pi).\end{aligned} \tag{43}$$

Comparing the real parts in (42) and (43) gives us a well-known Fourier series:

$$\sum_{n=1}^{\infty} \frac{\cos n\theta}{n^2} = \frac{\pi^2}{6} - \frac{\theta(2\pi - \theta)}{4} \quad (0 \leqq \theta \leqq 2\pi). \tag{44}$$

Conversely, if we compare the imaginary parts in (42) and (43), we obtain *Clausen's integral* (or *function*) $\mathrm{Cl}_2(\theta)$ defined by [*cf.* Equation (2) above]

$$\mathrm{Cl}_2(\theta) := \sum_{n=1}^{\infty} \frac{\sin n\theta}{n^2} = -\int_0^{\theta} \log\left[2\sin\left(\frac{1}{2}\eta\right)\right] d\eta. \tag{45}$$

Equation (43) can, therefore, be written in the form:

$$\mathrm{Li}_2\left(e^{i\theta}\right) = \frac{\pi^2}{6} - \frac{\theta(2\pi - \theta)}{4} + i\,\mathrm{Cl}_2(\theta) \quad (0 \leqq \theta \leqq 2\pi). \tag{46}$$

Some known properties and special values of the Clausen integral (or function) include the periodic properties:

$$\mathrm{Cl}_2(2n\pi \pm \theta) = \mathrm{Cl}_2(\pm\theta) = \pm\mathrm{Cl}_2(\theta), \tag{47}$$

which, for $n = 1$ and with θ replaced by $\pi + \theta$, yields

$$\mathrm{Cl}_2(\pi + \theta) = -\mathrm{Cl}_2(\pi - \theta). \tag{48}$$

From the series definition (45), it is obvious that

$$\mathrm{Cl}_2(n\pi) = 0 \quad (n \in \mathbb{Z}), \tag{49}$$

which, for $n = 1$, gives

$$\int_0^{\pi} \log\left(2\sin\frac{1}{2}\theta\right) d\theta = 0 \quad \text{and} \quad \int_0^{\pi/2} \log(\sin\theta)\, d\theta = -\frac{1}{2}\pi \log 2. \tag{50}$$

Setting $\theta = \frac{1}{2}\pi$ in the series definition (45), and using (48), we find that

$$\mathrm{Cl}_2\left(\frac{1}{2}\pi\right) = G = -\mathrm{Cl}_2\left(\frac{3}{2}\pi\right), \tag{51}$$

where G is the Catalan constant given in 1.4(16);

$$\frac{1}{2}\mathrm{Cl}_2(2\theta) = \mathrm{Cl}_2(\theta) - \mathrm{Cl}_2(\pi - \theta). \tag{52}$$

From (45), by employing integration by parts, we have

$$\mathrm{Cl}_2(\theta) = -\theta \log\left(\sin\frac{1}{2}\theta\right) + \int_0^\theta \frac{1}{2}\eta \cot\left(\frac{1}{2}\eta\right) d\eta. \tag{53}$$

If we substitute the known formula:

$$\cot\left(\frac{1}{2}\theta\right) = \frac{1}{\frac{1}{2}\theta} - \sum_{k=1}^{\infty} \frac{\theta}{k^2\pi^2 - \frac{1}{4}\theta^2} \tag{54}$$

into (53) and evaluate the resulting integral, we obtain

$$\mathrm{Cl}_2(\theta) = -\theta \log\left(\sin\frac{1}{2}\theta\right) + \theta + \sum_{n=1}^{\infty}\left[2\theta - 2n\pi \log\left(\frac{2n\pi + \theta}{2n\pi - \theta}\right)\right] \quad (0 \leqq \theta < 2\pi); \tag{55}$$

$$\mathrm{Ti}_2(\tan\theta) = \theta \log(\tan\theta) + \frac{1}{2}\mathrm{Cl}_2(2\theta) + \frac{1}{2}\mathrm{Cl}_2(\pi - 2\theta), \tag{56}$$

whose interesting special case when $\theta = \frac{1}{12}\pi$ becomes

$$\mathrm{Cl}_2\left(\frac{\pi}{6}\right) + \mathrm{Cl}_2\left(\frac{5\pi}{6}\right) = \frac{4}{3}G. \tag{57}$$

The Trilogarithm Function

The *Trilogarithm function* $\mathrm{Li}_3(z)$ is defined by

$$\begin{aligned}\mathrm{Li}_3(z) &:= \sum_{n=1}^{\infty} \frac{z^n}{n^3} \quad (|z| \leqq 1)\\ &= \int_0^z \frac{\mathrm{Li}_2(t)}{t}\, dt.\end{aligned} \tag{58}$$

An obvious special case of (58) is

$$\mathrm{Li}_3(1) = \zeta(3), \tag{59}$$

where $\zeta(s)$ is the Riemann Zeta function (see Section 2.3).

The following simple functional relationship for $\mathrm{Li}_3(x)$ would follow, if we divide both sides of (14) by x and then integrate each term:

$$\frac{1}{4}\mathrm{Li}_3\left(x^2\right) = \mathrm{Li}_3(x) + \mathrm{Li}_3(-x), \tag{60}$$

which, for $x = 1$, yields

$$\mathrm{Li}_3(-1) = -\frac{3}{4}\mathrm{Li}_3(1). \tag{61}$$

Similarly, we find from (4) and (9) that

$$\mathrm{Li}_3(-x) - \mathrm{Li}_3\left(-\frac{1}{x}\right) = -\frac{\pi^2}{6}\log x - \frac{1}{6}(\log x)^3 \tag{62}$$

and

$$\mathrm{Li}_3(y) - \mathrm{Li}_3\left(\frac{1}{y}\right) = \frac{\pi^2}{3}\log y - \frac{1}{6}(\log y)^3 - \frac{1}{2}i\pi(\log y)^2 \quad (y > 1), \tag{63}$$

respectively.

Landen's formula for $\mathrm{Li}_3(x)$ (see Lewin [751, p. 138]) is recalled here as follows:

$$\begin{aligned}\mathrm{Li}_3\left(\frac{x}{x-1}\right) &+ \mathrm{Li}_3(1-x) + \mathrm{Li}_3(x) = \mathrm{Li}_3(1) + \frac{\pi^2}{6}\log(1-x)\\ &- \frac{1}{2}\log x[\log(1-x)]^2 + \frac{1}{6}[\log(1-x)]^3 \quad (0 < x < 1),\end{aligned} \tag{64}$$

whose domain can be extended by using (63) with $y = 1 - x$:

$$\begin{aligned}\mathrm{Li}_3\left(\frac{x}{x-1}\right) &+ \mathrm{Li}_3\left(\frac{1}{1-x}\right) + \mathrm{Li}_3(x) = \mathrm{Li}_3(1) - \frac{\pi^2}{6}\log(1-x)\\ &- \frac{1}{2}\log(-x)[\log(1-x)]^2 + \frac{1}{3}[\log(1-x)]^3 \quad (x < 0).\end{aligned} \tag{65}$$

Taking $x = \frac{1}{2}$ in (64) or $x = -1$ in (65), we obtain

$$2\mathrm{Li}_3\left(\frac{1}{2}\right) + \mathrm{Li}_3(-1) = \mathrm{Li}_3(1) - \frac{\pi^2}{6}\log 2 + \frac{1}{3}(\log 2)^3,$$

which, upon eliminating $\mathrm{Li}_3(-1)$ with the aid of (61), yields

$$\mathrm{Li}_3\left(\frac{1}{2}\right)=\frac{7}{8}\mathrm{Li}_3(1)-\frac{\pi^2}{12}\log 2+\frac{1}{6}(\log 2)^3. \tag{66}$$

Setting $1-x=x/(1-x)$, we have

$$(1-x)^2=x \quad \text{or} \quad x^2-3x+1=0,$$

which gives the value:

$$x=\frac{3-\sqrt{5}}{2}=4\sin^2\left(\frac{\pi}{10}\right) \quad (x<1). \tag{67}$$

Now, writing $1-x$ for x in (60), and evaluating (64) with x given by (67), we obtain

$$\mathrm{Li}_3\left(\frac{3-\sqrt{5}}{2}\right)=\frac{4}{5}\mathrm{Li}_3(1)+\frac{\pi^2}{15}\log\left(\frac{3-\sqrt{5}}{2}\right)-\frac{1}{12}\left[\log\left(\frac{3-\sqrt{5}}{2}\right)\right]^3. \tag{68}$$

An interesting definite integral of $\mathrm{Li}_3(1)$ was given by Hjortnaes [559]:

$$\mathrm{Li}_3(1)=10\int_0^{\log\left(\frac{1+\sqrt{5}}{2}\right)} t^2\coth t\,dt, \tag{69}$$

which, upon writing $\coth t$ in exponential form and putting $y=1-e^{-2t}$, yields

$$\mathrm{Li}_3(1)=\frac{5}{4}\int_0^{\frac{\sqrt{5}-1}{2}}[\log(1-y)]^2\left(\frac{2}{y}+\frac{1}{1-y}\right)dy. \tag{70}$$

Integrating (70) by parts and expressing it in terms of trilogarithms, we arrive, once again, at (68).

The Polylogarithm Functions

The *Polylogarithm function* $\mathrm{Li}_n(z)$ is defined by

$$\begin{aligned}\mathrm{Li}_n(z)&:=\sum_{k=1}^{\infty}\frac{z^k}{k^n} \quad (|z|\leqq 1;\ n\in\mathbb{N}\setminus\{1\})\\ &=\int_0^z\frac{\mathrm{Li}_{n-1}(t)}{t}\,dt \quad (n\in\mathbb{N}\setminus\{1,2\}).\end{aligned} \tag{71}$$

Clearly, we have

$$\mathrm{Li}_n(1) = \zeta(n) \quad (n \in \mathbb{N} \setminus \{1\}), \tag{72}$$

in terms of the Riemann Zeta function $\zeta(s)$ (see Section 2.3). Setting $z = iy$ ($y \in \mathbb{R}$) in (71), we have

$$\mathrm{Li}_n(iy) = \frac{1}{2^n}\mathrm{Li}_n\left(-y^2\right) + i\,\mathrm{Ti}_n(y) \quad (y \in \mathbb{R}), \tag{73}$$

where $\mathrm{Ti}_n(x)$ is the *inverse tangent integral of order n* defined by

$$\mathrm{Ti}_n(x) := \sum_{k=1}^{\infty} (-1)^{k+1} \frac{x^{2k-1}}{(2k-1)^n} \quad (|x| \leqq 1). \tag{74}$$

It is obvious that

$$\mathrm{Ti}_n(x) = \int_0^x \frac{\mathrm{Ti}_{n-1}(t)}{t}\, dt, \tag{75}$$

which, in conjunction with (34) for $\mathrm{Ti}_2(x)$, extends the domain of $\mathrm{Ti}_n(x)$.

If $z = re^{i\theta}$, one defines

$$\mathrm{Li}_n\left(re^{i\theta}\right) := \mathrm{Li}_n(r,\theta) + i[\mathrm{Ti}_n(\varrho) - \mathrm{Ti}_n(\varrho, \tan\theta)], \tag{76}$$

where $\varrho = r\sin\theta/(1 - r\cos\theta)$. Then, it is readily seen that

$$\mathrm{Li}_n(r,\theta) = \int_0^r \frac{\mathrm{Li}_{n-1}(t,\theta)}{t}\, dt \tag{77}$$

and

$$\mathrm{Ti}_n(\varrho, \tan\theta) = \int_0^{\varrho} \left(\frac{\mathrm{Ti}_{n-1}(t)}{t+\tan\theta} + \frac{\mathrm{Ti}_{n-1}(t,\tan\theta)}{t} - \frac{\mathrm{Ti}_{n-1}(t,\tan\theta)}{t+\tan\theta}\right) dt. \tag{78}$$

The *generalized Clausen function* $\mathrm{Cl}_n(\theta)$ is defined by

$$\mathrm{Cl}_n(\theta) := \begin{cases} \displaystyle\sum_{k=1}^{\infty} \frac{\sin k\theta}{k^n} & (n \text{ is even}), \\ \displaystyle\sum_{k=1}^{\infty} \frac{\cos k\theta}{k^n} & (n \text{ is odd}). \end{cases} \tag{79}$$

From these definitions, we have

$$\mathrm{Cl}_{2n+1}(\theta) = \mathrm{Li}_{2n+1}(1) - \int_0^\theta \mathrm{Cl}_{2n}(t)\,dt \tag{80}$$

and

$$\mathrm{Cl}_{2n}(\theta) = \int_0^\theta \mathrm{Cl}_{2n-1}(t)\,dt. \tag{81}$$

The *Log-Sine integral* $\mathrm{Ls}_n(\theta)$ *of order* n is defined by

$$\mathrm{Ls}_n(\theta) := -\int_0^\theta \left(\log\left|2\sin\frac{1}{2}t\right|\right)^{n-1} dt \quad (n \in \mathbb{N}\setminus\{1\}), \tag{82}$$

whose special case *only* when $n = 2$ satisfies the relationship (*cf.* Equation (45)):

$$\mathrm{Ls}_2(\theta) = \mathrm{Cl}_2(\theta). \tag{83}$$

The *generalized Log-Sine integral of order n and index m* is defined by

$$\mathrm{Ls}_n^{(m)}(\theta) := -\int_0^\theta t^m \left(\log\left|2\sin\frac{1}{2}t\right|\right)^{n-m-1} dt. \tag{84}$$

The *associated Clausen function* $\mathrm{Gl}_n(\theta)$ *of order* n is defined by

$$\mathrm{Gl}_n(\theta) := \begin{cases} \displaystyle\sum_{k=1}^{\infty} \frac{\cos k\theta}{k^n} & (n \text{ is even}), \\[2ex] \displaystyle\sum_{k=1}^{\infty} \frac{\sin k\theta}{k^n} & (n \text{ is odd}), \end{cases} \tag{85}$$

which also satisfies the following obvious relationships analogous to (80) and (81):

$$\mathrm{Gl}_{2n}(\theta) = \mathrm{Li}_{2n}(1) - \int_0^\theta \mathrm{Gl}_{2n-1}(t)\,dt \tag{86}$$

and

$$\mathrm{Gl}_{2n+1}(\theta) = \int_0^\theta \mathrm{Gl}_{2n}(t)\,dt, \tag{87}$$

each of which was studied extensively by Glaisher [488].

Dividing both sides of (4) by x and integrating from 1 to x, we obtain (*cf.* Equation (62))

$$\mathrm{Li}_3(-x)-\mathrm{Li}_3\left(-\frac{1}{x}\right)=-\frac{1}{3!}(\log x)^3+2\,\mathrm{Li}_2(-1)\,\log x. \tag{88}$$

By repeating this process $n-2$ times, we finally arrive at

$$\begin{aligned}\mathrm{Li}_n(-x)+(-1)^n\,\mathrm{Li}_n\left(-\frac{1}{x}\right)&=-\frac{1}{n!}(\log x)^n\\&\quad+2\sum_{k=1}^{\left[\frac{1}{2}n\right]}\frac{(\log x)^{n-2k}}{(n-2k)!}\mathrm{Li}_{2k}(-1)\quad(n\in\mathbb{N}\setminus\{1\}),\end{aligned}\tag{89}$$

which was attributed to Jonquiére by Nielsen [861], although equivalent formulas were given much earlier by Spence and Kummer (see Lewin [751, p. 173]).

Taking $x=e^{i\pi}$ in (89), we find that

$$\left[1+(-1)^n\right]\mathrm{Li}_n(1)=-\frac{i^n\pi^n}{n!}+2\sum_{k=1}^{\left[\frac{1}{2}n\right]}\frac{i^{n-2k}\pi^{n-2k}}{(n-2k)!}\mathrm{Li}_{2k}(-1). \tag{90}$$

Since the left-hand side of (90) vanishes when n is odd, the evaluation of $\mathrm{Li}_{2n+1}(1)$ is not possible from (90). Yet, using the following known relationship between $\mathrm{Li}_n(1)$ and $\mathrm{Li}_n(-1)$:

$$\mathrm{Li}_n(-1)=-\left(1-2^{1-n}\right)\mathrm{Li}_n(1) \tag{91}$$

and writing $2n$ for n in (90), we readily get

$$\frac{2(-1)^{n-1}\left(2^{2n}-1\right)}{(2n)!}\mathrm{L}_{2n}=\frac{1}{(2n)!}-2\sum_{k=1}^{n-1}\frac{(-1)^{k-1}\left(2^{2k-1}-1\right)}{(2k)!\,(2n-2k)!}\mathrm{L}_{2k}, \tag{92}$$

where, for the sake of brevity,

$$\mathrm{L}_{2k}:=(2k)!\pi^{-2k}2^{1-2k}\,\mathrm{Li}_{2k}(1). \tag{93}$$

By using the generating function 1.6(2) of the Bernoulli numbers, one can also obtain

$$\mathrm{Li}_{2n}(-1)=(-1)^n\left(2^{2n-1}-1\right)B_{2n}\frac{\pi^{2n}}{(2n)!}\quad(n\in\mathbb{N}) \tag{94}$$

and

$$\mathrm{Li}_{2n}(1)=(-1)^{n+1}\frac{(2\pi)^{2n}}{2(2n)!}B_{2n}\quad(n\in\mathbb{N}), \tag{95}$$

which is precisely the same as 2.3(18).

Although there is no result like (94) when the order of the Polylogarithm function is odd, there is an analogous relation involving the Euler numbers E_n given in Section 1.7. Dividing (35) by x and integrating from 1 to x would yield

$$\mathrm{Ti}_3(x) + \mathrm{Ti}_3\left(\frac{1}{x}\right) = 2\mathrm{Ti}_3(1) + \frac{1}{4}\pi(\log x)^2 \quad (x > 0). \tag{96}$$

Repeated applications of this process lead us to *Spence's formula*:

$$\begin{aligned}\mathrm{Ti}_n(x) + (-1)^{n-1}\mathrm{Ti}_n\left(\frac{1}{x}\right) &= \frac{1}{2}\pi\frac{(\log x)^{n-1}}{(n-1)!} \\ &+ 2\sum_{k=1}^{\left[\frac{1}{2}(n-1)\right]}\frac{(\log x)^{n-2k-1}}{(n-2k-1)!}\mathrm{Ti}_{2k+1}(1) \quad (x>0).\end{aligned} \tag{97}$$

Next, we find from (73) that

$$\mathrm{Li}_{2n+1}(i) - \mathrm{Li}_{2n+1}(-i) = 2i\,\mathrm{Ti}_{2n+1}(1). \tag{98}$$

Setting $x = \exp\left(\frac{1}{2}\pi i\right)$ and writing $2n+1$ for n in (89), we obtain

$$\begin{aligned}&\mathrm{Li}_{2n+1}(-i) - \mathrm{Li}_{2n+1}(i) \\ &= (-1)^n i\left[-\frac{\pi^{2n+1}}{(2n+1)!\,2^{2n+1}} + 2\pi^{2n+1}\sum_{k=1}^{n}\frac{B_{2k}\left(2^{2k-1}-1\right)}{(2k)!\,(2n-2k+1)!\,2^{2n-2k+1}}\right].\end{aligned} \tag{99}$$

Consider

$$\frac{2e^{\frac{1}{2}z}}{e^z+1} = 2\frac{e^{\frac{1}{2}z}-e^{-\frac{1}{2}z}}{2}\left(\frac{z}{e^z-1} - \frac{1}{2}\frac{2z}{e^{2z}-1}\right), \tag{100}$$

which can be expanded by 1.6(2) and 1.6(40) in powers of z. Upon equating the coefficients of z^{2n}, we find that

$$\frac{E_{2n}}{(2n)!\,2^{2n+1}} = \left[\frac{1}{(2n+1)!\,2^{2n+1}} - 2\sum_{k=1}^{n}\frac{B_{2k}\left(2^{2k-1}-1\right)}{(2k)!\,(2n-2k+1)!\,2^{2n-2k+1}}\right]. \tag{101}$$

Now, a combination of (98), (99) and (101) leads us to the well-known Euler series:

$$\mathrm{Ti}_{2n+1}(1) = \sum_{k=1}^{\infty}(-1)^{k+1}\frac{1}{(2k-1)^{2n+1}} = \frac{(-1)^n\pi^{2n+1}}{(2n)!\,2^{2n+2}}E_{2n}, \tag{102}$$

which, in view of 1.6(59), gives

$$\mathrm{Ti}_3(1)=\frac{\pi^3}{32},\quad \mathrm{Ti}_5(1)=\frac{5\pi^5}{1536},\quad \mathrm{Ti}_7(1)=\frac{61\pi^7}{184320},\ \ldots. \tag{103}$$

Starting with the known formula for $\mathrm{Li}_2(x,\theta)$:

$$\mathrm{Li}_2(r,\theta)+\mathrm{Li}_2\left(\frac{1}{r},\theta\right)=2\,\mathrm{Gl}_2(\theta)-\frac{1}{2}(\log r)^2 \tag{104}$$

and following a procedure analogous to that in proving (89), one obtains

$$\begin{aligned}\mathrm{Li}_n(r,\theta)+(-1)^n\,\mathrm{Li}_n\left(\frac{1}{r},\theta\right)&=-\frac{1}{n!}(\log r)^n\\&\quad+2\sum_{k=1}^{\left[\frac{1}{2}n\right]}\frac{(\log r)^{n-2k}}{(n-2k)!}\mathrm{Gl}_{2k}(\theta)\quad (n\in\mathbb{N}\setminus\{1\}),\end{aligned} \tag{105}$$

which, upon setting $r=e^{i\pi}$, noting that $\mathrm{Li}_n(-r,\theta)=\mathrm{Li}_n(r,\pi-\theta)$ and writing $2n$ for n, gives

$$\mathrm{Gl}_{2n}(\pi-\theta)=\frac{(-1)^{n-1}\pi^{2n}}{2(2n)!}+\sum_{k=1}^{n}\frac{(-1)^{n-k}\pi^{2n-2k}}{(2n-2k)!}\mathrm{Gl}_{2k}(\theta). \tag{106}$$

The following relationship between $\mathrm{Gl}_{2n}(\theta)$ and $\mathrm{Gl}_{2n}(\pi-\theta)$ can readily be obtained from the series definition (85):

$$\mathrm{Gl}_{2n}(\pi-\theta)+\mathrm{Gl}_{2n}(\theta)=\frac{1}{2^{2n-1}}\mathrm{Gl}_{2n}(2\theta). \tag{107}$$

The generating function 1.7(1) for the Bernoulli polynomials can be rewritten in the form:

$$\frac{te^{xt}}{e^t-1}=1+t\sum_{n=1}^{\infty}\frac{2t\cos(2n\pi x)-4n\pi\sin(2n\pi x)}{t^2+4n^2\pi^2}. \tag{108}$$

Making use of the expansion:

$$\frac{1}{t^2+4n^2\pi^2}=\frac{1}{4n^2\pi^2}\sum_{k=0}^{\infty}(-1)^k\left(\frac{t}{2n\pi}\right)^{2k}\qquad (|t|<2n\pi;\ n\in\mathbb{N})$$

and rearranging the double series in (108) in powers of t, the coefficients are easily expressible in terms of $\mathrm{Gl}_n(2\pi x)$, defined by (85). Thus, employing 1.7(1) on the left-hand side of (108) and equating the coefficients of the same powers on both sides, we find that

$$\mathrm{Gl}_n(2\pi x) = (-1)^{1+\left[\frac{1}{2}n\right]} 2^{n-1} \pi^n \frac{B_n(x)}{n!} \quad (0 \leqq x \leqq 1;\ n \in \mathbb{N} \setminus \{1\}), \tag{109}$$

which gives us the following special cases:

$$\begin{aligned}
\mathrm{Gl}_2(\theta) &= \frac{1}{4}(\pi - \theta)^2 - \frac{\pi^2}{12}, \\
\mathrm{Gl}_3(\theta) &= \frac{1}{12}\theta(\pi - \theta)(2\pi - \theta), \\
\mathrm{Gl}_4(\theta) &= \frac{1}{90}\pi^4 - \frac{1}{12}\theta^2\left(\pi^2 - \pi\theta + \frac{1}{4}\theta^2\right), \\
\mathrm{Gl}_5(\theta) &= \frac{1}{720}\theta(\pi - \theta)(2\pi - \theta)\left(4\pi^2 + 6\pi\theta - 3\theta^2\right).
\end{aligned} \tag{110}$$

The Log-Sine Integrals

A recurrence relationship for $\mathrm{Ls}_n(\pi)$, defined by (82), is given by

$$\begin{aligned}
\frac{(-1)^m}{m!}\mathrm{Ls}_{m+2}(\pi) &= \pi\left(1 - 2^{-m}\right)\zeta(m+1) \\
&+ \sum_{k=2}^{m-1} \frac{(-1)^{k+1}}{k!}\left(1 - 2^{k-m}\right)\zeta(m-k+1)\,\mathrm{Ls}_{k+1}(\pi) \quad (m \in \mathbb{N}).
\end{aligned} \tag{111}$$

To derive the recurrence relation (111), we let

$$\begin{aligned}
I &:= \int_0^\pi \exp\left[x \log\left(2\sin\frac{1}{2}\theta\right)\right] d\theta \\
&= \sum_{n=0}^{\infty} \int_0^\pi \frac{x^n}{n!}\left[\log\left(2\sin\frac{1}{2}\theta\right)\right]^n d\theta \\
&= -\sum_{n=0}^{\infty} \frac{x^n}{n!}\mathrm{Ls}_{n+1}(\pi).
\end{aligned} \tag{112}$$

Alternatively, in view of 1.1(44), we have

$$I = \int_0^\pi \left[\exp\left\{ \log\left(2\sin\frac{1}{2}\theta \right) \right\} \right]^x d\theta$$

$$= \int_0^\pi 2^x \sin^x \left(\frac{1}{2}\theta \right) d\theta$$

$$= 2^x \sqrt{\pi} \, \frac{\Gamma\left(\frac{1}{2} + \frac{1}{2}x\right)}{\Gamma\left(1 + \frac{1}{2}x\right)}.$$

By the Legendre duplication formula 1.1(29) for Γ, we, thus, obtain

$$I = \pi\, y, \tag{113}$$

where, for convenience,

$$y := \frac{\Gamma(1+x)}{\left[\Gamma\left(1 + \frac{1}{2}x\right)\right]^2}.$$

Making use of the following notations:

$$D^n f(x) = \frac{d^n}{dx^n} f(x) = f^{(n)}(x) \quad \text{and} \quad D_0^n f(x) = \left. \frac{d^n}{dx^n} f(x) \right|_{x=0} = f^{(n)}(0),$$

differentiating (112) n times and setting $x = 0$, we find that

$$\mathrm{Ls}_{n+1}(\pi) = -D_0^n I = -\pi\, D_0^n y, \tag{114}$$

which, by virtue of 1.3(53), yields

$$D_0^n \log y = (-1)^n (n-1)! \left(1 - 2^{1-n}\right) \zeta(n) \quad (n \in \mathbb{N} \setminus \{1\}), \tag{115}$$

so that we have

$$\log y =: f(x) = \sum_{n=2}^{\infty} \frac{x^n}{n} (-1)^n \left(1 - 2^{1-n}\right) \zeta(n), \tag{116}$$

since

$$\log y|_{x=0} = 0 = D_0^1 \log y.$$

From (114), we get $\mathrm{Ls}_{n+1}(\pi) = -\pi\, D_0^n \exp(f(x))$. Let $y_m := D^m\, e^f$. Then,

$$y_{m+1} = D^{m+1}\, e^f = D^m \left(De^f\right) = D^m \left(f'\, e^f\right).$$

Hence, using Leibniz's rule for differentiation, we have

$$y_{m+1} = \sum_{k=0}^{m} \binom{m}{k} f^{(m-k+1)}\, y_k, \tag{117}$$

which, upon considering

$$y_m|_{x=0} = -\frac{1}{\pi}\, \mathrm{Ls}_{m+1}(\pi)$$

in (114), yields

$$\mathrm{Ls}_{m+2}(\pi) = -\pi\, f_0^{(m+1)} + \sum_{k=1}^{m} \binom{m}{k} f_0^{(m-k+1)}\, \mathrm{Ls}_{k+1}(\pi). \tag{118}$$

Now, setting

$$\mathrm{Ls}_2(\pi) = 0, \quad f_0^{'} = 0 \quad \text{and} \quad f_0^{(m)} = (-1)^m\, (m-1)! \left(1 - 2^{1-m}\right) \zeta(m) \quad (m \in \mathbb{N} \setminus \{1\})$$

in (118), we immediately arrive at the desired recurrence relation (111).

Some simple consequences of (111) are presented below:

$$\mathrm{Ls}_2(\pi) = -\int_0^{\pi} \log\left(2 \sin \frac{1}{2}\theta\right) d\theta = 0,$$

$$\mathrm{Ls}_3(\pi) = -\int_0^{\pi} \left[\log\left(2 \sin \frac{1}{2}\theta\right)\right]^2 d\theta = -\frac{1}{12}\pi^3,$$

$$\mathrm{Ls}_4(\pi) = -\int_0^{\pi} \left[\log\left(2 \sin \frac{1}{2}\theta\right)\right]^3 d\theta = \frac{3}{2}\pi\, \zeta(3),$$

$$\mathrm{Ls}_5(\pi) = -\int_0^{\pi} \left[\log\left(2 \sin \frac{1}{2}\theta\right)\right]^4 d\theta = -\frac{19}{240}\pi^5,$$

$$\mathrm{Ls}_6(\pi) = -\int_0^{\pi} \left[\log\left(2\sin\frac{1}{2}\theta\right)\right]^5 d\theta = \frac{45}{2}\pi\,\zeta(5) + \frac{5}{4}\pi^3\,\zeta(3),$$

$$\mathrm{Ls}_7(\pi) = -\int_0^{\pi} \left[\log\left(2\sin\frac{1}{2}\theta\right)\right]^6 d\theta = -\frac{45}{2}\pi\,\{\zeta(3)\}^2 - \frac{275}{1344}\pi^7,$$

$$\mathrm{Ls}_8(\pi) = -\int_0^{\pi} \left[\log\left(2\sin\frac{1}{2}\theta\right)\right]^7 d\theta = \frac{2835}{4}\pi\,\zeta(7) + \frac{315}{8}\pi^3\,\zeta(5) + \frac{133}{20}\pi^5\,\zeta(3),$$

$$\mathrm{Ls}_9(\pi) = -\int_0^{\pi} \left[\log\left(2\sin\frac{1}{2}\theta\right)\right]^8 d\theta$$

$$= -\frac{24177}{26880}\pi^9 - 1890\pi\,\zeta(3)\,\zeta(5) - \frac{105}{2}\pi^3\,\{\zeta(3)\}^2.$$

2.5 Hurwitz–Lerch Zeta Functions

The *Hurwitz–Lerch Zeta function* $\Phi(z,s,a)$ is defined by

$$\Phi(z,s,a) := \sum_{n=0}^{\infty} \frac{z^n}{(n+a)^s} \tag{1}$$

$$\left(a \in \mathbb{C}\setminus\mathbb{Z}_0^-;\ s \in \mathbb{C} \text{ when } |z|<1;\ \Re(s)>1 \text{ when } |z|=1\right),$$

which satisfies the obvious functional relation:

$$\Phi(z,s,a) = z^n\,\Phi(z,s,n+a) + \sum_{k=0}^{n-1} \frac{z^k}{(k+a)^s} \quad \left(n \in \mathbb{N};\ a \in \mathbb{C}\setminus\mathbb{Z}_0^-\right). \tag{2}$$

By writing the Eulerian integral 1.1(1) in the form:

$$\Gamma(z) = s^z \int_0^{\infty} e^{-st}\,t^{z-1}\,dt \quad (\Re(z)>0;\ \Re(s)>0), \tag{3}$$

we can deduce the following integral representation from (1):

$$\Phi(z,s,a) = \frac{1}{\Gamma(s)} \int_0^{\infty} \frac{t^{s-1}\,e^{-at}}{1-ze^{-t}}\,dt = \frac{1}{\Gamma(s)} \int_0^{\infty} \frac{t^{s-1}\,e^{-(a-1)t}}{e^t - z}\,dt \tag{4}$$

$$(\Re(a)>0;\ |z| \leqq 1,\ z \neq 1,\ \Re(s)>0;\ z=1,\ \Re(s)>1),$$

by noting that

$$\frac{z^n}{(n+a)^s} = \frac{1}{\Gamma(s)} \int_0^\infty e^{-at}\, t^{s-1} \left(ze^{-t}\right)^n dt \quad (\Re(a) > 0;\ \Re(s) > 0). \tag{5}$$

If use is made of the infinite-series version of Plana's summation formula 2.2(9) (*cf.* Lindelöf [769, p. 61]; see also Erdélyi *et al.* [421, p. 22]) and the definition (1), another definite integral representation of $\Phi(z,s,a)$ is obtained in the form:

$$\begin{aligned}\Phi(z,s,a) &= \frac{1}{2a^s} + \int_0^\infty \frac{z^t}{(t+a)^s}\, dt \\ &\quad - 2\int_0^\infty \sin\left[t\log z - s\tan^{-1}\left(\frac{t}{a}\right)\right]\left(t^2+a^2\right)^{-\frac{1}{2}s} \frac{dt}{e^{2\pi t}-1} \quad (\Re(a) > 0),\end{aligned} \tag{6}$$

which, for $z = 1$, immediately reduces to Hermite's formula 2.2(12) for $\zeta(s,a)$.

By setting $z = e^{i\theta}$ in (4) and using (3), we get *Lipschitz's formula*:

$$\begin{aligned}2\,\Gamma(s) \sum_{n=1}^\infty \frac{e^{in\theta}}{(n+a)^s} &= \int_0^\infty e^{-at}\, t^{s-1}\, \frac{e^{i\theta} - e^{-t}}{\cosh t - \cos\theta}\, dt \\ &(0 < \theta < 2\pi;\ \Re(a) > -1;\ \Re(s) > 0).\end{aligned} \tag{7}$$

Several contour integral and series representations of $\Phi(z,s,a)$ include (see Erdélyi *et al.* [421, pp. 28–31]):

$$\begin{aligned}\Phi(z,s,a) &= -\frac{\Gamma(1-s)}{2\pi i} \int_C \frac{(-t)^{s-1}\, e^{-at}}{1 - ze^{-t}}\, dt \\ &(\Re(a) > 0;\ |\arg(-t)| \leqq \pi),\end{aligned} \tag{8}$$

where the contour C is the Hankel loop of Theorem 2.5, which, obviously, does not enclose any pole of the integrand, that is, $t = \log z \pm 2n\pi i\ (n \in \mathbb{N}_0)$;

$$\begin{aligned}\Phi(z,s,a) &= \frac{\Gamma(1-s)}{z^a} \sum_{n=-\infty}^\infty (-\log z + 2n\pi i)^{s-1}\, e^{2n\pi i a} \\ &(0 < a \leqq 1;\ \Re(s) < 0;\ |\arg(-\log z + 2n\pi i)| \leqq \pi);\end{aligned} \tag{9}$$

Lerch's transformation formula:

$$\begin{aligned}\Phi(z,s,a) &= \frac{i(2\pi)^{s-1}\,\Gamma(1-s)}{z^a} \left\{ e^{-\frac{1}{2}i\pi s} L\left(-a, 1-s, \frac{\log z}{2\pi i}\right) \right. \\ &\quad \left. - e^{i\pi\left(\frac{1}{2}s+2a\right)} L\left(a, 1-s, 1 - \frac{\log z}{2\pi i}\right) \right\},\end{aligned} \tag{10}$$

where the so-called Lipschitz-Lerch transcendent $L(\xi, s, a)$ is defined by

$$\phi(\xi, a, s) := \sum_{n=0}^{\infty} \frac{e^{2n\pi i\xi}}{(n+a)^s} = \Phi\left(e^{2\pi i\xi}, s, a\right) =: L(\xi, s, a) \tag{11}$$
$$\left(a \in \mathbb{C} \setminus \mathbb{Z}_0^-;\ \Re(s) > 0 \quad \text{when} \quad \xi \in \mathbb{R} \setminus \mathbb{Z};\ \Re(s) > 1 \quad \text{when} \quad \xi \in \mathbb{Z}\right),$$

which was first studied by Rudolf Lipschitz (1832–1903) and Matyáš Lerch (1860–1922) in connection with Dirichlet's famous theorem on primes in arithmetic progressions.

Replacing s by $1-s$ in 2.2(6), we obtain another equivalent form of Hurwitz's formula 2.2(6):

$$\zeta(s,a) = 2(2\pi)^{s-1}\,\Gamma(1-s) \sum_{n=1}^{\infty} n^{s-1} \sin\left(2n\pi a + \frac{\pi s}{2}\right) \tag{12}$$
$$(\Re(s) < 0;\ 0 < a \leqq 1).$$

By applying the binomial theorem to $(-\log z \pm 2n\pi i)^{s-1}$ in (9) and making use of (12), 1.1(12) and 1.1(21), we readily obtain the following formulas (*cf.* Erdélyi *et al.* [421, p. 29, Eq. (8)]):

$$\Phi(z, s, a) = \frac{\Gamma(1-s)}{z^a} \left(\log \frac{1}{z}\right)^{s-1} + \frac{1}{z^a} \sum_{k=0}^{\infty} \zeta(s-k, a) \frac{(\log z)^k}{k!} \tag{13}$$
$$(|\log z| < 2\pi;\ s \notin \mathbb{N};\ a \notin \mathbb{Z}_0^-);$$

$$\Phi(z, m, a)$$
$$= \frac{1}{z^a} \left\{ \sum_{\substack{k=0 \\ k \neq m-1}}^{\infty} \zeta(m-k, a) \frac{(\log z)^k}{k!} + \frac{(\log z)^{m-1}}{(m-1)!} \left[\psi(m) - \psi(a) - \log\left(\log \frac{1}{z}\right)\right] \right\}$$
$$(m \in \mathbb{N} \setminus \{1\};\ |\log z| < 2\pi;\ a \notin \mathbb{Z}_0^-). \tag{14}$$

By setting $s = -m$ ($m \in \mathbb{N}$) in (13) and using 2.1(17), Φ can also be expressed in terms of the Bernoulli polynomials as follows:

$$\Phi(z, -m, a) = \frac{m!}{z^a} \left(\log \frac{1}{z}\right)^{-m-1} - \frac{1}{z^a} \sum_{k=0}^{\infty} \frac{B_{m+k+1}(a) (\log z)^k}{k!(m+k+1)} \tag{15}$$
$$(|\log z| < 2\pi).$$

The special case of (1) when $s = 1$ yields

$$\Phi(z,1,a) = \sum_{n=0}^{\infty} \frac{z^n}{n+a} = a^{-1}\,{}_2F_1(1,\, a;\, 1+a;\, z) \quad (|z| < 1), \tag{16}$$

and, more generally, we have

$$\Phi(z,k,a) = a^{-k}\,{}_{k+1}F_k(1,\, a,\, \ldots,\, a;\, a+1,\, \ldots, a+1;\, z) \quad (k \in \mathbb{N}_0), \tag{17}$$

where ${}_pF_q$ denotes the generalized hypergeometric function (see Section 1.5).

It follows from (13) and (16) that

$$\lim_{z\to 1} \left\{ (1-z)^{1-s}\,\Phi(z,s,a) \right\} = \Gamma(1-s) \quad (\Re(s) < 1) \tag{18}$$

and

$$\lim_{z\to 1} \left\{ \frac{\Phi(z,1,1)}{\log(1-z)} \right\} = -1. \tag{19}$$

The *Jonquière function* $F(z,s)$ is defined by

$$F(z,s) := \sum_{k=1}^{\infty} \frac{z^k}{k^s} = z\,\Phi(z,s,1), \tag{20}$$

whose many properties can be deduced from those of the Φ-function. For example, Lerch's transformation formula (10) readily yields *Jonquière's formula*:

$$F(z,s) + e^{is\pi}\, F\left(\frac{1}{z}, s\right) = \frac{(2\pi)^s}{\Gamma(s)} e^{\frac{1}{2}i\pi s}\, \zeta\left(1-s, \frac{\log z}{2\pi i}\right). \tag{21}$$

Moreover, we obtain

$$F(z,-m) = (-1)^{m+1}\, F\left(\frac{1}{z}, -m\right) \quad (m \in \mathbb{N}) \tag{22}$$

and

$$F(z,m) + (-1)^m\, F\left(\frac{1}{z}, m\right) = -\frac{2\pi i}{m!} B_m\left(\frac{\log z}{2\pi i}\right) \quad (m \in \mathbb{N} \setminus \{1\}), \tag{23}$$

both of which provide the analytic continuation of $F(z,s)$ in (20) outside its circle of convergence $|z| = 1$.

By taking $a = 1$ in (15) and using 1.7(5), we get

$$F(z, -m) = m! \left(\log \frac{1}{z}\right)^{-m-1} - \sum_{k=0}^{\infty} \frac{B_{m+k+1}}{k!(m+k+1)} (\log z)^k \tag{24}$$
$$(m \in \mathbb{N};\ |\log z| < 2\pi).$$

In addition to the Jonquière function $F(z, s)$ in (20), various special cases of $\Phi(z, s, a)$ are listed below:

$$\zeta(s) = \Phi(1, s, 1); \tag{25}$$

$$\zeta(s, a) = \Phi(1, s, a); \tag{26}$$

$$\mathrm{Li}_2(z) = z\Phi(z, 2, 1); \tag{27}$$

$$\mathrm{Li}_s(z) := \sum_{k=1}^{\infty} \frac{z^k}{k^s} = z\Phi(z, s, 1) = F(z, s) \tag{28}$$
$$(s \in \mathbb{C} \text{ and } |z| < 1;\ \Re(s) > 1 \text{ and } |z| = 1).$$

The Taylor Series Expansion of the Lipschitz-Lerch Transcendent $L(x, s, a)$

The function $L(x, s, a)$, defined by (11), was first studied by Lipschitz [770, 771] and Lerch [745] in connection with Dirichlet's famous theorem on primes in arithmetic progressions. For $x \in \mathbb{Z}$, (11) reduces immediately to the Hurwitz Zeta function $\zeta(s, a)$ (see Section 2.2). Since $L(x, s, a)$ is a special case of $\Phi(z, s, a)$, many properties of this function can be obtained from those of $\Phi(z, s, a)$.

The Lerch functional equation for $L(x, s, a)$ can be deduced from (10):

$$L(x, 1-s, a) = \frac{\Gamma(s)}{(2\pi)^s} \left\{ e^{\pi i\left(\frac{1}{2}s - 2ax\right)} L(-a, s, x) + e^{\pi i\left(-\frac{1}{2}s + 2a(1-x)\right)} L(a, s, 1-x) \right\}$$
$$(0 < x < 1;\ 0 < a < 1). \tag{29}$$

Several interesting proofs of (29) were given by Apostol [58] and Berndt [117].

The function $L(x, s, a)$ satisfies the differential-difference equations:

$$\frac{\partial L(x, s, a)}{\partial a} = -sL(x, s+1, a) \tag{30}$$

and

$$\frac{\partial L(x, s, a)}{\partial x} + 2\pi i a L(x, s, a) = 2\pi i L(x, s-1, a). \tag{31}$$

Klusch [673] gave a Taylor series expansion of the function $L(x,s,a+\xi)$ in the neighborhood of $\xi=0$ in the form: For fixed $a\in\mathbb{R}^+$,

$$\sum_{k=0}^{\infty}(-1)^k\frac{(s)_k}{k!}L(x,s+k,a)\,\xi^k=L(x,s,a+\xi)\quad(|\xi|<a),\tag{32}$$

which provides the analytic continuation of $L(x,s,a)$ to the whole complex s-plane. Klusch [673] also applied (32) to derive several summation formulas which will be investigated systematically in Chapter 3.

By setting $\xi=-t$ in (32), we get

$$\sum_{k=0}^{\infty}\frac{(s)_k}{k!}L(x,s+k,a)\,t^k=L(x,s,a-t)\quad(|t|<a),\tag{33}$$

which, upon replacing k by $k+2$ and s by $s-1$, dividing the resulting equation by t^2 and differentiating both sides with respect to t yields the following *corrected* version of one of Klusch's results (Klusch [673, p. 517, Eq. (3.3)]):

$$\begin{aligned}&\sum_{k=1}^{\infty}\frac{k}{(k+2)!}\,\frac{\Gamma(s+k+1)}{r\Gamma(s)}L(x,s+k+1,a)\,t^{k-1}\\&=t^{-2}\{L(x,s,a-t)+L(x,s,a)\}-\frac{2t^{-3}}{rs-1}\{L(x,s-1,a-t)-L(x,s-1,a)\}\\&\qquad\qquad(|t|<a;\ t\neq 0).\end{aligned}\tag{34}$$

Evaluation of $L(x,-n,a)$

Since $L(x,s,a)=\zeta(s,a)$ $(x\in\mathbb{Z})$, whose properties are studied in Section 2.2, in what follows, we assume that x is not an integer. By making use of the classical method presented in Section 2.1, $L(x,s,a)$ can be extended to the whole complex s-plane by means of the contour integral (see Apostol [58]):

$$I(x,s,a)=\frac{1}{2\pi i}\int_{C_1}\frac{z^{s-1}\,e^{az}}{1-e^{z+2\pi ix}}\,dz,\tag{35}$$

where the contour C_1 is a loop which begins at $-\infty$, encircles the origin once in the positive (counter-clockwise) direction and returns to $-\infty$. Since $I(x,s,a)$ is an entire function of s and

$$L(x,s,a)=\Gamma(1-s)\,I(x,s,a),\tag{36}$$

we can use (36) to continue $L(x,s,a)$ analytically to the whole complex s-plane as an entire function of s $(x\notin\mathbb{Z})$.

Apostol [58] computed recursively the values of $L(x,-n,a)$ $(n \in \mathbb{N}_0;\ x \notin \mathbb{Z})$ by applying (31) and (35). Thus, he obtained (*cf.* Apostol [58])

$$L(x,0,a) = \frac{1}{1-\exp(2\pi ix)} = \frac{i}{2}\cot(\pi x) + \frac{1}{2}, \tag{37}$$

$$L(x,-1,a) = \frac{a}{2}[i\cot(\pi x)+1] - \frac{1}{4}\csc^2(\pi x), \tag{38}$$

$$L(x,-2,a) = \frac{a^2}{2}\left[i\cot(\pi x)+\frac{1}{4}\right] - \frac{a}{2}\csc^2(\pi x) - \frac{i}{4}\cot(\pi x)\csc^2(\pi x) \tag{39}$$

and

$$L(x,-n,a) = -\frac{\mathcal{B}_{n+1}\left(a;e^{2\pi ix}\right)}{n+1} \qquad (n \in \mathbb{N}_0), \tag{40}$$

where the $\mathcal{B}_n(a;\alpha)$ are defined by 1.8(1).

2.6 Generalizations of the Hurwitz–Lerch Zeta Function

Lin and Srivastava [765] introduced and investigated the following generalization of the Hurwitz–Lerch Zeta function $\Phi(z,s,a)$, defined by 2.5(1):

$$\Phi_{\mu,\nu}^{(\rho,\sigma)}(z,s,a) := \sum_{n=0}^{\infty} \frac{(\mu)_{\rho n}}{(\nu)_{\sigma n}} \frac{z^n}{(n+a)^s}$$

$$\big(\mu \in \mathbb{C};\ a,\nu \in \mathbb{C}\setminus\mathbb{Z}_0^-;\ \rho,\sigma \in \mathbb{R}^+;\ \rho < \sigma \quad \text{when} \quad s,z \in \mathbb{C}; \tag{1}$$
$$\rho=\sigma \ \text{ and } \ s \in \mathbb{C} \ \text{ when } \ |z| < \delta := \rho^{-\rho}\,\sigma^{\sigma};\ \rho=\sigma \ \text{ and } \ \Re(s-\mu+\nu) > 1$$
$$\text{when} \quad |z| = \delta\big),$$

where $(\lambda)_n$ denotes the Pochhammer symbol, defined by 1.1(5).

Clearly, we have

$$\Phi_{\nu,\nu}^{(\sigma,\sigma)}(z,s,a) = \Phi_{\mu,\nu}^{(0,0)}(z,s,a) = \Phi(z,s,a) \tag{2}$$

and

$$\Phi_{\mu,1}^{(1,1)}(z,s,a) = \Phi_{\mu}^{*}(z,s,a) := \sum_{n=0}^{\infty} \frac{(\mu)_n}{n!} \frac{z^n}{(n+a)^s}, \tag{3}$$

where, as also noted by Lin and Srivastava [765], $\Phi_{\mu}^{*}(z,s,a)$ is a generalization of the Hurwitz–Lerch Zeta function considered by Goyal and Laddha [504, p. 100, Eq. (1.5)].

For further results involving these classes of generalized Hurwitz–Lerch Zeta functions, see the recent works by Garg *et al.* [467] and Lin *et al.* [767]. Recently, among other things, Choi *et al.* [273] presented and investigated the following multiple-series generalization of the Hurwitz–Lerch Zeta function $\Phi(z,s,a)$, by calling it the multiple (or, simply, n-tuple) Hurwitz–Lerch Zeta function $\Phi_n(z,s,a)$, defined by

$$\Phi_n(z,s,a) := \sum_{m_1,\ldots,m_n=0}^{\infty} \frac{z^{m_1+\cdots+m_n}}{(m_1+\cdots+m_n+a)^s} \tag{4}$$
$$\left(a \in \mathbb{C}\setminus\mathbb{Z}_0^-;\ s \in \mathbb{C} \text{ when } |z|<1;\ \Re(s)>n \text{ when } |z|=1\right).$$

In its special case when $z=1$, the definition (4) yields the following familiar multiple (or, simply, n-tuple) Hurwitz Zeta function $\zeta_n(s,a)$ given in 2.1(2):

$$\Phi_n(1,s,a) = \zeta_n(s,a)\left(\Re(s)>n;\ a \in \mathbb{C}\setminus\mathbb{Z}_0^-\right). \tag{5}$$

The special case of (4) when $n=1$ reduces to the Hurwitz–Lerch Zeta function $\Phi(z,s,a)$, defined by 2.5(1), as follows:

$$\Phi_1(z,s,a) = \Phi(z,s,a). \tag{6}$$

Furthermore, when $n=1$ and $z=1$, the definition (4) yields the Hurwitz (or generalized) Zeta function $\zeta(s,a)$, defined by 2.2(1):

$$\Phi_1(1,s,a) = \zeta(s,a) \qquad \left(a \in \mathbb{C}\setminus\mathbb{Z}_0^-;\ \Re(s)>1\right). \tag{7}$$

See also a recent investigation by Srivastava and Attiya [1092], using the Hurwitz–Lerch Zeta function in *Geometric Function Theory*.

We recall several interesting properties of the multiple Hurwitz–Lerch Zeta function $\Phi_n(z,s,a)$ in (4) investigated by Choi *et al.* [267, 273].

First of all, the multiple Hurwitz–Lerch Zeta function $\Phi_n(z,s,a)$ satisfies the following functional relation:

$$\Phi_n(z,s,a) = z^{\ell}\,\Phi_n(z,s,\ell+a) + \sum_{\substack{m_1,\ldots,m_n=0 \\ m_1+\cdots+m_n \leqq \ell-1}} \frac{z^{m_1+\cdots+m_n}}{(m_1+\cdots+m_n+a)^s} \tag{8}$$
$$(\ell \in \mathbb{N};\ a \in \mathbb{C}\setminus\mathbb{Z}_0^-),$$

which, for $n=1$, yields a known functional relation for $\Phi(z,s,a)$ given by 2.5(2).

It is easy to derive the following summation formula:

$$\sum_{k=0}^{\infty} \frac{(s)_k}{k!} \Phi_n(z, s+k, a) t^k = \Phi_n(z, s, a-t) \qquad (|t| < |a|; \ s \neq 1), \tag{9}$$

where $(\lambda)_m$ denotes the Pochhammer symbol defined by 1.1(5). Indeed, we see that

$$\begin{aligned}
\Phi_n(z, s, a-t) &= \sum_{m_1, \ldots, m_n \geqq 0} \frac{z^{m_1+\cdots+m_n}}{(m_1+\cdots+m_n+a-t)^s} \\
&= \sum_{m_1, \ldots, m_n \geqq 0} \frac{z^{m_1+\cdots+m_n}}{(m_1+\cdots+m_n+a)^s} \left(1 - \frac{t}{m_1+\cdots+m_n+a}\right)^{-s} \\
&= \sum_{m_1, \ldots, m_n \geqq 0} \frac{z^{m_1+\cdots+m_n}}{(m_1+\cdots+m_n+a)^s} \sum_{k=0}^{\infty} \frac{(s)_k}{k!} \left(\frac{t}{m_1+\cdots+m_n+a}\right)^k \\
&\qquad (|t| < |m_1+\cdots+m_n+a|; \ m_j \in \mathbb{N}_0; \ j = 1, \ldots, n) \\
&= \sum_{k=0}^{\infty} \frac{(s)_k}{k!} \left(\sum_{m_1, \ldots, m_n \geqq 0} \frac{z^{m_1+\cdots+m_n}}{(m_1+\cdots+m_n+a)^{s+k}} \right) t^k \\
&= \sum_{k=0}^{\infty} \frac{(s)_k}{k!} \Phi_n(z, s+k, a) t^k \qquad (|t| < |a|; \ s \neq 1),
\end{aligned}$$

which proves (9). The interchange of the order of summation can be justified by the absolute convergence of the series involved.

Now, from the Eulerian integral form of the Gamma function Γ (see Section 1.1), we have

$$\Gamma(s) = u^s \int_0^{\infty} e^{-ut} t^{s-1} \, dt \qquad \big(\Re(s) > 0; \ \Re(u) > 0\big). \tag{10}$$

By setting

$$u = m_1 + \cdots + m_n + a \qquad \big((m_1, \ldots, m_n) \in \mathbb{N}_0{}^n\big),$$

multiplying each side of the resulting equation by $z^{m_1+\cdots+m_n}$ and summing up for each m_j from 0 to ∞, $(j = 1, \ldots, n)$, we can get the following integral representation for

$\Phi_n(z,s,a)$:

$$\Phi_n(z,s,a) = \frac{1}{\Gamma(s)} \int_0^\infty \frac{t^{s-1}\, e^{-at}}{\left(1 - z e^{-t}\right)^n}\, dt = \frac{1}{\Gamma(s)} \int_0^\infty \frac{t^{s-1}\, e^{-(a-n)t}}{(e^t - z)^n}\, dt \tag{11}$$

$$\big(n \in \mathbb{N};\ \Re(a) > 0;\ |z| \leqq 1\ (z \neq 1) \text{ and } \Re(s) > 0;\ z = 1 \text{ and } \Re(s) > 1\big).$$

In its special case when $n = 1$, (11) yields a known integral representation for the Hurwitz–Lerch Zeta function $\Phi(z,s,a)$ (see 2.5(4)).

We can get a particular infinite-series version of Plana's summation formula 2.2(9) in the form:

$$\sum_{k=0}^{\infty} f(k) = \frac{1}{2} f(0) + \int_0^\infty f(\tau)\, d\tau - 2 \int_0^\infty \frac{q(0,t)}{e^{2\pi t} - 1}\, dt, \tag{12}$$

where $f(z)$ is a bounded analytic function in $0 \leqq \Re(z) < \infty$, the integral

$$\int_0^\infty f(\tau)\, d\tau$$

is assumed to be convergent, and

$$q(n,t) \to 0 \qquad \text{as} \qquad n \to \infty \qquad \text{or} \qquad t \to \infty.$$

If we apply (12) to the function

$$f(u) := \frac{z^{u+m_1+\cdots+m_{n-1}}}{(u + m_1 + \cdots + m_n + a)^s},$$

then we get

$$\sum_{k=0}^{\infty} \frac{z^{k+m_1+\cdots+m_{n-1}}}{(k + m_1 + \cdots + m_{n-1} + a)^s} = \frac{z^{m_1+\cdots+m_{n-1}}}{2\,(m_1 + \cdots + m_{n-1} + a)^s}$$

$$+ \int_0^\infty \frac{z^{\tau+m_1+\cdots+m_{n-1}}}{(\tau + m_1 + \cdots + m_{n-1} + a)^s}\, d\tau$$

$$- 2 \int_0^\infty z^{m_1+\cdots+m_{n-1}} \left[(m_1 + \cdots + m_{n-1} + a)^2 + t^2\right]^{-\frac{s}{2}}$$

$$\cdot \sin\left[t \log z - s \arctan\left(\frac{t}{m_1 + \cdots + m_{n-1} + a}\right)\right] \frac{dt}{e^{2\pi t} - 1}.$$

Now, summing up for each m_j from 0 to ∞, $(j = 1, \ldots, n-1)$, we finally obtain the following integral representation for $\Phi_n(z,s,a)$:

$$\begin{aligned}\Phi_n(z,s,a) &= \frac{1}{2}\,\Phi_{n-1}(z,s,a) + \int_0^\infty z^t\,\Phi_{n-1}(z,s,a+t)\,dt \\ &- 2\sum_{m_1,\ldots,m_{n-1}=0}^{\infty} z^{m_1+\cdots+m_{n-1}} \int_0^\infty \left[(m_1+\cdots+m_{n-1}+a)^2+t^2\right]^{-\frac{s}{2}} \\ &\cdot \sin\left[t\log z - s\arctan\left(\frac{t}{m_1+\cdots+m_{n-1}+a}\right)\right]\frac{dt}{e^{2\pi t}-1} \qquad (n\in\mathbb{N}),\end{aligned} \tag{13}$$

provided that the involved integrals and series converge.

The special case of (13) when $n = 1$ yields the known integral representation for $\Phi(z,s,a)$ (see 2.5(6)).

It is not difficult to deduce the following contour integral representation of $\Phi_n(z,s,a)$, by applying a procedure that is used to get a contour integral representation of the multiple Hurwitz Zeta function $\zeta_n(s,a)$ (see, for details, Section 2.1):

$$\begin{aligned}\Phi_n(z,s,a) &= -\frac{\Gamma(1-s)}{2\pi i}\int_C \frac{(-t)^{s-1}\,e^{-at}}{\left(1-ze^{-t}\right)^n}\,dt \\ &\left(n\in\mathbb{N};\ \Re(a)>0;\ |\arg(-t)| \leqq \pi\right),\end{aligned} \tag{14}$$

where the contour C is the Hankel loop given as in 2.1(7).

The special cases of (14) when $z = 1$ and $n = 1$ would immediately yield the known integral representations of the multiple Hurwitz Zeta function $\zeta_n(s,a)$ in 2.1(8) and the Hurwitz–Lerch Zeta function $\Phi(z,s,a)$ in 2.5(8), respectively.

The special case of $\Phi_n(z,s,a)$ in (14) when

$$s = -\ell \qquad (\ell \in \mathbb{N}_0)$$

can be expressed in terms of the Apostol-Bernoulli polynomials $B_k^{(\alpha)}(a;\lambda)$ of order α, defined by means of the following generating functions (see Section 1.8):

$$\Phi_n(z,-\ell,a) = (-1)^n\,\frac{\ell!}{(n+\ell)!}\,B_{n+\ell}^{(n)}(a;z) \quad (\ell\in\mathbb{N}_0). \tag{15}$$

Indeed, upon setting

$$s = -\ell \qquad (\ell \in \mathbb{N}_0)$$

in (14), we find, by using 1.8(13), that

$$\begin{aligned}\Phi_n(z,-\ell,a) &= -\frac{\Gamma(1+\ell)}{2\pi i}\int_C \frac{(-t)^{-\ell-1}\,e^{-at}}{\left(1-ze^{-t}\right)^n}\,dt \\ &= (-1)^{\ell}\,\ell!\,\operatorname*{Res}_{t=0} t^{-n-\ell-1}\left(\frac{t}{1-ze^{-t}}\right)^n e^{-at} \\ &= (-1)^{\ell}\,\ell!\,\operatorname*{Res}_{t=0} t^{-n-\ell-1}\sum_{k=0}^{\infty} B_k^{(n)}(a;z)\,\frac{(-1)^k t^k}{k!} \\ &= (-1)^n\,\frac{\ell!}{(n+\ell)!}\,B_{n+\ell}^{(n)}(a;z).\end{aligned}$$

The special case of (15) when $z=1$, together with 1.8(14), would reduce immediately to the known formula (see 2.1(16)).

The multiple Hurwitz–Lerch Zeta function $\Phi_n(z,s,a)$ can, thus, be expressed as follows, as a simple series (see 2.1(19)):

$$\Phi_n(z,s,a) = \sum_{m=0}^{\infty}\binom{m+n-1}{n-1}\frac{z^m}{(m+a)^s} \tag{16}$$

$$\left(a\in\mathbb{C}\setminus\mathbb{Z}_0^-;\ s\in\mathbb{C}\ \text{when}\ |z|<1;\ \Re(s)>n\ \text{when}\ |z|=1\right).$$

We find from (16), as in getting 2.1(21), that $\Phi_n(z,s,a)$ is expressible as a finite combination of the Hurwitz–Lerch Zeta function $\Phi(z,s,a)$ with polynomial coefficients in a as follows:

$$\Phi_n(z,s,a) = \sum_{j=0}^{n-1} p_{n,j}(a)\,\Phi(z,s-j,a), \tag{17}$$

where

$$p_{n,j}(a) = \frac{1}{(n-1)!}\sum_{\ell=j}^{n-1}(-1)^{n+1-j}\binom{\ell}{j}s(n,\ell+1)\,a^{\ell-j}. \tag{18}$$

We now find $p_{n,j}(a)$ in (18) as a polynomial in a of degree $n-1-j$ with rational coefficients.

Since

$$\sum_{\ell=0}^{\infty} f(\ell) = \sum_{j=0}^{k-1}\sum_{\ell=0}^{\infty} f(k\ell+j)\quad (k\in\mathbb{N}), \tag{19}$$

we get

$$\Phi_n(z,s,a) = k^{-s} \sum_{j=0}^{k-1} \Phi_n\left(z^k, s, \frac{a+j}{k}\right) z^j \quad (k \in \mathbb{N}). \tag{20}$$

Thus, by combining (11) and (20), we immediately obtain the following sum-integral representation:

$$\Phi_n(z,s,a) = \sum_{j=0}^{k-1} \frac{z^j}{\Gamma(s)} \int_0^\infty \frac{t^{s-1}\, e^{-(a+j)t}}{\left(1 - z^k\, e^{-kt}\right)^n}\, dt \tag{21}$$

$$\left(k,\, n \in \mathbb{N};\ \Re(a) > 0;\ \Re(s) > 0 \text{ when } |z| \leqq 1\ (z \neq 1);\ \Re(s) > 1 \text{ when } z = 1\right),$$

which is a straightforward generalization of the following sum-integral representation given by Lin and Srivastava [765, p. 727, Eq. (7)]:

$$\Phi(z,s,a) = \sum_{j=0}^{k-1} \frac{z^j}{\Gamma(s)} \int_0^\infty \frac{t^{s-1}\, e^{-(a+j)t}}{\left(1 - z^k\, e^{-kt}\right)}\, dt \tag{22}$$

$$\left(k \in \mathbb{N};\ \Re(a) > 0;\ \Re(s) > 0 \text{ when } |z| \leqq 1\ (z \neq 1);\ \Re(s) > 1 \text{ when } z = 1\right).$$

For further special cases of (21) or (22), just as noted also by Lin and Srivastava [765, p. 726], see Yen *et al.* [1244, p. 100, Theorem] and Nishimoto *et al.* [864, p. 94, Theorem 4].

In view of (17), we can deduce some properties of the multiple Hurwitz–Lerch Zeta function $\Phi_n(z,s,a)$ from those (see, e.g., [447]) of the Hurwitz–Lerch Zeta function $\Phi(z,s,a)$. Combining the integral formula 2.5(6) for $\Phi(z,s,a)$ with (17), we obtain another sum-integral representation for $\Phi_n(z,s,a)$:

$$\begin{aligned}
&\Phi_n(z,s,a) \\
&= \frac{(-1)^{n+1}}{2a^s\,(n-1)!} \sum_{j=0}^{n-1} (-1)^j \sum_{\ell=j}^{n-1} \binom{\ell}{j} s(n,\ell+1)\, a^\ell + \sum_{j=0}^{n-1} p_{n,j}(a) \int_0^\infty \frac{z^t}{(t+a)^{s-j}}\, dt \\
&\quad - 2 \sum_{j=0}^{n-1} p_{n,j}(a) \int_0^\infty \sin\left[t \log z + (j-s) \tan^{-1}\left(\frac{t}{a}\right)\right] \left(t^2 + a^2\right)^{\frac{1}{2}(j-s)} \frac{dt}{e^{2\pi t} - 1}
\end{aligned}$$

$$\left(n \in \mathbb{N};\ \Re(a) > 0;\ s \in \mathbb{C}\right). \tag{23}$$

By applying Apostol's formula 2.5(40) (see [58, p. 164]; see also [467, p. 806, Eq. (19)]):

$$\phi(\xi, a, 1-\ell) = \Phi\left(e^{2\pi i\xi}, 1-\ell, a\right) = -\frac{\mathcal{B}_\ell\left(a; e^{2\pi i\xi}\right)}{\ell} \qquad (\ell \in \mathbb{N};\ \xi \in \mathbb{R}) \tag{24}$$

to (17), we obtain

$$\Phi_n\left(e^{2\pi i\xi}, 1-\ell, a\right) = -\sum_{j=0}^{n-1} p_{n,j}(a)\,\frac{\mathcal{B}_{\ell+j}\left(a; e^{2\pi i\xi}\right)}{\ell+j} \qquad (n,\ \ell \in \mathbb{N}). \tag{25}$$

The following known summation formula for $\Phi(z, s, a)$, expressed in terms of the Hurwitz (or generalized) Zeta function $\zeta(s, a)$ (see [788, p. 298, Eq. (38)]):

$$\Phi\left(e^{2\pi i\frac{p}{q}}, s, a\right) = q^{-s}\sum_{\ell=1}^{q} \zeta\left(s, \frac{a+\ell-1}{q}\right)\exp\left(\frac{2(\ell-1)p\pi i}{q}\right) \qquad (p \in \mathbb{Z};\ q \in \mathbb{N}), \tag{26}$$

when applied to (17), would readily yield

$$\Phi_n\left(e^{2\pi i\frac{p}{q}}, s, a\right) = q^{-s}\sum_{j=0}^{n-1} p_{n,j}(a)\,q^j \sum_{\ell=1}^{q} \zeta\left(s-j, \frac{a+\ell-1}{q}\right)\exp\left(\frac{2(\ell-1)p\pi i}{q}\right)$$
$$(n,\ q \in \mathbb{N};\ p \in \mathbb{Z}). \tag{27}$$

Lin and Srivastava [765, p. 729, Eq. (17)] presented the following integral representation for $\Phi^*_\mu(z, s, a)$:

$$\Phi^*_\mu(z, s, a) = \frac{1}{\Gamma(s)}\int_0^\infty \frac{t^{s-1}\,e^{-at}}{\left(1-ze^{-t}\right)^\mu}\,dt \tag{28}$$
$$\left(\Re(a) > 0;\ \Re(s) > 0 \ \text{when}\ |z| \leqq 1\ (z \neq 1);\ \Re(s) > 1\ \text{when}\ z = 1\right),$$

which was given earlier by Goyal and Laddha [504, p. 100, Eq. (1.6)], together with the seemingly unnecessary constraint $\mu \geqq 1$ (see, for details, [765, p. 729]). Here, in view of (11) and (28), it is seen that

$$\Phi_n(z, s, a) = \Phi^*_n(z, s, a) \quad (n \in \mathbb{N}). \tag{29}$$

Lin and Srivastava [765] also recalled the *Riemann-Liouville fractional derivative operator* $\mathcal{D}_z^{\mu}$, defined by (*cf.* [422, p. 181 *et seq.*]; see also [896] and [648])

$$\mathcal{D}_z^{\mu}\{f(z)\} := \begin{cases} \dfrac{1}{\Gamma(-\mu)} \displaystyle\int_0^z (z-t)^{-\mu-1} f(t)\,dt & (\Re(\mu) < 0) \\[2ex] \dfrac{d^m}{dz^m}\, \mathcal{D}_z^{\mu-m}\{f(z)\} & (m-1 \leqq \Re(\mu) < m \ (m \in \mathbb{N})), \end{cases} \tag{30}$$

to, first, notice that $\Phi_{\mu}^{*}(z,s,a)$ is essentially a Riemann-Liouville fractional derivative of the classical Hurwitz–Lerch function $\Phi(z,s,a)$ of order $\mu - 1$ as follows:

$$\Phi_{\mu}^{*}(z,s,a) = \frac{1}{\Gamma(\mu)}\, \mathcal{D}_z^{\mu-1}\left\{z^{\mu-1}\,\Phi(z,s,a)\right\} \quad (\Re(\mu) > 0). \tag{31}$$

By using their remarkable observation (31), together with the Eulerian integrals of the first and second kind, Hankel's integral representation for the Gamma function [421, p. 13, Eq. 1.6(2)] and the theory of the Mellin-Barnes contour integration (*cf.* [421, p. 49]), Lin and Srivastava [765] presented various integral representations for $\Phi_{\mu}^{*}(z,s,a)$. In view of (29), as well as for the sake of completeness, we recall the known integral representations for $\Phi_{\mu}^{*}(z,s,a)$ given by Lin and Srivastava (see [765, Eqs. (27), (28), (34), and (37)]) in their following *special* forms for $\Phi_n(z,s,a)$:

$$\Phi_n(z,s,a) = \frac{1}{(n-1)!} \int_0^{\infty} t^{n-1} e^{-t}\, \Phi_{n,1}^{(0,1)}(zt,s,a)\,dt \quad (n \in \mathbb{N}), \tag{32}$$

$$\Phi_n(z,s,a) = \frac{1}{(n-1)!} \int_0^{1} \left(\log\frac{1}{t}\right)^{n-1} \Phi_{n,1}^{(0,1)}\left(z\log\frac{1}{t},s,a\right) dt \quad (n \in \mathbb{N}), \tag{33}$$

$$\Phi_n(z,s,a) = \frac{1}{2\pi i} \int_{-\infty}^{(+0)} w^{-1} e^{w}\, \Phi_{n,1}^{(0,1)}\left(zw^{-1},s,a\right) dw \quad (n \in \mathbb{N};\ |\arg(w)| \leqq \pi), \tag{34}$$

where the contour of integration in the complex w-plane is *Hankel's loop* (*cf.*, e.g., [1225, p. 245]), which starts from the point at $-\infty$ along the lower side of the negative real axis, encircles the origin once in the positive (counter-clockwise) direction and then returns to the point at $-\infty$ along the upper side of the negative real axis.

$$\Phi_n(z,s,a) = \frac{1}{2\pi i} \int_{\tau-i\infty}^{\tau+i\infty} w^{-1} e^{w}\, \Phi_{n,1}^{(1,0)}\left(zw^{-1},s,a\right) dw \quad \left(n \in \mathbb{N};\ \tau \in \mathbb{R}^{+}\right), \tag{35}$$

where the contour of integration is a Mellin-Barnes contour (*cf.* [421, p. 49]).

Garg *et al.* [467, p. 809, Theorem 2; p. 811, Theorem 3] gave transformation formulas for $\Phi_n^*(z,s,a)$ and Lin *et al.* [767, p. 825, Eqs. (39) and (40)] presented several expansion and transformation formulas for the general case $\Phi_\mu^*(z,s,a)$. By means of (29), we choose to rewrite these known results in their *special* cases applicable to $\Phi_n(z,s,a)$ as follows:

$$\Phi_n(z,s,a)=\frac{i\,z^{-a}\,\Gamma(1-s)}{(n-1)!}\sum_{k=0}^{n-1}\binom{n-1}{k}B_{n-k-1}^{(n)}$$
$$\cdot\sum_{j=0}^{k}(-1)^{n-k+j-1}\binom{k}{j}\binom{s-1}{j}j!\,(-a)^{k-j}\,(2\pi)^{s-j-1}$$
$$\cdot\left[\exp\left(-\frac{1}{2}(s-j)\pi i\right)\Phi\left(e^{-2\pi ai},1-s+j,\frac{\log z}{2\pi i}\right)\right.$$
$$\left.-\exp\left[\left(2a+\frac{1}{2}(s-j)\right)\pi i\right]\Phi\left(e^{2\pi ai},1-s+j,1-\frac{\log z}{2\pi i}\right)\right]\qquad(n\in\mathbb{N}),\tag{36}$$

provided that each side of (36) exists;

$$\Phi_n(z,s,a)=i\,z^{-a}\,\Gamma(1-s)\sum_{j=0}^{n-1}\frac{(j-a+1)_{n-j-1}}{j!(n-j-1)!}\sum_{k=0}^{j}\binom{j-1}{k-1}(1-s)_kB_{j-k}^{(j)}(2\pi)^{s-k-1}$$
$$\cdot\left[\exp\left(-\frac{1}{2}(s-k)\pi i\right)\Phi\left(e^{-2\pi ai},1-s+k,\frac{\log z}{2\pi i}\right)\right.$$
$$\left.-\exp\left[\left(2a+\frac{1}{2}(s-k)\right)\pi i\right]\Phi\left(e^{2\pi ai},1-s+k,1-\frac{\log z}{2\pi i}\right)\right]\qquad(n\in\mathbb{N}),\tag{37}$$

provided that each side of (37) exists;

$$\Phi_n(z,s,a)=i\,z^{-a}\,\Gamma(1-s)\sum_{j=0}^{n-1}\frac{(j-a+1)_{n-j-1}}{j!\,(n-j-1)!}$$
$$\cdot\sum_{k=0}^{j}\binom{j-1}{k-1}(1-s)_k\,B_{j-k}^{(j)}\cdot(2\pi)^{s-k-1}$$
$$\cdot\left[\exp\left(-\frac{1}{2}(s-k)\pi i\right)\Phi\left(e^{-2\pi ai},1-s+k,\frac{\log z}{2\pi i}\right)\right.$$
$$\left.-\exp\left[\left(2a+\frac{1}{2}(s-k)\right)\pi i\right]\Phi\left(e^{2\pi ai},1-s+k,1-\frac{\log z}{2\pi i}\right)\right]\qquad(n\in\mathbb{N}),\tag{38}$$

and

$$\Phi_n\left(e^{2\pi i\xi}, n-s, \frac{p}{q}\right) = \frac{i\,\Gamma(1-n+s)}{(n-1)!}\sum_{j=0}^{n-1}\binom{n-1}{j}\left(j-\frac{p}{q}+1\right)_{n-j-1}$$

$$\cdot\sum_{k=0}^{j}\binom{j-1}{k-1}(1-n+s)_k\, B_{j-k}^{(j)}\cdot(2\pi q)^{n-s-k-1}$$

$$\cdot\left[\sum_{r=1}^{q}\zeta\left(1-n+s+k, \frac{\xi+r-1}{q}\right)\exp\left[-\left(\frac{1}{2}(n-s-k)+\frac{2(\xi+r-1)p}{q}\right)\pi i\right]\right.$$

$$\left.-\sum_{r=1}^{q}\zeta\left(1-n+s+k, \frac{r-\xi}{q}\right)\exp\left[\left(\frac{1}{2}(n-s-k)+\frac{2(r-\xi)p}{q}\right)\pi i\right]\right]$$

$$(p\in\mathbb{Z};\ q, n\in\mathbb{N}). \tag{39}$$

Garg *et al.* [467, p. 813, Eqs. (51) and (52)] applied the definition 2.5(11) and the summation formula (26) to (36) and (37), by setting

$$z = e^{2\pi i\xi} \qquad \text{and} \qquad a = \frac{p}{q} \qquad (p\in\mathbb{Z};\ q\in\mathbb{N};\ \xi\in\mathbb{R}),$$

and, thereby, expressed the function:

$$\Phi_n^*\left(e^{2\pi i\xi}, n-s, \frac{p}{q}\right)$$

in terms of the Hurwitz Zeta function $\zeta(s,a)$ and the generalized Apostol-Bernoulli polynomials $\mathcal{B}_k^{(j)}(x;\lambda)$. We rewrite these known results of Garg *et al.* [467, Eqs. (51) and (52)] in their following *special* forms for $\Phi_n\left(e^{2\pi i\xi}, n-s, \frac{p}{q}\right)$):

$$\Phi_n\left(e^{2\pi i\xi}, n-s, \frac{p}{q}\right) = \frac{i\,(-1)^{n-1}\,\Gamma(1-n+s)}{(n-1)!}\sum_{k=0}^{n-1}\binom{n-1}{k}B_{n-k-1}^{(n)}\,q^{n-s-k-1}$$

$$\cdot\sum_{j=0}^{k}\binom{k}{j}\binom{n-s-1}{j}j!\,p^{k-j}\,(2\pi)^{n-s-j-1}$$

$$\cdot\left[\sum_{r=1}^{q}\zeta\left(1-n+s+j, \frac{\xi+r-1}{q}\right)\exp\left[-\left(\frac{1}{2}(n-s-j)+\frac{2(\xi+r-1)p}{q}\right)\pi i\right]\right.$$

$$\left.-\sum_{r=1}^{q}\zeta\left(1-n+s+j, \frac{r-\xi}{q}\right)\exp\left[\left(\frac{1}{2}(n-s-j)+\frac{2(r-\xi)p}{q}\right)\pi i\right]\right]$$

$$(n\in\mathbb{N}) \tag{40}$$

and

$$\Phi_n\left(e^{2\pi i\xi}, n-s, \frac{p}{q}\right) = \frac{i\,\Gamma(1-n+s)}{(n-1)!}\sum_{j=0}^{n-1}\binom{n-1}{j}\left(j-\frac{p}{q}+1\right)_{n-j-1}$$

$$\cdot\sum_{k=0}^{j}\binom{j-1}{k-1}(1-n+s)_k\, B_{j-k}^{(j)}\cdot(2\pi q)^{n-s-k-1}$$

$$\cdot\left[\sum_{r=1}^{q}\zeta\left(1-n+s+k, \frac{\xi+r-1}{q}\right)\exp\left[-\left(\frac{1}{2}\,(n-s-k)+\frac{2(\xi+r-1)p}{q}\right)\pi i\right]\right.$$

$$\left.-\sum_{r=1}^{q}\zeta\left(1-n+s+k, \frac{r-\xi}{q}\right)\exp\left[\left(\frac{1}{2}\,(n-s-k)+\frac{2(r-\xi)p}{q}\right)\pi i\right]\right]$$

$$(n\in\mathbb{N}). \tag{41}$$

In a recent paper, Srivastava *et al.* [1107] introduced and investigated *further* generalizations of the above-defined Lin-Srivastava model $\Phi_{\mu,\nu}^{(\rho,\sigma)}(z,s,a)$, *first*, in the following form:

$$\Phi_{\lambda,\mu;\nu}^{(\rho,\sigma,\kappa)}(z,s,a) := \sum_{n=0}^{\infty}\frac{(\lambda)_{\rho n}(\mu)_{\sigma n}}{(\nu)_{\kappa n}\cdot n!}\,\frac{z^n}{(n+a)^s}$$

$$\big(\lambda,\mu\in\mathbb{C};\ a,\nu\in\mathbb{C}\setminus\mathbb{Z}_0^-;\ \rho,\sigma,\kappa\in\mathbb{R}^+;\ \kappa-\rho-\sigma>-1 \text{ when } s,z\in\mathbb{C}; \tag{42}$$

$$\kappa-\rho-\sigma=-1 \text{ and } s\in\mathbb{C} \text{ when } |z|<\delta^*:=\rho^{-\rho}\,\sigma^{-\sigma}\,\kappa^{\kappa};$$

$$\kappa-\rho-\sigma=-1 \text{ and } \Re(s+\nu-\lambda-\mu)>1 \text{ when } |z|=\delta^*\big)$$

and, *then*, in a substantially more general form given by

$$\Phi_{\lambda_1,\ldots,\lambda_p;\mu_1,\ldots,\mu_q}^{(\rho_1,\ldots,\rho_p,\sigma_1,\ldots,\sigma_q)}(z,s,a) = \sum_{n=0}^{\infty}\frac{\prod_{j=1}^{p}(\lambda_j)_{n\rho_j}}{n!\,\prod_{j=1}^{q}(\mu_j)_{n\sigma_j}}\,\frac{z^n}{(n+a)^s}$$

$$\Big(p,q\in\mathbb{N}_0;\ \lambda_j\in\mathbb{C}\,(j=1,\ldots,p);\ a,\mu_j\in\mathbb{C}\setminus Z_0^-\ (j=1,\ldots,q);$$

$$\rho_j,\sigma_k\in\mathbb{R}^+\ (j=1,\ldots,p;\ k=1,\ldots,q);\ \Delta>-1 \text{ when } s,z\in\mathbb{C};$$

$$\Delta=-1 \text{ and } s\in\mathbb{C} \text{ when } |z|<\nabla;\ \Delta=-1 \text{ and } \Re(\Xi)>\frac{1}{2} \text{ when } |z|=\nabla\Big), \tag{43}$$

where, for convenience,

$$\nabla := \left(\prod_{j=1}^{p} \rho_j^{-\rho_j}\right)\left(\prod_{j=1}^{q} \sigma_j^{\sigma_j}\right),$$

$$\Delta := \sum_{j=1}^{q} \sigma_j - \sum_{j=1}^{p} \rho_j$$

and

$$\Xi := s + \sum_{j=1}^{q} \mu_j - \sum_{j=1}^{p} \lambda_j + \frac{p-q}{2}.$$

The special case of the definition (43) when $p-1=q=1$ would obviously correspond to the above-investigated (Lin-Srivastava) generalized Hurwitz–Lerch Zeta function $\Phi_{\lambda,\mu;\nu}^{(\rho,\sigma,\kappa)}(z,s,a)$, defined by (42).

The following *further* special case of the generalized Hurwitz–Lerch Zeta function $\Phi_{\lambda,\mu;\nu}^{(\rho,\sigma,\kappa)}(z,s,a)$ (which corresponds to the definition (43) with $p-2=q=1$) when $\rho=\sigma=\kappa=1$:

$$\Phi_{\lambda,\mu;\nu}(z,s,a) := \sum_{n=0}^{\infty} \frac{(\lambda)_n(\mu)_n}{(\nu)_n \cdot n!} \frac{z^n}{(n+a)^s} =: \Phi_{\lambda,\mu;\nu}^{(1,1,1)}(z,s,a) \tag{44}$$

$$\big(\lambda, \mu \in \mathbb{C};\ \nu, a \in \mathbb{C}\setminus\mathbb{Z}_0^-;\ s \in \mathbb{C} \text{ when } |z|<1;\ \Re(s+\nu-\lambda-\mu)>1 \text{ when } |z|=1\big)$$

was investigated earlier by Garg *et al.* [466, p. 313, Eq. (1.7)].

For various interesting properties and results involving the families of generalized Hurwitz–Lerch Zeta functions:

$$\Phi_{\lambda,\mu;\nu}^{(\rho,\sigma,\kappa)}(z,s,a) \quad \text{and} \quad \Phi_{\lambda_1,\ldots,\lambda_p;\mu_1,\ldots,\mu_q}^{(\rho_1,\ldots,\rho_p,\sigma_1,\ldots,\sigma_q)}(z,s,a),$$

which are defined by (42) and (43), respectively, as well as for the numerous special cases and consequences of each of these general models, we choose to refer the interested reader to the above-mentioned paper by Srivastava *et al.* [1107], as well as to a subsequent investigation by Srivastava *et al.* [1106] in which several two-sided bounding inequalities for the extended Hurwitz–Lerch Zeta function

$$\Phi_{\lambda_1,\ldots,\lambda_p;\mu_1,\ldots,\mu_q}^{(\rho_1,\ldots,\rho_p,\sigma_1,\ldots,\sigma_q)}(z,s,a)$$

were given (see also Problems 51, 52 and 53 at the end of this chapter).

2.7 Analytic Continuations of Multiple Zeta Functions

The multiple Hurwitz zeta function $\zeta_n(s,a)$ ($n \in \mathbb{N}$), defined by 2.1(2), is obviously a generalization of the Riemann zeta function $\zeta(s)$ given in 2.3(1). Another very useful and widely-investigated generalization of $\zeta(s)$ is the multiple zeta functions $\zeta_d(s_1,\ldots,s_d)$ of depth d ($d \in \mathbb{N}$), defined by

$$\zeta_d(s_1,\ldots,s_d) := \sum_{0<n_1<n_2<\cdots<n_d} \frac{1}{{n_1}^{s_1}\,{n_2}^{s_2}\cdots{n_d}^{s_d}} \tag{1}$$

$$\left((s_1,\ldots,s_d) \in \mathbb{C}^d;\ \Re(s_d) > 1;\ \sum_{j=1}^{d} \Re(s_j) > d\right),$$

which is often referred to as Euler-Riemann-Zagier zeta functions of depth d.

If $(s_1,\ldots,s_d) \in \mathbb{N}^d$, $\zeta_d(s_1,\ldots,s_d)$ is also called the multiple zeta values (or Euler-Zagier sums) (see [1246]). Since these values have turned out to be connected with a variety of research subjects, such as topology, theoretical physics and the theory of mixed Tate motives, an extensive research has been made of those values (see, e.g., [151], [182], [734], [1245], [1246], [1247]). Ohno [872] presented a unified algebraic relationship among the multiple zeta values that generalizes the so-called *sum formula* (or *sum conjecture*) (see [560]) and another remarkable identity, referred to as the *duality* theorem (see [1246]). Here, we introduce two totally different methods of analytic continuation of $\zeta_d(s_1,\ldots,s_d)$, defined by (1), which were given by Zhao [1260] and Akiyama *et al.* [16]. An analytic continuation of $\zeta_2(s_1,s_2)$ was proven by, for example, Apostol and Vu [68], Atkinson [78], Motohashi [849] and Katsurada and Matsumoto [639]. Arakawa and Kaneko [70] gave the analytic continuation of $\zeta_d(s_1,\ldots,s_d)$ by considering a function of one variable s_d, while $s_1,\ldots,s_{d-1}$ are positive integers.

Generalized Functions of Gel'fand and Shilov

We remark in proceeding this subsection that most of the arguments given here are due to Zhao [1260].

Theorem 2.7 *The infinite sum in* (1) *converges absolutely when*

$$\Re(s_d) > 1 \quad \text{and} \quad \sum_{j=1}^{d} \Re(s_j) > d.$$

Proof. The case $d=1$ is obvious, since $\zeta(s_1)$ becomes the Riemann zeta function, defined by 2.3(1), and converges absolutely for $\Re(s_1) > 1$. So, let us assume $d=2$ and write $\sigma_i = \Re(s_i)$. If $\sigma_1 > 1$, then, being also $\sigma_2 > 1$, the result is clear:

$$|\zeta_2(s_1,s_2)| \leqq \sum_{0<n_1<n_2} \frac{1}{{n_1}^{\sigma_1}\,{n_2}^{\sigma_2}} \leqq \sum_{n_1=1}^{\infty} \frac{1}{{n_1}^{\sigma_1}} \cdot \sum_{n_2=1}^{\infty} \frac{1}{{n_2}^{\sigma_1}} < \infty.$$

If $\sigma_1 = 1$, then the theorem is proven by using the following well-known asymptotic formula (see 1.2(68)):

$$\sum_{n_1=1}^{n} \frac{1}{n_1} = \log n + O(1) \quad (n \to \infty). \tag{2}$$

Since $\sigma_2 > 1$, write $\sigma_2 = \delta + \mu$ for some $\delta > 1$ and $\mu > 0$. Then, we have, as $n \to \infty$,

$$\begin{aligned}\sum_{n_1=1}^{n} \frac{1}{n_1 \cdot n^{\sigma_2}} &= \frac{1}{n^{\sigma_2}} \sum_{n_1=1}^{n} \frac{1}{n_1} = \frac{1}{n^{\sigma_2}} (\log n + O(1)) \\ &= \frac{1}{n^{\delta}} \cdot \frac{1}{n^{\mu}} [\log n + O(1)] \leqq \frac{L}{n^{\delta}}\end{aligned}$$

for some $L > 0$, the second term in the third equality tending to as $n \to \infty$. We, therefore, find that

$$|\zeta_2(s_1, s_2)| \leqq L \sum_{n=1}^{\infty} \frac{1}{n^{\delta}} < \infty.$$

If $\sigma_1 < 1$, then one has the following estimate:

$$\begin{aligned}\sum_{n_1=1}^{n-1} \frac{1}{{n_1}^{\sigma_1} \cdot n^{\sigma_2}} &= \sum_{k=1}^{[\log_2 n]} \sum_{\frac{n}{2^k} \leqq n_1 < \frac{n}{2^{k-1}}} \frac{1}{{n_1}^{\sigma_1}\, n^{\sigma_2}} \\ &< 2^{\max\{0,\, -\sigma_1\}} \sum_{k=1}^{\infty} \frac{n/2^k}{\left(n/2^k\right)^{\sigma_1} n^{\sigma_2}} \\ &= \frac{2^{\max\{0,\, -\sigma_1\}}}{n^{\sigma_1+\sigma_2-1}} \sum_{k=1}^{\infty} \frac{1}{2^{(1-\sigma_1)k}} \leqq \frac{M}{n^{\sigma_1+\sigma_2-1}}\end{aligned}$$

for some $M > 0$, being $1 - \sigma_1 > 0$. So, in this case, being $\sigma_1 + \sigma_2 - 1 > 1$,

$$|\zeta_2(s_1, s_2)| \leqq M \sum_{n=1}^{\infty} \frac{1}{n^{\sigma_1+\sigma_2-1}} < \infty.$$

The general case is seen to be easily verified by following the argument of case $d = 2$. ∎

Preliminary on Generalized Functions

For any $n \in \mathbb{N}$, Zhao [1260] introduces the space $\mathcal{K}$ of test functions consisting of all the complex-valued smooth functions on $\mathbb{R}^n$, which decrease to zero faster than any negative power of

$$|x| = \sqrt{\sum_{i=1}^{n} x_i^2} \quad \text{as } |x| \to \infty \quad \left(x = (x_1, \ldots, x_n) \in \mathbb{R}^n\right)$$

(for details of test functions, see [478, Chapter 1]). Throughout this subsection, we use the abbreviations $x = (x_1, \ldots, x_n) \in \mathbb{R}^n$ and $s = (s_1, \ldots, s_n) \in \mathbb{C}^n$. We say that a generalized function $g(x; s)$ (for this definition, see [478, Chapter 1]) is entire in s, if the inner product $< g(x; s), \varphi(x) >$ is an entire function of s for any test function $\varphi(x)$.

As an example of the generalized functions, let us suppose, for $x \in \mathbb{R}$, that

$$x_+^s = \begin{cases} x^s & (x > 0) \\ 0 & (x \leqq 0), \end{cases} \tag{3}$$

whose value on a test function $\varphi(x)$ is given by

$$< x_+^s, \varphi(x) > := \int_0^\infty x^s \varphi(x)\, dx.$$

The following lemmas play dominant rôles in this subsection.

Lemma 2.8 ([478, p. 48, Eq. (3)]) *We have, for any test function* φ,

$$\begin{aligned} \int_0^\infty x^{s-1} \varphi(x)\, dx = \int_0^1 x^{s-1} \left[\varphi(x) - \sum_{j=0}^{n} \frac{\varphi^{(j)}(0)}{j!} x^j \right] dx \\ + \int_1^\infty x^{s-1} \varphi(x)\, dx + \sum_{j=0}^{n} \frac{\varphi^{(j)}(0)}{j!\,(s+j)} \\ (\Re(s) > -n-1;\ s \neq -1, -2, \ldots, -n;\ n \in \mathbb{N}_0). \end{aligned} \tag{4}$$

As an easy consequence of Lemma 2.8, one finds the following result.

Lemma 2.9 ([478, p. 57]) *The generalized function* $x_+^{s-1}/\Gamma(s)$ *has an analytic continuation to an entire function in* s*, such that*

$$\left. \frac{x_+^{s-1}}{\Gamma(s)} \right|_{s=-n} = \delta^{(n)}(x) \quad (n \in \mathbb{N}_0),$$

where δ *is the delta function (see [478, Section 1.3]) and, for any test function* φ,

$$< \delta^{(n)}(x), \varphi(x) > := (-1)^n \delta^{(n)}(0) \quad (n \in \mathbb{N}_0).$$

Lemma 2.10 *The generalized function*

$$f(x; s, u) = \frac{(1-x)_+^{s-1}\, x_+^{u-1}}{\Gamma(s)\,\Gamma(u)}$$

can be extended to an entire function in complex variables s and u.

Proof. We need to show that, for any test function φ on [0, 1], the function of two complex variables s and u defined by

$$\int_0^1 \frac{(1-x)^{s-1}\, x^{u-1}}{\Gamma(s)\,\Gamma(u)}\,\varphi(x)\,dx \quad (\Re(u) > 0;\ \Re(s) > 0),$$

can be analytically continued to an entire function on all of $\mathbb{C}^2$. This follows easily from Lemma 2.8, since the interval [0, 1] can be separated into $[0, 1/2] \cup [1/2, 1]$. ■

Remark 1 The Riemann zeta function $\zeta(s)$, defined by 2.3(1), can be obtained by the Mellin-transformation (see 2.3(4)):

$$\zeta(s) = \frac{1}{\Gamma(s)} \int_0^\infty \frac{x^{s-2} \cdot x}{e^x - 1}\,dx = \frac{1}{s-1} \int_0^\infty \frac{x^{s-2}}{\Gamma(s-1)} \cdot \frac{x}{e^x - 1}\,dx \quad (\Re(s) > 1). \tag{5}$$

In view of Lemma 2.9, one finds that $(s-1)\,\zeta(s)$ can be considered as the value of the generalized function $x^{s-2}/\,\Gamma(s-1)$ (which is entire) on the test function $x/(e^x - 1)$. In this way, one immediately recovers the analytic continuation of $\zeta(s)$. Zhao [1260] used this idea to give an analytic continuation the multiple zeta functions of any depth as follows:

Theorem 2.11 *The multiple zeta function* $\zeta_d(s_1, \ldots, s_d)$ *of depth* d $(d \in \mathbb{N})$ *defined by* (1), *can be continued meromorphically to the whole space* $\mathbb{C}^d$ *with possible simple poles at* $s_d = 1$ *and* $s_d(j) := s_j + \cdots + s_d = d - j + 2 - \ell$ $(j, \ell \in \mathbb{N};\ 1 \leqq j < d)$.

Proof. If $d = 1$, then $\zeta(s_1)$ is the Riemann zeta function, defined by 2.3(1). So, one may assume $d \in \mathbb{N} \setminus \{1\}$. By the identities

$$\begin{aligned}
&\sum_{0<n_1<n_2<\cdots<n_d} e^{-n_1 t_1 - n_2 t_2 - \cdots - n_d t_d} \\
&= \sum_{j=1,\, n_j \in \mathbb{N}}^{d} e^{-n_1 t_1 - (n_1+n_2) t_2 - \cdots - (n_1+\cdots+n_d) t_d} \\
&= \prod_{j=1}^{d} \sum_{n_j \in \mathbb{N}} e^{-n_j (t_j + \cdots + t_d)} = \prod_{j=1}^{d} \left(e^{t_j + \cdots + t_d} - 1\right)^{-1}
\end{aligned}$$

and

$$\int_0^\infty e^{-\alpha t} t^{s-1}\, dt = \frac{\Gamma(s)}{\alpha^s} \quad (\Re(s) > 0;\ \Re(\alpha) > 0), \tag{6}$$

it is easy to find that

$$\Gamma(s_1)\cdots\Gamma(s_d)\,\zeta_d(s_1,\ldots,s_d) = \int_0^\infty\int_0^\infty\cdots\int_0^\infty \frac{t_1^{s_1-1}\cdots t_d^{s_d-1}\,dt_1\cdots dt_d}{\prod_{j=1}^d \left(e^{t_j+\cdots+t_d}-1\right)}. \tag{7}$$

Now, put $x_{d+1} := 0$ and make the change of variables

$$x_1\cdots x_j = t_j + \cdots + t_d, \quad \text{if and only if} \quad t_j = x_1\cdots x_j\left(1 - x_{j+1}\right) \tag{8}$$
$$(j \in \mathbb{N};\ 1 \leqq j \leqq d).$$

It is not difficult to get the Jacobian of (8):

$$\frac{\partial(t_1,\ldots,t_d)}{\partial(x_1,\ldots,x_d)} = \det\begin{bmatrix} \frac{\partial t_1}{\partial x_1} & \frac{\partial t_1}{\partial x_2} \cdots & \frac{\partial t_1}{\partial x_d} \\ \vdots & \vdots \ \ \ldots & \vdots \\ \frac{\partial t_d}{\partial x_1} & \frac{\partial t_d}{\partial x_2} \cdots & \frac{\partial t_d}{\partial x_d} \end{bmatrix} = x_1^{d-1}\cdot x_2^{d-2}\cdots x_{d-2}^2\cdot x_{d-1}. \tag{9}$$

It follows from (8) and (9) that

$$\Gamma(s_1)\cdots\Gamma(s_d)\,\zeta_d(s_1,\ldots,s_d)$$
$$= \int_0^1\cdots\int_0^1\int_0^\infty \prod_{j=1}^d x_j^{s_d(j)-d+j-2} \prod_{j=2}^d \left(1-x_j\right)^{s_{j-1}-1} \varphi(x)\,dx_1\cdots dx_d, \tag{10}$$

where the function $\varphi(x)$ is given by

$$\varphi(x) := \frac{x_1^d\cdot x_2^{d-1}\cdots x_{d-1}^2\cdot x_d}{\prod_{j=1}^d \left(e^{x_1\cdots x_j}-1\right)}, \tag{11}$$

which is seen to be a test function in $\mathcal{K}$. Letting $u_j := s_d(j) - d + j - 1$, we can apply the generalized function

$$\frac{x_{1+}^{u_1-1}}{\Gamma(u_1)}\cdot\prod_{j=2}^d \frac{\left(1-x_j\right)_+^{s_{j-1}-1}\cdot x_{j+}^{u_j-1}}{\Gamma\left(s_{j-1}\right)\Gamma\left(u_j\right)}$$

on $\varphi(x)$ and get exactly $(s_d-1)\,\zeta_d\,(s_1,\ldots,s_d)/\prod_{j=1}^{d-1}\Gamma\left(u_j\right)$, since $\Gamma\,(s_d)=(s_d-1)\,\Gamma\,(u_d)$.

By Lemma 2.9 and Lemma 2.10, all of the following generalized functions

$$f_1(x;s)=\frac{x_{1+}^{u_1-1}}{\Gamma\,(u_1)}\quad\text{and}\quad f_j(x;s)=\frac{\left(1-x_j\right)_+^{s_{j-1}-1}\cdot x_{j+}^{u_j-1}}{\Gamma\left(s_{j-1}\right)\Gamma\left(u_j\right)}\qquad (j\in\mathbb{N};\ 2\leqq j\leqq d)$$

can be extended to entire functions in s. Furthermore, f_j depends only on x_j as a function of x. Hence, one can set the entire function

$$\xi\,(s_1,\ldots,s_d):=\left\langle\prod_{j=1}^{d}f_j(x;s),\,\varphi(x)\right\rangle \tag{12}$$

on $\mathbb{C}^d$ and define an analytic continuation of $\zeta_d\,(s_1,\ldots,s_d)$ on $\mathbb{C}^d$ by

$$\zeta_d\,(s_1,\ldots,s_d):=\frac{\prod_{j=1}^{d-1}\Gamma\left(u_j\right)}{s_d-1}\,\xi\,(s_1,\ldots,s_d). \tag{13}$$

This completes the proof of the theorem, because Gamma functions have simple poles only at nonpositive integers (see 1.1(10) and 1.1(12)). ■

Zhao [1260, Theorem 6 and Theorem 7] evaluates the residues of $\zeta_d\,(s_1,\ldots,s_d)$ at the simple poles given in Theorem 2.11 as follows:

Theorem 2.12 *The residue of $\zeta_d\,(s_1,\ldots,s_d)$ at $s_d=1$ is 1 or $\zeta_{d-1}\,(s_1,\ldots,s_{d-1})$, according to whether $d=1$ or $d\in\mathbb{N}\setminus\{1\}$.*

For depth $d\in\mathbb{N}\setminus\{1\}$ and $i,\,\ell\in\mathbb{N}$ with $1\leqq i\leqq d-1$, the residue of $\zeta_d\,(s_1,\ldots,s_d)$ on the hyperplane

$$s_d(i)=d-i+2-\ell$$

in $\mathbb{C}^d$ is equal to

$$\zeta_{i-1}\,(s_1,\ldots,s_{i-1})\sum_{\substack{a_d(i+1)=\ell-1\\ a_{i+1},\ldots,a_d\geqq 0}}\left(\prod_{j=i+1}^{d}\frac{B_{a_j}\,\Gamma\left(a_d(j)+u_j\right)}{a_j!\,\Gamma\left(a_d(j+1)+u_j+1\right)}\right), \tag{14}$$

where $\zeta\,(s_0):=1$, $a_d(j):=a_j+\cdots+a_d$, $a_d(d+1):=0$, and $u_j:=s_d(j)-d+j-1$.

Proof. The case $d=1$ is obvious, since $\zeta\,(s_1)$ is the Riemann zeta function. So, we may assume $d\in\mathbb{N}\setminus\{1\}$. Let $\varphi(x)$ and $\xi\,(s_1,\ldots,s_d)$ be as in (11) and (12). In view of (13), we find that

$$\begin{aligned}\operatorname*{Res}_{s_d=1} \zeta_d(s_1,\ldots,s_d) &= \lim_{s_d\to 1} (s_d-1)\,\zeta_d(s_1,\ldots,s_d)\\ &= \left\{\lim_{s_d\to 1}\prod_{j=1}^{d-1}\Gamma\left(u_j\right)\right\}\left\{\lim_{s_d\to 1}\xi(s_1,\ldots,s_d)\right\},\end{aligned}$$

where the two involved limits are readily seen to be

$$\lim_{s_d\to 1}\prod_{j=1}^{d-1}\Gamma\left(u_j\right) = \Gamma(s_{d-1}-1)\cdot\prod_{j=1}^{d-2}\Gamma(s_{d-1}(j)-d+j)$$

and

$$\lim_{s_d\to 1}\xi(s_1,\ldots,s_d) = \frac{\xi(s_1,\ldots,s_{d-1})}{\Gamma(s_{d-1})}.$$

Now, it is easy to see that

$$\begin{aligned}\operatorname*{Res}_{s_d=1} \zeta_d(s_1,\ldots,s_d) = &\frac{\prod_{j=1}^{d-2}\Gamma(s_{d-1}(j)-d+j)}{s_{d-1}-1}\\ &\cdot\xi(s_1,\ldots,s_{d-1}) = \zeta_{d-1}(s_1,\ldots,s_{d-1}).\end{aligned}$$

The proof of the remaining part, that is (14), is left to an interested reader. ■

Remark 2 By the power series definition, one easily sees that

$$\zeta_2(0,s_2) = \zeta(s_2-1)-\zeta(s_2), \tag{15}$$

whereas it is known (see 3.3(35)) that

$$\zeta_2(s_1,s_2)+\zeta_2(s_2,s_1)+\zeta(s_1+s_2) = \zeta(s_1)\,\zeta(s_2). \tag{16}$$

Setting $s_2=0$ in (16) and using (15) yields

$$\zeta_2(s_1,0) = \zeta(s_1)\,\zeta(0)-\zeta(s_1-1). \tag{17}$$

Now, setting $s_2=0$ in (15) and $s_1=0$ in (17), together with 2.3(10), gives

$$\lim_{s_2\to 0}\lim_{s_1\to 0}\zeta_2(s_1,s_2) = \frac{5}{12} \quad\text{and}\quad \lim_{s_1\to 0}\lim_{s_2\to 0}\zeta_2(s_1,s_2) = \frac{1}{3}. \tag{18}$$

It is observed that the limits of a function at some point along different routes in the multivariable space may be different.

From (10) and the identity (use 1.7(2))

$$\frac{x^2 y}{(e^x-1)(e^{xy}-1)} = 1 - \frac{1}{2}(1+y)x + \frac{1}{12}\left(y^2+3y+1\right)x^2 + O\left(x^3\right) \quad (x \to 0), \tag{19}$$

we readily obtain (see Zhao [1260, p. 1280]) that

$$\zeta_2(s_1, s_2) = \frac{4s_1+5s_2}{12(s_1+s_2)} + \mathcal{R}(s_1, s_2), \tag{20}$$

where $\mathcal{R}(s_1, s_2)$ represents an analytic function of (s_1, s_2) in a neighborhood of $(0, 0)$ and $\lim_{(s_1,s_2)\to(0,0)} \mathcal{R}(s_1, s_2) = 0$. It is seen that (20) also gives the same limit values as in (18).

Euler-Maclaurin Summation Formula

We begin by remarking that most of this subsection is due to Akiyama *et al.* [16] and recalling another version of Euler-Maclaurin summation formula (*cf.* [3, p. 806] and 1.3(68)): Let $N \in \mathbb{N}$ and $a, b \in \mathbb{R}$ with $a < b$. Suppose that $f \in C^N([a, b])$ denotes (as usual) the set of functions having continuous derivatives of Nth order on the interval $[a, b]$. Then, we have

$$\begin{aligned}\sum_{a<n\leq b} f(n) &= \int_a^b f(x)dx + \sum_{k=1}^{K} \frac{(-1)^k}{k!}\widetilde{B}_k(b)f^{(k-1)}(b) \\ &\quad - \sum_{k=1}^{K} \frac{(-1)^k}{k!}\widetilde{B}_k(a)f^{(k-1)}(a) - \frac{(-1)^K}{K!}\int_a^b \widetilde{B}_K(x)f^{(K)}(x)dx,\end{aligned} \tag{21}$$

where $\widetilde{B}_k(x)$ is the periodic Bernoulli polynomial, defined by $\widetilde{B}_k(x) := B_k(x-[x])$ ($[x]$ denotes the greatest integer $\leqq x$).

Setting $a = 1$, $b = m$ and $K = \ell + 1$ in (21) yields

$$\begin{aligned}\sum_{n=1}^{m} f(n) &= \int_1^m f(x)\,dx + \frac{1}{2}\left(f(1)+f(m)\right) + \sum_{k=1}^{\ell} \frac{B_{k+1}}{(k+1)!}\left(f^{(k)}(m) - f^{(k)}(1)\right) \\ &\quad - \frac{(-1)^{\ell+1}}{(\ell+1)!}\int_1^m \widetilde{B}_{\ell+1}(x)f^{(\ell+1)}(x)\,dx,\end{aligned} \tag{22}$$

where $f \in C^{\ell+1}([1, m])$, $m \in \mathbb{N}$ and $\ell \in \mathbb{N}_0$.

Consider $f(x) = x^{-s}$ in (22). We get, by subtracting the resulting identity of $\sum_{n=1}^{m} n^{-s}$ from that of $\sum_{n=1}^{\infty} n^{-s} = \zeta(s)$, the following formula:

$$\phi_\ell(m,s) = \sum_{n=1}^{m} \frac{1}{n^s} - \left\{ \frac{m^{1-s}-1}{1-s} + \frac{1}{2m^s} - \sum_{k=1}^{\ell} \frac{(s)_k\, a_k}{m^{s+k}} + \zeta(s) - \frac{1}{s-1} \right\}, \tag{23}$$

where $(s)_k$ is the Pochhammer symbol given in 1.1(5), $a_k := B_{k+1}/(k+1)!$, and

$$\begin{aligned} \phi_\ell(m,s) :&= \frac{(s)_{\ell+1}}{(\ell+1)!} \int_m^\infty \widetilde{B}_{\ell+1}(x)\, x^{-s-\ell-1}\, dx \\ &= O\left(m^{-\Re(s)-\ell}\right), \end{aligned} \tag{24}$$

where, for example, see Hardy [537, p. 332, 13.10.3]. We rewrite (23) in the following equivalent form:

$$\sum_{n=m+1}^{\infty} \frac{1}{n^s} = -\phi_\ell(m,s) + \frac{m^{1-s}}{s-1} - \frac{1}{2m^s} + \sum_{k=1}^{\ell} \frac{(s)_k\, a_k}{m^{s+k}} \qquad (\Re(s) > 1). \tag{25}$$

Consider the following multiple Zeta function in two variables:

$$\zeta_2(s_1, s_2) = \sum_{0<n_1<n_2} \frac{1}{n_1^{s_1}\, n_2^{s_2}} \qquad (\Re(s_i) > 1;\, i = 1, 2).$$

We make use of (25) to get

$$\begin{aligned} \zeta_2(s_1, s_2) &= \sum_{n_1=1}^{\infty} \frac{1}{n_1^{s_1}} \sum_{n_2=n_1+1}^{\infty} \frac{1}{n_2^{s_2}} \\ &= \sum_{n_1=1}^{\infty} \frac{1}{n_1^{s_1}} \left\{ -\phi_\ell(n_1, s_2) + \frac{n_1^{1-s_2}}{s_2-1} - \frac{1}{2n_1^{s_2}} + \sum_{k=1}^{\ell} \frac{(s_2)_k\, a_k}{n_1^{s_2+k}} \right\} \\ &= \frac{\zeta(s_1+s_2-1)}{s_2-1} - \frac{\zeta(s_1+s_2)}{2} \\ &\quad + \sum_{k=1}^{\ell} (s_2)_k\, a_k\, \zeta(s_1+s_2+k) - \sum_{n_1=1}^{\infty} \frac{\phi_\ell(n_1, s_2)}{n_1^{s_1}} \end{aligned} \tag{26}$$

for $\Re(s_i) > 1$ $(i = 1, 2)$. The terms on the right hand side, except the last one, have meromorphic continuations. The last sum is absolutely convergent and, hence, holomorphic in $\Re(s_1 + s_2 + \ell - 1) > 0$. Thus, we now have a meromorphic continuation of $\zeta_2(s_1, s_2)$ to $\mathbb{C}^2$, which is holomorphic in

$$\left\{ (s_1, s_2) \in \mathbb{C}^2 \mid s_2 \neq 1,\ s_1 + s_2 \notin \{2, 1, 0, -2, -4, -6, \ldots\} \right\}.$$

It is easy to see that this manipulation can be employed to a multiple zeta function with d variables. Indeed, we have

$$\begin{aligned}
&\zeta_d(s_1,\ldots,s_d)\\
&=\sum_{n_1=1}^{\infty}\frac{1}{n_1^{s_1}}\sum_{n_2=n_1+1}^{\infty}\frac{1}{n_2^{s_2}}\cdots\sum_{n_{d-1}=n_{d-2}+1}^{\infty}\frac{1}{n_{d-1}^{s_{d-1}}}\sum_{n_d=n_{d-1}+1}^{\infty}\frac{1}{n_d^{s_d}}\\
&=\sum_{n_1=1}^{\infty}\frac{1}{n_1^{s_1}}\sum_{n_2=n_1+1}^{\infty}\frac{1}{n_2^{s_2}}\cdots\sum_{n_{d-1}=n_{d-2}+1}^{\infty}\frac{1}{n_{d-1}^{s_{d-1}}}\\
&\quad\cdot\left\{-\phi_\ell(n_{d-1},s_d)+\frac{n_{d-1}^{1-s_d}}{s_d-1}-\frac{1}{2n_{d-1}^{s_d}}+\sum_{k=1}^{\ell}\frac{(s_d)_k\,a_k}{n_{d-1}^{s_d+k}}\right\} \qquad (27)\\
&=\frac{\zeta_{d-1}(s_1,\ldots,s_{d-2},s_{d-1}+s_d-1)}{s_d-1}-\frac{\zeta_{d-1}(s_1,\ldots,s_{d-2},s_{d-1}+s_d)}{2}\\
&\quad+\sum_{k=1}^{\ell}(s_d)_k\,a_k\,\zeta_{d-1}(s_1,\ldots,s_{d-2},s_{d-1}+s_d+k)\\
&\quad-\sum_{0<n_1<\cdots<n_{d-1}}\frac{\phi_\ell(n_{d-1},s_d)}{n_1^{s_1}\cdots n_{d-1}^{s_{d-1}}}\qquad(\Re(s_i)>1;\ i=1,\,\ldots,\,d).
\end{aligned}$$

Since

$$\sum_{0<n_1<\cdots<n_{d-1}}\left|\frac{\phi_\ell(n_{d-1},s_d)}{n_1^{s_1}\cdots n_{d-1}^{s_{d-1}}}\right|\leqq\sum_{n_{d-1}}\frac{n_{d-1}^{-\ell-\Re(s_d)+d-2}}{n_{d-1}^{L}}$$

with

$$L:=\Re(s_{d-1})+\sum_{\substack{1\leqq i\leqq d-2\\ \Re(s_i)\leqq 0}}\Re(s_i),$$

the last summation is absolutely convergent in the set

$$\ell-d+1+\Re(s_d)+\Re(s_{d-1})+\sum_{\substack{1\leqq i\leqq d-2\\ \Re(s_i)\leqq 0}}\Re(s_i)>0. \qquad (28)$$

Since ℓ can be taken arbitrarily large, we get an analytic continuation of $\zeta_d(s_1,\ldots,s_d)$ to $\mathbb{C}^d$. Now, we consider the set singularities. It is seen that the singular part of $\zeta_2(s_1,s_2)$ may be *formally* given as follows:

$$\frac{\zeta(s_1+s_2-1)}{s_2-1}+\sum_{k_1\geqq 0}\frac{a_{k_1}(s_2)_{k_1}}{s_1+s_2+k_1-1},$$

since, by analytic continuation, $\zeta(s)$ can be written as $\zeta(s) = 1/(s-1) + \tau(s)$, where $\tau(s)$ is an entire function. We find from this expression that

$$s_2 = 1, \quad s_1 + s_2 \in \{2,\, 1,\, 0,\, -2,\, -4,\, -6,\, \ldots\}$$

forms the set of all singularities of $\zeta_2(s_1, s_2)$. For the case $\zeta_3(s_1, s_2, s_3)$, by using the singular part of ζ_2, we see that singularities lie on

$$s_3 = 1, \quad s_2 + s_3 \in \{2,\, 1,\, 0,\, -2,\, -4,\, -6,\, \ldots\}$$

and

$$s_1 + s_2 + s_3 \in \{3,\, 2,\, 1,\, 0,\, -1,\, -2,\, -3,\, \ldots\}.$$

To show these are all singularities of $\zeta_3(s_1, s_2, s_3)$, it suffices to prove that no singularities, defined by one of the above equations, will identically vanish. This can be shown by changing variables:

$$u_1 = s_1, \quad u_2 = s_2 + s_3, \quad u_3 = s_3.$$

In fact, it is seen that the singular part of $\zeta_3(u_1, u_2 - u_3, u_3)$ is given by

$$\frac{1}{u_3 - 1}\,\zeta_2(u_1, u_2 - 1) + \sum_{k_2 \geqq 0} a_{k_2}\,(u_3)_{k_2}\,\zeta_2(u_1, u_2 + k_2).$$

It is found from this expression that the singularities of $\zeta_2(u_1, u_2 + k_2)$ are summed with functions of u_3 of different degrees. Thus, these singularities, as a weighted sum by another variable u_3, will not vanish identically. Similarly, we can prove Theorem 2.13 below.

Theorem 2.13 *The multiple zeta function* $\zeta_d(s_1, \ldots, s_d)$ *continues meromorphically to* $\mathbb{C}^d$ *and has singularities on*

$$s_d = 1, \quad s_{d-1} + s_d = 2,\, 1,\, 0,\, -2,\, -4,\, \ldots,$$

and

$$\sum_{i=1}^{j} s_{d-i+1} \in \mathbb{Z}_{\leqq j} \quad (j = 3,\, 4,\, \ldots,\, d),$$

where $\mathbb{Z}_{\leqq j}$ *is the set of integers less than or equal to* j.

We conclude this section by remarking the following facts: Akiyama *et al.* [16] defines the multiple zeta values at nonpositive integers by

$$\zeta_d(-r_1, \ldots, -r_d) := \lim_{s_1 \to -r_1} \cdots \lim_{s_d \to -r_d} \zeta_d(s_1, \ldots, s_d) \quad \left(r_j \in \mathbb{N}_0;\; j = 1, \ldots, d\right). \tag{29}$$

If, for simplicity, we write

$$(s)_{-1} := \frac{1}{s-1},$$

then it follows from (24), (27) and (29) that

$$\zeta_d(-r_1, \ldots, -r_d) = \sum_{k=-1}^{r_d} (-r_d)_k \, a_k \, \zeta_{d-1}(-r_1, \ldots, -r_{d-2}, -r_{d-1} - r_d + k), \quad (30)$$

where we have used the fact that, in view of (24),

$$\phi_\ell(n_{d-1}, -r_d) = 0 \qquad (\ell \geqq r_d).$$

Akiyama *et al.* [16] and Akiyama and Tanigawa [17] made a basic use of (30) to give the multiple zeta values at nonpositive integers as asserted by Theorem 2.13 below.

Theorem 2.14 [17, p. 330, Theorem 2]

$$\zeta_d(-n, 0, \ldots, 0) = \frac{(-1)^n}{d!} \sum_{j=1}^{d} s(d,j) \frac{j B_{n+j}}{n+j} \qquad (n \in \mathbb{N}_0), \quad (31)$$

where $s(d,j)$ denotes the Stirling numbers of the first kind (see Section 1.6) and B_j the Bernoulli numbers (see Section 1.7).

Problems

1. Show that

$$\begin{aligned}\zeta'\left(-2k+1, \frac{p}{q}\right) &= \frac{1}{2k}\left[\psi(2k) - \log(2\pi q)\right] B_{2k}\left(\frac{p}{q}\right) - \frac{q^{-2k}}{2k}\left[\psi(2k) - \log(2\pi)\right] B_{2k} \\ &+ \frac{(-1)^{k+1}\pi}{(2\pi q)^{2k}} \sum_{n=1}^{q-1} \sin\left(\frac{2\pi p n}{q}\right) \psi^{(2k-1)}\left(\frac{n}{q}\right) \\ &+ \frac{(-1)^{k+1} 2 \cdot (2k-1)!}{(2\pi q)^{2k}} \sum_{n=1}^{q-1} \cos\left(\frac{2\pi p n}{q}\right) \zeta'\left(2k, \frac{n}{q}\right) \\ &+ \frac{\zeta'(-2k+1)}{q^{2k}} \qquad (p < q;\ p, q, k \in \mathbb{N}).\end{aligned}$$

(Miller and Adamchik [829, p. 203, Proposition 1])

2. Prove that there exist constants D_k, defined by

$$\log D_k := \lim_{n \to \infty} \left(\sum_{m=1}^{n} m^k \log m - p(n,k) \right) \qquad (k \in \mathbb{N}_0),$$

where the definition of $p(n,k)$ in Adamchik [6, p. 198, Eq. (20)] is *corrected* here as follows:

$$p(n,k) := \frac{n^k}{2}\log n + \frac{n^{k+1}}{k+1}\left(\log n - \frac{1}{k+1}\right) + k!\sum_{j=1}^{k}\frac{n^{k-j}B_{j+1}}{(j+1)!\,(k-j)!}\left[\log n + \left(1-\delta_{kj}\right)\sum_{\ell=1}^{j}\frac{1}{k-\ell+1}\right]$$

and δ_{kj} is the Kronecker symbol defined by 1.6(25).

(Bendersky [114, pp. 273–275]; Adamchik [6, p. 198])

3. For the constants D_k $(k \in \mathbb{N}_0)$, defined in Problem 2, show that

$$D_0 = (2\pi)^{\frac{1}{2}} \quad D_1 = A, \quad D_2 = B \quad \text{and} \quad D_3 = C$$

and

$$\log D_k = \frac{B_{k+1}H_k}{k+1} - \zeta'(-k) \quad (k \in \mathbb{N}_0),$$

where B_n are the Bernoulli numbers and H_n are the harmonic numbers, and the mathematical constants A, B, and C are given as in Section 1.4.

(Adamchik [6, pp. 198–199])

4. Prove that

$$\int_0^{\pi}\left(\log\left[\left(2\sin\tfrac{1}{2}t\right)\left(2\cos\tfrac{1}{2}t\right)^{p}\right]\right)^3 dt = -2\pi\,\zeta(3)\left[\tfrac{3}{4}\left(1+p^3\right) - \tfrac{3}{8}\left(p+p^2\right)\right].$$

(Lewin [752, p. 220])

5. The Barnes double Zeta function $\zeta_2(v;\alpha,w)$ is the meromorphic continuation of the series defined by

$$\zeta_2(v;\alpha,w) := \sum_{m=0}^{\infty}\sum_{n=0}^{\infty}(\alpha+m+nw)^{-v} \quad (\alpha > 0;\ w > 0;\ v \in \mathbb{C}).$$

The functions $\rho_2(1,w)$ and $\Gamma_2(\alpha,(1,w))$ are defined by

$$\log \rho_2(1,w) := -\lim_{\alpha\to 0}\left[\zeta_2'(0;\alpha,w) + \log\alpha\right]$$

and

$$\Gamma_2(\alpha,(1,w)) := \rho_2(1,w)\exp\left[\zeta_2'(0;\alpha,w)\right],$$

respectively.

(a) Prove that, for any $N \in \mathbb{N}$ and $\Re(v) > -N+1$ $(v \in \mathbb{C}\setminus[\mathbb{Z}_0^- \cup \{1,2\}])$,

$$\zeta_2(v;\alpha,w) = \zeta(v,\alpha) + \frac{\zeta(v-1)}{v-1}w^{1-v} + \sum_{n=0}^{N-1}\binom{-v}{n}\zeta(-n,\alpha)\zeta(v+n)w^{-v-n} + R_N(v;\alpha,w)$$

with the estimate:

$$R_N(v;\alpha,w) = O\left(w^{-\Re(v)-N}\right),$$

where the O-constant depends on v, N and α.

(b) Show also that, for $N \in \mathbb{N} \setminus \{1, 2\}$,

$$\begin{aligned}\zeta_2'(0;\alpha,w) = &-\frac{1}{12} w \log w + \left[\frac{1}{12} - \zeta'(-1)\right] w \\ &+ \frac{1}{2}\left(\frac{1}{2} - \alpha\right)\log w + \log\Gamma(\alpha) + \left(\frac{1}{2}\alpha - \frac{3}{4}\right)\log(2\pi) \\ &+ \zeta(-1,\alpha)\, w^{-1}(\log w - \gamma) \\ &+ \sum_{n=2}^{N-1} \frac{(-1)^n}{n} \zeta(-n,\alpha)\,\zeta(n)\, w^{-n} + R_N'(0;\alpha,w)\end{aligned}$$

and

$$\begin{aligned}\log\Gamma_2(\alpha,(1,w)) = &\frac{1}{2}\alpha\log\left(2\pi w^{-1}\right) + \log\Gamma(\alpha) \\ &+ [\zeta(-1,\alpha) - \zeta(-1)]\, w^{-1}(\log w - \gamma) \\ &+ \sum_{n=2}^{N-1} \frac{(-1)^n}{n} [\zeta(-n,\alpha) - \zeta(n)]\,\zeta(n)\, w^{-n} + S_N(\alpha,(1,w))\end{aligned}$$

with the estimates:

$$R_N'(0;\alpha,w) = O\left[w^{-N}(|\log w| + 1)\right]$$

and

$$S_N(\alpha,(1,w)) = O\left(w^{-N}(|\log w| + 1)\right],$$

where the O-constants depend on N and α and the prime in $\zeta_2'(v;\alpha,w)$ denotes differentiation with respect to v.

(Matsumoto [804, 805])

6. For the Bernoulli numbers and polynomials, show that

(a)

$$B_{2n} = 4n(-1)^{n+1} \int_0^\infty \frac{x^{2n-1}}{e^{2\pi x} - 1}\, dx \quad (n \in \mathbb{N});$$

(b)

$$|B_{2n}(x)| \leq |B_{2n}| \quad (n \in \mathbb{N};\ 0 \leqq x \leqq 1).$$

7. Show that

$$\zeta(s) = \frac{1}{\left(1 - 2^{1-s}\right)\Gamma(s)} \int_0^\infty \frac{x^{s-1}}{e^x + 1}\, dx \quad (\Re(s) > 0;\ s \neq 1).$$

(See, e.g., Chen [240, p. 263])

8. Prove that

$$\int_{m}^{m+n+\frac{1}{2}} \log \Gamma(t)\,dt = -\sum_{k=1}^{m-1} k \log k + \frac{1}{2} \sum_{k=1}^{m+n-1} (2k+1) \log (2k+1)$$
$$- \frac{1}{2}(m+n)^2 (1+\log 2) + \frac{m(m-1)}{2} - \frac{3}{2}\zeta'(-1)$$
$$+ \left(-\frac{3}{2}+\frac{1}{2}\log 2\right)\zeta(-1) + \left(n+\frac{1}{2}\right)\log\sqrt{2\pi} \quad (m \in \mathbb{N};\ n \in \mathbb{N}_0).$$

(Elizalde *et al.* [412, p. 19])

9. Prove that

$$\int_0^z t^n\,\psi(t)\,dt = (-1)^{n-1}\,\zeta'(-n) + \frac{(-1)^n}{n+1} B_{n+1} H_n$$
$$- \sum_{k=0}^{n} (-1)^k \binom{n}{k} \frac{z^{n-k}}{k+1} B_{k+1}(z) H_k + \sum_{k=0}^{n} (-1)^k \binom{n}{k} z^{n-k}\,\zeta'(-k,z)$$
$$(n \in \mathbb{N}_0;\ \Re(z) > 0).$$

(Adamchik [6, p. 197])

10. Prove that

$$\frac{z^a}{\Gamma(1-s)}\,\Phi(z,s,a) = \left(\log\frac{1}{z}\right)^{s-1} + \sum_{k=0}^{\infty} \binom{s-1}{2k} (\log z)^{2k} \frac{\zeta(s-2k,a)}{\Gamma(2k+1-s)}$$
$$- \sum_{k=0}^{\infty} \binom{s-1}{2k+1} (\log z)^{2k+1} \frac{\zeta(s-2k-1,a)}{\Gamma(2k+2-s)} \quad (|\log z| < 2\pi;\ s \notin \mathbb{N};\ a \notin \mathbb{Z}_0^-).$$

Also deduce the relatively more familiar result 2.5(13).

11. Prove that

$$E_{2n-1}(0) = \frac{4(-1)^n}{(2\pi)^{2n}}(2n-1)!\left(2^{2n}-1\right)\zeta(2n) \quad (n \in \mathbb{N}),$$

where $E_n(x)$ denotes the Euler polynomials (see Section 1.6).

(*Cf.* Srivastava [1084, p. 390])

12. Let $f : \mathbb{N} \times \mathbb{N} \to \mathbb{R}$ be a function. For a fixed $n \in \mathbb{N}$, let

$$T_n := \{(k,j) \in \mathbb{N} \times \mathbb{N} \mid 1 \leqq k, j \leqq n;\ k+j \geqq n+1\}$$

and

$$S_n := \sum_{(x,y)\in T_n} f(x,y),$$

the sum being taken over all pairs $(x,y) \in T_n$. Show that

$$S_n = \sum_{k=1}^{n} f(k,k) + \sum_{k=2}^{n}\sum_{j=1}^{k-1} \{f(j,k) + f(k,j) - f(j,k-j)\}.$$

(Sitaramachandrarao and Sivaramsarma [1035, p. 603])

13. Prove that

$$2\sum_{q=1}^{\infty}\frac{1}{q^n}\left(\sum_{\substack{k=1\\(k,q)=1}}^{q-1}\frac{1}{k}\right)=n+2-\frac{1}{\zeta(n+1)}\sum_{j=1}^{n-2}\zeta(n-j)\,\zeta(j+1)\quad(n\in\mathbb{N}\setminus\{1\}),$$

where an empty sum is understood (as usual) to be nil and $(k,q)=1$ denotes that k and q are relatively prime.

(*Cf.* Equation (54); see also Sitaramachandrarao and Sivaramsarma [1035, p. 602])

14. Prove that

$$\sum_{k=1}^{\infty}\frac{H_k}{k^{2n+1}}=\frac{1}{2}\sum_{j=2}^{2n}(-1)^j\,\zeta(j)\,\zeta(2n-j+2)\quad(n\in\mathbb{N}),$$

and

$$\sum_{k=1}^{\infty}\frac{H_k}{k^n}=\left(1+\frac{n}{2}\right)\zeta(n+1)-\frac{1}{2}\sum_{j=2}^{n-1}\zeta(j)\,\zeta(n-j+1)\quad(n\in\mathbb{N}\setminus\{1\}),$$

where H_k are the harmonic numbers defined by 3.2(36).

(Georghiou and Philippou [479, p. 29])

15. For bounded maps $f,\ g:\mathbb{N}\to\mathbb{C}$, let

$$S_{f,g}(u,v;w)=\sum_{r,k=1}^{\infty}\frac{f(r+k)\,g(k)}{r^u\,k^u\,(r+k)^w}\quad\left(u,\ v\in\mathbb{N};\ w\in\mathbb{R}^+\right)$$

and

$$C_{f,g}(x,y)=\sum_{r=2}^{\infty}\frac{f(r)}{r^x}\sum_{k=1}^{r-1}\frac{g(k)+g(r-k)}{k^y}\quad(x>1;\ y\in\mathbb{N}).$$

Show that

$$\begin{aligned}S_{f,g}(u,v;w)+S_{f,g}(v,u;w)&=\sum_{j=0}^{u-1}\binom{v+j-1}{v-1}C_{f,g}(v+w+j,u-j)\\&\quad+\sum_{j=0}^{v-1}\binom{u+j-1}{u-1}C_{f,g}(u+w+j,v-j).\end{aligned}$$

(Subbarao and Sitaramachandrarao [1133, p. 246])

16. Prove that

$$\sum_{r,k=1}^{\infty}\frac{(-1)^{r+k}}{rk\,(r+k)}=\frac{1}{4}\,\zeta(3)$$

and

$$\sum_{r,k=1}^{\infty}\frac{(-1)^{k-1}}{rk\,(r+k)}=\frac{5}{8}\,\zeta(3).$$

(Subbarao and Sitaramachandrarao [1133, p. 247])

17. Prove that

$$\sum_{k=1}^{\infty} \frac{H_k^{(2)}}{k^{2n+1}} = \zeta(2)\,\zeta(2n+1) - \frac{(n+2)(2n+1)}{2}\zeta(2n+3)$$
$$+2\sum_{j=2}^{n+1}(j-1)\zeta(2j-1)\,\zeta(2n+4-2j) \quad (n \in \mathbb{N}),$$

where $H_n^{(m)}$ denotes the generalized harmonic number, defined by

$$H_0^{(m)} = 0 \quad \text{and} \quad H_n^{(m)} = \sum_{\ell=1}^{n} \ell^{-m} \quad (m,\, n \in \mathbb{N}).$$

(Georghiou and Philippou [479, p. 35])

18. Prove that

$$\sum_{m,\,n=0}^{\infty}{}' \frac{1}{m^6\left(m^2+n^2\right)} = \frac{13\pi^8}{28350} \quad \text{and} \quad \sum_{m,\,n=0}^{\infty}{}' \frac{1}{m^2\left(m^2-mn+n^2\right)} = \frac{\sqrt{3}\pi^4}{30},$$

where the prime denotes that the summations are taken over all pairs of integers, except $(0,0)$.

(Smart [1042, p. 10])

19. Prove that

$$\sum_{k=1}^{\infty} \frac{1}{k}\left(H_{2k-1} - H_k - \log 2\right) = (\log 2)^2 - \frac{\pi^2}{6}.$$

(Knuth [679, p. 138])

20. Let $a,\, b,\, c \in \mathbb{R}$ with $a > 0$ and $d = b^2 - 4ac < 0$. The Epstein Zeta function is given by

$$Z(s) := \frac{1}{2}\sum{}' \left(am^2 + bmn + cn^2\right)^{-s} \quad (\Re(s) > 1),$$

where the prime denotes that the summation is to be taken over all pairs (m,n) of integers other than the pair $(0,0)$. Let $k \in \mathbb{R}^+$ be a number satisfying

$$k^2 = \frac{c}{a} - \left(\frac{b}{2a}\right)^2 \quad (a,\, b,\, c \in \mathbb{R};\ a > 0)$$

and suppose that the Bessel function $K_\nu(z)$ is defined by

$$K_\nu(z) = \frac{1}{2}\int_0^{\infty} e^{-\frac{z}{2}\left(u+\frac{1}{u}\right)} u^{\nu-1}\,du = \frac{1}{2}\int_{-\infty}^{\infty} e^{-z\cosh t} e^{\nu t}\,dt$$
$$= \int_0^{\infty} e^{-z\cosh t}\cosh(\nu t)\,dt \quad \left(\nu \in \mathbb{C};\ |\arg z| < \frac{\pi}{2}\right).$$

Also, let

$$\sigma_\nu(n) := \sum_{d|n} d^\nu = \sum_{d|n} \left(\frac{n}{d}\right)^\nu \qquad (\nu \in \mathbb{C};\ n \in \mathbb{N}).$$

Show that

$$a^s Z(s) = \zeta(2s) + k^{1-2s}\,\zeta(2s-1)\,\frac{\Gamma\left(s-\frac{1}{2}\right)\Gamma\left(\frac{1}{2}\right)}{\Gamma(s)} + \frac{\pi^s}{\Gamma(s)} k^{\frac{1}{2}-s} H(s) \qquad (\Re(s) > 1),$$

where

$$H(s) = 4\sum_{n=1}^{\infty} n^{s-\frac{1}{2}}\,\sigma_{1-2s}(n)\,\cos\left(\frac{n\pi b}{a}\right) K_{s-\frac{1}{2}}(2\pi kn).$$

Show, also, that $H(s)$ is an entire function of s, such that

$$H(s) = H(1-s).$$

(Bateman and Grosswald [101, p. 366])

21. Show that

$$(cz+d)^{2n} A\left(\frac{az+b}{cz+d}, -2n\right) = A(z,-2n) + \frac{1}{2}\left\{1-(cz+d)^{2n}\right\}\zeta(2n+1)$$
$$+ \frac{(-1)^n (2\pi)^{2n+1}\, i}{2\,(2n+2)!} \sum_{j=1}^{c}\sum_{k=0}^{2n+2} \binom{2n+2}{k} B_k\left(\frac{j}{c}\right) \bar{B}_{2n-k+2}\left(\frac{jd}{c}\right) \{-(cz+d)\}^{k-1}$$
$$(c,\ n \in \mathbb{N}),$$

where $A(z,-2n)$ denotes a Lambert series in the variable $e^{2\pi iz}$, defined by

$$A(z,-2n) := \sum_{k=1}^{\infty} k^{-2n-1}\,\frac{e^{2\pi ikz}}{1-e^{2\pi ikz}}$$

and $\bar{B}_n(x) := B_n(x-[x])$ ($[x]$ being, as usual, the greatest integer in x).

(*Cf.* Berndt [119, p. 505]; see also Apostol [57, p. 153])

22. Show that

$$2^{k-s}\left\{\zeta\left(s,\frac{a}{2}\right) - \zeta\left(s,\frac{a+1}{2}\right)\right\} = \sum_{\ell=1}^{k}(-1)^{\ell-1}\sum_{j=0}^{\ell-1}(-1)^j\binom{k}{j}(\ell-1-j+a)^{-s}$$
$$+\sum_{n=0}^{\infty}(-1)^{n+k}\sum_{j=0}^{k}(-1)^j\binom{k}{j}(n+k-j+a)^{-s} \qquad (k \in \mathbb{N};\ \Re(s) > -k),$$

where $\zeta(s,a)$ $(a > 0)$ is the Hurwitz (or generalized) Zeta function, defined by 2.2(1).

(Balakrishnan [91, p. 205])

23. Show that

$$\int_0^{\frac{\pi}{3}} \sum_{k=0}^{m-2} \frac{(-1)^k}{2k+1} \binom{2m-2}{2k} \left(\frac{\theta}{2}\right)^{2k+1} \left\{\log\left(2\sin\frac{\theta}{2}\right)\right\}^{2m-2k-2} d\theta$$

$$= \frac{(-)^m \pi^{2m}}{8m(2m-1)} \left[\left(\frac{1}{6}\right)^{2m-1} - 2\left(1 - \frac{1}{2^{2m-1}}\right) B_{2m}\right] \quad (m \in \mathbb{N} \setminus \{1\})$$

and

$$\int_0^{\frac{\pi}{3}} \sum_{k=0}^{m-1} (-1)^k \binom{2m}{2k} \left(\frac{\theta}{2}\right)^{2k} \left\{\log\left(2\sin\frac{\theta}{2}\right)\right\}^{2m-2k} d\theta$$

$$= \frac{(-)^m \pi^{2m+1}}{2^{2m+2}} \left(E_{2m} - \frac{1}{(2m+1)\,3^{2m}}\right) \quad (m \in \mathbb{N}),$$

where B_n and E_n are the Bernoulli and Euler numbers, respectively, given in Section 1.7.

(Zhang and Williams [1253, p. 282])

24. Show that

$$\sum_{n=1}^{\infty} \frac{1}{n^{m+2}\binom{2n}{n}} = \frac{(-1)^m 2^m}{m!} \int_0^{\frac{\pi}{3}} \theta \left\{\log\left(2\sin\frac{\theta}{2}\right)\right\}^m d\theta \quad (m \in \mathbb{N}_0)$$

and

$$\sum_{n=1}^{\infty} \frac{(-1)^{n-1}}{n^{m+2}\binom{2n}{n}} = \frac{(-1)^m 2^m}{m!} \int_0^{2\log\tau} \theta \left\{\log\left(2\sinh\frac{\theta}{2}\right)\right\}^m d\theta \quad (m \in \mathbb{N}_0).$$

(Zhang and Williams [1253, pp. 286 and 288])

25. Show that

$$\int_0^{\frac{\pi}{3}} \left[\left\{\log\left(2\sin\frac{\theta}{2}\right)\right\}^4 - \frac{3\theta^2}{2}\left\{\log\left(2\sin\frac{\theta}{2}\right)\right\}^2\right] d\theta = \frac{253\pi^5}{2^3 \cdot 3^4 \cdot 5}$$

and

$$\int_0^{\frac{\pi}{3}} \left[\theta\left\{\log\left(2\sin\frac{\theta}{2}\right)\right\}^4 - \frac{\theta^3}{2}\log\left(2\sin\frac{\theta}{2}\right)\right] d\theta = \frac{313\pi^6}{2^4 \cdot 3^6 \cdot 5 \cdot 7}.$$

(Zucker [1267, pp. 98 and 99])

26. Prove each of the following claims:

(a) For all $a_1, a_2, \ldots$

$$\sum_{k=1}^{\infty} \frac{a_1 a_2 \cdots a_{k-1}}{(x+a_1)\cdots(x+a_k)} = \frac{1}{x}.$$

(b)

$$\zeta(3) =: \sum_{n=1}^{\infty} \frac{1}{n^3} = \frac{5}{2} \sum_{n=1}^{\infty} \frac{(-1)^{n-1}}{n^3 \binom{2n}{n}}.$$

(c) Consider the recursion:

$$n^3 u_n + (n-1)^3 u_{n-2} = \left(34n^3 - 51n^2 + 27n - 5\right) u_{n-1} \quad (n \in \mathbb{N} \setminus \{1\}).$$

Let $\{b_n\}$ be the sequence defined by $b_0 = 1$, $b_1 = 5$ and $b_n = u_n$ for all $n \in \mathbb{N} \setminus \{1\}$; then the b_n are all integers.

Let $\{a_n\}$ be the sequence defined by $a_0 = 0$, $a_1 = 6$ and $a_n = u_n$ for all $n \in \mathbb{N} \setminus \{1\}$; then the a_n are rational numbers with denominator dividing $2[1, 2, \ldots, n]^3$ (here, $[1, 2, \ldots, n]$ is the lowest common multiple of $1, \ldots, n$).

(d) $a_n/b_n \to \zeta(3)$ as $n \to \infty$; indeed, the convergence is so fast as to prove that $\zeta(3)$ cannot be rational. To be precise, for all integers p, q with q sufficiently large relative to $\epsilon > 0$,

$$\left|\zeta(3) - \frac{p}{q}\right| > \frac{1}{q^{\theta+\epsilon}} \quad \text{and} \quad \theta = 13.41782\cdots,$$

which proves the irrationality of $\zeta(3)$.

(*Cf.* van der Poorten [1180, p. 195]; see also Apéry [56])

27. Prove the following inequalities: For $n, p \in \mathbb{N}$,

$$0 \leq \left(p + \frac{1}{2}\right) \sum_{k=1}^{n} \frac{1}{k^{2p}} - \sum_{j=1}^{p-1} \left(\sum_{k=1}^{n} \frac{1}{k^{2p-2j}}\right) \left(\sum_{k=1}^{n} \frac{1}{k^{2j}}\right) < \frac{8}{n}$$

and

$$\left|\sum_{k=1}^{n} \frac{1}{(2k-1)^2} - 2\left(\sum_{k=1}^{n} \frac{(-1)^{k+1}}{2k-1}\right)^2\right| < \frac{6}{n}.$$

(Hovstad [571, p. 93])

28. Show that the first inequality of Problem 27 implies that

$$\sum_{k=1}^{\infty} \frac{1}{k^{2p}} = \mathcal{R}_p \left(\sum_{k=1}^{\infty} \frac{1}{k^2}\right)^p \quad (p \in \mathbb{N}),$$

where $\mathcal{R}_p$ is a rational number, satisfying the recurrence formula:

$$\mathcal{R}_1 = 1 \quad \text{and} \quad \left(p + \frac{1}{2}\right) \mathcal{R}_p = \sum_{j=1}^{p-1} \mathcal{R}_j \mathcal{R}_{p-j} \quad (p \in \mathbb{N} \setminus \{1\}).$$

(Hovstad [571, p. 93])

29. Prove that $\zeta(s)$ has the factorization

$$\zeta(s) = \frac{e^{bs}}{2(s-1)\Gamma\left(1+\frac{1}{2}s\right)} \prod_{\rho} \left[\left(1-\frac{s}{\rho}\right)e^{\frac{s}{\rho}}\right],$$

where

$$b = -1 - \frac{\gamma}{2} + \log(2\pi);$$

γ is the Euler-Mascheroni constant, and the product is taken over all the so-called non-trivial zeros of $\zeta(s)$.

(*Cf.* Titchmarsh [1151, pp. 30–31]; see also Melzak [821, p. 111])

30. Prove that

$$\zeta\left(\frac{1}{2}, a+1\right) = \sum_{n=0}^{\infty} \frac{(2n)!}{2^{2n}(n!)^2} \zeta\left(n+\frac{1}{2}\right)(-a)^n \quad (|a|<1)$$

and

$$\zeta\left(\frac{1}{2}, na\right) = n^{-\frac{1}{2}} \sum_{k=0}^{n-1} \zeta\left(\frac{1}{2}, a+\frac{k}{n}\right) \quad (n \in \mathbb{N}).$$

(Powell [911, p. 117])

31. Let

$$(s-1)\zeta(s,a) = 1 + \sum_{n=0}^{\infty} \gamma_n(a)(s-1)^{n+1} \quad (0 < a \leq 1;\; n \in \mathbb{N}_0).$$

Show that

$$\gamma_n(a) = \frac{(-1)^n}{n!} \lim_{m\to\infty} \left(\sum_{k=0}^{m} \frac{\{\log(k+a)\}^n}{k+a} - \frac{\{\log(m+a)\}^{n+1}}{n+1}\right),$$

which, for $a=1$, gives the coefficients in the Laurent expansion of $\zeta(s)$ about $s=1$ (*cf.* Equation 2.3(7)).

(Berndt [118, p. 152])

32. Prove that

$$\limsup_{t\to\infty} \frac{|\zeta(1+it)|}{\log(\log t)} \geq e^{\gamma},$$

where γ is the Euler-Mascheroni constant defined by 1.1(3).

(Titchmarsh [1150, p. 79])

33. Prove that

$$\mathrm{Li}_m(e^z) = \sum_{\substack{n=0 \\ n\neq m-1}}^{\infty} \zeta(m-n)\frac{z^n}{n!} + \left(1+\frac{1}{2}+\cdots+\frac{1}{m-1} - \log(-z)\right)\frac{z^{m-1}}{(m-1)!} \quad (m\in\mathbb{N};\; |z|<2\pi).$$

(Cohen *et al.* [336, p. 26])

34. Prove that

$$\zeta(5) = \frac{5}{2}\sum_{n=1}^{\infty}\left(\sum_{k=1}^{n-1}\frac{1}{k^2} - \frac{4}{5}\frac{1}{n^2}\right)\frac{(-1)^n}{n^3\binom{2n}{n}}.$$

(van der Poorten [1181, p. 274])

35. Let K be a normal extension field of degree n over the rational number field $\mathbb{Q}$. Denote by O_K the integer ring of K. Let $I(K)$ be the set of all nonzero ideals of O_K, $N_{\mathfrak{a}}$ the absolute norm of an ideal $\mathfrak{a} \in I(K)$ and $\mathrm{Tr}\,\alpha$ the trace of $\alpha \in K$ over $\mathbb{Q}$. We assume that $[K : \mathbb{Q}] = n > 1$, let

$$O_{K,0} = \{\alpha \in O_K \mid \mathrm{Tr}\,\alpha = 0\};$$
$$TK = \min\{\mathrm{Tr}\,\alpha \mid \alpha \in O_K,\ \mathrm{Tr}\,\alpha > 0\}.$$

For $\mathfrak{a} \in I(K)$ and $\mathfrak{a}_0 = O_{K,0} \cap \mathfrak{a}$, define

$$T(\mathfrak{a}) = \mathrm{Min}\left\{\frac{\mathrm{Tr}\alpha}{T(K)} \mid \alpha \in \mathfrak{a},\ \mathrm{Tr}\alpha > 0\right\};$$
$$N_0(\mathfrak{a}) = \#\{O_{K,0}/\mathfrak{a}_0\},$$

where $O_{K,0}/\mathfrak{a}_0$ is the quotient of $\mathbb{Z}$-module $O_{K,0}$ by submodule $\mathfrak{a}_0$. For any $x \in \mathbb{R}^+$, let

$$j_K(x) = \#\{\mathfrak{a} \in I(K) \mid N\mathfrak{a} \le x\};$$
$$j_{K,0}(x) = \#\{\mathfrak{a} \in I(K) \mid N_0(\mathfrak{a}) \le x\}$$
$$= \#\left\{\mathfrak{a} \in I(K) \mid \frac{N\mathfrak{a}}{T(\mathfrak{a})} \le x\right\},$$

and $q \equiv 1 \pmod{n}$. Suppose that K is the subfield of the qth cyclotomic field C_q, such that $K/\mathbb{Q}$ is a tamely ramified cyclic extension of degree n. Show that

$$\zeta(2k+1) = \frac{2q^n - q^{n-1} - q}{q^n - 1}\,\zeta(2k)\lim_{x\to+\infty}\frac{j_k(x)}{j_{K,0}(x)}.$$

(Lan [728, p. 273])

36. Prove that

$$\sum_{n=1}^{\infty}\frac{(H_n)^3}{n^5} - \frac{11}{4}\sum_{n=1}^{\infty}\frac{H_n^{(2)}}{n^6} = \frac{469}{32}\zeta(8) - 16\,\zeta(3)\,\zeta(5) + \frac{3}{2}\zeta(2)\,\{\zeta(3)\}^2$$

and

$$\sum_{n=1}^{\infty}\frac{(H_n)^3}{n^7} - \frac{13}{4}\sum_{n=1}^{\infty}\frac{H_n^{(2)}}{n^8} = \frac{561}{20}\zeta(10) - \frac{47}{4}\{\zeta(5)\}^2$$
$$- \frac{49}{2}\zeta(7)\,\zeta(3) + 3\,\zeta(2)\,\zeta(3)\,\zeta(5) + \frac{15}{4}\{\zeta(3)\}^2\,\zeta(4).$$

(Flajolet and Salvy [454, p. 27])

37. Prove that, for an odd weight $m = p + q$,

$$\sum_{n=1}^{\infty} \frac{H^{(p)}(n)}{n^q} = \zeta(m)\left(\frac{1}{2} - \frac{(-1)^p}{2}\binom{m-1}{p} - \frac{(-1)^p}{2}\binom{m-1}{q}\right)$$
$$+ \frac{1-(-1)^p}{2}\,\zeta(p)\,\zeta(q) + (-1)^p \sum_{k=1}^{[p/2]} \binom{m-2k-1}{q-1}\zeta(2k)\,\zeta(m-2k)$$
$$+ (-1)^p \sum_{k=1}^{[q/2]} \binom{m-2k-1}{p-1}\zeta(2k)\,\zeta(m-2k),$$

where $\zeta(1)$ should be interpreted as 0 wherever it occurs.

(*Cf.* Borwein *et al.* [146, p. 278]; see also Flajolet and Salvy [454, p. 22])

38. Prove that

$$\sum_{n=1}^{\infty} \sum_{k=1}^{2n-1} \frac{(-1)^{n+k}}{n^2\,k} = \pi\,G - \frac{27}{16}\zeta(3)$$

and

$$\sum_{n=1}^{\infty} \sum_{k=1}^{2n-1} \frac{(-1)^{n-1}}{n^2\,k} = \pi\,G - \frac{29}{16}\zeta(3),$$

where G denotes the Catalan constant defined by 1.3(16).

(Sitaramachandrarao [1034, p. 13])

39. Prove that, for $m \in \mathbb{N}$ and $\Re(a) > 0$,

$$\zeta'(-m,a) = \left.\frac{\partial}{\partial z}\zeta(z,a)\right|_{z=-m}$$
$$= \frac{1}{m+1}\,a^{m+1}\log a - \frac{1}{(m+1)^2}\,a^{m+1} - \frac{1}{2}\,a^m \log a$$
$$+ \frac{m}{12}\,a^{m-1}\log a + \frac{1}{12}\,a^{m-1} + \sum_{k=1}^{\infty} \alpha_{2k}\,a^{-(2k-m+1)},$$

where

$$\alpha_{2k} := \begin{cases} \dfrac{B_{2k+2}}{2k+2}\left[\dbinom{m}{2k+1}\log a + \displaystyle\sum_{j=0}^{2k}\binom{m}{j}\frac{(-1)^j}{2k-j+1}\right] & (2k \leqq m-1), \\[2ex] \dfrac{B_{2k+2}}{2k+2}\displaystyle\sum_{j=0}^{m}\binom{m}{j}\frac{(-1)^j}{2k-j+1} & (2k \geqq m). \end{cases}$$

(Elizalde [410, p. 349])

40. Define constants Φ_k, for $k > 1$ an odd integer, by

$$\Phi_k = -\frac{2}{\pi}\sum_{\substack{d=1\\ d\,\text{odd}}}^{k-2} (-1)^{\frac{d-1}{2}}\,\frac{\pi^d}{d!}\,\zeta(k-d+1).$$

Show that, for real $r \geq 1$ and $s > 1$,

$$\zeta(r,s) = -\frac{1}{2}\zeta(r+s) + \sum_{\substack{k=3 \\ k\,\text{odd}}}^{\infty} \Phi_k \sum_{\substack{j=0 \\ j\,\text{even}}}^{k-1} \binom{k}{j} \eta(r-j)\,\eta(s-k+j),$$

where the Eta function is defined by

$$\eta(s) := \left(1 - 2^{1-s}\right)\zeta(s).$$

(Crandall and Buhler [344, p. 279])

41. Let $f(s)$ be a function defined by the Dirichlet series as follows:

$$f(s) := \sum_{n=2}^{\infty} n^{-s} \sum_{k<n} k^{-1} \qquad (\Re(s) = \sigma > 1).$$

Show that $f(s)$ is analytic in the whole complex s-plane, except at simple poles $s = 0$ and $s = 1 - 2a$ $(a \in \mathbb{N})$ with residues given by

$$\operatorname*{Res}_{s=0} f(s) = -\frac{1}{2}$$

and

$$\operatorname*{Res}_{s=1-2a} f(s) = -\frac{B_{2a}}{2a} \qquad (a \in \mathbb{N}),$$

and a pole of order 2 at $s = 1$ with residue given by

$$\operatorname*{Res}_{s=1} f(s) = \gamma,$$

where B_{2a} are the Bernoulli numbers, defined by 1.6(2), and γ is the Euler-Mascheroni constant, defined by 1.1(3).

(Matsuoka [813, p. 399])

42. By using the definitions and restrictions as in Problem 20 of Chapter 2 with an *additional* condition $c > 0$, show that

$$Z(s) = \zeta(2s)\,a^{-s} + \frac{2^{2s-1}\,a^{s-1}\,\sqrt{\pi}}{\Gamma(s)\,(-d)^{s-\frac{1}{2}}}\zeta(2s-1)\,\Gamma\left(s - \frac{1}{2}\right) + Q(s),$$

where

$$Q(s) = \frac{2\pi^{s}\cdot 2^{s-\frac{1}{2}}}{\sqrt{a}\,\Gamma(s)\cdot(-d)^{\frac{s}{2}-\frac{1}{4}}} \sum_{n=1}^{\infty} n^{s-\frac{1}{2}}\,\sigma_{1-2s}(n)\cos\left(\frac{n\pi b}{a}\right)$$
$$\cdot \int_0^{\infty} u^{s-\frac{3}{2}} \exp\left(-\frac{\pi n\sqrt{-d}}{2a}\left(u + u^{-1}\right)\right) du.$$

(Selberg and Chowla [1017, p. 87])

43. Let $r, s \in \mathbb{N}_0$ with $r > s$. Show that

$$\int_0^1 \int_0^1 \frac{x^r y^s}{1-xy}\,dxdy$$

is a rational number whose denominator is a divisor of d_r^2, where d_r denotes the lowest common multiple of $1, 2, \ldots, r$.

(Beukers [126, p. 268])

44. Prove that

$$\zeta(3) = \frac{1}{4}\sum_{n=1}^{\infty}(-1)^{n-1}\frac{56n^2-32n+5}{(2n-1)^2}\frac{1}{\binom{3n}{n}\binom{2n}{n}n^3}$$

and

$$\zeta(3) = \sum_{n=0}^{\infty}\frac{(-1)^n}{72\binom{4n}{n}\binom{3n}{n}}\frac{5265n^4+13878n^3+13761n^2+6120n++1040}{(4n+1)(4n+3)(n+1)(3n+1)^2(3n+2)^2}.$$

(Amdeberhan [34, p. 2])

45. Prove that

$$\sum_{n=1}^{\infty}\frac{(-1)^n}{n^2}\sum_{k=1}^{n}\frac{1}{k^2} = -\frac{1}{6}\left[24\,\mathrm{Li}_4\left(\frac{1}{2}\right)+21\,\zeta(3)\log 2+(\log 2)^4\right] + \frac{\pi^2}{6}(\log 2)^2+\frac{17}{480}\pi^4,$$

where $\mathrm{Li}_4(z)$ denotes the Tetralogarithm (see Section 2.4).

(Daudé *et al.* [367, p. 421])

46. Prove that

$$\sum_{n=1}^{\infty}(-1)^n\frac{(\log(2n+1))^2}{2n+1} = \frac{\pi}{4}\left\{2\left[\zeta''\left(0,\tfrac{1}{4}\right)-\zeta''\left(0,\tfrac{3}{4}\right)\right] -4[\gamma+\log(2\pi)]\log\left(\frac{\Gamma\left(\frac{1}{4}\right)}{\Gamma\left(\frac{3}{4}\right)}\right)+[\gamma+\log(2\pi)]^2+\frac{\pi^2}{12}\right\}$$

and

$$\int_0^{\infty}\frac{(\log t)^2}{\cosh t}dt = \frac{\pi}{2}\left\{2\left[\zeta''\left(0,\tfrac{1}{4}\right)-\zeta''\left(0,\tfrac{3}{4}\right)\right] -4\log(2\pi)\log\left(\frac{\Gamma\left(\frac{1}{4}\right)}{\Gamma\left(\frac{3}{4}\right)}\right)+[\log(2\pi)]^2+\frac{1}{4}\pi^2\right\}.$$

(Shail [1021])

47. For an integer $a \geqq -1$, prove the following asymptotic expansion:

$$\begin{aligned} G_a(x) &:= \sum_{n \geqq a+1}^{\infty} \zeta(n-a) \frac{(-x)^n}{n!} \\ &= -\frac{(-x)^{a+1}}{(a+1)!} \{\log x - \psi(a+2) + \gamma\} \\ &\quad + \sum_{n=0}^{a} (-x)^n \left(\sum_{n=0}^{n} \frac{(a-k)!}{(a+1)!\,(n-k)!} - \frac{1}{n!\,(a+1-n)} \right) + O\left(\frac{1}{x}\right) \quad (x \to \infty). \end{aligned}$$

(Buschman and Srivastava [195, p. 296])

48. For any multi-index $\mathbf{k} = (k_1, k_2, \ldots, k_r))$ $(k_i \in \mathbb{N})$, the *weight* $\mathrm{wt}(\mathbf{k})$ and *depth* $\mathrm{dep}(\mathbf{k})$ of $\mathbf{k}$ are defined by $|\mathbf{k}| = k_1 + k_2 + \cdots + k_r$ and r, respectively. The height of the index $\mathbf{k}$ is also defined by $\mathrm{ht}\,\mathbf{k} = \#\{j \mid k_j \geqq 2\}$.

Denote, by $I(k, r)$, the set of multi-indices $\mathbf{k}$ of weight k and depth r and, by $I_0(k, r)$, the subset of $I(k, r)$ with *admissible* indices, that is indices with the additional requirement that $k_1 \geqq 2$. For $(k_1, \ldots, k_r) \in I_0(k, r)$, the multiple zeta value (MZV) and the non-strict multiple zeta value (MZSV) can often be defined, respectively, as follows:

$$\zeta(k_1, k_2, \ldots, k_r) := \sum_{n_1 > \cdots > n_r > 0} \frac{1}{n_1^{k_1} \cdots n_r^{k_r}}$$

and

$$\zeta^*(k_1, k_2, \ldots, k_r) := \sum_{n_1 \geqq \cdots \geqq n_r \geqq 1} \frac{1}{n_1^{k_1} \cdots n_r^{k_r}}.$$

Prove the following formulas:

(a) Sum Formula. For $r < k$ $(r, k \in \mathbb{N})$, there hold

$$\sum_{\mathbf{k} \in I_0(k, r)} \zeta(\mathbf{k}) = \zeta(k) \quad \text{and} \quad \sum_{\mathbf{k} \in I_0(k, r)} \zeta^*(\mathbf{k}) = \binom{k-1}{r-1} \zeta(k).$$

(b) Cyclic Sum Formula. For $(k_1, \ldots, k_r) \in I_0(k, r)$,

$$\sum_{i=1}^{r} \sum_{j=0}^{k_i - 2} \zeta^*(k_i - j, k_{i+1}, \ldots, k_r, k_1, \ldots, k_{i-1}, j+1) = k\,\zeta(k+1),$$

where the empty sum means zero.

(Ohno and Okuda [873, p. 3030])

49. Kamano [624] investigated the following multiple zeta function:

$$\zeta_n(s_1, \ldots, s_n; a) = \sum_{0 \leqq m_1 < \cdots < m_n} \frac{1}{(m_1 + a)^{s_1} \cdots (m_n + a)^{s_n}} \tag{a}$$

$$\left(a > 0;\ (m_1, \ldots, m_n) \in \mathbb{Z}^n;\ (s_1, \ldots, s_n) \in \mathbb{C}^n\right),$$

where $\mathbb{Z}$ denotes the set of integers. The special case $n = 1$ of $\zeta_n(s_1, \ldots, s_n; a)$ in (a) reduces to the Hurwitz (or generalized) zeta function $\zeta(s, a)$. Also $\zeta_n(s_1, \ldots, s_n; 1)$

becomes the Euler-Zagier multiple zeta function denoted by $\zeta_n(s_1, \ldots, s_n)$. Matsumoto [806] proved the analytic continuation of a more general class of multiple zeta functions, including (a) as a special case. Kamano [624] presented three kinds of limiting values of $\zeta_n(s_1, \ldots, s_n; a)$ in (a) at nonpositive integers.

Show that

$$\zeta_n'(0; a) = \frac{(-1)^{n-1}}{(n-1)!} \prod_{k=1}^{n-1} \left(k + a - \frac{1}{2}\right) \log \frac{\Gamma(a)}{\sqrt{2\pi}} \qquad (n \in \mathbb{N}),$$

where $\zeta_n'(s; a) = \frac{\partial}{\partial s} \zeta_n(s; a)$, $\zeta_n(s; a) := \zeta_n(s, \ldots, s; a)$ in (a) and an empty sum is understood to be nil.

(Kamano [624, Theorem 3])

50. Show that

$$z\gamma(z) = \sum_{k=2}^{\infty} (-1)^k \frac{\mathrm{Li}(z)}{k} \qquad (|z| \leqq 1),$$

where $\gamma(z)$ denotes the generalized-Euler-constant function, defined by

$$\gamma(z) = \sum_{n=1}^{\infty} z^{n-1} \left(\frac{1}{n} - \log \frac{n+1}{n}\right) \qquad (|z| \leqq 1)$$

$$= \int_0^1 \int_0^1 \frac{1-x}{(1-xyz)(-\log xy)} dxdy \qquad (\mathbb{C} \setminus [1, \infty)).$$

(Sondow and Hadjicostas [1053, Theorem 1])

51. Let $\{\alpha_n\}_{n \in \mathbb{N}_0}$ be a positive sequence, such that the following infinite series:

$$\sum_{n=0}^{\infty} \mathrm{e}^{-\alpha_n t}$$

converges for any $t \in \mathbb{R}^+$. Then, for the generalized Hurwitz–Lerch Zeta function

$$\Phi_{\lambda_1, \ldots, \lambda_p; \mu_1, \ldots, \mu_q}^{(\rho_1, \ldots, \rho_p, \sigma_1, \ldots, \sigma_q)}(z, s, a),$$

defined by 2.6(44), show that

$$\Phi_{\lambda_1, \ldots, \lambda_p; \mu_1, \ldots, \mu_q}^{(\rho_1, \ldots, \rho_p, \sigma_1, \ldots, \sigma_q)}(z, s, a) = \frac{1}{\Gamma(s)} \sum_{n=0}^{\infty} \int_0^{\infty} t^{s-1}\, \mathrm{e}^{-(a-\alpha_0+\alpha_n)t} \left(1 - \mathrm{e}^{-(\alpha_{n+1}-\alpha_n)t}\right)$$

$$\cdot {}_p\Psi_q^* \left[\begin{matrix} (\lambda_1, \rho_1), \ldots, (\lambda_p, \rho_p); \\ (\mu_1, \sigma_1), \ldots, (\mu_q, \sigma_q); \end{matrix} \; z\mathrm{e}^{-t} \right] \mathrm{d}t \qquad \left(\min\{\Re(a), \Re(s)\} > 0\right),$$

provided that each member exists, ${}_p\Psi_q^*$ being the Fox-Wright hypergeometric function defined in Problem 64 (Chapter 1).

(Srivastava *et al.* [1107])

52. For the generalized Hurwitz–Lerch Zeta function

$$\Phi_{\lambda_1,\ldots,\lambda_p;\mu_1,\ldots,\mu_q}^{(\rho_1,\ldots,\rho_p,\sigma_1,\ldots,\sigma_q)}(z,s,a),$$

defined by 2.6(44), derive the following extension of the fractional derivative formulas, such as the one given by 2.6(32):

$$\mathcal{D}_z^{\nu-\tau}\left\{z^{\nu-1}\,\Phi_{\lambda_1,\ldots,\lambda_p;\mu_1,\ldots,\mu_q}^{(\rho_1,\ldots,\rho_p,\sigma_1,\ldots,\sigma_q)}(z^{\kappa},s,a)\right\}$$
$$=\frac{\Gamma(\nu)}{\Gamma(\tau)}\,z^{\tau-1}\,\Phi_{\lambda_1,\ldots,\lambda_p,\nu;\mu_1,\ldots,\mu_q,\tau}^{(\rho_1,\ldots,\rho_p,\kappa,\sigma_1,\ldots,\sigma_q,\kappa)}(z^{\kappa},s,a)\qquad\big(\Re(\nu)>0;\ \kappa>0\big).$$

(Srivastava *et al.* [1107])

53. Show that each of the following integral representations holds true:

$$\Phi_{\lambda_1,\ldots,\lambda_p;\mu_1,\ldots,\mu_q}^{(\rho_1,\ldots,\rho_p,\sigma_1,\ldots,\sigma_q)}(z,s,a)=\frac{1}{\Gamma(s)}\int_0^\infty t^{s-1}\,\mathrm{e}^{-at}\,{}_p\Psi_q^*\left[\begin{matrix}(\lambda_1,\rho_1),\ldots,(\lambda_p,\rho_p);\\ \\(\mu_1,\sigma_1),\ldots,(\mu_q,\sigma_q);\end{matrix}\ z\mathrm{e}^{-t}\right]dt$$
$$\big(\min\{\Re(a),\Re(s)\}>0\big)$$

and

$$\Phi_{\lambda_1,\ldots,\lambda_p;\mu_1,\ldots,\mu_q}^{(\rho_1,\ldots,\rho_p,\sigma_1,\ldots,\sigma_q)}(z,s,a)=\frac{\prod_{j=1}^{q}\Gamma(\mu_j)}{\prod_{j=1}^{p}\Gamma(\lambda_j)}$$
$$\cdot\frac{1}{2\pi\mathrm{i}}\int_{\mathcal{L}}\frac{\Gamma(-\xi)\,\{\Gamma(\xi+a)\}^s\prod_{j=1}^{p}\Gamma(\lambda_j+\rho_j\xi)}{\{\Gamma(\xi+a+1)\}^s\prod_{j=1}^{q}\Gamma(\mu_j+\sigma_j\xi)}(-z)^{\xi}\,d\xi\qquad\big(|\arg(-z)|<\pi\big),$$

where the path of integration $\mathcal{L}$ is a Mellin-Barnes type contour in the complex ξ-plane, which starts at the point $-\mathrm{i}\infty$ and terminates at the point $\mathrm{i}\infty$ $\big(\mathrm{i}:=\sqrt{-1}\big)$ with indentations, if necessary, in such a manner as to separate the poles of $\Gamma(-\xi)$ from the poles of $\Gamma\left(\lambda_j+\rho_j\xi\right)$ $(j=1,\ldots,p)$.

(Srivastava *et al.* [1107])

54. For a given sequence $\{a_n\}_{n\in\mathbb{N}_0}$, let the operator Δ be defined by

$$\Delta^0 a_n:=a_n\quad\text{and}\quad\Delta^j a_n:=\Delta^{j-1}a_n-\Delta^{j-1}a_{n+1}=\sum_{k=0}^{j}(-1)^k\binom{j}{k}a_{n+k}\qquad(j\in\mathbb{N}).$$

Then derive the following analytic continuation formula for the Riemann Zeta function $\zeta(s)$ for all $s\in\mathbb{C}\setminus\{1\}$:

$$\zeta(s)=\frac{1}{1-2^{1-s}}\sum_{j=0}^{\infty}\frac{\Delta^j 1^{-s}}{2^{j+1}}.$$

Also, deduce the following special case, which was used in Hardy's celebrated proof of the Riemann functional equation 2.3(11):

$$\zeta(s) = \frac{1}{1-2^{1-s}}\left[\frac{1}{2}+\frac{1}{2}\sum_{n=1}^{\infty}(-1)^{n-1}\left(\frac{1}{n^s}-\frac{1}{(n+1)^s}\right)\right] \quad \left(\Re(s) > -1;\ s \neq 1\right).$$

(*Cf.* Sondow [1048])

55. For the *extended Fermi-Dirac function* $\Theta_\nu(s;x)$, defined in terms of the *Weyl fractional integral operator* $\mathcal{W}_{x+}^{s}$ by

$$\Theta_\nu(s;x) := \mathcal{W}_{x+}^{s}[\vartheta(t;\nu)] = \frac{1}{\Gamma(s)}\int_0^\infty t^{s-1}\,\vartheta(x+t;\nu)dt$$

$$= \frac{1}{\Gamma(s)}\int_x^\infty (t-x)^{s-1}\,\vartheta(t;\nu)dt$$

$$\left(\vartheta(t;\nu) := \frac{e^{-\nu t}}{e^t+1};\ \Re(s) > 0;\ x \geqq 0;\ \Re(\nu) > -1\right),$$

show that

$$\Theta_\nu(\mu+\nu;x) = \frac{1}{\Gamma(\mu)}\int_0^\infty t^{\mu-1}\,\Theta_\nu(\nu;x+t)dt = \frac{1}{\Gamma(\nu)}\int_0^\infty t^{\nu-1}\,\Theta_\nu(\mu;x+t)dt$$

$$\left(\min\{\Re(\mu),\Re(\nu)\} > 0;\ \Re(\nu) > -1\right)$$

and

$$\Theta_\nu(s;x) = \sum_{n=0}^{\infty}\Theta_\nu(s-n;0)\,\frac{(-x)^n}{n!} \qquad \left(s \neq n+1\ \ (n \in \mathbb{N}_0);\ x \geqq 0;\ \Re(\nu) > -1\right),$$

provided that each member of these assertions exist.

(See, for details, Srivastava *et al.* [1093])

56. For the *extended Bose-Einstein function* $\Theta_\nu(s;x)$, defined in terms of the *Weyl fractional integral operator* $\mathcal{W}_{x+}^{s}$, by

$$\Psi_\nu(s;x) := \mathcal{W}_{x+}^{s}[\psi(t;\nu)] = \frac{1}{\Gamma(s)}\int_0^\infty t^{s-1}\,\psi(x+t;\nu)dt$$

$$= \frac{1}{\Gamma(s)}\int_x^\infty (t-x)^{s-1}\,\psi(t;\nu)dt$$

$$\left(\psi(t;\nu) := \frac{e^{-\nu t}}{e^t-1};\ \Re(s) > 1;\ x \geqq 0;\ \Re(\nu) > -1\right),$$

show that

$$\Psi_\nu(s;x) = q^{-s}\sum_{j=1}^{q}\Psi_{\frac{\nu+j-q}{q}}(s;qx) \qquad \left(\Re(s)>1;\ x\geqq 0;\ \Re(\nu)>-1\right).$$

(See, for details, Srivastava *et al.* [1093])

57. For the *extended Fermi-Dirac function* $\Theta_\nu(s;x)$ and *extended Bose-Einstein function* $\Theta_\nu(s;x)$ (see Problems 55 and 56 above), derive each of the following identities:

$$\Theta_{2\nu}(s;x) = \Psi_{2\nu}(s;x) - 2^{1-s}\,\Psi_\nu(s,2x) \qquad \left(\Re(s)>1;\ x\geqq 0;\ \Re(\nu)>-1\right),$$

$$\Theta_{\nu+1}(s,x) = 2^{-s}\left[\Psi_{\frac{\nu}{2}}(s;2x) - \Psi_{\frac{\nu+1}{2}}(s;2x)\right] \qquad \left(\Re(s)>1;\ x\geqq 0;\ \Re(\nu)>-1\right)$$

and

$$\Theta_{\nu+1}(s;x) + \Theta_\nu(s;x) = (\nu+1)^{-s}\,e^{-(\nu+1)x} \qquad \left(\Re(s)>0;\ x\geqq 0;\ \Re(\nu)>-1\right).$$

(See, for details, Srivastava *et al.* [1093])

58. For the Apostol-Genocchi polynomials $\mathcal{G}_n^{(\alpha)}(x;\lambda)$ $(\lambda\in\mathbb{C})$, defined by 1.8(58), derive each of the following explicit series representations:

$$\mathcal{G}_n\left(\frac{p}{q};e^{2\pi i\xi}\right) = \frac{2\cdot n!}{(2q\pi)^n}\left\{\sum_{j=1}^{q}\zeta\left(n,\frac{2\xi+2j-1}{2q}\right)\exp\left[\left(\frac{n}{2}-\frac{(2\xi+2j-1)p}{q}\right)\pi i\right]\right.$$

$$\left.+\sum_{j=1}^{q}\zeta\left(n,\frac{2j-2\xi-1}{2q}\right)\exp\left[\left(-\frac{n}{2}+\frac{(2j-2\xi-1)p}{q}\right)\pi i\right]\right\}$$

$$\left(n\in\mathbb{N}\setminus\{1\};\ p\in\mathbb{Z};\ q\in\mathbb{N};\ \xi\in\mathbb{R}\right),$$

$$\mathcal{G}_n^{(\alpha)}\left(\frac{p}{q};e^{2\pi i\xi}\right) = \frac{2n}{e^{2\pi i\xi}+1}\,\mathcal{G}_{n-1}^{(\alpha-1)}\left(e^{2\pi i\xi}\right) + \sum_{k=2}^{n}\frac{2\cdot k!}{(2q\pi)^k}\binom{n}{k}\mathcal{G}_{n-k}^{(\alpha-1)}(e^{2\pi i\xi})$$

$$\cdot\left\{\sum_{j=1}^{q}\zeta\left(k,\frac{2\xi+2j-1}{2q}\right)\exp\left[\left(\frac{k}{2}-\frac{(2\xi+2j-1)p}{q}\right)\pi i\right]\right.$$

$$\left.+\sum_{j=1}^{q}\zeta\left(k,\frac{2j-2\xi-1}{2q}\right)\exp\left[\left(-\frac{k}{2}+\frac{(2j-2\xi-1)p}{q}\right)\pi i\right]\right\}$$

$$\left(n\in\mathbb{N}\setminus\{1\};\ p\in\mathbb{Z};\ q\in\mathbb{N};\ \xi\in\mathbb{R}\setminus\Lambda\ \left(\Lambda:=\left\{k+\frac{1}{2}:k\in\mathbb{Z}\right\}\right);\ \alpha\in\mathbb{C}\right)$$

and

$$\mathcal{G}_n^{(\alpha)}\left(\frac{p}{q}; e^{2\pi i}\right) = -\frac{n\,(-2)^{\alpha}}{e^{2\pi i\xi}+1}\mathcal{B}_{n-1}^{(\alpha-1)}\left(-e^{2\pi i\xi}\right) - \sum_{k=2}^{n}\frac{k!}{(2q\pi)^k}\binom{n}{k}\mathcal{B}_{n-k}^{(\alpha-1)}\left(-e^{2\pi i\xi}\right)$$

$$\cdot\left\{\sum_{j=1}^{q}\zeta\left(k, \frac{2\xi+2j-1}{2q}\right)\exp\left[\left(\frac{k}{2}-\frac{(2\xi+2j-1)p}{q}\right)\pi i\right]\right.$$

$$\left.+\sum_{j=1}^{q}\zeta\left(k, \frac{2j-2\xi-1}{2q}\right)\exp\left[\left(-\frac{k}{2}+\frac{(2j-2\xi-1)p}{q}\right)\pi i\right]\right\}$$

$$\left(n\in\mathbb{N}\setminus\{1\};\ p\in\mathbb{Z};\ q\in\mathbb{N};\ \xi\in\mathbb{R}\setminus\Lambda\ \left(\Lambda := \left\{k+\frac{1}{2} : k\in\mathbb{Z}\right\}\right);\ \alpha\in\mathbb{C}\right)$$

in terms of the Hurwitz (or generalized) Zeta function $\zeta(s,a)$.

(See Luo and Srivastava [791, p. 5713, Theorem 4; p. 5715, Theorems 6 and 7])

3 Series Involving Zeta Functions

The main purpose of this chapter is to present a rather extensive collection of closed-form sums of series involving the Zeta functions. Many of these summation formulas will find their applications in Chapter 5 (especially Section 5.3) in the evaluations of the determinants of the Laplacians for the n-dimensional sphere $\mathbf{S}^n$ with the standard metric.

We begin this chapter by presenting an interesting historical introduction to the remarkably widely investigated subject of closed-form summation of series involving the Zeta functions. Among the various methods and techniques used in the vast literature on the subject, we give here reasonably detailed accounts of those using the binomial theorem, generating functions, multiple Gamma functions and hypergeometric identities. The last section (Section 3.6) deals with other methods, based (for example) on the Weierstrass canonical product form for the Gamma function and higher-order derivatives of the Gamma function.

3.1 Historical Introduction

A rather classical (over two centuries old) theorem of Christian Goldbach (1690–1764), which was stated in a letter dated 1729 from Goldbach to Daniel Bernoulli (1700–1782), was revived in 1986 by Shallit and Zikan [1022] as the following problem:

$$\sum_{\omega \in \mathcal{S}} (\omega - 1)^{-1} = 1, \tag{1}$$

where $\mathcal{S}$ denotes the set of all nontrivial integer kth powers, that is,

$$\mathcal{S} := \left\{ n^k \mid n, k \in \mathbb{N} \setminus \{1\} \right\}. \tag{2}$$

In terms of the Riemann Zeta function $\zeta(s)$, defined by 2.3(1), Goldbach's theorem (1) assumes the elegant form (*cf.* Shallit and Zikan [1022, p. 403]):

$$\sum_{k=2}^{\infty} \{\zeta(k) - 1\} = 1 \tag{3}$$

Zeta and q-Zeta Functions and Associated Series and Integrals. DOI: 10.1016/B978-0-12-385218-2.00003-7

or, equivalently,

$$\sum_{k=2}^{\infty} \mathcal{F}(\zeta(k)) = 1, \tag{4}$$

where $\mathcal{F}(x) := x - [x]$ denotes the *fractional* part of $x \in \mathbb{R}$. As a matter of fact, it is fairly straightforward to observe also that

$$\sum_{k=2}^{\infty} (-1)^k \mathcal{F}(\zeta(k)) = \frac{1}{2}, \tag{5}$$

$$\sum_{k=1}^{\infty} \mathcal{F}(\zeta(2k)) = \frac{3}{4} \quad \text{and} \quad \sum_{k=1}^{\infty} \mathcal{F}(\zeta(2k+1)) = \frac{1}{4}. \tag{6}$$

Another remarkable result involving the Riemann's ζ-function is the following series representation for $\zeta(3)$:

$$\zeta(3) = \frac{\pi^2}{7} \left\{ 1 - 4 \sum_{k=1}^{\infty} \frac{\zeta(2k)}{(2k+1)(2k+2)\,2^{2k}} \right\} \tag{7}$$

or, equivalently,

$$\zeta(3) = -\frac{4\pi^2}{7} \sum_{k=0}^{\infty} \frac{\zeta(2k)}{(2k+1)(2k+2)\,2^{2k}}, \tag{8}$$

since $\zeta(0) = -\frac{1}{2}$ (see Equation 2.3(10)). The series representation (7) is contained in a 1772 paper by Leonhard Euler (1707–1783) (see, e.g., Ayoub [81, pp. 1084–1085]). It was rediscovered by Ramaswami [966] and (more recently) by Ewell [436]. (See also Srivastava [1072, p. 7, Equation (2.23)]), where Euler's result (7) was reproduced *actually* from the work of Ramaswami [966]). Numerous further series representations for $\zeta(3)$, which are analogous to (7) or (8), can be found in the works of Wilton [1233], Zhang and Williams [1250, 1251], Cvijović and Klinowski [351], Chen and Srivastava [252], Srivastava [1082–1084, 1086], Srivastava and Tsumura [1111, 1112] and others (*cf.*, e.g., Tsumura [1169, p. 384] and Ewell [438, p. 1004]); see also Berndt [122]).

A considerably large variety of methods were used in the aforementioned works, as well as in the works of (among others) Jensen [607], Dinghas [385], Srivastava [1068, 1069, 1072, 1073], Klusch [673], Choi *et al.* [308], Choi and Srivastava [288, 289, 292, 294, 295], Choi and Seo [285] and Kanemitsu *et al.* [628], dealing with summation of series involving the Riemann ζ-function and its such extensions as the (Hurwitz's) generalized Zeta function $\zeta(s,a)$, defined by 2.2(1).

3.2 Use of the Binomial Theorem

Landau's formula (*cf.* Landau [729, p. 274, Eq. (3)]; see also Titchmarsh [1151, p. 33, Eq. (2.14.1)]):

$$\zeta(s) = 1 + \frac{1}{s-1} - \sum_{k=1}^{\infty} \frac{(s)_k}{(k+1)!}\{\zeta(s+k)-1\} \tag{1}$$

is capable of providing the analytic continuation of $\zeta(s)$ over the whole complex s-plane; here, $(s)_k$ denotes the Pochhammer symbol, defined by 1.1(5).

Another formula, which can also be used in a similar way, is attributed to Ramaswami [966, p. 166] (see also Titchmarsh [1151, p. 33, Eq. (2.14.2)]):

$$(1-2^{1-s})\zeta(s) = \sum_{k=1}^{\infty} \frac{(s)_k}{k!}\frac{\zeta(s+k)}{2^{s+k}}. \tag{2}$$

Motivated by these well-known results (1) and (2) in the theory of the Riemann zeta function $\zeta(s)$, Singh and Verma [1033] derived the following infinite series involving $\zeta(s)$:

$$\zeta(s) = \frac{1}{2} + \frac{1}{s-1} + \frac{1}{2}\sum_{k=1}^{\infty}(-1)^{k-1}\frac{k\cdot(s)_{k+1}}{(k+2)!}\zeta(s+k+1) \quad (\Re(s)<1) \tag{3}$$

and

$$\zeta(s) = 1 + \frac{1}{2^{s+1}}\frac{s+3}{s-1} + \frac{1}{2}\sum_{k=1}^{\infty}(-1)^{k-1}\frac{k\cdot(s)_{k+1}}{(k+2)!}\{\zeta(s+k+1)-1\}. \tag{4}$$

The proofs of (3) and (4) by Singh and Verma [1033, Sections 2 and 3] depend rather heavily on the integral representation (see Titchmarsh [1151, p. 14, Eq. (2.1.4)]):

$$\zeta(s) = \frac{1}{2} + \frac{1}{s-1} - s\int_1^{\infty}\left(x-[x]-\frac{1}{2}\right)\frac{dx}{x^{s+1}} \quad (\Re(s)>-1). \tag{5}$$

Srivastava [1069] gave relatively simple proofs of (3) and (4), *without* using the integral representation (5). He was, thus, led naturally to an interesting unification (and generalization) of (3) and (4), involving the Hurwitz (or generalized) Zeta function $\zeta(s,a)$ (see Section 2.2). The elementary techniques employed in Srivastava [1069] are shown to apply also to the derivation of numerous other results for $\zeta(s,a)$, including, for example, some useful analogues of (1) and (2).

The derivation of Srivastava's unification (and generalization) of (3) and (4) is based simply upon the familiar binomial expansion (see Equation 1.5(22)):

$$\sum_{k=0}^{\infty} \frac{(\lambda)_k}{k!} z^k = (1-z)^{-\lambda} \quad (|z| < 1). \tag{6}$$

Indeed, it follows readily from (6) and the definition 2.2(1) of the generalized Zeta function $\zeta(s,a)$ that (*cf.*, e.g., Wilton [1233, p. 90, Eq. (1)])

$$\sum_{k=0}^{\infty} \frac{(\lambda)_k}{k!} \zeta(\lambda+k,a)\, t^k = \zeta(\lambda, a-t) \quad (|t| < |a|), \tag{7}$$

which holds true, by the principle of analytic continuation of $\zeta(s,a)$, for all values of $\lambda \neq 1$. Observe that (7) is the special case of 2.5(33) when $x \in \mathbb{Z}$.

Now, replace the summation index in (7) by $k+2$, set $\lambda = s-1$ and divide both sides of the resulting equation by t^2. We, thus, observe from (7) that

$$\begin{aligned}\sum_{k=0}^{\infty} \frac{(s-1)_{k+2}}{(k+2)!} \zeta(s+k+1,a) t^k &= \{\zeta(s-1,a-t) - \zeta(s-1,a)\} t^{-2} \\ &\quad - (s-1)\zeta(s,a) t^{-1} \quad (0 < |t| < |a|).\end{aligned} \tag{8}$$

Differentiating both sides of (8) with respect to t and noticing from the definition 1.1(5) and 2.2(18) that

$$(s-1)_{k+2} = (s-1)(s)_{k+1} \quad \text{and} \quad \frac{\partial}{\partial t}\{\zeta(s-1,a-t)\} = (s-1)\zeta(s,a-t), \tag{9}$$

we have

$$\begin{aligned}\sum_{k=1}^{\infty} \frac{k\cdot(s)_{k+1}}{(k+2)!} \zeta(s+k+1,a) t^{k-1} &= \{\zeta(s,a-t) + \zeta(s,a)\} t^{-2} \\ &- \frac{2}{s-1}\{\zeta(s-1,a-t) - \zeta(s-1,a)\} t^{-3} \quad (0 < |t| < |a|),\end{aligned} \tag{10}$$

which can also be obtained as the special case of 2.5(34) when $x \in \mathbb{Z}$.

For $t = -1$, (10) readily yields the desired unification (and generalization) of (3) and (4) in the form (see Srivastava [1069, p. 49, Eq. (2.6)]):

$$\zeta(s,a) = a^{-s}\left(\frac{1}{2} + \frac{a}{s-1}\right) + \frac{1}{2}\sum_{k=1}^{\infty} (-1)^{k-1} \frac{k\cdot(s)_{k+1}}{(k+2)!} \zeta(s+k+1,a), \tag{11}$$

provided that the series converges.

In the special case of (11) when $a = 1$, the series converges, if $\Re(s) < 1$, and we immediately obtain (3). Furthermore, in view of 2.3(2) and the special case of 2.3(9) when $n = 1$, the formula (11) with $a = 2$ is precisely the same as the known result (4).

It may be of interest to remark here that alternative proofs of the well-known results (1) and (2), based upon the integral representation 2.3(30), were given by Menon [823]. As a matter of fact, both (1) and (2) can also be deduced fairly easily from (7).

Replacing the summation index n in (7) by $n+1$ and setting $\lambda = s-1$, we have

$$\begin{aligned}\sum_{k=0}^{\infty} &\frac{(s)_k}{(k+1)!}\zeta(s+k,a)t^{k+1} \\ &= \frac{1}{s-1}\{\zeta(s-1,a-t)-\zeta(s-1,a)\} \quad (|t|<|a|),\end{aligned} \tag{12}$$

where we have made use of the first identity in (9).

By virtue of the definition 2.2(1), (12) with $t = 1$ assumes the form:

$$\zeta(s,a) = \frac{(a-1)^{1-s}}{s-1} - \sum_{k=1}^{\infty}\frac{(s)_k}{(k+1)!}\zeta(s+k,a), \tag{13}$$

which, in view of 2.3(9) when $n = 1$, yields Landau's formula (1) for $a = 2$. Conversely, (12) with $t = \frac{1}{2}$ (and s replaced by $s+1$) or (7) with $t = \frac{1}{2}$ (and $\lambda = s$) similarly yields

$$\zeta(s,2a-1) - 2^{1-s}\zeta(s,a) = \sum_{k=1}^{\infty}\frac{(s)_k}{k!}\frac{\zeta(s+k,a)}{2^{s+k}}, \tag{14}$$

which leads us immediately to Ramaswami's formula (2) upon setting $a = 1$.

For $t = -1$, (12) yields

$$\zeta(s,a) = \frac{a^{1-s}}{s-1} + \sum_{k=1}^{\infty}(-1)^{k-1}\frac{(s)_k}{(k+1)!}\zeta(s+k,a), \tag{15}$$

and (12) with $t = -\frac{1}{2}$ (and s replaced by $s+1$) or (7) with $t = -\frac{1}{2}$ (and $\lambda = s$) gives

$$\zeta(s,2a) - 2^{1-s}\zeta(s,a) = -\sum_{k=1}^{\infty}(-1)^{k-1}\frac{(s)_k}{k!}\frac{\zeta(s+k,a)}{2^{s+k}}. \tag{16}$$

Formulas (15) with $a = 2$ and (16) with $a = 1$ provide interesting analogues of (1) and (2), respectively; in fact, this indicated analogue of (2) [that is, (16) with $a = 1$] was also given by Ramaswami [966, p. 166]. It is not difficult to deduce (11) as a natural consequence of (15).

Numerous other consequences of the general results (7), (11) and (12) can be deduced by suitably specializing the parameter t in a manner detailed above.

In terms of the binomial coefficient 1.1(40), (6) and (7) with $a = 2$ are written in the following equivalent forms:

$$\sum_{k=0}^{\infty} \binom{\lambda+k-1}{k} t^k = (1-t)^{-\lambda} \quad (|t| < 1) \tag{17}$$

and

$$\sum_{k=0}^{\infty} \binom{\lambda+k-1}{k} \{\zeta(\lambda+k)-1\} t^k = \zeta(\lambda, 2-t) \quad (|t| < 2), \tag{18}$$

respectively. This last identity (18) is, in fact, equivalent to (*cf.* Ramanujan [964, p. 78, Eq. (15)]); Apostol [61, p. 240, Eq. (7)])

$$\sum_{k=0}^{\infty} \binom{\lambda+k-1}{k} \zeta(\lambda+k) t^k = \zeta(\lambda, 1-t) \quad (|t| < 1). \tag{19}$$

For fixed $\lambda \neq 1$, the series in (18) and (19) converge absolutely for $|t| < 2$ and $|t| < 1$, respectively. Thus, by the principle of analytic continuation, the formulas (18) and (19) are valid for all values of $\lambda \neq 1$.

Formula (18) provides a unification (and generalization) of 3.1(4) and 3.1(5) and, indeed, also of a fairly large number of other summation formulas scattered in the literature. For example, in view of the relationships 2.3(1) and 2.2(4), (18) with $t = 1$ gives us (*cf.* Hansen [531, p. 356, Eq. (54.4.1)])

$$\sum_{k=1}^{\infty} \binom{\lambda+k-1}{k} \{\zeta(\lambda+k)-1\} = 1, \tag{20}$$

which generalizes 3.1(4), and a special case of (18) when $t = -1$ yields (*cf.* Hansen [531, p. 356, Eq. (54.4.2)])

$$\sum_{k=1}^{\infty} (-1)^{k-1} \binom{\lambda+k-1}{k} \{\zeta(\lambda+k)-1\} = 2^{-\lambda}, \tag{21}$$

which generalizes 3.1(5).

Several additional consequences of the general summation formulas (18) and (19) are worthy of note. First, replace the summation index k in (18) by $k+1$ and set $\lambda = s-1$, so that

$$\sum_{k=0}^{\infty} \binom{s+k-1}{k+1} \{\zeta(s+k)-1\} t^{k+1} = \zeta(s-1, 2-t) - \zeta(s-1) + 1 \quad (|t| < 2), \tag{22}$$

which, for $t = 1$, reduces immediately to the following alternative form of (20)

$$\sum_{k=0}^{\infty} \binom{s+k-1}{k+1} \{\zeta(s+k)-1\} = 1. \tag{23}$$

Now, it follows from the definition 1.1(20) of the binomial coefficient that

$$\binom{s+k-1}{k+1} = \frac{(s-1)(s)_k}{(k+1)!} \quad (k \in \mathbb{N}_0). \tag{24}$$

Thus, the formula (23) can easily be rewritten as Landau's formula (1).

For $t = -1$, (22) readily yields (see Srivastava [1073, p. 131, Eq. (2.4)]):

$$\zeta(s) = 1 + \frac{1}{2^{s-1}} \frac{1}{s-1} + \sum_{k=1}^{\infty} (-1)^{k-1} \frac{(s)_k}{(k+1)!} \{\zeta(s+k)-1\}, \tag{25}$$

which provides an interesting companion of Landau's formula (1).

Setting $t = \frac{1}{2}$ in (22) and making use of 2.2(4) with $a = \frac{1}{2}$ and $n = 1$, we obtain another series representation for $\zeta(s)$ (see Srivastava [1072, p. 5, Eq. (2.13)]):

$$\zeta(s) = \frac{2^s - 1}{2^s - 2} + \frac{1}{2^s - 2} \sum_{k=1}^{\infty} \frac{(s)_k}{k! 2^k} \{\zeta(s+k)-1\}. \tag{26}$$

In their special cases when $s = 2$, (1) and (25) reduce simply to the summation formulas (4) and (5), respectively, whereas (26) similarly yields the elegant sum:

$$\sum_{k=2}^{\infty} \frac{k-1}{2^k} \{\zeta(k)-1\} = \frac{\pi^2}{8} - 1. \tag{27}$$

Next, we turn to the summation formula (19), which (for $\lambda = s - 1$ and with k replaced by $k+1$) assumes the form:

$$\sum_{k=0}^{\infty} \binom{s+k-1}{k+1} \zeta(s+k) t^{k+1} = \zeta(s-1, 1-t) - \zeta(s-1) \quad (|t| < 1). \tag{28}$$

In view of the identities in 2.3(1) and (24), a special case of (28) when $t = \frac{1}{2}$ readily yields Ramaswami's result (2).

In case we add (19) to itself (with t replaced by $-t$), we obtain the summation formula (*cf.* Hansen [531, p. 357, Eq. (54.6.3)]):

$$\sum_{k=0}^{\infty} \binom{\lambda+2k-1}{2k} \zeta(\lambda+2k) t^{2k} = \frac{1}{2} \{\zeta(\lambda-1, 1-t) + \zeta(\lambda, 1+t)\} \quad (|t| < 1), \tag{29}$$

whereas a similar subtraction yields

$$\sum_{k=0}^{\infty} \binom{\lambda+2k}{2k+1} \zeta(\lambda+2k+1) t^{2k+1} = \frac{1}{2}\{\zeta(\lambda, 1-t) - \zeta(\lambda, 1+t)\} \quad (|t| < 1). \tag{30}$$

Various interesting special cases of (29) and (30) are given in the literature. In particular, the special cases of (29) when $t = \frac{1}{2}$, $t = \frac{1}{3}$ and $t = \frac{1}{6}$ were considered by Ramaswami [966, p. 167, Eqs. (1), (3), and (4)], who also gave a special case of (30) when $t = \frac{1}{2}$ (*cf.* Ramaswami [966, p. 167, Eq. (2)]), and by Apostol [61], who proved various generalizations of Ramaswami's results.

By assigning suitable numerical values to the variable s in some of the aforementioned special cases of (29) and (30), Ramaswami [966] also evaluated a number of special sums, including various formulas in Section 3.4 and 3.1(7). Wilton [1233, p. 92] showed that

$$\sum_{k=1}^{\infty} \frac{(2k-1)!}{(2k+3)!} \frac{\zeta(2k)}{2^{2k}} = \frac{\zeta(3)}{2\pi^2} + \frac{1}{12}\log\pi - \frac{11}{72} \tag{31}$$

and

$$\sum_{k=1}^{\infty} \frac{(2k)!}{(2k+3)!} \frac{\zeta(2k+1)}{2^{2k+1}} = \frac{3}{8} - \frac{1}{6}\log\pi - \frac{1}{4}\gamma + \frac{1}{12}\log 2 + \frac{\zeta'(2)}{\pi^2}. \tag{32}$$

Many closed-form evaluations of series involving the Zeta function were investigated systematically by Srivastava [1072], who gave a rather detailed discussion about their derivations and relationships (if any) among them. As a matter of fact, most of the series identities considered by Srivastava [1072] will be included in our extensive list given in Section 3.4.

We record, here, the following identity, involving the Bernoulli polynomials:

$$\frac{B_{n+1}(a+t)}{n+1} = \sum_{k=0}^{n} \binom{n}{k} \frac{B_{k+1}(a)}{k+1} t^{n-k} + \frac{t^{n+1}}{n+1} \quad (n \in \mathbb{N}_0), \tag{33}$$

which is, in fact, an immediate consequence of the known formula 1.7(13).

It follows from 2.2(15) that the Laurent series expansion of $\zeta(s,a)$ at $s = 1$ is of the form

$$\zeta(s,a) = \frac{1}{s-1} - \psi(a) + \sum_{n=1}^{\infty} c_n (s-1)^n, \tag{34}$$

where the coefficients c_n are constants to be determined and the Psi (or Digamma) function $\psi(z)$ is given in Section 1.3.

Start with the following known identity for the Hurwitz-Lerch Zeta function $\Phi(z, s, a)$, defined by 2.5(1) (see 2.5(33) and 2.6(9)):

$$\sum_{k=0}^{\infty} \frac{(s)_k}{k!} \Phi(z, s+k, a)\, t^k = \Phi(z, s, a-t) \quad (|t| < |a|;\ s \neq 1), \tag{35}$$

where $(\lambda)_n$ denotes the Pochhammer symbol, defined by 1.1(5). The general series identity, which is to be proven in this section, is contained in Theorem 3.1 below (see [297]).

Theorem 3.1 *For every nonnegative integer n,*

$$\begin{aligned} \sum_{k=2}^{\infty} \Phi(z, k, a) \frac{t^{n+k}}{(k)_{n+1}} &= \frac{(-1)^n}{n!} [\Phi'(z, -n, a-t) - \Phi'(z, -n, a)] \\ &+ \sum_{k=1}^{n} \frac{(-1)^{n+k}}{n!} \binom{n}{k} \big[(H_n - H_{n-k})\, \Phi(z, k-n, a) - \Phi'(z, k-n, a)\big]\, t^k \\ &+ [H_n L_1(z, a) - L_2(z, a)] \frac{t^{n+1}}{(n+1)!} \quad (|t| < |a|;\ |z| < 1;\ n \in \mathbb{N}_0), \end{aligned} \tag{36}$$

where H_n denotes the harmonic numbers, defined by

$$H_n := \sum_{j=1}^{n} \frac{1}{j}, \tag{37}$$

it being understood (as elsewhere in this paper) that an empty sum is nil,

$$\Phi'(z, s, a) = \frac{\partial}{\partial s} \Phi(z, s, a),$$

and $L_1(z, a)$ and $L_2(z, a)$ are defined by

$$L_1(z, a) := \lim_{s \to -n} \{(s+n)\, \Phi(z, s+n+1, a)\} \tag{38}$$

and

$$L_2(z, a) := \lim_{s \to -n} \big\{\Phi(z, s+n+1, a) + (s+n)\, \Phi'(z, s+n+1, a)\big\}. \tag{39}$$

Before proceeding to prove Theorem 3.1, we show that the assertion (36) of Theorem 3.1 is equivalent to that of Theorem 3.2 below, by making use of the elementary identity:

$$\frac{1}{(k)_{n+1}} = \frac{1}{k(k+1)\cdots(k+n)} := \sum_{j=0}^{n} \frac{A_j^n}{k+j}, \tag{40}$$

where

$$\mathcal{A}_j^n = \frac{(-1)^j}{n!}\binom{n}{j} \quad (0 \leqq j \leqq n;\ j,\, n \in \mathbb{N}_0);$$

the combinatorial identities:

$$\sum_{j=0}^{n} \frac{(-1)^j}{j+1}\binom{n}{j} = \frac{1}{n+1} \quad (n \in \mathbb{N}_0), \tag{41}$$

$$\sum_{j=1}^{n} \frac{(-1)^{j+1}}{j+1}\binom{n}{j} H_j = \frac{H_n}{n+1} \quad (n \in \mathbb{N}_0), \tag{42}$$

and

$$\sum_{j=n-k+1}^{n} \frac{(-1)^{j+1}}{j-(n-k)}\binom{n}{j} = (-1)^{n+k}\binom{n}{k}(H_n - H_{n-k}) \quad (0 \leqq k \leqq n;\ n,\, k \in \mathbb{N}_0), \tag{43}$$

the special case $k = n$ of which is recorded in [505, p. 5, Entry 0.155]; the familiar result:

$$\sum_{j=k}^{n} (-1)^j \binom{j}{k}\binom{n}{j} = 0 \quad (0 \leqq k \leqq n-1;\ k \in \mathbb{N}_0;\ n \in \mathbb{N}), \tag{44}$$

and some rather simple manipulations, using such elementary series identities as the ones involved in

$$\sum_{j=0}^{n}\sum_{k=0}^{j-1} A_{j,k} = \sum_{k=0}^{n-1}\sum_{j=k+1}^{n} A_{j,k} \tag{45}$$

and

$$\sum_{j=0}^{n}\sum_{k=0}^{j} A_{j,k} = \sum_{k=0}^{n}\sum_{j=k}^{n} A_{j,k}, \tag{46}$$

where $\{A_{j,k}\}\,(j,k \in \mathbb{N}_0)$ is a suitably bounded double sequence.

Theorem 3.2 *For every nonnegative integer n,*

$$\sum_{k=2}^{\infty} \frac{\Phi(z,k,a)}{k+n}\, t^{k+n} = \sum_{k=0}^{n} \binom{n}{k} \Phi'(z,-k,a-t)\, t^{n-k} - \sum_{k=0}^{n-1} \frac{\Phi(z,-k,a)}{n-k}\, t^{n-k}$$
$$- [H_n L_1(z,a) + L_2(z,a)]\frac{t^{n+1}}{n+1} - \Phi'(z,-n,a) \quad (|t| < |a|;\ |z| < 1;\ n \in \mathbb{N}_0), \tag{47}$$

where $L_1(z,a)$ *and* $L_2(z,a)$ *are given as in* (38) *and* (39), *respectively.*

It follows easily from the definitions 1.1(19) and 1.3(1) that

$$\frac{d}{dz}\{(z)_n\} = (z)_n[\psi(z+n) - \psi(z)]. \tag{48}$$

Proof of Theorem 3.1 Upon transposing the first $n+2$ terms from $k=0$ to $k=n+1$ in (35) to the right-hand side, if we divide both sides of the resulting equation by $s+n$, we get

$$\sum_{k=n+2}^{\infty} (s)_n\,(s+n+1)_{k-n-1}\,\Phi(z,s+k,\,a)\,\frac{t^k}{k!} = \frac{g_n(z,s,t,a)}{s+n} \tag{49}$$

$$(|t| < |a|;\ s \neq 1;\ n \in \mathbb{N}_0),$$

where, for convenience,

$$g_n(z,s,t,a) := \Phi(z,s,\,a-t) - \sum_{k=0}^{n+1} \frac{(s)_k}{k!}\,\Phi(z,s+k,\,a)\,t^k. \tag{50}$$

Now we shall show that

$$\lim_{s\to -n} g_n(z,s,t,a) = 0. \tag{51}$$

Indeed, we observe that

$$\begin{aligned}\Phi(z,-n,\,a-t) &= \sum_{k=0}^{\infty} (k+a-t)^n\, z^k \\ &= \sum_{k=0}^{\infty}\left\{\sum_{j=0}^{n}\binom{n}{j}(k+a)^{n-j}\,(-t)^j\right\} z^k \\ &= \sum_{j=0}^{n}\binom{n}{j}(-t)^j \sum_{k=0}^{\infty}\frac{z^k}{(k+a)^{-n+j}} \\ &= \sum_{j=0}^{n}\frac{(-n)_j}{j!}\,\Phi(z,-n+j,\,a)\,t^j,\end{aligned}$$

from which (51) follows easily. Thus, by l'Hôpital's rule, we have

$$\lim_{s\to -n}\frac{g_n(z,s,t,a)}{s+n} = \lim_{s\to -n}\frac{\partial}{\partial s}\{g_n(z,s,t,a)\}. \tag{52}$$

Next, by appealing to (50) and (48), we find that

$$\begin{aligned}\frac{\partial}{\partial s}\{g_n(z,s,t,a)\} &= \Phi'(s,\,a-t) - \Phi'(s,\,a)\\ &\quad -\sum_{k=1}^{n} h(z,s,a,k)\frac{t^k}{k!} - h(z,s,a,n+1)\frac{t^{n+1}}{(n+1)!},\end{aligned} \tag{53}$$

where, for convenience,

$$h(z,s,a,k) := (s)_k\,[\{\psi(s+k)-\psi(s)\}\,\Phi(z,s+k,\,a) + \Phi'(z,s+k,\,a)]. \tag{54}$$

In view of 1.3(7), (54) yields

$$\begin{aligned}\lim_{s\to -n} h(z,s,a,k) &= -(-n)_k\big[(H_n - H_{n-k})\,\Phi(z,k-n,\,a) - \Phi'(z,k-n,\,a)\big]\\ &\qquad (k=1,\ldots,n)\end{aligned} \tag{55}$$

and

$$\begin{aligned}&\lim_{s\to -n} h(z,s,a,n+1)\\ &\quad = \lim_{s\to -n}\left[\left(\sum_{j=0}^{n}\frac{(s)_n}{s+j}\right)(s+n)\,\Phi(z,s+n+1,\,a) + (s)_{n+1}\,\Phi'(z,s+n+1,\,a)\right],\end{aligned} \tag{56}$$

which, upon writing

$$\sum_{j=0}^{n}\frac{(s)_n}{s+j} = \sum_{j=0}^{n-1}\frac{(s)_n}{s+j} + \frac{(s)_n}{s+n},$$

reduces to the following form:

$$\lim_{s\to -n} h(z,s,a,n+1) = (-n)_n\,[-H_n\,L_1(z,a) + L_2(z,a)], \tag{57}$$

where $L_1(z,a)$ and $L_2(z,a)$ are given by (38) and (39), respectively.

Conversely, upon taking the limit on the left-side of (49) as $s\to -n$, we have

$$\begin{aligned}&\lim_{s\to -n}\sum_{k=n+2}^{\infty}(s)_n\,(s+n+1)_{k-n-1}\,\Phi(z,s+k,a)\frac{t^k}{k!}\\ &\quad = (-n)_n\sum_{k=2}^{\infty}\Phi(z,k,\,a)\,\frac{t^{k+n}}{(k)_{n+1}}.\end{aligned} \tag{58}$$

Finally, upon taking the limit on both sides of (49) as $s\to -n$ and considering (52), (53), (54), (57) and (58), we are led immediately to the desired series identity (36). This evidently completes our proof of Theorem 3.1.

Applications of Theorems 3.1 and 3.2

We first note that $\Phi(z,s,a)$ is an entire function of s when $|z|<1$, converges for $\Re(s)>0$ when $|z|=1$ and $z\neq 1$ and reduces to $\zeta(s,a)$ when $z=1$. So, when $|z|\leqq 1$ and $z\neq 1$, we have

$$L_1(z,a)=0 \quad \text{and} \quad L_2(z,a)=\Phi(z,1,a)=a^{-1}\,{}_2F_1(1,\,a;\,1+a;\,z). \tag{59}$$

It follows from 2.5(26) that

$$L_1(1,a)=\lim_{s\to -n}(s+n)\,\zeta(s+n+1,a)=1, \tag{60}$$

since $\zeta(s,a)$ has a simple pole at $s=1$ with residue 1. From (34), we get

$$L_2(1,a)=-\psi(a). \tag{61}$$

Upon setting $z=1$ in (36) and (47) and considering 2.5(26), (60) and (61), we have certain families of series associated with the generalized (or Hurwitz) Zeta function $\zeta(s,a)$:

Corollary 3.3 *For every nonnegative integer n,*

$$\begin{aligned}\sum_{k=2}^{\infty}\frac{\zeta(k,\,a)}{(k)_{n+1}}\,t^{n+k} &= \frac{(-1)^n}{n!}\,[\zeta'(-n,\,a-t)-\zeta'(-n,\,a)]\\ &+\sum_{k=1}^{n}\frac{(-1)^{n+k}}{n!}\binom{n}{k}\big[(H_n-H_{n-k})\,\zeta(k-n,\,a)-\zeta'(k-n,\,a)\big]\,t^k\\ &+[H_n+\psi(a)]\,\frac{t^{n+1}}{(n+1)!}\qquad (|t|<|a|;\ n\in\mathbb{N}_0),\end{aligned} \tag{62}$$

where H_n denotes the harmonic numbers defined by (37), it being understood that

$$\zeta'(s,a)=\frac{\partial}{\partial s}\,\zeta(s,a).$$

Corollary 3.4 *For every nonnegative integer n,*

$$\begin{aligned}\sum_{k=2}^{\infty}\frac{\zeta(k,a)}{k+n}\,t^{k+n} &= \sum_{k=0}^{n}\binom{n}{k}\zeta'(-k,a-t)\,t^{n-k}-\sum_{k=0}^{n-1}\frac{\zeta(-k,a)}{n-k}\,t^{n-k}\\ &+[\psi(a)-H_n]\,\frac{t^{n+1}}{n+1}-\zeta'(-n,a)\qquad (|t|<|a|;\ n\in\mathbb{N}_0),\end{aligned} \tag{63}$$

where H_n denotes the harmonic numbers defined by (37), it being understood that

$$\zeta'(s,a)=\frac{\partial}{\partial s}\zeta(s,a).$$

Infinite sums of the type occurring in (62) can also be evaluated, in a *markedly different* manner, in terms of such higher transcendental functions as the *multiple* Gamma functions (see, for details, [628, p. 10, Theorem 3.1]).

The series identities (62) and (63) can include a number of formulas involving various classes of earlier closed-form evaluations of series associated with the Zeta function as its special cases. For example, we can get the following (easily derivable) consequence of the summation formula (63):

$$\sum_{k=1}^{\infty}\frac{\zeta(2k,a)}{k+n}t^{2k+2n}=\sum_{k=0}^{2n}\binom{2n}{k}\left[\zeta'(-k,a-t)+(-1)^k\zeta'(-k,a+t)\right]t^{2n-k}$$
$$-\sum_{\ell=0}^{n-1}\frac{\zeta(-2\ell,a)}{n-\ell}t^{2n-2\ell}-2\zeta'(-2n,a)\quad(n\in\mathbb{N}_0;\ |t|<|a|);\tag{64}$$

$$\sum_{k=1}^{\infty}\frac{\zeta(2k,a)}{2k+2n+1}t^{2k+2n+1}$$
$$=\frac{1}{2}\sum_{k=0}^{2n+1}\binom{2n+1}{k}\left[\zeta'(-k,a-t)+(-1)^k\zeta'(-k,a+t)\right]t^{2n+1-k}\tag{65}$$
$$-\sum_{\ell=0}^{n}\frac{\zeta(-2\ell,a)}{2n+1-2\ell}t^{2n+1-2\ell}\quad(n\in\mathbb{N}_0;\ |t|<|a|);$$

$$\sum_{k=1}^{\infty}\frac{\zeta(2k+1,a)}{2k+2n+1}t^{2k+2n+1}$$
$$=\frac{1}{2}\sum_{k=0}^{2n}\binom{2n}{k}\left[\zeta'(-k,a-t)-(-1)^k\zeta'(-k,a+t)\right]t^{2n-k}$$
$$-\sum_{\ell=1}^{n}\frac{\zeta(1-2\ell,a)}{2n-2\ell+1}t^{2n-2\ell+1}-\frac{t^{2n+1}}{2n+1}[\psi(2n+1)-\psi(a)+\gamma]\tag{66}$$
$$(n\in\mathbb{N}_0;\ |t|<|a|);$$

$$\sum_{k=1}^{\infty}\frac{\zeta(2k+1,a)}{k+n+1}t^{2k+2n+2}$$
$$=\sum_{k=0}^{2n+1}\binom{2n+1}{k}\left[\zeta'(-k,a-t)-(-1)^k\zeta'(-k,a+t)\right]t^{2n+1-k}$$
$$-\sum_{\ell=1}^{n}\frac{\zeta(1-2\ell,a)}{n-\ell+1}t^{2n+2-2\ell}-\frac{t^{2n+2}}{n+1}[\psi(2n+2)-\psi(a)+\gamma]\tag{67}$$
$$-2\zeta'(-2n-1,a)\quad(n\in\mathbb{N}_0;\ |t|<|a|).$$

Setting $a = 1$ and $t = -1$ in (63) and using various identities given in the previous chapters, for example, 2.2(4) and 1.3(7), we obtain

$$\sum_{k=2}^{\infty} \frac{(-1)^k}{k+n} \zeta(k) = \sum_{\ell=0}^{n-1} (-1)^\ell \binom{n}{\ell} \zeta'(-\ell) - \sum_{\ell=0}^{n-1} \frac{(-1)^\ell}{rn-\ell} \zeta(-\ell) + \frac{1}{n+1} (\gamma + H_n) \quad (n \in \mathbb{N}_0). \tag{68}$$

Likewise, setting $a = 1$ and $t = 1/2$ in (63), we get

$$\sum_{k=2}^{\infty} \frac{\zeta(k)}{(k+n)\,2^k} = \sum_{\ell=0}^{n-1} \left[\binom{n}{\ell} \log 2 - \frac{2^\ell}{n-\ell} \right] \zeta(-\ell) + \sum_{\ell=0}^{n-1} \binom{n}{\ell} \left(1 - 2^\ell\right) \zeta'(-\ell) - \frac{\gamma + H_n}{2(n+1)} + \zeta(-n) \log 2 + \left(1 - 2^{n+1}\right) \zeta'(-n) \quad (n \in \mathbb{N}_0). \tag{69}$$

The special cases of (64) and (65) when $a = 1$ and $t = 1/2$ are written here:

$$\sum_{k=1}^{\infty} \frac{\zeta(2k)}{(k+n)\,2^{2k}} = \frac{1}{2n} - \log 2 + (-1)^n (2n)! \frac{1 - 2^{2n+1}}{(2\pi)^{2n}} \zeta(2n+1) + \sum_{k=1}^{n-1} (-1)^k \binom{2n}{2k} \frac{(2k)!}{(2\pi)^{2k}} \left(1 - 2^{2k}\right) \zeta(2k+1) \quad (n \in \mathbb{N}). \tag{70}$$

$$\sum_{k=1}^{\infty} \frac{\zeta(2k)}{(2k+2n-1)\,2^{2k}} = \frac{1}{2(2n-1)} - \frac{1}{2} \log 2 + \frac{1}{2} \sum_{k=1}^{n-1} (-1)^k \binom{2n-1}{2k} \frac{(2k)!}{(2\pi)^{2k}} \left(1 - 2^{2k}\right) \zeta(2k+1) \quad (n \in \mathbb{N}), \tag{71}$$

the special case $n = 1$ of which yields 3.4(519).

Setting $a = 1$ and $t = 1/2$ in (66), we get

$$\sum_{k=1}^{\infty} \frac{\zeta(2k+1)}{(2k+2n+1)\,2^{2k}} = \sum_{\ell=1}^{n} \frac{2^{2\ell-1} B_{2\ell}}{\ell(2n-2\ell+1)} - \log 2 \sum_{\ell=1}^{n} \binom{2n}{2\ell-1} \frac{B_{2\ell}}{\ell} + 2 \sum_{\ell=1}^{n} \binom{2n}{2\ell-1} \left(1 - 2^{2\ell-1}\right) \zeta'(1-2\ell) - \frac{\gamma + H_{2n}}{2n+1} + \log 2 \sum_{\ell=0}^{2n} (-1)^\ell \binom{2n}{\ell} \quad (n \in \mathbb{N}_0), \tag{72}$$

where we have assumed that

$$\sum_{\ell=0}^{2n}(-1)^{\ell}\binom{2n}{\ell}:=1 \quad (n=0).$$

Of course, it is true that

$$\sum_{\ell=0}^{2n}(-1)^{\ell}\binom{2n}{\ell}=0 \quad (n\in\mathbb{N}). \tag{73}$$

Setting $a=1$ and $t=1/2$ in (67), we obtain

$$\begin{aligned}\sum_{k=1}^{\infty}\frac{\zeta(2k+1)}{(k+n+1)\,2^{2k}} &= -2\log 2\sum_{\ell=0}^{n}\binom{2n+1}{2\ell+1}\frac{B_{2\ell+2}}{\ell+1}-\frac{\gamma+H_{2n+1}}{n+1}\\ &+\sum_{\ell=1}^{n}\frac{2^{2\ell-1}}{\ell(n-\ell+1)}B_{2\ell}+4\sum_{\ell=0}^{n-1}\left(1-2^{2\ell+1}\right)\binom{2n+1}{2\ell+1}\zeta'(-2\ell-1)\\ &+4\left(1-2^{2n+2}\right)\zeta'(-2n-1) \quad (n\in\mathbb{N}_0).\end{aligned} \tag{74}$$

If we set $n=1$ in (74), we get

$$\sum_{k=1}^{\infty}\frac{\zeta(2k+1)}{(k+2)\cdot 2^{2k}}=-\frac{2}{3}-\frac{\gamma}{2}-\frac{29}{30}\log 2+12\log A+30\,\log C. \tag{75}$$

Upon setting $a=1$ in (65) and then $t=\frac{1}{2}$ in the resulting equation, if we use the various identities for $\zeta(s)$ and $\zeta(s,a)$ given in Chapter 2, in particular, together with the known identity 2.3(22), we obtain (see [256, p. 262, Eq. (14)])

$$\begin{aligned}\sum_{k=1}^{\infty}\frac{\zeta(2k)}{(2k+2\ell-1)\,2^{2k}} &= \frac{1}{2(2\ell-1)}-\frac{1}{2}\log 2\\ &+\frac{1}{2}\sum_{k=1}^{\ell-1}(-1)^{k}\binom{2\ell-1}{2k}\frac{(2k)!}{(2\pi)^{2k}}\left(1-2^{2k}\right)\zeta(2k+1) \quad (\ell\in\mathbb{N}),\end{aligned} \tag{76}$$

whose special case $\ell=1$ is also recorded in 3.4(519).

As a consequence, the following integral can be evaluated as a finite series involving the Riemann Zeta function (see [256, Eq. (18)]):

$$\begin{aligned}\int_{0}^{\frac{\pi}{2}}\theta^{2\ell}\cot\theta\,d\theta &= \left(\frac{\pi}{2}\right)^{2\ell}\left[\log 2+(-1)^{\ell}(2\ell)!\,\frac{2^{2\ell+1}-1}{(2\pi)^{2\ell}}\zeta(2\ell+1)\right.\\ &\left.+\sum_{k=1}^{\ell-1}(-1)^{k+1}\binom{2\ell}{2k}\frac{(2k)!}{(2\pi)^{2k}}\left(1-2^{2k}\right)\zeta(2k+1)\right] \quad (\ell\in\mathbb{N}).\end{aligned} \tag{77}$$

We note that the integral in (77) was evaluated earlier as an infinite series involving the Riemann Zeta function (see [505, p. 428, Entry 3.748.2]; also see [256]).

We deduce another interesting identity by suitably combining the special cases of (62) when $a = 1$ and $a = 2$. By applying 2.2(4), 1.3(4) and 1.3(7), we, thus, find that

$$\sum_{k=1}^{\infty} \frac{t^{n+k}}{(k)_{n+1}} = \frac{(-1)^{n+1}}{n!}(1-t)^n \log(1-t) + \sum_{k=1}^{n} \frac{(-1)^{n+k}}{n!} \binom{n}{k} (H_n - H_{n-k})\, t^k \quad (|t| < 1;\ n \in \mathbb{N}_0), \tag{78}$$

which, in the special case when $n = 2$, is a known result recorded (for example) by Hansen [531, p. 37, Entry (5.7.40); p. 74, Entry (5.16.26)].

We shall deal with other interesting applications of Corollary 3.3 in Section 3.4.

3.3 Use of Generating Functions

Adamchik and Srivastava [11] proposed and developed a novel method of evaluation of the sums of series involving Zeta functions by using generating functions. The key ingredients in their approach happen to include the familiar integral representations 2.2(21) and 2.3(30). This method has already been implemented in *Mathematica* (Version 3.0).

Consider the sum:

$$\Omega(a) = \sum_{k=1}^{\infty} f(k)\, \zeta(k+1, a) \quad (\Re(a) > 0), \tag{1}$$

where the sequence $\{f(n)\}_{n=1}^{\infty}$ is assumed to possess a generating function:

$$F(t) = \sum_{k=1}^{\infty} f(k)\, \frac{t^k}{k!} \tag{2}$$

and

$$f(n) = O\left(\frac{1}{n}\right) \quad (n \to \infty). \tag{3}$$

Upon replacing the Hurwitz (or generalized) Zeta function in (1) by its integral representation given by 2.2(21) with $s = k + 1$, if we invert the order of summation and integration, we obtain

$$\Omega(a) = \int_0^{\infty} F(t)\, \frac{e^{-(a-1)t}}{e^t - 1}\, dt, \tag{4}$$

where we have also made use of the generating function (2).

Thus, the problem of summation of series of the type (1) has been reduced formally to that of integration in (4). Although the integral in (4) appears to be fairly involved for symbolic integration, it may be possible to reduce it to 2.2(21) or 2.3(30) (or another known integral), especially when $F(t)$ is a power, exponential, trigonometric or hyperbolic function.

The first example (Proposition 3.5 below) would illustrate how this technique actually works (see Adamchik and Srivastava [11, p. 135, Proposition 1]):

Proposition 3.5 *Let n be a positive integer. Then,*

$$\sum_{k=1}^{\infty} \frac{(-1)^k}{k} \{\zeta(nk) - 1\} = \log\left(\prod_{j=0}^{n-1} \Gamma\left(2 - (-1)^{(2j+1)/n}\right) \right). \tag{5}$$

Remark 1 In Proposition 3.5, as also in Equations (7), (8) and (10) below, it is tacitly assumed that

$$-1 = e^{\log(-1)},$$

where $\log z$ denotes the principal branch of the logarithmic function in the complex z-plane for which

$$-\pi < \arg(z) \leqq \pi \quad (z \neq 0).$$

Proof. Denote, for convenience, the left-hand side of the summation formula (5) by $\Theta(n)$. Then, in view of the special case of 2.3(9) when $n = 1$, we can apply the integral representation 2.2(21) with $s = nk$ and $a = 2$. Upon inverting the order of summation and integration, which can be justified by the absolute convergence of the series and the integral involved, we, thus, find that

$$\Theta(n) = n \int_0^{\infty} \frac{dt}{te^t(e^t - 1)} \sum_{k=1}^{\infty} \frac{(-t^n)^k}{\Gamma(nk+1)}. \tag{6}$$

By recognizing the series in (6) as a trigonometric function of order n (see, for details, Erdélyi *et al.* [422, Section 18.2]) or, alternatively, by using an easily derivable special case of a known result in Hansen [531, p. 207, Entry (10.49.1)], we have

$$\sum_{k=1}^{\infty} \frac{(-t^n)^k}{\Gamma(nk+1)} = -1 + \frac{1}{n} \sum_{j=1}^{n} \exp\left(t(-1)^{(2j+1)/n}\right). \tag{7}$$

Substituting from (7) into (6) and inverting the order of summation and integration once again, we obtain

$$\Theta(n) = \lim_{r \to 1-} \left\{ -n \int_0^{\infty} \frac{dt}{t^r e^t(e^t - 1)} + \sum_{j=1}^{n} \int_0^{\infty} \frac{\exp\left(-\left[1 - (-1)^{(2j+1)/n}\right] t\right)}{t^r(e^t - 1)} dt \right\}, \tag{8}$$

where the parameter $r < 1$ is inserted with a view to providing convergence of the integrals at their lower terminal $t = 0$.

Finally, we evaluate each integral in (8) by means of 2.2(21) and proceed to the limit as $r \to 1-$. Indeed, by making use of the known behavior of $\zeta(1-r,a)$ near $r = 1$ for fixed a (*cf.*, e.g., Erdélyi *et al.* [421, p. 26]):

$$\zeta(1-r,a) \sim \frac{1}{2} - a + (1-r)\log\left(\frac{\Gamma(a)}{\sqrt{2\pi}}\right) \quad (r \to 1-;\ a \text{ fixed}), \tag{9}$$

we arrive at the right-hand side of the assertion (5).

In precisely the same manner, we can prove a *mild* generalization of Proposition 3.5, which we state here as (see Adamchik and Srivastava [11, p. 136, Proposition 2]) ■

Proposition 3.6 *Let n be a positive integer. Then,*

$$\sum_{k=1}^{\infty} \frac{(-1)^k}{k} \zeta(nk,a) = -n \log \Gamma(a) + \log\left(\prod_{j=0}^{n-1} \Gamma\left(a - (-1)^{(2j+1)/n}\right)\right) \quad \left(\Re(a) \geqq 1\right). \tag{10}$$

Next we prove (see Adamchik and Srivastava [11, p. 136, Proposition 3])

Proposition 3.7 *Let n be a positive integer. Then, in terms of the Bernoulli numbers* B_n, *defined by* 1.7(2), *and the Stirling numbers* $S(n,k)$, *defined by* 1.6(14),

$$\sum_{k=2}^{\infty} (-1)^k \{\zeta(k)-1\}\, k^n = -1 + \frac{1-2^{n+1}}{n+1} B_{n+1} - \sum_{k=1}^{n} (-1)^k k!\, \zeta(k+1) S(n+1,k+1). \tag{11}$$

Proof. Making use of the integral representation 2.2(21) with $a = 2$, if we invert the order of summation and integration and then evaluate the resulting integral and sum, we find that

$$\begin{aligned}\Lambda(n) &:= \sum_{k=2}^{\infty} (-1)^k \{\zeta(k)-1\}\, k^n \\ &= \lim_{r \to 1} \left(r \frac{d}{dr}\right)^n \{r[\psi(r+2) - \psi(2)]\}.\end{aligned} \tag{12}$$

Since

$$\left(r \frac{d}{dr}\right)^r \{r f(r)\} = \sum_{k=0}^{n} S(n+1,k+1) f^{(k)}(r)\, r^{k+1} \tag{13}$$

and (*cf.* Equation 1.2(53))

$$\psi^{(k)}(3) = (-1)^k\, k!\, \left\{1 + 2^{-k-1} - \zeta(k+1)\right\} \quad (k \in \mathbb{N}), \tag{14}$$

we find from (12) that

$$\begin{aligned}\Lambda(n) = -1 &- \sum_{k=1}^{n} (-1)^k\, k!\, \zeta(k+1)\, S(n+1,k+1) \\ &+ \sum_{k=0}^{n} (-1)^k\, k!\, 2^{-k-1} S(n+1,k+1).\end{aligned} \tag{15}$$

To complete the proof of Proposition 3.7, we, thus, need to show that

$$\sum_{k=0}^{n} (-1)^k\, k!\, 2^{-k-1}\, S(n+1,k+1) = \frac{1-2^{n+1}}{n+1} B_{n+1} \tag{16}$$

in terms of the Bernoulli numbers B_n, defined by 1.7(2). In fact, it is known that (Hansen [531, p. 351, Entry (52.2.36)])

$$\sum_{k=1}^{n} (-1)^k\, k!\, 2^{-k}\, S(n,k) = \frac{2}{n+1}\left(1-2^{n+1}\right) B_{n+1}, \tag{17}$$

which, in view of the recurrence relation 1.6(18) with k replaced by $k+1$, yields the desired identity (16).

Similarly, we can prove (see Adamchik and Srivastava [11, p. 137, Proposition 4]) ∎

Proposition 3.8 *Let n be a positive integer. Then,*

$$\sum_{k=2}^{\infty} \{\zeta(k) - 1\}\, k^n = 1 + \sum_{k=1}^{n} k!\, \zeta(k+1)\, S(n+1,k+1). \tag{18}$$

The following list provides further summation formulas, involving series of Zeta functions, which can be derived by applying the foregoing technique (see Adamchik and Srivastava [11, pp. 138–139]):

$$\sum_{k=2}^{\infty} \frac{k^2}{k+1} \{\zeta(k) - 1\} = \frac{3}{2} - \frac{\gamma}{2} + \frac{\pi^2}{6} - \frac{1}{2}\log(2\pi); \tag{19}$$

$$\sum_{k=2}^{\infty} \frac{k^2}{k+1} \{\zeta(2k+1) - 1\} = \frac{9}{16} - \gamma + \log 2 - \frac{1}{2}\zeta(3); \tag{20}$$

$$\sum_{k=1}^{\infty}\{\zeta(4k)-1\}=\frac{7}{8}-\frac{\pi}{4}\coth\pi; \tag{21}$$

$$\sum_{k=1}^{\infty}(-1)^k\{\zeta(4k)-1\}=1+\frac{\pi}{2\sqrt{2}}\,\frac{\sin\left(\pi\sqrt{2}\right)+\sinh\left(\pi\sqrt{2}\right)}{\cos\left(\pi\sqrt{2}\right)-\cosh\left(\pi\sqrt{2}\right)}; \tag{22}$$

$$\sum_{k=1}^{\infty}\{\zeta(4k)-1\}\,z^{4k}=\frac{3z^4-1}{2(z^4-1)}-\frac{\pi z}{4}\{\cot(\pi z)+\coth(\pi z)\}\quad(|z|<2), \tag{23}$$

which contains both (21) and (22) as limiting cases;

$$\sum_{k=1}^{\infty}\{\zeta(2k)-1\}\sin k=-\frac{1}{2}\cot\left(\frac{1}{2}\right)$$
$$+\frac{\pi}{2}\,\frac{\sin\left(\frac{1}{2}\right)\sin\left(2\pi\cos\left(\frac{1}{2}\right)\right)-\cos\left(\frac{1}{2}\right)\sinh\left(2\pi\sin\left(\frac{1}{2}\right)\right)}{\cos\left(2\pi\cos\left(\frac{1}{2}\right)\right)-\cosh\left(2\pi\sin\left(\frac{1}{2}\right)\right)}; \tag{24}$$

$$\sum_{k=1}^{\infty}\binom{p+k}{k}\zeta(p+k+1,a)\,z^k=\frac{(-1)^p}{p!}\left\{\psi^{(p)}(a)-\psi^{(p)}(a-z)\right\}$$
$$(p\in\mathbb{N};\ \Re(a)>0;\ |z|<|a|); \tag{25}$$

$$\sum_{k=2}^{\infty}\frac{k}{2^k}\zeta\left(k+1,\frac{3}{4}\right)=8G \tag{26}$$

and

$$\sum_{k=2}^{\infty}\frac{k}{2^k}\zeta\left(k+1,\frac{5}{4}\right)=8(1-G), \tag{27}$$

where G denotes Catalan's constant, defined by 1.3(16).

Remark 2 Since the right-hand side of the relationship 1.3(53) (with $n!$ replaced by $\Gamma(n+1)$) is well-defined for $n\in\mathbb{C}\setminus\{-1\}$, the summation formula (25) may be put in a slightly more general form (*cf.* Wilton [1233]; see also Srivastava [1072, p. 137, Equation (6.6)]):

$$\sum_{k=0}^{\infty}\binom{s+k-1}{k}\zeta(s+k,a)\,z^k=\zeta(s,a-z)\quad\left(s\in\mathbb{C}\setminus\{1\};\ a\neq\mathbb{Z}_0^-;\ |z|<|a|\right), \tag{28}$$

which, in view of 1.1(20), is equivalent to 3.2(7) for $\lambda=s$ and $t=z$.

The foregoing method has been implemented in *Mathematica* (Version 3.0), as we remarked at the beginning of this section.

Series Involving Polygamma Functions

The generalized harmonic numbers $H_n^{(s)}$, defined by (*cf.* Graham *et al.* [507])

$$H_n^{(s)} := \sum_{k=1}^{n} \frac{1}{k^s} \quad (n \in \mathbb{N},\ s \in \mathbb{C}), \tag{29}$$

which, by virtue of 2.3(9), can be written in the form:

$$H_n^{(s)} = \zeta(s) - \zeta(s, n+1) \quad (\Re(s) > 1;\ n \in \mathbb{N}). \tag{30}$$

In view of the relationships 1.2(53) and (30), the foregoing techniques can be applied also to series involving Polygamma functions and generalized harmonic numbers. We first state (see Adamchik and Srivastava [11, p. 139, Proposition 5])

Proposition 3.9 *Let p be a positive integer. Then, in terms of the function* $\Phi(z,s,a)$, *defined by* 2.5(1),

$$\begin{aligned}\sum_{k=1}^{\infty} \psi^{(p)}(a+k)\, z^k &= \frac{(-1)^{p+1}\, p!}{a^{p+1}} + \frac{\psi^{(p)}(a+1)}{1-z} \\ &+ \frac{(-1)^p\, p!\, z^2}{1-z}\, \Phi(z, p+1, a+1) \quad \left(|z| < 1;\ a \neq \mathbb{Z}_0^- \setminus \{0\}\right).\end{aligned} \tag{31}$$

Proof. Denoting, for convenience, the left-hand side of the summation formula (31) by $\Xi(z)$, it is not difficult to find from 1.3(53) and 2.2(21) that

$$\begin{aligned}\Xi(z) &= (-1)^{p+1} \int_0^{\infty} \frac{t^p\, e^{-(a-1)t}}{e^t - 1} \left(\sum_{k=1}^{\infty} z^k\, e^{-kt} \right) dt \\ &= - \int_0^1 \frac{\tau^{a-1} (\log \tau)^p}{(1-\tau)(1-z\tau)}\, d\tau \quad (\Re(a) > 0).\end{aligned}$$

Upon evaluating this last integral and waiving the restriction on the parameter a by appealing to the principle of analytic continuation, we complete the proof of Proposition 3.9.

Next, we turn to a family of *linear* harmonic sums:

$$S_{p,q} := \sum_{n=1}^{\infty} \frac{H_n^{(p)}}{n^q}, \tag{32}$$

which were discussed extensively by Flajolet and Salvy [454]. By applying (30), 2.2(21) and 2.3(30), it is not difficult to prove (see Adamchik and Srivastava [11, p. 140, Proposition 6]) ■

Proposition 3.10 *Let $\mathcal{S}_{p,q}$ be defined by* (32). *Then,*

$$\mathcal{S}_{p,q} = \zeta(p)\zeta(q) + \frac{(-1)^p}{(p-1)!} \int_0^1 (\log t)^{p-1} \operatorname{Li}_q(t) \frac{dt}{1-t} \tag{33}$$

and

$$\mathcal{S}_{p,q} = \zeta(p+q) - \frac{(-1)^q}{(q-1)!} \int_0^1 (\log t)^{q-1} \operatorname{Li}_p(t) \frac{dt}{1-t}. \tag{34}$$

Remark 3 In view of the following symmetry relation in Flajolet and Salvy [454]:

$$\mathcal{S}_{p,q} + \mathcal{S}_{q,p} = \zeta(p)\zeta(q) + \zeta(p+q), \tag{35}$$

the integral representations (33) and (34) are essentially the same. Although, in general, the integrals occurring in (33) and (34) cannot be evaluated in closed forms, many interesting particular cases of linear harmonic sums would follow from Proposition 3.10.

Series Involving Polylogarithm Functions

We shall derive, among other results, an integral representation for the Khintchine constant K_0, which arises in the measure theory of continued fractions. Every positive irrational number μ can, indeed, be written uniquely as a simple continued fraction as follows:

$$\mu = a_0 + \frac{a_1}{a_2+}\frac{a_3}{a_4+}\cdots\frac{a_n}{a_{n+1}+}\cdots, \tag{36}$$

that is, with a_0 a non-negative integer and with all other a_j $(j \in \mathbb{N})$ positive integers. The Gauss-Kuz'min distribution (*cf.*, e.g., Khintchine [647]) predicts that the density of occurrence of some chosen positive integer k in the continued fraction (36) of a random real number is given by

$$\operatorname{Prob}\{a_n = k\} = -\log_2\left(1 - \frac{1}{(k+1)^2}\right). \tag{37}$$

And, making use of the Gauss-Kuz'min distribution involving (37), Khintchine [647] showed that, for almost all irrational numbers, the limiting geometric mean of the positive integer elements a_j ($j \in \mathbb{N}$) of the relevant continued fraction exists and equals

$$\begin{aligned} K_0 &:= \prod_{k=1}^{\infty} \left\{ 1 + \frac{1}{k(k+2)} \right\}^{\log_2 k} \\ &= \prod_{k=1}^{\infty} \left\{ k^{\log_2 \left(1+\frac{1}{k(k+2)}\right)} \right\}. \end{aligned} \tag{38}$$

An interesting explicit representation of the Khintchine constant K_0 in terms of Polylogarithm functions was proven recently by Bailey *et al.* [85, p. 422]:

$$\log(K_0)\log 2 = (\log 2)^2 + \mathrm{Li}_2\left(-\frac{1}{2}\right) + \frac{1}{2}\sum_{n=2}^{\infty}(-1)^n \mathrm{Li}_2\left(\frac{4}{n^2}\right). \tag{39}$$

If we set

$$\mathcal{L}(n) := \sum_{n=2}^{\infty}(-1)^n \mathrm{Li}_2\left(\frac{4}{n^2}\right), \tag{40}$$

replace the Dilogarithm function by its series representation given by 2.4(3), change the order of summation and evaluate the inner sum, we shall obtain

$$\mathcal{L}(n) = 2\sum_{k=1}^{\infty}\frac{\zeta(2k)}{k^2} - \sum_{k=1}^{\infty}\{\zeta(2k)-1\}\frac{4^k}{k^2}. \tag{41}$$

It seems very unlikely that the sums occurring in (41) can be evaluated in terms of well-known functions. Nevertheless, by noting that

$$\sum_{k=1}^{\infty}\frac{t^k}{k^2}\zeta(2k) = \int_0^1 \frac{dt}{t}\sum_{k=1}^{\infty}\frac{t^k}{k}\zeta(2k) \tag{42}$$

and evaluating the inner sum by the method illustrated in the preceding sections, we find that

$$\sum_{k=1}^{\infty}\frac{t^k}{k^2}\zeta(2k) = \log\left(\pi\sqrt{t}\csc\left(\pi\sqrt{t}\right)\right). \tag{43}$$

Combining (41), (42) and (43), we have

$$\mathcal{L}(n) := \sum_{n=2}^{\infty}(-1)^n \mathrm{Li}_2\left(\frac{4}{n^2}\right) = \int_0^1 \frac{dt}{t}\log\left(\frac{\pi\sqrt{t}\cot\left(\pi\sqrt{t}\right)}{1-4t}\right), \tag{44}$$

which leads us immediately to the following integral representation for the Khintchine constant K_0 (see Adamchik and Srivastava [11, p. 141, Eq. (4.10)]):

$$\log(K_0)\log 2 = \frac{\pi^2}{12} + \frac{(\log 2)^2}{2} + \int_0^{\pi} \log\left(t\left|\cot t\right|\right) \frac{dt}{t}. \tag{45}$$

Other sums involving the Polylogarithm function, which were also evaluated by Adamchik and Srivastava [11], are given below.

$$\sum_{k=1}^{\infty} \left(-\frac{1}{2}\right)^k \operatorname{Li}_k\left(z^2\right) = 1 - z^{-1}\operatorname{arctanh} z; \tag{46}$$

$$\sum_{k=1}^{\infty} \left(\frac{1}{2}\right)^k \operatorname{Li}_k\left(z^2\right) = \frac{1}{2} z \log\left(\frac{1+z}{1-z}\right). \tag{47}$$

Since $\operatorname{arctanh}\left(\frac{1}{2}\right) = \log 3$, both (46) and (47) can be expressed in terms of $\log 3$ when $z = \frac{1}{2}$.

3.4 Use of Multiple Gamma Functions

We provide a rather extensive list of evaluations of series involving the Zeta functions, by making use of the Gamma and multiple Gamma functions. Although our list in this section includes some of the series identities considered in previous sections or earlier works, we choose to present, here, all of the identities evaluated by the use of multiple Gamma functions for the sake of completeness.

Evaluation by Using the Gamma Function

Upon setting $n = 0$ in 3.2(35) and applying 2.2(17), we obtain the familiar result (*cf.* Whittaker and Watson [1225, p. 276], Hansen [531, p. 358, Entry (54.11.1)], and Srivastava [1072, p. 18]):

$$\sum_{k=2}^{\infty} \zeta(k,a)\frac{t^k}{k} = \log\Gamma(a-t) - \log\Gamma(a) + t\psi(a) \quad (|t| < |a|), \tag{1}$$

which, upon replacing t by $-t$, yields

$$\sum_{k=2}^{\infty} (-1)^k \zeta(k,a)\frac{t^k}{k} = \log\Gamma(a+t) - \log\Gamma(a) - t\psi(a) \quad (|t| < |a|). \tag{2}$$

By adding and subtracting, we find from (1) and (2) that

$$\sum_{k=1}^{\infty} \zeta(2k,a)\frac{t^{2k}}{k} = \log\Gamma(a+t) + \log\Gamma(a-t) - 2\log\Gamma(a) \quad (|t| < |a|); \tag{3}$$

$$\sum_{k=1}^{\infty} \zeta(2k+1,a)\frac{t^{2k+1}}{2k+1} = \frac{1}{2}\{\log\Gamma(a-t) - \log\Gamma(a+t)\} + t\psi(a) \quad (|t| < |a|). \tag{4}$$

Differentiating both sides of (1), (2), (3) and (4) with respect to t, we have

$$\sum_{k=2}^{\infty} \zeta(k,a)t^{k-1} = -\psi(a-t) + \psi(a) \quad (|t| < |a|); \tag{5}$$

$$\sum_{k=2}^{\infty} (-1)^k \zeta(k,a)t^{k-1} = \psi(a+t) - \psi(a) \quad (|t| < |a|); \tag{6}$$

$$\sum_{k=1}^{\infty} \zeta(2k,a)t^{2k-1} = \frac{1}{2}\{\psi(a+t) - \psi(a-t)\} \quad (|t| < |a|); \tag{7}$$

$$\sum_{k=1}^{\infty} \zeta(2k+1,a)t^{2k} = -\frac{1}{2}\{\psi(a+t) + \psi(a-t)\} + \psi(a) \quad (|t| < |a|). \tag{8}$$

Setting $a = 1$ in (1) through (8) and using 1.3(4) and 2.3(2), we obtain

$$\sum_{k=2}^{\infty} \zeta(k)\frac{t^k}{k} = \log\Gamma(1-t) - \gamma t \quad (|t| < 1); \tag{9}$$

$$\sum_{k=2}^{\infty} (-1)^k \zeta(k)\frac{t^k}{k} = \log\Gamma(1+t) + \gamma t \quad (|t| < 1); \tag{10}$$

$$\sum_{k=1}^{\infty} \zeta(2k)\frac{t^{2k}}{k} = \log\Gamma(1+t) + \log\Gamma(1-t) \quad (|t| < 1); \tag{11}$$

$$\sum_{k=1}^{\infty} \zeta(2k+1)\frac{t^{2k+1}}{2k+1} = \frac{1}{2}\{\log\Gamma(1-t) - \log\Gamma(1+t)\} - \gamma t \quad (|t| < 1); \tag{12}$$

$$\sum_{k=2}^{\infty} \zeta(k)t^{k-1} = -\psi(1-t) - \gamma \quad (|t| < 1); \tag{13}$$

$$\sum_{k=2}^{\infty}(-1)^k\zeta(k)t^{k-1} = \psi(1+t)+\gamma \quad (|t|<1); \tag{14}$$

$$\sum_{k=1}^{\infty}\zeta(2k)t^{2k-1} = \frac{1}{2}\{\psi(1+t)-\psi(1-t)\} \quad (|t|<1); \tag{15}$$

$$\sum_{k=1}^{\infty}\zeta(2k+1)t^{2k} = -\frac{1}{2}\{\psi(1+t)+\psi(1-t)\}-\gamma \quad (|t|<1). \tag{16}$$

Equations (11) and (15) can, in view of 1.1(12) and 1.3(9), be written in their respective equivalent forms:

$$\sum_{k=1}^{\infty}\zeta(2k)\frac{t^{2k}}{k} = \log\left(\frac{\pi t}{\sin \pi t}\right) \quad (|t|<1); \tag{17}$$

$$\sum_{k=1}^{\infty}\zeta(2k)t^{2k-1} = -\frac{\pi}{2}\cot(\pi t)+\frac{1}{2t} \quad (|t|<1). \tag{18}$$

If we use some elementary trigonometric identities, we obtain the following trigonometric evaluations:

$$\begin{aligned}
&\sin\left(\frac{\pi}{8}\right) = \sin\left(\frac{7\pi}{8}\right) = \frac{1}{2}\sqrt{2-\sqrt{2}};\\
&\sin\left(\frac{3\pi}{8}\right) = \sin\left(\frac{5\pi}{8}\right) = \frac{1}{2}\sqrt{2+\sqrt{2}};
\end{aligned} \tag{19}$$

$$\begin{aligned}
&\sin\left(\frac{\pi}{5}\right) = \sin\left(\frac{4\pi}{5}\right) = \cos\left(\frac{3\pi}{10}\right) = \frac{\sqrt{5-\sqrt{5}}}{2\sqrt{2}};\\
&\sin\left(\frac{2\pi}{5}\right) = \sin\left(\frac{3\pi}{5}\right) = \cos\left(\frac{\pi}{10}\right) = \frac{\sqrt{5+\sqrt{5}}}{2\sqrt{2}};
\end{aligned} \tag{20}$$

$$\begin{aligned}
&\cos\left(\frac{\pi}{5}\right) = -\cos\left(\frac{4\pi}{5}\right) = \sin\left(\frac{3\pi}{10}\right) = \frac{\sqrt{5}+1}{4};\\
&\cos\left(\frac{2\pi}{5}\right) = -\cos\left(\frac{3\pi}{5}\right) = \sin\left(\frac{\pi}{10}\right) = \frac{\sqrt{5}-1}{4};
\end{aligned} \tag{21}$$

$$\begin{aligned}
&\cot\left(\frac{\pi}{5}\right) = -\cot\left(\frac{4\pi}{5}\right) = \sqrt{1+\frac{2}{5}\sqrt{5}}; \quad \cot\left(\frac{2\pi}{5}\right) = -\cot\left(\frac{3\pi}{5}\right) = \sqrt{1-\frac{2}{5}\sqrt{5}};\\
&\cot\left(\frac{\pi}{10}\right) = -\cot\left(\frac{9\pi}{10}\right) = \sqrt{5+2\sqrt{5}}; \quad \cot\left(\frac{3\pi}{10}\right) = -\cot\left(\frac{7\pi}{10}\right) = \sqrt{5-2\sqrt{5}}.
\end{aligned} \tag{22}$$

Taking the limit as $t \to 1$ on both sides of (10), we have the well-known result:

$$\sum_{k=2}^{\infty}(-1)^k \frac{\zeta(k)}{k} = \gamma. \tag{23}$$

For suitable special values of the argument t, we can deduce the following series identities from (9) through (18), by making use of such evaluations as (19) through (22), as well as the ψ-function values listed in Section 1.3.

$$\sum_{k=2}^{\infty} \frac{\zeta(k)}{k \cdot 2^k} = \frac{1}{2}\log \pi - \frac{\gamma}{2}; \tag{24}$$

$$\sum_{k=2}^{\infty}(-1)^k \frac{\zeta(k)}{k \cdot 2^k} = \frac{\gamma}{2} + \frac{1}{2}\log \pi - \log 2; \tag{25}$$

$$\sum_{k=1}^{\infty} \frac{\zeta(2k)}{k \cdot 2^{2k}} = \log\left(\frac{\pi}{2}\right); \tag{26}$$

$$\sum_{k=1}^{\infty} \frac{\zeta(2k+1)}{(2k+1)2^{2k}} = \log 2 - \gamma; \tag{27}$$

$$\sum_{k=1}^{\infty} \frac{\zeta(2k)}{k \cdot 3^{2k}} = \log\left(\frac{2\pi}{3\sqrt{3}}\right); \tag{28}$$

$$\sum_{k=1}^{\infty} \frac{\zeta(2k)}{k}\left(\frac{2}{3}\right)^{2k} = \log\left(\frac{4\pi}{3\sqrt{3}}\right); \tag{29}$$

$$\sum_{k=1}^{\infty} \frac{\zeta(2k)}{k \cdot 2^{4k}} = \log\left(\frac{\pi}{2\sqrt{2}}\right); \tag{30}$$

$$\sum_{k=1}^{\infty} \frac{\zeta(2k)}{k}\left(\frac{3}{4}\right)^{2k} = \log\left(\frac{3\pi}{2\sqrt{2}}\right); \tag{31}$$

$$\sum_{k=1}^{\infty} \frac{\zeta(2k)}{k \cdot 5^{2k}} = \log\left(\frac{2\sqrt{2}\pi}{5\sqrt{5-\sqrt{5}}}\right); \tag{32}$$

$$\sum_{k=1}^{\infty} \frac{\zeta(2k)}{k}\left(\frac{2}{5}\right)^{2k} = \log\left(\frac{4\sqrt{2}\pi}{5\sqrt{5+\sqrt{5}}}\right); \tag{33}$$

$$\sum_{k=1}^{\infty} \frac{\zeta(2k)}{k}\left(\frac{3}{5}\right)^{2k} = \log\left(\frac{6\sqrt{2}\pi}{5\sqrt{5+\sqrt{5}}}\right); \tag{34}$$

$$\sum_{k=1}^{\infty} \frac{\zeta(2k)}{k}\left(\frac{4}{5}\right)^{2k} = \log\left(\frac{8\sqrt{2}\pi}{5\sqrt{5+\sqrt{5}}}\right); \tag{35}$$

$$\sum_{k=1}^{\infty} \frac{\zeta(2k)}{k \cdot 10^{2k}} = \log\left\{\frac{\left(\sqrt{5}+1\right)\pi}{10}\right\}; \tag{36}$$

$$\sum_{k=1}^{\infty} \frac{\zeta(2k)}{k} \left(\frac{3}{10}\right)^{2k} = \log \left\{ \frac{3\left(\sqrt{5}-1\right)\pi}{10} \right\}; \tag{37}$$

$$\sum_{k=1}^{\infty} \frac{\zeta(2k)}{k} \left(\frac{7}{10}\right)^{2k} = \log \left\{ \frac{7\left(\sqrt{5}-1\right)\pi}{10} \right\}; \tag{38}$$

$$\sum_{k=1}^{\infty} \frac{\zeta(2k)}{k} \left(\frac{9}{10}\right)^{2k} = \log \left\{ \frac{9\left(\sqrt{5}+1\right)\pi}{10} \right\}; \tag{39}$$

$$\sum_{k=1}^{\infty} \frac{\zeta(2k)}{k \cdot 6^{2k}} = \log \left(\frac{\pi}{3}\right); \tag{40}$$

$$\sum_{k=1}^{\infty} \frac{\zeta(2k)}{k} \left(\frac{5}{6}\right)^{2k} = \log \left(\frac{5\pi}{3}\right); \tag{41}$$

$$\sum_{k=1}^{\infty} \frac{\zeta(2k)}{k \cdot 2^{6k}} = \log \left(\frac{\pi}{4\sqrt{2-\sqrt{2}}}\right); \tag{42}$$

$$\sum_{k=1}^{\infty} \frac{\zeta(2k)}{k} \left(\frac{3}{8}\right)^{2k} = \log \left(\frac{3\pi}{4\sqrt{2+\sqrt{2}}}\right); \tag{43}$$

$$\sum_{k=1}^{\infty} \frac{\zeta(2k)}{k} \left(\frac{5}{8}\right)^{2k} = \log \left(\frac{5\pi}{4\sqrt{2+\sqrt{2}}}\right); \tag{44}$$

$$\sum_{k=1}^{\infty} \frac{\zeta(2k)}{k} \left(\frac{7}{8}\right)^{2k} = \log \left(\frac{7\pi}{8}\sqrt{4+2\sqrt{2}}\right); \tag{45}$$

$$\sum_{k=1}^{\infty} \frac{\zeta(2k)}{k \cdot 12^{2k}} = \log \left(\frac{\pi\left(\sqrt{3}+1\right)}{6\sqrt{2}}\right); \tag{46}$$

$$\sum_{k=1}^{\infty} \frac{\zeta(2k)}{k} \left(\frac{5}{12}\right)^{2k} = \log \left(\frac{5\pi\left(\sqrt{3}-1\right)}{6\sqrt{2}}\right); \tag{47}$$

$$\sum_{k=1}^{\infty} \frac{\zeta(2k)}{k} \left(\frac{7}{12}\right)^{2k} = \log \left(\frac{7\pi\left(\sqrt{3}-1\right)}{6\sqrt{2}}\right); \tag{48}$$

$$\sum_{k=1}^{\infty} \frac{\zeta(2k)}{k} \left(\frac{11}{12}\right)^{2k} = \log \left(\frac{11\pi\left(\sqrt{3}+1\right)}{6\sqrt{2}}\right); \tag{49}$$

$$\sum_{k=2}^{\infty} (-1)^k \frac{\zeta(k)}{2^k} = 1 - \log 2; \tag{50}$$

$$\sum_{k=2}^{\infty} \frac{\zeta(k)}{2^k} = \log 2; \tag{51}$$

$$\sum_{k=1}^{\infty} \frac{\zeta(2k)}{2^{2k-1}} = 1; \tag{52}$$

$$\sum_{k=1}^{\infty} \frac{\zeta(2k+1)}{2^{2k}} = 2\log 2 - 1; \tag{53}$$

$$\sum_{k=2}^{\infty} (-1)^k \frac{\zeta(k)}{3^k} = 1 - \frac{1}{18}\pi\sqrt{3} - \frac{1}{2}\log 3; \tag{54}$$

$$\sum_{k=2}^{\infty} \frac{\zeta(k)}{3^k} = \frac{1}{2}\log 3 - \frac{1}{18}\pi\sqrt{3}; \tag{55}$$

$$\sum_{k=1}^{\infty} \frac{\zeta(2k)}{3^{2k}} = \frac{1}{2} - \frac{1}{18}\pi\sqrt{3}; \tag{56}$$

$$\sum_{k=1}^{\infty} \frac{\zeta(2k+1)}{3^{2k}} = \frac{3}{2}(\log 3 - 1); \tag{57}$$

$$\sum_{k=2}^{\infty} (-1)^k \zeta(k) \left(\frac{2}{3}\right)^k = 1 + \frac{1}{9}\pi\sqrt{3} - \log 3; \tag{58}$$

$$\sum_{k=2}^{\infty} \zeta(k) \left(\frac{2}{3}\right)^k = \frac{1}{9}\pi\sqrt{3} + \log 3; \tag{59}$$

$$\sum_{k=1}^{\infty} \zeta(2k) \left(\frac{2}{3}\right)^{2k} = \frac{1}{2} + \frac{1}{9}\pi\sqrt{3}; \tag{60}$$

$$\sum_{k=1}^{\infty} \zeta(2k+1) \left(\frac{2}{3}\right)^{2k} = \frac{3}{2}\log 3 - \frac{3}{4}; \tag{61}$$

$$\sum_{k=2}^{\infty} (-1)^k \frac{\zeta(k)}{2^{2k}} = 1 - \frac{1}{8}\pi - \frac{3}{4}\log 2; \tag{62}$$

$$\sum_{k=2}^{\infty} \frac{\zeta(k)}{2^{2k}} = \frac{3}{4}\log 2 - \frac{1}{8}\pi; \tag{63}$$

$$\sum_{k=1}^{\infty} \frac{\zeta(2k)}{2^{4k}} = \frac{1}{2} - \frac{1}{8}\pi; \tag{64}$$

$$\sum_{k=1}^{\infty} \frac{\zeta(2k+1)}{2^{4k}} = 3\log 2 - 2; \tag{65}$$

$$\sum_{k=2}^{\infty} (-1)^k \zeta(k) \left(\frac{3}{4}\right)^k = 1 + \frac{3}{8}\pi - \frac{9}{4}\log 2; \tag{66}$$

$$\sum_{k=2}^{\infty}\zeta(k)\left(\frac{3}{4}\right)^{k}=\frac{3}{8}\pi+\frac{9}{4}\log 2; \tag{67}$$

$$\sum_{k=1}^{\infty}\zeta(2k)\left(\frac{3}{4}\right)^{2k}=\frac{1}{2}+\frac{3}{8}\pi; \tag{68}$$

$$\sum_{k=1}^{\infty}\zeta(2k+1)\left(\frac{3}{4}\right)^{2k}=3\log 2-\frac{2}{3}; \tag{69}$$

$$\sum_{k=2}^{\infty}(-1)^{k}\frac{\zeta(k)}{5^{k}}=1-\frac{\pi}{10}\sqrt{1+\frac{2}{5}\sqrt{5}}-\frac{1}{4}\log 5-\frac{\sqrt{5}}{10}\log\frac{1+\sqrt{5}}{2}; \tag{70}$$

$$\sum_{k=2}^{\infty}\frac{\zeta(k)}{5^{k}}=-\frac{\pi}{10}\sqrt{1+\frac{2}{5}\sqrt{5}}+\frac{1}{4}\log 5+\frac{\sqrt{5}}{10}\log\frac{1+\sqrt{5}}{2}; \tag{71}$$

$$\sum_{k=1}^{\infty}\frac{\zeta(2k)}{5^{2k}}=\frac{1}{2}-\frac{\pi}{10}\sqrt{1+\frac{2}{5}\sqrt{5}}; \tag{72}$$

$$\sum_{k=1}^{\infty}\frac{\zeta(2k+1)}{5^{2k}}=-\frac{5}{2}+\frac{5}{4}\log 5+\frac{\sqrt{5}}{2}\log\frac{1+\sqrt{5}}{2}; \tag{73}$$

$$\sum_{k=2}^{\infty}(-1)^{k}\zeta(k)\left(\frac{2}{5}\right)^{k}=1-\frac{\pi}{5}\sqrt{1-\frac{2}{5}\sqrt{5}}-\frac{1}{2}\log 5+\frac{\sqrt{5}}{5}\log\frac{1+\sqrt{5}}{2}; \tag{74}$$

$$\sum_{k=2}^{\infty}\zeta(k)\left(\frac{2}{5}\right)^{k}=-\frac{\pi}{5}\sqrt{1-\frac{2}{5}\sqrt{5}}+\frac{1}{2}\log 5-\frac{\sqrt{5}}{5}\log\frac{1+\sqrt{5}}{2}; \tag{75}$$

$$\sum_{k=1}^{\infty}\zeta(2k)\left(\frac{2}{5}\right)^{2k}=\frac{1}{2}-\frac{\pi}{5}\sqrt{1-\frac{2}{5}\sqrt{5}}; \tag{76}$$

$$\sum_{k=1}^{\infty}\zeta(2k+1)\left(\frac{2}{5}\right)^{2k}=-\frac{5}{4}+\frac{5}{4}\log 5-\frac{\sqrt{5}}{2}\log\frac{1+\sqrt{5}}{2}; \tag{77}$$

$$\sum_{k=2}^{\infty}(-1)^{k}\zeta(k)\left(\frac{3}{5}\right)^{k}=1+\frac{3\pi}{10}\sqrt{1-\frac{2}{5}\sqrt{5}}-\frac{3}{4}\log 5+\frac{3\sqrt{5}}{10}\log\frac{1+\sqrt{5}}{2}; \tag{78}$$

$$\sum_{k=2}^{\infty}\zeta(k)\left(\frac{3}{5}\right)^{k}=\frac{3\pi}{10}\sqrt{1-\frac{2}{5}\sqrt{5}}+\frac{3}{4}\log 5-\frac{3\sqrt{5}}{10}\log\frac{1+\sqrt{5}}{2}; \tag{79}$$

$$\sum_{k=1}^{\infty}\zeta(2k)\left(\frac{3}{5}\right)^{2k}=\frac{1}{2}+\frac{3\pi}{10}\sqrt{1-\frac{2}{5}\sqrt{5}}; \tag{80}$$

$$\sum_{k=1}^{\infty}\zeta(2k+1)\left(\frac{3}{5}\right)^{2k}=-\frac{5}{6}+\frac{5}{4}\log 5-\frac{\sqrt{5}}{2}\log\frac{1+\sqrt{5}}{2}; \tag{81}$$

$$\sum_{k=2}^{\infty}(-1)^{k}\zeta(k)\left(\frac{4}{5}\right)^{k}=1+\frac{2\pi}{5}\sqrt{1+\frac{2}{5}\sqrt{5}}-\log 5-\frac{2\sqrt{5}}{5}\log\frac{1+\sqrt{5}}{2}; \tag{82}$$

$$\sum_{k=2}^{\infty}\zeta(k)\left(\frac{4}{5}\right)^{k}=\frac{2\pi}{5}\sqrt{1+\frac{2}{5}\sqrt{5}}+\log 5+\frac{2\sqrt{5}}{5}\log\frac{1+\sqrt{5}}{2}; \tag{83}$$

$$\sum_{k=1}^{\infty}\zeta(2k)\left(\frac{4}{5}\right)^{2k}=\frac{1}{2}+\frac{2\pi}{5}\sqrt{1+\frac{2}{5}\sqrt{5}}; \tag{84}$$

$$\sum_{k=1}^{\infty}\zeta(2k+1)\left(\frac{4}{5}\right)^{2k}=-\frac{5}{8}+\frac{5}{4}\log 5+\frac{\sqrt{5}}{2}\log\frac{1+\sqrt{5}}{2}; \tag{85}$$

$$\sum_{k=2}^{\infty}(-1)^{k}\frac{\zeta(k)}{6^{k}}=1-\frac{1}{12}\pi\sqrt{3}-\frac{1}{4}\log 3-\frac{1}{3}\log 2; \tag{86}$$

$$\sum_{k=2}^{\infty}\frac{\zeta(k)}{6^{k}}=-\frac{1}{12}\pi\sqrt{3}+\frac{1}{4}\log 3+\frac{1}{3}\log 2; \tag{87}$$

$$\sum_{k=1}^{\infty}\frac{\zeta(2k)}{6^{2k}}=\frac{1}{2}-\frac{1}{12}\pi\sqrt{3}; \tag{88}$$

$$\sum_{k=1}^{\infty}\frac{\zeta(2k+1)}{6^{2k}}=-3+\frac{3}{2}\log 3+2\log 2; \tag{89}$$

$$\sum_{k=2}^{\infty}(-1)^{k}\zeta(k)\left(\frac{5}{6}\right)^{k}=1+\frac{5}{12}\pi\sqrt{3}-\frac{5}{4}\log 3-\frac{5}{3}\log 2; \tag{90}$$

$$\sum_{k=2}^{\infty}\zeta(k)\left(\frac{5}{6}\right)^{k}=\frac{5}{12}\pi\sqrt{3}+\frac{5}{4}\log 3+\frac{5}{3}\log 2; \tag{91}$$

$$\sum_{k=1}^{\infty}\zeta(2k)\left(\frac{5}{6}\right)^{2k}=\frac{1}{2}+\frac{5}{12}\pi\sqrt{3}; \tag{92}$$

$$\sum_{k=1}^{\infty}\zeta(2k+1)\left(\frac{5}{6}\right)^{2k}=-\frac{3}{5}+\frac{3}{2}\log 3+2\log 2; \tag{93}$$

$$\sum_{k=2}^{\infty}(-1)^{k}\frac{\zeta(k)}{2^{3k}}=1-\frac{\pi}{16}\left(\sqrt{2}+1\right)-\frac{1}{2}\log 2+\frac{\sqrt{2}}{16}\log\left(3-2\sqrt{2}\right); \tag{94}$$

$$\sum_{k=2}^{\infty}\frac{\zeta(k)}{2^{3k}}=-\frac{\pi}{16}\left(\sqrt{2}+1\right)+\frac{1}{2}\log 2-\frac{\sqrt{2}}{16}\log\left(3-2\sqrt{2}\right); \tag{95}$$

$$\sum_{k=1}^{\infty}\frac{\zeta(2k)}{2^{6k}}=\frac{1}{2}-\frac{\pi}{16}\left(\sqrt{2}+1\right); \tag{96}$$

$$\sum_{k=1}^{\infty}\frac{\zeta(2k+1)}{2^{6k}}=-4+4\log 2-\frac{\sqrt{2}}{2}\log\left(3-2\sqrt{2}\right); \tag{97}$$

$$\sum_{k=2}^{\infty}(-1)^{k}\zeta(k)\left(\frac{3}{8}\right)^{k}=1-\frac{3\pi}{16}\left(\sqrt{2}-1\right)-\frac{3}{2}\log 2-\frac{3\sqrt{2}}{16}\log\left(3-2\sqrt{2}\right); \tag{98}$$

$$\sum_{k=2}^{\infty} \zeta(k) \left(\frac{3}{8}\right)^k = -\frac{3\pi}{16}\left(\sqrt{2}-1\right) + \frac{3}{2}\log 2 + \frac{3\sqrt{2}}{16}\log\left(3-2\sqrt{2}\right); \tag{99}$$

$$\sum_{k=1}^{\infty} \zeta(2k) \left(\frac{3}{8}\right)^{2k} = \frac{1}{2} - \frac{3\pi}{16}\left(\sqrt{2}-1\right); \tag{100}$$

$$\sum_{k=1}^{\infty} \zeta(2k+1) \left(\frac{3}{8}\right)^{2k} = -\frac{4}{3} + 4\log 2 + \frac{\sqrt{2}}{2}\log\left(3-2\sqrt{2}\right); \tag{101}$$

$$\sum_{k=2}^{\infty} (-1)^k \zeta(k) \left(\frac{5}{8}\right)^k = 1 + \frac{5\pi}{16}\left(\sqrt{2}-1\right) - \frac{5}{2}\log 2 - \frac{5\sqrt{2}}{16}\log\left(3-2\sqrt{2}\right); \tag{102}$$

$$\sum_{k=2}^{\infty} \zeta(k) \left(\frac{5}{8}\right)^k = \frac{5\pi}{16}\left(\sqrt{2}-1\right) + \frac{5}{2}\log 2 + \frac{5\sqrt{2}}{16}\log\left(3-2\sqrt{2}\right); \tag{103}$$

$$\sum_{k=1}^{\infty} \zeta(2k) \left(\frac{5}{8}\right)^{2k} = \frac{1}{2} + \frac{5\pi}{16}\left(\sqrt{2}-1\right); \tag{104}$$

$$\sum_{k=1}^{\infty} \zeta(2k+1) \left(\frac{5}{8}\right)^{2k} = -\frac{4}{5} + 4\log 2 + \frac{\sqrt{2}}{2}\log\left(3-2\sqrt{2}\right); \tag{105}$$

$$\sum_{k=2}^{\infty} (-1)^k \zeta(k) \left(\frac{7}{8}\right)^k = 1 + \frac{7\pi}{16}\left(\sqrt{2}+1\right) - \frac{7}{2}\log 2 + \frac{7\sqrt{2}}{16}\log\left(3-2\sqrt{2}\right); \tag{106}$$

$$\sum_{k=2}^{\infty} \zeta(k) \left(\frac{7}{8}\right)^k = \frac{7\pi}{16}\left(\sqrt{2}+1\right) + \frac{7}{2}\log 2 - \frac{7\sqrt{2}}{16}\log\left(3-2\sqrt{2}\right); \tag{107}$$

$$\sum_{k=1}^{\infty} \zeta(2k) \left(\frac{7}{8}\right)^{2k} = \frac{1}{2} + \frac{7\pi}{16}\left(\sqrt{2}+1\right); \tag{108}$$

$$\sum_{k=1}^{\infty} \zeta(2k+1) \left(\frac{7}{8}\right)^{2k} = -\frac{4}{7} + 4\log 2 - \frac{\sqrt{2}}{2}\log\left(3-2\sqrt{2}\right); \tag{109}$$

$$\sum_{k=2}^{\infty} (-1)^k \frac{\zeta(k)}{10^k} = 1 - \frac{\pi}{20}\sqrt{5+2\sqrt{5}} - \frac{1}{5}\log 2 - \frac{1}{8}\log 5 - \frac{\sqrt{5}}{20}\log\left(2+\sqrt{5}\right); \tag{110}$$

$$\sum_{k=2}^{\infty} \frac{\zeta(k)}{10^k} = -\frac{\pi}{20}\sqrt{5+2\sqrt{5}} + \frac{1}{5}\log 2 + \frac{1}{8}\log 5 + \frac{\sqrt{5}}{20}\log\left(2+\sqrt{5}\right); \tag{111}$$

$$\sum_{k=1}^{\infty} \frac{\zeta(2k)}{10^{2k}} = \frac{1}{2} - \frac{\pi}{20}\sqrt{5+2\sqrt{5}}; \tag{112}$$

$$\sum_{k=1}^{\infty} \frac{\zeta(2k+1)}{10^{2k}} = -5 + 2\log 2 + \frac{5}{4}\log 5 + \frac{\sqrt{5}}{2}\log\left(2+\sqrt{5}\right); \tag{113}$$

$$\sum_{k=2}^{\infty}(-1)^k\zeta(k)\left(\frac{3}{10}\right)^k = 1-\frac{3\pi}{20}\sqrt{5-2\sqrt{5}}-\frac{3}{5}\log 2 - \frac{3}{8}\log 5+\frac{3\sqrt{5}}{20}\log\left(2+\sqrt{5}\right); \tag{114}$$

$$\sum_{k=2}^{\infty}\zeta(k)\left(\frac{3}{10}\right)^k = -\frac{3\pi}{20}\sqrt{5-2\sqrt{5}}+\frac{3}{5}\log 2+\frac{3}{8}\log 5-\frac{3\sqrt{5}}{20}\log\left(2+\sqrt{5}\right); \tag{115}$$

$$\sum_{k=1}^{\infty}\zeta(2k)\left(\frac{3}{10}\right)^{2k} = \frac{1}{2}-\frac{3\pi}{20}\sqrt{5-2\sqrt{5}}; \tag{116}$$

$$\sum_{k=1}^{\infty}\zeta(2k+1)\left(\frac{3}{10}\right)^{2k} = -\frac{5}{3}+2\log 2+\frac{5}{4}\log 5-\frac{\sqrt{5}}{2}\log\left(2+\sqrt{5}\right); \tag{117}$$

$$\sum_{k=2}^{\infty}(-1)^k\zeta(k)\left(\frac{7}{10}\right)^k = 1+\frac{7\pi}{20}\sqrt{5-2\sqrt{5}}-\frac{7}{5}\log 2 - \frac{7}{8}\log 5+\frac{7\sqrt{5}}{20}\log\left(2+\sqrt{5}\right); \tag{118}$$

$$\sum_{k=2}^{\infty}\zeta(k)\left(\frac{7}{10}\right)^k = \frac{7\pi}{20}\sqrt{5-2\sqrt{5}}+\frac{7}{5}\log 2+\frac{7}{8}\log 5-\frac{7\sqrt{5}}{20}\log\left(2+\sqrt{5}\right); \tag{119}$$

$$\sum_{k=1}^{\infty}\zeta(2k)\left(\frac{7}{10}\right)^{2k} = \frac{1}{2}+\frac{7\pi}{20}\sqrt{5-2\sqrt{5}}; \tag{120}$$

$$\sum_{k=1}^{\infty}\zeta(2k+1)\left(\frac{7}{10}\right)^{2k} = -\frac{5}{7}+2\log 2+\frac{5}{4}\log 5-\frac{\sqrt{5}}{2}\log\left(2+\sqrt{5}\right); \tag{121}$$

$$\sum_{k=2}^{\infty}(-1)^k\zeta(k)\left(\frac{9}{10}\right)^k = 1+\frac{9\pi}{20}\sqrt{5+2\sqrt{5}}-\frac{9}{5}\log 2 - \frac{9}{8}\log 5-\frac{9\sqrt{5}}{20}\log\left(2+\sqrt{5}\right); \tag{122}$$

$$\sum_{k=2}^{\infty}\zeta(k)\left(\frac{9}{10}\right)^k = \frac{9\pi}{20}\sqrt{5+2\sqrt{5}}+\frac{9}{5}\log 2+\frac{9}{8}\log 5+\frac{9\sqrt{5}}{20}\log\left(2+\sqrt{5}\right); \tag{123}$$

$$\sum_{k=1}^{\infty}\zeta(2k)\left(\frac{9}{10}\right)^{2k} = \frac{1}{2}+\frac{9\pi}{20}\sqrt{5+2\sqrt{5}}; \tag{124}$$

$$\sum_{k=1}^{\infty}\zeta(2k+1)\left(\frac{9}{10}\right)^{2k} = -\frac{5}{9}+2\log 2+\frac{5}{4}\log 5+\frac{\sqrt{5}}{2}\log\left(2+\sqrt{5}\right); \tag{125}$$

$$\sum_{k=2}^{\infty}(-1)^k\frac{\zeta(k)}{12^k}=1-\frac{\pi}{24}\left(2+\sqrt{3}\right)+\frac{\sqrt{3}}{24}\log\left(7-4\sqrt{3}\right)-\frac{1}{8}\log 3+\frac{1}{4}\log 2; \tag{126}$$

$$\sum_{k=2}^{\infty}\frac{\zeta(k)}{12^k}=-\frac{\pi}{24}\left(2+\sqrt{3}\right)-\frac{\sqrt{3}}{24}\log\left(7-4\sqrt{3}\right)+\frac{1}{8}\log 3+\frac{1}{4}\log 2; \tag{127}$$

$$\sum_{k=1}^{\infty}\frac{\zeta(2k)}{12^{2k}}=\frac{1}{2}-\frac{\pi}{24}\left(2+\sqrt{3}\right); \tag{128}$$

$$\sum_{k=1}^{\infty}\frac{\zeta(2k+1)}{12^{2k}}=-6-\frac{1}{2}\sqrt{3}\log\left(7-4\sqrt{3}\right)+\frac{3}{2}\log 3+3\log 2; \tag{129}$$

$$\sum_{k=2}^{\infty}(-1)^k\zeta(k)\left(\frac{5}{12}\right)^k$$
$$=1-\frac{5\pi}{24}\left(2-\sqrt{3}\right)-\frac{5\sqrt{3}}{24}\log\left(7-4\sqrt{3}\right)-\frac{5}{8}\log 3-\frac{5}{4}\log 2; \tag{130}$$

$$\sum_{k=2}^{\infty}\zeta(k)\left(\frac{5}{12}\right)^k$$
$$=-\frac{5\pi}{24}\left(2-\sqrt{3}\right)+\frac{5\sqrt{3}}{24}\log\left(7-4\sqrt{3}\right)+\frac{5}{8}\log 3+\frac{5}{4}\log 2; \tag{131}$$

$$\sum_{k=1}^{\infty}\zeta(2k)\left(\frac{5}{12}\right)^{2k}=\frac{1}{2}-\frac{5\pi}{24}\left(2-\sqrt{3}\right); \tag{132}$$

$$\sum_{k=1}^{\infty}\zeta(2k+1)\left(\frac{5}{12}\right)^{2k}=-\frac{6}{5}+\frac{\sqrt{3}}{2}\log\left(7-4\sqrt{3}\right)+\frac{3}{2}\log 3+3\log 2; \tag{133}$$

$$\sum_{k=2}^{\infty}(-1)^k\zeta(k)\left(\frac{7}{12}\right)^k$$
$$=1+\frac{7\pi}{24}\left(2-\sqrt{3}\right)-\frac{7\sqrt{3}}{24}\log\left(7-4\sqrt{3}\right)-\frac{7}{8}\log 3-\frac{7}{4}\log 2; \tag{134}$$

$$\sum_{k=2}^{\infty}\zeta(k)\left(\frac{7}{12}\right)^k$$
$$=\frac{7\pi}{24}\left(2-\sqrt{3}\right)+\frac{7\sqrt{3}}{24}\log\left(7-4\sqrt{3}\right)+\frac{7}{8}\log 3+\frac{7}{4}\log 2; \tag{135}$$

$$\sum_{k=1}^{\infty}\zeta(2k)\left(\frac{7}{12}\right)^{2k}=\frac{1}{2}+\frac{7\pi}{24}\left(2-\sqrt{3}\right); \tag{136}$$

$$\sum_{k=1}^{\infty}\zeta(2k+1)\left(\frac{7}{12}\right)^{2k}=-\frac{6}{7}+\frac{\sqrt{3}}{2}\log\left(7-4\sqrt{3}\right)+\frac{3}{2}\log 3+3\log 2; \tag{137}$$

$$\sum_{k=2}^{\infty}(-1)^k\zeta(k)\left(\frac{11}{12}\right)^k$$
$$= 1+\frac{11\pi}{24}\left(2+\sqrt{3}\right)+\frac{11\sqrt{3}}{24}\log\left(7-4\sqrt{3}\right)-\frac{11}{8}\log 3-\frac{11}{4}\log 2; \tag{138}$$

$$\sum_{k=2}^{\infty}\zeta(k)\left(\frac{11}{12}\right)^k$$
$$= \frac{11\pi}{24}\left(2+\sqrt{3}\right)-\frac{11\sqrt{3}}{24}\log\left(7-4\sqrt{3}\right)+\frac{11}{8}\log 3+\frac{11}{4}\log 2; \tag{139}$$

$$\sum_{k=1}^{\infty}\zeta(2k)\left(\frac{11}{12}\right)^{2k} = \frac{1}{2}+\frac{11\pi}{24}\left(2+\sqrt{3}\right); \tag{140}$$

$$\sum_{k=1}^{\infty}\zeta(2k+1)\left(\frac{11}{12}\right)^{2k} = -\frac{6}{11}-\frac{\sqrt{3}}{2}\log\left(7-4\sqrt{3}\right)+\frac{3}{2}\log 3+3\log 2. \tag{141}$$

In view of the various identities given in previous sections and the special case of 2.3(9) when $n = 1$, by setting $a = 2$ in (1) through (8), we obtain the following additional series identities:

$$\sum_{k=2}^{\infty}\{\zeta(k)-1\}\frac{t^k}{k} = \log\Gamma(2-t)+(1-\gamma)t \quad (|t|<2); \tag{142}$$

$$\sum_{k=2}^{\infty}(-1)^k\{\zeta(k)-1\}\frac{t^k}{k} = \log\Gamma(2+t)+(\gamma-1)t \quad (|t|<2); \tag{143}$$

$$\sum_{k=1}^{\infty}\{\zeta(2k)-1\}\frac{t^{2k}}{k} = \log\Gamma(2+t)+\log\Gamma(2-t) \quad (|t|<2); \tag{144}$$

$$\sum_{k=1}^{\infty}\{\zeta(2k+1)-1\}\frac{t^{2k+1}}{2k+1}$$
$$= \frac{1}{2}\{\log\Gamma(2-t)-\log\Gamma(2+t)\}+(1-\gamma)t \quad (|t|<2); \tag{145}$$

$$\sum_{k=2}^{\infty}\{\zeta(k)-1\}t^{k-1} = -\psi(2-t)+1-\gamma \quad (|t|<2); \tag{146}$$

$$\sum_{k=2}^{\infty}(-1)^k(\zeta(k)-1)t^{k-1} = \psi(2+t)+\gamma-1 \quad (|t|<2); \tag{147}$$

$$\sum_{k=1}^{\infty}\{\zeta(2k)-1\}t^{2k-1} = \frac{1}{2}\{\psi(2+t)-\psi(2-t)\} \quad (|t|<2); \tag{148}$$

$$\sum_{k=1}^{\infty}\{\zeta(2k+1)-1\}t^{2k} = -\frac{1}{2}\{\psi(2+t)+\psi(2-t)\}+1-\gamma \quad (|t|<2); \tag{149}$$

$$\sum_{k=1}^{\infty}\{\zeta(2k)-1\}\frac{t^{2k}}{k}=\log\left(\frac{\pi t(1-t^2)}{\sin\pi t}\right)\quad(|t|<2); \tag{150}$$

$$\sum_{k=1}^{\infty}\{\zeta(2k)-1\}t^{2k-1}=-\frac{\pi}{2}\cot(\pi t)+\frac{3t^2-1}{2t(t^2-1)}\quad(|t|<2). \tag{151}$$

Similarly, as above, setting various suitable arguments in (142) through (151), we readily obtain the following evaluations: for instance, by taking the limit in (143) as $t\to 2$, we obtain

$$\sum_{k=2}^{\infty}(-1)^k\{\zeta(k)-1\}\frac{2^k}{k}=2\gamma-2+\log 6; \tag{152}$$

$$\sum_{k=2}^{\infty}(-1)^k\frac{\zeta(k)-1}{k\cdot 2^k}=\frac{\gamma}{2}-\frac{1}{2}+\log\left(\frac{3\sqrt{\pi}}{4}\right); \tag{153}$$

$$\sum_{k=2}^{\infty}\frac{\zeta(k)-1}{k\cdot 2^k}=\frac{1}{2}-\frac{\gamma}{2}+\log\left(\frac{\sqrt{\pi}}{2}\right); \tag{154}$$

$$\sum_{k=1}^{\infty}\frac{\zeta(2k)-1}{k\cdot 2^{2k}}=\log\left(\frac{3\pi}{8}\right); \tag{155}$$

$$\sum_{k=1}^{\infty}\frac{\zeta(2k+1)-1}{(2k+1)2^{2k}}=1-\gamma+\log\left(\frac{2}{3}\right); \tag{156}$$

$$\sum_{k=2}^{\infty}(-1)^k\frac{\zeta(k)-1}{k}=\gamma-1+\log 2; \tag{157}$$

$$\sum_{k=2}^{\infty}\frac{\zeta(k)-1}{k}=1-\gamma; \tag{158}$$

$$\sum_{k=1}^{\infty}\frac{\zeta(2k)-1}{k}=\log 2; \tag{159}$$

$$\sum_{k=1}^{\infty}\frac{\zeta(2k+1)-1}{2k+1}=1-\gamma-\frac{1}{2}\log 2; \tag{160}$$

$$\sum_{k=2}^{\infty}(-1)^k\frac{\zeta(k)-1}{k}\left(\frac{3}{2}\right)^k=\frac{3\gamma}{2}-\frac{3}{2}+\log\left(\frac{15\sqrt{\pi}}{8}\right); \tag{161}$$

$$\sum_{k=2}^{\infty}\frac{\zeta(k)-1}{k}\left(\frac{3}{2}\right)^k=\frac{3}{2}-\frac{3\gamma}{2}+\frac{1}{2}\log\pi; \tag{162}$$

$$\sum_{k=1}^{\infty}\frac{\zeta(2k)-1}{k}\left(\frac{3}{2}\right)^{2k}=\log\left(\frac{15\pi}{8}\right); \tag{163}$$

$$\sum_{k=1}^{\infty} \frac{\zeta(2k+1)-1}{2k+1} \left(\frac{3}{2}\right)^{2k} = 1 - \gamma + \frac{1}{3} \log\left(\frac{8}{15}\right); \tag{164}$$

$$\sum_{k=1}^{\infty} \frac{\zeta(2k)-1}{k \cdot 3^{2k}} = \log\left(\frac{16\pi}{27\sqrt{3}}\right); \tag{165}$$

$$\sum_{k=1}^{\infty} \frac{\zeta(2k)-1}{k} \left(\frac{2}{3}\right)^{2k} = \log\left(\frac{20\pi}{27\sqrt{3}}\right); \tag{166}$$

$$\sum_{k=1}^{\infty} \frac{\zeta(2k)-1}{k} \left(\frac{4}{3}\right)^{2k} = \log\left(\frac{56\pi}{27\sqrt{3}}\right); \tag{167}$$

$$\sum_{k=1}^{\infty} \frac{\zeta(2k)-1}{k} \left(\frac{5}{3}\right)^{2k} = \log\left(\frac{160\pi}{27\sqrt{3}}\right); \tag{168}$$

$$\sum_{k=1}^{\infty} \frac{\zeta(2k)-1}{k \cdot 2^{4k}} = \log\left(\frac{15\pi}{32\sqrt{2}}\right); \tag{169}$$

$$\sum_{k=1}^{\infty} \frac{\zeta(2k)-1}{k} \left(\frac{3}{4}\right)^{2k} = \log\left(\frac{21\pi}{32\sqrt{2}}\right); \tag{170}$$

$$\sum_{k=1}^{\infty} \frac{\zeta(2k)-1}{k} \left(\frac{5}{4}\right)^{2k} = \log\left(\frac{45\pi}{32\sqrt{2}}\right); \tag{171}$$

$$\sum_{k=1}^{\infty} \frac{\zeta(2k)-1}{k} \left(\frac{7}{4}\right)^{2k} = \log\left(\frac{231\pi}{32\sqrt{2}}\right); \tag{172}$$

$$\sum_{k=1}^{\infty} \frac{\zeta(2k)-1}{k \cdot 5^{2k}} = \log\left(\frac{48\sqrt{2}\,\pi}{125\sqrt{5-\sqrt{5}}}\right); \tag{173}$$

$$\sum_{k=1}^{\infty} \frac{\zeta(2k)-1}{k} \left(\frac{2}{5}\right)^{2k} = \log\left(\frac{84\sqrt{2}\,\pi}{125\sqrt{5+\sqrt{5}}}\right); \tag{174}$$

$$\sum_{k=1}^{\infty} \frac{\zeta(2k)-1}{k} \left(\frac{3}{5}\right)^{2k} = \log\left(\frac{96\sqrt{2}\,\pi}{125\sqrt{5+\sqrt{5}}}\right); \tag{175}$$

$$\sum_{k=1}^{\infty} \frac{\zeta(2k)-1}{k} \left(\frac{4}{5}\right)^{2k} = \log\left(\frac{72\sqrt{2}\,\pi}{125\sqrt{5-\sqrt{5}}}\right); \tag{176}$$

$$\sum_{k=1}^{\infty} \frac{\zeta(2k)-1}{k} \left(\frac{6}{5}\right)^{2k} = \log\left(\frac{132\sqrt{2}\,\pi}{125\sqrt{5-\sqrt{5}}}\right); \tag{177}$$

$$\sum_{k=1}^{\infty} \frac{\zeta(2k)-1}{k} \left(\frac{7}{5}\right)^{2k} = \log\left(\frac{336\sqrt{2}\,\pi}{125\sqrt{5+\sqrt{5}}}\right); \tag{178}$$

$$\sum_{k=1}^{\infty} \frac{\zeta(2k)-1}{k}\left(\frac{8}{5}\right)^{2k} = \log\left(\frac{624\sqrt{2}\pi}{125\sqrt{5+\sqrt{5}}}\right); \tag{179}$$

$$\sum_{k=1}^{\infty} \frac{\zeta(2k)-1}{k}\left(\frac{9}{5}\right)^{2k} = \log\left(\frac{1008\sqrt{2}\pi}{125\sqrt{5-\sqrt{5}}}\right); \tag{180}$$

$$\sum_{k=1}^{\infty} \frac{\zeta(2k)-1}{k\cdot 10^{2k}} = \log\left\{\frac{99\left(\sqrt{5}+1\right)\pi}{1000}\right\}; \tag{181}$$

$$\sum_{k=1}^{\infty} \frac{\zeta(2k)-1}{k}\left(\frac{3}{10}\right)^{2k} = \log\left\{\frac{273\left(\sqrt{5}-1\right)\pi}{1000}\right\}; \tag{182}$$

$$\sum_{k=1}^{\infty} \frac{\zeta(2k)-1}{k}\left(\frac{7}{10}\right)^{2k} = \log\left\{\frac{357\left(\sqrt{5}-1\right)\pi}{1000}\right\}; \tag{183}$$

$$\sum_{k=1}^{\infty} \frac{\zeta(2k)-1}{k}\left(\frac{9}{10}\right)^{2k} = \log\left\{\frac{171\left(\sqrt{5}+1\right)\pi}{1000}\right\}; \tag{184}$$

$$\sum_{k=1}^{\infty} \frac{\zeta(2k)-1}{k}\left(\frac{11}{10}\right)^{2k} = \log\left\{\frac{231\left(\sqrt{5}+1\right)\pi}{1000}\right\}; \tag{185}$$

$$\sum_{k=1}^{\infty} \frac{\zeta(2k)-1}{k}\left(\frac{13}{10}\right)^{2k} = \log\left\{\frac{897\left(\sqrt{5}-1\right)\pi}{1000}\right\}; \tag{186}$$

$$\sum_{k=1}^{\infty} \frac{\zeta(2k)-1}{k}\left(\frac{17}{10}\right)^{2k} = \log\left\{\frac{3213\left(\sqrt{5}-1\right)\pi}{1000}\right\}; \tag{187}$$

$$\sum_{k=1}^{\infty} \frac{\zeta(2k)-1}{k}\left(\frac{19}{10}\right)^{2k} = \log\left\{\frac{4959\left(\sqrt{5}+1\right)\pi}{1000}\right\}; \tag{188}$$

$$\sum_{k=1}^{\infty} \frac{\zeta(2k)-1}{k\cdot 6^{2k}} = \log\left(\frac{35\pi}{108}\right); \tag{189}$$

$$\sum_{k=1}^{\infty} \frac{\zeta(2k)-1}{k}\left(\frac{5}{6}\right)^{2k} = \log\left(\frac{55\pi}{108}\right); \tag{190}$$

$$\sum_{k=1}^{\infty} \frac{\zeta(2k)-1}{k}\left(\frac{7}{6}\right)^{2k} = \log\left(\frac{91\pi}{108}\right); \tag{191}$$

$$\sum_{k=1}^{\infty} \frac{\zeta(2k)-1}{k}\left(\frac{11}{6}\right)^{2k} = \log\left(\frac{935\pi}{108}\right); \tag{192}$$

$$\sum_{k=1}^{\infty} \frac{\zeta(2k)-1}{k\cdot 2^{6k}} = \log\left(\frac{63\pi}{256\sqrt{2-\sqrt{2}}}\right); \tag{193}$$

$$\sum_{k=1}^{\infty} \frac{\zeta(2k)-1}{k}\left(\frac{3}{8}\right)^{2k} = \log\left(\frac{165\pi}{256\sqrt{2+\sqrt{2}}}\right); \tag{194}$$

$$\sum_{k=1}^{\infty} \frac{\zeta(2k)-1}{k}\left(\frac{5}{8}\right)^{2k} = \log\left(\frac{195\pi}{256\sqrt{2+\sqrt{2}}}\right); \tag{195}$$

$$\sum_{k=1}^{\infty} \frac{\zeta(2k)-1}{k}\left(\frac{7}{8}\right)^{2k} = \log\left(\frac{105\pi}{256\sqrt{2-\sqrt{2}}}\right); \tag{196}$$

$$\sum_{k=1}^{\infty} \frac{\zeta(2k)-1}{k}\left(\frac{9}{8}\right)^{2k} = \log\left(\frac{153\pi}{256\sqrt{2-\sqrt{2}}}\right); \tag{197}$$

$$\sum_{k=1}^{\infty} \frac{\zeta(2k)-1}{k}\left(\frac{11}{8}\right)^{2k} = \log\left(\frac{627\pi}{256\sqrt{2+\sqrt{2}}}\right); \tag{198}$$

$$\sum_{k=1}^{\infty} \frac{\zeta(2k)-1}{k}\left(\frac{13}{8}\right)^{2k} = \log\left(\frac{1365\pi}{256\sqrt{2+\sqrt{2}}}\right); \tag{199}$$

$$\sum_{k=1}^{\infty} \frac{\zeta(2k)-1}{k}\left(\frac{15}{8}\right)^{2k} = \log\left(\frac{2415\pi}{256\sqrt{2-\sqrt{2}}}\right); \tag{200}$$

$$\sum_{k=1}^{\infty} \frac{\zeta(2k)-1}{k\cdot 12^{2k}} = \log\left(\frac{143\pi(\sqrt{3}+1)}{864\sqrt{2}}\right); \tag{201}$$

$$\sum_{k=1}^{\infty} \frac{\zeta(2k)-1}{k}\left(\frac{5}{12}\right)^{2k} = \log\left(\frac{595\pi(\sqrt{3}-1)}{864\sqrt{2}}\right); \tag{202}$$

$$\sum_{k=1}^{\infty} \frac{\zeta(2k)-1}{k}\left(\frac{7}{12}\right)^{2k} = \log\left(\frac{665\pi(\sqrt{3}-1)}{864\sqrt{2}}\right); \tag{203}$$

$$\sum_{k=1}^{\infty} \frac{\zeta(2k)-1}{k}\left(\frac{11}{12}\right)^{2k} = \log\left(\frac{253\pi(\sqrt{3}+1)}{864\sqrt{2}}\right); \tag{204}$$

$$\sum_{k=1}^{\infty} \frac{\zeta(2k)-1}{k}\left(\frac{13}{12}\right)^{2k} = \log\left(\frac{325\pi(\sqrt{3}+1)}{864\sqrt{2}}\right); \tag{205}$$

$$\sum_{k=1}^{\infty} \frac{\zeta(2k)-1}{k}\left(\frac{17}{12}\right)^{2k} = \log\left(\frac{2465\pi(\sqrt{3}-1)}{864\sqrt{2}}\right); \tag{206}$$

$$\sum_{k=1}^{\infty} \frac{\zeta(2k)-1}{k}\left(\frac{19}{12}\right)^{2k} = \log\left(\frac{4123\pi(\sqrt{3}-1)}{864\sqrt{2}}\right); \tag{207}$$

$$\sum_{k=1}^{\infty} \frac{\zeta(2k)-1}{k} \left(\frac{23}{12}\right)^{2k} = \log\left(\frac{8855\pi(\sqrt{3}+1)}{864\sqrt{2}}\right); \tag{208}$$

$$\sum_{k=2}^{\infty} (-1)^k \frac{\zeta(k)-1}{2^k} = \frac{5}{6} - \log 2; \tag{209}$$

$$\sum_{k=2}^{\infty} \frac{\zeta(k)-1}{2^k} = \log 2 - \frac{1}{2}; \tag{210}$$

$$\sum_{k=1}^{\infty} \frac{\zeta(2k)-1}{2^{2k}} = \frac{1}{6}; \tag{211}$$

$$\sum_{k=1}^{\infty} \frac{\zeta(2k+1)-1}{2^{2k}} = -\frac{4}{3} + 2\log 2; \tag{212}$$

$$\sum_{k=2}^{\infty} (-1)^k \{\zeta(k)-1\} = \frac{1}{2}; \tag{213}$$

$$\sum_{k=2}^{\infty} \{\zeta(k)-1\} = 1; \tag{214}$$

$$\sum_{k=1}^{\infty} \{\zeta(2k)-1\} = \frac{3}{4}; \tag{215}$$

$$\sum_{k=1}^{\infty} \{\zeta(2k+1)-1\} = \frac{1}{4}; \tag{216}$$

$$\sum_{k=2}^{\infty} (-1)^k \{\zeta(k)-1\} \left(\frac{3}{2}\right)^k = \frac{31}{10} - 3\log 2; \tag{217}$$

$$\sum_{k=2}^{\infty} \{\zeta(k)-1\} \left(\frac{3}{2}\right)^k = \frac{3}{2} + 3\log 2; \tag{218}$$

$$\sum_{k=1}^{\infty} \{\zeta(2k)-1\} \left(\frac{3}{2}\right)^{2k} = \frac{23}{10}; \tag{219}$$

$$\sum_{k=1}^{\infty} \{\zeta(2k+1)-1\} \left(\frac{3}{2}\right)^{2k} = -\frac{8}{15} + 2\log 2; \tag{220}$$

$$\sum_{k=2}^{\infty} (-1)^k \frac{\zeta(k)-1}{3^k} = \frac{11}{12} - \frac{\pi}{18}\sqrt{3} - \frac{1}{2}\log 3; \tag{221}$$

$$\sum_{k=2}^{\infty} \frac{\zeta(k)-1}{3^k} = -\frac{1}{6} - \frac{\pi}{18}\sqrt{3} + \frac{1}{2}\log 3; \tag{222}$$

$$\sum_{k=1}^{\infty} \frac{\zeta(2k)-1}{3^{2k}} = \frac{3}{8} - \frac{1}{18}\pi\sqrt{3}; \tag{223}$$

$$\sum_{k=1}^{\infty} \frac{\zeta(2k+1)-1}{3^{2k}} = -\frac{13}{8} + \frac{3}{2}\log 3; \tag{224}$$

$$\sum_{k=2}^{\infty} (-1)^k (\zeta(k)-1) \left(\frac{2}{3}\right)^k = \frac{11}{15} + \frac{1}{9}\pi\sqrt{3} - \log 3; \tag{225}$$

$$\sum_{k=2}^{\infty} \{\zeta(k)-1\} \left(\frac{2}{3}\right)^k = -\frac{4}{3} + \frac{1}{9}\pi\sqrt{3} + \log 3; \tag{226}$$

$$\sum_{k=1}^{\infty} \{\zeta(2k)-1\} \left(\frac{2}{3}\right)^{2k} = -\frac{3}{10} + \frac{1}{9}\pi\sqrt{3}; \tag{227}$$

$$\sum_{k=1}^{\infty} \{\zeta(2k+1)-1\} \left(\frac{2}{3}\right)^{2k} = -\frac{31}{20} + \frac{3}{2}\log 3; \tag{228}$$

$$\sum_{k=2}^{\infty} (-1)^k \{\zeta(k)-1\} \left(\frac{4}{3}\right)^k = \frac{89}{21} - \frac{2}{9}\pi\sqrt{3} - 2\log 3; \tag{229}$$

$$\sum_{k=2}^{\infty} \{\zeta(k)-1\} \left(\frac{4}{3}\right)^k = \frac{4}{3} - \frac{2}{9}\pi\sqrt{3} + 2\log 3; \tag{230}$$

$$\sum_{k=1}^{\infty} \{\zeta(2k)-1\} \left(\frac{4}{3}\right)^{2k} = \frac{39}{14} - \frac{2}{9}\pi\sqrt{3}; \tag{231}$$

$$\sum_{k=1}^{\infty} \{\zeta(2k+1)-1\} \left(\frac{4}{3}\right)^{2k} = -\frac{61}{56} + \frac{3}{2}\log 3; \tag{232}$$

$$\sum_{k=2}^{\infty} (-1)^k \{\zeta(k)-1\} \left(\frac{5}{3}\right)^k = \frac{59}{24} + \frac{5}{18}\pi\sqrt{3} - \frac{5}{2}\log 3; \tag{233}$$

$$\sum_{k=2}^{\infty} \{\zeta(k)-1\} \left(\frac{5}{3}\right)^k = \frac{5}{3} + \frac{5}{18}\pi\sqrt{3} + \frac{5}{2}\log 3; \tag{234}$$

$$\sum_{k=1}^{\infty} \{\zeta(2k)-1\} \left(\frac{5}{3}\right)^{2k} = \frac{33}{16} + \frac{5}{18}\pi\sqrt{3}; \tag{235}$$

$$\sum_{k=1}^{\infty} \{\zeta(2k+1)-1\} \left(\frac{5}{3}\right)^{2k} = -\frac{19}{80} + \frac{3}{2}\log 3; \tag{236}$$

$$\sum_{k=2}^{\infty} (-1)^k \frac{\zeta(k)-1}{2^{2k}} = \frac{19}{20} - \frac{1}{8}\pi - \frac{3}{4}\log 2; \tag{237}$$

$$\sum_{k=2}^{\infty} \frac{\zeta(k)-1}{2^{2k}} = -\frac{1}{12} - \frac{1}{8}\pi + \frac{3}{4}\log 2; \tag{238}$$

$$\sum_{k=1}^{\infty} \frac{\zeta(2k)-1}{2^{4k}} = \frac{13}{30} - \frac{1}{8}\pi; \tag{239}$$

$$\sum_{k=1}^{\infty} \frac{\zeta(2k+1)-1}{2^{4k}} = -\frac{31}{15} + 3\log 2; \tag{240}$$

$$\sum_{k=2}^{\infty} (-1)^k\{\zeta(k)-1\}\left(\frac{3}{4}\right)^k = \frac{19}{28} + \frac{3}{8}\pi - \frac{9}{4}\log 2; \tag{241}$$

$$\sum_{k=2}^{\infty} \{\zeta(k)-1\}\left(\frac{3}{4}\right)^k = -\frac{9}{4} + \frac{3}{8}\pi + \frac{9}{4}\log 2; \tag{242}$$

$$\sum_{k=1}^{\infty} \{\zeta(2k)-1\}\left(\frac{3}{4}\right)^{2k} = -\frac{11}{14} + \frac{3}{8}\pi; \tag{243}$$

$$\sum_{k=1}^{\infty} \{\zeta(2k+1)-1\}\left(\frac{3}{4}\right)^{2k} = -\frac{41}{21} + 3\log 2; \tag{244}$$

$$\sum_{k=2}^{\infty} (-1)^k\{\zeta(k)-1\}\left(\frac{5}{4}\right)^k = \frac{191}{36} - \frac{5}{8}\pi - \frac{15}{4}\log 2; \tag{245}$$

$$\sum_{k=2}^{\infty} \{\zeta(k)-1\}\left(\frac{5}{4}\right)^k = \frac{5}{4} - \frac{5}{8}\pi + \frac{15}{4}\log 2; \tag{246}$$

$$\sum_{k=1}^{\infty} \{\zeta(2k)-1\}\left(\frac{5}{4}\right)^{2k} = \frac{59}{18} - \frac{5}{8}\pi; \tag{247}$$

$$\sum_{k=1}^{\infty} \{\zeta(2k+1)-1\}\left(\frac{5}{4}\right)^{2k} = -\frac{73}{45} + 3\log 2; \tag{248}$$

$$\sum_{k=2}^{\infty} (-1)^k\{\zeta(k)-1\}\left(\frac{7}{4}\right)^k = \frac{293}{132} + \frac{7}{8}\pi - \frac{21}{4}\log 2; \tag{249}$$

$$\sum_{k=2}^{\infty} \{\zeta(k)-1\}\left(\frac{7}{4}\right)^k = \frac{7}{4} + \frac{7}{8}\pi + \frac{21}{4}\log 2; \tag{250}$$

$$\sum_{k=1}^{\infty} \{\zeta(2k)-1\}\left(\frac{7}{4}\right)^{2k} = \frac{131}{66} + \frac{7}{8}\pi; \tag{251}$$

$$\sum_{k=1}^{\infty} \{\zeta(2k+1)-1\}\left(\frac{7}{4}\right)^{2k} = -\frac{31}{231} + 3\log 2; \tag{252}$$

$$\sum_{k=2}^{\infty} (-1)^k \frac{\zeta(k)-1}{5^k} = \frac{29}{30} - \frac{\pi}{10}\sqrt{1+\frac{2}{5}\sqrt{5}} - \frac{1}{4}\log 5 - \frac{\sqrt{5}}{10}\log\frac{1+\sqrt{5}}{2}; \tag{253}$$

$$\sum_{k=2}^{\infty} \frac{\zeta(k)-1}{5^k} = -\frac{1}{20} - \frac{\pi}{10}\sqrt{1+\frac{2}{5}\sqrt{5}} + \frac{1}{4}\log 5 + \frac{\sqrt{5}}{10}\log\frac{1+\sqrt{5}}{2}; \tag{254}$$

$$\sum_{k=1}^{\infty} \frac{\zeta(2k)-1}{5^{2k}} = \frac{11}{24} - \frac{\pi}{10}\sqrt{1+\frac{2}{5}\sqrt{5}}; \tag{255}$$

$$\sum_{k=1}^{\infty} \frac{\zeta(2k+1)-1}{5^{2k}} = -\frac{61}{24} + \frac{5}{4}\log 5 + \frac{\sqrt{5}}{2}\log\frac{1+\sqrt{5}}{2}; \tag{256}$$

$$\sum_{k=2}^{\infty} (-1)^k \{\zeta(k)-1\} \left(\frac{2}{5}\right)^k = \frac{31}{35} - \frac{\pi}{5}\sqrt{1-\frac{2}{5}\sqrt{5}} - \frac{1}{2}\log 5 + \frac{\sqrt{5}}{5}\log\frac{1+\sqrt{5}}{2}; \tag{257}$$

$$\sum_{k=2}^{\infty} \{\zeta(k)-1\} \left(\frac{2}{5}\right)^k = -\frac{4}{15} - \frac{\pi}{5}\sqrt{1-\frac{2}{5}\sqrt{5}} + \frac{1}{2}\log 5 - \frac{\sqrt{5}}{5}\log\frac{1+\sqrt{5}}{2}; \tag{258}$$

$$\sum_{k=1}^{\infty} \{\zeta(2k)-1\} \left(\frac{2}{5}\right)^{2k} = \frac{13}{42} - \frac{\pi}{5}\sqrt{1-\frac{2}{5}\sqrt{5}}; \tag{259}$$

$$\sum_{k=1}^{\infty} \{\zeta(2k+1)-1\} \left(\frac{2}{5}\right)^{2k} = -\frac{121}{84} + \frac{5}{4}\log 5 - \frac{\sqrt{5}}{2}\log\frac{1+\sqrt{5}}{2}; \tag{260}$$

$$\sum_{k=2}^{\infty} (-1)^k \{\zeta(k)-1\} \left(\frac{3}{5}\right)^k = \frac{31}{40} + \frac{3\pi}{10}\sqrt{1-\frac{2}{5}\sqrt{5}} - \frac{3}{4}\log 5 + \frac{3\sqrt{5}}{10}\log\frac{1+\sqrt{5}}{2}; \tag{261}$$

$$\sum_{k=2}^{\infty} \{\zeta(k)-1\} \left(\frac{3}{5}\right)^k = -\frac{9}{10} + \frac{3\pi}{10}\sqrt{1-\frac{2}{5}\sqrt{5}} + \frac{3}{4}\log 5 - \frac{3\sqrt{5}}{10}\log\frac{1+\sqrt{5}}{2}; \tag{262}$$

$$\sum_{k=1}^{\infty} \{\zeta(2k)-1\} \left(\frac{3}{5}\right)^{2k} = -\frac{1}{16} + \frac{3\pi}{10}\sqrt{1-\frac{2}{5}\sqrt{5}}; \tag{263}$$

$$\sum_{k=1}^{\infty} \{\zeta(2k+1)-1\} \left(\frac{3}{5}\right)^{2k} = -\frac{67}{48} + \frac{5}{4}\log 5 - \frac{\sqrt{5}}{2}\log\frac{1+\sqrt{5}}{2}; \tag{264}$$

$$\sum_{k=2}^{\infty} (-1)^k \{\zeta(k)-1\} \left(\frac{4}{5}\right)^k = \frac{29}{45} + \frac{2\pi}{5}\sqrt{1+\frac{2}{5}\sqrt{5}} - \log 5 - \frac{2\sqrt{5}}{5}\log\frac{1+\sqrt{5}}{2}; \tag{265}$$

$$\sum_{k=2}^{\infty}\{\zeta(k)-1\}\left(\frac{4}{5}\right)^{k} = -\frac{16}{5}+\frac{2\pi}{5}\sqrt{1+\frac{2}{5}\sqrt{5}}+\log 5+\frac{2\sqrt{5}}{5}\log\frac{1+\sqrt{5}}{2}; \tag{266}$$

$$\sum_{k=1}^{\infty}\{\zeta(2k)-1\}\left(\frac{4}{5}\right)^{2k} = -\frac{23}{18}+\frac{2\pi}{5}\sqrt{1+\frac{2}{5}\sqrt{5}}; \tag{267}$$

$$\sum_{k=1}^{\infty}\{\zeta(2k+1)-1\}\left(\frac{4}{5}\right)^{2k} = -\frac{173}{72}+\frac{5}{4}\log 5+\frac{\sqrt{5}}{2}\log\frac{1+\sqrt{5}}{2}; \tag{268}$$

$$\sum_{k=2}^{\infty}(-1)^k\{\zeta(k)-1\}\left(\frac{6}{5}\right)^{k} = \frac{349}{55}-\frac{3\pi}{5}\sqrt{1+\frac{2}{5}\sqrt{5}}-\frac{3}{2}\log 5-\frac{3\sqrt{5}}{5}\log\frac{1+\sqrt{5}}{2}; \tag{269}$$

$$\sum_{k=2}^{\infty}\{\zeta(k)-1\}\left(\frac{6}{5}\right)^{k} = \frac{6}{5}-\frac{3\pi}{5}\sqrt{1+\frac{2}{5}\sqrt{5}}+\frac{3}{2}\log 5+\frac{3\sqrt{5}}{5}\log\frac{1+\sqrt{5}}{2}; \tag{270}$$

$$\sum_{k=1}^{\infty}\{\zeta(2k)-1\}\left(\frac{6}{5}\right)^{2k} = \frac{83}{22}-\frac{3\pi}{5}\sqrt{1+\frac{2}{5}\sqrt{5}}; \tag{271}$$

$$\sum_{k=1}^{\infty}\{\zeta(2k+1)-1\}\left(\frac{6}{5}\right)^{2k} = -\frac{283}{132}+\frac{5}{4}\log 5+\frac{\sqrt{5}}{2}\log\frac{1+\sqrt{5}}{2}; \tag{272}$$

$$\sum_{k=2}^{\infty}(-1)^k\{\zeta(k)-1\}\left(\frac{7}{5}\right)^{k} = \frac{221}{60}-\frac{7\pi}{10}\sqrt{1-\frac{2}{5}\sqrt{5}}-\frac{7}{4}\log 5+\frac{7\sqrt{5}}{10}\log\frac{1+\sqrt{5}}{2}; \tag{273}$$

$$\sum_{k=2}^{\infty}\{\zeta(k)-1\}\left(\frac{7}{5}\right)^{k} = \frac{7}{5}-\frac{7\pi}{10}\sqrt{1-\frac{2}{5}\sqrt{5}}+\frac{7}{4}\log 5-\frac{7\sqrt{5}}{10}\log\frac{1+\sqrt{5}}{2}; \tag{274}$$

$$\sum_{k=1}^{\infty}\{\zeta(2k)-1\}\left(\frac{7}{5}\right)^{2k} = \frac{61}{24}-\frac{7\pi}{10}\sqrt{1-\frac{2}{5}\sqrt{5}}; \tag{275}$$

$$\sum_{k=1}^{\infty}\{\zeta(2k+1)-1\}\left(\frac{7}{5}\right)^{2k} = -\frac{137}{168}+\frac{5}{4}\log 5-\frac{\sqrt{5}}{2}\log\frac{1+\sqrt{5}}{2}; \tag{276}$$

$$\sum_{k=2}^{\infty}(-1)^k\{\zeta(k)-1\}\left(\frac{8}{5}\right)^k = \frac{523}{195}+\frac{4\pi}{5}\sqrt{1-\frac{2}{5}\sqrt{5}}-2\log 5+\frac{4\sqrt{5}}{5}\log\frac{1+\sqrt{5}}{2}; \tag{277}$$

$$\sum_{k=2}^{\infty}\{\zeta(k)-1\}\left(\frac{8}{5}\right)^k = \frac{8}{5}+\frac{4\pi}{5}\sqrt{1-\frac{2}{5}\sqrt{5}}+2\log 5-\frac{4\sqrt{5}}{5}\log\frac{1+\sqrt{5}}{2}; \tag{278}$$

$$\sum_{k=1}^{\infty}\{\zeta(2k)-1\}\left(\frac{8}{5}\right)^{2k} = \frac{167}{78}+\frac{4\pi}{5}\sqrt{1-\frac{2}{5}\sqrt{5}}; \tag{279}$$

$$\sum_{k=1}^{\infty}\{\zeta(2k+1)-1\}\left(\frac{8}{5}\right)^{2k} = -\frac{211}{624}+\frac{5}{4}\log 5-\frac{\sqrt{5}}{2}\log\frac{1+\sqrt{5}}{2}; \tag{280}$$

$$\sum_{k=2}^{\infty}(-1)^k\{\zeta(k)-1\}\left(\frac{9}{5}\right)^k = \frac{293}{140}+\frac{9\pi}{10}\sqrt{1+\frac{2}{5}\sqrt{5}}-\frac{9}{4}\log 5-\frac{9\sqrt{5}}{10}\log\frac{1+\sqrt{5}}{2}; \tag{281}$$

$$\sum_{k=2}^{\infty}\{\zeta(k)-1\}\left(\frac{9}{5}\right)^k = \frac{9}{5}+\frac{9\pi}{10}\sqrt{1+\frac{2}{5}\sqrt{5}}+\frac{9}{4}\log 5+\frac{9\sqrt{5}}{10}\log\frac{1+\sqrt{5}}{2}; \tag{282}$$

$$\sum_{k=1}^{\infty}\{\zeta(2k)-1\}\left(\frac{9}{5}\right)^{2k} = \frac{109}{56}+\frac{9\pi}{10}\sqrt{1+\frac{2}{5}\sqrt{5}}; \tag{283}$$

$$\sum_{k=1}^{\infty}\{\zeta(2k+1)-1\}\left(\frac{9}{5}\right)^{2k} = -\frac{41}{504}+\frac{5}{4}\log 5+\frac{\sqrt{5}}{2}\log\frac{1+\sqrt{5}}{2}; \tag{284}$$

$$\sum_{k=2}^{\infty}(-1)^k\frac{\zeta(k)-1}{6^k} = \frac{41}{42}-\frac{\pi}{12}\sqrt{3}-\frac{1}{4}\log 3-\frac{1}{3}\log 2; \tag{285}$$

$$\sum_{k=2}^{\infty}\frac{\zeta(k)-1}{6^k} = -\frac{1}{30}-\frac{\pi}{12}\sqrt{3}+\frac{1}{4}\log 3+\frac{1}{3}\log 2; \tag{286}$$

$$\sum_{k=1}^{\infty}\frac{\zeta(2k)-1}{6^{2k}} = \frac{33}{70}-\frac{\pi}{12}\sqrt{3}; \tag{287}$$

$$\sum_{k=1}^{\infty}\frac{\zeta(2k+1)-1}{6^{2k}} = -\frac{106}{35}+\frac{3}{2}\log 3+2\log 2; \tag{288}$$

$$\sum_{k=2}^{\infty}(-1)^k\{\zeta(k)-1\}\left(\frac{5}{6}\right)^k=\frac{41}{66}+\frac{5\pi}{12}\sqrt{3}-\frac{5}{4}\log 3-\frac{5}{3}\log 2; \tag{289}$$

$$\sum_{k=2}^{\infty}\{\zeta(k)-1\}\left(\frac{5}{6}\right)^k=-\frac{25}{6}+\frac{5\pi}{12}\sqrt{3}+\frac{5}{4}\log 3+\frac{5}{3}\log 2; \tag{290}$$

$$\sum_{k=1}^{\infty}\{\zeta(2k)-1\}\left(\frac{5}{6}\right)^{2k}=-\frac{107}{66}+\frac{5\pi}{12}\sqrt{3}; \tag{291}$$

$$\sum_{k=1}^{\infty}\{\zeta(2k+1)-1\}\left(\frac{5}{6}\right)^{2k}=-\frac{158}{55}+\frac{3}{2}\log 3+2\log 2; \tag{292}$$

$$\sum_{k=2}^{\infty}(-1)^k\{\zeta(k)-1\}\left(\frac{7}{6}\right)^k=\frac{575}{78}-\frac{7\pi}{12}\sqrt{3}-\frac{7}{4}\log 3-\frac{7}{3}\log 2; \tag{293}$$

$$\sum_{k=2}^{\infty}\{\zeta(k)-1\}\left(\frac{7}{6}\right)^k=\frac{7}{6}-\frac{7\pi}{12}\sqrt{3}+\frac{7}{4}\log 3+\frac{7}{3}\log 2; \tag{294}$$

$$\sum_{k=1}^{\infty}\{\zeta(2k)-1\}\left(\frac{7}{6}\right)^{2k}=\frac{111}{26}-\frac{7\pi}{12}\sqrt{3}; \tag{295}$$

$$\sum_{k=1}^{\infty}\{\zeta(2k+1)-1\}\left(\frac{7}{6}\right)^{2k}=-\frac{242}{91}+\frac{3}{2}\log 3+2\log 2; \tag{296}$$

$$\sum_{k=2}^{\infty}(-1)^k\{\zeta(k)-1\}\left(\frac{11}{6}\right)^k=\frac{11297}{5610}+\frac{11\pi}{12}\sqrt{3}-\frac{11}{4}\log 3-\frac{11}{3}\log 2; \tag{297}$$

$$\sum_{k=2}^{\infty}\{\zeta(k)-1\}\left(\frac{11}{6}\right)^k=\frac{11}{6}+\frac{11\pi}{12}\sqrt{3}+\frac{11}{4}\log 3+\frac{11}{3}\log 2; \tag{298}$$

$$\sum_{k=1}^{\infty}\{\zeta(2k)-1\}\left(\frac{11}{6}\right)^{2k}=\frac{3597}{1870}+\frac{11\pi}{12}\sqrt{3}; \tag{299}$$

$$\sum_{k=1}^{\infty}\{\zeta(2k+1)-1\}\left(\frac{11}{6}\right)^{2k}=-\frac{46}{935}+\frac{3}{2}\log 3+2\log 2; \tag{300}$$

$$\sum_{k=2}^{\infty}(-1)^k\frac{\zeta(k)-1}{2^{3k}}=\frac{71}{72}-\frac{\pi}{16}\left(\sqrt{2}+1\right)-\frac{1}{2}\log 2+\frac{\sqrt{2}}{16}\log\left(3-2\sqrt{2}\right); \tag{301}$$

$$\sum_{k=2}^{\infty}\frac{\zeta(k)-1}{2^{3k}}=-\frac{1}{56}-\frac{\pi}{16}\left(\sqrt{2}+1\right)+\frac{1}{2}\log 2-\frac{\sqrt{2}}{16}\log\left(3-2\sqrt{2}\right); \tag{302}$$

$$\sum_{k=1}^{\infty}\frac{\zeta(2k)-1}{2^{6k}}=\frac{61}{126}-\frac{\pi}{16}\left(\sqrt{2}+1\right); \tag{303}$$

$$\sum_{k=1}^{\infty} \frac{\zeta(2k+1)-1}{2^{6k}} = -\frac{253}{63} + 4\log 2 - \frac{\sqrt{2}}{2}\log(3-2\sqrt{2}); \tag{304}$$

$$\sum_{k=2}^{\infty} (-1)^k \{\zeta(k)-1\} \left(\frac{3}{8}\right)^k = \frac{79}{88} - \frac{3\pi}{16}\left(\sqrt{2}-1\right) - \frac{3}{2}\log 2 - \frac{3\sqrt{2}}{16}\log\left(3-2\sqrt{2}\right); \tag{305}$$

$$\sum_{k=2}^{\infty} \{\zeta(k)-1\} \left(\frac{3}{8}\right)^k = -\frac{9}{40} - \frac{3\pi}{16}\left(\sqrt{2}-1\right) + \frac{3}{2}\log 2 + \frac{3}{16}\sqrt{2}\log\left(3-2\sqrt{2}\right); \tag{306}$$

$$\sum_{k=1}^{\infty} \{\zeta(2k)-1\} \left(\frac{3}{8}\right)^{2k} = \frac{37}{110} - \frac{3\pi}{16}\left(\sqrt{2}-1\right); \tag{307}$$

$$\sum_{k=1}^{\infty} \{\zeta(2k+1)-1\} \left(\frac{3}{8}\right)^{2k} = -\frac{247}{165} + 4\log 2 + \frac{\sqrt{2}}{2}\log\left(3-2\sqrt{2}\right); \tag{308}$$

$$\sum_{k=2}^{\infty} (-1)^k \{\zeta(k)-1\} \left(\frac{5}{8}\right)^k = \frac{79}{104} + \frac{5\pi}{16}(\sqrt{2}-1) - \frac{5}{2}\log 2 - \frac{5\sqrt{2}}{16}\log\left(3-2\sqrt{2}\right); \tag{309}$$

$$\sum_{k=2}^{\infty} \{\zeta(k)-1\} \left(\frac{5}{8}\right)^k = -\frac{25}{24} + \frac{5\pi}{16}(\sqrt{2}-1) + \frac{5}{2}\log 2 + \frac{5\sqrt{2}}{16}\log\left(3-2\sqrt{2}\right); \tag{310}$$

$$\sum_{k=1}^{\infty} \{\zeta(2k)-1\} \left(\frac{5}{8}\right)^{2k} = -\frac{55}{78} + \frac{5\pi}{16}\left(\sqrt{2}-1\right); \tag{311}$$

$$\sum_{k=1}^{\infty} \{\zeta(2k+1)-1\} \left(\frac{5}{8}\right)^{2k} = -\frac{281}{195} + 4\log 2 + \frac{\sqrt{2}}{2}\log\left(3-2\sqrt{2}\right); \tag{312}$$

$$\sum_{k=2}^{\infty} (-1)^k \{\zeta(k)-1\} \left(\frac{7}{8}\right)^k = \frac{71}{120} + \frac{7\pi}{16}\left(\sqrt{2}+1\right) - \frac{7}{2}\log 2 + \frac{7\sqrt{2}}{16}\log\left(3-2\sqrt{2}\right); \tag{313}$$

$$\sum_{k=2}^{\infty} \{\zeta(k)-1\} \left(\frac{7}{8}\right)^k = -\frac{49}{8} + \frac{7\pi}{16}\left(\sqrt{2}+1\right) + \frac{7}{2}\log 2 - \frac{7\sqrt{2}}{16}\log\left(3-2\sqrt{2}\right); \tag{314}$$

$$\sum_{k=1}^{\infty}\{\zeta(2k)-1\}\left(\frac{7}{8}\right)^{2k}=-\frac{83}{30}+\frac{7\pi}{16}\left(\sqrt{2}+1\right); \tag{315}$$

$$\sum_{k=1}^{\infty}\{\zeta(2k+1)-1\}\left(\frac{7}{8}\right)^{2k}=-\frac{403}{105}+4\log 2-\frac{\sqrt{2}}{2}\log\left(3-2\sqrt{2}\right); \tag{316}$$

$$\sum_{k=2}^{\infty}(-1)^k\{\zeta(k)-1\}\left(\frac{9}{8}\right)^{k}$$
$$=\frac{1279}{136}-\frac{9\pi}{16}\left(\sqrt{2}+1\right)-\frac{9}{2}\log 2+\frac{9\sqrt{2}}{16}\log\left(3-2\sqrt{2}\right); \tag{317}$$

$$\sum_{k=2}^{\infty}\{\zeta(k)-1\}\left(\frac{9}{8}\right)^{k}$$
$$=\frac{9}{8}-\frac{9\pi}{16}\left(\sqrt{2}+1\right)+\frac{9}{2}\log 2-\frac{9\sqrt{2}}{16}\log\left(3-2\sqrt{2}\right); \tag{318}$$

$$\sum_{k=1}^{\infty}\{\zeta(2k)-1\}\left(\frac{9}{8}\right)^{2k}=\frac{179}{34}-\frac{9\pi}{16}\left(\sqrt{2}+1\right); \tag{319}$$

$$\sum_{k=1}^{\infty}\{\zeta(2k+1)-1\}\left(\frac{9}{8}\right)^{2k}=-\frac{563}{153}+4\log 2-\frac{\sqrt{2}}{2}\log\left(3-2\sqrt{2}\right); \tag{320}$$

$$\sum_{k=2}^{\infty}(-1)^k\{\zeta(k)-1\}\left(\frac{11}{8}\right)^{k}$$
$$=\frac{1765}{456}-\frac{11\pi}{16}\left(\sqrt{2}-1\right)-\frac{11}{2}\log 2-\frac{11\sqrt{2}}{16}\log\left(3-2\sqrt{2}\right); \tag{321}$$

$$\sum_{k=2}^{\infty}\{\zeta(k)-1\}\left(\frac{11}{8}\right)^{k}$$
$$=\frac{11}{8}-\frac{11\pi}{16}\left(\sqrt{2}-1\right)+\frac{11}{2}\log 2+\frac{11\sqrt{2}}{16}\log\left(3-2\sqrt{2}\right); \tag{322}$$

$$\sum_{k=1}^{\infty}\{\zeta(2k)-1\}\left(\frac{11}{8}\right)^{2k}=\frac{299}{114}-\frac{11\pi}{16}\left(\sqrt{2}-1\right); \tag{323}$$

$$\sum_{k=1}^{\infty}\{\zeta(2k+1)-1\}\left(\frac{11}{8}\right)^{2k}=-\frac{569}{627}+4\log 2+\frac{\sqrt{2}}{2}\log\left(3-2\sqrt{2}\right); \tag{324}$$

$$\sum_{k=2}^{\infty}(-1)^k\{\zeta(k)-1\}\left(\frac{13}{8}\right)^{k}$$
$$=\frac{2179}{840}+\frac{13\pi}{16}\left(\sqrt{2}-1\right)-\frac{13}{2}\log 2-\frac{13\sqrt{2}}{16}\log\left(3-2\sqrt{2}\right); \tag{325}$$

$$\sum_{k=2}^{\infty}\{\zeta(k)-1\}\left(\frac{13}{8}\right)^{k} = \frac{13}{8}+\frac{13\pi}{16}\left(\sqrt{2}-1\right)+\frac{13}{2}\log 2+\frac{13\sqrt{2}}{16}\log\left(3-2\sqrt{2}\right); \tag{326}$$

$$\sum_{k=1}^{\infty}\{\zeta(2k)-1\}\left(\frac{13}{8}\right)^{2k}=\frac{443}{210}+\frac{13\pi}{16}\left(\sqrt{2}-1\right); \tag{327}$$

$$\sum_{k=1}^{\infty}\{\zeta(2k+1)-1\}\left(\frac{13}{8}\right)^{2k}=-\frac{407}{1365}+4\log 2+\frac{\sqrt{2}}{2}\log\left(3-2\sqrt{2}\right); \tag{328}$$

$$\sum_{k=2}^{\infty}(-1)^{k}\{\zeta(k)-1\}\left(\frac{15}{8}\right)^{k} = \frac{2473}{1288}+\frac{15\pi}{16}\left(\sqrt{2}+1\right)-\frac{15}{2}\log 2+\frac{15\sqrt{2}}{16}\log\left(3-2\sqrt{2}\right); \tag{329}$$

$$\sum_{k=2}^{\infty}\{\zeta(k)-1\}\left(\frac{15}{8}\right)^{k} = \frac{15}{8}+\frac{15\pi}{16}\left(\sqrt{2}+1\right)+\frac{15}{2}\log 2-\frac{15\sqrt{2}}{16}\log\left(3-2\sqrt{2}\right); \tag{330}$$

$$\sum_{k=1}^{\infty}\{\zeta(2k)-1\}\left(\frac{15}{8}\right)^{2k}=\frac{611}{322}+\frac{15\pi}{16}\left(\sqrt{2}+1\right); \tag{331}$$

$$\sum_{k=1}^{\infty}\{\zeta(2k+1)-1\}\left(\frac{15}{8}\right)^{2k}=-\frac{29}{2415}+4\log 2-\frac{\sqrt{2}}{2}\log\left(3-2\sqrt{2}\right); \tag{332}$$

$$\sum_{k=2}^{\infty}(-1)^{k}\frac{\zeta(k)-1}{10^{k}} = \frac{109}{110}-\frac{\pi}{20}\sqrt{5+2\sqrt{5}}-\frac{1}{5}\log 2-\frac{1}{8}\log 5-\frac{\sqrt{5}}{20}\log\left(2+\sqrt{5}\right); \tag{333}$$

$$\sum_{k=2}^{\infty}\frac{\zeta(k)-1}{10^{k}} = -\frac{1}{90}-\frac{\pi}{20}\sqrt{5+2\sqrt{5}}+\frac{1}{5}\log 2+\frac{1}{8}\log 5+\frac{\sqrt{5}}{20}\log\left(2+\sqrt{5}\right); \tag{334}$$

$$\sum_{k=1}^{\infty}\frac{\zeta(2k)-1}{10^{2k}}=\frac{97}{198}-\frac{\pi}{20}\sqrt{5+2\sqrt{5}}; \tag{335}$$

$$\sum_{k=1}^{\infty}\frac{\zeta(2k+1)-1}{10^{2k}}=-\frac{496}{99}+2\log 2+\frac{5}{4}\log 5+\frac{\sqrt{5}}{2}\log\left(2+\sqrt{5}\right); \tag{336}$$

$$\sum_{k=2}^{\infty}(-1)^k\{\zeta(k)-1\}\left(\frac{3}{10}\right)^k = \frac{121}{130}-\frac{3\pi}{20}\sqrt{5-2\sqrt{5}}-\frac{3}{5}\log 2-\frac{3}{8}\log 5+\frac{3\sqrt{5}}{20}\log\left(2+\sqrt{5}\right); \tag{337}$$

$$\sum_{k=2}^{\infty}\{\zeta(k)-1\}\left(\frac{3}{10}\right)^k = -\frac{9}{70}-\frac{3\pi}{20}\sqrt{5-2\sqrt{5}}+\frac{3}{5}\log 2+\frac{3}{8}\log 5-\frac{3\sqrt{5}}{20}\log\left(2+\sqrt{5}\right); \tag{338}$$

$$\sum_{k=1}^{\infty}\{\zeta(2k)-1\}\left(\frac{3}{10}\right)^{2k} = \frac{73}{182}-\frac{3\pi}{20}\sqrt{5-2\sqrt{5}}; \tag{339}$$

$$\sum_{k=1}^{\infty}\{\zeta(2k+1)-1\}\left(\frac{3}{10}\right)^{2k} = -\frac{482}{273}+2\log 2+\frac{5}{4}\log 5-\frac{\sqrt{5}}{2}\log\left(2+\sqrt{5}\right); \tag{340}$$

$$\sum_{k=2}^{\infty}(-1)^k\{\zeta(k)-1\}\left(\frac{7}{10}\right)^k = \frac{121}{170}+\frac{7\pi}{20}\sqrt{5-2\sqrt{5}} -\frac{7}{5}\log 2-\frac{7}{8}\log 5+\frac{7\sqrt{5}}{20}\log\left(2+\sqrt{5}\right); \tag{341}$$

$$\sum_{k=2}^{\infty}\{\zeta(k)-1\}\left(\frac{7}{10}\right)^k = -\frac{49}{30}+\frac{7\pi}{20}\sqrt{5-2\sqrt{5}}+\frac{7}{5}\log 2+\frac{7}{8}\log 5-\frac{7\sqrt{5}}{20}\log\left(2+\sqrt{5}\right); \tag{342}$$

$$\sum_{k=1}^{\infty}\{\zeta(2k)-1\}\left(\frac{7}{10}\right)^{2k} = -\frac{47}{102}+\frac{7\pi}{20}\sqrt{5-2\sqrt{5}}; \tag{343}$$

$$\sum_{k=1}^{\infty}\{\zeta(2k+1)-1\}\left(\frac{7}{10}\right)^{2k} = -\frac{598}{357}+2\log 2+\frac{5}{4}\log 5-\frac{\sqrt{5}}{2}\log\left(2+\sqrt{5}\right); \tag{344}$$

$$\sum_{k=2}^{\infty}(-1)^k\{\zeta(k)-1\}\left(\frac{9}{10}\right)^k = \frac{91}{190}+\frac{9\pi}{20}\sqrt{5+2\sqrt{5}} -\frac{9}{5}\log 2-\frac{9}{8}\log 5-\frac{9\sqrt{5}}{20}\log\left(2+\sqrt{5}\right); \tag{345}$$

$$\sum_{k=2}^{\infty}\{\zeta(k)-1\}\left(\frac{9}{10}\right)^{k} = -\frac{81}{10}+\frac{9\pi}{20}\sqrt{5+2\sqrt{5}}+\frac{9}{5}\log 2+\frac{9}{8}\log 5+\frac{9\sqrt{5}}{20}\log\left(2+\sqrt{5}\right); \tag{346}$$

$$\sum_{k=1}^{\infty}\{\zeta(2k)-1\}\left(\frac{9}{10}\right)^{2k} = -\frac{143}{38}+\frac{9\pi}{20}\sqrt{5+2\sqrt{5}}; \tag{347}$$

$$\sum_{k=1}^{\infty}\{\zeta(2k+1)-1\}\left(\frac{9}{10}\right)^{2k} = -\frac{824}{171}+2\log 2+\frac{5}{4}\log 5+\frac{\sqrt{5}}{2}\log\left(2+\sqrt{5}\right); \tag{348}$$

$$\sum_{k=2}^{\infty}(-1)^k\{\zeta(k)-1\}\left(\frac{11}{10}\right)^{k} = \frac{2399}{210}-\frac{11\pi}{20}\sqrt{5+2\sqrt{5}} -\frac{11}{5}\log 2-\frac{11}{8}\log 5-\frac{11\sqrt{5}}{20}\log\left(2+\sqrt{5}\right); \tag{349}$$

$$\sum_{k=2}^{\infty}\{\zeta(k)-1\}\left(\frac{11}{10}\right)^{k} = \frac{11}{10}-\frac{11\pi}{20}\sqrt{5+2\sqrt{5}} +\frac{11}{5}\log 2+\frac{11}{8}\log 5+\frac{11\sqrt{5}}{20}\log\left(2+\sqrt{5}\right); \tag{350}$$

$$\sum_{k=1}^{\infty}\{\zeta(2k)-1\}\left(\frac{11}{10}\right)^{2k} = \frac{263}{42}-\frac{11\pi}{20}\sqrt{5+2\sqrt{5}}; \tag{351}$$

$$\sum_{k=1}^{\infty}\{\zeta(2k+1)-1\}\left(\frac{11}{10}\right)^{2k} = -\frac{1084}{231}+2\log 2+\frac{5}{4}\log 5+\frac{\sqrt{5}}{2}\log\left(2+\sqrt{5}\right); \tag{352}$$

$$\sum_{k=2}^{\infty}(-1)^k\{\zeta(k)-1\}\left(\frac{13}{10}\right)^{k} = \frac{3173}{690}-\frac{13\pi}{20}\sqrt{5-2\sqrt{5}} -\frac{13}{5}\log 2-\frac{13}{8}\log 5+\frac{13\sqrt{5}}{20}\log\left(2+\sqrt{5}\right); \tag{353}$$

$$\sum_{k=2}^{\infty}\{\zeta(k)-1\}\left(\frac{13}{10}\right)^{k} = \frac{13}{10}-\frac{13\pi}{20}\sqrt{5-2\sqrt{5}} +\frac{13}{5}\log 2+\frac{13}{8}\log 5-\frac{13\sqrt{5}}{20}\log\left(2+\sqrt{5}\right); \tag{354}$$

$$\sum_{k=1}^{\infty}\{\zeta(2k)-1\}\left(\frac{13}{10}\right)^{2k} = \frac{407}{138}-\frac{13\pi}{20}\sqrt{5-2\sqrt{5}}; \tag{355}$$

$$\sum_{k=1}^{\infty}\{\zeta(2k+1)-1\}\left(\frac{13}{10}\right)^{2k} = -\frac{1138}{897}+2\log 2+\frac{5}{4}\log 5-\frac{\sqrt{5}}{2}\log\left(2+\sqrt{5}\right); \tag{356}$$

$$\sum_{k=2}^{\infty}(-1)^k\{\zeta(k)-1\}\left(\frac{17}{10}\right)^{k} = \frac{4457}{1890}+\frac{17\pi}{20}\sqrt{5-2\sqrt{5}} -\frac{17}{5}\log 2-\frac{17}{8}\log 5+\frac{17\sqrt{5}}{20}\log\left(2+\sqrt{5}\right); \tag{357}$$

$$\sum_{k=2}^{\infty}\{\zeta(k)-1\}\left(\frac{17}{10}\right)^{k} = \frac{17}{10}+\frac{17\pi}{20}\sqrt{5-2\sqrt{5}} +\frac{17}{5}\log 2+\frac{17}{8}\log 5-\frac{17\sqrt{5}}{20}\log\left(2+\sqrt{5}\right); \tag{358}$$

$$\sum_{k=1}^{\infty}\{\zeta(2k)-1\}\left(\frac{17}{10}\right)^{2k} = \frac{767}{378}+\frac{17\pi}{20}\sqrt{5-2\sqrt{5}}; \tag{359}$$

$$\sum_{k=1}^{\infty}\{\zeta(2k+1)-1\}\left(\frac{17}{10}\right)^{2k} = -\frac{622}{3213}+2\log 2+\frac{5}{4}\log 5-\frac{\sqrt{5}}{2}\log\left(2+\sqrt{5}\right); \tag{360}$$

$$\sum_{k=2}^{\infty}(-1)^k\{\zeta(k)-1\}\left(\frac{19}{10}\right)^{k} = \frac{4871}{261}+\frac{19\pi}{20}\sqrt{5+2\sqrt{5}} -\frac{19}{5}\log 2-\frac{19}{8}\log 5-\frac{19\sqrt{5}}{20}\log\left(2+\sqrt{5}\right); \tag{361}$$

$$\sum_{k=2}^{\infty}\{\zeta(k)-1\}\left(\frac{19}{10}\right)^{k} = \frac{19}{20}+\frac{19\pi}{20}\sqrt{5+2\sqrt{5}} +\frac{19}{5}\log 2+\frac{19}{8}\log 5+\frac{19\sqrt{5}}{20}\log\left(2+\sqrt{5}\right); \tag{362}$$

$$\sum_{k=1}^{\infty}\{\zeta(2k)-1\}\left(\frac{19}{10}\right)^{2k} = \frac{983}{522}+\frac{19\pi}{20}\sqrt{5+2\sqrt{5}}; \tag{363}$$

$$\sum_{k=1}^{\infty}\{\zeta(2k+1)-1\}\left(\frac{19}{10}\right)^{2k} = \frac{44}{4959}+2\log 2+\frac{5}{4}\log 5+\frac{\sqrt{5}}{2}\log\left(2+\sqrt{5}\right); \tag{364}$$

$$\sum_{k=2}^{\infty}(-1)^k\frac{\zeta(k)-1}{12^k}$$
$$=\frac{155}{156}-\frac{\pi}{24}(2+\sqrt{3})+\frac{\sqrt{3}}{24}\log\left(7-4\sqrt{3}\right)-\frac{1}{8}\log 3-\frac{1}{4}\log 2; \tag{365}$$

$$\sum_{k=2}^{\infty}\frac{\zeta(k)-1}{12^k}$$
$$=-\frac{1}{132}-\frac{\pi}{r24}\left(2+\sqrt{3}\right)-\frac{\sqrt{3}}{24}\log\left(7-4\sqrt{3}\right)+\frac{1}{8}\log 3+\frac{1}{4}\log 2; \tag{366}$$

$$\sum_{k=1}^{\infty}\frac{\zeta(2k)-1}{12^{2k}}=\frac{141}{286}-\frac{\pi}{24}\left(2+\sqrt{3}\right); \tag{367}$$

$$\sum_{k=1}^{\infty}\frac{\zeta(2k+1)-1}{12^{2k}}=-\frac{859}{143}-\frac{\sqrt{3}}{2}\log\left(7-4\sqrt{3}\right)+\frac{3}{2}\log 3+3\log 2; \tag{368}$$

$$\sum_{k=2}^{\infty}(-1)^k\{\zeta(k)-1\}\left(\frac{5}{12}\right)^k$$
$$=\frac{179}{204}-\frac{5\pi}{24}\left(2-\sqrt{3}\right)-\frac{5\sqrt{3}}{24}\log\left(7-4\sqrt{3}\right)-\frac{5}{8}\log 3-\frac{5}{4}\log 2; \tag{369}$$

$$\sum_{k=2}^{\infty}\{\zeta(k)-1\}\left(\frac{5}{12}\right)^k$$
$$=-\frac{25}{84}-\frac{5\pi}{24}\left(2-\sqrt{3}\right)+\frac{5\sqrt{3}}{24}\log\left(7-4\sqrt{3}\right)+\frac{5}{8}\log 3+\frac{5}{4}\log 2; \tag{370}$$

$$\sum_{k=1}^{\infty}\{\zeta(2k)-1\}\left(\frac{5}{12}\right)^{2k}=\frac{69}{238}-\frac{5\pi}{24}\left(2-\sqrt{3}\right); \tag{371}$$

$$\sum_{k=1}^{\infty}\{\zeta(2k+1)-1\}\left(\frac{5}{12}\right)^{2k}=-\frac{839}{595}+\frac{\sqrt{3}}{2}\log\left(7-4\sqrt{3}\right)+\frac{3}{2}\log 3+3\log 2; \tag{372}$$

$$\sum_{k=2}^{\infty}(-1)^k\{\zeta(k)-1\}\left(\frac{7}{12}\right)^k$$
$$=\frac{179}{228}+\frac{7\pi}{24}\left(2-\sqrt{3}\right)-\frac{7\sqrt{3}}{24}\log\left(7-4\sqrt{3}\right)-\frac{7}{8}\log 3-\frac{7}{4}\log 2; \tag{373}$$

$$\sum_{k=2}^{\infty}\{\zeta(k)-1\}\left(\frac{7}{12}\right)^k$$
$$=-\frac{49}{60}+\frac{7\pi}{24}\left(2-\sqrt{3}\right)+\frac{7\sqrt{3}}{24}\log\left(7-4\sqrt{3}\right)+\frac{7}{8}\log 3+\frac{7}{4}\log 2; \tag{374}$$

$$\sum_{k=1}^{\infty}\{\zeta(2k)-1\}\left(\frac{7}{12}\right)^{2k}=-\frac{3}{190}+\frac{7\pi}{24}\left(2-\sqrt{3}\right); \tag{375}$$

$$\sum_{k=1}^{\infty}\{\zeta(2k+1)-1\}\left(\frac{7}{12}\right)^{2k}=-\frac{913}{665}+\frac{\sqrt{3}}{2}\log\left(7-4\sqrt{3}\right)+\frac{3}{2}\log 3+3\log 2; \tag{376}$$

$$\sum_{k=2}^{\infty}(-1)^k\{\zeta(k)-1\}\left(\frac{11}{12}\right)^{k}$$
$$=\frac{155}{276}+\frac{11\pi}{24}\left(2+\sqrt{3}\right)+\frac{11\sqrt{3}}{24}\log\left(7-4\sqrt{3}\right)-\frac{11}{8}\log 3-\frac{11}{4}\log 2; \tag{377}$$

$$\sum_{k=2}^{\infty}\{\zeta(k)-1\}\left(\frac{11}{12}\right)^{k}$$
$$=-\frac{121}{12}+\frac{11\pi}{24}\left(2+\sqrt{3}\right)-\frac{11\sqrt{3}}{24}\log\left(7-4\sqrt{3}\right)+\frac{11}{8}\log 3+\frac{11}{4}\log 2; \tag{378}$$

$$\sum_{k=1}^{\infty}\{\zeta(2k)-1\}\left(\frac{11}{12}\right)^{2k}=-\frac{219}{46}+\frac{11\pi}{24}\left(2+\sqrt{3}\right); \tag{379}$$

$$\sum_{k=1}^{\infty}\{\zeta(2k+1)-1\}\left(\frac{11}{12}\right)^{2k}$$
$$=-\frac{1469}{253}-\frac{\sqrt{3}}{2}\log\left(7-4\sqrt{3}\right)+\frac{3}{2}\log 3+3\log 2; \tag{380}$$

$$\sum_{k=2}^{\infty}(-1)^k\{\zeta(k)-1\}\left(\frac{13}{12}\right)^{k}$$
$$=\frac{4031}{300}-\frac{13\pi}{24}(2+\sqrt{3})+\frac{13\sqrt{3}}{24}\log\left(7-4\sqrt{3}\right)-\frac{13}{8}\log 3-\frac{13}{4}\log 2; \tag{381}$$

$$\sum_{k=2}^{\infty}\{\zeta(k)-1\}\left(\frac{13}{12}\right)^{k}$$
$$=\frac{13}{12}-\frac{13\pi}{24}\left(2+\sqrt{3}\right)-\frac{13\sqrt{3}}{24}\log\left(7-4\sqrt{3}\right)+\frac{13}{8}\log 3+\frac{13}{4}\log 2; \tag{382}$$

$$\sum_{k=1}^{\infty}\{\zeta(2k)-1\}\left(\frac{13}{12}\right)^{2k}=\frac{363}{50}-\frac{13\pi}{24}\left(2+\sqrt{3}\right); \tag{383}$$

$$\sum_{k=1}^{\infty}\{\zeta(2k+1)-1\}\left(\frac{13}{12}\right)^{2k}=-\frac{1853}{325}-\frac{\sqrt{3}}{2}\log\left(7-4\sqrt{3}\right)+\frac{3}{2}\log 3+3\log 2; \tag{384}$$

$$\sum_{k=2}^{\infty}(-1)^k\{\zeta(k)-1\}\left(\frac{17}{12}\right)^k = \frac{6211}{1740}-\frac{17\pi}{24}\left(2-\sqrt{3}\right)-\frac{17\sqrt{3}}{24}\log\left(7-4\sqrt{3}\right)-\frac{17}{8}\log 3-\frac{17}{4}\log 2; \tag{385}$$

$$\sum_{k=2}^{\infty}\{\zeta(k)-1\}\left(\frac{17}{12}\right)^k = \frac{17}{12}-\frac{17\pi}{24}(2-\sqrt{3})+\frac{17\sqrt{3}}{24}\log\left(7-4\sqrt{3}\right)+\frac{17}{8}\log 3+\frac{17}{4}\log 2; \tag{386}$$

$$\sum_{k=1}^{\infty}\{\zeta(2k)-1\}\left(\frac{17}{12}\right)^{2k} = \frac{723}{290}-\frac{17\pi}{24}\left(2-\sqrt{3}\right); \tag{387}$$

$$\sum_{k=1}^{\infty}\{\zeta(2k+1)-1\}\left(\frac{17}{12}\right)^{2k} = -\frac{1873}{2465}+\frac{\sqrt{3}}{2}\log\left(7-4\sqrt{3}\right)+\frac{3}{2}\log 3+3\log 2; \tag{388}$$

$$\sum_{k=2}^{\infty}(-1)^k\{\zeta(k)-1\}\left(\frac{19}{12}\right)^k = \frac{7145}{2604}+\frac{19\pi}{24}(2-\sqrt{3})-\frac{19\sqrt{3}}{24}\log\left(7-4\sqrt{3}\right)-\frac{19}{8}\log 3-\frac{19}{4}\log 2; \tag{389}$$

$$\sum_{k=2}^{\infty}\{\zeta(k)-1\}\left(\frac{19}{12}\right)^k = \frac{19}{12}+\frac{19\pi}{24}\left(2-\sqrt{3}\right)+\frac{19\sqrt{3}}{24}\log\left(7-4\sqrt{3}\right)+\frac{19}{8}\log 3+\frac{19}{4}\log 2; \tag{390}$$

$$\sum_{k=1}^{\infty}\{\zeta(2k)-1\}\left(\frac{19}{12}\right)^{2k} = \frac{939}{434}+\frac{19\pi}{24}\left(2-\sqrt{3}\right); \tag{391}$$

$$\sum_{k=1}^{\infty}\{\zeta(2k+1)-1\}\left(\frac{19}{12}\right)^{2k} = -\frac{1511}{4123}+\frac{\sqrt{3}}{2}\log\left(7-4\sqrt{3}\right)+\frac{3}{2}\log 3+3\log 2; \tag{392}$$

$$\sum_{k=2}^{\infty}(-1)^k\{\zeta(k)-1\}\left(\frac{23}{12}\right)^k = \frac{8461}{4620}+\frac{23\pi}{24}(2+\sqrt{3})+\frac{23\sqrt{3}}{24}\log\left(7-4\sqrt{3}\right)-\frac{23}{8}\log 3-\frac{23}{4}\log 2; \tag{393}$$

$$\sum_{k=2}^{\infty}\{\zeta(k)-1\}\left(\frac{23}{12}\right)^k = \frac{23}{12}+\frac{23\pi}{24}(2+\sqrt{3})-\frac{23\sqrt{3}}{24}\log\left(7-4\sqrt{3}\right)+\frac{23}{8}\log 3+\frac{23}{4}\log 2; \tag{394}$$

$$\sum_{k=1}^{\infty}\{\zeta(2k)-1\}\left(\frac{23}{12}\right)^{2k}=\frac{1443}{770}+\frac{23\pi}{24}\left(2+\sqrt{3}\right); \tag{395}$$

$$\sum_{k=1}^{\infty}\{\zeta(2k+1)-1\}\left(\frac{23}{12}\right)^{2k}=\frac{197}{8855}-\frac{\sqrt{3}}{2}\log\left(7-4\sqrt{3}\right)+\frac{3}{2}\log 3+3\log 2. \tag{396}$$

Setting $a=\frac{1}{2}$ in (1) through (8) and using 1.1(14), 2.3(2) and appropriate special values of the ψ-function listed in Section 1.3, we obtain

$$\sum_{k=2}^{\infty}(2^k-1)\zeta(k)\frac{t^k}{k}=\log\Gamma\left(\frac{1}{2}-t\right)-(\gamma+2\log 2)t-\frac{1}{2}\log\pi \quad \left(|t|<\frac{1}{2}\right); \tag{397}$$

$$\sum_{k=2}^{\infty}(-1)^k(2^k-1)\zeta(k)\frac{t^k}{k}=\log\Gamma\left(\frac{1}{2}+t\right)+(\gamma+2\log 2)t-\frac{1}{2}\log\pi \quad \left(|t|<\frac{1}{2}\right); \tag{398}$$

$$\sum_{k=1}^{\infty}\left(2^{2k}-1\right)\zeta(2k)\frac{t^{2k}}{k}=\log\Gamma\left(\frac{1}{2}+t\right)+\log\Gamma\left(\frac{1}{2}-t\right)-\log\pi \quad \left(|t|<\frac{1}{2}\right); \tag{399}$$

$$\begin{aligned}\sum_{k=1}^{\infty}&\left(2^{2k+1}-1\right)\zeta(2k+1)\frac{t^{2k+1}}{2k+1}\\&=\frac{1}{2}\left\{\log\Gamma\left(\frac{1}{2}-t\right)-\log\Gamma\left(\frac{1}{2}+t\right)\right\}-(\gamma+2\log 2)t \quad \left(|t|<\frac{1}{2}\right);\end{aligned} \tag{400}$$

$$\sum_{k=2}^{\infty}(2^k-1)\zeta(k)t^{k-1}=-\psi\left(\frac{1}{2}-t\right)-\gamma-2\log 2 \quad \left(|t|<\frac{1}{2}\right); \tag{401}$$

$$\sum_{k=2}^{\infty}(-1)^k(2^k-1)\zeta(k)t^{k-1}=\psi\left(\frac{1}{2}+t\right)+\gamma+2\log 2 \quad \left(|t|<\frac{1}{2}\right); \tag{402}$$

$$\sum_{k=1}^{\infty}\left(2^{2k}-1\right)\zeta(2k)t^{2k-1}=\frac{1}{2}\left\{\psi\left(\frac{1}{2}+t\right)-\psi\left(\frac{1}{2}-t\right)\right\} \quad \left(|t|<\frac{1}{2}\right); \tag{403}$$

$$\begin{aligned}\sum_{k=1}^{\infty}&\left(2^{2k+1}-1\right)\zeta(2k+1)t^{2k}\\&=-\frac{1}{2}\left\{\psi\left(\frac{1}{2}+t\right)+\psi\left(\frac{1}{2}-t\right)\right\}-\gamma-2\log 2 \quad \left(|t|<\frac{1}{2}\right).\end{aligned} \tag{404}$$

For $z = t + \frac{1}{2}$, 1.1(12) yields

$$\Gamma\left(\frac{1}{2}+t\right)\Gamma\left(\frac{1}{2}-t\right) = \frac{\pi}{\cos(\pi t)}, \tag{405}$$

which can be used to rewrite the series identity (399) in its equivalent form:

$$\sum_{k=1}^{\infty}\left(2^{2k}-1\right)\zeta(2k)\frac{t^{2k}}{k} = -\log\cos(\pi t) \qquad \left(|t| < \frac{1}{2}\right). \tag{406}$$

Furthermore, in view of 1.3(11), the series identity (403) can be rewritten in its equivalent form:

$$\sum_{k=1}^{\infty}\left(2^{2k}-1\right)\zeta(2k)t^{2k-1} = \frac{\pi}{2}\tan(\pi t) \qquad \left(|t| < \frac{1}{2}\right). \tag{407}$$

For suitable special values of the argument t, we can easily deduce the following series identities from (397) through (404), (406) and (407):

$$\sum_{k=2}^{\infty}(-1)^k\left(1-\frac{1}{2^k}\right)\frac{\zeta(k)}{k} = \frac{1}{2}\gamma - \frac{1}{2}\log\pi + \log 2; \tag{408}$$

$$\sum_{k=1}^{\infty}\frac{2^{2k}-1}{k\cdot 3^{2k}}\zeta(2k) = \log 2; \tag{409}$$

$$\sum_{k=1}^{\infty}\frac{2^{2k}-1}{k\cdot 4^{2k}}\zeta(2k) = \frac{1}{2}\log 2; \tag{410}$$

$$\sum_{k=1}^{\infty}\frac{2^{2k}-1}{k\cdot 6^{2k}}\zeta(2k) = \log 2 - \frac{1}{2}\log 3; \tag{411}$$

$$\sum_{k=1}^{\infty}\frac{2^{2k}-1}{k\cdot 8^{2k}}\zeta(2k) = \frac{1}{2}\log\left(4-2\sqrt{2}\right); \tag{412}$$

$$\sum_{k=1}^{\infty}\frac{2^{2k}-1}{k}\zeta(2k)\left(\frac{3}{8}\right)^{2k} = \frac{1}{2}\log\left(4+2\sqrt{2}\right); \tag{413}$$

$$\sum_{k=1}^{\infty}\frac{2^{2k}-1}{k\cdot 12^{2k}}\zeta(2k) = \log\left(\sqrt{6}-\sqrt{2}\right); \tag{414}$$

$$\sum_{k=1}^{\infty}\frac{2^{2k}-1}{k}\zeta(2k)\left(\frac{5}{12}\right)^{2k} = \log\left(\sqrt{6}+\sqrt{2}\right); \tag{415}$$

$$\sum_{k=2}^{\infty}(-1)^k\left(1-\frac{1}{2^k}\right)\zeta(k) = \log 2; \tag{416}$$

$$\sum_{k=2}^{\infty}(-1)^k \frac{2^k-1}{3^k}\zeta(k) = \frac{\pi}{6}\sqrt{3} - \frac{1}{2}\log 3; \tag{417}$$

$$\sum_{k=2}^{\infty}\frac{2^k-1}{3^k}\zeta(k) = \frac{\pi}{6}\sqrt{3} + \frac{1}{2}\log 3; \tag{418}$$

$$\sum_{k=1}^{\infty}\frac{2^{2k}-1}{3^{2k}}\zeta(2k) = \frac{\pi}{6}\sqrt{3}; \tag{419}$$

$$\sum_{k=1}^{\infty}\frac{2^{2k+1}-1}{3^{2k+1}}\zeta(2k+1) = \frac{1}{2}\log 3; \tag{420}$$

$$\sum_{k=2}^{\infty}(-1)^k \frac{2^k-1}{4^k}\zeta(k) = \frac{\pi}{8} - \frac{1}{4}\log 2; \tag{421}$$

$$\sum_{k=2}^{\infty}\frac{2^k-1}{4^k}\zeta(k) = \frac{\pi}{8} + \frac{1}{4}\log 2; \tag{422}$$

$$\sum_{k=1}^{\infty}\frac{2^{2k}-1}{4^{2k}}\zeta(2k) = \frac{\pi}{8}; \tag{423}$$

$$\sum_{k=1}^{\infty}\frac{2^{2k+1}-1}{4^{2k}}\zeta(2k+1) = \log 2; \tag{424}$$

$$\sum_{k=2}^{\infty}(-1)^k \frac{2^k-1}{5^k}\zeta(k) = \frac{\pi}{10}\sqrt{5-2\sqrt{5}} - \frac{1}{4}\log 5 + \frac{\sqrt{5}}{10}\log\left(2+\sqrt{5}\right); \tag{425}$$

$$\sum_{k=2}^{\infty}\frac{2^k-1}{5^k}\zeta(k) = \frac{\pi}{10}\sqrt{5-2\sqrt{5}} + \frac{1}{4}\log 5 - \frac{\sqrt{5}}{10}\log\left(2+\sqrt{5}\right); \tag{426}$$

$$\sum_{k=1}^{\infty}\frac{2^{2k}-1}{5^{2k}}\zeta(2k) = \frac{\pi}{10}\sqrt{5-2\sqrt{5}}; \tag{427}$$

$$\sum_{k=1}^{\infty}\frac{2^{2k+1}-1}{5^{2k}}\zeta(2k+1) = \frac{5}{4}\log 5 - \frac{\sqrt{5}}{2}\log\left(2+\sqrt{5}\right); \tag{428}$$

$$\sum_{k=2}^{\infty}(-1)^k\left(2^k-1\right)\zeta(k)\left(\frac{2}{5}\right)^k = \frac{\pi}{5}\sqrt{5+2\sqrt{5}} - \frac{1}{2}\log 5 - \frac{\sqrt{5}}{5}\log\left(2+\sqrt{5}\right); \tag{429}$$

$$\sum_{k=2}^{\infty}\left(2^k-1\right)\zeta(k)\left(\frac{2}{5}\right)^k=\frac{\pi}{5}\sqrt{5+2\sqrt{5}}+\frac{1}{2}\log 5+\frac{\sqrt{5}}{5}\log\left(2+\sqrt{5}\right); \tag{430}$$

$$\sum_{k=1}^{\infty}\left(2^{2k}-1\right)\zeta(2k)\left(\frac{2}{5}\right)^{2k}=\frac{\pi}{5}\sqrt{5+2\sqrt{5}}; \tag{431}$$

$$\sum_{k=1}^{\infty}\left(2^{2k+1}-1\right)\zeta(2k+1)\left(\frac{2}{5}\right)^{2k}=\frac{5}{4}\log 5+\frac{\sqrt{5}}{2}\log\left(2+\sqrt{5}\right); \tag{432}$$

$$\sum_{k=2}^{\infty}(-1)^k\frac{2^k-1}{6^k}\zeta(k)=\frac{\pi}{36}\sqrt{3}-\frac{1}{4}\log 3+\frac{1}{3}\log 2; \tag{433}$$

$$\sum_{k=2}^{\infty}\frac{2^k-1}{6^k}\zeta(k)=\frac{\pi}{36}\sqrt{3}+\frac{1}{4}\log 3-\frac{1}{3}\log 2; \tag{434}$$

$$\sum_{k=1}^{\infty}\frac{2^{2k}-1}{k\cdot 5^{2k}}\zeta(2k)=-\log\frac{\sqrt{5}+1}{4}; \tag{435}$$

$$\sum_{k=1}^{\infty}\frac{2^{2k}-1}{k}\zeta(2k)\left(\frac{2}{5}\right)^{2k}=-\log\frac{\sqrt{5}-1}{4}; \tag{436}$$

$$\sum_{k=1}^{\infty}\frac{2^{2k}-1}{k\cdot 10^{2k}}\zeta(2k)=\log\frac{2\sqrt{2}}{\sqrt{5+\sqrt{5}}}; \tag{437}$$

$$\sum_{k=1}^{\infty}\frac{2^{2k}-1}{k}\zeta(2k)\left(\frac{3}{10}\right)^{2k}=\log\frac{2\sqrt{2}}{\sqrt{5-\sqrt{5}}}; \tag{438}$$

$$\sum_{k=1}^{\infty}\frac{2^{2k}-1}{6^{2k}}\zeta(2k)=\frac{\pi}{36}\sqrt{3}; \tag{439}$$

$$\sum_{k=1}^{\infty}\frac{2^{2k+1}-1}{6^{2k+1}}\zeta(2k+1)=\frac{1}{4}\log 3-\frac{1}{3}\log 2; \tag{440}$$

$$\sum_{k=2}^{\infty}(-1)^k\frac{2^k-1}{8^k}\zeta(k)=\frac{\pi}{16}\left(\sqrt{2}-1\right)-\frac{\sqrt{2}}{16}\log\left(3-2\sqrt{2}\right)-\frac{1}{4}\log 2; \tag{441}$$

$$\sum_{k=2}^{\infty}\frac{2^k-1}{8^k}\zeta(k)=\frac{\pi}{16}\left(\sqrt{2}-1\right)+\frac{\sqrt{2}}{16}\log\left(3-2\sqrt{2}\right)+\frac{1}{4}\log 2; \tag{442}$$

$$\sum_{k=1}^{\infty}\frac{2^{2k}-1}{8^{2k}}\zeta(2k)=\frac{\pi}{16}\left(\sqrt{2}-1\right); \tag{443}$$

$$\sum_{k=1}^{\infty} \frac{2^{2k+1}-1}{8^{2k+1}} \zeta(2k+1) = \frac{1}{4}\log 2 + \frac{\sqrt{2}}{16}\log\left(3-2\sqrt{2}\right); \tag{444}$$

$$\sum_{k=2}^{\infty} (-1)^k \left(2^k-1\right) \zeta(k) \left(\frac{3}{8}\right)^k = \frac{3\pi}{16}\left(\sqrt{2}+1\right) - \frac{3}{4}\log 2 + \frac{3\sqrt{2}}{16}\log\left(3-2\sqrt{2}\right); \tag{445}$$

$$\sum_{k=2}^{\infty} \left(2^k-1\right) \zeta(k) \left(\frac{3}{8}\right)^k = \frac{3\pi}{16}\left(\sqrt{2}+1\right) + \frac{3}{4}\log 2 - \frac{3\sqrt{2}}{16}\log\left(3-2\sqrt{2}\right); \tag{446}$$

$$\sum_{k=1}^{\infty} \left(2^{2k}-1\right) \zeta(2k) \left(\frac{3}{8}\right)^{2k} = \frac{3\pi}{16}\left(\sqrt{2}+1\right); \tag{447}$$

$$\sum_{k=1}^{\infty} \left(2^{2k+1}-1\right) \zeta(2k+1) \left(\frac{3}{8}\right)^{2k} = 2\log 2 - \frac{\sqrt{2}}{2}\log\left(3-2\sqrt{2}\right); \tag{448}$$

$$\sum_{k=2}^{\infty} (-1)^k \frac{2^k-1}{10^k} \zeta(k) = \frac{\pi}{20}\sqrt{1-\frac{2}{5}\sqrt{5}} + \frac{1}{5}\log 2 - \frac{1}{8}\log 5 + \frac{\sqrt{5}}{20}\log\frac{1+\sqrt{5}}{2}; \tag{449}$$

$$\sum_{k=2}^{\infty} \frac{2^k-1}{10^k} \zeta(k) = \frac{\pi}{20}\sqrt{1-\frac{2}{5}\sqrt{5}} - \frac{1}{5}\log 2 + \frac{1}{8}\log 5 - \frac{\sqrt{5}}{20}\log\frac{1+\sqrt{5}}{2}; \tag{450}$$

$$\sum_{k=1}^{\infty} \frac{2^{2k}-1}{10^{2k}} \zeta(2k) = \frac{\pi}{20}\sqrt{1-\frac{2}{5}\sqrt{5}}; \tag{451}$$

$$\sum_{k=1}^{\infty} \frac{2^{2k+1}-1}{10^{2k}} \zeta(2k+1) = -2\log 2 + \frac{5}{4}\log 5 - \frac{\sqrt{5}}{2}\log\frac{1+\sqrt{5}}{2}; \tag{452}$$

$$\sum_{k=2}^{\infty} (-1)^k \left(2^k-1\right) \zeta(k) \left(\frac{3}{10}\right)^k = \frac{3\pi}{20}\sqrt{1+\frac{2}{5}\sqrt{5}} + \frac{3}{5}\log 2 - \frac{3}{8}\log 5 - \frac{3\sqrt{5}}{20}\log\frac{1+\sqrt{5}}{2}; \tag{453}$$

$$\sum_{k=2}^{\infty} \left(2^k-1\right) \zeta(k) \left(\frac{3}{10}\right)^k = \frac{3\pi}{20}\sqrt{1+\frac{2}{5}\sqrt{5}} - \frac{3}{5}\log 2 + \frac{3}{8}\log 5 + \frac{3\sqrt{5}}{20}\log\frac{1+\sqrt{5}}{2}; \tag{454}$$

$$\sum_{k=1}^{\infty}\left(2^{2k}-1\right)\zeta(2k)\left(\frac{3}{10}\right)^{2k}=\frac{3\pi}{20}\sqrt{1+\frac{2}{5}\sqrt{5}}; \tag{455}$$

$$\sum_{k=1}^{\infty}\left(2^{2k+1}-1\right)\zeta(2k+1)\left(\frac{3}{10}\right)^{2k}=-2\log 2+\frac{5}{4}\log 5+\frac{\sqrt{5}}{2}\log\frac{1+\sqrt{5}}{2}; \tag{456}$$

$$\sum_{k=2}^{\infty}(-1)^k\frac{2^k-1}{12^k}\zeta(k)=\frac{\pi}{24}(2-\sqrt{3})-\frac{\sqrt{3}}{24}\log(7-4\sqrt{3})-\frac{1}{8}\log 3-\frac{1}{12}\log 2; \tag{457}$$

$$\sum_{k=2}^{\infty}\frac{2^k-1}{12^k}\zeta(k)=\frac{\pi}{24}(2-\sqrt{3})+\frac{\sqrt{3}}{24}\log\left(7-4\sqrt{3}\right)+\frac{1}{8}\log 3+\frac{1}{12}\log 2; \tag{458}$$

$$\sum_{k=1}^{\infty}\frac{2^{2k}-1}{12^{2k}}\zeta(2k)=\frac{\pi}{24}\left(2-\sqrt{3}\right); \tag{459}$$

$$\sum_{k=1}^{\infty}\frac{2^{2k+1}-1}{12^{2k+1}}\zeta(2k+1)=\frac{\sqrt{3}}{24}\log\left(7-4\sqrt{3}\right)+\frac{1}{8}\log 3+\frac{1}{12}\log 2; \tag{460}$$

$$\sum_{k=2}^{\infty}(-1)^k\left(2^k-1\right)\zeta(k)\left(\frac{5}{12}\right)^{k}=\frac{5\pi}{24}\left(2+\sqrt{3}\right)+\frac{5\sqrt{3}}{24}\log\left(7-4\sqrt{3}\right)-\frac{5}{8}\log 3-\frac{5}{12}\log 2; \tag{461}$$

$$\sum_{k=2}^{\infty}\left(2^k-1\right)\zeta(k)(512)^k=\frac{5\pi}{24}\left(2+\sqrt{3}\right)-\frac{5\sqrt{3}}{24}\log\left(7-4\sqrt{3}\right)+\frac{5}{8}\log 3+\frac{5}{12}\log 2; \tag{462}$$

$$\sum_{k=1}^{\infty}\left(2^{2k}-1\right)\zeta(2k)\left(\frac{5}{12}\right)^{2k}=\frac{5\pi}{24}\left(2+\sqrt{3}\right); \tag{463}$$

$$\sum_{k=1}^{\infty}\left(2^{2k+1}-1\right)\zeta(2k+1)\left(\frac{5}{12}\right)^{2k}=-\frac{\sqrt{3}}{2}\log\left(7-4\sqrt{3}\right)+\frac{3}{2}\log 3+\log 2. \tag{464}$$

Evaluation by Using the Double and Triple Gamma Functions

We first derive some useful integral expressions for $\log\Gamma(z+a)$, $\log G(z+a)$ and $\log\Gamma_3(z+a)$. Setting $z=t+a-1$ in 1.1(2) and 1.4(3) and taking the logarithmic derivatives of the resulting equations, we obtain

$$\sum_{k=1}^{\infty}\left(\frac{1}{t+a-1+k}-\frac{1}{k}\right)=-\frac{\Gamma'(t+a)}{\Gamma(t+a)}-\gamma \tag{465}$$

and

$$\sum_{k=1}^{\infty}\left(\frac{k}{t+a-1+k}-1+\frac{t+a-1}{k}\right)=\frac{G'(t+a)}{G(t+a)}-\frac{1}{2}\log(2\pi)+\frac{1}{2}+(1+\gamma)(t+a-1), \tag{466}$$

respectively.

Now, we set $z=t+a-1$ in 1.4(90) and take the logarithmic derivative of the resulting equation. We, thus, find that

$$\begin{aligned}\frac{\Gamma_3'(t+a)}{\Gamma_3(t+a)}&=\left(\frac{3}{8}-\frac{1}{4}\log(2\pi)-\log A\right)\\&+\frac{1}{2}\left(\gamma+\log(2\pi)+\frac{1}{2}\right)(t+a-1)-\frac{1}{2}\left(\gamma+\frac{\pi^2}{6}+\frac{3}{2}\right)(t+a-1)^2\\&+\frac{t+a-1}{2}\left[\sum_{k=1}^{\infty}\left(\frac{k}{t+a-1+k}-1+\frac{t+a-1}{rk}\right)\right.\\&\left.+\sum_{k=1}^{\infty}\left(\frac{1}{t+a-1+k}-\frac{1}{k}\right)+\frac{\pi^2}{6}(t+a-1)\right].\end{aligned} \tag{467}$$

A combination of (465) and (466) yields

$$\frac{G'(t+a)}{G(t+a)}=\frac{1}{2}+\frac{1}{2}\log(2\pi)-a-t+(t+a-1)\frac{\Gamma'(t+a)}{\Gamma(t+a)}. \tag{468}$$

Equation (467), by virtue of (465) and (466), becomes

$$\begin{aligned}\frac{\Gamma_3'(t+a)}{\Gamma_3(t+a)}&=\left(\frac{3}{8}-\frac{1}{4}\log(2\pi)-\log A\right)\\&+\left(\frac{1}{2}+\frac{1}{4}\log(2\pi)\right)(t+a-1)-\frac{(t+a-1)^2}{4}\\&+\frac{1}{2}(t+a-1)\frac{G'(t+a)}{G(t+a)}-\frac{1}{2}(t+a-1)\frac{\Gamma'(t+a)}{\Gamma(t+a)}.\end{aligned} \tag{469}$$

Integrating both sides of (468) from $t=0$ to $t=z$, we have

$$\begin{aligned}&\int_0^z \log\Gamma(t+a)dt\\&=\left(\frac{1}{2}+\frac{1}{2}\log(2\pi)-a\right)z-\frac{z^2}{2}+(z+a-1)\log\Gamma(z+a)\\&\quad-\log G(z+a)+(1-a)\log\Gamma(a)+\log G(a),\end{aligned} \tag{470}$$

which is due to Barnes [94, p. 283] and, in the special case when $a = 1$, reduces at once to Alexeiewsky's theorem 1.4(42).

Integrating both sides of (469) with respect to t from $t = 0$ to $t = z$ with the aid of Barnes's integral formula (470), we obtain

$$\begin{aligned}\int_0^z \log G(t+a)dt = &\left[\frac{1}{2}(a-1)\log(2\pi) - 2\log A - \frac{a^2}{2} + a - \frac{1}{4}\right]z \\ &+ \frac{1}{4}[\log(2\pi) + 2 - 2a]z^2 - \frac{1}{6}z^3 \\ &+ (z+a-2)\log G(z+a) - 2\log\Gamma_3(z+a) \\ &+ (2-a)\log G(a) + 2\log\Gamma_3(a).\end{aligned} \tag{471}$$

In addition to 1.4(73) and 1.4(77), we can also evaluate each of the following integrals by direct use of the triple Gamma function Γ_3 in 1.4(90):

$$\int_0^1 \log\Gamma_3(t+1)dt = -\frac{1}{24}\log(2\pi) + \frac{3}{2}\log B \tag{472}$$

and

$$\begin{aligned}\int_0^{\frac{1}{2}} \log\Gamma_3(t+1)dt = &-\frac{1}{256} - \frac{29}{1920}\log 2 - \frac{1}{48}\log\pi \\ &+ \frac{1}{16}\log A + \frac{3}{4}\log B + \frac{15}{16}\log C.\end{aligned} \tag{473}$$

For example, to evaluate the integral in (473), we take the logarithms on both sides of the equation 1.4(90) and integrate the resulting equation from $t = 0$ to $t = \frac{1}{2}$. We then obtain

$$\begin{aligned}\int_0^{\frac{1}{2}} \log\Gamma_3(t+1)dt = &\frac{37}{768} + \frac{\gamma}{128} - \frac{1}{48}\log(2\pi) - \frac{1}{8}\log A \\ &+ \frac{1}{16}\lim_{n\to\infty} S_n,\end{aligned} \tag{474}$$

where

$$\begin{aligned}S_n := \sum_{k=1}^{n}\Big[&-\Big\{(2k+1)^3\log(2k+1) + (2k)^3\log(2k)\Big\} \\ &+ \{(2k+1)\log(2k+1) + (2k)\log(2k)\} \\ &+ (16k^3 + 12k^2 + 2k)\log(2k) + (4k^2 + 5k) + \frac{5}{6} - \frac{1}{8k}\Big],\end{aligned}$$

which immediately yields

$$\begin{aligned}S_n &= -\sum_{k=1}^{2n+1} k^3 \log k + \sum_{k=1}^{2n+1} k \log k + 16\sum_{k=1}^{n} k^3 \log k + 12\sum_{k=1}^{n} k^2 \log k \\ &+ 2\sum_{k=1}^{n} k \log k + 16 \log 2 \sum_{k=1}^{n} k^3 + (4 + 12\log 2)\sum_{k=1}^{n} k^2 \\ &+ (5 + 2\log 2)\sum_{k=1}^{n} k - \frac{1}{8}\sum_{k=1}^{n} \frac{1}{k} + \frac{5}{6}n.\end{aligned}$$

Using 1.1(3), 1.3(2), 1.3(69) and 1.3(70), we have

$$\begin{aligned}S_n &= \frac{11}{120}\log 2 - \frac{1}{8}\gamma + 3\log A + 12\log B + 15\log C \\ &- \left(4n^4 + 12n^3 + 11n^2 + 3n - \frac{11}{120}\right)\log\left(1 + \frac{1}{2n}\right) \\ &+ 2n^3 + \frac{11}{2}n^2 + \frac{25}{6}n - \frac{13}{48} + O\left(\frac{1}{n}\right) \quad (n \to \infty),\end{aligned}$$

which, by means of the asymptotic formula 1.3(76), yields

$$\lim_{n\to\infty} S_n = -\frac{5}{6} + \frac{11}{120}\log 2 - \frac{\gamma}{8} + 3\log A + 12\log B + 15\log C. \tag{475}$$

Finally, by substituting (475) into (474), we obtain the desired formula (473).
We shall show that integrals of the forms:

$$\int_0^z t^k \psi(t+a)dt \ (k \in \mathbb{N})$$

can be expressed in terms of multiple Gamma functions.

First of all, integrating by parts with the aid of (470), we obtain

$$\begin{aligned}\int_0^z t\psi(t+a)dt &= \frac{1}{2}[\,2a - 1 - \log(2\pi)\,]z + \frac{z^2}{2} + (1-a)\log\Gamma(z+a) \\ &+ \log G(z+a) + (a-1)\log\Gamma(a) - \log G(a).\end{aligned} \tag{476}$$

Conversely, integrating by parts with the aid of (470) and (471), we obtain

$$2\int_0^z t\log\Gamma(t+a)dt = \left(\frac{1}{4}-\frac{1}{2}a+\frac{1}{2}a^2-2\log A\right)z+\left(\frac{1}{2}\log(2\pi)-\frac{a}{2}+\frac{1}{4}\right)z^2$$
$$-\frac{z^3}{2}+(z^2-(a-1)^2)\log\Gamma(z+a)+(2a-3)\log G(z+a)$$
$$-2\log\Gamma_3(z+a)+(a-1)^2\log\Gamma(a)+(3-2a)\log G(a)+2\log\Gamma_3(a). \tag{477}$$

Next, integrating by parts with the aid of (477), we obtain

$$\int_0^z t^2\psi(t+a)dt = \frac{a-1}{2}\{1-2a+\log(2\pi)\}z+\{1-\log(2\pi)\}\frac{z^2}{4}+\frac{z^3}{3}$$
$$+(a-1)^2\log\Gamma(a+z)+(z-a+1)\log G(a+z)$$
$$-(a-1)^2\log\Gamma(a)+(a-1)\log G(a)-\int_0^z \log G(t+a)\,dt$$
$$=\left(-\frac{1}{4}+\frac{1}{2}a-\frac{1}{2}a^2+2\log A\right)z$$
$$+\left(-\frac{1}{2}\log(2\pi)+\frac{a}{2}-\frac{1}{4}\right)z^2+\frac{z^3}{2}+(a-1)^2\log\Gamma(z+a)$$
$$+(3-2a)\log G(z+a)+2\log\Gamma_3(z+a)-(a-1)^2\log\Gamma(a)$$
$$+(2a-3)\log G(a)-2\log\Gamma_3(a). \tag{478}$$

Furthermore, integrating by parts with the aid of (471), we obtain

$$2\int_0^z t\log G(t+a)dt = (2-a)\left(-\frac{1}{4}+\frac{1}{2}(a-1)\log(2\pi)-2\log A-\frac{a^2}{2}+a\right)z$$
$$+\frac{1}{2}\left(\frac{7}{4}+\frac{1}{2}\log(2\pi)-2\log A+\frac{a^2}{2}-2a\right)z^2$$
$$+\frac{1}{6}(\log(2\pi)-a)z^3-\frac{1}{8}z^4+(z^2-a^2+4a-4)\log G(z+a)$$
$$+2(a-2-z)\log\Gamma_3(z+a)+(a-2)^2\log G(a)$$
$$+2(2-a)\log\Gamma_3(a)+2\int_0^z \log\Gamma_3(t+a)dt. \tag{479}$$

Finally, integrating by parts with the aid of (477) and (479), we obtain

$$
\begin{aligned}
3\int_0^z t^2 \log\Gamma(t+a)\,dt
&= \left[-a^2+\frac{3}{2}a-\frac{1}{4}+2(2a-3)\log A-\frac{1}{2}(a^2-3a+2)\log(2\pi)\right]z \\
&+\left[\frac{9}{8}-\frac{7}{4}a+\frac{3}{4}a^2+\frac{1}{4}(3-2a)\log(2\pi)-\log A\right]z^2 \\
&+\frac{1}{3}[\log(2\pi)-1]z^3-\frac{3}{8}z^4+\left\{z^3+(a-1)^3\right\}\log\Gamma(z+a) \\
&-(3a^2-9a+7)\log G(z+a)+2(2a-3-z)\log\Gamma_3(z+a)+(1-a)^3\log\Gamma(a) \\
&+(3a^2-9a+7)\log G(a)+2(3-2a)\log\Gamma_3(a)+2\int_0^z \log\Gamma_3(t+a)\,dt
\end{aligned}
\tag{480}
$$

and

$$
\begin{aligned}
\int_0^z t^3\psi(t+a)\,dt
&= \left[a^2-\frac{3}{2}a+\frac{1}{4}+2(3-2a)\log A+\frac{1}{2}(a^2-3a+2)\log(2\pi)\right]z \\
&+\left[-\frac{3}{4}a^2+\frac{7}{4}a-\frac{9}{8}+\frac{1}{4}(2a-3)\log(2\pi)+\log A\right]z^2 \\
&+\frac{1}{3}[1-\log(2\pi)]z^3+\frac{3}{8}z^4+(1-a)^3\log\Gamma(z+a) \\
&+(3a^2-9a+7)\log G(z+a)+2(z-2a+3)\log\Gamma_3(z+a)+(a-1)^3\log\Gamma(a) \\
&-(3a^2-9a+7)\log G(a)+2(2a-3)\log\Gamma_3(a)-2\int_0^z \log\Gamma_3(t+a)\,dt.
\end{aligned}
\tag{481}
$$

By taking logarithms on both sides of 1.4(3) and using the Maclaurin series of $\log(1+x)$, we are led to the following identity (*cf.* Srivastava [1072, p. 13, Eq. (5.2)]):

$$
\sum_{k=2}^{\infty}(-1)^k\frac{\zeta(k)}{k+1}z^{k+1}=[1-\log(2\pi)]\frac{z}{2}+(1+\gamma)\frac{z^2}{2}+\log G(z+1)\quad (|z|<1).
\tag{482}
$$

Letting $z \to 1$ in (482), in view of 1.3(6), we obtain

$$\sum_{k=2}^{\infty}(-1)^k \frac{\zeta(k)}{k+1} = 1 + \frac{\gamma}{2} - \frac{1}{2}\log(2\pi), \tag{483}$$

which was proved by Suryanarayana [1138] and, again, by Singh and Verma [1033, p. 3, Section 4].

If we set $z = \frac{1}{2}$ in (482) and make use of 1.4(6) and 1.3(8), we get

$$\sum_{k=2}^{\infty}(-1)^k \frac{\zeta(k)}{k+1} 2^{-k} = 1 + \frac{\gamma}{4} - \frac{5}{12}\log 2 - 3\log A. \tag{484}$$

Multiplying each member of (5) through (8) by t, integrating both sides of the resulting equations from $t = 0$ to $t = z$ and using (476), we obtain

$$\begin{aligned}\sum_{k=2}^{\infty}\zeta(k,a)\frac{z^{k+1}}{k+1} &= \frac{\psi(a)}{2}z^2 + z\log\Gamma(a-z) + \int_0^{-z}\log\Gamma(a+t)\,dt \\ &= [2a-1-\log(2\pi)]\frac{z}{2} + [\psi(a)-1]\frac{z^2}{2} + (a-1)\log\Gamma(a-z) \\ &\quad - \log G(a-z) + (1-a)\log\Gamma(a) + \log G(a) \quad (|z|<|a|);\end{aligned} \tag{485}$$

$$\begin{aligned}\sum_{k=2}^{\infty}(-1)^k\zeta(k,a)\frac{z^{k+1}}{k+1} &= -\frac{\psi(a)}{2}z^2 + z\log\Gamma(a+z) - \int_0^{z}\log\Gamma(a+t)\,dt \\ &= [2a-1-\log(2\pi)]\frac{z}{2} + [1-\psi(a)]\frac{z^2}{2} + (1-a)\log\Gamma(a+z) \\ &\quad + \log G(a+z) + (a-1)\log\Gamma(a) - \log G(a) \quad (|z|<|a|),\end{aligned} \tag{486}$$

which, in the special case when $a = 1$, corresponds to (482);

$$\begin{aligned}\sum_{k=1}^{\infty}\zeta(2k,a)\frac{z^{2k+1}}{2k+1} &= \frac{z}{2}\log\Gamma(a+z)\Gamma(a-z) - \frac{1}{2}\int_{-z}^{z}\log\Gamma(a+t)\,dt \\ &= \frac{1}{2}\left\{[2a-1-\log(2\pi)]z + (1-a)\log\frac{\Gamma(a+z)}{\Gamma(a-z)} + \log\frac{G(a+z)}{G(a-z)}\right\} \\ &\qquad (|z|<|a|);\end{aligned} \tag{487}$$

$$\sum_{k=1}^{\infty} \zeta(2k+1,a) \frac{z^{2k+2}}{k+1}$$

$$= \psi(a)z^2 + z \log \frac{\Gamma(a-z)}{\Gamma(a+z)} + \int_0^z \log \Gamma(a+t)\,dt + \int_0^{-z} \log \Gamma(a+t)\,dt \tag{488}$$

$$= [\psi(a)-1]z^2 + (a-1)\log \Gamma(a+z)\Gamma(a-z) - \log G(a+z)G(a-z)$$
$$+ 2(1-a)\log \Gamma(a) + 2\log G(a) \quad (|z| < |a|).$$

Taking $a = 1$ in (485), (487) and (488) and using 1.4(6), 2.3(2) and 1.3(4), we obtain

$$\sum_{k=2}^{\infty} \zeta(k) \frac{z^{k+1}}{k+1} = [1 - \log(2\pi)]\frac{z}{2} - \frac{1+\gamma}{2} z^2 - \log G(1-z) \quad (|z| < 1); \tag{489}$$

$$\sum_{k=1}^{\infty} \zeta(2k) \frac{z^{2k+1}}{2k+1} = \frac{1}{2}\left\{[1 - \log(2\pi)]z + \log \frac{G(1+z)}{G(1-z)}\right\} \quad (|z| < 1); \tag{490}$$

$$\sum_{k=1}^{\infty} \zeta(2k+1) \frac{z^{2k+2}}{k+1} = -(1+\gamma)z^2 - \log G(1+z)G(1-z) \quad (|z| < 1). \tag{491}$$

Setting $z = \frac{1}{2}$ in (489), (490) and (491) and using 1.4(6) and 1.4(8), we obtain

$$\sum_{k=2}^{\infty} \frac{\zeta(k)}{(k+1)\cdot 2^k} = -\frac{\gamma}{4} - \frac{7}{12}\log 2 + 3\log A; \tag{492}$$

$$\sum_{k=1}^{\infty} \frac{\zeta(2k)}{(2k+1)\cdot 2^{2k}} = \frac{1}{2} - \frac{1}{2}\log 2; \tag{493}$$

$$\sum_{k=1}^{\infty} \frac{\zeta(2k+1)}{(k+1)\cdot 2^{2k}} = -2 - \gamma - \frac{1}{3}\log 2 + 12\log A. \tag{494}$$

Taking $a = 2$ in (485) through (488) and using 1.4(6), 2.3(2) and 1.3(50), we obtain

$$\sum_{k=2}^{\infty} \{\zeta(k) - 1\} \frac{z^{k+1}}{k+1} = [3 - \log(2\pi)]\frac{z}{2} - \frac{\gamma}{2} z^2 + \log \frac{\Gamma(2-z)}{G(2-z)} \quad (|z| < 2); \tag{495}$$

$$\sum_{k=2}^{\infty} (-1)^k \{\zeta(k) - 1\} \frac{z^{k+1}}{k+1} = [3 - \log(2\pi)]\frac{z}{2} + \frac{\gamma}{2} z^2 + \log \frac{G(2+z)}{\Gamma(2+z)} \quad (|z| < 2); \tag{496}$$

$$\sum_{k=1}^{\infty}\{\zeta(2k)-1\}\frac{z^{2k+1}}{2k+1}=\frac{1}{2}\left\{[3-\log(2\pi)]z+\log\frac{G(2+z)\Gamma(2-z)}{G(2-z)\Gamma(2+z)}\right\} \quad (|z|<2); \tag{497}$$

$$\sum_{k=1}^{\infty}\{\zeta(2k+1)-1\}\frac{z^{2k+2}}{k+1}=-\gamma z^2+\log\frac{\Gamma(2+z)\Gamma(2-z)}{G(2+z)G(2-z)} \quad (|z|<2). \tag{498}$$

Taking the limit as $z \to 2$ on both sides of (496) and using 1.4(6), we obtain

$$\sum_{k=2}^{\infty}(-1)^k\{\zeta(k)-1\}\frac{2^{k+1}}{k+1}=3+2\gamma-\log(6\pi). \tag{499}$$

Setting $z=1$, $z=\frac{1}{2}$ and $z=\frac{3}{2}$ in (495) through (498) and using 1.4(6) and 1.4(8), we obtain

$$\sum_{k=2}^{\infty}\frac{\zeta(k)-1}{k+1}=\frac{3}{2}-\frac{\gamma}{2}-\frac{1}{2}\log(2\pi); \tag{500}$$

$$\sum_{k=2}^{\infty}\frac{(-1)^k}{k+1}\{\zeta(k)-1\}=\frac{3}{2}+\frac{\gamma}{2}-\frac{1}{2}\log(8\pi); \tag{501}$$

$$\sum_{k=1}^{\infty}\frac{\zeta(2k)-1}{2k+1}=\frac{3}{2}-\frac{1}{2}\log(4\pi); \tag{502}$$

$$\sum_{k=1}^{\infty}\frac{\zeta(2k+1)-1}{k+1}=-\gamma+\log 2; \tag{503}$$

$$\sum_{k=2}^{\infty}\frac{\zeta(k)-1}{(k+1)\cdot 2^k}=\frac{5}{4}-\frac{\gamma}{4}+\log\left(2^{-\frac{31}{12}}\cdot A^3\right); \tag{504}$$

$$\sum_{k=2}^{\infty}(-1)^k\frac{\zeta(k)-1}{(k+1)\cdot 2^k}=\frac{7}{4}+\frac{\gamma}{4}+\log\left(2^{\frac{19}{12}}\cdot 3^{-2}\cdot A^{-3}\right); \tag{505}$$

$$\sum_{k=1}^{\infty}\frac{\zeta(2k)-1}{(2k+1)\cdot 2^{2k}}=\frac{3}{2}+\log\left(2^{-\frac{1}{2}}\cdot 3^{-1}\right); \tag{506}$$

$$\sum_{k=1}^{\infty}\frac{\zeta(2k+1)-1}{(k+1)\cdot 2^{2k}}=-1-\gamma+\log\left(2^{-\frac{25}{3}}\cdot 3^4\cdot A^{12}\right); \tag{507}$$

$$\sum_{k=2}^{\infty}\frac{\zeta(k)-1}{k+1}\left(\frac{3}{2}\right)^k=\frac{17}{12}-\frac{3\gamma}{4}+\log\left(2^{-\mathit{frac}1936}\cdot A\right); \tag{508}$$

$$\sum_{k=2}^{\infty}(-1)^k\frac{\zeta(k)-1}{k+1}\left(\frac{3}{2}\right)^k=\frac{19}{12}+\frac{3\gamma}{4}+\log\left(2^{-\frac{17}{36}}\cdot 5^{-\frac{2}{3}}\cdot A^{-1}\right); \tag{509}$$

$$\sum_{k=1}^{\infty} \frac{\zeta(2k)-1}{2k+1}\left(\frac{3}{2}\right)^{2k} = \frac{3}{2} + \log\left(2^{-\frac{1}{2}} \cdot 5^{-\frac{1}{3}}\right); \tag{510}$$

$$\sum_{k=1}^{\infty} \frac{\zeta(2k+1)-1}{k+1}\left(\frac{3}{2}\right)^{2k} = -\frac{1}{9} - \gamma + \log\left(2^{-\frac{1}{27}} \cdot 5^{\frac{4}{9}} \cdot A^{\frac{4}{3}}\right). \tag{511}$$

Taking $a = \frac{1}{2}$ in (485) through (488) and using 1.4(8), 2.3(2) and 1.3(51), we obtain

$$\begin{aligned}&\sum_{k=2}^{\infty} \left(2^k - 1\right)\zeta(k)\frac{z^{k+1}}{k+1} \\ &\quad = \frac{1}{8} - \frac{z}{2}\log(2\pi) - (1+\gamma+2\log 2)\frac{z^2}{2} - \frac{1}{2}\log\Gamma\left(\frac{1}{2}-z\right) \\ &\quad\quad - \log G\left(\frac{1}{2}-z\right) + \log\left(2^{\frac{1}{24}} \cdot A^{-\frac{3}{2}}\right) \quad \left(|z| < \frac{1}{2}\right);\end{aligned} \tag{512}$$

$$\begin{aligned}&\sum_{k=2}^{\infty} (-1)^k \left(2^k - 1\right)\zeta(k)\frac{z^{k+1}}{k+1} \\ &\quad = -\frac{1}{8} - \frac{z}{2}\log(2\pi) + (1+\gamma+2\log 2)\frac{z^2}{2} + \frac{1}{2}\log\Gamma\left(\frac{1}{2}+z\right) \\ &\quad\quad + \log G\left(\frac{1}{2}+z\right) + \log\left(2^{-\frac{1}{24}} \cdot A^{\frac{3}{2}}\right) \quad \left(|z| < \frac{1}{2}\right);\end{aligned} \tag{513}$$

$$\begin{aligned}&\sum_{k=1}^{\infty} \left(2^{2k} - 1\right)\zeta(2k)\frac{z^{2k+1}}{2k+1} \\ &\quad = -\frac{z}{2}\log(2\pi) + \frac{1}{4}\log\frac{\Gamma\left(\frac{1}{2}+z\right)}{\Gamma\left(\frac{1}{2}-z\right)} + \frac{1}{2}\log\frac{G\left(\frac{1}{2}+z\right)}{G\left(\frac{1}{2}-z\right)} \quad \left(|z| < \frac{1}{2}\right);\end{aligned} \tag{514}$$

$$\begin{aligned}&\sum_{k=1}^{\infty} \left(2^{2k+1} - 1\right)\zeta(2k+1)\frac{z^{2k+2}}{k+1} \\ &\quad = \frac{1}{4} - (1+\gamma+2\log 2)z^2 - \frac{1}{2}\log\Gamma\left(\frac{1}{2}+z\right)\Gamma\left(\frac{1}{2}-z\right) \\ &\quad\quad - \log G\left(\frac{1}{2}+z\right)G\left(\frac{1}{2}-z\right) + \log\left(2^{\frac{1}{12}} \cdot A^{-3}\right) \quad \left(|z| < \frac{1}{2}\right).\end{aligned} \tag{515}$$

Taking the limit in (513) as $z \to \frac{1}{2}$ and using 1.3(6), we obtain

$$\sum_{k=2}^{\infty} (-1)^k \frac{2^k - 1}{(k+1)\cdot 2^k}\zeta(k) = \frac{\gamma}{4} + \log\left(2^{-\frac{1}{12}} \cdot \pi^{-\frac{1}{2}} \cdot A^3\right). \tag{516}$$

Multiplying each member of (5) through (8) by t^2, integrating both sides of the resulting equations from $t = 0$ to $t = z$ and using (478), we obtain

$$\begin{aligned}\sum_{k=2}^{\infty} \zeta(k,a) \frac{z^{k+2}}{k+2} &= \frac{1-a}{2}[1-2a+\log(2\pi)]z + [1-\log(2\pi)]\frac{z^2}{4} + [\psi(a)-1]\frac{z^3}{3} \\ &\quad + (a-1)^2 \log\Gamma(a-z) - (z+a-1)\log G(a-z) - (a-1)^2\log\Gamma(a) \\ &\quad + (a-1)\log G(a) - \int_0^{-z} \log G(t+a)\,dt \quad (|z| < |a|);\end{aligned} \tag{517}$$

$$\begin{aligned}\sum_{k=2}^{\infty} (-1)^k \zeta(k,a) \frac{z^{k+2}}{k+2} &= \frac{a-1}{2}[1-2a+\log(2\pi)]z + [1-\log(2\pi)]\frac{z^2}{4} + [1-\psi(a)]\frac{z^3}{3} \\ &\quad + (a-1)^2 \log\Gamma(a+z) + (z-a+1)\log G(a+z) - (a-1)^2\log\Gamma(a) \\ &\quad + (a-1)\log G(a) - \int_0^{z} \log G(t+a)\,dt \quad (|z| < |a|);\end{aligned} \tag{518}$$

$$\begin{aligned}\sum_{k=1}^{\infty} \zeta(2k,a) \frac{z^{2k+2}}{k+1} &= 2(a-1)\log G(a) - 2(a-1)^2\log\Gamma(a) + [1-\log(2\pi)]\frac{z^2}{2} \\ &\quad + (a-1)^2\log\Gamma(a+z)\Gamma(a-z) + (z-a+1)\log G(a+z) \\ &\quad - (z+a-1)\log G(a-z) - \int_0^{z} \log G(t+a)\,dt \\ &\quad - \int_0^{-z} \log G(t+a)\,dt \quad (|z| < |a|);\end{aligned} \tag{519}$$

$$\sum_{k=1}^{\infty} \zeta(2k+1,a) \frac{z^{2k+3}}{2k+3} = \frac{1-a}{2}[1-2a+\log(2\pi)]z + [\psi(a)-1]\frac{z^3}{3} - \frac{1}{2}(a-1)^2\log\frac{\Gamma(a+z)}{\Gamma(a-z)}$$

$$-\frac{1}{2}(z-a+1)\log G(a+z)-\frac{1}{2}(z+a-1)\log G(a-z) \tag{520}$$

$$+\frac{1}{2}\left\{\int_0^z \log G(t+a)\,dt-\int_0^{-z}\log G(t+a)\,dt\right\} \quad (|z|<|a|).$$

If we use 1.4(6), 1.4(55), 1.4(56), 1.4(73) and 1.4(77), we readily find that

$$\int_0^{-1}\log\Gamma(t+1)\,dt=-\frac{1}{2}\log(2\pi); \tag{521}$$

$$\int_0^{-1}\log G(t+1)\,dt=-\frac{1}{12}+\frac{1}{4}\log(2\pi)+2\log A; \tag{522}$$

$$\int_0^{-\frac{1}{2}}\log\Gamma(t+1)\,dt=-\frac{7}{24}\log 2-\frac{1}{4}\log\pi+\frac{3}{2}\log A; \tag{523}$$

$$\int_0^{-\frac{1}{2}}\log G(t+1)\,dt=-\frac{1}{24}+\frac{1}{12}\log 2+\frac{1}{16}\log\pi+\frac{1}{4}\log A-\frac{7}{4}\log B. \tag{524}$$

Taking $a=1$ in (517) through (520) and using 1.4(6), 2.3(2) and 1.3(4), we obtain

$$\sum_{k=2}^{\infty}\zeta(k)\frac{z^{k+2}}{k+2}$$

$$=[1-\log(2\pi)]\frac{z^2}{4}-(1+\gamma)\frac{z^3 3}{-}z\log G(1-z) \tag{525}$$

$$-\int_0^{-z}\log G(t+1)\,dt \quad (|z|<1);$$

$$\sum_{k=2}^{\infty}(-1)^k\zeta(k)\frac{z^{k+2}}{k+2}$$

$$=[1-\log(2\pi)]\frac{z^2}{4}+(1+\gamma)\frac{z^3}{3}+z\log G(1+z) \tag{526}$$

$$-\int_0^{z}\log G(t+1)\,dt \quad (|z|<1);$$

$$\begin{aligned}\sum_{k=1}^{\infty} \zeta(2k) \frac{z^{2k+2}}{k+1} &= [1-\log(2\pi)]\frac{z^2}{2} + z\log\frac{G(1+z)}{G(1-z)} \\ &\quad - \int_0^z \log G(t+1)\,dt - \int_0^{-z} \log G(t+1)\,dt \quad (|z|<1);\end{aligned} \tag{527}$$

$$\begin{aligned}\sum_{k=1}^{\infty} \zeta(2k+1) \frac{z^{2k+3}}{2k+3} &= -(1+\gamma)\frac{z^3}{3} - \frac{1}{2} z\log G(1+z)G(1-z) \\ &\quad + \frac{1}{2}\left\{\int_0^z \log G(t+1)\,dt - \int_0^{-z} \log G(t+1)\,dt\right\} \quad (|z|<1).\end{aligned} \tag{528}$$

Taking the limit in (526) as $z \to 1$ and using 1.4(6) and 1.4(73), we obtain

$$\sum_{k=2}^{\infty} (-1)^k \frac{\zeta(k)}{k+2} = \frac{1}{2} + \frac{\gamma}{3} + \log\left(2^{-\frac{1}{2}} \cdot \pi^{-\frac{1}{2}} \cdot A^2\right). \tag{529}$$

If we set $z = \frac{1}{2}$ in (525) through (528), by virtue of 1.4(6), 1.4(8), 1.4(77) and (524), we get

$$\sum_{k=2}^{\infty} \frac{\zeta(k)}{(k+2)\cdot 2^k} = -\frac{\gamma}{6} + \log\left(2^{-\frac{2}{3}} \cdot A^2 \cdot B^7\right); \tag{530}$$

$$\sum_{k=2}^{\infty} (-1)^k \frac{\zeta(k)}{(k+2)\cdot 2^k} = \frac{1}{2} + \frac{\gamma}{6} + \log\left(2^{-\frac{1}{3}} \cdot A^{-2} \cdot B^7\right); \tag{531}$$

$$\sum_{k=1}^{\infty} \frac{\zeta(2k)}{(k+1)\cdot 2^{2k}} = \frac{1}{2} + \log\left(2^{-1} \cdot B^{14}\right); \tag{532}$$

$$\sum_{k=1}^{\infty} \frac{\zeta(2k+1)}{(2k+3)\cdot 2^{2k}} = -\frac{1}{2} - \frac{\gamma}{3} + \log\left(2^{-\frac{1}{3}} \cdot A^4\right). \tag{533}$$

Now, we recall the following integral formula (*cf.* Choi *et al.* [308, p. 386]):

$$\begin{aligned}\int_0^1 \log G(t+a)\,dt &= \frac{a(1-a)}{2} + \frac{a}{2}\log(2\pi) + a\log\Gamma(a) - \log G(a+1) \\ &\quad + \log\left(2^{-\frac{1}{4}} \cdot \pi^{-\frac{1}{4}} \cdot e^{\frac{1}{12}} \cdot A^{-2}\right).\end{aligned} \tag{534}$$

In view of 1.4(6), (534) with $a = 2$ yields

$$\int_0^1 \log G(t+2)\,dt = -\frac{11}{12} + \frac{3}{4}\log(2\pi) - 2\log A. \tag{535}$$

By setting $a = 2$ in (517) through (520) and using 1.4(6), 2.3(2) and 1.3(50), we find that

$$\begin{aligned}\sum_{k=2}^{\infty}\{\zeta(k)-1\}\frac{z^{k+2}}{k+2}\\ &= [3-\log(2\pi)]\frac{z}{2} + [1-\log(2\pi)]\frac{z^2}{4} - \frac{\gamma}{3}z^3 + \log\Gamma(2-z)\\ &\quad - (z+1)\log G(2-z) - \int_0^{-z}\log G(t+2)\,dt \quad (|z|<2);\end{aligned} \tag{536}$$

$$\begin{aligned}\sum_{k=2}^{\infty}(-1)^k\{\zeta(k)-1\}\frac{z^{k+2}}{k+2}\\ &= [\log(2\pi)-3]\frac{z}{2} + [1-\log(2\pi)]\frac{z^2}{4} + \frac{\gamma}{3}z^3 + \log\Gamma(2+z)\\ &\quad + (z-1)\log G(2+z) - \int_0^{z}\log G(t+2)\,dt \quad (|z|<2);\end{aligned} \tag{537}$$

$$\begin{aligned}\sum_{k=1}^{\infty}\{\zeta(2k)-1\}\frac{z^{2k+2}}{k+1}\\ &= [1-\log(2\pi)]\frac{z^2}{2} + \log\Gamma(2+z)\Gamma(2-z) + (z-1)\log G(2+z)\\ &\quad - (z+1)\log G(2-z) - \int_0^{z}\log G(t+2)\,dt\\ &\quad - \int_0^{-z}\log G(t+2)\,dt \quad (|z|<2);\end{aligned} \tag{538}$$

$$\begin{aligned}\sum_{k=1}^{\infty}\{\zeta(2k+1)-1\}\frac{z^{2k+3}}{2k+3}\\ &= [3-\log(2\pi)]\frac{z}{2} - \frac{\gamma}{3}z^3 - \frac{1}{2}\log\frac{\Gamma(2+z)}{\Gamma(2-z)}\end{aligned}$$

$$-\frac{1}{2}(z-1)\log G(2+z)-\frac{1}{2}(z+1)\log G(2-z)$$
$$+\frac{1}{2}\left\{\int_0^z \log G(t+2)\,dt-\int_0^{-z}\log G(t+2)\,dt\right\}\qquad(|z|<2). \tag{539}$$

Taking the limit in (537) as $z\to 2$ and using 1.4(6), 1.4(55) and (535), we obtain

$$\sum_{k=2}^{\infty}(-1)^k(\zeta(k)-1)\frac{2^k}{k+2}=\frac{11}{24}+\frac{2\gamma}{3}+\log\left(3^{\frac{1}{4}}\cdot\pi^{-\frac{1}{2}}\cdot A\right). \tag{540}$$

If use is made of 1.4(6), 1.4(56), 1.4(73), 1.4(77), (524) and (535), the following integral formulas are readily obtained:

$$\int_0^{\frac{3}{2}}\log G(t+2)\,dt$$
$$=\int_0^1\log G(t+2)\,dt+\int_0^{\frac{1}{2}}\log G(t+2)\,dt+\int_0^{\frac{1}{2}}\log\Gamma(t+2)\,dt \tag{541}$$
$$=-\frac{19}{8}-\frac{31}{24}\log 2+\frac{3}{2}\log 3+\frac{21}{16}\log\pi+\frac{3}{4}\log A-\frac{7}{4}\log B;$$

$$\int_0^{-\frac{3}{2}}\log G(t+2)\,dt$$
$$=-\int_0^1\log G(t+1)\,dt+\int_0^{-\frac{1}{2}}\log G(t+1)\,dt \tag{542}$$
$$=-\frac{1}{8}-\frac{1}{6}\log 2-\frac{3}{16}\log\pi+\frac{9}{4}\log A-\frac{7}{4}\log B.$$

If we set $z=1$, $z=\frac{1}{2}$, and $z=\frac{3}{2}$ in (536) through (539), in view of 1.4(6), 1.4(8), 1.4(56), 1.4(73), 1.4(77), (523), (524), (535) and (541), we get

$$\sum_{k=2}^{\infty}\frac{\zeta(k)-1}{k+2}=\frac{11}{6}-\frac{\gamma}{3}+\log\left(2^{-\frac{1}{2}}\cdot\pi^{-\frac{1}{2}}\cdot A^{-2}\right); \tag{543}$$

$$\sum_{k=2}^{\infty}(-1)^k\frac{\zeta(k)-1}{k+2}=-\frac{13\ \gamma}{+\ 3}+\log\left(2^{\frac{1}{2}}\cdot\pi^{-\frac{1}{2}}\cdot A^{2}\right); \tag{544}$$

$$\sum_{k=1}^{\infty}\frac{\zeta(2k)-1}{k+1}=\frac{3}{2}-\log\pi; \tag{545}$$

$$\sum_{k=1}^{\infty}\frac{\zeta(2k+1)-1}{2k+3}=1312-\frac{\gamma}{3}+\log\left(2^{-\frac{1}{2}}\cdot A^{-2}\right); \tag{546}$$

$$\sum_{k=2}^{\infty}\frac{\zeta(k)-1}{(k+2)\cdot 2^k}=\frac{8}{3}-\frac{\gamma}{6}+\log\left(2^{-\frac{14}{3}}\cdot A^2\cdot B^7\right); \tag{547}$$

$$\sum_{k=2}^{\infty}(-1)^k\frac{\zeta(k)-1}{(k+2)\cdot 2^k}=-\frac{7}{6}+\frac{\gamma}{6}+\log\left(2^{-\frac{13}{3}}\cdot 3^4\cdot A^{-2}\cdot B^7\right); \tag{548}$$

$$\sum_{k=1}^{\infty}\frac{\zeta(2k)-1}{(k+1)\cdot 2^{2k}}=\frac{3}{2}+\log\left(2^{-9}\cdot 3^4\cdot B^{14}\right); \tag{549}$$

$$\sum_{k=1}^{\infty}\frac{\zeta(2k+1)-1}{(2k+3)\cdot 2^{2k}}=\frac{23}{6}-\frac{\gamma}{3}+\log\left(2^{\frac{1}{2}}\cdot 3^{-4}\cdot A^4\right); \tag{550}$$

$$\sum_{k=2}^{\infty}\frac{\zeta(k)-1}{k+2}\left(\frac{3}{2}\right)^k=\frac{7}{6}-\frac{\gamma}{2}+\log\left(2^{-\frac{5}{9}}\cdot A^{\frac{2}{3}}\cdot B^{\frac{7}{9}}\right); \tag{551}$$

$$\sum_{k=2}^{\infty}(-1)^k\frac{\zeta(k)-1}{k+2}\left(\frac{3}{2}\right)^k=\frac{1}{3}+\frac{\gamma}{2}+\log\left(2^{-\frac{4}{3}}\cdot 5^{\frac{4}{9}}\cdot A^{-\frac{2}{3}}\cdot B^{\frac{7}{9}}\right); \tag{552}$$

$$\sum_{k=1}^{\infty}\frac{\zeta(2k)-1}{k+1}\left(\frac{3}{2}\right)^{2k}=\frac{3}{2}+\log\left(2^{-\frac{17}{9}}\cdot 5^{\frac{4}{9}}\cdot B^{\frac{14}{9}}\right); \tag{553}$$

$$\sum_{k=1}^{\infty}\frac{\zeta(2k+1)-1}{2k+3}\left(\frac{3}{2}\right)^{2k}=\frac{5}{18}-\frac{\gamma}{3}+\log\left(2^{\frac{7}{27}}\cdot 5^{-\frac{4}{27}}\cdot A^{\frac{4}{9}}\right). \tag{554}$$

Taking $a=\frac{1}{2}$ in (517) through (520) and using 1.4(8), 2.3(2) and 1.3(51), we obtain

$$\begin{aligned}\sum_{k=2}^{\infty}&\left(2^k-1\right)\zeta(k)\frac{z^{k+2}}{k+2}\\&=\log\left(2^{-\frac{1}{48}}\cdot e^{-\frac{1}{16}}\cdot A^{\frac{3}{4}}\right)+\frac{z}{4}\log(2\pi)+[1-\log(2\pi)]\frac{z^2}{4}\\&\quad-(1+\gamma+2\log 2)\frac{z^3}{3}+\frac{1}{4}\log\Gamma\left(\frac{1}{2}-z\right)+\left(\frac{1}{2}-z\right)\log G\left(\frac{1}{2}-z\right)\\&\quad-\int_0^{-z}\log G\left(t+\frac{1}{2}\right)dt\quad\left(|z|<\frac{1}{2}\right);\end{aligned} \tag{555}$$

$$\begin{aligned}\sum_{k=2}^{\infty}&(-1)^k\left(2^k-1\right)\zeta(k)\frac{z^{k+2}}{k+2}\\&=\log\left(2^{-\frac{1}{48}}\cdot e^{-\frac{1}{16}}\cdot A^{\frac{3}{4}}\right)-\frac{z}{4}\log(2\pi)+[1-\log(2\pi)]\frac{z^2}{4}\end{aligned}$$

$$+(1+\gamma+2\log 2)\frac{z^3}{3}+\frac{1}{4}\log\Gamma\left(\frac{1}{2}+z\right)+\left(\frac{1}{2}+z\right)\log G\left(\frac{1}{2}+z\right)$$
$$-\int_0^z \log G\left(t+\frac{1}{2}\right)dt \quad \left(|z|<\frac{1}{2}\right); \tag{556}$$

$$\sum_{k=1}^{\infty}\left(2^{2k}-1\right)\zeta(2k)\frac{z^{2k+2}}{k+1}$$
$$=\log\left(2^{-\frac{1}{24}}\cdot e^{-\frac{1}{8}}\cdot A^{\frac{3}{2}}\right)+[1-\log(2\pi)]\frac{z^2}{2}+\frac{1}{4}\log\Gamma\left(\frac{1}{2}+z\right)\Gamma\left(\frac{1}{2}-z\right)$$
$$+\left(\frac{1}{2}+z\right)\log G\left(\frac{1}{2}+z\right)+\left(\frac{1}{2}-z\right)\log G\left(\frac{1}{2}-z\right) \tag{557}$$
$$-\int_0^z \log G\left(t+\frac{1}{2}\right)dt-\int_0^{-z}\log G\left(t+\frac{1}{2}\right)dt \quad (|z|<\frac{1}{2});$$

$$\sum_{k=1}^{\infty}\left(2^{2k+1}-1\right)\zeta(2k+1)\frac{z^{2k+3}}{2k+3}$$
$$=\frac{z}{4}\log(2\pi)-(1+\gamma+2\log 2)\frac{z^3}{3}-\frac{1}{8}\log\frac{\Gamma\left(\frac{1}{2}+z\right)}{\Gamma\left(\frac{1}{2}-z\right)}$$
$$-\frac{1}{2}\left(\frac{1}{2}+z\right)\log G\left(\frac{1}{2}+z\right)+\frac{1}{2}\left(\frac{1}{2}-z\right)\log G\left(\frac{1}{2}-z\right) \tag{558}$$
$$+\frac{1}{r2}\left\{\int_0^z \log G\left(t+\frac{1}{2}\right)dt-\int_0^{-z}\log G\left(t+\frac{1}{2}\right)dt\right\} \quad \left(|z|<\frac{1}{2}\right).$$

By taking the limit in (556) as $z\to\frac{1}{2}$ and using 1.4(6), 1.4(55), 1.4(56), 1.4(73) and 1.4(77), we obtain

$$\sum_{k=2}^{\infty}(-1)^k\frac{2^k-1}{(k+2)\cdot 2^k}\zeta(k)=\frac{\gamma}{6}+\log\left(2^{\frac{23}{6}}\cdot\pi^{-\frac{1}{2}}\cdot A^4\cdot B^{-7}\right). \tag{559}$$

Integrating both sides of (1) through (4) from $t=0$ to $t=z$, in view of (470), we find that

$$\sum_{k=2}^{\infty}\zeta(k,a)\frac{z^{k+1}}{k(k+1)}$$
$$=-z\log\Gamma(a)+\psi(a)\frac{z^2}{2}-\int_0^{-z}\log\Gamma(a+t)\,dt$$

$$= -z\log\Gamma(a) + \frac{\psi(a)}{2}z^2 + \int_0^z \log\Gamma(a-t)\,dt$$

$$= (a-1)\log\Gamma(a) - \log G(a) + \left(\frac{1}{2} + \frac{1}{2}\log(2\pi) - a - \log\Gamma(a)\right)z \tag{560}$$

$$+ [1+\psi(a)]\frac{z^2}{2} + (z-a+1)\log\Gamma(a-z) + \log G(a-z) \quad (|z| < |a|);$$

$$\sum_{k=2}^{\infty}(-1)^k\zeta(k,a)\frac{z^{k+1}}{k(k+1)} = -z\log\Gamma(a) - \psi(a)\frac{z^2}{2} + \int_0^z \log\Gamma(a+t)\,dt$$

$$= (1-a)\log\Gamma(a) + \log G(a) + \left(\frac{1}{2} + \frac{1}{2}\log(2\pi) - a - \log\Gamma(a)\right)z \tag{561}$$

$$- [1+\psi(a)]\frac{z^2}{2} + (z+a-1)\log\Gamma(z+a) - \log G(z+a) \quad (|z| < |a|);$$

$$\sum_{k=1}^{\infty}\zeta(2k,a)\frac{z^{2k+1}}{k(2k+1)}$$

$$= -2z\log\Gamma(a) + \int_{-z}^{z} \log\Gamma(a+t)\,dt$$

$$= -2z\log\Gamma(a) + \int_0^z \{\log\Gamma(a+t) + \log\Gamma(a-t)\}\,dt \tag{562}$$

$$= [1 - 2a + \log(2\pi) - 2\log\Gamma(a)]z + (z+a-1)\log\Gamma(a+z)$$

$$+ (z-a+1)\log\Gamma(a-z) + \log\frac{G(a-z)}{G(a+z)} \quad (|z| < |a|);$$

$$\sum_{k=1}^{\infty}\zeta(2k+1,a)\frac{z^{2k+2}}{(k+1)(2k+1)}$$

$$= \psi(a)z^2 - \int_0^z \log\Gamma(a+t)\,dt - \int_0^{-z} \log\Gamma(a+t)\,dt$$

$$= \psi(a)z^2 + \int_0^z \{\log\Gamma(a-t) - \log\Gamma(a+t)\}\,dt \tag{563}$$

$$= 2(a-1)\log\Gamma(a) - 2\log G(a) + [1+\psi(a)]z^2 + (1-a+z)\log\Gamma(a-z)$$

$$+ (1-a-z)\log\Gamma(a+z) + \log G(a-z)G(a+z) \quad (|z| < |a|).$$

If we set $a = 1$ in (560) through (563), by virtue of 1.4(6), 2.3(2) and 1.3(4), we get

$$\sum_{k=2}^{\infty} \zeta(k) \frac{z^{k+1}}{k(k+1)} = [\log(2\pi) - 1] \frac{z}{r2} + (1-\gamma)\frac{z^2}{2} + z \log \Gamma(1-z) + \log G(1-z) \quad (|z| < 1); \tag{564}$$

$$\sum_{k=2}^{\infty} (-1)^k \zeta(k) \frac{z^{k+1}}{k(k+1)} = [\log(2\pi) - 1] \frac{z}{2} + (\gamma - 1)\frac{z^2}{2} + z \log \Gamma(z+1) - \log G(z+1) \quad (|z| < 1); \tag{565}$$

$$\sum_{k=1}^{\infty} \zeta(2k) \frac{z^{2k+1}}{k(2k+1)} = [\log(2\pi) - 1] z + z \log \Gamma(1-z)\Gamma(1+z) + \log \frac{G(1-z)}{G(1+z)} \quad (|z| < 1); \tag{566}$$

$$\sum_{k=1}^{\infty} \zeta(2k+1) \frac{z^{2k+2}}{(k+1)(2k+1)} = (1-\gamma) z^2 + z \log \frac{\Gamma(1-z)}{\Gamma(1+z)} + \log G(1-z) G(1+z) \quad (|z| < 1). \tag{567}$$

Taking the limit in (565) as $z \to 1$ and using 1.3(6), we obtain

$$\sum_{k=2}^{\infty} (-1)^k \frac{\zeta(k)}{k(k+1)} = -1 + \frac{\gamma}{2} + \frac{1}{2} \log(2\pi). \tag{568}$$

Setting $z = \frac{1}{2}$ in (564) through (567) and using 1.4(6) and 1.4(8), we obtain

$$\sum_{k=2}^{\infty} \frac{\zeta(k)}{k(k+1) \cdot 2^k} = -\frac{\gamma}{4} + \log\left(2^{\frac{7}{12}} \cdot \pi^{\frac{1}{2}} \cdot A^{-3}\right); \tag{569}$$

$$\sum_{k=2}^{\infty} (-1)^k \frac{\zeta(k)}{k(k+1) \cdot 2^k} = -1 + \frac{\gamma}{4} + \log\left(2^{-\frac{7}{12}} \cdot \pi^{\frac{1}{2}} \cdot A^{3}\right); \tag{570}$$

$$\sum_{k=1}^{\infty} \frac{\zeta(2k)}{k(2k+1)\cdot 2^{2k}} = -1 + \log \pi; \tag{571}$$

$$\sum_{k=1}^{\infty} \frac{\zeta(2k+1)}{(k+1)(2k+1)\cdot 2^{2k}} = 2 - \gamma + \log\left(2^{\frac{7}{3}} \cdot A^{-12}\right). \tag{572}$$

By putting $a = 2$ in (560) through (563) and using 1.4(6), 2.3(2) and 1.3(50), we obtain

$$\begin{aligned}\sum_{k=2}^{\infty} \{\zeta(k) - 1\} \frac{z^{k+1}}{k(k+1)} \\ &= [\log(2\pi) - 3]\frac{z}{2} + (2-\gamma)\frac{z^2}{2} + (z-1)\log\Gamma(2-z) \\ &\quad + \log G(2-z) \quad (|z| < 2);\end{aligned} \tag{573}$$

$$\begin{aligned}\sum_{k=2}^{\infty} (-1)^k \{\zeta(k) - 1\} \frac{z^{k+1}}{k(k+1)} \\ &= [\log(2\pi) - 3]\frac{z}{2} + (\gamma - 2)\frac{z^2}{2} + (z+1)\log\Gamma(z+2) \\ &\quad - \log G(z+2) \quad (|z| < 2);\end{aligned} \tag{574}$$

$$\begin{aligned}\sum_{k=1}^{\infty} \{\zeta(2k) - 1\} \frac{z^{2k+1}}{k(2k+1)} \\ &= [\log(2\pi) - 3]z + (z+1)\log\Gamma(2+z) + (z-1)\log\Gamma(2-z) \\ &\quad + \log\frac{G(2-z)}{G(2+z)} \quad (|z| < 2);\end{aligned} \tag{575}$$

$$\begin{aligned}\sum_{k=1}^{\infty} \{\zeta(2k+1) - 1\} \frac{z^{2k+2}}{(k+1)(2k+1)} \\ &= (2-\gamma)z^2 + (z-1)\log\Gamma(2-z) - (z+1)\log\Gamma(2+z) \\ &\quad + \log G(2-z)G(2+z) \quad (|z| < 2).\end{aligned} \tag{576}$$

Taking the limit in (574) as $z \to 2$ and using 1.4(6), we have

$$\sum_{k=2}^{\infty} (-1)^k \frac{\{\zeta(k) - 1\}2^k}{k(k+1)} = -\frac{7}{2} + \gamma + \log\left(2^{\frac{3}{2}} \cdot 3^{\frac{3}{2}} \cdot \pi^{\frac{1}{2}}\right). \tag{577}$$

If we set $z = 1$, $z = \frac{1}{2}$ and $z = \frac{3}{2}$ in (573) through (576), in view of 1.4(6) and 1.4(8), we obtain

$$\sum_{k=2}^{\infty} \frac{\zeta(k) - 1}{k(k+1)} = -\frac{1}{2} - \frac{\gamma}{2} + \log\left(2^{\frac{1}{2}} \cdot \pi^{\frac{1}{2}}\right); \tag{578}$$

$$\sum_{k=2}^{\infty}(-1)^k \frac{\zeta(k)-1}{k(k+1)} = -\frac{5}{2}+\frac{\gamma}{2}+\log\left(2^{\frac{5}{2}}\cdot\pi^{\frac{1}{2}}\right); \tag{579}$$

$$\sum_{k=1}^{\infty}\frac{\zeta(2k)-1}{k(2k+1)} = -3+\log(8\pi); \tag{580}$$

$$\sum_{k=1}^{\infty}\frac{\zeta(2k+1)-1}{(k+1)(2k+1)} = 2-\gamma-2\log 2; \tag{581}$$

$$\sum_{k=2}^{\infty}\frac{\zeta(k)-1}{k(k+1)\cdot 2^k} = -\frac{3}{4}-\frac{\gamma}{4}+\log\left(2^{\frac{19}{12}}\cdot\pi^{\frac{1}{2}}\cdot A^{-3}\right); \tag{582}$$

$$\sum_{k=2}^{\infty}(-1)^k \frac{\zeta(k)-1}{k(k+1)\cdot 2^k} = -\frac{9}{4}+\frac{\gamma}{4}+\log\left(2^{-\frac{43}{12}}\cdot 3^3\cdot\pi^{\frac{1}{2}}\cdot A^{3}\right); \tag{583}$$

$$\sum_{k=1}^{\infty}\frac{\zeta(2k)-1}{k(2k+1)\cdot 2^{2k}} = -3+\log\left(2^{-2}\cdot 3^3\cdot\pi\right); \tag{584}$$

$$\sum_{k=1}^{\infty}\frac{\zeta(2k+1)-1}{(k+1)(2k+1)\cdot 2^{2k}} = 3-\gamma+\log\left(2^{\frac{31}{3}}\cdot 3^{-6}\cdot A^{-12}\right); \tag{585}$$

$$\sum_{k=2}^{\infty}\frac{\zeta(k)-1}{k(k+1)}\left(\frac{3}{2}\right)^k = \frac{1}{12}-\frac{3\gamma}{4}+\log\left(2^{\frac{19}{36}}\cdot\pi^{\frac{1}{2}}\cdot A^{-1}\right); \tag{586}$$

$$\sum_{k=2}^{\infty}(-1)^k\frac{\zeta(k)-1}{k(k+1)}\left(\frac{3}{2}\right)^k = -\frac{37}{12}+\frac{3\gamma}{4}+\log\left(2^{-\frac{91}{36}}\cdot 3\cdot 5^{\frac{5}{3}}\cdot\pi^{\frac{1}{2}}\cdot A\right); \tag{587}$$

$$\sum_{k=1}^{\infty}\frac{\zeta(2k)-1}{k(2k+1)}\left(\frac{3}{2}\right)^{2k} = -3+\log\left(2^{-2}\cdot 3\cdot 5^{\frac{5}{3}}\cdot\pi\right); \tag{588}$$

$$\sum_{k=1}^{\infty}\frac{\zeta(2k+1)-1}{(k+1)(2k+1)}\left(\frac{3}{2}\right)^{2k} = \frac{19}{9}-\gamma+\log\left(2^{\frac{55}{27}}\cdot 3^{-\frac{2}{3}}\cdot 5^{-\frac{10}{9}}\cdot A^{-\frac{4}{3}}\right). \tag{589}$$

If we take $a=\frac{1}{2}$ in (560) through (563), by virtue 1.4(8), 2.3(2) and 1.3(51), we obtain

$$\begin{aligned}\sum_{k=2}^{\infty}\left(2^k-1\right)\zeta(k)\frac{z^{k+1}}{k(k+1)}\\ &= \log\left(2^{-\frac{1}{24}}\cdot e^{-\frac{1}{8}}\cdot A^{\frac{3}{2}}\right)+\frac{z}{2}\log 2+(1-\gamma-2\log 2)\frac{z^2}{2}\\ &\quad+\left(\frac{1}{2}+z\right)\log\Gamma\left(\frac{1}{2}-z\right)+\log G\left(\frac{1}{2}-z\right) \quad \left(|z|<\frac{1}{2}\right);\end{aligned} \tag{590}$$

$$\sum_{k=2}^{\infty}(-1)^k\left(2^k-1\right)\zeta(k)\frac{z^{k+1}}{k(k+1)}$$
$$= \log\left(2^{\frac{1}{24}}\cdot e^{\frac{1}{8}}\cdot A^{-\frac{3}{2}}\right)+\frac{z}{2}\log 2+(\gamma+2\log 2-1)\frac{z^2}{2}$$
$$+\left(z-\frac{1}{2}\right)\log\Gamma\left(z+\frac{1}{2}\right)-\log G\left(z+\frac{1}{2}\right)\quad\left(|z|<\frac{1}{2}\right); \tag{591}$$

$$\sum_{k=1}^{\infty}\left(2^{2k}-1\right)\zeta(2k)\frac{z^{2k+1}}{k(2k+1)}$$
$$= z\log 2+\left(z-\frac{1}{2}\right)\log\Gamma\left(\frac{1}{2}+z\right)+\left(\frac{1}{2}+z\right)\log\Gamma\left(\frac{1}{2}-z\right) \tag{592}$$
$$+\log\frac{G\left(\frac{1}{2}-z\right)}{G\left(\frac{1}{2}+z\right)}\quad\left(|z|<\frac{1}{2}\right);$$

$$\sum_{k=1}^{\infty}\left(2^{2k+1}-1\right)\zeta(2k+1)\frac{z^{2k+2}}{(k+1)(2k+1)}$$
$$= \log\left(2^{-\frac{1}{12}}\cdot e^{-\frac{1}{4}}\cdot A^{3}\right)+(1-\gamma-2\log 2)z^2$$
$$+\left(\frac{1}{2}+z\right)\log\Gamma\left(\frac{1}{2}-z\right)+\left(\frac{1}{2}-z\right)\log\Gamma\left(\frac{1}{2}+z\right) \tag{593}$$
$$+\log G\left(\frac{1}{2}-z\right)G\left(\frac{1}{2}+z\right)\quad\left(|z|<\frac{1}{2}\right).$$

Taking the limit in (591) as $z\to\frac{1}{2}$ and using 1.3(6), we obtain

$$\sum_{k=2}^{\infty}(-1)^k\frac{2^k-1}{k(k+1)\cdot 2^k}\zeta(k)=\frac{\gamma}{4}+\log\left(2^{\frac{13}{12}}\cdot A^{-3}\right). \tag{594}$$

By setting $z=t$ in (560) through (563), integrating both sides of the resulting equations from $t=0$ to $t=z$ and using (470) and (478), we obtain

$$\sum_{k=2}^{\infty}\zeta(k,a)k(k+1)(k+2)z^{k+2}$$
$$= \frac{1}{2}(a-1)\log G(a)-\frac{1}{2}(a-1)^2\log\Gamma(a)$$
$$+\left(\frac{1}{4}(a-1)[2a-1-\log(2\pi)]+(a-1)\log\Gamma(a)-\log G(a)\right)z \tag{595}$$
$$+[5-8a+3\log(2\pi)-4\log\Gamma(a)]\frac{z^2}{8}+[2+\psi(a)]\frac{z^3}{6}$$

$$+\frac{1}{2}(z-a+1)^2\log\Gamma(a-z)+\frac{1}{2}(z-a+1)\log G(a-z)$$
$$-\frac{1}{2}\int_0^{-z}\log G(t+a)\,dt \quad (|z|<|a|);$$

$$\sum_{k=2}^{\infty}(-1)^k\frac{\zeta(k,a)}{k(k+1)(k+2)}z^{k+2}$$
$$=\frac{1}{2}(a-1)\log G(a)-\frac{1}{2}(a-1)^2\log\Gamma(a)$$
$$+\left(\frac{1}{4}(a-1)[1-2a+\log(2\pi)]+(1-a)\log\Gamma(a)+\log G(a)\right)z$$
$$+[5-8a+3\log(2\pi)-4\log\Gamma(a)]\frac{z^2}{8}-[2+\psi(a)]\frac{z^3}{6} \tag{596}$$
$$+\frac{1}{2}(z+a-1)^2\log\Gamma(z+a)-\frac{1}{2}(z+a-1)\log G(z+a)$$
$$-\frac{1}{2}\int_0^{z}\log G(t+a)\,dt \quad (|z|<|a|);$$

$$\sum_{k=1}^{\infty}\frac{\zeta(2k,a)}{k(k+1)(2k+1)}z^{2k+2}=2(a-1)\log G(a)$$
$$-2(a-1)^2\log\Gamma(a)+[5-8a+3\log(2\pi)-4\log\Gamma(a)]\frac{z^2}{2}$$
$$+(z+a-1)^2\log\Gamma(a+z)+(z-a+1)^2\log\Gamma(a-z) \tag{597}$$
$$+(1-a-z)\log G(a+z)+(1-a+z)\log G(a-z)$$
$$-\int_0^{z}\log G(t+a)\,dt-\int_0^{-z}\log G(t+a)\,dt \quad (|z|<|a|);$$

$$\sum_{k=1}^{\infty}\frac{\zeta(2k+1,a)}{(k+1)(2k+1)(2k+3)}z^{2k+3}$$
$$=\left(2(a-1)\log\Gamma(a)-2\log G(a)+\frac{1}{2}(a-1)[2a-1-\log(2\pi)]\right)z$$
$$+[2+\psi(a)]\frac{z^3}{3}+\frac{1}{2}(z-a+1)^2\log\Gamma(a-z)-\frac{1}{2}(z+a-1)^2\log\Gamma(a+z)$$
$$+\frac{1}{2}(z-a+1)\log G(a-z)+\frac{1}{2}(z+a-1)\log G(a+z)$$
$$+\frac{1}{2}\int_0^{z}\log G(t+a)\,dt-\frac{1}{2}\int_0^{-z}\log G(t+a)\,dt \quad (|z|<|a|). \tag{598}$$

Taking $a = 1$ in (595) through (598) and using 1.4(6), 2.3(2) and 1.3(4), we obtain

$$\sum_{k=2}^{\infty} \frac{\zeta(k)}{k(k+1)(k+2)} z^{k+2}$$
$$= [\log(2\pi) - 1]\frac{3z^2}{8} + (2-\gamma)\frac{z^3}{6} + \frac{z^2}{2}\log\Gamma(1-z) \tag{599}$$
$$+ \frac{z}{2}\log G(1-z) - \frac{1}{2}\int_0^{-z} \log G(t+1)\,dt \quad (|z| < 1);$$

$$\sum_{k=2}^{\infty} (-1)^k \frac{\zeta(k)}{k(k+1)(k+2)} z^{k+2}$$
$$= [\log(2\pi) - 1]\frac{3z^2}{8} + (\gamma - 2)\frac{z^3}{6} + \frac{z^2}{2}\log\Gamma(z+1) \tag{600}$$
$$- \frac{z}{2}\log G(z+1) - \frac{1}{2}\int_0^{z} \log G(t+1)\,dt \quad (|z| < 1);$$

$$\sum_{k=1}^{\infty} \frac{\zeta(2k)}{k(k+1)(2k+1)} z^{2k+2}$$
$$= [\log(2\pi) - 1]\frac{3z^2}{2} + z^2\log\Gamma(1+z)\Gamma(1-z) + z\log\frac{G(1-z)}{G(1+z)} \tag{601}$$
$$- \int_0^{z} \log G(t+1)\,dt - \int_0^{-z} \log G(t+1)\,dt \quad (|z| < 1);$$

$$\sum_{k=1}^{\infty} \frac{\zeta(2k+1)}{(k+1)(2k+1)(2k+3)} z^{2k+3}$$
$$= (2-\gamma)\frac{z^3}{3} + \frac{z^2}{2}\log\frac{\Gamma(1-z)}{\Gamma(1+z)} + \frac{z}{2}\log G(1-z)G(1+z) \tag{602}$$
$$+ \frac{1}{2}\int_0^{z} \log G(t+1)\,dt - \frac{1}{2}\int_0^{-z} \log G(t+1)\,dt \quad (|z| < 1).$$

Taking the limit in (600) as $z \to 1$ and using 1.4(6) and 1.4(73), we obtain

$$\sum_{k=2}^{\infty} (-1)^k \frac{\zeta(k)}{k(k+1)(k+2)} = -\frac{3}{4} + \frac{\gamma}{6} + \log\left(2^{\frac{1}{4}} \cdot \pi^{\frac{1}{4}} \cdot A\right). \tag{603}$$

If we set $z = \frac{1}{2}$ in (599) through (602), in view of 1.4(6), 1.4(8), 1.4(77) and (524), we get

$$\sum_{k=2}^{\infty} \frac{\zeta(k)}{k(k+1)(k+2)\cdot 2^k} = -\frac{\gamma}{12} + \log\left(2^{\frac{1}{4}} \cdot \pi^{\frac{1}{4}} \cdot A^{-2} \cdot B^{\frac{7}{2}}\right); \tag{604}$$

$$\sum_{k=2}^{\infty} (-1)^k \frac{\zeta(k)}{k(k+1)(k+2)\cdot 2^k} = -\frac{3}{4} + \frac{\gamma}{12} + \log\left(2^{-\frac{1}{4}} \cdot \pi^{\frac{1}{4}} \cdot A^{2} \cdot B^{\frac{7}{2}}\right); \tag{605}$$

$$\sum_{k=1}^{\infty} \frac{\zeta(2k)}{k(k+1)(2k+1)\cdot 2^{2k}} = -\frac{3}{2} + \log\left(\pi \cdot B^{14}\right); \tag{606}$$

$$\sum_{k=1}^{\infty} \frac{\zeta(2k+1)}{(k+1)(2k+1)(2k+3)\cdot 2^{2k}} = \frac{3}{2} - \frac{\gamma}{3} + \log\left(2 \cdot A^{-8}\right). \tag{607}$$

Taking $a = 2$ in (595) through (598) and using 1.4(6), 2.3(2) and 1.3(50), we have

$$\begin{aligned} &\sum_{k=2}^{\infty} \frac{\zeta(k)-1}{k(k+1)(k+2)} z^{k+2} \\ &\quad = [3 - \log(2\pi)]z4 + [3\log(2\pi) - 11]z^2 8 + (3-\gamma)z^3 6 \\ &\qquad + 12(z-1)^2 \log\Gamma(2-z) + 12(z-1)\log G(2-z) \\ &\qquad - 12\int_0^{-z} \log G(t+2)\,dt \quad (|z| < 2); \end{aligned} \tag{608}$$

$$\begin{aligned} &\sum_{k=2}^{\infty} (-1)^k \frac{\zeta(k)-1}{k(k+1)(k+2)} z^{k+2} \\ &\quad = [\log(2\pi) - 3]\frac{z}{4} + [3\log(2\pi) - 11]\frac{z^2}{8} + (\gamma - 3)\frac{z^3}{6} \\ &\qquad + \frac{1}{2}(z+1)^2 \log\Gamma(z+2) - \frac{1}{2}(z+1)\log G(z+2) \\ &\qquad - \frac{1}{2}\int_0^{z} \log G(t+2)\,dt \quad (|z| < 2); \end{aligned} \tag{609}$$

$$\begin{aligned} &\sum_{k=1}^{\infty} \frac{\zeta(2k)-1}{k(k+1)(2k+1)} z^{2k+2} \\ &\quad = [3\log(2\pi) - 11]\frac{z^2}{2} + (z+1)^2 \log\Gamma(2+z) \end{aligned} \tag{610}$$

$$+(z-1)^2\log\Gamma(2-z)-(z+1)\log G(z+2)+(z-1)\log G(2-z)$$
$$-\int_0^z \log G(t+2)\,dt-\int_0^{-z}\log G(t+2)\,dt \quad (|z|<2);$$

$$\sum_{k=1}^{\infty}\frac{\zeta(2k+1)-1}{(k+1)(2k+1)(2k+3)}z^{2k+3}$$
$$=[3-\log(2\pi)]\frac{z}{2}+(3-\gamma)\frac{z^3}{3}+\frac{1}{2}(z-1)^2\log\Gamma(2-z)$$
$$-\frac{1}{2}(z+1)^2\log\Gamma(2+z)+\frac{1}{r2}(z-1)\log G(2-z)+\frac{1}{2}(z+1)\log G(2+z) \tag{611}$$
$$+\frac{1}{2}\int_0^z \log G(t+2)\,dt-\frac{1}{2}\int_0^{-z}\log G(t+2)\,dt \quad (|z|<2).$$

Taking the limit in (609) as $z\to 2$ and using 1.4(6) and (535), we obtain

$$\sum_{k=2}^{\infty}(-1)^k\frac{\zeta(k)-1}{k(k+1)(k+2)}2^k=-\frac{109}{48}+\frac{\gamma}{3}+\log\left(2^{\frac{3}{4}}\cdot 3^{\frac{9}{8}}\cdot\pi^{\frac{1}{4}}\cdot A^{\frac{1}{2}}\right). \tag{612}$$

If we set $z=1$, $z=\frac{1}{2}$ and $z=\frac{3}{2}$ in (608) through (611), by virtue of 1.4(6), 1.4(8), 1.4(73), (523), (524), (535), (541) and (542), we get

$$\sum_{k=2}^{\infty}\frac{\zeta(k)-1}{k(k+1)(k+2)}=-\frac{1}{12}-\frac{\gamma}{6}+\log\left(2^{\frac{1}{4}}\cdot\pi^{\frac{1}{4}}\cdot A^{-1}\right); \tag{613}$$

$$\sum_{k=2}^{\infty}(-1)^k\frac{\zeta(k)-1}{k(k+1)(k+2)}=-\frac{13}{6}+\frac{\gamma}{6}+\log\left(2^{\frac{9}{4}}\cdot\pi^{\frac{1}{4}}\cdot A\right); \tag{614}$$

$$\sum_{k=1}^{\infty}\frac{\zeta(2k)-1}{k(k+1)(2k+1)}=-\frac{9}{2}+5\log 2+\log\pi; \tag{615}$$

$$\sum_{k=1}^{\infty}\frac{\zeta(2k+1)-1}{(k+1)(2k+1)(2k+3)}=\frac{25}{12}-\frac{\gamma}{3}+\log\left(2^{-2}\cdot A^{-2}\right); \tag{616}$$

$$\sum_{k=2}^{\infty}\frac{\zeta(k)-1}{k(k+1)(k+2)\cdot 2^k}=\frac{1}{3}-\frac{\gamma}{12}+\log\left(2^{-\frac{1}{4}}\cdot\pi^{\frac{1}{4}}\cdot A^{-2}\cdot B^{\frac{7}{2}}\right); \tag{617}$$

$$\sum_{k=2}^{\infty}(-1)^k\frac{\zeta(k)-1}{k(k+1)(k+2)\cdot 2^k}=-\frac{31}{12}+\frac{\gamma}{12}+\log\left(2^{-\frac{19}{4}}\cdot 3^{\frac{9}{2}}\cdot\pi^{\frac{1}{4}}\cdot A^{2}\cdot B^{\frac{7}{2}}\right); \tag{618}$$

$$\sum_{k=1}^{\infty}\frac{\zeta(2k)-1}{k(k+1)(2k+1)\cdot 2^{2k}}=-\frac{9}{2}+\log\left(2^{-10}\cdot 3^{9}\cdot\pi\cdot B^{14}\right); \tag{619}$$

$$\sum_{k=1}^{\infty}\frac{\zeta(2k+1)-1}{(k+1)(2k+1)(2k+3)\cdot 2^{2k}}=\frac{35}{6}-\frac{\gamma}{3}+\log\left(2^{\frac{59}{6}}\cdot 3^{-9}\cdot A^{-8}\right); \tag{620}$$

$$\sum_{k=2}^{\infty}\frac{\zeta(k)-1}{k(k+1)(k+2)}\left(\frac{3}{2}\right)^{k}=-\frac{1}{12}-\frac{\gamma}{4}+\log\left(2^{\frac{1}{4}}\cdot\pi^{\frac{1}{4}}\cdot A^{-\frac{2}{3}}\cdot B^{\frac{7}{18}}\right); \tag{621}$$

$$\begin{aligned}&\sum_{k=2}^{\infty}(-1)^{k}\frac{\zeta(k)-1}{k(k+1)(k+2)}\left(\frac{3}{2}\right)^{k}\\&\quad=-\frac{13}{6}+\frac{\gamma}{4}+\log\left(2^{-\frac{61}{36}}\cdot 3^{\frac{1}{2}}\cdot 5^{\frac{25}{18}}\cdot\pi^{\frac{1}{4}}\cdot A^{\frac{2}{3}}\cdot B^{\frac{7}{18}}\right);\end{aligned} \tag{622}$$

$$\sum_{k=1}^{\infty}\frac{\zeta(2k)-1}{k(k+1)(2k+1)}\left(\frac{3}{2}\right)^{2k}=-\frac{9}{2}+\log\left(2^{-\frac{26}{9}}\cdot 3\cdot 5^{\frac{25}{9}}\cdot\pi\cdot B^{\frac{14}{9}}\right); \tag{623}$$

$$\sum_{k=1}^{\infty}\frac{\zeta(2k+1)-1}{(k+1)(2k+1)(2k+3)}\left(\frac{3}{2}\right)^{2k}=\frac{25}{18}-\frac{\gamma}{3}+\log\left(2^{\frac{35}{27}}\cdot 3^{-\frac{1}{3}}\cdot 5^{-\frac{25}{27}}\cdot A^{-\frac{8}{9}}\right). \tag{624}$$

Taking $a=\frac{1}{2}$ in (595) through (598) and using 1.4(6), 2.3(2) and 1.3(51), we get

$$\begin{aligned}&\sum_{k=2}^{\infty}\frac{(2^{k}-1)\zeta(k)}{k(k+1)(k+2)}z^{k+2}=\log\left(2^{-\frac{1}{96}}\cdot e^{-\frac{1}{32}}\cdot A^{\frac{3}{8}}\right)\\&\quad+\left(\frac{1}{12}\log 2+\frac{1}{8}\log\pi+\frac{3}{2}\log A-\frac{1}{8}\right)z+[1+3\log 2+\log\pi]\frac{z^{2}}{8}\\&\quad+[2-\gamma-2\log 2]\frac{z^{3}}{6}+\frac{1}{2}\left(z+\frac{1}{2}\right)^{2}\log\Gamma\left(\frac{1}{2}-z\right)\\&\quad+\frac{1}{2}\left(\frac{1}{2}+z\right)\log G\left(\frac{1}{2}-z\right)-\frac{1}{2}\int_{0}^{-z}\log G\left(t+\frac{1}{2}\right)dt\qquad\left(|z|<\frac{1}{2}\right);\end{aligned} \tag{625}$$

$$\begin{aligned}&\sum_{k=2}^{\infty}(-1)^{k}\frac{(2^{k}-1)\zeta(k)}{k(k+1)(k+2)}z^{k+2}=\log\left(2^{-\frac{1}{96}}\cdot e^{-\frac{1}{32}}\cdot A^{\frac{3}{8}}\right)\\&\quad+\left(\frac{1}{8}-\frac{1}{12}\log 2-\frac{1}{8}\log\pi-\frac{3}{2}\log A\right)z+[1+3\log 2+\log\pi]\frac{z^{2}}{r8}\\&\quad+[\gamma-2+2\log 2]\frac{z^{3}}{6}+\frac{1}{2}\left(z-\frac{1}{2}\right)^{2}\log\Gamma\left(z+\frac{1}{2}\right)\\&\quad+\frac{1}{2}\left(\frac{1}{2}-z\right)\log G\left(\frac{1}{2}+z\right)-\frac{1}{2}\int_{0}^{z}\log G\left(t+\frac{1}{2}\right)dt\qquad\left(|z|<\frac{1}{2}\right);\end{aligned} \tag{626}$$

$$\sum_{k=1}^{\infty} \frac{(2^{2k}-1)\zeta(2k)}{k(k+1)(2k+1)} z^{2k+2}$$
$$= \log\left(2^{-\frac{1}{24}} \cdot e^{-\frac{1}{8}} \cdot A^{\frac{3}{2}}\right) + [1 + 3\log 2 + \log \pi]\frac{z^2}{2}$$
$$+ \left(\frac{1}{2} - z\right)^2 \log\Gamma\left(\frac{1}{2} + z\right) + \left(\frac{1}{2} + z\right)^2 \log\Gamma\left(\frac{1}{2} - z\right) \tag{627}$$
$$+ \left(\frac{1}{2} - z\right) \log G\left(\frac{1}{2} + z\right) + \left(\frac{1}{2} + z\right) \log G\left(\frac{1}{2} - z\right)$$
$$- \int_0^z \log G\left(t + \frac{1}{2}\right) dt - \int_0^{-z} \log G\left(t + \frac{1}{2}\right) dt \quad \left(|z| < \frac{1}{2}\right);$$

$$\sum_{k=1}^{\infty} \frac{(2^{2k+1}-1)\zeta(2k+1)}{(k+1)(2k+1)(2k+3)} z^{2k+3}$$
$$= \left(\frac{1}{6}\log 2 + \frac{1}{4}\log \pi + 3\log A - \frac{1}{4}\right) z + [2 - \gamma - 2\log 2]\frac{z^3}{3}$$
$$+ \frac{1}{2}\left(\frac{1}{2} + z\right)^2 \log\Gamma\left(\frac{1}{2} - z\right) - \frac{1}{2}\left(\frac{1}{2} - z\right)^2 \log\Gamma\left(\frac{1}{2} + z\right) \tag{628}$$
$$+ \frac{1}{2}\left(\frac{1}{2} + z\right) \log G\left(\frac{1}{2} - z\right) + \frac{1}{2}\left(z - \frac{1}{2}\right) \log G\left(z + \frac{1}{2}\right)$$
$$+ \frac{1}{2}\int_0^z \log G\left(t + \frac{1}{2}\right) dt - \frac{1}{2}\int_0^{-z} \log G\left(t + \frac{1}{2}\right) dt \quad \left(|z| < \frac{1}{2}\right).$$

By using 1.4(6), 1.4(55), 1.4(56), 1.4(73) and 1.4(77), we can show that

$$\int_0^{\frac{1}{2}} \log G\left(t + \frac{1}{2}\right) dt = -\frac{1}{2} + \frac{1}{2}\log 2$$
$$+ \int_0^1 \log G(t+1)\, dt - \int_0^1 \log\Gamma(t+1)\, dt \tag{629}$$
$$- \int_0^{\frac{1}{2}} \log G(t+1)\, dt + \int_0^{\frac{1}{2}} \log\Gamma(t+1)\, dt$$
$$= \frac{1}{24} - \frac{13}{12}\log 2 - \frac{1}{16}\log \pi - \frac{1}{4}\log A + \frac{7}{4}\log B.$$

Taking the limit in (626) as $z \to \frac{1}{2}$ and applying 1.4(6) and (629), we obtain

$$\sum_{k=2}^{\infty}(-1)^k \frac{(2^k-1)\zeta(k)}{k(k+1)(k+2)\cdot 2^k} = \frac{\gamma}{12} + \log\left(2^{\frac{5}{2}}\cdot A^{-1}\cdot B^{-\frac{7}{2}}\right). \tag{630}$$

By setting $z = t$ in (485) through (488), integrating both sides of the resulting equations from $t = 0$ to $t = z$ and using (470), we obtain

$$\begin{aligned}\sum_{k=2}^{\infty} \frac{\zeta(k,a)}{(k+1)(k+2)} z^{k+2} &= (1-a)^2 \log\Gamma(a) + (1-a)\log G(a) \\ &+ \left(\frac{1}{2}(a-1)[1+\log(2\pi)] + a(1-a) + (1-a)\log\Gamma(a) + \log G(a)\right) z \\ &+ [4a-3-\log(2\pi)]\frac{z^2}{4} + [\psi(a)-1]\frac{z^3}{6} \\ &+ (a-1)(z-a+1)\log\Gamma(a-z) + (a-1)\log G(a-z) \\ &+ \int_0^{-z} \log G(t+a)\,dt \quad (|z|<|a|);\end{aligned} \tag{631}$$

$$\begin{aligned}\sum_{k=2}^{\infty}(-1)^k \frac{\zeta(k,a)}{(k+1)(k+2)} z^{k+2} &= (1-a)^2 \log\Gamma(a) + (1-a)\log G(a) \\ &+ \left(\frac{1}{2}(1-a)[1+\log(2\pi)] + a(a-1) + (a-1)\log\Gamma(a) - \log G(a)\right) z \\ &+ [4a-3-\log(2\pi)]\frac{z^2}{4} + [1-\psi(a)]\frac{z^3}{6} \\ &+ (1-a)(z+a-1)\log\Gamma(z+a) + (a-1)\log G(z+a) \\ &+ \int_0^{z} \log G(t+a)\,dt \quad (|z|<|a|);\end{aligned} \tag{632}$$

$$\begin{aligned}\sum_{k=1}^{\infty} \frac{\zeta(2k,a)}{(k+1)(2k+1)} z^{2k+2} &= 2(1-a)^2 \log\Gamma(a) + 2(1-a)\log G(a) \\ &+ [4a-3-\log(2\pi)]\frac{z^2}{2} + (1-a)(z+a-1)\log\Gamma(a+z) \\ &+ (a-1)(z-a+1)\log\Gamma(a-z) + (a-1)\log G(a+z)G(a-z) \\ &+ \int_0^{z} \log G(t+a)\,dt + \int_0^{-z} \log G(t+a)\,dt \quad (|z|<|a|);\end{aligned} \tag{633}$$

$$\sum_{k=1}^{\infty} \frac{\zeta(2k+1,a)}{(k+1)(2k+3)} z^{2k+3}$$
$$= \{(a-1)[1+\log(2\pi)] + 2a(1-a) + 2(1-a)\log\Gamma(a) + 2\log G(a)\} z$$
$$+ [\psi(a)-1]\frac{z^3}{3} + (a-1)(z+a-1)\log\Gamma(a+z) \tag{634}$$
$$+ (a-1)(z-a+1)\log\Gamma(a-z) + (a-1)\log\frac{G(a-z)}{G(a+z)}$$
$$- \int_0^z \log G(t+a)\,dt + \int_0^{-z} \log G(t+a)\,dt \quad (|z| < |a|).$$

Setting $a = 1$ in (631) through (634) and using 1.4(6), 2.3(2) and 1.3(4), we obtain

$$\sum_{k=2}^{\infty} \frac{\zeta(k)}{(k+1)(k+2)} z^{k+2}$$
$$= [1-\log(2\pi)]\frac{z^2}{4} - (1+\gamma)\frac{z^3}{6} + \int_0^{-z} \log G(t+1)\,dt \quad (|z| < 1); \tag{635}$$

$$\sum_{k=2}^{\infty} (-1)^k \frac{\zeta(k)}{(k+1)(k+2)} z^{k+2}$$
$$= [1-\log(2\pi)]\frac{z^2}{4} + (1+\gamma)\frac{z^3}{6} + \int_0^{z} \log G(t+1)\,dt \quad (|z| < 1); \tag{636}$$

$$\sum_{k=1}^{\infty} \frac{\zeta(2k)}{(k+1)(2k+1)} z^{2k+2}$$
$$= [1-\log(2\pi)]\frac{z^2}{2} + \int_0^{z} \log G(t+1)\,dt + \int_0^{-z} \log G(t+1)\,dt \quad (|z| < 1); \tag{637}$$

$$\sum_{k=1}^{\infty} \frac{\zeta(2k+1)}{(k+1)(2k+3)} z^{2k+3}$$
$$= -(1+\gamma)\frac{z^3}{3} - \int_0^{z} \log G(t+1)\,dt + \int_0^{-z} \log G(t+1)\,dt \quad (|z| < 1); \tag{638}$$

Taking the limit in (636) as $z \to 1$ and using 1.4(6) and 1.4(73), we obtain

$$\sum_{k=2}^{\infty}(-1)^k \frac{\zeta(k)}{(k+1)(k+2)} = \frac{1}{2} + \frac{\gamma}{6} - 2\log A. \tag{639}$$

Setting $z = \frac{1}{2}$ in (635) through (638), and using 1.4(6), 1.4(8), 1.4(77), and (524), we obtain

$$\sum_{k=2}^{\infty} \frac{\zeta(k)}{(k+1)(k+2)\cdot 2^k} = -\frac{\gamma}{12} + \log\left(2^{\frac{1}{12}} \cdot A \cdot B^{-7}\right); \tag{640}$$

$$\sum_{k=2}^{\infty}(-1)^k \frac{\zeta(k)}{(k+1)(k+2)\cdot 2^k} = \frac{1}{2} + \frac{\gamma}{12} + \log\left(2^{-\frac{1}{12}} \cdot A^{-1} \cdot B^{-7}\right); \tag{641}$$

$$\sum_{k=1}^{\infty} \frac{\zeta(2k)}{(k+1)(2k+1)\cdot 2^{2k}} = \frac{1}{2} - 14\log B; \tag{642}$$

$$\sum_{k=1}^{\infty} \frac{\zeta(2k+1)}{(k+1)(2k+3)\cdot 2^{2k}} = -1 - \frac{\gamma}{3} + \log\left(2^{\frac{1}{3}} \cdot A^4\right). \tag{643}$$

If we take $a = 2$ in (631) through (634), in view of 1.4(6), 2.3(2) and 1.3(50), we get

$$\begin{aligned}\sum_{k=2}^{\infty} \frac{\zeta(k)-1}{(k+1)(k+2)} z^{k+2} &= [\log(2\pi) - 3]\frac{z}{2} + [5 - \log(2\pi)]\frac{z^2}{4} - \frac{\gamma}{6}z^3 + (z-1)\log\Gamma(2-z) \\ &\quad + \log G(2-z) + \int_0^{-z} \log G(t+2)\,dt \quad (|z| < 2);\end{aligned} \tag{644}$$

$$\begin{aligned}\sum_{k=2}^{\infty}(-1)^k \frac{\zeta(k)-1}{(k+1)(k+2)} z^{k+2} &= [3 - \log(2\pi)]\frac{z}{2} + [5 - \log(2\pi)]\frac{z^2}{4} + \frac{\gamma}{6}z^3 - (z+1)\log\Gamma(z+2) \\ &\quad + \log G(z+2) + \int_0^{z} \log G(t+2)\,dt \quad (|z| < 2);\end{aligned} \tag{645}$$

$$\sum_{k=1}^{\infty} \zeta(2k) - 1(k+1)(2k+1)z^{2k+2}$$
$$= [5 - \log(2\pi)]\frac{z^2}{2} - (z+1)\log\Gamma(2+z)$$
$$+ (z-1)\log\Gamma(2-z) + \log G(2+z)G(2-z) \tag{646}$$
$$+ \int_0^z \log G(t+2)\,dt + \int_0^{-z} \log G(t+2)\,dt \quad (|z| < 2);$$

$$\sum_{k=1}^{\infty} \frac{\zeta(2k+1)-1}{(k+1)(2k+3)} z^{2k+3}$$
$$= [\log(2\pi) - 3]z - \frac{\gamma}{3}z^3 + (z-1)\log\Gamma(2-z) + (z+1)\log\Gamma(2+z) \tag{647}$$
$$+ \log\frac{G(2-z)}{G(2+z)} - \int_0^z \log G(t+2)\,dt + \int_0^{-z} \log G(t+2)\,dt \quad (|z| < 2).$$

Taking the limit in (645) as $z \to 2$ and using 1.4(6), 1.4(73) and (535), we obtain

$$\sum_{k=2}^{\infty} (-1)^k \frac{(\zeta(k)-1)2^k}{(k+1)(k+2)} = \frac{25}{24} + \frac{\gamma}{3} + \log\left(3^{-\frac{3}{4}} \cdot A^{-1}\right). \tag{648}$$

Setting $z = \frac{1}{2}$, $z = 1$ and $z = \frac{3}{2}$ in (644) through (647) and using 1.4(6), 1.4(8), 1.4(73), 1.4(77), (524), (535) and (541), we get

$$\sum_{k=2}^{\infty} \frac{\zeta(k)-1}{(k+1)(k+2)\cdot 2^k} = -\frac{17}{12} - \frac{\gamma}{12} + \log\left(2^{\frac{25}{12}} \cdot A \cdot B^{-7}\right); \tag{649}$$

$$\sum_{k=2}^{\infty} (-1)^k \frac{\zeta(k)-1}{(k+1)(k+2)\cdot 2^k} = \frac{35}{12} + \frac{\gamma}{12} + \log\left(2^{\frac{71}{12}} \cdot 3^{-6} \cdot A^{-1} \cdot B^{-7}\right); \tag{650}$$

$$\sum_{k=1}^{\infty} \frac{\zeta(2k)-1}{(k+1)(2k+1)\cdot 2^{2k}} = \frac{3}{2} + \log\left(2^8 \cdot 3^{-6} \cdot B^{-14}\right); \tag{651}$$

$$\sum_{k=1}^{\infty} \frac{\zeta(2k+1)-1}{(k+1)(2k+3)\cdot 2^{2k}} = -\frac{26}{3} - \frac{\gamma}{3} + \log\left(2^{-\frac{28}{3}} \cdot 3^{12} \cdot A^4\right); \tag{652}$$

$$\sum_{k=2}^{\infty} \frac{\zeta(k)-1}{(k+1)(k+2)} = -\frac{1}{3} - \frac{\gamma}{6} + 2\log A; \tag{653}$$

$$\sum_{k=2}^{\infty} (-1)^k \frac{\zeta(k)-1}{(k+1)(k+2)} = \frac{11}{6} + \frac{\gamma}{6} - 2\log(2A); \tag{654}$$

$$\sum_{k=1}^{\infty} \frac{\zeta(2k)-1}{(k+1)(2k+1)} = \frac{3}{2} - 2\log 2; \tag{655}$$

$$\sum_{k=1}^{\infty} \frac{\zeta(2k+1)-1}{(k+1)(2k+3)} = -\frac{13}{6} - \frac{\gamma}{3} + \log\left(2^2 \cdot A^4\right); \tag{656}$$

$$\sum_{k=2}^{\infty} \frac{\zeta(k)-1}{(k+1)(k+2)} \left(\frac{3}{2}\right)^k = \frac{1}{4} - \frac{\gamma}{4} + \log\left(2^{\frac{1}{36}} \cdot A^{\frac{1}{3}} \cdot B^{-\frac{7}{9}}\right); \tag{657}$$

$$\sum_{k=2}^{\infty} (-1)^k \frac{\zeta(k)-1}{(k+1)(k+2)} \left(\frac{3}{2}\right)^k = \frac{5}{4} + \frac{\gamma}{4} + \log\left(2^{\frac{31}{36}} \cdot 5^{-\frac{10}{9}} \cdot A^{-\frac{1}{3}} \cdot B^{-\frac{7}{9}}\right); \tag{658}$$

$$\sum_{k=1}^{\infty} \frac{\zeta(2k)-1}{(k+1)(2k+1)} \left(\frac{3}{2}\right)^{2k} = \frac{3}{2} + \log\left(2^{\frac{8}{9}} \cdot 5^{-\frac{10}{9}} \cdot B^{-\frac{14}{9}}\right); \tag{659}$$

$$\sum_{k=1}^{\infty} \frac{\zeta(2k+1)-1}{(k+1)(2k+3)} \left(\frac{3}{2}\right)^{2k} = -\frac{2}{3} - \frac{\gamma}{3} + \log\left(2^{-\frac{5}{9}} \cdot 5^{\frac{20}{27}} \cdot A^{\frac{4}{9}}\right). \tag{660}$$

Taking $a = \frac{1}{2}$ in (631) through (634) and using 1.4(8), 2.3(2) and 1.3(51), we obtain

$$\begin{aligned}
&\sum_{k=2}^{\infty} \frac{(2^k-1)\zeta(k)}{(k+1)(k+2)} z^{k+2} \\
&\quad = \log\left(2^{\frac{1}{48}} \cdot e^{\frac{1}{16}} \cdot A^{-\frac{3}{4}}\right) - \left(\frac{5}{12}\log 2 + \frac{1}{2}\log\pi - \frac{1}{4} + 3\log A\right)\frac{z}{2} \\
&\qquad - [1+\log(2\pi)]\frac{z^2}{4} - (1+\gamma+2\log 2)\frac{z^3}{6} - \frac{1}{2}\left(\frac{1}{2}+z\right)\log\Gamma\left(\frac{1}{2}-z\right) \\
&\qquad - \frac{1}{2}\log G\left(\frac{1}{2}-z\right) + \int_0^{-z} \log G\left(t+\frac{1}{2}\right) dt \quad \left(|z| < \frac{1}{2}\right);
\end{aligned} \tag{661}$$

$$\begin{aligned}
&\sum_{k=2}^{\infty} (-1)^k \frac{(2^k-1)\zeta(k)}{(k+1)(k+2)} z^{k+2} \\
&\quad = \log\left(2^{\frac{1}{48}} \cdot e^{\frac{1}{16}} \cdot A^{-\frac{3}{4}}\right) + \left(\frac{5}{12}\log 2 + \frac{1}{2}\log\pi - \frac{1}{4} + 3\log A\right)\frac{z}{2} \\
&\qquad - [1+\log(2\pi)]\frac{z^2}{4} + (1+\gamma+2\log 2)\frac{z^3}{6} + \frac{1}{2}\left(z-\frac{1}{2}\right)\log\Gamma\left(z+\frac{1}{2}\right) \\
&\qquad - \frac{1}{2}\log G\left(z+\frac{1}{2}\right) + \int_0^{z} \log G\left(t+\frac{1}{2}\right) dt \quad \left(|z| < \frac{1}{2}\right);
\end{aligned} \tag{662}$$

$$\sum_{k=1}^{\infty} \frac{(2^{2k}-1)\zeta(2k)}{(k+1)(2k+1)} z^{2k+2}$$
$$= \log\left(2^{\frac{1}{24}} \cdot e^{\frac{1}{8}} \cdot A^{-\frac{3}{2}}\right) - [1+\log(2\pi)]\frac{z^2}{2} + \frac{1}{2}\left(z-\frac{1}{2}\right)\log\Gamma\left(z+\frac{1}{2}\right)$$
$$-\frac{1}{2}\left(\frac{1}{2}+z\right)\log\Gamma\left(\frac{1}{2}-z\right) - \frac{1}{2}\log G\left(\frac{1}{2}+z\right)G\left(\frac{1}{2}-z\right) \tag{663}$$
$$+\int_0^z \log G(t+12)\,dt + \int_0^{-z} \log G\left(t+\frac{1}{2}\right)dt \quad \left(|z|<\frac{1}{2}\right);$$

$$\sum_{k=1}^{\infty} \frac{(2^{2k+1}-1)\zeta(2k+1)}{(k+1)(2k+3)} z^{2k+3}$$
$$= -\left(\frac{5}{12}\log 2 + \frac{1}{2}\log\pi - \frac{1}{4} + 3\log A\right)z - (1+\gamma+2\log 2)\frac{z^3}{3}$$
$$+\frac{1}{2}\left(\frac{1}{2}-z\right)\log\Gamma\left(\frac{1}{2}+z\right) - \frac{1}{2}\left(\frac{1}{2}+z\right)\log\Gamma\left(\frac{1}{2}-z\right) + \frac{1}{2}\log\frac{G\left(\frac{1}{2}+z\right)}{G\left(\frac{1}{2}-z\right)}$$
$$-\int_0^z \log G\left(t+\frac{1}{2}\right)dt + \int_0^{-z} \log G\left(t+\frac{1}{2}\right)dt \quad \left(|z|<\frac{1}{2}\right). \tag{664}$$

Taking the limit in (662) as $z \to \frac{1}{2}$ and using 1.4(6) and (629), we obtain

$$\sum_{k=2}^{\infty} (-1)^k \frac{(2^k-1)\zeta(k)}{(k+1)(k+2)\cdot 2^k} = \frac{\gamma}{12} + \log\left(2^{-\frac{47}{12}} \cdot A^{-1} \cdot B^7\right). \tag{665}$$

Evaluation in Terms of Catalan's Constant G

We now evaluate series associated with the Zeta function in terms of Catalan's constant G and $\Gamma\left(\frac{1}{4}\right)$, by using some of the formulas we have already developed in this section.

Setting $z = \frac{1}{4}$ and $z = \frac{3}{4}$ in (482) and using 1.3(6), (1.12), 1.3(23) and 1.3(24), we obtain

$$\sum_{k=2}^{\infty} (-1)^k \frac{\zeta(k)}{(k+1)\cdot 2^{2k}} = 1 + \gamma 8 - \frac{G}{\pi} + \log\left[2^{-\frac{1}{2}} \cdot \pi^{-\frac{1}{2}} \cdot A^{-\frac{9}{2}} \cdot \Gamma\left(\frac{1}{4}\right)\right]; \tag{666}$$

$$\sum_{k=2}^{\infty} (-1)^k \frac{\zeta(k)}{k+1}\left(\frac{3}{4}\right)^k = 1 + \frac{3\gamma}{8} + \frac{G}{3\pi} + \log\left[\pi^{\frac{1}{2}} \cdot A^{-\frac{3}{2}} \cdot \left\{\Gamma\left(\frac{1}{4}\right)\right\}^{-1}\right]. \tag{667}$$

Setting $z=\frac{1}{4}$ and $z=\frac{3}{4}$ in (489), (490) and (491) and using 1.4(6), 1.1(12), 1.4(23) and 1.4(24), we obtain

$$\sum_{k=2}^{\infty}\frac{\zeta(k)}{(k+1)\cdot 2^{2k}}=-\frac{\gamma}{8}-\frac{G}{\pi}+\log\left[\pi^{\frac{1}{2}}\cdot A^{\frac{9}{2}}\cdot\left\{\Gamma\left(\frac{1}{4}\right)\right\}^{-1}\right]; \tag{668}$$

$$\sum_{k=2}^{\infty}\frac{\zeta(k)}{rk+1}\left(\frac{3}{4}\right)^{k}=-\frac{3\gamma}{8}+\frac{G}{3\pi}+\log\left[2^{-\frac{1}{2}}\cdot\pi^{-\frac{1}{2}}\cdot A^{\frac{3}{2}}\cdot\Gamma\left(\frac{1}{4}\right)\right]; \tag{669}$$

$$\sum_{k=1}^{\infty}\frac{\zeta(2k)}{(2k+1)\cdot 2^{4k}}=\frac{1}{2}-\frac{G}{\pi}-\frac{1}{4}\log 2; \tag{670}$$

$$\sum_{k=1}^{\infty}\frac{\zeta(2k)}{2k+1}\left(\frac{3}{4}\right)^{2k}=\frac{1}{2}+\frac{G}{3\pi}-\frac{1}{4}\log 2; \tag{671}$$

$$\sum_{k=1}^{\infty}\frac{\zeta(2k+1)}{(k+1)\cdot 2^{4k}}=-4-\gamma+\log\left[2^{2}\cdot\pi^{4}\cdot A^{36}\cdot\left\{\Gamma\left(\frac{1}{4}\right)\right\}^{-8}\right]; \tag{672}$$

$$\sum_{k=1}^{\infty}\frac{\zeta(2k+1)}{k+1}\left(\frac{3}{4}\right)^{2k}=-43-\gamma+\log\left[2^{-\frac{2}{3}}\cdot\pi^{-\frac{4}{3}}\cdot A^{4}\cdot\left\{\Gamma\left(\frac{1}{4}\right)\right\}^{\frac{8}{3}}\right]. \tag{673}$$

If we set $z=\frac{1}{4}$, $z=\frac{3}{4}$, $z=\frac{5}{4}$ and $z=\frac{7}{4}$ in (495) to (498), in view of 1.4(6), 1.1(12), 1.4(23) and 1.4(24), we get

$$\sum_{k=2}^{\infty}(-1)^{k}\frac{\zeta(k)-1}{(k+1)\cdot 2^{2k}}=\frac{15}{8}+\frac{\gamma}{8}-\frac{G}{\pi}+\log\left[2^{\frac{15}{2}}\cdot 5^{-4}\cdot\pi^{-\frac{1}{2}}\cdot A^{-\frac{9}{2}}\cdot\Gamma\left(\frac{1}{4}\right)\right]; \tag{674}$$

$$\begin{aligned}&\sum_{k=2}^{\infty}(-1)^{k}\frac{\zeta(k)-1}{k+1}\left(\frac{3}{4}\right)^{k}\\&=\frac{13}{8}+\frac{3\gamma}{8}+\frac{G}{3\pi}+\log\left[2^{\frac{8}{3}}\cdot 7^{-\frac{4}{3}}\cdot\pi^{\frac{1}{2}}\cdot A^{-\frac{3}{2}}\cdot\left\{\Gamma\left(\frac{1}{4}\right)\right\}^{-1}\right];\end{aligned} \tag{675}$$

$$\sum_{k=2}^{\infty}\frac{\zeta(k)-1}{(k+1)\cdot 2^{2k}}=\frac{9}{8}-\frac{\gamma}{8}-\frac{G}{\pi}+\log\left[2^{-8}\cdot 3^{4}\cdot\pi^{\frac{1}{2}}\cdot A^{\frac{9}{2}}\cdot\left\{\Gamma\left(\frac{1}{4}\right)\right\}^{-1}\right]; \tag{676}$$

$$\sum_{k=2}^{\infty}\frac{\zeta(k)-1}{k+1}\left(\frac{3}{4}\right)^{k}=118-\frac{3\gamma}{8}+\frac{G}{3\pi}+\log\left[2^{-\frac{19}{6}}\cdot\pi^{-\frac{1}{2}}\cdot A^{\frac{3}{2}}\cdot\Gamma\left(\frac{1}{4}\right)\right]; \tag{677}$$

$$\sum_{k=1}^{\infty}\frac{\zeta(2k)-1}{(2k+1)\cdot 2^{4k}}=\frac{3}{2}-\frac{G}{\pi}+\log\left[2^{-\frac{1}{4}}\cdot 3^{2}\cdot 5^{-2}\right]; \tag{678}$$

$$\sum_{k=1}^{\infty}\frac{\zeta(2k)-1}{2k+1}\left(\frac{3}{4}\right)^{2k}=\frac{3}{2}+\frac{G}{3\pi}+\log\left[2^{-\frac{1}{4}}\cdot 7^{-\frac{2}{3}}\right]; \tag{679}$$

$$\sum_{k=1}^{\infty}\frac{\zeta(2k+1)-1}{(k+1)\cdot 2^{4k}}=-3-\gamma+\log\left[2^{-62}\cdot 3^{16}\cdot 5^{16}\cdot\pi^{4}\cdot A^{36}\cdot\left\{\Gamma\left(\frac{1}{4}\right)\right\}^{-8}\right]; \tag{680}$$

$$\sum_{k=1}^{\infty}\frac{\zeta(2k+1)-1}{k+1}\left(\frac{3}{4}\right)^{2k}=-\frac{1}{3}-\gamma+\log\left[2^{-\frac{70}{9}}\cdot 7^{\frac{16}{9}}\cdot\pi^{-\frac{4}{3}}\cdot A^{4}\cdot\left\{\Gamma\left(\frac{1}{4}\right)\right\}^{\frac{8}{3}}\right]; \tag{681}$$

$$\begin{aligned}&\sum_{k=2}^{\infty}(-1)^{k}\frac{\zeta(k)-1}{k+1}\left(\frac{5}{4}\right)^{k}\\&=\frac{63}{40}+\frac{5\gamma}{8}-\frac{G}{5\pi}+\log\left[2^{-\frac{1}{2}}\cdot\pi^{-\frac{1}{2}}\cdot 3^{-\frac{8}{5}}\cdot A^{-\frac{9}{10}}\cdot\Gamma\left(\frac{1}{4}\right)\right];\end{aligned} \tag{682}$$

$$\begin{aligned}&\sum_{k=2}^{\infty}(-1)^{k}\frac{\zeta(k)-1}{k+1}\left(\frac{7}{4}\right)^{k}\\&=\frac{87}{56}+\frac{7\gamma}{8}+\frac{G}{7\pi}+\log\left[3^{\frac{4}{7}}\cdot 11^{-\frac{4}{7}}\cdot\pi^{\frac{1}{2}}\cdot A^{-\frac{9}{14}}\cdot\left\{\Gamma\left(\frac{1}{4}\right)\right\}^{-1}\right];\end{aligned} \tag{683}$$

$$\sum_{k=2}^{\infty}\frac{\zeta(k)-1}{k+1}\left(\frac{5}{4}\right)^{k}=\frac{57}{40}-\frac{5\gamma}{8}-\frac{G}{5\pi}+\log\left[\pi^{\frac{1}{2}}\cdot A^{\frac{9}{10}}\cdot\left\{\Gamma\left(\frac{1}{4}\right)\right\}^{-1}\right]; \tag{684}$$

$$\sum_{k=2}^{\infty}\frac{\zeta(k)-1}{k+1}\left(\frac{7}{4}\right)^{k}=\frac{81}{56}-\frac{7\gamma}{8}+\frac{G}{7\pi}+\log\left[2^{-\frac{1}{2}}\cdot\pi^{-\frac{1}{2}}\cdot A^{\frac{9}{14}}\cdot\Gamma\left(\frac{1}{4}\right)\right]; \tag{685}$$

$$\sum_{k=1}^{\infty}\frac{\zeta(2k)-1}{2k+1}\left(\frac{5}{4}\right)^{2k}=\frac{3}{2}-\frac{G}{5\pi}+\log\left[2^{-\frac{1}{4}}\cdot 3^{-\frac{4}{5}}\right]; \tag{686}$$

$$\sum_{k=1}^{\infty}\frac{\zeta(2k)-1}{2k+1}\left(\frac{7}{4}\right)^{2k}=\frac{3}{2}+\frac{G}{7\pi}+\log\left[2^{-\frac{1}{4}}\cdot 3^{\frac{2}{7}}\cdot 11^{-\frac{2}{7}}\right]; \tag{687}$$

$$\begin{aligned}&\sum_{k=1}^{\infty}\frac{\zeta(2k+1)-1}{k+1}\left(\frac{5}{4}\right)^{2k}\\&=-\frac{3}{25}-\gamma+\log\left[2^{\frac{2}{5}}\cdot 3^{\frac{32}{25}}\cdot\pi^{\frac{4}{5}}\cdot A^{\frac{36}{25}}\cdot\left\{\Gamma\left(\frac{1}{4}\right)\right\}^{-\frac{8}{5}}\right];\end{aligned} \tag{688}$$

$$\begin{aligned}&\sum_{k=1}^{\infty}\frac{\zeta(2k+1)-1}{k+1}\left(\frac{7}{4}\right)^{2k}\\&=-\frac{3}{49}-\gamma+\log\left[2^{-\frac{2}{7}}\cdot 3^{-\frac{16}{49}}\cdot 11^{\frac{16}{49}}\cdot\pi^{-\frac{4}{7}}\cdot A^{\frac{36}{49}}\cdot\left\{\Gamma\left(\frac{1}{4}\right)\right\}^{\frac{8}{7}}\right].\end{aligned} \tag{689}$$

Setting $z = \frac{1}{4}$ in (512) through (515) and using 1.4(6), 1.1(12), 1.4(23) and 1.4(24), we obtain

$$\sum_{k=2}^{\infty} \frac{(2^k-1)\zeta(k)}{(k+1)\cdot 2^{2k}} = -\frac{\gamma}{8} + \frac{G}{\pi} + \log\left[2^{-\frac{7}{12}} \cdot \pi^{-\frac{1}{2}} \cdot A^{-\frac{3}{2}} \cdot \Gamma\left(\frac{1}{4}\right)\right]; \tag{690}$$

$$\sum_{k=2}^{\infty} (-1)^k \frac{(2^k-1)\zeta(k)}{(k+1)\cdot 2^{2k}} = \frac{\gamma}{8} + \frac{G}{\pi} + \log\left[2^{\frac{1}{12}} \cdot \pi^{\frac{1}{2}} \cdot A^{\frac{3}{2}} \cdot \left\{\Gamma\left(\frac{1}{4}\right)\right\}^{-1}\right]; \tag{691}$$

$$\sum_{k=1}^{\infty} \frac{(2^{2k}-1)\zeta(2k)}{(2k+1)\cdot 2^{4k}} = \frac{G}{\pi} - \frac{1}{4}\log 2; \tag{692}$$

$$\sum_{k=1}^{\infty} \frac{(2^{2k+1}-1)\zeta(2k+1)}{(k+1)\cdot 2^{4k}} = -\gamma + \log\left[2^{-\frac{8}{3}} \cdot \pi^{-4} \cdot A^{-12} \cdot \left\{\Gamma\left(\frac{1}{4}\right)\right\}^{8}\right]. \tag{693}$$

If we set $z = \frac{1}{4}$ and $z = \frac{3}{4}$ in (564) through (567), by virtue of 1.4(6), 1.1(12), 1.4(23) and 1.4(24), we obtain

$$\sum_{k=2}^{\infty} (-1)^k \frac{\zeta(k)}{k(k+1)\cdot 2^{2k}} = -1 + \frac{\gamma}{8} + \frac{G}{\pi} + \log\left[2^{-\frac{3}{2}} \cdot \pi^{\frac{1}{2}} \cdot A^{\frac{9}{2}}\right]; \tag{694}$$

$$\sum_{k=2}^{\infty} (-1)^k \frac{\zeta(k)}{k(k+1)} \left(\frac{3}{4}\right)^k = -1 + \frac{3\gamma}{8} - \frac{G}{3\pi} + \log\left[2^{-\frac{3}{2}} \cdot 3 \cdot \pi^{\frac{1}{2}} \cdot A^{\frac{3}{2}}\right]; \tag{695}$$

$$\sum_{k=2}^{\infty} \frac{\zeta(k)}{k(k+1)\cdot 2^{2k}} = -\frac{\gamma}{8} + \frac{G}{\pi} + \log\left[2^{\frac{1}{2}} \cdot \pi^{\frac{1}{2}} \cdot A^{-\frac{9}{2}}\right]; \tag{696}$$

$$\sum_{k=2}^{\infty} \frac{\zeta(k)}{k(k+1)} \left(\frac{3}{4}\right)^k = -\frac{3\gamma}{8} - \frac{G}{3\pi} + \log\left[2^{\frac{1}{2}} \cdot \pi^{\frac{1}{2}} \cdot A^{-\frac{3}{2}}\right]; \tag{697}$$

$$\sum_{k=1}^{\infty} \frac{\zeta(2k)}{k(2k+1)\cdot 2^{4k}} = -1 + \frac{2G}{\pi} + \log\left(\frac{\pi}{2}\right); \tag{698}$$

$$\sum_{k=1}^{\infty} \frac{\zeta(2k)}{k(2k+1)} \left(\frac{3}{4}\right)^{2k} = -1 - \frac{2G}{3\pi} + \log\left(\frac{3\pi}{2}\right); \tag{699}$$

$$\sum_{k=1}^{\infty} \frac{\zeta(2k+1)}{(k+1)(2k+1)\cdot 2^{4k}} = 4 - \gamma + \log\left[2^{8} \cdot A^{-36}\right]; \tag{700}$$

$$\sum_{k=1}^{\infty} \frac{\zeta(2k+1)}{(k+1)(2k+1)} \left(\frac{3}{4}\right)^{2k} = \frac{4}{3} - \gamma + \log\left[2^{\frac{8}{3}} \cdot 3^{-\frac{4}{3}} \cdot A^{-4}\right]. \tag{701}$$

Setting $z=\frac{1}{4}$, $z=\frac{3}{4}$, $z=\frac{5}{4}$ and $z=\frac{7}{4}$ in (573) through (576) and using 1.4(6), 1.1(12), 1.4(23) and 1.4(24), we get

$$\sum_{k=2}^{\infty}(-1)^k\frac{\zeta(k)-1}{k(k+1)\cdot 2^{2k}}=-\frac{17}{8}+\frac{\gamma}{8}+\frac{G}{\pi}+\log\left[2^{-\frac{23}{2}}\cdot 5^5\cdot\pi^{\frac{1}{2}}\cdot A^{\frac{9}{2}}\right]; \tag{702}$$

$$\sum_{k=2}^{\infty}(-1)^k\frac{\zeta(k)-1}{k(k+1)}\left(\frac{3}{4}\right)^k=-\frac{19}{8}+\frac{3\gamma}{8}-\frac{G}{3\pi}+\log\left[2^{-\frac{37}{6}}\cdot 3\cdot 7^{\frac{7}{3}}\cdot\pi^{\frac{1}{2}}\cdot A^{\frac{3}{2}}\right]; \tag{703}$$

$$\sum_{k=2}^{\infty}(-1)^k\frac{\zeta(k)-1}{k(k+1)}\left(\frac{5}{4}\right)^k=-\frac{113}{40}+\frac{5\gamma}{8}+\frac{G}{5\pi}+\log\left[2^{-\frac{11}{2}}\cdot 3^{\frac{18}{5}}\cdot 5\cdot\pi^{\frac{1}{2}}\cdot A^{\frac{9}{10}}\right]; \tag{704}$$

$$\sum_{k=2}^{\infty}(-1)^k\frac{\zeta(k)-1}{k(k+1)}\left(\frac{7}{4}\right)^k$$
$$=-\frac{185}{56}+\frac{7\gamma}{8}-\frac{G}{7\pi}+\log\left[2^{-\frac{11}{2}}\cdot 3^{\frac{3}{7}}\cdot 7\cdot 11^{\frac{11}{7}}\cdot\pi^{\frac{1}{2}}\cdot A^{\frac{9}{14}}\right]; \tag{705}$$

$$\sum_{k=2}^{\infty}\frac{\zeta(k)-1}{k(k+1)\cdot 2^{2k}}=-\frac{7}{8}-\frac{\gamma}{8}+\frac{G\pi}{+}\log\left[2^{\frac{13}{2}}\cdot 3^{-3}\cdot\pi^{\frac{1}{2}}\cdot A^{-\frac{9}{2}}\right]; \tag{706}$$

$$\sum_{k=2}^{\infty}\frac{\zeta(k)-1}{k(k+1)}\left(\frac{3}{4}\right)^k=-\frac{5}{8}-\frac{3\gamma}{8}-\frac{G}{3\pi}+\log\left[2^{\frac{7}{6}}\cdot\pi^{\frac{1}{2}}\cdot A^{-\frac{3}{2}}\right]; \tag{707}$$

$$\sum_{k=2}^{\infty}\frac{\zeta(k)-1}{k(k+1)}\left(\frac{5}{4}\right)^k=-\frac{7}{40}-\frac{5\gamma}{8}+\frac{G}{5\pi}+\log\left[2^{\frac{1}{2}}\cdot\pi^{\frac{1}{2}}\cdot A^{-\frac{9}{10}}\right]; \tag{708}$$

$$\sum_{k=2}^{\infty}\frac{\zeta(k)-1}{k(k+1)}\left(\frac{7}{4}\right)^k=\frac{17}{56}-\frac{7\gamma}{8}-\frac{G}{7\pi}+\log\left[2^{\frac{1}{2}}\cdot\pi^{\frac{1}{2}}\cdot A^{-\frac{9}{14}}\right]; \tag{709}$$

$$\sum_{k=1}^{\infty}\frac{\zeta(2k)-1}{k(2k+1)\cdot 2^{4k}}=-3+\frac{2G}{\pi}+\log\left[2^{-5}\cdot 3^{-3}\cdot 5^5\cdot\pi\right]; \tag{710}$$

$$\sum_{k=1}^{\infty}\frac{\zeta(2k)-1}{k(2k+1)}\left(\frac{3}{4}\right)^{2k}=-3-\frac{2G}{3\pi}+\log\left[2^{-5}\cdot 3\cdot 7^{\frac{7}{3}}\cdot\pi\right]; \tag{711}$$

$$\sum_{k=1}^{\infty}\frac{\zeta(2k)-1}{k(2k+1)}\left(\frac{5}{4}\right)^{2k}=-3+\frac{2G}{5\pi}+\log\left[2^{-5}\cdot 3^{\frac{18}{5}}\cdot 5\cdot\pi\right]; \tag{712}$$

$$\sum_{k=1}^{\infty}\frac{\zeta(2k)-1}{k(2k+1)}\left(\frac{7}{4}\right)^{2k}=-3-\frac{2G}{7\pi}+\log\left[2^{-5}\cdot 3^{\frac{3}{7}}\cdot 7\cdot 11^{\frac{11}{7}}\cdot\pi\right]; \tag{713}$$

$$\sum_{k=1}^{\infty}\frac{\zeta(2k+1)-1}{(k+1)(2k+1)\cdot 2^{4k}}=5-\gamma+\log\left[2^{72}\cdot 3^{-12}\cdot 5^{-20}\cdot A^{-36}\right]; \tag{714}$$

$$\sum_{k=1}^{\infty} \frac{\zeta(2k+1)-1}{(k+1)(2k+1)} \left(\frac{3}{4}\right)^{2k} = \frac{7}{3} - \gamma + \log\left[2^{\frac{88}{9}} \cdot 3^{-\frac{4}{3}} \cdot 7^{-\frac{28}{9}} \cdot A^{-4}\right]; \tag{715}$$

$$\sum_{k=1}^{\infty} \frac{\zeta(2k+1)-1}{(k+1)(2k+1)} \left(\frac{5}{4}\right)^{2k} = \frac{53}{25} - \gamma + \log\left[2^{\frac{24}{5}} \cdot 3^{-\frac{72}{25}} \cdot 5^{-\frac{4}{5}} \cdot A^{-\frac{36}{25}}\right]; \tag{716}$$

$$\sum_{k=1}^{\infty} \frac{\zeta(2k+1)-1}{(k+1)(2k+1)} \left(\frac{7}{4}\right)^{2k} = \frac{101}{49} - \gamma + \log\left[2^{\frac{24}{7}} \cdot 3^{-\frac{12}{49}} \cdot 7^{-\frac{4}{7}} \cdot 11^{-\frac{44}{49}} \cdot A^{-\frac{36}{49}}\right]. \tag{717}$$

Setting $z = \frac{1}{4}$ in (590) through (593) and using 1.4(6), 1.1(12), 1.4(23) and 1.4(24), we obtain

$$\sum_{k=2}^{\infty} \frac{(2^k-1)\zeta(k)}{k(k+1)\cdot 2^{2k}} = -\frac{\gamma}{8} - \frac{G}{\pi} + \log\left[2^{\frac{1}{12}} \cdot A^3 2\right]; \tag{718}$$

$$\sum_{k=2}^{\infty} (-1)^k \frac{(2^k-1)\zeta(k)}{k(k+1)\cdot 2^{2k}} = \frac{\gamma}{8} - \frac{G}{\pi} + \log\left[2^{\frac{11}{12}} \cdot A^{-\frac{3}{2}}\right]; \tag{719}$$

$$\sum_{k=1}^{\infty} \frac{(2^{2k}-1)\zeta(2k)}{k(2k+1)\cdot 2^{4k}} = -\frac{2G}{\pi} + \log 2; \tag{720}$$

$$\sum_{k=1}^{\infty} \frac{(2^{2k+1}-1)\zeta(2k+1)}{(k+1)(2k+1)\cdot 2^{4k}} = -\gamma + \log\left[2^{-\frac{10}{3}} \cdot A^{12}\right]. \tag{721}$$

Further Evaluation by Using the Triple Gamma Function

We, first, rewrite the left-hand side of (488) as follows:

$$\begin{aligned}\sum_{n=2}^{\infty} \zeta(2n-1,a) z^{2n} n &= [\psi(a)-1]z^2 + (a-1)\log[\Gamma(a+z)\Gamma(a-z)] \\ &\quad - \log[G(a+z)G(a-z)] + 2(1-a)\log\Gamma(a) + 2\log G(a) \quad (|z|<|a|).\end{aligned} \tag{722}$$

By differentiating both sides of (4) with respect to t and multiplying the resulting equation by t^3, if we integrate the resulting equation with respect to t from $t=0$ to $t=z$, we obtain

$$\sum_{n=3}^{\infty} \zeta(2n-3,a) \frac{z^{2n}}{n} = -\int_0^z t^3 \psi(a+t)dt - \int_0^{-z} t^3 \psi(a+t)dt + \frac{\psi(a)}{2} z^4 \quad (|z|<|a|), \tag{723}$$

which, by virtue of (481), yields

$$\begin{aligned}
&\sum_{n=3}^{\infty} \zeta(2n-3,a)\frac{z^{2n}}{n} \\
&\quad = \left[\frac{3}{2}a^2 - \frac{7}{2}a + \frac{9}{4} + \left(\frac{3}{2} - a\right)\log(2\pi) - 2\log A\right] z^2 \\
&\quad + \frac{1}{4}[2\psi(a) - 3]z^4 + (a-1)^3 \log[\Gamma(a+z)\Gamma(a-z)] \\
&\quad - (3a^2 - 9a + 7)\log[G(a+z)G(a-z)] \\
&\quad - 2(z - 2a + 3)\log\Gamma_3(a+z) + 2(z + 2a - 3)\log\Gamma_3(a-z) \\
&\quad + 2(1-a)^3 \log\Gamma(a) + 2(3a^2 - 9a + 7)\log G(a) \\
&\quad + 4(3-2a)\log\Gamma_3(a) + 2\int_0^z \log\Gamma_3(t+a)dt \\
&\quad + 2\int_0^{-z} \log\Gamma_3(t+a)dt \quad (|z| < |a|).
\end{aligned} \tag{724}$$

Setting $a = 2$ in (722) and applying 1.3(50) and 2.3(9), we obtain

$$\begin{aligned}
&\sum_{n=3}^{\infty} \{\zeta(2n-1) - 1\}\frac{z^{2n}}{n} = -\gamma z^2 + \frac{1}{2}[1 - \zeta(3)]z^4 \\
&\quad + \log[\Gamma(2+z)\Gamma(2-z)] - \log[G(2+z)G(2-z)] \quad (|z| < 2).
\end{aligned} \tag{725}$$

Setting $z = \frac{3}{2}$ in (725) and making use of 1.4(8) and 2.3(9), we obtain

$$\sum_{n=3}^{\infty} \frac{1}{n}\left(\frac{3}{2}\right)^{2n} \{\zeta(2n-1) - 1\} = \frac{73}{32} - \frac{9\gamma}{4} + 3\log A - \frac{81}{32}\zeta(3) + \log\left(2^{-\frac{1}{12}} \cdot 5\right). \tag{726}$$

Setting $a = 4$ in (722) and applying 1.3(50), 2.3(9) and 2.3(12), we obtain

$$\begin{aligned}
&\sum_{n=3}^{\infty} \left(\zeta(2n-1) - 1 - \frac{1}{2^{2n-1}} - \frac{1}{3^{2n-1}}\right)\frac{z^{2n}}{n} = \left(\frac{5}{6} - \gamma\right) z^2 + \frac{1}{2}\left(\frac{251}{216} - \zeta(3)\right) z^4 \\
&\quad + 3\log[\Gamma(4+z)\Gamma(4-z)] - \log[G(4+z)G(4-z)] + \log\left(2^{-4} \cdot 3^{-6}\right) \quad (|z| < 4),
\end{aligned} \tag{727}$$

which, for $z = 3$, yields

$$\sum_{n=3}^{\infty} \frac{3^{2n}}{n}\left(\zeta(2n-1) - 1 - \frac{1}{2^{2n-1}} - \frac{1}{3^{2n-1}}\right) = \frac{873}{16} - 9\gamma - \frac{81}{2}\zeta(3) + \log\left(3^{-3}\cdot 5^2\right). \tag{728}$$

Setting $a = 4$ in (724) and applying 1.3(50) and 2.3(9), we obtain

$$\begin{aligned}
\sum_{n=3}^{\infty}&\left(\zeta(2n-3) - 1 - \frac{1}{2^{2n-3}} - \frac{1}{3^{2n-3}}\right)\frac{z^{2n}}{n} = \left(\frac{49}{4} - \frac{5}{2}\log(2\pi) - 2\log A\right)z^2 \\
&+ 12\left(\frac{1}{3} - \gamma\right)z^4 + 27\log[\Gamma(4+z)\Gamma(4-z)] - 19\log[G(4+z)G(4-z)] \\
&+ 2(5-z)\log\Gamma_3(4+z) + 2(5+z)\log\Gamma_3(4-z) + \log\left(2^{-16}3^{-54}\right) \\
&+ 2\int_0^z \log\Gamma_3(t+4)dt + 2\int_0^{-z} \log\Gamma_3(t+4)dt \quad (|z| < 4).
\end{aligned} \tag{729}$$

Using 1.3(39), 1.4(73) and (472), we find that

$$\begin{aligned}
\int_0^3 \log\Gamma_3(t+4)\,dt &= 10\int_0^1 \log(t+1)\,dt + 4\int_0^1 \log(t+2)\,dt + \int_0^1 \log(t+3)\,dt \\
&\quad + 19\int_0^1 \log\Gamma(t+1)\,dt + 12\int_0^1 \log G(t+1)\,dt + 3\int_0^1 \log\Gamma_3(t+1)\,dt \\
&= -33 + \frac{259}{8}\log 2 + 9\log 3 + \frac{99}{8}\log\pi - 24\log A + \frac{9}{2}\log B
\end{aligned} \tag{730}$$

and

$$\begin{aligned}
&\int_0^{-3} \log\Gamma_3(t+4)\,dt \\
&\quad = -\int_0^1 \log\Gamma(t+1)\,dt - 3\int_0^1 \log G(t+1)\,dt - 3\int_0^1 \log\Gamma_3(t+1)\,dt \\
&\quad = \frac{3}{4} - \frac{9}{8}\log(2\pi) + 6\log A - \frac{9}{2}\log B.
\end{aligned} \tag{731}$$

If we set $z = 3$ in (729) and make use of (730) and (731), we obtain

$$\sum_{n=3}^{\infty} \frac{3^{2n}}{n} \left(\zeta(2n-3) - 1 - \frac{1}{2^{2n-3}} - \frac{1}{3^{2n-3}} \right) = \frac{237}{4} - \frac{81}{2}\gamma - 54 \log A + \log\left(2^{12} \cdot 3^{-27} \cdot 5^{8}\right). \tag{732}$$

Setting $a = 2$ in (724) and applying 1.3(50) and 2.3(9), we obtain

$$\begin{aligned} \sum_{n=3}^{\infty} [\zeta(2n-3) - 1] \frac{z^{2n}}{n} &= \left(\frac{5}{4} - \frac{1}{2}\log(2\pi) - 2\log A \right) z^2 \\ &- \frac{1}{4}(1+2\gamma) z^4 + \log[\Gamma(2+z)\Gamma(2-z)] - \log[G(2+z)G(2-z)] \\ &+ 2(1-z)\log\Gamma_3(2+z) + 2(1+z)\log\Gamma_3(2-z) \\ &+ 2\int_0^z \log\Gamma_3(t+2)dt + 2\int_0^{-z} \log\Gamma_3(t+2)dt \quad (|z| < 2). \end{aligned} \tag{733}$$

Making use of 1.3(39), 1.4(55), 1.4(73), 1.4(77), (472) and (473), we find that

$$\begin{aligned} \int_0^{\frac{3}{2}} \log\Gamma_3(t+2)\,dt &= \int_0^1 \log G(t+1)\,dt + \int_0^1 \log\Gamma_3(t+1)\,dt \\ &+ \int_0^{\frac{12}{\log}} \Gamma(t+1)\,dt + 2\int_0^{\frac{12}{\log}} G(t+1)\,dt + \int_0^{\frac{12}{\log}} \Gamma_3(t+1)\,dt \\ &= -\frac{259}{768} - \frac{29}{1920}\log 2 + \frac{9}{16}\log\pi - \frac{15}{16}\log A - \frac{5}{4}\log B + \frac{15}{16}\log C \end{aligned} \tag{734}$$

and

$$\begin{aligned} \int_0^{-\frac{3}{2}} \log\Gamma_3(t+2)\,dt &= -2\int_0^1 \log\Gamma_3(t+1)\,dt + \int_0^{\frac{12}{\log}} \Gamma_3(t+1)\,dt \\ &+ \int_0^1 \log G(t+1)\,dt - \int_0^{\frac{12}{\log}} G(t+1)\,dt - \int_0^1 \log\Gamma(t+1)\,dt \\ &+ \int_0^{\frac{12}{\log}} \Gamma(t+1)\,dt - \int_0^{-\frac{1}{2}} \log(t+1)\,dt \\ &= \frac{29}{768} - \frac{29}{1920}\log 2 - \frac{3}{16}\log A - \frac{1}{2}\log B + \frac{15}{16}\log C. \end{aligned} \tag{735}$$

Setting $z = \frac{3}{2}$ in (733) and applying 1.4(8), 1.4(96), (734) and (735), we obtain

$$\begin{aligned}&\sum_{n=3}^{\infty} \frac{\zeta(2n-3)-1}{n}\left(\frac{3}{2}\right)^{2n}\\&= -\frac{1796}{-}\frac{81}{32}\gamma + \frac{27}{4}\log A + \frac{15}{4}\log C + \log\left(2^{-\frac{269}{480}}\cdot 5\right).\end{aligned} \tag{736}$$

Applications of Corollary 3.3

Upon setting $n = 1, 2, 3$ and 4 in 3.2(62), if we make use of the appropriate identities given in previous sections, we obtain the following known or new formulas for closed-form evaluations of several families of series involving the Hurwitz (or generalized) Zeta function $\zeta(s, a)$:

$$\begin{aligned}&\sum_{k=2}^{\infty} \frac{\zeta(k,a)}{k(k+1)} t^{k+1} = \zeta'(-1,a) - \zeta'(-1,a-t)\\&+\left(\frac{1}{2} - a - \log\Gamma(a) + \frac{1}{2}\log(2\pi)\right) t + [1+\psi(a)]\,\frac{t^2}{2} \quad (|t| < |a|);\end{aligned} \tag{737}$$

$$\begin{aligned}&\sum_{k=2}^{\infty} \frac{\zeta(k,a)}{k(k+1)(k+2)} t^{k+2}\\&= \frac{1}{2}[\zeta'(-2,a-t) - \zeta'(-2,a)] + \left[\frac{1}{4}\left(a^2 - a + \frac{1}{6}\right) + \zeta'(-1,a)\right] t\\&+\left(\frac{3}{2} - 3a + \log(2\pi) - 2\log\Gamma(a)\right)\frac{t^2}{2} + [3+2\psi(a)]\frac{t^3}{12} \quad (|t| < |a|);\end{aligned} \tag{738}$$

$$\begin{aligned}&\sum_{k=2}^{\infty} \frac{\zeta(k,a)}{k(k+1)(k+2)(k+3)} t^{k+3} = \frac{1}{6}[\zeta'(-3,a) - \zeta'(-3,a-t)]\\&-\left[\frac{1}{3}\left(a^3 - \frac{3}{2}a^2 + \frac{1}{2}a\right) + 3\zeta'(-2,a)\right]\frac{t}{6}\\&+\left[\frac{5}{2}\left(a^2 - a + \frac{1}{6}\right) + 6\zeta'(-1,a)\right]\frac{t^2}{12}\\&+\left(\frac{11}{2} - 11a + 3\log(2\pi) - 6\log\Gamma(a)\right)\frac{t^3}{36}\\&+[11+6\psi(a)]\,\frac{t^4}{144} \quad (|t| < |a|);\end{aligned} \tag{739}$$

$$\begin{aligned}&\sum_{k=2}^{\infty} \frac{\zeta(k,a)}{k(k+1)(k+2)(k+3)(k+4)} t^{k+4} = \frac{1}{24}[\zeta'(-4,a-t) - \zeta'(-4,a)]\\&+\left[\frac{1}{4}\left(a^4 - 2a^3 + a^2 - \frac{1}{30}\right) + 4\zeta'(-3,a)\right]\frac{t}{24}\end{aligned}$$

$$- \left[\frac{7}{3} \left(a^3 - \frac{3}{2} a^2 + \frac{1}{2} a \right) + 12 \zeta'(-2, a) \right] \frac{t^2}{48}$$
$$+ \left[13 \left(a^2 - a + \frac{1}{6} \right) + 24 \zeta'(-1, a) \right] \frac{t^3}{144}$$
$$+ [25 - 50a + 12 \log(2\pi) - 24 \log \Gamma(a)] \frac{t^4}{576}$$
$$+ [25 + 12 \psi(a)] \frac{t^5}{1440} \quad (|t| < |a|). \tag{740}$$

Setting $a = 1$ in (740) and applying the formulas presented in this and previous sections, we readily have

$$\begin{aligned} \sum_{k=2}^{\infty} \frac{\zeta(k)}{k(k+1)(k+2)(k+3)(k+4)} t^{k+4} &= \frac{1}{24} \left(\zeta'(-4, 1-t) - \frac{3\zeta(5)}{4\pi^4} \right) \\ &- \left(\frac{5}{72} + 4 \log C \right) \frac{t}{24} + \frac{\zeta(3)}{16\pi^2} t^2 + \left(\frac{25}{6} - 24 \log A \right) \frac{t^3}{144} \\ &+ [12 \log(2\pi) - 25] \frac{t^4}{576} + (25 - 12\gamma) \frac{t^5}{1440} \quad (|t| < 1), \end{aligned} \tag{741}$$

which, for $t = \frac{1}{2}$, yields

$$\begin{aligned} &\sum_{k=2}^{\infty} \frac{\zeta(k)}{k(k+1)(k+2)(k+3)(k+4) \cdot 2^k} \\ &= \frac{1}{48} \log(2\pi) - \frac{\gamma}{240} - \frac{1}{3} \log A - \frac{4}{3} \log C + \frac{\zeta(3)}{4\pi^2} - \frac{31\zeta(5)}{32\pi^4}. \end{aligned} \tag{742}$$

Similarly, we obtain

$$\begin{aligned} &\sum_{k=2}^{\infty} \frac{\zeta(k)}{k(k+1)(k+2)(k+3) \cdot 2^k} \\ &= \frac{59}{720} \log 2 + \frac{1}{12} \log \pi - \frac{\gamma}{48} - \log A - \frac{5}{2} \log C + \frac{\zeta(3)}{2\pi^2}; \end{aligned} \tag{743}$$

$$\begin{aligned} &\sum_{k=2}^{\infty} \frac{\zeta(k)}{k(k+1)(k+2) \cdot 2^k} \\ &= -\frac{3}{8} + \frac{1}{2} \log(2\pi) - \frac{\gamma}{12} - 2 \log A + \frac{7\zeta(3)}{8\pi^2}. \end{aligned} \tag{744}$$

We remark in passing that the various series identities presented here are potentially useful in deriving further identities for series involving the Zeta and related functions.

For example, if we replace t in (737) by $-t$, we get

$$\sum_{k=2}^{\infty}(-1)^k \frac{\zeta(k,a)}{k(k+1)} t^{k+1} \\ = \zeta'(-1,a-t) - \zeta'(-1,a) + \left(\frac{1}{2} - a - \log\Gamma(a) + \frac{1}{2}\log(2\pi)\right) t \\ - [1+\psi(a)]\,\frac{t^2}{2} \qquad (|t| < |a|). \tag{745}$$

Furthermore, by making use of 2.2(17) and 2.2(33), the series identity (745) can be written in an equivalent form involving the double Gamma function Γ_2 (*cf.* Equation (561)):

$$\sum_{k=2}^{\infty}(-1)^k \frac{\zeta(k,a)}{k(k+1)} t^{k+1} \\ = (1-a)\log\Gamma(a) - \log\Gamma_2(a) + (t+a-1)\log\Gamma(a+t) + \log\Gamma_2(a+t) \\ + \left(\frac{1}{2} - a + \frac{1}{2}\log(2\pi) - \log\Gamma(a)\right) t - [\psi(a)+1]\,\frac{t^2}{2} \quad (|t| < |a|), \tag{746}$$

which, in the form (561), was proven (in a markedly different way) by Choi and Srivastava [288, p. 108, Eq. (2.14)].

3.5 Use of Hypergeometric Identities

Al-Saqabi *et al.* [21] presented a systematic account of several interesting infinite series expressed in terms of the Psi (or Digamma) function. Aular de Durán *et al.* [79] examined rather systematically the sums of numerous interesting families of infinite series with or without the use of fractional calculus (see also [1237]). Shen [1024] investigated the connections between the Stirling numbers $s(n,k)$ of the first kind and the Riemann Zeta function $\zeta(n)$ by means of the Gauss summation formula 1.5(7) for the hypergeometric series. We show how the sums of certain interesting classes of infinite series can be evaluated by analyzing three known identities involving hypergeometric series. Our presentation in this section is based essentially on the works of Shen [1024], Choi and Seo [282] Choi and Srivastava [290] and Choi *et al.* [309].

In addition to Gauss's summation formula 1.5(7), we, first, recall some well-known summation formulas for hypergeometric series.

Gauss's formula:

$${}_2F_1\left(a, b; \frac{1}{2}(a+b+1); \frac{1}{2}\right) = \frac{\Gamma\left(\frac{1}{2}\right)\Gamma\left(\frac{1}{2}+\frac{1}{2}a+\frac{1}{2}b\right)}{\Gamma\left(\frac{1}{2}+\frac{1}{2}a\right)\Gamma\left(\frac{1}{2}+\frac{1}{2}b\right)} \tag{1}$$

$$(a+b \neq -1, -3, -5, \ldots).$$

Kummer's formulas:

$$ {}_2F_1\left(a,\, 1-a;\, c;\, \frac{1}{2}\right) = \frac{2^{1-c}\Gamma(c)\Gamma\left(\frac{1}{2}\right)}{\Gamma\left(\frac{1}{2}c+\frac{1}{2}a\right)\Gamma\left(\frac{1}{2}c-\frac{1}{2}a+\frac{1}{2}\right)} \quad \left(c \notin \mathbb{Z}_0^-\right), \tag{2} $$

$$ {}_2F_1(a,\, b;\, 1+a-b;\, -1) = \frac{\Gamma(1+a-b)\Gamma\left(1+\frac{1}{2}a\right)}{\Gamma\left(1+\frac{1}{2}a-b\right)\Gamma(1+a)} \tag{3} $$

$$ \left(\Re(b) < 1;\; 1+a-b \notin \mathbb{Z}_0^-\right). $$

Dixon's theorem:

$$ {}_3F_2\left[\begin{matrix} a,\, b,\, c; \\ 1+a-b,\, 1+a-c; \end{matrix}\; 1\right] $$

$$ = \frac{\Gamma\left(1+\frac{1}{2}a\right)\Gamma(1+a-b)\Gamma(1+a-c)\Gamma\left(1+\frac{1}{2}a-b-c\right)}{\Gamma(1+a)\Gamma\left(1+\frac{1}{2}a-b\right)\Gamma\left(1+\frac{1}{2}a-c\right)\Gamma(1+a-b-c)} \tag{4} $$

$$ \left(\Re\left(\frac{1}{2}a-b-c\right) > -1\right). $$

Watson's theorem:

$$ {}_3F_2\left[\begin{matrix} a,\, b,\, c; \\ \frac{1}{2}+\frac{1}{2}a+12b,\, 2c; \end{matrix}\; 1\right] $$

$$ = \frac{\Gamma\left(\frac{1}{2}\right)\Gamma\left(c+\frac{1}{2}\right)\Gamma\left(\frac{1}{2}+\frac{1}{2}a+\frac{1}{2}b\right)\Gamma\left(\frac{1}{2}-\frac{1}{2}a-\frac{1}{2}b+c\right)}{\Gamma\left(\frac{1}{2}+\frac{1}{2}a\right)\Gamma\left(\frac{1}{2}+\frac{1}{2}b\right)\Gamma\left(\frac{1}{2}-\frac{1}{2}a+c\right)\Gamma\left(\frac{1}{2}-\frac{1}{2}b+c\right)}. \tag{5} $$

Whipple's theorem:

$$ {}_3F_2\,[a,\, b,\, c;\, d,\, e;\, 1] $$

$$ = \pi 2^{1-2c} \frac{\Gamma(d)\Gamma(e)}{\Gamma\left(\frac{1}{2}a+\frac{1}{2}e\right)\Gamma\left(\frac{1}{2}a+\frac{1}{2}d\right)\Gamma\left(\frac{1}{2}b+\frac{1}{2}e\right)\Gamma\left(\frac{1}{2}b+\frac{1}{2}d\right)} \tag{6} $$

$$ (a+b=1;\; d+e=1+2c). $$

Series Derivable from Gauss's Summation Formula 1.4(7)

Consider a as a variable in 1.5(7) and write

$$ f(z) := 1 + \sum_{n=1}^{\infty} \frac{(b)_n\,(z)_n}{(c)_n\, n!} = \frac{\Gamma(c)\,\Gamma(c-b-z)}{\Gamma(c-b)\,\Gamma(c-z)}. \tag{7} $$

By making use of the definition and properties of the Stirling numbers $s(n,k)$ of the first kind, we find that

$$f(z) = 1 + \sum_{n=1}^{\infty} \frac{(b)_n}{(c)_n} \left\{ \frac{z}{n} + \frac{H_{n-1}}{n} z^2 + \frac{H_{n-1}^2 - H_{n-1}^{(2)}}{2n} z^3 \right. \\ \left. + \frac{H_{n-1}^3 - 3H_{n-1}H_{n-1}^{(2)} + 2H_{n-1}^{(3)}}{6n} z^4 + \cdots \right\}. \tag{8}$$

Conversely, we let

$$f(z) = \frac{\Gamma(c)\,\Gamma(c-b-z)}{\Gamma(c-b)\,\Gamma(c-z)} := 1 + \sum_{n=1}^{\infty} a_n z^n, \tag{9}$$

where the coefficients a_n are to be determined. Differentiating (9) logarithmically, we obtain

$$\frac{f'(z)}{f(z)} := \sum_{n=0}^{\infty} c_n z^n,$$

which, in view of 1.3(1) and 1.3(53), yields

$$c_n = \begin{cases} \psi(c) - \psi(c-b) & (n = 0), \\ \zeta(n+1, c-b) - \zeta(n+1, c) & (n \in \mathbb{N}), \end{cases} \tag{10}$$

where $\psi(z)$ and $\zeta(s,a)$ denote the Digamma (or Psi) and Hurwitz (or generalized) Zeta functions, respectively (see Sections 1.3 and 2.2). Observe that

$$a_0 = f(0) = 1 \quad \text{and} \quad c_0 = f'(0)f(0) = a_1.$$

Since

$$\sum_{n=0}^{\infty} (n+1)a_{n+1} z^n = f'(z) = f(z) \cdot \frac{f'(z)}{f(z)} = \sum_{n=0}^{\infty} \left(\sum_{k=0}^{n} a_{n-k}\, c_k \right) z^n,$$

we have the following recursion formula for a_n:

$$(n+1)a_{n+1} = \sum_{k=0}^{n} a_{n-k}\, c_k \quad (n \in \mathbb{N}). \tag{11}$$

Now, by comparing the power series in (8) and (9), we get

$$\sum_{n=1}^{\infty} \frac{(b)_n}{n\,(c)_n} = \psi(c) - \psi(c-b) \quad \left(\Re(c-b) > 0;\ c \notin \mathbb{Z}_0^-\right), \tag{12}$$

which is a well-known (rather classical) result;

$$\sum_{n=1}^{\infty} \frac{(b)_n}{(c)_n} \frac{H_{n-1}}{n} = \frac{1}{2}\Big[\{\psi(c) - \psi(c-b)\}^2 + \zeta(2, c-b) - \zeta(2, c)\Big], \tag{13}$$

which is the *corrected* version of a result in Shen [1024, p. 1397, Eq. (31)]):

$$\sum_{n=1}^{\infty} \frac{1}{n} \frac{(b)_n}{(c)_n} \left\{H_{n-1}^2 - H_{n-1}^{(2)}\right\} = \frac{1}{3}[\psi(c) - \psi(c-b)]^3$$
$$+ [\psi(c) - \psi(c-b)][\zeta(2, c-b) - \zeta(2, c)] + \frac{2}{3}[\zeta(3, c-b) - \zeta(3, c)]. \tag{14}$$

If use is made of 1.3(3), the classical result (12) is rewritten in its equivalent form:

$$\sum_{n=1}^{\infty} \frac{(b)_n}{n\,(c)_n} = \sum_{n=0}^{\infty} \left(\frac{1}{n+c-b} - \frac{1}{n+c}\right) \quad \left(\Re(c-b) > 0;\ c \notin \mathbb{Z}_0^-\right). \tag{15}$$

Choose $c = 1$ and $b = z$ in (15) and expand both sides as power series about $z = 0$. We, thus, obtain

$$\sum_{k=1}^{\infty} \left(\sum_{n=k}^{\infty} \frac{(-1)^{n+k} s(n,k)}{n \cdot n!}\right) z^k = \sum_{n=1}^{\infty} \frac{(z)_n}{n \cdot n!} = \sum_{n=1}^{\infty} \left(\frac{1}{n-z} - \frac{1}{n}\right) = \sum_{k=1}^{\infty} \zeta(k+1)\, z^k,$$

which, upon equating the coefficients of z^k, yields an interesting relation between the Riemann Zeta function and the Stirling numbers of the first kind:

$$\zeta(k+1) = \sum_{n=k}^{\infty} \frac{(-1)^{n+k} s(n,k)}{n \cdot n!} \quad (k \in \mathbb{N}). \tag{16}$$

Similarly, we can rewrite (13) in the following form:

$$\sum_{n=1}^{\infty} \frac{(b)_n}{(c)_n} \frac{H_{n-1}}{n} = \frac{1}{2}\left[\left\{\sum_{n=0}^{\infty} \left(\frac{1}{n+c-b} - \frac{1}{n+c}\right)\right\}^2\right.$$
$$\left. + \sum_{n=0}^{\infty} \left(\frac{1}{(n+c-b)^2} - 1(n+c)^2\right)\right], \tag{17}$$

which, upon choosing $c = 1$ and $b = z$ and expanding both sides as power series about $z = 0$, yields

$$\left(\sum_{n=1}^{\infty} \frac{H_{n-1}}{n^2}\right) z + \left\{\sum_{n=1}^{\infty} \left(\frac{H_{n-1}}{n}\right)^2\right\} z^2 + \cdots$$
$$= \zeta(3)\, z + \frac{1}{2}\Big[\{\zeta(2)\}^2 + 3\,\zeta(4)\Big] z^2 + \cdots.$$

Therefore, we have (see Shen [1024, p. 1398])

$$\sum_{n=1}^{\infty} \frac{H_{n-1}}{n^2} = \zeta(3), \tag{18}$$

which is an obvious special case of 2.3(54) when $n = 2$:

$$\sum_{n=1}^{\infty} \left(\frac{H_{n-1}}{n}\right)^2 = \frac{1}{2}\left[\{\zeta(2)\}^2 + 3\zeta(4)\right] = \frac{11}{4}\zeta(4). \tag{19}$$

If we set ($b = 1$ and $c = 2$) and ($b = \frac{1}{2}$ and $c = 1$) in (13) and (14) and use 1.3(50), 1.3(51), 2.3(2) and 2.3(9), we also obtain the following interesting formulas:

$$\sum_{n=1}^{\infty} \frac{H_{n-1}}{n(n+1)} = 1; \tag{20}$$

$$\sum_{n=1}^{\infty} \frac{H_{n-1}^2 - H_{n-1}^{(2)}}{n(n+1)} = 2; \tag{21}$$

$$\sum_{n=1}^{\infty} \frac{\left(\frac{1}{2}\right)_n}{n \cdot n!} H_{n-1} = \zeta(2) + 2(\log 2)^2; \tag{22}$$

$$\sum_{n=1}^{\infty} \frac{\left(\frac{1}{2}\right)_n}{n \cdot n!} \left\{H_{n-1}^2 - H_{n-1}^{(2)}\right\} = 4\zeta(3) + 4\zeta(2)\log 2 + \frac{8}{3}(\log 2)^3. \tag{23}$$

Series Derivable from Kummer's Formula (3)

Consider a as a variable in (3) and apply the same technique as above. We readily obtain

$$\sum_{n=1}^{\infty} \frac{(-1)^n (b)_n}{n(1-b)_n} = \frac{1}{2}[\gamma + \psi(1-b)], \tag{24}$$

which is, in fact, an immediate consequence of a known result recorded by Aular de Durán *et al.* [79, p. 746, Eq. (2.9)] with $z = 1 - b$;

$$\begin{aligned} &\sum_{n=1}^{\infty} \frac{(-1)^n (b)_n}{n(1-b)_n} \left(H_{n-1} + \sum_{k=1}^{n} \frac{1}{b-k}\right) \\ &\quad = \frac{1}{8}[\gamma + \psi(1-b)]^2 + \frac{3}{8}[\zeta(2, 1-b) - \zeta(2)]; \end{aligned} \tag{25}$$

$$\sum_{n=1}^{\infty} \frac{(-1)^n (b)_n}{n(1-b)_n} \left[\left(\sum_{k=1}^{n} \frac{1}{b-k} + H_{n-1} \right)^2 + \sum_{k=1}^{n} \frac{1}{(b-k)^2} - H_{n-1}^{(2)} \right]$$
$$= \frac{1}{24}[\gamma + \psi(1-b)]^3 + \frac{3}{8}[\gamma + \psi(1-b)][\zeta(2, 1-b) - \zeta(2)] + \frac{7}{12}[\zeta(3) - \zeta(3, 1-b)]. \tag{26}$$

By setting $b = \frac{1}{2}$ in (25) and applying some elementary identities given in Sections 1.3 and 2.3, we obtain

$$\frac{1}{2}(\log 2)^2 + \frac{3}{4}\zeta(2) = \sum_{n=1}^{\infty} \frac{(-1)^n}{n} \left\{ H_{n-1} - \sum_{k=0}^{n-1} \frac{2}{2k+1} \right\}$$
$$= \sum_{n=1}^{\infty} \frac{(-1)^n}{n} \left\{ -2 + \sum_{k=1}^{n-1} \frac{1}{k(2k+1)} \right\}$$

from which, by using the Maclaurin series of $\log(1+x)$, we obtain

$$\sum_{n=1}^{\infty} \frac{(-1)^n}{n} \sum_{k=1}^{n-1} \frac{1}{k(2k+1)} = \frac{1}{2}(\log 2)^2 - 2\log 2 + \frac{3}{4}\zeta(2). \tag{27}$$

Setting $b = \frac{2}{3}$ in (25) yields

$$\sum_{n=1}^{\infty} \frac{(-1)^n}{n} \cdot \left(\frac{2}{3}\right)_n \left(\frac{1}{3}\right)_n \left\{ -3 + \sum_{k=1}^{n-1} \frac{1}{k(3k+1)} \right\}$$
$$= \frac{1}{32} \left(\frac{\sqrt{3}}{3}\pi + 3\log 3 \right)^2 + \frac{3}{8}[\zeta(2, \tfrac{1}{3}) - \zeta(2)]. \tag{28}$$

Finally, by setting $b = \frac{1}{2}$ in (26), we obtain the following series evaluation:

$$\sum_{n=1}^{\infty} \frac{(-1)^{n+1}}{n} \left[\left(\frac{2}{2n-1} + \sum_{k=1}^{n-1} \frac{1}{k(2k-1)} \right)^2 + \sum_{k=1}^{n} \frac{4}{(2k-1)^2} - H_{n-1}^{(2)} \right]$$
$$= \frac{1}{3}(\log 2)^3 + \frac{3}{2}\zeta(2)\log 2 + \frac{7}{2}\zeta(3). \tag{29}$$

Next, by considering b as a variable in (3), we also show that the sums of a different class of infinite series can be evaluated. For $b = z$, we find from (3) that

$$f(z) := 1 + \sum_{n=1}^{\infty} \frac{(-1)^n (a)_n}{n!} \cdot \frac{(z)_n}{(1+a-z)_n} = \frac{\Gamma\left(1+\frac{1}{2}a\right)}{\Gamma(1+a)} \cdot \frac{\Gamma(1+a-z)}{\Gamma\left(1+\frac{1}{2}a-z\right)}. \tag{30}$$

Now, since

$$\sum_{k=0}^{n}\sum_{\ell=0}^{k} B_{k,\ell} = \sum_{\ell=0}^{n}\sum_{k=\ell}^{n} B_{k,\ell},$$

we can apply 1.5(5) to obtain

$$\begin{aligned}(1+a-z)_n &= \sum_{k=0}^{n}(-1)^{n+k}s(n,k)(1+a-z)^k\\ &= \sum_{k=0}^{n}(-1)^{n+k}s(n,k)\left\{\sum_{\ell=0}^{k}\binom{k}{\ell}(1+a)^{k-\ell}\cdot(-1)^{\ell}\cdot z^{\ell}\right\}\\ &= \sum_{\ell=0}^{n}\left[\sum_{k=\ell}^{n}(-1)^{n+k}s(n,k)\binom{k}{\ell}(-1)^{\ell}\cdot(1+a)^{k-\ell}\right]z^{\ell}.\end{aligned} \tag{31}$$

Expanding the first member of (30) in terms of the Stirling numbers $s(n,k)$ and using 1.6(9), 1.6(11) and (31), we have

$$f(z) = 1+\sum_{n=1}^{\infty}(-1)^n(a)_n rn!\left(\alpha_1 z+\alpha_2 z^2+\alpha_3 z^3+\cdots\right), \tag{32}$$

where

$$\begin{aligned}\alpha_1 &= \frac{(n-1)!}{(1+a)_n};\\ \alpha_2 &= \frac{(n-1)!}{(1+a)_n}(H_{n-1}+\psi(1+a+n)-\psi(1+a));\\ \alpha_3 &= \frac{(n-1)!}{2(1+a)_n}\Big[-H_{n-1}^{(2)}+(H_{n-1}+\psi(1+a+n)-\psi(1+a))^2\\ &\qquad +\zeta(2,1+a)-\zeta(2,1+a+n)\Big]\end{aligned}$$

and so on.

Now, using the same technique as above, we obtain

$$\sum_{n=1}^{\infty}\frac{(-1)^n a}{n(n+a)} = \psi\left(1+\frac{1}{2}a\right)-\psi(1+a), \tag{33}$$

which was recorded by Al-Saqabi *et al.* [21, p. 366, Eq. (2.7)], who commented that (33) is essentially the same as the known result (*cf.* Hansen [531, p. 102, Eq. (6.1.14)]):

$$\sum_{n=1}^{\infty}\frac{(-1)^n}{n(\lambda+\mu n)} = \frac{1}{\lambda}\left\{\beta\left(1+\frac{\lambda}{\mu}\right)-\log 2\right\}, \tag{34}$$

where $\beta(z)$ is defined by

$$\beta(z) := \frac{1}{2}\left\{\psi\left(\frac{1}{2}+\frac{1}{2}z\right)-\psi\left(\frac{1}{2}z\right)\right\} = \sum_{N=0}^{\infty}\frac{(-1)^N}{z+N}; \tag{35}$$

$$\begin{aligned}&\sum_{n=1}^{\infty}\frac{(-1)^n a}{n(n+a)}\left\{H_{n-1}+\sum_{k=1}^{n}\frac{1}{k+a}\right\}\\&\quad=\frac{1}{2}\left[\left\{\psi\left(1+\frac{1}{2}a\right)-\psi(1+a)\right\}^2+\zeta(2,1+a)-\zeta\left(2,1+\frac{1}{2}a\right)\right];\end{aligned} \tag{36}$$

$$\begin{aligned}&\sum_{n=1}^{\infty}\frac{(-1)^n a}{n(n+1)}\left\{\sum_{k=1}^{n}\frac{1}{(k+a)^2}-H_{n-1}^{(2)}+\left(\sum_{k=1}^{n}\frac{1}{k+a}+H_{n-1}\right)^2\right\}\\&\quad=\frac{1}{3}\left\{\psi\left(1+\frac{1}{2}a\right)-\psi(1+a)\right\}\left[\left\{\psi\left(1+\frac{1}{2}a\right)-\psi(1+a)\right\}^2\right.\\&\qquad\left.+3\left\{\zeta(2,1+a)-\zeta\left(2,1+\frac{1}{2}a\right)\right\}\right]+\frac{2}{3}\left\{\zeta(3,1+a)-\zeta\left(3,1+\frac{1}{2}a\right)\right\}\\&\qquad\qquad\big(a\notin\mathbb{Z}_0^-\setminus\{0\}\big).\end{aligned} \tag{37}$$

Setting $a=1$ in (36) and (37) and simplifying the resulting equations with the aid of various known (or easily derivable) identities, including the following (*cf.*, e.g., Hansen [531, p. 116, Entry (6.2.14)] for $m-2=n=1$ and $m=n-2=1$):

$$\begin{aligned}\sum_{n=1}^{\infty}\frac{(-1)^{n+1}}{n(n+1)^3}&=2\log 2-3+\frac{1}{2}\zeta(2)+\frac{3}{4}\zeta(3);\\\sum_{n=1}^{\infty}\frac{(-1)^{n+1}}{n^3(n+1)}&=2\log 2-1-\frac{1}{2}\zeta(2)+\frac{3}{4}\zeta(3);\\\sum_{n=1}^{\infty}\frac{(-1)^{n+1}}{n^2(n+1)^2}H_n&=4\sum_{n=1}^{\infty}\frac{(-1)^n}{n}H_n+\zeta(2)+\frac{3}{4}\zeta(3),\end{aligned} \tag{38}$$

we readily obtain

$$\sum_{n=1}^{\infty}\frac{(-1)^n}{n(n+1)}H_n=(\log 2)^2-\frac{1}{2}\zeta(2), \tag{39}$$

which reduces immediately to

$$\sum_{n=1}^{\infty}\frac{(-1)^{n+1}}{n}H_n=\frac{1}{2}\left[\zeta(2)-(\log 2)^2\right]. \tag{40}$$

We can also readily obtain the following identity:

$$\sum_{n=1}^{\infty} \frac{(-1)^n}{n(n+1)} H_n^2 = -\frac{7}{9} - \frac{2}{3} (\log 2)^3 + \zeta(2) \log 2 - \frac{1}{3} \zeta(3). \tag{41}$$

Series Derivable from Other Hypergeometric Summation Formulas

Consider a as a variable in (2) and apply the same technique as above. We, thus, find that

$$\sum_{n=1}^{\infty} \frac{(n-1)!}{(c)_n \cdot 2^n} = \frac{1}{2} \left\{ \psi\left(12c + \frac{1}{2}\right) - \psi\left(\frac{1}{2}c\right) \right\}; \tag{42}$$

$$\sum_{n=1}^{\infty} \frac{(n-1)!}{n\,(c)_n \cdot 2^n} = \frac{1}{8} \left[\zeta\left(2, \frac{1}{2}c\right) + \zeta\left(2, \frac{1}{2}c + \frac{1}{2}\right) - \left\{ \psi\left(\frac{1}{2}c + \frac{1}{2}\right) - \psi\left(\frac{1}{2}c\right) \right\}^2 \right], \tag{43}$$

which, upon setting $c = 1$, 2 and 3, yields

$$\sum_{n=1}^{\infty} \frac{1}{n^2 \cdot 2^n} = \frac{1}{2} \left[\zeta(2) - (\log 2)^2 \right], \tag{44}$$

which may be compared with (40):

$$\sum_{n=1}^{\infty} \frac{1}{n^2 (n+1) \cdot 2^{n-1}} = \zeta(2) - 1 - (\log 2 - 1)^2; \tag{45}$$

$$\sum_{n=1}^{\infty} 1n^2 (n+1)(n+2) \cdot 2^{n-1} = \frac{1}{2} \zeta(2) - \frac{58}{-} \frac{1}{8} (2 \log 2 - 1)^2. \tag{46}$$

By setting $c = z$ in (2) and 1.5(7), multiplying the resulting equations by z and making use of the same technique as above, we obtain the following summation formula:

$$\sum_{n=1}^{\infty} \frac{\Gamma(a+n)\,\Gamma(1-a+n)}{n!\,(n-1)!\,2^n} = \frac{1}{\sqrt{\pi}} \Gamma\left(\frac{1}{2} + \frac{1}{2}a\right) \Gamma\left(1 - \frac{1}{2}a\right), \tag{47}$$

which, after some elementary manipulation, becomes

$$ {}_2F_1\left(1+a, 2-a; 2; \frac{1}{2}\right) = \frac{2}{\sqrt{\pi}} \frac{\Gamma\left(\frac{1}{2} + \frac{1}{2}a\right) \Gamma\left(1 - \frac{1}{2}a\right)}{\Gamma(1+a)\,\Gamma(2-a)}. \tag{48}$$

By using Legendre's duplication formula 1.1(29), (48) is rewritten in its *equivalent* form:

$$ {}_2F_1\left(1+a,\, 2-a;\, 2;\, \frac{1}{2}\right) = \frac{\sqrt{\pi}}{\Gamma\left(1+\frac{1}{2}a\right)\Gamma\left(\frac{3}{r2}-\frac{1}{2}a\right)}, \tag{49} $$

which can be shown to be a special case of Kummer's summation formula (2) in view of Euler's formula 1.5(21):

$$ \sum_{n=1}^{\infty} \frac{\Gamma(a+n)\Gamma(1-a+n)}{n!\,(n-1)!\,2^n} H_{n-1} = \Gamma(a)\Gamma(1-a) \tag{50} $$
$$ + \frac{1}{\sqrt{\pi}}\Gamma\left(\frac{1}{2}+\frac{1}{2}a\right)\Gamma\left(1-\frac{1}{2}a\right)\left\{\log 2+\gamma+\frac{1}{2}\psi\left(\frac{1}{2}a\right)+\frac{1}{2}\psi\left(\frac{1}{2}-\frac{1}{2}a\right)\right\}; $$

$$ \sum_{n=1}^{\infty} \frac{\Gamma(a+n)\Gamma(1-a+n)}{n!\,(n-1)!\,2^n}\left\{H_{n-1}^2+H_{n-1}^{(2)}\right\} $$
$$ = \frac{1}{\sqrt{\pi}}\Gamma\left(\frac{1}{2}+\frac{1}{2}a\right)\Gamma\left(1-\frac{1}{2}a\right)\left[\left\{\log 2+\gamma+\frac{1}{2}\psi\left(\frac{1}{2}a\right)+\frac{1}{2}\psi\left(\frac{1}{2}-\frac{1}{2}a\right)\right\}^2\right. $$
$$ \left. +\zeta(2)-\frac{1}{4}\left\{\zeta\left(2,\,\frac{1}{2}a\right)+\zeta\left(2,\,\frac{1}{2}-\frac{1}{2}a\right)\right\}\right]; \tag{51} $$

$$ \sum_{n=1}^{\infty} \frac{(a)_n(b)_n}{(n-1)!\,n!} = \frac{\Gamma(-a-b)}{\Gamma(-a)\Gamma(-b)} \qquad (\Re(a+b)<0), \tag{52} $$

which, after some elementary manipulation, is seen to be the same as the following special case of Gauss's summation theorem 1.5(7):

$$ {}_2F_1(a+1,\, b+1;\, 2;\, 1) = \frac{\Gamma(-a-b)}{\Gamma(1-a)\Gamma(1-b)} \qquad (\Re(a+b)<0); \tag{53} $$

$$ \sum_{n=1}^{\infty} \frac{(a)_n(b)_n}{(n-1)!\,n!} H_{n-1} = 1+\frac{\Gamma(-a-b)}{\Gamma(-a)\Gamma(-b)}[\gamma+\psi(-a)+\psi(-b)-\psi(-a-b)]; \tag{54} $$

$$ \sum_{n=1}^{\infty} \frac{(a)_n(b)_n}{(n-1)!\,n!}\left\{H_{n-1}^2+H_{n-1}^{(2)}\right\} = \frac{\Gamma(-a-b)}{\Gamma(-a)\Gamma(-b)} $$
$$ \cdot\left[\{\gamma+\psi(-a)+\psi(-b)-\psi(-a-b)\}^2+\zeta(2)\right. \tag{55} $$
$$ \left. +\zeta(2,\,-a-b)-\zeta(2,\,-a)-\zeta(2,\,-b)\right]. $$

Setting $a = \frac{1}{2}$ in (47), (50) and (51) and simplifying with the aid of various known identities involving Gamma and Psi functions, we obtain

$$\sum_{n=1}^{\infty} \frac{\{(2n)!\}^2}{(n!)^3 (n-1)!\, 2^{5n}} = \frac{1}{\pi\sqrt{\pi}} \left\{\Gamma\left(\tfrac{3}{4}\right)\right\}^2; \tag{56}$$

$$\sum_{n=1}^{\infty} \frac{\{(2n)!\}^2}{(n!)^3 (n-1)!\, 2^{5n}} H_{n-1} = 1 - \frac{1}{\pi\sqrt{\pi}} \left\{\Gamma\left(\tfrac{3}{4}\right)\right\}^2 \left(\frac{\pi}{2} + 2\log 2\right); \tag{57}$$

$$\begin{aligned} &\sum_{n=1}^{\infty} \frac{\{(2n)!\}^2}{(n!)^3\, (n-1)!\, 2^{5n}} \left\{H_{n-1}^2 + H_{n-1}^{(2)}\right\} \\ &= \frac{1}{\pi\sqrt{\pi}} \left\{\Gamma\left(\tfrac{3}{4}\right)\right\}^2 \left[\left(\frac{\pi}{2} + 2\log 2\right)^2 + \zeta(2) - \frac{1}{2}\zeta\left(2, \tfrac{1}{4}\right)\right]. \end{aligned} \tag{58}$$

In view of Euler's formula 1.5(21), the summation formula (56) can be deduced directly from (2), by setting $a = \frac{1}{2}$ and $c = 2$. Moreover, (56) follows readily from the special case of the identity (1) when $a = b = \frac{3}{2}$.

Considering (1) just as above, we obtain

$$\sum_{n=1}^{\infty} \frac{(b)_n}{n \cdot 2^{n-1} \left(\frac{1}{2} + \frac{1}{2}b\right)_n} = \psi\left(\frac{1}{2} + \frac{1}{2}b\right) - \psi\left(\frac{1}{2}\right), \tag{59}$$

which is precisely the same as the aforementioned known result recorded (with correction) by Al-Saqabi *et al.* [21, p. 362, Eq. (1.7)]:

$$\begin{aligned} &\sum_{n=1}^{\infty} \frac{(b)_n}{n \cdot 2^{n-1} \left(\frac{1}{2} + \frac{1}{2}b\right)_n} \left(2H_{n-1} - \sum_{k=0}^{n-1} \frac{1}{k + \frac{1}{2} + \frac{1}{2}b}\right) \\ &= \frac{1}{4}\left\{\psi\left(\frac{1}{2} + \frac{1}{2}b\right) - \psi\left(\frac{1}{2}\right)\right\}^2 + \frac{1}{4}\left\{\zeta\left(2, \frac{1}{2} + \frac{1}{2}b\right) - \zeta\left(2, \frac{1}{2}\right)\right\}. \end{aligned} \tag{60}$$

If we consider a as a variable in Dixon's theorem (4), similarly as above, we obtain

$$\sum_{n=1}^{\infty} \frac{(b)_n (c)_n}{n (1-b)_n (1-c)_n} = \frac{1}{2}\{\gamma + \psi(1-b) + \psi(1-c) - \psi(1-b-c)\}, \tag{61}$$

which is an immediate consequence of the special case $a = 0$ of a known result recorded by Srivastava [1079, p. 82, Eq. (3.2)]:

$$\sum_{n=1}^{\infty} \frac{(b)_n (c)_n}{n(1-b)_n(1-c)_n} \left[\sum_{k=1}^{n} \left(\frac{1}{k} - \frac{1}{k-b} - \frac{1}{k-c} \right) - \frac{1}{n} \right]$$
$$= \frac{1}{8} [\gamma + \psi(1-b) + \psi(1-c) - \psi(1-b-c)]^2$$
$$+ 38 [\zeta(2, 1-b) + \zeta(2, 1-c) - \zeta(2) - \zeta(2, 1-b-c)]; \tag{62}$$

$$\sum_{n=1}^{\infty} \frac{(b)_n (c)_n}{n(1-b)_n(1-c)_n} \left[\left\{ \sum_{k=1}^{n} \left(\frac{1}{k-b} + \frac{1}{k-c} - \frac{1}{k} \right) + \frac{1}{n} \right\}^2 \right.$$
$$\left. + \sum_{k=1}^{n} \left\{ \frac{1}{(k-b)^2} + \frac{1}{(k-c)^2} - \frac{1}{k^2} \right\} + \frac{1}{n^2} \right]$$
$$= \frac{1}{24} [\gamma + \psi(1-b) + \psi(1-c) - \psi(1-b-c)]$$
$$\cdot [\{\gamma + \psi(1-b) + \psi(1-c) - \psi(1-b-c)\}^2$$
$$+ 3\{\zeta(2, 1-b) + \zeta(2, 1-c) - \zeta(2) - \zeta(2, 1-b-c)\}]$$
$$+ \frac{7}{12} [\zeta(3) - \zeta(3, 1-b) - \zeta(3, 1-c) + \zeta(3, 1-b-c)]. \tag{63}$$

Further Summation Formulas Related to Generalized Harmonic Numbers

As shown in preceding subsections, by employing the univariate series expansion of classical hypergeometric formulas, Shen [1024] and Choi and Srivastava [290, 298] investigated the evaluation of infinite series related to generalized harmonic numbers. More summation formulas have been systematically derived by Chu [315] and Chu and de Donno [316], who developed fully this approach to the multivariate case. For example, Chu [315] investigated the hypergeometric summation theorem due to Gauss:

$${}_2F_1 \left[\begin{matrix} x, \; y \\ 1-z \end{matrix} \middle| \, 1 \right] = \frac{\Gamma(1-z)\Gamma(1-x-y-z)}{\Gamma(1-x-z)\Gamma(1-y-z)}, \tag{64}$$

as well as Kummer's and Dixon's summation theorems. The hypergeometric series of (64) may be expanded in terms of the partial sums of the Riemann Zeta function through the symmetric functions generated by the following finite products (*cf.* [793]):

$$\prod_{k=1}^{n} \left(1 + \frac{x}{k} \right) = 1 + xH_n + \frac{x^2}{2} \left(H_n^2 - H_n^{(2)} \right) + \frac{x^3}{6} \left(H_n^3 - 3H_n H_n^{(2)} + 2H_n^{(3)} \right) + \cdots, \tag{65}$$

$$\prod_{k=1}^{n}\left(1-\frac{x}{k}\right)^{-1}=1+xH_n+\frac{x^2}{2}\left(H_n^2+H_n^{(2)}\right) \\ +\frac{x^3}{6}\left(H_n^3+3\,H_n\,H_n^{(2)}+2\,H_n^{(3)}\right)+\cdots, \tag{66}$$

$$\prod_{k=1}^{n}\left(1+\frac{y}{2k-1}\right)=1+yO_n+\frac{y^2}{2}\left(O_n^2-O_n^{(2)}\right) \\ +\frac{y^3}{6}\left(O_n^3-3\,O_n\,O_n^{(2)}+2\,O_n^{(3)}\right)+\cdots, \tag{67}$$

$$\prod_{k=1}^{n}\left(1-\frac{y}{2k-1}\right)^{-1}=1+yO_n+\frac{y^2}{2}\left(O_n^2+O_n^{(2)}\right) \\ +\frac{y^3}{6}\left(O_n^3+3\,O_n\,O_n^{(2)}+2\,O_n^{(3)}\right)+\cdots, \tag{68}$$

where O_n's are defined by

$$O_n=\sum_{k=1}^{n}\frac{1}{2k-1} \quad \text{and} \quad O_n^{(r)}=\sum_{k=1}^{n}\frac{1}{(2k-1)^r} \quad (r\in\mathbb{N}\setminus\{1\}). \tag{69}$$

Conversely, the right-hand side of (64) may be expanded as multivariate formal power series, by means of the following expansions:

$$\Gamma(1-z)=\exp\left(\sum_{k=1}^{\infty}\frac{\sigma_k}{k}z^k\right) \quad [\sigma_1=\gamma;\ \sigma_k=\zeta(k)\ (k\in\mathbb{N}\setminus\{1\})] \tag{70}$$

and

$$\Gamma\left(\frac{1}{2}-z\right)=\sqrt{\pi}\exp\left(\sum_{k=1}^{\infty}\frac{\tau_k}{k}z^k\right) \\ [\tau_1=\gamma+2\log 2;\quad \tau_k=(2^k-1)\,\zeta(k)\ (k\in\mathbb{N}\setminus\{1\})]. \tag{71}$$

Then, term-by-term comparison of the coefficients from two power series may yield an infinite number of summation formulas.

Here, we choose to record some summation formulas involving harmonic numbers and other related numbers presented very recently by Zheng [1264], who, motivated by the work of Chu [315], established further numerous summation formulas expressing infinite series related to generalized harmonic numbers in terms of the Riemann

Zeta function $\zeta(m)$ with $m = 5, 6, 7$.

$$\sum_{n=1}^{\infty} \frac{H_{n-1}^3 - 3H_{n-1}H_{n-1}^{(2)} + 2H_{n-1}^{(3)}}{n^2} = 6\,\zeta(5). \tag{72}$$

$$\sum_{n=1}^{\infty} \left(\frac{8H_n^2 + 2H_{n-1}^{(2)}}{n^3} - \frac{H_{n-1} + H_n}{n^4} \right) = 12\,\zeta(5). \tag{73}$$

$$\sum_{n=1}^{\infty} \left(\frac{H_n^{(2)}}{n^3} + \frac{3H_n}{n^4} \right) = \frac{9}{2}\,\zeta(5). \tag{74}$$

$$\sum_{n=1}^{\infty} \frac{\left\{ (H_{n-1} + H_n)^2 - H_{n-1}^{(2)} + H_n^{(2)} \right\} O_n}{n^2} = 31\,\zeta(5). \tag{75}$$

$$\sum_{n=1}^{\infty} \frac{(H_{n-1} + H_n)^2\, O_n}{n^2} = \frac{7\pi^2}{12}\,\zeta(3) + \frac{93}{4}\,\zeta(5). \tag{76}$$

$$\sum_{n=1}^{\infty} \frac{(H_{n-1} + H_n)\, O_n^2}{n^2} = \frac{93}{8}\,\zeta(5). \tag{77}$$

$$\sum_{n=1}^{\infty} \frac{2O_n^3 + O_n^{(3)}}{n^2} = \frac{93}{8}\,\zeta(5). \tag{78}$$

$$\sum_{n=1}^{\infty} \frac{O_n}{n^4} = -\frac{7\pi^2}{12}\,\zeta(3) + \frac{31}{4}\,\zeta(5). \tag{79}$$

$$\sum_{n=1}^{\infty} \frac{\left(H_{n-1}^2 + H_n^2\right) O_n}{n^2} = \frac{31}{2}\,\zeta(5). \tag{80}$$

$$\sum_{n=1}^{\infty} \frac{H_{n-1} H_n O_n}{n^2} = \frac{7\pi^2}{24}\,\zeta(3) + \frac{31}{8}\,\zeta(5). \tag{81}$$

$$\sum_{n=1}^{\infty} \frac{(O_{n-1} + O_n)^3}{(2n-1)^2} = \frac{\pi^4}{16}\log 2 + \pi^2 \log^2 2 + \frac{7\pi^2}{32}\,\zeta(3). \tag{82}$$

$$\sum_{n=1}^{\infty} \frac{O_{n-1} + O_n}{(2n-1)^4} = \frac{\pi^4}{48}\log 2 - \frac{\pi^2}{32}\,\zeta(3). \tag{83}$$

$$\sum_{n=1}^{\infty} \frac{O_{n-1}^{(3)} + O_n^{(3)}}{(2n-1)^2} = \frac{\pi^2}{8}\,\zeta(3). \tag{84}$$

$$\sum_{n=1}^{\infty} \frac{(O_{n-1}+O_n)\left\{(O_{n-1}+O_n)^2 - O_{n-1}^{(2)} + O_n^{(2)}\right\}}{(2n-1)^2} = \frac{\pi^4}{12}\log 2 + \pi^2 \log^3 2 + \frac{3\pi^2}{16}\zeta(3). \tag{85}$$

$$\sum_{n=1}^{\infty} \frac{(O_{n-1}+O_n)\left\{(O_{n-1}+O_n)^2 - 3\left(O_{n-1}^{(2)} - O_n^{(2)}\right)\right\} + 2\left(O_{n-1}^{(3)} + O_n^{(2)}\right)}{(2n-1)^2} = \frac{\pi^4}{8}\log 2 + \pi^2 \log^3 2 + \frac{3\pi^2}{8}\zeta(3). \tag{86}$$

$$\sum_{n=1}^{\infty} \frac{(O_{n-1}+O_n)^4}{(2n-1)^2} = \frac{\pi^6}{90} + \frac{\pi^4}{4}\log^2 2 + 2\pi^2 \log^4 2 + \frac{7\pi^2}{4}\log 2\zeta(3). \tag{87}$$

3.6 Other Methods and their Applications

The Weierstrass Canonical Product Form for the Gamma Function

Starting from the Weierstrass canonical product formula 1.1(2), Zhang and Williams [1250] proved the following well-known identity (*cf.* Hansen [531, p. 46, Entry (54.3.4)]):

$$\sum_{k=1}^{\infty} \zeta(2k+1)\, t^{2k+1} = -\frac{1}{2}[1 + \pi t \cot(\pi t)] - t[\psi(t) + \gamma] \quad (|t| < 1), \tag{1}$$

which may be compared with 3.4(16).

Now, for the generalized Euler constant $\gamma(r,n)$, defined by

$$\gamma(r,n) := \lim_{m\to\infty} \left\{ \sum_{\substack{\ell=1 \\ \ell\equiv r (\mathrm{mod}\, n)}}^{m} \left(\frac{1}{\ell} - \frac{1}{n}\log m\right) \right\} \quad (r \in \mathbb{Z};\ n \in \mathbb{N}), \tag{2}$$

so that

$$\gamma(0,1) = \gamma, \tag{3}$$

Lehmer [741, Theorem 7] showed that

$$\gamma(r,n) = -\frac{1}{n}\left[\psi\left(\frac{r}{n}\right) + \log n\right] \quad (0 < r \leqq n;\ r, n \in \mathbb{N}). \tag{4}$$

A combination of (1) and (4) yields

$$\sum_{k=1}^{\infty} \zeta(2k+1)\left(\frac{r}{n}\right)^{2k} = \log n - \frac{\pi}{2}\cot\left(\frac{r\pi}{n}\right) + n\gamma(r,n) - \gamma - \frac{n}{2r} \tag{5}$$
$$(0 < r \leq n;\ r, n \in \mathbb{N}),$$

which was given by Zhang and Williams [1250, Corollary 4], who also considered its many special cases, included in Section 3.4, by evaluating $\gamma(r,n)$.

Next, we rewrite (1) as

$$\sum_{k=1}^{\infty} \zeta(2k+1)\, t^{2k} = \frac{1}{2t} - \frac{\pi}{2}\cot(\pi t) - \psi(t+1) - \gamma \quad (|t| < 1), \tag{6}$$

which, upon integrating and differentiating, yields the well-known result (see Hansen [531, p. 356, Entry (54.5.8)]; Zhang and Williams [1250, Theorems 5 and 6]):

$$\sum_{k=1}^{\infty} \frac{\zeta(2k+1)}{2k+1} t^{2k+1} = \frac{1}{2}\log\left(\frac{\pi t}{\sin \pi t}\right) - \log\Gamma(1+t) - \gamma t \quad (0 < |t| < 1); \tag{7}$$

$$\sum_{k=1}^{\infty} 2k\,\zeta(2k+1)\, t^{2k-1} = \frac{1}{2}\left(\frac{\pi^2 t}{\sin^2 \pi t} + \frac{1}{t^2}\right) - \psi'(t) \quad (0 < |t| < 1); \tag{8}$$

$$\begin{aligned}\sum_{k=1}^{\infty} &(2k+1)\,\zeta(2k+1)\, t^{2k} \\ &= \frac{1}{2}\left[\frac{\pi^2 t}{\sin^2 \pi t} - \pi\cot(\pi t)\right] - \psi(t) - \gamma - \psi'(t)\, t \quad (0 < |t| < 1).\end{aligned} \tag{9}$$

The special cases of (8) and (9) when $t = \frac{1}{2}$ yield

$$\sum_{k=1}^{\infty} \frac{2k\,\zeta(2k+1)}{2^{2k}} = 1; \tag{10}$$

$$\sum_{k=1}^{\infty} \frac{(2k+1)\,\zeta(2k+1)}{2^{2k+1}} = \log 2. \tag{11}$$

By replacing t in 3.4(18) by it, we obtain

$$\sum_{k=1}^{\infty} (-1)^{k-1}\,\zeta(2k)\, t^{2k-1} = \frac{\pi}{2}\coth(\pi t) - \frac{1}{2t} \quad (0 < |t| < 1); \tag{12}$$

$$\sum_{k=1}^{\infty} (-1)^{k-1}\{\zeta(2k) - 1\}\, t^{2k-1} = \frac{\pi}{2}\coth(\pi t) - \frac{3t^2+1}{2t\left(t^2+1\right)} \quad (0 < |t| \leq 1). \tag{13}$$

The special cases of (12) and (13) when $t = \frac{1}{2}$ and $t = 1$ give us the following sums:

$$\sum_{k=1}^{\infty} \frac{(-1)^{k-1}\,\zeta(2k)}{2^{2k}} = \frac{\pi}{4}\coth\left(\frac{\pi}{2}\right) - \frac{1}{2}; \tag{14}$$

$$\sum_{k=1}^{\infty} (-1)^{k-1}\,\{\zeta(2k) - 1\}\,t^{2k-1} = \frac{\pi}{2}\coth\pi - 1. \tag{15}$$

By integrating (12) (or, alternatively, replacing t in 3.4(17) by it), we find that

$$\sum_{k=1}^{\infty} \frac{(-1)^{k-1}\,\zeta(2k)}{k}\,t^{2k} = \log\left(\frac{\sinh \pi t}{\pi t}\right) \qquad (0 < |t| \leq 1), \tag{16}$$

which, for $t = 1$ and $t = \frac{1}{2}$, yields

$$\sum_{k=1}^{\infty} \frac{(-1)^{k-1}\,\zeta(2k)}{k} = \log\left(\frac{\sinh \pi}{\pi}\right); \tag{17}$$

$$\sum_{k=1}^{\infty} \frac{(-1)^{k-1}\,\zeta(2k)}{k \cdot 2^{2k}} = \log\left(\frac{2\sinh \frac{\pi}{2}}{\pi}\right). \tag{18}$$

Evaluation by Using Infinite Products

A remarkable infinite product formula was recently posed as a problem by Wilf [1228]. Subsequently, Choi and Seo [282] proved Wilf's formula, as well as three other analogous product formulas. Choi *et al.* [275], as a sequel to these earlier works, presented several general infinite product formulas, which include, as their special cases, the aforementioned product formulas of Wilf [1228] and Choi and Seo [282]. Some other related results, series involving the Riemann Zeta function, are also considered briefly.

In 1997, Wilf [1228] posed as a problem the following elegant infinite product formula, which contains some of the most important mathematical constants, such as π, e and the Euler-Mascheroni constant γ, defined by 1.1(3):

$$\prod_{j=1}^{\infty} \left\{ e^{-\frac{1}{j}} \left(1 + \frac{1}{j} + \frac{1}{2j^2}\right) \right\} = \frac{e^{\frac{\pi}{2}} + e^{-\frac{\pi}{2}}}{\pi\, e^{\gamma}}. \tag{19}$$

Subsequently, Choi and Seo [282] proved (19), as well as three other similar product formulas, by making use of well-known infinite product formulas for circular and hyperbolic functions and the familiar Stirling formula 1.1(32).

Choi *et al.* [275] presented the following two general infinite product formulas, which include Wilf's formula (19) and other similar formulas in [282] as their special cases:

$$\prod_{j=1}^{\infty}\left\{e^{-\frac{1}{j}}\left(1+\frac{1}{j}+\frac{\alpha^2+\frac{1}{4}}{j^2}\right)\right\}=\frac{2\left(e^{\alpha\pi}+e^{-\alpha\pi}\right)}{\left(4\alpha^2+1\right)\pi\, e^{\gamma}}$$

$$\left(\alpha\in\mathbb{C};\ \alpha\neq\pm\frac{1}{2}i;\ i=\sqrt{-1}\right) \tag{20}$$

and

$$\prod_{j=1}^{\infty}\left\{e^{-\frac{2}{j}}\left(1+\frac{2}{j}+\frac{\beta^2+1}{j^2}\right)\right\}=\frac{e^{\beta\pi}-e^{-\beta\pi}}{2\beta\left(\beta^2+1\right)\pi\, e^{2\gamma}} \quad (\beta\in\mathbb{C}\setminus\{0\};\ \beta\neq\pm i). \tag{21}$$

Indeed, we begin by letting

$$\mathcal{A}(p,q):=\prod_{j=1}^{\infty}\left\{e^{-\frac{p}{j}}\left(1+\frac{p}{j}+\frac{q}{j^2}\right)\right\} \quad (p,q\in\mathbb{C};\ \Re(p)>0) \tag{22}$$

and take logarithms on both sides of (22). Then, using the Euler-Mascheroni constant γ given in 1.1(3), we obtain

$$\log\mathcal{A}(p,q)=-p\gamma+\lim_{n\to\infty}\left[-p\log n+\sum_{j=1}^{n}\log\left(j^2+pj+q\right)-2\sum_{j=1}^{n}\log j\right],$$

which, upon applying Stirling's formula 1.1(52) to the last summation, yields

$$\begin{aligned}\log\mathcal{A}(p,q)&=-p\gamma-\log(2\pi)\\&\quad+\lim_{n\to\infty}\left[-(2n+p+1)\log n+2n+\sum_{j=1}^{n}\log\left(j^2+pj+q\right)\right].\end{aligned} \tag{23}$$

Now, we recall the infinite product formula of $\cosh z$:

$$\cosh z=\prod_{j=1}^{\infty}\left\{\left(1+\frac{4z^2}{(2j-1)^2\pi^2}\right)\right\}=\frac{e^z+e^{-z}}{2}. \tag{24}$$

If we set $z=\alpha\pi$ in (24), take logarithms on both sides of the resulting equation and use the following version of Stirling's formula 1.1(32):

$$\sum_{j=1}^{n}\log(2j-1)\sim\left(n+\frac{1}{2}\right)\log 2-n+n\log n \quad (n\to\infty),$$

we find that

$$\log\left(e^{\alpha\pi}+e^{-\alpha\pi}\right)=\lim_{n\to\infty}\left[\sum_{j=1}^{n}\log\left\{(2j-1)^2+4\alpha^2\right\}+2n-2n\log(2n)\right]. \tag{25}$$

Combining (23) and (25), we obtain

$$\begin{aligned}\log\mathcal{A}(p,q)=-p\gamma-\log(2\pi)+\log\left(e^{\alpha\pi}+e^{-\alpha\pi}\right)\\ +\lim_{n\to\infty}\left[-(p+1)\log n+\sum_{j=1}^{n}\log\left(4j^2+4pj+4q\right)\right.\\ \left.-\sum_{j=1}^{n}\log\left\{(2j-1)^2+4\alpha^2\right\}\right].\end{aligned} \tag{26}$$

In a similar manner, if we consider the infinite product formula of $\sinh z$, instead of $\cosh z$ in (24), we also have

$$\begin{aligned}\log\mathcal{A}(p,q)=-p\gamma-\log(2\beta\pi)+\log\left(e^{\beta\pi}-e^{-\beta\pi}\right)\\ +\lim_{n\to\infty}\left[-p\log n+\sum_{j=1}^{n}\log\left(j^2+pj+q\right)-\sum_{j=1}^{n}\log\left(j^2+\beta^2\right)\right].\end{aligned} \tag{27}$$

Finally, by setting

$$(p,q)=\left(1,\alpha^2+14\right)\quad\text{and}\quad(p,q)=\left(2,\beta^2+1\right)$$

in (26) and (27), respectively, we arrive at the desired formulas (20) and (21).

We note that the special case of (20) when $\alpha=12$ leads to Wilf's formula (19). Other interesting special cases of (20) and (21) include all of the aforementioned results of Choi and Seo [282].

Taking logarithms on both sides of (20) and (21) and using the Riemann Zeta function $\zeta(s)$, defined by 2.3(1), together with the Maclaurin expansion of $\log(1+x)$, we obtain the following interesting closed-form evaluation of two families of series involving the Riemann Zeta function:

$$\begin{aligned}\sum_{n=2}^{\infty}\frac{(-1)^{n+1}}{n}\sum_{\ell=0}^{n}\binom{n}{\ell}\left(\alpha^2+\frac{1}{4}\right)^{\ell}\left\{\zeta(n+\ell)-\sum_{k=1}^{j-1}\frac{1}{k^{n+\ell}}\right\}\\ =\sum_{k=1}^{j-1}\left[\frac{1}{k}+\left(\alpha^2+\frac{1}{4}\right)\frac{1}{k^2}-\log\left(1+\frac{1}{k}+\frac{\alpha^2+\frac{1}{4}}{k^2}\right)\right]\end{aligned} \tag{28}$$

$$
-\left(\alpha^2+\frac{1}{4}\right)\zeta(2)+\log\left(\frac{2\left(e^{\alpha\pi}+e^{-\alpha\pi}\right)}{\left(4\alpha^2+1\right)\pi\, e^{\gamma}}\right)
$$

$$
\left(j-1\in\mathbb{N};\ \alpha^2<j^2-j-\frac{1}{4};\ \alpha\in\mathbb{R}\right);
$$

$$
\sum_{n=2}^{\infty}(-1)^{n+1}\frac{2^n}{n}\sum_{\ell=0}^{n}\binom{n}{\ell}\left(\frac{\beta^2+1}{2}\right)^{\ell}\left\{\zeta(n+\ell)-\sum_{k=1}^{j-1}\frac{1}{k^{n+\ell}}\right\}
$$

$$
=\sum_{k=1}^{j-1}\left[\frac{2}{k}+\frac{\beta^2+1}{k^2}-\log\left(1+\frac{2}{k}+\frac{\beta^2+1}{k^2}\right)\right] \tag{29}
$$

$$
-\left(\beta^2+1\right)\zeta(2)+\log\left(\frac{e^{\beta\pi}-e^{-\beta\pi}}{2\beta\left(\beta^2+1\right)\pi\, e^{2\gamma}}\right)
$$

$$
\left(j-2\in\mathbb{N};\ \beta^2<j^2-2j-1;\ \beta\in\mathbb{R}\right).
$$

The summation formulas (28) and (29), which we have presented here, are obviously different from the various classes of series involving Zeta functions, which we considered elsewhere in this chapter.

Higher-Order Derivatives of the Gamma Function

Here, we present explicit formulas for the evaluation of higher-order derivatives of the familiar Gamma function and consider several applications of these explicit formulas, including evaluation of some families of definite integrals (see Choi and Srivastava [293]).

It follows from 2.3(19) that

$$
\begin{aligned}
&\{\zeta(2)\}^2=\frac{5}{2}\zeta(4),\quad \zeta(2)\zeta(4)=\frac{7}{6}\zeta(6),\\
&\zeta(2)\zeta(6)-\frac{5}{3}\zeta(8),\quad \{\zeta(4)\}^2-\frac{7}{6}\zeta(8),\\
&\zeta(2)\zeta(8)=\frac{33}{20}\zeta(10),\quad \zeta(4)\zeta(6)=\frac{11}{10}\zeta(10).
\end{aligned} \tag{30}
$$

In addition to the Maclaurin series expansion 3.4(10) for $\log\Gamma(1+z)$, it is also known that (*cf.*, e.g., Luke [779, Vol. I, p. 27])

$$
\Gamma(1+z)=\sum_{n=0}^{\infty}a_n z^n \quad (|z|<1), \tag{31}
$$

where

$$
a_0=1 \quad \text{and} \quad n\,a_n=-\gamma a_{n-1}+\sum_{k=2}^{n}(-1)^k a_{n-k}\,\zeta(k). \tag{32}
$$

We, first, evaluate some higher-order derivatives of the Gamma function $\Gamma(z)$ at $z=1$ and $z=\frac{1}{2}$. Upon replacing z by $z-1$ in (31), we readily obtain

$$a_n = \frac{\Gamma^{(n)}(1)}{n!} \quad (n \in \mathbb{N}_0). \tag{33}$$

Formula (32) and (33), together, immediately yield the recursion formula:

$$\Gamma^{(n+1)}(1) = -\gamma\,\Gamma^{(n)}(1) + n! \sum_{k=1}^{n} \frac{(-1)^{k+1}}{(n-k)!} \zeta(k+1)\,\Gamma^{(n-k)}(1) \quad (n \in \mathbb{N}_0), \tag{34}$$

which can alternatively be derived directly from the definition 1.3(52) with $n=1$ (and z replaced by $z+1$) by applying such results as the relationship 1.3(53) with $z=1$.

Similarly, we obtain another recursion formula:

$$\begin{aligned}\Gamma^{(n+1)}\left(\frac{1}{2}\right) = -\delta\,\Gamma^{(n)}\left(\frac{1}{2}\right)& \\ + n! \sum_{k=1}^{n} \frac{(-1)^{k+1}}{(n-k)!} &\left(2^{k+1}-1\right) \zeta(k+1)\,\Gamma^{(n-k)}\left(\frac{1}{2}\right) \\ &(n \in \mathbb{N}_0;\ \delta := \gamma + 2\log 2).\end{aligned} \tag{35}$$

Now, using the formulas in earlier sections, we compute $\Gamma^{(n)}(1)$ explicitly for the first ten derivatives:

$$\begin{aligned}\Gamma'(1) &= -\gamma = \psi(1);\\ \Gamma^{(2)}(1) &= \gamma^2 + \zeta(2);\\ \Gamma^{(3)}(1) &= -\gamma^3 - 3\gamma\zeta(2) - 2\zeta(3),\end{aligned}$$

which is the *corrected* version of a formula given by Campbell [208, p. 25]:

$$\begin{aligned}\Gamma^{(4)}(1) &= \gamma^4 + 6\gamma^2\zeta(2) + 8\gamma\zeta(3) + \frac{27}{2}\zeta(4);\\ \Gamma^{(5)}(1) &= -\gamma^5 - 10\gamma^3\zeta(2) - 20\gamma^2\zeta(3)\\ &\quad - \frac{135}{2}\gamma\zeta(4) - 20\zeta(2)\zeta(3) - 24\zeta(5);\\ \Gamma^{(6)}(1) &= \gamma^6 + 15\gamma^4\zeta(2) + 40\gamma^3\zeta(3) + \frac{405}{2}\gamma^2\zeta(4)\\ &\quad + 120\gamma\,\zeta(2)\zeta(3) + 144\gamma\,\zeta(5) + 40\{\zeta(3)\}^2 + \frac{2745}{8}\zeta(6);\\ \Gamma^{(7)}(1) &= -\gamma^7 - 21\gamma^5\zeta(2) - 70\gamma^4\zeta(3) - \frac{945}{2}\gamma^3\zeta(4)\\ &\quad - 420\gamma^2\zeta(2)\zeta(3) - 504\gamma^2\zeta(5) - 280\gamma\{\zeta(3)\}^2 - \frac{19215}{8}\gamma\zeta(6)\\ &\quad - 945\zeta(3)\zeta(4) - 504\zeta(2)\zeta(5) - 720\zeta(7);\end{aligned}$$

$$\begin{aligned}\Gamma^{(8)}(1) &= \gamma^8 + 28\gamma^6\zeta(2) + 112\gamma^5\zeta(3) + 945\gamma^4\zeta(4) \\ &+ 1120\gamma^3\zeta(2)\zeta(3) + 1344\gamma^3\zeta(5) + 1120\gamma^2\{\zeta(3)\}^2 \\ &+ \frac{19215}{2}\gamma^2\zeta(6) + 7560\gamma\zeta(3)\zeta(4) + 4032\gamma\zeta(2)\zeta(5) + 5760\gamma\zeta(7) \\ &+ 1120\zeta(2)\{\zeta(3)\}^2 + 2688\zeta(3)\zeta(5) + \frac{132405}{8}\zeta(8);\end{aligned}$$

$$\begin{aligned}\Gamma^{(9)}(1) &= -\gamma^9 - 36\gamma^7\zeta(2) - 168\gamma^6\zeta(3) - 1701\gamma^5\zeta(4) - 2520\gamma^4\zeta(2)\zeta(3) \\ &- 3024\gamma^4\zeta(5) - 3360\gamma^3\{\zeta(3)\}^2 - \frac{57645}{2}\gamma^3\zeta(6) - 34020\gamma^2\zeta(3)\zeta(4) \\ &- 18144\gamma^2\zeta(2)\zeta(5) - 25920\gamma^2\zeta(7) - 10080\gamma^2\zeta(2)\{\zeta(3)\}^2 \\ &- 24192\gamma\zeta(3)\zeta(5) - \frac{1191645}{8}\gamma\zeta(8) - 57645\zeta(3)\zeta(6) \\ &- 40824\zeta(4)\zeta(5) - 25920\zeta(2)\zeta(7) - 2240\{\zeta(3)\}^3 - 40320\zeta(9);\end{aligned}$$

$$\begin{aligned}\Gamma^{(10)}(1) &= \gamma^{10} + 45\gamma^8\zeta(2) + 240\gamma^7\zeta(3) + 2835\gamma^6\zeta(4) + 5040\gamma^5\zeta(2)\zeta(3) \\ &+ 6048\gamma^5\zeta(5) + 8400\gamma^4\{\zeta(3)\}^2 + \frac{288225}{4}\gamma^4\zeta(6) \\ &+ 113400\gamma^3\zeta(3)\zeta(4) + 60480\gamma^3\zeta(2)\zeta(5) + 86400\gamma^3\zeta(7) \\ &+ 50400\gamma^2\zeta(2)\{\zeta(3)\}^2 + 120952\gamma^2\zeta(3)\zeta(5) + \frac{5958185}{8}\gamma^2\zeta(8) \\ &+ 576450\gamma\zeta(3)\zeta(6) + 408240\gamma\zeta(4)\zeta(5) + 259200\gamma\zeta(2)\zeta(7) \\ &+ 22400\gamma\{\zeta(3)\}^3 + 403200\gamma\zeta(9) + 113400\{\zeta(3)\}^2\zeta(4) \\ &+ 120960\zeta(2)\zeta(3)\zeta(5) + 72576\{\zeta(5)\}^2 + 172800\zeta(3)\zeta(7) \\ &+ \frac{42329385}{32}\zeta(10).\end{aligned}$$

The corresponding problem for $\Gamma^{(n)}\left(\frac{1}{2}\right)$ $(n = 1, \ldots, 10)$ yields

$$\Gamma'\left(\frac{1}{2}\right) = -\delta\sqrt{\pi},$$

$$\frac{1}{\sqrt{\pi}}\Gamma^{(2)}\left(\frac{1}{2}\right) = \delta^2 + 3\zeta(2),$$

$$\frac{1}{\sqrt{\pi}}\Gamma^{(3)}\left(\frac{1}{2}\right) = -\delta^3 - 9\delta\zeta(2) - 14\zeta(3),$$

all three of which are also recorded by Campbell [208, p. 25];

$$\frac{1}{\sqrt{\pi}}\Gamma^{(4)}\left(\frac{1}{2}\right) = \delta^4 + 18\delta^2\zeta(2) + 56\delta\zeta(3) + \frac{315}{2}\zeta(4);$$

$$\frac{1}{\sqrt{\pi}}\Gamma^{(5)}\left(\frac{1}{2}\right) = -\delta^5 - 30\delta^3\zeta(2) - 140\delta^2\zeta(3) - \frac{1575}{2}\delta\zeta(4) - 420\zeta(2)\zeta(3) - 744\zeta(5);$$

$$\frac{1}{\sqrt{\pi}}\Gamma^{(6)}\left(\frac{1}{2}\right) = \delta^6 + 45\delta^4\zeta(2) + 280\delta^3\zeta(3) + \frac{4725}{2}\delta^2\zeta(4) + 2520\delta\zeta(2)\zeta(3) + 4464\delta\zeta(5) + 1960\{\zeta(3)\}^2 + \frac{131355}{8}\zeta(6);$$

$$\frac{1}{\sqrt{\pi}}\Gamma^{(7)}\left(\frac{1}{2}\right) = -\delta^7 - 63\delta^5\zeta(2) - 490\delta^4\zeta(3) - \frac{11025}{2}\delta^3\zeta(4) - 8820\delta^2\zeta(2)\zeta(3) - 15624\delta^2\zeta(5) - 13720\delta\{\zeta(3)\}^2 - \frac{919485}{8}\delta\zeta(6) - 77175\zeta(3)\zeta(4) - 46872\zeta(2)\zeta(5) - 91440\zeta(7);$$

$$\frac{1}{\sqrt{\pi}}\Gamma^{(8)}\left(\frac{1}{2}\right) = \delta^8 + 84\delta^6\zeta(2) + 784\delta^5\zeta(3) + 11025\delta^4\zeta(4) + 23520\delta^3\zeta(2)\zeta(3) + 41664\delta^3\zeta(5) + 54880\delta^2\{\zeta(3)\}^2 + \frac{919485}{2}\delta^2\zeta(6) + 617400\delta\zeta(3)\zeta(4) + 374976\delta\zeta(2)\zeta(5) + 731520\delta\zeta(7) + 164640\zeta(2)\{\zeta(3)\}^2 + 583296\zeta(3)\zeta(5) + \frac{25859925}{8}\zeta(8);$$

$$\frac{1}{\sqrt{\pi}}\Gamma^{(9)}\left(\frac{1}{2}\right) = -\delta^9 - 108\,\delta^7\,\zeta(2) - 1176\,\delta^6\,\zeta(3) - 19845\,\delta^5\,\zeta(4) - 52920\,\delta^4\,\zeta(2)\zeta(3) - 93744\delta^4\zeta(5) - 164640\,\delta^3\{\zeta(3)\}^2 - \frac{2758455}{2}\delta^3\,\zeta(6) - 2778300\,\delta^2\,\zeta(3)\zeta(4) - 1687392\,\delta^2\,\zeta(2)\zeta(5) - 3291840\,\delta^2\,\zeta(7) - 1481760\,\delta\,\zeta(2)\,\{\zeta(3)\}^2 - 5249664\,\delta\,\zeta(3)\zeta(5) - \frac{232739325}{8}\,\delta\,\zeta(8) - 17688510\,\zeta(3)\zeta(6) - 14764680\,\zeta(4)\zeta(5) - 768320\,\{\zeta(3)\}^3 - 9875520\,\zeta(2)\zeta(7) - 20603520\,\zeta(9);$$

$$\frac{1}{\sqrt{\pi}}\Gamma^{(10)}\left(\frac{1}{2}\right) = \delta^{10} + 135\,\delta^8\,\zeta(2) + 1680\,\delta^7\,\zeta(3) + 33075\,\delta^6\,\zeta(4) + 105840\delta^5\zeta(2)\zeta(3) + 187488\delta^5\zeta(5) + 411600\delta^4\{\zeta(3)\}^2 + \frac{13792275}{4}\delta^4\,\zeta(6) + 9261000\,\delta^3\,\zeta(3)\,\zeta(4) + 5624640\,\delta^3\,\zeta(2)\zeta(5) + 10972800\,\delta^3\,\zeta(7) + 7408800\,\delta^2\,\zeta(2)\,\{\zeta(3)\}^2 + 26248320\,\delta^2\,\zeta(3)\zeta(5) + \frac{1163696625}{8}\,\delta^2\,\zeta(8) + 191471175\,\delta\,\zeta(3)\zeta(6) + 147646800\,\delta\,\zeta(4)\zeta(5) + 7683200\,\delta\,\{\zeta(3)\}^3$$

$$+ 98755200\,\delta\,\zeta(2)\zeta(7) + 206035200\,\delta\,\zeta(9) + 64827000\,\{\zeta(3)\}^2\,\zeta(4)$$
$$+ 78744960\,\zeta(2)\zeta(3)\zeta(5) + 153619200\,\zeta(3)\zeta(7)$$
$$+ 69745536\,\{\zeta(5)\}^2 + \frac{33466588155}{32}\,\zeta(10),$$

δ being defined already with (35).

For applications of our evaluations for $\Gamma^{(n)}(1)$ and $\Gamma^{(n)}\left(\frac{1}{2}\right)$, by replacing z by $z+1$ in 1.1(1), differentiating both sides of the resulting equation n times and then setting $z=0$, we obtain

$$\Gamma^{(n)}(1) = \int_0^\infty e^{-t}\,(\log t)^n\,dt \quad (n \in \mathbb{N}_0). \tag{36}$$

Similarly, we have

$$\Gamma^{(n)}\left(\frac{1}{2}\right) = \int_0^\infty \frac{e^{-t}}{\sqrt{t}}\,(\log t)^n\,dt \quad (n \in \mathbb{N}_0). \tag{37}$$

Both (36) and (37) are recorded by Campbell [208, p. 25].

Making use of 1.1(8), instead of 1.1(1) (or, alternatively, by setting $t = \log(1/\tau)$ in (36) and (37)), we obtain

$$\Gamma^{(n)}(1) = \int_0^1 \left[\log\left(\log\frac{1}{t}\right)\right]^n dt \quad (n \in \mathbb{N}_0) \tag{38}$$

and

$$\Gamma^{(n)}\left(\frac{1}{2}\right) = \int_0^1 \left(\log\frac{1}{t}\right)^{-\frac{1}{2}} \left[\log\left(\log\frac{1}{t}\right)\right]^n dt \quad (n \in \mathbb{N}_0). \tag{39}$$

By using the computations presented here, each of the integral formulas (36), (37), (38) and (39) can be expressed explicitly in terms of γ and $\zeta(n)$, at least for $n = 1, \ldots, 10$.

Differentiating both sides of 1.1(46) n times, by employing the Leibniz rule for differentiating the Γ-product and setting $p = \frac{1}{2}$ in the resulting equation, we obtain

$$\begin{aligned}\int_0^\infty \frac{(\log t)^n}{(1+t)\sqrt{t}}dt &:= I(n)\\ &= \sum_{k=0}^{n}(-1)^k \binom{n}{k}\,\Gamma^{(n-k)}\left(\frac{1}{2}\right)\Gamma^{(k)}\left(\frac{1}{2}\right) \quad (n \in \mathbb{N}_0).\end{aligned} \tag{40}$$

It is interesting to observe that

$$\int_0^\infty \frac{(\log t)^{2n+1}}{(1+t)\sqrt{t}}dt = 0 \quad (n \in \mathbb{N}_0), \tag{41}$$

which can also be shown as follows, by separating the sum on the right-hand side of (40) into two parts:

$$\begin{aligned} I(2n+1) &= \left(\sum_{k=0}^{n} + \sum_{k=n+1}^{2n+1}\right)(-1)^k \binom{2n+1}{k} \Gamma^{(2n+1-k)}\left(\frac{1}{2}\right)\Gamma^{(k)}\left(\frac{1}{2}\right) \\ &= \sum_{k=0}^{n}(-1)^k \binom{2n+1}{k} \Gamma^{(2n+1-k)}\left(\frac{1}{2}\right)\Gamma^{(k)}\left(\frac{1}{2}\right) \\ &\quad + \sum_{k=0}^{n}(-1)^{n+1-k} \binom{2n+1}{n-k} \Gamma^{(n-k)}\left(\frac{1}{2}\right)\Gamma^{(n+1+k)}\left(\frac{1}{2}\right), \end{aligned}$$

which, upon reversing the order of the latter sum, yields

$$\begin{aligned} I(2n+1) &= \left\{1+(-1)^{2n+1}\right\} \sum_{k=0}^{n}(-1)^k \binom{2n+1}{k} \Gamma^{(2n+1-k)}\left(\frac{1}{2}\right)\Gamma^{(k)}\left(\frac{1}{2}\right) \\ &= 0. \end{aligned} \tag{42}$$

Conversely, in the case of even integers, we have

$$\begin{aligned} I(2n) &= \int_0^\infty \frac{(\log t)^{2n}}{(1+t)\sqrt{t}}dt \\ &= 2\sum_{k=0}^{n-1}(-1)^k \binom{2n}{k} \Gamma^{(2n-k)}\left(\frac{1}{2}\right)\Gamma^{(k)}\left(\frac{1}{2}\right) \\ &\quad + (-1)^n \binom{2n}{n} \left[\Gamma^{(n)}\left(\frac{1}{2}\right)\right]^2 \quad (n \in \mathbb{N}_0), \end{aligned} \tag{43}$$

which, in view of the results already presented here, readily yields the following special cases:

$$\int_0^\infty \frac{(\log t)^2}{(1+t)\sqrt{t}}dt = \pi^3, \tag{44}$$

$$\int_0^\infty \frac{(\log t)^4}{(1+t)\sqrt{t}}dt = 5\pi^5, \tag{45}$$

$$\int_0^\infty \frac{(\log t)^6}{(1+t)\sqrt{t}}dt = 61\pi^7, \tag{46}$$

$$\int_0^\infty \frac{(\log t)^8}{(1+t)\sqrt{t}}dt = 1385\pi^9, \tag{47}$$

and

$$\int_0^\infty \frac{(\log t)^{10}}{(1+t)\sqrt{t}}dt = 50521\pi^{11}. \tag{48}$$

We conclude by remarking that the evaluations given here may find further applications involving (for example) computation and evaluation of several families of definite integrals.

3.7 Applications of Series Involving the Zeta Function

The Multiple Gamma Functions

The theory of multiple Gamma functions was applied in several earlier works to evaluate some families of series involving the Riemann Zeta function, as well as to compute the determinants of the Laplacians (see Section 3.4 and Chapter 5).

Recently, Choi *et al.* [269] addressed the converse problem and applied various (known or new) formulas for series associated with the Zeta and related functions with a view to developing the corresponding theory of multiple Gamma functions.

We begin by mainly using 1.4(86) to derive the explicit forms of $\Gamma_n(z)$ like 1.4(3) and 1.4(91). For example, we have

$$\Gamma_4(1+z) = \exp\left(d_1 z + d_2 z^2 + d_3 z^3 + d_4 z^4\right) \cdot \prod_{k=1}^{\infty}\left\{\left(1+\frac{z}{k}\right)^{-\binom{k+2}{3}} \exp\left[\binom{k+2}{3}\left(\frac{z}{k} - \frac{z^2}{2k^2} + \frac{z^3}{3k^3} - \frac{z^4}{4k^4}\right)\right]\right\}, \tag{1}$$

where

$$d_1 = \frac{7}{24} - \log A - \frac{1}{2}\log B - \frac{1}{6}\log(2\pi), \quad d_2 = -\frac{1}{144} + \frac{\gamma}{6} + \frac{1}{4}\log(2\pi) + \frac{1}{2}\log A,$$

$$d_3 = -\frac{2}{9} - \frac{\gamma}{6} - \frac{1}{12}\log(2\pi) - \frac{\pi^2}{54}, \quad d_4 = \frac{11}{144} + \frac{\gamma}{24} + \frac{\pi^2}{48} + \frac{\zeta(3)}{12};$$

$$\Gamma_5(1+z) = \exp\left(e_1 z + e_2 z^2 + e_3 z^3 + e_4 z^4 + e_5 z^5\right) \cdot \prod_{k=1}^{\infty}\left\{\left(1+\frac{z}{k}\right)^{-\binom{k+3}{4}} \exp\left[\binom{k+3}{4}\left(\frac{z}{k} - \frac{z^2}{2k^2} + \frac{z^3}{3k^3} - \frac{z^4}{4k^4} + \frac{z^5}{5k^5}\right)\right]\right\}, \tag{2}$$

where

$$e_1 = \frac{469}{2^6 \cdot 3^2 \cdot 5} - \frac{1}{8}\log(2\pi) + \frac{11}{12}\zeta'(-1) - \frac{3\zeta(3)}{16\pi^2} + \frac{1}{6}\zeta'(-3) + \frac{1}{20}\zeta(4) - \frac{1}{20}\zeta(5),$$

$$e_2 = \frac{\gamma}{8} + \frac{11}{48}\log(2\pi) - \frac{3}{4}\zeta'(-1) + \frac{\zeta(3)}{16\pi^2},$$

$$e_3 = -\frac{161}{2^5 \cdot 3^3} - \frac{11}{2^3 \cdot 3^2}\gamma - \frac{1}{8}\log(2\pi) + \frac{1}{6}\zeta'(-1) - \frac{1}{12}\zeta(2),$$

$$e_4 = \frac{7}{64} + \frac{1}{16}\gamma + \frac{1}{48}\log(2\pi) + \frac{11}{96}\zeta(2) + \frac{1}{16}\zeta(3),$$

$$e_5 = -\frac{5}{2^5 \cdot 3^2} - \frac{1}{2^3 \cdot 3 \cdot 5}\gamma - \frac{1}{20}\zeta(2) - \frac{11}{120}\zeta(3) - \frac{1}{20}\zeta(4).$$

By virtue of 1.4(82), to get the constants d_j's and e_j's involved in (1) and (2), respectively, it is indispensable to evaluate $A_n(1)$ explicitly. To do this, using 1.4(83) and 1.4(8.6), we find that

$$A_n(z) = \sum_{k=1}^{\infty} \binom{n+k-2}{n-1} \left[\sum_{\ell=1}^{n} \frac{(-1)^{n-\ell}}{\ell} \left(\frac{z}{k}\right)^{\ell} + (-1)^n \log\left(1 + \frac{z}{k}\right) \right]$$

$$= \sum_{\ell=n+1}^{\infty} \frac{(-1)^{n+\ell-1}}{\ell} z^{\ell} \sum_{k=1}^{\infty} \binom{n+k-2}{n-1} \frac{1}{k^{\ell}},$$

which, by means of the expansion:

$$\binom{n+k-2}{n-1} = \frac{1}{(n-1)!} \sum_{j=0}^{n-1} (-1)^{n-1+j} s(n-1,j)\, k^j,$$

immediately yields

$$A_n(z) = \frac{1}{(n-1)!} \sum_{j=0}^{n-1} (-1)^j s(n-1,j) \sum_{\ell=n+1}^{\infty} \frac{(-1)^{\ell}}{\ell} \zeta(\ell - j) z^{\ell}, \tag{3}$$

where $s(n,k)$ denote the Stirling numbers of the first kind given in Section 1.5.

Now, we can apply 3.2(63) to (3), and we get

$$A_n(z) = \frac{1}{(n-1)!} \sum_{j=0}^{n-1} s(n-1,j) \left[\sum_{k=0}^{j} (-1)^k \binom{j}{k} \zeta'(-k, 1+z)\, z^{j-k} \right.$$

$$+ (-1)^{j+1}\zeta'(-j) - \sum_{\ell=0}^{j-1} (-1)^{\ell} \frac{\zeta(-\ell)}{j-\ell} z^{j-\ell} + \frac{z^{j+1}}{j+1}\left(H_j + \gamma\right) \tag{4}$$

$$\left. - \sum_{k=2}^{n-j} (-1)^k \frac{\zeta(k)}{k+j} z^{k+j} \right],$$

the special case $z = 1$ of which is given here for convenience:

$$\begin{aligned} A_n(1) = \frac{1}{(n-1)!} \sum_{j=0}^{n-1} s(n-1,j) \Bigg[\sum_{k=0}^{j-1} (-1)^k \binom{j}{k} \zeta'(-k) \\ - \sum_{k=0}^{j-1} \frac{(-1)^k}{j-k} \zeta(-k) + \frac{1}{j+1} (H_j + 2\gamma) - \sum_{k=2}^{n-j} \frac{(-1)^k}{k+j} \zeta(k) \Bigg] \quad (n \in \mathbb{N}). \end{aligned} \tag{5}$$

By using various identities given in this and the foregoing chapters, some special cases of (5) are explicitly obtained here as follows:

$$\begin{aligned} &A_1(1) = \gamma, \quad A_2(1) = 1 + \frac{\gamma}{2} - \frac{1}{2}\log(2\pi), \quad A_3(1) = -\frac{1}{4} - \frac{\gamma}{12} + \frac{\pi^2}{36} + \log A, \\ &A_4(1) = \frac{5}{36} + \frac{\gamma}{24} + \frac{\pi^2}{432} + \frac{\zeta(3)}{12} - \frac{1}{2}\log A - \frac{1}{2}\log B, \\ &A_5(1) = -\frac{37}{540} - \frac{19}{720}\gamma + \frac{3}{160}\zeta(2) + \frac{7}{240}\zeta(3) \\ &\qquad + \frac{1}{20}\zeta(5) - \frac{1}{3}\zeta'(-1) + \frac{\zeta(3)}{8\pi^2} - \frac{1}{6}\zeta'(-3). \end{aligned} \tag{6}$$

Next, by taking logarithms on each side of 1.4(3), 1.4(91) and (1), we obtain

$$\log \Gamma_2(1+z) = \frac{1}{2}z - \frac{1}{2}z\log(2\pi) + \frac{1}{2}(\gamma+1)z^2 - \sum_{n=2}^{\infty} \frac{(-1)^n}{n+1} \zeta(n) z^{n+1},$$

$$\begin{aligned} \log \Gamma_3(1+z) = \left[\frac{3}{8} - \frac{1}{4}\log(2\pi) - \log A \right] z + \left[\frac{\gamma}{4} + \frac{1}{4}\log(2\pi) + \frac{1}{8} \right] z^2 \\ - \left(\frac{1}{4} + \frac{\gamma}{6} \right) z^3 - \frac{1}{2} \sum_{n=2}^{\infty} \frac{(-1)^n}{n+1} \zeta(n) z^{n+1} + \frac{1}{2} \sum_{n=2}^{\infty} \frac{(-1)^n}{n+2} \zeta(n) z^{n+2}, \end{aligned}$$

and

$$\begin{aligned} \log \Gamma_4(1+z) = &\left[\frac{7}{24} - \log A - \frac{1}{2}\log B - \frac{1}{6}\log(2\pi) \right] z \\ &+ \frac{1}{2}\left[-\frac{1}{72} + \frac{\gamma}{3} + \frac{1}{2}\log(2\pi) + \log A \right] z^2 \\ &- \frac{1}{6}\left[\frac{4}{3} + \gamma + \frac{1}{2}\log(2\pi) \right] z^3 \\ &+ \frac{1}{24}\left(\frac{11}{6} + \gamma \right) z^4 - \frac{1}{6} \sum_{n=2}^{\infty} \frac{(-1)^n}{n+3} \zeta(n) z^{n+3} \\ &+ \frac{1}{2} \sum_{n=2}^{\infty} \frac{(-1)^n}{n+2} \zeta(n) z^{n+2} - \frac{1}{3} \sum_{n=2}^{\infty} \frac{(-1)^n}{n+1} \zeta(n) z^{n+1}, \end{aligned}$$

which, upon employing the special case $a = 1$ of 3.2(63), can be expressed in terms of the Zeta functions as follows:

$$\log \Gamma_2(1+z) = -\frac{1}{12} + \log A - z \log \Gamma(1+z) + \zeta'(-1, 1+z), \tag{7}$$

$$\begin{aligned} \log \Gamma_3(1+z) &= -\frac{1}{24} + \frac{1}{2} \log A + \frac{\zeta(3)}{8\pi^2} + \left(\frac{1}{12} - \log A\right) zn \\ &+ \frac{1}{2}(z^2 - z) \log \Gamma(1+z) + \left(\frac{1}{2} - z\right) \zeta'(-1, 1+z) + \frac{1}{2} \zeta'(-2, 1+z), \end{aligned} \tag{8}$$

and

$$\begin{aligned} \log \Gamma_4(1+z) &= -\frac{1}{36} + \frac{1}{3} \log A + \frac{\zeta(3)}{8\pi^2} - \frac{1}{6}\zeta'(-3) + \left(\frac{1}{12} - \log A - \frac{\zeta(3)}{8\pi^2}\right) z \\ &+ \left(-\frac{1}{24} + \frac{1}{2} \log A\right) z^2 - \frac{1}{6}(z^3 - 3z^2 + 2z) \log \Gamma(1+z) \\ &+ \left(\frac{1}{2} z^2 - z + \frac{1}{3}\right) \zeta'(-1, 1+z) + \frac{1}{6} \zeta'(-3, 1+z), \end{aligned} \tag{9}$$

where we have also made use of 7.1(24) and the known identity 2.3(22).

Analogous to the classical result $\Gamma\left(\frac{1}{2}\right) = \pi^{\frac{1}{2}}$, setting $z = -\frac{1}{2}$ in (7), (8) and (9), we obtain

$$\Gamma_2\left(\frac{1}{2}\right) = 2^{-\frac{1}{24}} \cdot \pi^{\frac{1}{4}} \cdot e^{-\frac{1}{8}} \cdot A^{\frac{3}{2}}, \tag{10}$$

$$\Gamma_3\left(\frac{1}{2}\right) = 2^{-\frac{1}{24}} \cdot \pi^{\frac{3}{16}} \cdot \exp\left[-\frac{1}{8} + \frac{7\zeta(3)}{32\pi^2}\right] \cdot A^{\frac{3}{2}}, \tag{11}$$

and

$$\Gamma_4\left(\frac{1}{2}\right) = 2^{-\frac{229}{5760}} \cdot \pi^{\frac{5}{32}} \cdot e^{-\frac{265}{2304}} \cdot A^{\frac{23}{16}} \cdot B^{\frac{3}{4}} \cdot C^{\frac{5}{16}}. \tag{12}$$

It is noted that (10) and (11) are the same as 1.4(18) and 1.4(97), respectively.

By integrating the very three equations before (7) from 0 to z and using the special case $a = 1$ of 3.2(63), we get

$$\begin{aligned} \int_0^z \log \Gamma_2(1+t)\, dt &= \frac{\zeta(3)}{4\pi^2} - \left[\frac{1}{12} + \zeta'(-1)\right] z - \frac{1}{4} z^2 \log(2\pi) + \frac{1}{6} z^3 \\ &- z\zeta'(-1, 1+z) + \zeta'(-2, 1+z), \end{aligned} \tag{13}$$

$$\int_0^z \log\Gamma_3(1+t)\,dt = \frac{\zeta(3)}{8\pi^2} - \frac{1}{2}\zeta'(-3) + \left[\frac{\zeta(3)}{8\pi^2} - \frac{1}{12} + \frac{1}{2}\log A\right] z$$

$$+ \left[\frac{1}{24} - \frac{1}{8}\log(2\pi) - \frac{1}{2}\log A\right] z^2 + \frac{1}{12}[1+\log(2\pi)]z^3 - \frac{1}{24}z^4 - \frac{z}{2}\zeta'(-1,1+z)$$

$$+ \frac{1}{2}z^2\zeta'(-1,1+z) + \frac{1}{2}\zeta'(-2,1+z) - z\zeta'(-2,1+z) + \frac{1}{2}\zeta'(-3,1+z), \tag{14}$$

and

$$\int_0^z \log\Gamma_4(1+t)\,dt$$

$$= \frac{\zeta(3)}{12\pi^2} - \frac{1}{2}\zeta'(-3) - \frac{\zeta(5)}{8\pi^4} - \left[\frac{19}{720} + \frac{1}{3}\zeta'(-1) - \frac{\zeta(3)}{8\pi^2} + \frac{1}{6}\zeta'(-3)\right] z$$

$$+ \left[\frac{1}{2}\zeta'(-1) - \frac{\zeta(3)}{16\pi^2} - \frac{1}{12}\log(2\pi)\right] z^2 + \left[\frac{1}{18} + \frac{1}{12}\log(2\pi) - \frac{1}{6}\zeta'(-1)\right] z^3 \tag{15}$$

$$- \frac{1}{24}\left[1 + \frac{1}{2}\log(2\pi)\right] z^4 + \frac{1}{120}z^5 - \frac{1}{6}\left(z^3 - 3z^2 + 2z\right)\zeta'(-1,1+z)$$

$$+ \left(\frac{1}{2}z^2 - z + \frac{1}{3}\right)\zeta'(-2,1+z) - \frac{1}{2}(z-1)\zeta'(-3,1+z) + \frac{1}{6}\zeta'(-4,1+z).$$

Upon setting $z = 1$ and $z = \frac{1}{2}$ in (13), (14) and (15) and using some previously recorded identities, we have

$$\int_0^1 \log\Gamma_2(1+t)\,dt = -\frac{1}{12} - \frac{1}{4}\log(2\pi) + 2\log A, \tag{16}$$

$$\int_0^1 \log\Gamma_3(1+t)\,dt = \frac{3\zeta(3)}{8\pi^2} - \frac{1}{24}\log(2\pi), \tag{17}$$

$$\int_0^1 \log\Gamma_4(1+t)\,dt = \frac{13}{2160} + \frac{3\zeta(3)}{16\pi^2} - \frac{1}{48}\log(2\pi) + \frac{2}{3}\log C, \tag{18}$$

$$\int_0^{\frac{1}{2}} \log\Gamma_2(1+t)\,dt = -\frac{1}{24}(1+\log 2) - \frac{1}{16}\log\pi + \frac{1}{4}\log A + \frac{7\zeta(3)}{16\pi^2}, \tag{19}$$

$$\int_0^{\frac{1}{2}} \log \Gamma_3(1+t)\,dt = -\frac{1}{256} - \frac{29}{1920}\log 2 - \frac{1}{48}\log \pi + \frac{1}{16}\log A + \frac{3}{4}\log B + \frac{15}{16}\log C, \tag{20}$$

and

$$\int_0^{\frac{1}{2}} \log \Gamma_4(1+t)\,dt = \frac{73}{69120} - \frac{17}{1920}\log 2 - \frac{3}{256}\log \pi + \frac{1}{32}\log A + \frac{77}{96}\log C + \frac{47\zeta(3)}{384\pi^2} - \frac{31\zeta(5)}{128\pi^4}. \tag{21}$$

Barnes [94, p. 283] expressed $\log \Gamma_2(z+a)$ as an integral of $\log \Gamma(t+a)$ as follows:

$$\int_0^z \log \Gamma(t+a)\,dt = \frac{1}{2}\left[\log(2\pi) + 1 - 2a\right] z - \frac{z^2}{2} + (z+a-1)\log \Gamma(z+a) + \log \Gamma_2(z+a) + (1-a)\log \Gamma(a) - \log \Gamma_2(a). \tag{22}$$

Similarly, we have

$$\begin{aligned}\int_0^z \log \Gamma_2(t+a)dt &= \left[\frac{1}{4} - \frac{1}{2}(a-1)\log(2\pi) + 2\log A + \frac{a^2}{2} - a\right] z \\ &+ \frac{1}{4}[2a - 2 - \log(2\pi)]z^2 + \frac{1}{6}z^3 + (z+a-2)\log \Gamma_2(z+a) \\ &+ 2\log \Gamma_3(z+a) + (2-a)\log \Gamma_2(a) - 2\log \Gamma_3(a),\end{aligned} \tag{23}$$

$$\begin{aligned}&\int_0^z \log \Gamma_3(t+a)\,dt \\ &= \left[\frac{1}{8} - \frac{5}{6}a + \frac{3}{4}a^2 - \frac{1}{6}a^3 + (3-2a)\log A + \frac{3}{2}\log B + \frac{1}{4}(a-1)(a-2)\log(2\pi)\right] z \\ &+ \left[-\frac{1}{4}a^2 + \frac{3}{4}a - \frac{5}{12} + \left(\frac{1}{4}a - \frac{3}{8}\right)\log(2\pi) - \log A\right] z^2 + \left[\frac{1}{4} - \frac{1}{6}a + \frac{1}{12}\log(2\pi)\right] z^3 \\ &- \frac{1}{24}z^4 + (z+a-3)\log \Gamma_3(z+a) + (3-a)\log \Gamma_3(a) + 3\log \Gamma_4(z+a) - 3\log \Gamma_4(a),\end{aligned} \tag{24}$$

and

$$\begin{aligned}\int_0^z \log \Gamma_4(t+a)\,dt &= \beta_1 z + \beta_2 z^2 + \beta_3 z^3 + \beta_4 z^4 + \beta_5 z^5 - 4\log \Gamma_5(a) \\ &+ (z+a-4)\log \Gamma_4(z+a) + (4-a)\log \Gamma_4(a) + 4\log \Gamma_5(z+a),\end{aligned} \tag{25}$$

where

$$\beta_1 = \frac{1937}{2160} + \frac{19}{90}\gamma - a + \frac{11}{12}a^2 - \frac{1}{3}a^3 + \frac{1}{24}a^4$$
$$- \frac{1}{12}\left(a^3 - 6a^2 + 11a - 6\right)\log(2\pi) - \left(a^2 - 4a + 1\right)\zeta'(-1)$$
$$+ \frac{1}{8}(1-3a)\frac{\zeta(3)}{\pi^2} + \frac{2}{3}\zeta'(-3) - \frac{3}{20}\zeta(2) - \frac{7}{30}\zeta(3) - \frac{1}{5}\zeta(4) - \frac{1}{5}\zeta(5),$$
$$\beta_2 = -\frac{1}{2} + \frac{11}{12}a - \frac{1}{2}a^2 + \frac{1}{12}a^3 - \left(\frac{1}{8}a^2 - \frac{1}{2}a + \frac{11}{24}\right)\log(2\pi)$$
$$+ (2-a)\,\zeta'(-1) - \frac{3\zeta(3)}{16\pi^2},$$
$$\beta_3 = \frac{11}{36} - \frac{1}{3}a + \frac{1}{12}a^2 + \frac{1}{12}(2-a)\log(2\pi) - \frac{1}{3}\zeta'(-1),$$
$$\beta_4 = -\frac{1}{12} + \frac{1}{24}a - \frac{1}{48}\log(2\pi), \quad \text{and} \quad \beta_5 = \frac{1}{120}.$$

To derive (25), for example, by setting $z = t + a$ in 1.1(2) and $z = t + a - 1$ in 1.4(3), 1.4(91), (1) and (2) and taking the logarithmic derivatives on each side of the resulting equations, we find that

$$\frac{\Gamma'(t+a)}{\Gamma(t+a)} = -\gamma - \sum_{k=1}^{\infty}\left(\frac{1}{t+a-1+k} - \frac{1}{k}\right),$$
$$\frac{\Gamma_2'(t+a)}{\Gamma_2(t+a)} = \frac{1}{2} - \frac{1}{2}\log(2\pi) + (1+\gamma)(t+a-1)$$
$$+ \sum_{k=1}^{\infty} k\left(-\frac{1}{t+a-1+k} + \frac{1}{k} - \frac{t+a-1}{k^2}\right),$$
$$\frac{\Gamma_3'(t+a)}{\Gamma_3(t+a)} = c_1 + 2c_2\,(t+a-1) + 3c_3\,(t+a-1)^2$$
$$+ \sum_{k=1}^{\infty}\frac{1}{2}k(k+1)\left(-\frac{1}{t+a-1+k} + \frac{1}{k} - \frac{t+a-1}{k^2} + \frac{(t+a-1)^2}{k^3}\right),$$

and

$$\frac{\Gamma_4'(t+a)}{\Gamma_4(t+a)} = d_1 + 2d_2\,(t+a-1) + 3d_3\,(t+a-1)^2 + 4d_4\,(t+a-1)^3$$
$$+ \sum_{k=1}^{\infty}\binom{k+2}{3}\left[-\frac{1}{t+a-1+k} + \frac{1}{k} - \frac{t+a-1}{k^2}\right.$$
$$\left. + \frac{(t+a-1)^2}{k^3} - \frac{(t+a-1)^3}{k^4}\right],$$

all of which can be applied in conjunction with

$$\frac{\Gamma_5'(t+a)}{\Gamma_5(t+a)} = e_1 + 2e_2(t+a-1) + 3e_3(t+a-1)^2 + 4e_4(t+a-1)^3$$
$$+5e_5(t+a-1)^4 + \sum_{k=1}^{\infty}\binom{k+3}{4}\left[-\frac{1}{t+a-1+k} + \frac{1}{k} - \frac{t+a-1}{k^2}\right.$$
$$\left. + \frac{(t+a-1)^2}{k^3} - \frac{(t+a-1)^3}{k^4} + \frac{(t+a-1)^4}{k^5}\right]$$

to obtain

$$\frac{\Gamma_5'(t+a)}{\Gamma_5(t+a)} = e_1 + \omega_1(t+a-1) + \omega_2(t+a-1)^2 + \omega_3(t+a-1)^3$$
$$+\omega_4(t+a-1)^4 - \frac{1}{4}(t+a-1)\frac{\Gamma'(t+a)}{\Gamma(t+a)} - \frac{1}{4}(t+a-1)\frac{\Gamma_2'(t+a)}{\Gamma_2(t+a)} \tag{26}$$
$$-\frac{1}{4}(t+a-1)\frac{\Gamma_3'(t+a)}{\Gamma_3(t+a)} - \frac{1}{4}(t+a-1)\frac{\Gamma_4'(t+a)}{\Gamma_4(t+a)},$$

where

$$\omega_1 = \frac{1}{4} + \frac{11}{48}\log(2\pi) - \zeta'(-1) + \frac{3\zeta(3)}{32\pi^2}, \quad \omega_2 = -\frac{11}{48} - \frac{1}{8}\log(2\pi) + \frac{1}{4}\zeta'(-1),$$
$$\omega_3 = \frac{1}{12} + \frac{1}{48}\log(2\pi), \quad \text{and} \quad \omega_4 = -\frac{1}{96}.$$

Finally, upon integrating both sides of (26) from $t=0$ to $t=z$ and using (22), (23) and (24), we are led to the desired identity (25).

The foregoing methods can also be applied with a view to deriving explicit expressions for the multiple Gamma functions Γ_n of higher order and various other related identities.

Mathieu Series

Almost twelve decades ago, Mathieu investigated an interesting series $S(r)$ in the study of elasticity of solid bodies. Since then, many authors have studied various problems arising from the Mathieu series $S(r)$ in various diverse ways. Here, we present a relationship between the Mathieu series $S(r)$ and certain series involving the Zeta functions (see [303]). By means of this relationship, we then express the Mathieu series $S(r)$ in terms of the Trigamma function $\psi'(z)$ or (equivalently) the Hurwitz (or generalized) Zeta function $\zeta(s,a)$. Accordingly, various interesting properties of $S(r)$ can be obtained from those of $\psi'(z)$ and $\zeta(s,a)$. Among other results, certain integral representations of $S(r)$ are deduced here, by using the aforementioned relationships among $S(r)$, $\psi'(z)$ and $\zeta(s,a)$.

Émile Leonard Mathieu (1835–1890) [803] investigated the following infinite series:

$$S(r) = \sum_{n=1}^{\infty} \frac{2n}{\left(n^2 + r^2\right)^2} \qquad \left(r \in \mathbb{R}^+\right) \tag{27}$$

in the study of elasticity of solid bodies (see also [418]), $\mathbb{R}^+$ being (as usual) the set of positive real numbers. Pogány *et al.* [902] introduced an *alternating* version of the Mathieu series (1.1) as follows:

$$\tilde{S}(r) = \sum_{n=1}^{\infty} (-1)^{n-1} \frac{2n}{\left(n^2 + r^2\right)^2} \qquad \left(r \in \mathbb{R}^+\right). \tag{28}$$

Since the time of Mathieu, many authors (see, e.g., [30], [376], [403], [407], [418], [897], [898], [899], [900], [901], [902], [903], [904], [925], [1110], [1155], [1156], [1157], [1158], [1159], [1160] and [1161]; see also the references cited in each of these works) have investigated various problems arising from the Mathieu series (27) and its extensions and generalizations in various diverse ways. In particular, Pogány *et al.* [902] presented the following integral representations of the Mathieu series (27) and the alternating Mathieu series (28):

$$S(r) = \frac{1}{r} \int_0^{\infty} \frac{t \sin(rt)}{e^t - 1}\, dt \tag{29}$$

and

$$\tilde{S}(r) = \frac{1}{r} \int_0^{\infty} \frac{t \sin(rt)}{e^t + 1}\, dt. \tag{30}$$

Srivastava and Tomovski [1110], conversely, defined the following five-parameter family of generalized Mathieu series:

$$S_{\mu}^{(\alpha,\beta)}(r; \mathbf{a}) = S_{\mu}^{(\alpha,\beta)}(r; \{a_k\}) = \sum_{n=1}^{\infty} \frac{2\, a_n^{\beta}}{\left(a_n^{\alpha} + r^2\right)^{\mu}} \qquad \left(r, \alpha, \beta, \mu \in \mathbb{R}^+\right), \tag{31}$$

where it is *tacitly* assumed that the *positive* sequence

$$\mathbf{a} := \{a_k\}_{k=1}^{\infty} \qquad \left(\lim_{k \to \infty} a_k = \infty\right) \tag{32}$$

is so chosen (and then the positive parameters α, β and μ) are so constrained that the infinite series in (31) converges, that is, that the following auxiliary series:

$$\sum_{n=1}^{\infty} \frac{1}{a_n^{\mu\alpha - \beta}}$$

is convergent.

Several authors have considered some interesting variants and special cases of the generalized Mathieu series (31) (see, e.g., [897], [899], [902] and [925]). And, very recently, Elezović *et al.* [407] derived several results involving the Laplace, Fourier and Mellin transforms of various functions belonging to the family of generalized Mathieu series (31), including, for example, the Laplace transforms of $S(r)$ and $\tilde{S}(r)$ as given below:

$$\mathcal{L}\big(S(r)\big)(x) = \int_0^\infty \left(\frac{t}{e^t - 1}\right) \arctan\left(\frac{t}{x}\right) dt \tag{33}$$

and

$$\mathcal{L}\big(\tilde{S}(r)\big)(x) = \int_0^\infty \left(\frac{t}{e^t + 1}\right) \arctan\left(\frac{t}{x}\right) dt. \tag{34}$$

The Dirichlet Eta function (or the alternating Riemann Zeta function) $\eta(s)$ is defined by

$$\eta(s) = \sum_{n=1}^\infty \frac{(-1)^{n-1}}{n^s} \qquad \big(\Re(s) > 0\big). \tag{35}$$

It is easy to find from 2.3(1) and (35) that

$$\eta(s) = \left(1 - 2^{1-s}\right) \zeta(s) \qquad \big(\Re(s) > 1\big). \tag{36}$$

We express the Mathieu series in (27) and the alternating Mathieu series in (28) as sums involving the Riemann Zeta function, which can also be evaluated in terms of the Trigamma function $\psi'(z)$ and so the Hurwitz Zeta function $\zeta(s,a)$ as follows:

Each of the following series representations holds true:

$$S(r) = 2\sum_{k=1}^\infty (-1)^{k-1}\, k\, \zeta(2k+1)\, r^{2(k-1)} \qquad (|r| < 1) \tag{37}$$

and

$$\tilde{S}(r) = 2\sum_{k=1}^\infty (-1)^{k-1}\, k\, \eta(2k+1)\, r^{2(k-1)} \qquad (|r| < 1), \tag{38}$$

where $\zeta(s)$ *and* $\eta(s)$ *are the Riemann Zeta function and the Dirichlet Eta function given in* 2.3(1) *and* (35), *respectively.*

Proof. By applying the following elementary identity:

$$\frac{1}{(1+z)^2} = \sum_{k=0}^{\infty} (-1)^k (k+1) z^k \qquad (|z| < 1) \tag{39}$$

to (27), we obtain

$$\begin{aligned} S(r) &= \sum_{n=1}^{\infty} \frac{2n}{n^4} \frac{1}{\{1+(r/n)^2\}^2} \\ &= \sum_{n=1}^{\infty} \frac{2n}{n^4} \sum_{k=0}^{\infty} (-1)^k (k+1) \frac{r^{2k}}{n^{2k}} \qquad \left(\left|\frac{r}{n}\right| < 1;\ n \in \mathbb{N}\right) \\ &= 2 \sum_{k=0}^{\infty} (-1)^k (k+1) r^{2k} \sum_{n=1}^{\infty} \frac{1}{n^{2k+3}} \\ &= 2 \sum_{k=0}^{\infty} (-1)^k (k+1) \zeta(2k+3) r^{2k}, \end{aligned}$$

where the rearrangement of the sums in the penultimate step is guaranteed by the absolute convergence of the double series involved.

Finally, upon replacing the summation index k by $k-1$ in the last expression, we are led to the assertion (37). Similarly, we can prove the assertion (38).

By using (37) and (38), we obtain:

Each of the following relationships holds true:

$$S(r) = \frac{i}{2r} \left[\psi'(1+ir) - \psi'(1-ir)\right] \qquad \left(0 < |r| < 1;\ i = \sqrt{-1}\right), \tag{40}$$

$$S(r) = \frac{i}{2r} \left[\zeta(2,\, 1+ir) - \zeta(2,\, 1-ir)\right] \qquad \left(0 < |r| < 1;\ i = \sqrt{-1}\right) \tag{41}$$

and

$$\tilde{S}(r) = S(r) - \frac{1}{4} S\left(\frac{r}{2}\right) \qquad (|r| < 1), \tag{42}$$

where $\psi'(z)$ and $\zeta(s,a)$ are the Trigamma function and the Hurwitz (or generalized) Zeta function given in 1.3(52) *and* 2.1(1), *respectively.* ■

Proof. We recall the following known identity (see [1094, p. 160, Eq. (16)]):

$$\sum_{k=1}^{\infty} \zeta(2k+1) t^{2k} = -\frac{1}{2} \left[\psi(1+t) + \psi(1-t)\right] - \gamma \qquad (|t| < 1), \tag{43}$$

where γ denotes the Euler-Mascheroni constant given in 1.1(3) (also see Section 1.2).

Differentiating each side of (43) with respect to t and dividing the resulting identity by t, we get

$$2\sum_{k=1}^{\infty} k\,\zeta(2k+1)\,t^{2(k-1)} = -\frac{1}{2t}\left[\psi'(1+t)-\psi'(1-t)\right] \qquad (0<|t|<1). \quad (44)$$

By setting $t = ir$ $\left(i=\sqrt{-1}\right)$ in (44), and making use of (37) we prove the assertion (40). Similarly, by applying 1.3(53) to (40), we obtain (41). Furthermore, by combining (37), (38) and (36), we can prove (42).

It is noted that, in view of the relationship in (42), several properties of the alternating Mathieu series $\tilde{S}(r)$ can easily be obtained from the corresponding properties of the Mathieu series $S(r)$.

First of all, we analytically continue the Mathieu series $S(r)$ from $r \in \mathbb{R}^+$ to the complex plane $\mathbb{C}$ as follows:

The Mathieu series $S(r)$ is analytic on the punctured complex plane

$$\mathbb{C}\setminus\left\{z : z = \pm ni \quad \left(n\in\mathbb{N};\ i=\sqrt{-1}\right)\right\}.$$

■

Proof. It is obvious that each term $2n/(n^2+r^2)$ in the series (1.1) defining the $S(r)$ is analytic on

$$\mathbb{C}\setminus\left\{z : z = \pm ni \quad \left(n\in\mathbb{N};\ i=\sqrt{-1}\right)\right\}.$$

Since the series for $S(r)$ is seen to be uniformly convergent on any compact subset $\mathbb{D}$ in

$$\mathbb{C}\setminus\left\{z : z = \pm ni \quad \left(n\in\mathbb{N};\ i=\sqrt{-1}\right)\right\},$$

$S(r)$ is analytic on $\mathbb{D}$. Now, by the well-known principle due to Weierstrass (see, e.g., [1025, p. 88, Theorem]), the Mathieu series $S(r)$ is analytic on

$$\mathbb{C}\setminus\left\{z : z = \pm ni \quad \left(n\in\mathbb{N};\ i=\sqrt{-1}\right)\right\}.$$

By means of the relationships given in (40), (41) and (42) and by making use of several known integral representations of $\zeta(s,a)$ and $\psi(z)$, for example, 2.2(21), 2.2(25), 1.3(18), 1.3(24) and another known integral formula for $\psi(z)$ (see [1225, p. 261, Example 19]):

$$\psi(z) = \log z - \int_0^1 \frac{1-t+\log t}{(1-t)\log t}\,t^{z-1}\,dt \qquad \left(\Re(z)>0\right), \quad (45)$$

we present various integral representations of $S(r)$ and $\tilde{S}(r)$, including (29) and (30) (see [303]).

Each of the following representations holds true:

$$S(r) = \frac{1}{(1+r^2)^2} + \frac{1}{1+r^2} + \frac{i}{2r}\sum_{k=2}^{n}\left(\frac{1}{(1+ir)^{k+1}} - \frac{1}{(1-ir)^{k+1}}\right)B_k$$
$$+\frac{1}{r}\int_0^\infty \left(\frac{1}{e^t-1} - \sum_{k=0}^{n}\frac{B_k}{k!}t^{k-1}\right) t\,e^{-t}\sin(rt)\,dt \tag{46}$$
$$\left(\Re(r) > 0;\ n \in \mathbb{N}_0;\ i = \sqrt{-1}\right),$$

where B_k denotes the Bernoulli numbers and an empty sum is (as usual) understood to be nil;

$$S(r) = \frac{1}{r}\int_0^1 \frac{\log t}{1-t}\sin(r\log t)\,dt \quad \left(\Re(r) > 0\right); \tag{47}$$

$$S(r) = \frac{1}{r}\int_0^\infty \frac{\log(1+t)}{t(1+t)}\sin[r\log(1+t)]\,dt \quad \left(\Re(r) > 0\right); \tag{48}$$

$$S(r) = \frac{1}{1+r^2} + \frac{1}{r}\int_0^1 \frac{1-t+\log t}{1-t}\sin(r\log t)\,dt \quad \left(\Re(r) > 0\right); \tag{49}$$

$$S(r) = \frac{1}{(1+r^2)^2} + \frac{1}{1+r^2} + 4\int_0^\infty \frac{1+r^2-t^2}{(1+t^2-r^2)^2 - 4r^2}\,\frac{t\,dt}{e^{2\pi t}-1} \quad \left(\Re(r) > 0\right). \tag{50}$$

■

Proof. By applying 2.2(25) and the third integral representation in 2.2(21) to (41), we obtain (46) and (47), respectively. Differentiating the right-hand side of each of the integral representations 1.3(18), (45) and 1.3(24) with respect to z under the sign of integration (which can be validated by means of a known result [1225, p. 74, Corollary]) and applying each of the resulting integral representations of the Trigamma function $\psi'(z)$ to (40), we obtain (48) to (49), respectively.

It is **noted** that formula (29) can be derived by employing the first integral representation of 2.2(21) in the identity (41) and the application of (29) to the relationship (42) yields (30).

It is also **noted** that, even though the identities (46) to (50) are proven under the condition that $0 < |r| < 1$, (in view of observation above) each of these identities holds true, by the principle of analytic continuation, throughout the extended region given by $\Re(r) > 0$.

As a by-product of the main identities (46) to (50), we obtain a variety of interesting identities, some of which are stated below.

Each of the following results holds true:

$$S(r) = \frac{1}{\left(1+r^2\right)^2} + \frac{1}{1+r^2} + \frac{1}{r}\int_0^\infty \left(\frac{1}{e^t-1} - \frac{1}{t} + \frac{1}{2}\right) t\, e^{-t} \sin(rt)\, dt \tag{51}$$

$$\left(\Re(r) > 0\right),$$

$$S(r) = \frac{1}{1+r^2} + O\left(\frac{1}{r^4}\right) \quad (r \to \infty;\ r \in \mathbb{R}^+) \tag{52}$$

and

$$\tilde{S}(r) = \frac{1}{2r}\int_0^1 \frac{\log t}{1-t} \sin\left(\frac{r}{2}\log t\right)\left[4\cos\left(\frac{r}{2}\log t\right) - 1\right] dt \quad \left(\Re(r) > 0\right). \tag{53}$$

■

Proof. The equation (51) is a special case of (46) when $n = 1$. By making use of the generating function of Bernoulli numbers (see 1.7(2) and 1.7(7)):

$$\frac{t}{e^t-1} - 1 + \frac{t}{2} = \sum_{k=1}^{\infty} B_{2k}\frac{t^{2k}}{(2k)!} \quad (|t| < 2\pi) \tag{54}$$

and the following entry in the readily available tables of Fourier sine transforms (see, e.g., [421, p. 72, Entry 2.4(3)]):

$$\int_0^\infty t^2 e^{-t} \sin(rt)\, dt = 2r\left(\frac{1}{1+r^2}\right)^3 \left(3 - r^2\right) \quad (r > 0), \tag{55}$$

in conjunction with (51), we obtain (52). Furthermore, if the various already-developed integral representations of $S(r)$ are applied to the relationship (42), several integral representations of $\tilde{S}(r)$ can be obtained. For example, by applying (47) to (42), we obtain the last identity (53).

Each of the following identities holds true:

$$\begin{aligned} T(r) :=& \sum_{n=1}^{\infty} \frac{8n}{\left(n^2+r^2\right)^3} \\ =& \frac{i}{2r^3}\left[\psi'(1+ir) - \psi'(1-ir)\right] + \frac{1}{2r^2}\left[\psi^{(2)}(1+ir) + \psi^{(2)}(1-ir)\right] \\ =& \frac{i}{2r^3}\left[\zeta(2,\, 1+ir) - \zeta(2,\, 1-ir)\right] - \frac{1}{r^2}\left[\zeta(3,\, 1+ir) + \zeta(3,\, 1-ir)\right] \end{aligned} \tag{56}$$

$$\left(0 < |r| < 1;\, i = \sqrt{-1}\right)$$

and

$$T(r) = \frac{2\left(3+r^2\right)}{\left(1+r^2\right)^3} + \frac{1}{r^3} \int_0^{\infty} \left(\frac{1}{e^t - 1} - \frac{1}{t} + \frac{1}{2} \right)$$
$$\cdot t e^{-t} \left[\sin(rt) - rt \cos(rt) \right] dt \quad \left(\Re(r) > 0 \right). \tag{57}$$

■

Proof. Upon differentiating (27) and (40) with respect to r, if we combine the resulting identities and make use of 1.3(53), we find a new series $T(r)$, which is expressed in terms of the Polygamma functions and the Hurwitz Zeta function as in (56). By applying (51) to (56), we obtain (57). ■

Problems

1. Evaluate the following special values of the G-function given in Section 1.4:

$$G\left(\tfrac{1}{3}\right) = \exp\left[\frac{1}{9} + \frac{\pi\sqrt{3}}{54} - \frac{\sqrt{3}\,\zeta\left(2, \frac{1}{3}\right)}{36\pi} \right] \cdot 3^{\frac{1}{72}} \cdot A^{-\frac{4}{3}} \cdot \left\{ \Gamma\left(\tfrac{1}{3}\right) \right\}^{-\frac{2}{3}};$$

$$G\left(\tfrac{2}{3}\right) = \exp\left[\frac{1}{9} - \frac{\pi\sqrt{3}}{54} + \frac{\sqrt{3}\,\zeta\left(2, \frac{1}{3}\right)}{36\pi} \right] \cdot 2^{-\frac{1}{3}} \cdot 3^{\frac{13}{72}} \cdot \pi^{-\frac{1}{3}} \cdot A^{-\frac{4}{3}} \cdot \left\{ \Gamma\left(\tfrac{1}{3}\right) \right\}^{\frac{1}{3}};$$

$$G\left(\tfrac{1}{6}\right) = \exp\left[\frac{5}{72} + \frac{\pi\sqrt{3}}{36} - \frac{\sqrt{3}\,\zeta\left(2, \frac{1}{3}\right)}{24\pi} \right] \cdot 2^{\frac{19}{72}} \cdot 3^{-\frac{61}{144}} \cdot \pi^{\frac{5}{12}} \cdot A^{-\frac{5}{6}} \cdot \left\{ \Gamma\left(\tfrac{1}{3}\right) \right\}^{-\frac{5}{3}};$$

$$G\left(\tfrac{5}{6}\right) = \exp\left[\frac{5}{72} - \frac{\pi\sqrt{3}}{36} + \frac{\sqrt{3}\,\zeta\left(2, \frac{1}{3}\right)}{24\pi} \right] \cdot 2^{-\frac{17}{72}} \cdot 3^{\frac{11}{144}} \pi^{-\frac{1}{4}} \cdot A^{-\frac{5}{6}} \cdot \left\{ \Gamma\left(\tfrac{1}{3}\right) \right\}^{\frac{1}{3}}.$$

(Gosper [498]; Choi and Srivastava [289, 294])

2. Applying the results of Problem 1, derive the following closed-form evaluations of series involving the Riemann Zeta function:

$$\sum_{k=2}^{\infty} (-1)^k \frac{\zeta(k)}{(k+1)\cdot 3^k} = 1 + \frac{\gamma}{6} + \frac{\pi\sqrt{3}}{18} - \frac{\sqrt{3}}{12\pi} \zeta\left(2, \frac{1}{3}\right)$$
$$+ \log\left(2^{-\frac{1}{2}} \cdot 3^{\frac{1}{24}} \cdot \pi^{-\frac{1}{2}} \cdot A^{-4} \cdot \Gamma\left(\tfrac{1}{3}\right) \right);$$

$$\sum_{k=2}^{\infty} (-1)^k \frac{\zeta(k) - 1}{(k+1)\cdot 3^k} = \frac{11}{6} + \frac{\gamma}{6} + \frac{\pi\sqrt{3}}{18} - \frac{\sqrt{3}}{12\pi} \zeta\left(2, \frac{1}{3}\right)$$
$$+ \log\left(2^{-\frac{13}{2}} \cdot 3^{\frac{73}{24}} \cdot \pi^{-\frac{1}{2}} \cdot A^{-4} \cdot \Gamma\left(\tfrac{1}{3}\right) \right);$$

$$\sum_{k=2}^{\infty} \frac{\zeta(k)}{(k+1)\cdot 3^k} = -\frac{\gamma}{6} + \frac{\pi\sqrt{3}}{18} - \frac{\sqrt{3}12\pi}{\zeta}\left(2, \frac{1}{3}\right)$$
$$+\log\left(2^{\frac{1}{2}} \cdot 3^{-\frac{13}{24}} \cdot \pi^{\frac{1}{2}} \cdot A^4 \cdot \left\{\Gamma\left(\tfrac{1}{3}\right)\right\}^{-1}\right);$$

$$\sum_{k=2}^{\infty} \frac{\zeta(k)-1}{(k+1)\cdot 3^k} = \frac{7}{6} - \frac{\gamma}{6} + \frac{\pi\sqrt{3}}{18} - \frac{\sqrt{3}}{12\pi}\zeta\left(2, \frac{1}{3}\right)$$
$$+\log\left(2^{\frac{7}{2}} \cdot 3^{-\frac{85}{24}} \cdot \pi^{\frac{1}{2}} \cdot A^4 \cdot \left\{\Gamma\left(\tfrac{1}{3}\right)\right\}^{-1}\right);$$

$$\sum_{k=1}^{\infty} \frac{\zeta(2k)}{(2k+1)\cdot 3^{2k}} = \frac{1}{2} + \frac{\pi\sqrt{3}}{18} - \frac{\sqrt{3}}{12\pi}\zeta\left(2, \frac{1}{3}\right) - \frac{1}{4}\log 3;$$

$$\sum_{k=1}^{\infty} \frac{\zeta(2k)-1}{(2k+1)\cdot 3^{2k}} = \frac{11}{6} + \frac{\pi\sqrt{3}}{18} - \frac{\sqrt{3}}{12\pi}\zeta\left(2, \frac{1}{3}\right) + \log\left(2^{-\frac{3}{2}} \cdot 3^{-\frac{1}{4}}\right);$$

$$\sum_{k=1}^{\infty} \frac{\zeta(2k+1)}{(k+1)\cdot 3^{2k}} = -3 - \gamma + \log\left(2^3 \cdot 3^{-\frac{7}{4}} \cdot \pi^3 \cdot A^{24} \cdot \left\{\Gamma\left(\tfrac{1}{3}\right)\right\}^{-6}\right);$$

$$\sum_{k=1}^{\infty} \frac{\zeta(2k+1)-1}{(k+1)\cdot 3^{2k}} = -2 - \gamma + \log\left(2^{30} \cdot 3^{-\frac{79}{4}} \cdot \pi^3 \cdot A^{24} \cdot \left\{\Gamma\left(\tfrac{1}{3}\right)\right\}^{-6}\right);$$

$$\sum_{k=2}^{\infty} (-1)^k \frac{\zeta(k)}{k+1}\left(\frac{2}{3}\right)^k = 1 + \frac{\gamma}{3} - \frac{\pi\sqrt{3}}{36} + \frac{\sqrt{3}}{24\pi}\zeta\left(2, \frac{1}{3}\right)$$
$$+\log\left(2^{\frac{1}{2}} \cdot 3^{-\frac{23}{48}} \cdot \pi^{\frac{1}{2}} \cdot A^{-2} \cdot \left\{\Gamma\left(\tfrac{1}{3}\right)\right\}^{-1}\right);$$

$$\sum_{k=2}^{\infty} (-1)^k \frac{\zeta(k)-1}{k+1}\left(\frac{2}{3}\right)^k = \frac{5}{3} + \frac{\gamma}{3} - \frac{\pi\sqrt{3}}{36} + \frac{\sqrt{3}}{24\pi}\zeta\left(2, \frac{1}{3}\right)$$
$$+\log\left(2^{\frac{1}{2}} \cdot 3^{\frac{49}{48}} \cdot 5^{-\frac{3}{2}} \cdot \pi^{\frac{1}{2}} \cdot A^{-2} \cdot \left\{\Gamma\left(\tfrac{1}{3}\right)\right\}^{-1}\right);$$

$$\sum_{k=2}^{\infty} \frac{\zeta(k)}{k+1}\left(\frac{2}{3}\right)^k = -\frac{\gamma}{3} - \frac{\pi\sqrt{3}}{36} + \frac{\sqrt{3}}{24\pi}\zeta\left(2, \frac{1}{3}\right)$$
$$+\log\left(2^{-\frac{1}{2}} \cdot 3^{-\frac{1}{48}} \cdot \pi^{-\frac{1}{2}} \cdot A^2 \cdot \Gamma\left(\tfrac{1}{3}\right)\right);$$

$$\sum_{k=2}^{\infty} \frac{\zeta(k)-1}{k+1}\left(\frac{2}{3}\right)^k = \frac{4}{3} - \frac{\gamma}{3} - \frac{\pi\sqrt{3}}{36} + \frac{\sqrt{3}}{24\pi}\zeta\left(2, \frac{1}{3}\right)$$
$$+\log\left(2^{-\frac{1}{2}} \cdot 3^{-\frac{73}{48}} \cdot \pi^{-\frac{1}{2}} \cdot A^2 \cdot \Gamma\left(\tfrac{1}{3}\right)\right);$$

$$\sum_{k=1}^{\infty} \frac{\zeta(2k)}{2k+1}\left(\frac{2}{3}\right)^{2k} = \frac{1}{2} - \frac{\pi\sqrt{3}}{36} + \frac{\sqrt{3}}{24\pi}\zeta\left(2, \frac{1}{3}\right) - \frac{1}{4}\log 3;$$

$$\sum_{k=1}^{\infty} \frac{\zeta(2k)-1}{2k+1}\left(\frac{2}{3}\right)^{2k} = \frac{3}{2} - \frac{\pi\sqrt{3}}{36} + \frac{\sqrt{3}}{24\pi}\zeta\left(2, \frac{1}{3}\right) + \log\left(3^{-\frac{1}{4}} \cdot 5^{-\frac{3}{4}}\right);$$

$$\sum_{k=1}^{\infty} \frac{\zeta(2k+1)}{k+1} \left(\frac{2}{3}\right)^{2k} = -\frac{3}{2} - \gamma + \log\left(2^{-\frac{3}{2}} \cdot 3^{\frac{11}{16}} \cdot \pi^{-\frac{3}{2}} \cdot A^6 \cdot \left\{\Gamma\left(\tfrac{1}{3}\right)\right\}^3\right);$$

$$\sum_{k=1}^{\infty} \frac{\zeta(2k+1)-1}{k+1} \left(\frac{2}{3}\right)^{2k} = -\frac{1}{2} - \gamma + \log\left(2^{-\frac{3}{2}} \cdot 3^{-\frac{61}{16}} \cdot 5^{\frac{9}{4}} \cdot \pi^{-\frac{3}{2}} \cdot A^6 \cdot \left\{\Gamma\left(\tfrac{1}{3}\right)\right\}^3\right);$$

$$\sum_{k=2}^{\infty} (-1)^k \frac{\zeta(k)}{(k+1)\cdot 6^k} = 1 + \frac{\gamma}{12} + \frac{\pi\sqrt{3}}{6} - \frac{\sqrt{3}}{4\pi}\zeta\left(2, \frac{1}{3}\right)$$
$$+ \log\left(2^{-\frac{11}{12}} \cdot 3^{\frac{11}{24}} \cdot \pi^{-1} \cdot A^{-5} \cdot \left\{\Gamma\left(\tfrac{1}{3}\right)\right\}^2\right);$$

$$\sum_{k=2}^{\infty} (-1)^k \frac{\zeta(k)-1}{(k+1)\cdot 6^k} = \frac{23}{12} + \frac{\gamma}{12} + \frac{\pi\sqrt{3}}{6} - \frac{\sqrt{3}}{4\pi}\zeta\left(2, \frac{1}{3}\right)$$
$$+ \log\left(2^{\frac{61}{12}} \cdot 3^{\frac{155}{24}} \cdot 7^{-6} \cdot \pi^{-1} \cdot A^{-5} \cdot \left\{\Gamma\left(\tfrac{1}{3}\right)\right\}^2\right);$$

$$\sum_{k=2}^{\infty} \frac{\zeta(k)}{(k+1)\cdot 6^k} = -\frac{\gamma}{12} + \frac{\pi\sqrt{3}}{6} - \frac{\sqrt{3}}{4\pi}\zeta\left(2, \frac{1}{3}\right)$$
$$+ \log\left(2^{\frac{11}{12}} \cdot 3^{-\frac{11}{24}} \cdot \pi \cdot A^5 \cdot \left\{\Gamma\left(\tfrac{1}{3}\right)\right\}^{-2}\right);$$

$$\sum_{k=2}^{\infty} \frac{\zeta(k)-1}{(k+1)\cdot 6^k} = \frac{13}{12} - \frac{\gamma}{12} + \frac{\pi\sqrt{3}}{6} - \frac{\sqrt{3}}{4\pi}\zeta\left(2, \frac{1}{3}\right)$$
$$+ \log\left(2^{-\frac{61}{12}} \cdot 3^{-\frac{155}{24}} \cdot 5^6 \cdot \pi \cdot A^5 \cdot \left\{\Gamma\left(\tfrac{1}{3}\right)\right\}^{-2}\right);$$

$$\sum_{k=1}^{\infty} \frac{\zeta(2k)}{(2k+1)\cdot 6^{2k}} = \frac{1}{2} + \frac{\pi\sqrt{3}}{6} - \frac{\sqrt{3}}{4\pi}\zeta\left(2, \frac{1}{3}\right);$$

$$\sum_{k=1}^{\infty} \frac{\zeta(2k)-1}{(2k+1)\cdot 6^{2k}} = \frac{3}{2} + \frac{\pi\sqrt{3}}{6} - \frac{\sqrt{3}}{4\pi}\zeta\left(2, \frac{1}{3}\right) + \log\left(5^3 \cdot 7^{-3}\right);$$

$$\sum_{k=1}^{\infty} \frac{\zeta(2k+1)}{(k+1)\cdot 6^{2k}} = -6 - \gamma + \log\left(2^{11} \cdot 3^{-\frac{11}{2}} \cdot \pi^{12} \cdot A^{60} \cdot \left\{\Gamma\left(\tfrac{1}{3}\right)\right\}^{-24}\right);$$

$$\sum_{k=1}^{\infty} \frac{\zeta(2k+1)-1}{(k+1)\cdot 6^{2k}} = -5 - \gamma$$
$$+ \log\left(2^{-61} \cdot 3^{-\frac{155}{2}} \cdot 5^{36} \cdot 7^{36} \cdot \pi^{12} \cdot A^{60} \cdot \left\{\Gamma\left(\tfrac{1}{3}\right)\right\}^{-24}\right);$$

$$\sum_{k=2}^{\infty} (-1)^k \frac{\zeta(k)}{k+1} \left(\frac{5}{6}\right)^k = 1 + \frac{5}{12}\gamma - \frac{\pi\sqrt{3}}{30} + \frac{\sqrt{3}}{20\pi}\zeta\left(2, \frac{1}{3}\right)$$
$$+ \log\left(2^{\frac{49}{60}} \cdot 3^{-\frac{61}{120}} \cdot \pi \cdot A^{-1} \cdot \left\{\Gamma\left(\tfrac{1}{3}\right)\right\}^{-2}\right);$$

$$\sum_{k=2}^{\infty}(-1)^k\frac{\zeta(k)-1}{k+1}\left(\frac{5}{6}\right)^k=\frac{19}{12}+\frac{5}{12}\gamma-\frac{\pi\sqrt{3}}{30}+\frac{\sqrt{3}}{20\pi}\zeta\left(2,\frac{1}{3}\right)$$
$$+\log\left(2^{\frac{121}{60}}\cdot 3^{\frac{83}{120}}\cdot 11^{-\frac{6}{5}}\cdot\pi\cdot A^{-1}\cdot\left\{\Gamma\left(\tfrac{1}{3}\right)\right\}^{-2}\right);$$

$$\sum_{k=2}^{\infty}\frac{\zeta(k)}{k+1}\left(\frac{5}{6}\right)^k=-\frac{5}{12}\gamma-\frac{\pi\sqrt{3}}{30}+\frac{\sqrt{3}}{20\pi}\zeta\left(2,\frac{1}{3}\right)$$
$$+\log\left(2^{-\frac{49}{60}}\cdot 3^{\frac{61}{120}}\cdot\pi^{-1}\cdot A\cdot\left\{\Gamma\left(\tfrac{1}{3}\right)\right\}^{2}\right);$$

$$\sum_{k=2}^{\infty}\frac{\zeta(k)-1}{k+1}\left(\frac{5}{6}\right)^k=\frac{17}{12}-\frac{5}{12}\gamma-\frac{\pi\sqrt{3}}{30}+\frac{\sqrt{3}}{20\pi}\zeta\left(2,\frac{1}{3}\right)$$
$$+\log\left(2^{-\frac{121}{60}}\cdot 3^{-\frac{83}{120}}\cdot\pi^{-1}\cdot A\cdot\left\{\Gamma\left(\tfrac{1}{3}\right)\right\}^{2}\right);$$

$$\sum_{k=1}^{\infty}\frac{\zeta(2k)}{2k+1}\left(\frac{5}{6}\right)^{2k}=\frac{1}{2}-\frac{\pi\sqrt{3}}{30}+\frac{\sqrt{3}}{20\pi}\zeta\left(2,\frac{1}{3}\right);$$

$$\sum_{k=1}^{\infty}\frac{\zeta(2k)-1}{2k+1}\left(\frac{5}{6}\right)^{2k}=\frac{3}{2}-\frac{\pi\sqrt{3}}{30}+\frac{\sqrt{3}}{20\pi}\zeta\left(2,\frac{1}{3}\right)-\frac{3}{5}\log 11;$$

$$\sum_{k=1}^{\infty}\frac{\zeta(2k+1)}{k+1}\left(\frac{5}{6}\right)^{2k}=-\frac{6}{5}-\gamma+\log\left(2^{-\frac{49}{25}}\cdot 3^{\frac{61}{50}}\cdot\pi^{-\frac{12}{5}}\cdot A^{\frac{12}{5}}\cdot\left\{\Gamma\left(\tfrac{1}{3}\right)\right\}^{\frac{24}{5}}\right);$$

$$\sum_{k=1}^{\infty}\frac{\zeta(2k+1)-1}{k+1}\left(\frac{5}{6}\right)^{2k}=-\frac{1}{5}-\gamma$$
$$+\log\left(2^{-\frac{121}{25}}\cdot 3^{-\frac{83}{50}}\cdot 11^{\frac{36}{25}}\cdot\pi^{-\frac{12}{5}}\cdot A^{\frac{12}{5}}\cdot\left\{\Gamma\left(\tfrac{1}{3}\right)\right\}^{\frac{24}{5}}\right);$$

$$\sum_{k=2}^{\infty}(-1)^k\frac{\zeta(k)-1}{k+1}\left(\frac{4}{3}\right)^k=\frac{19}{12}+\frac{2}{3}\gamma+\frac{\pi\sqrt{3}}{72}-\frac{\sqrt{3}}{48\pi}\zeta\left(2,\frac{1}{3}\right)$$
$$+\log\left(2^{-\frac{1}{2}}\cdot 3^{\frac{1}{96}}\cdot 7^{-\frac{3}{4}}\cdot\pi^{-\frac{1}{2}}\cdot A^{-1}\cdot\Gamma\left(\tfrac{1}{3}\right)\right);$$

$$\sum_{k=2}^{\infty}\frac{\zeta(k)-1}{k+1}\left(\frac{4}{3}\right)^k=\frac{17}{12}-\frac{2}{3}\gamma+\frac{\pi\sqrt{3}}{72}-\frac{\sqrt{3}}{48\pi}\zeta\left(2,\frac{1}{3}\right)$$
$$+\log\left(2^{\frac{1}{2}}\cdot 3^{-\frac{49}{96}}\cdot\pi^{\frac{1}{2}}\cdot A\cdot\left\{\Gamma\left(\tfrac{1}{3}\right)\right\}^{-1}\right);$$

$$\sum_{k=1}^{\infty}\frac{\zeta(2k)-1}{2k+1}\left(\frac{4}{3}\right)^{2k}=\frac{3}{2}+\frac{\pi\sqrt{3}}{72}-\frac{\sqrt{3}}{48\pi}\zeta\left(2,\frac{1}{3}\right)+\log\left(3^{-\frac{1}{4}}\cdot 7^{-\frac{3}{8}}\right);$$

$$\sum_{k=1}^{\infty}\frac{\zeta(2k+1)-1}{k+1}\left(\frac{4}{3}\right)^{2k}=-\frac{1}{8}-\gamma+\log\left(2^{\frac{3}{4}}\cdot 3^{-\frac{25}{64}}\cdot 7^{\frac{9}{16}}\cdot\pi^{\frac{3}{4}}\cdot A^{\frac{3}{2}}\cdot\left\{\Gamma\left(\tfrac{1}{3}\right)\right\}^{-\frac{3}{2}}\right);$$

$$\sum_{k=2}^{\infty}(-1)^k\frac{\zeta(k)-1}{k+1}\left(\frac{5}{3}\right)^k = \frac{47}{30}+\frac{5}{6}\gamma-\frac{\pi\sqrt{3}}{90}+\frac{\sqrt{3}}{60\pi}\zeta\left(2,\frac{1}{3}\right) + \log\left(2^{-\frac{7}{10}}\cdot 3^{-\frac{59}{120}}\cdot\pi^{\frac{1}{2}}\cdot A^{-\frac{4}{5}}\cdot\left\{\Gamma\left(\tfrac{1}{3}\right)\right\}^{-1}\right);$$

$$\sum_{k=2}^{\infty}\frac{\zeta(k)-1}{k+1}\left(\frac{5}{3}\right)^k = \frac{43}{30}-\frac{5}{6}\gamma-\frac{\pi\sqrt{3}}{90}+\frac{\sqrt{3}}{60\pi}\zeta\left(2,\frac{1}{3}\right) + \log\left(2^{-\frac{1}{2}}\cdot 3^{-\frac{1}{120}}\cdot\pi^{-\frac{1}{2}}\cdot A^{\frac{4}{5}}\cdot\Gamma\left(\tfrac{1}{3}\right)\right);$$

$$\sum_{k=1}^{\infty}\frac{\zeta(2k)-1}{2k+1}\left(\frac{5}{3}\right)^{2k} = \frac{3}{2}-\frac{\pi\sqrt{3}}{90}+\frac{\sqrt{3}}{60\pi}\zeta\left(2,\frac{1}{3}\right)+\log\left(2^{-\frac{3}{5}}\cdot 3^{-\frac{1}{4}}\right);$$

$$\sum_{k=1}^{\infty}\frac{\zeta(2k+1)-1}{k+1}\left(\frac{5}{3}\right)^{2k} = -\frac{2}{25}-\gamma+\log\left(2^{\frac{3}{25}}\cdot 3^{\frac{29}{100}}\cdot\pi^{-\frac{3}{5}}\cdot A^{\frac{24}{25}}\cdot\left\{\Gamma\left(\tfrac{1}{3}\right)\right\}^{\frac{6}{5}}\right);$$

$$\sum_{k=2}^{\infty}(-1)^k\frac{\zeta(k)-1}{k+1}\left(\frac{7}{6}\right)^k = \frac{131}{84}+\frac{7}{12}\gamma+\frac{\pi\sqrt{3}}{42}-\frac{\sqrt{3}}{28\pi}\zeta\left(2,\frac{1}{3}\right) + \log\left(2^{-\frac{71}{84}}\cdot 3^{\frac{83}{168}}\cdot 13^{-\frac{6}{7}}\cdot\pi^{-1}\cdot A^{-\frac{5}{7}}\cdot\left\{\Gamma\left(\tfrac{1}{3}\right)\right\}^{2}\right);$$

$$\sum_{k=2}^{\infty}\frac{\zeta(k)-1}{k+1}\left(\frac{7}{6}\right)^k = \frac{121}{84}-\frac{7}{12}\gamma+\frac{\pi\sqrt{3}}{42}-\frac{\sqrt{3}}{28\pi}\zeta\left(2,\frac{1}{3}\right) + \log\left(2^{\frac{71}{84}}\cdot 3^{-\frac{83}{168}}\cdot\pi\cdot A^{\frac{5}{7}}\cdot\left\{\Gamma\left(\tfrac{1}{3}\right)\right\}^{-2}\right);$$

$$\sum_{k=1}^{\infty}\frac{\zeta(2k)-1}{2k+1}\left(\frac{7}{6}\right)^{2k} = \frac{3}{2}+\frac{\pi\sqrt{3}}{42}-\frac{\sqrt{3}}{28\pi}\zeta\left(2,\frac{1}{3}\right)-\frac{3}{7}\log 13;$$

$$\sum_{k=1}^{\infty}\frac{\zeta(2k+1)-1}{k+1}\left(\frac{7}{6}\right)^{2k} = -\frac{5}{49}-\gamma+\log\left(2^{\frac{71}{49}}\cdot 3^{-\frac{83}{98}}\cdot\pi^{\frac{12}{7}}\cdot A^{\frac{60}{49}}\cdot\left\{\Gamma\left(\tfrac{1}{3}\right)\right\}^{-\frac{24}{7}}\right);$$

$$\sum_{k=2}^{\infty}(-1)^k\frac{\zeta(k)-1}{k+1}\left(\frac{11}{6}\right)^k = \frac{203}{132}+\frac{11}{12}\gamma-\frac{\pi\sqrt{3}}{66}+\frac{\sqrt{3}}{44\pi}\zeta\left(2,\frac{1}{3}\right) + \log\left(2^{\frac{109}{132}}\cdot 3^{-\frac{133}{264}}\cdot 5^{\frac{6}{11}}\cdot 17^{-\frac{6}{11}}\cdot\pi\cdot A^{-\frac{5}{11}}\cdot\left\{\Gamma\left(\tfrac{1}{3}\right)\right\}^{-2}\right);$$

$$\sum_{k=2}^{\infty}\frac{\zeta(k)-1}{k+1}\left(\frac{11}{6}\right)^k = \frac{193}{132}-\frac{11}{12}\gamma-\frac{\pi\sqrt{3}}{66}+\frac{\sqrt{3}}{44\pi}\zeta\left(2,\frac{1}{3}\right) + \log\left(2^{-\frac{109}{132}}\cdot 3^{\frac{133}{264}}\cdot\pi^{-1}\cdot A^{\frac{5}{11}}\cdot\left\{\Gamma\left(\tfrac{1}{3}\right)\right\}^{2}\right);$$

$$\sum_{k=1}^{\infty}\frac{\zeta(2k)-1}{2k+1}\left(\frac{11}{6}\right)^{2k} = \frac{3}{2}-\frac{\pi\sqrt{3}}{66}+\frac{\sqrt{3}}{44\pi}\zeta\left(2,\frac{1}{3}\right)+\log\left(5^{\frac{3}{11}}\cdot 17^{-\frac{3}{11}}\right);$$

$$\sum_{k=1}^{\infty} \frac{\zeta(2k+1)-1}{k+1} \left(\frac{11}{6}\right)^{2k} = -\frac{5}{121} - \gamma + \log\left(2^{-\frac{109}{121}} \cdot 3^{\frac{133}{242}} \cdot 5^{-\frac{36}{121}} \cdot 17^{\frac{36}{121}} \cdot \pi^{-\frac{12}{11}} \cdot A^{\frac{60}{121}} \cdot \left\{\Gamma\left(\tfrac{1}{3}\right)\right\}^{\frac{24}{11}}\right).$$

(Choi and Srivastava [289, 294])

3. Prove that

$$\sum_{m=2}^{\infty} \frac{\zeta(m,a)}{m+\lambda} z^{m+\lambda} = \sum_{k=0}^{\lambda} \binom{\lambda}{k} \zeta'(-k, a-z)\, z^{\lambda-k} - \zeta'(-\lambda, a) - \sum_{\ell=0}^{\lambda-1} \frac{\zeta(-\ell, a)}{\lambda-\ell} z^{\lambda-\ell} - \frac{z^{\lambda+1}}{\lambda+1} [\psi(\lambda+1) - \psi(a) + \gamma] \quad (\lambda \in \mathbb{N}_0;\ |z| < |a|).$$

(Kanemitsu *et al.* [628])

4. Prove the following identity:

$$\sum_{k=2}^{\infty} \frac{(-1)^k}{(k+1)(k+2)} \zeta(k) = \frac{\gamma}{6} + \frac{1}{2} + \frac{5}{18} \log 2 + \frac{1}{3} \log \pi - \frac{4}{3} \int_0^{\frac{1}{2}} \log \Gamma(t)\, dt,$$

where γ denotes the Euler-Mascheroni constant defined by 1.1(2).

(Janos and Srivastava [606])

5. Prove the following companion of the summation formulas deduced in Problems 46 and 47 (Chapter 1):

$$\zeta(k-1) = \sum_{n=k}^{\infty} \frac{(-1)^{n+k}}{(n-1)(n-2)\cdot(n-1)!} s(n,k).$$

(*cf.* Jordan [614, p. 339]; see also Equation 3.5(16) *and* Hansen [531, p. 348])

6. For $\Phi(z,s,a)$ defined by 2.5(1), derive the following generalization of the expansion formula 3.2(7):

$$\sum_{n=0}^{\infty} \frac{(\lambda)_n}{n!} \Phi(z, \lambda+n, a)\, t^n = \Phi(z, \lambda, a-t) \quad (|t| < |a|;\ \lambda \neq 1).$$

(*cf.* Equation 2.5(33))

7. For every nonnegative integer ℓ, prove the following formula:

$$\sum_{k=2}^{\infty} \Phi_n(z,k,a) \frac{t^{k+\ell}}{k+\ell} = \sum_{k=0}^{\ell} \binom{\ell}{k} \Phi_n'(z,-k,a-k)\, t^{\ell-k} - \sum_{k=0}^{\ell-1} \Phi_n(z,-k,a) \frac{t^{\ell-k}}{\ell-k} - \left[H_\ell L_{n,1}(z,a) + L_{n,2}(z,a)\right] \frac{t^{\ell+1}}{\ell+1} - \Phi_n'(z,-\ell,a)$$

$$(|t| < a;\ |z| < 1;\ \ell \in \mathbb{N}_0),$$

where $L_{n,1}(z,a)$ and $L_{n,2}(z,a)$ are given, respectively, as

$$L_{n,1}(z,a) := \lim_{s\to -\ell}\{(s+\ell)\Phi_n(z,s+\ell+1,a)\}$$

and

$$L_{n,2}(z,a) := \lim_{s\to -\ell}\{\Phi_n(z,s+\ell+1,a)+(s+l)\Phi'_n(z,s+\ell+1,a)\}.$$

(See Choi *et al.* [273, Theorem 2])

8. Show that

$$\sum_{n=1}^{\infty}(-1)^{n-1}\frac{\left(H_{n-1}^2+H_n^2\right)O_n}{n}=\frac{\pi^4}{96}+\frac{\pi^2}{8}\log^2 2-\frac{7}{4}\log 2\,\zeta(3).$$

(See Zheng [1264])

9. Show that, for $\lambda,\ \mu\in\mathbb{N}_0$,

$$\sum_{k=0}^{n}\binom{n+\mu n}{k}\binom{n+\lambda n}{n-k}H_{\lambda n+k}$$
$$=\binom{2n+\lambda n+\mu n}{n}\left(H_{\lambda n+n}+H_{\lambda n+\mu n+n}-H_{\lambda n+\mu n+2n}\right).$$

(See Chu and de Donno [316, Theorem 1])

10. Show that

$$\sum_{k=1}^{\infty}\frac{\left(2^{2k-1}-1\right)}{(2p)^{2k}\,(m+k)}\,\zeta(2k)$$
$$=-\frac{1}{4m}-\gamma(p)+(-1)^m(2m)!\left(1-2^{-2m-1}\right)\left(\frac{p}{\pi}\right)^{2m}\zeta(2m+1)$$
$$-\frac{(2m)!}{2p}\sum_{k=1}^{m}\frac{(-1)^k(2\pi)^{-2k}}{(2m-2k)!}\sum_{\ell=0}^{p-1}\cos\left(\frac{(2\ell+1)\pi}{p}\right)\zeta\left(2k+1,\frac{2\ell+1}{2p}\right)$$
$$-\frac{\pi}{p}(2m)!\sum_{k=1}^{m}\frac{(-1)^k(2\pi)^{-2k}}{(2m-2k+1)!}\sum_{\ell=0}^{p-1}\sin\left(\frac{(2\ell+1)\pi}{p}\right)\zeta\left(2k,\frac{2\ell+1}{2p}\right)$$
$$(m\in\mathbb{N};\ p\in\mathbb{N}\setminus\{1\})$$

and

$$\sum_{k=1}^{\infty}\frac{\left(2^{2k-1}-1\right)}{(2p)^{2k}\,(2m+2k+1)}\,\zeta(2k)=-\frac{1}{4(2m+1)}-\frac{1}{2}\gamma(p)$$
$$+\frac{(2m+1)!}{8p\pi}\sum_{k=0}^{m}\frac{(-1)^k(2\pi)^{-2k}}{(2m-2k)!}\sum_{\ell=0}^{p-1}\sin\left(\frac{(2\ell+1)\pi}{p}\right)\zeta\left(2k+2,\frac{2\ell+1}{2p}\right)$$
$$-\frac{(2m+1)!}{4p}\sum_{k=1}^{m}\frac{(-1)^k(2\pi)^{-2k}}{(2m-2k+1)!}\sum_{\ell=0}^{p-1}\cos\left(\frac{(2\ell+1)\pi}{p}\right)\zeta\left(2k+1,\frac{2\ell+1}{2p}\right)$$
$$(m\in\mathbb{N}_0;\ p\in\mathbb{N}\setminus\{1\}),$$

where $\gamma(p)$ is given by

$$\gamma(p)=\sum_{j=0}^{\infty}\sum_{\ell=0}^{p-1}\frac{1}{2pj+2\ell+1}\cos\left(\frac{(2\ell+1)\pi}{p}\right).$$

(Cho *et al.* [257, Equations (40) and (41)])

11. Show that

$$\sum_{j=0}^{n}(1-5jH_j+5jH_{n-j})\binom{n}{j}^5=(-1)^n\sum_{j=0}^{n}\binom{n}{j}^2\binom{n+j}{j},$$

where H_n is the harmonic number.

(Paule and Schneider [890, Eq. (5)])

12. For $n\in\mathbb{N}$, define

$$F(n)=\sum_{j=1}^{n}\binom{n+j}{j}^2\binom{n}{j}^2(1+2jH_{n+j}+2jH_{n-j}-4jH_j).$$

Show that $F(n)=0$.

(Ahlgren and Ono [14, Theorem 7])

13. A regularized version of Riemann Zeta function is

$$\zeta(s)-\frac{1}{s-1}=\sum_{n=0}^{\infty}(-1)^n b_n\binom{s}{n}.$$

Show that

$$b_n=n(1-\gamma-H_{n-1})-\frac{1}{2}+\sum_{k=2}^{\infty}\binom{n}{k}(-1)^k\zeta(k).$$

(Flajolet and Vepstas [455, Eq. (2)])

14. Show that

$$\int_0^t\log(\sin\theta)\,d\theta=t(\log t-1)-\sum_{k=1}^{\infty}\frac{(-1)^{k+1}B_{2k}}{4k(2k+1)!}(2t)^{2k+1}\qquad(|t|<\pi).$$

(Monegato and Strozzi [839, Eq. (17)])

15. Show that

$$\sum_{k=0}^{n}\left(\sum_{j=0}^{k}\binom{n}{j}\right)^3=\frac{n}{2}8^n+8^n-\frac{3n}{4}2^n\binom{2n}{n}.$$

(Zhang and Wang [1256, Eq. (1)])

16. Let $\{\Omega_n\}_{n\in\mathbb{N}_0}$ be a suitably bounded sequence of complex numbers. Also let the parameters λ and μ be constrained by

$$\Re(\lambda-\mu)>0\qquad(\lambda\notin\mathbb{Z}_0^-).$$

Then show that

$$\sum_{n=1}^{\infty} \frac{(\nu)_n}{n(\lambda)_n} \sum_{k=0}^{\infty} \frac{\Omega_k}{(\lambda+n)_k} \frac{z^k}{k!} = \sum_{k=0}^{\infty} \frac{\Omega_k}{(\lambda)_k} [\psi(\lambda+k) - \psi(\lambda-\nu+k)] \frac{z^k}{k!}$$

and

$$\sum_{n=1}^{\infty} \frac{(\nu)_n}{n(\lambda)_n} \sum_{k=0}^{\infty} \frac{(\nu+n)_k \, \Omega_k}{(\lambda+n)_k} \frac{z^k}{k!} = \sum_{k=0}^{\infty} \frac{(\nu)_k \, \Omega_k}{(\lambda)_k} [\psi(\lambda+k) - \psi(\lambda-\nu)] \frac{z^k}{k!},$$

provided that the series involved converge absolutely.

(Nishimoto and Srivastava [863, p. 104]; see also 3.5(12) and R. Srivastava [1115] for *further* results of these classes)

17. Derive the following hypergeometric identity:

$${}_8F_7\left[\begin{array}{r} 1+a, 2+\frac{1}{2}a, 1+b, 1+c, 2+2a-b-c+N, 1-N, 1, 1; \\ 2+a, 1+\frac{1}{2}a, 2+a-b, 2+a-c, 1+b+c-a-N, 2+a+N, 2; \end{array} 1\right]$$

$$= \frac{(1+a)(1+a-b)(1+a-c)(b+c-a-N)(1+a+N)}{(2+a)bc(1+2a-b-c+N)N}$$
$$\cdot \big[\psi(1+a) - \psi(1+a+N) + \psi(1+a-b+N) - \psi(1+a-b)$$
$$+ \psi(1+a-c+N) - \psi(1+a-c)$$
$$+ \psi(1+a-b-c) - \psi(1+a-b-c+N)\big] \quad (N \in \mathbb{N}),$$

provided that no zeros appear in the denominator of either member.

(Srivastava [1079, p. 80, Eq. (1.3)])

18. Prove the following identity:

$$\sum_{k=1}^{n} \frac{\binom{n}{k}\binom{m}{k}}{k\binom{n+k}{k}\binom{m+k}{k}} \left(H_{k-1}^2 + H_k^2\right) = \frac{1}{12} e_0^3 + \frac{1}{4} e_0 e_1 + \frac{1}{6} e_2,$$

where $m, n \in \mathbb{N};\ n \leqq m$ and

$$e_\ell := H_n^{(\ell+1)} + H_m^{(\ell+1)} - H_{n+m}^{(\ell+1)} \quad (\ell \in \mathbb{N}_0).$$

(Choi [266, Eq. (3.20)])

4 Evaluations and Series Representations

Based essentially on some of the most recent works on the subject by Srivastava [1084, 1086], Srivastava and Tsumura [1111] and others, this chapter aims at investigating rather systematically several interesting evaluations and representations of the Riemann Zeta function $\zeta(s)$ when $s \in \mathbb{N} \setminus \{1\}$. We begin by presenting some of the methods of evaluation of $\zeta(2n)$ $(n \in \mathbb{N})$. We then proceed to develop various families of rapidly convergent series representations for $\zeta(2n+1)$ $(n \in \mathbb{N})$. Finally, in one of the many computationally useful special cases considered here, we observe that $\zeta(3)$ can be represented by means of a series that converges much more rapidly than that in Euler's celebrated formula 3.1(8), as well as the series 4.2(2) used by Apéry [56] in his proof of the irrationality of $\zeta(3)$. Symbolic and numeric computations using *Mathematica* (Version 4.0) for Linux show, among other things, that only 50 terms of this series are capable of producing an accuracy of seven decimal places.

4.1 Evaluation of $\zeta(2n)$

The solution of the so-called *Basler problem* (*cf*., e.g., Spiess [1061, p. 66]):

$$\zeta(2) = \sum_{k=1}^{\infty} \frac{1}{k^2} = \frac{\pi^2}{6} \tag{1}$$

was first found in 1736 by Leonhard Euler (1707–1783), although Jakob Bernoulli (1654–1705) and Johann Bernoulli (1667–1748) did their utmost to sum the series in (1). In fact, the former of these Bernoulli brothers did not live to see the solution of the problem, and the solution became known to the latter soon after Euler found it (see, for details, Knopp [676, p. 238]).

The Basler problem (1) has been solved in the mathematical literature in many different ways. In addition to numerous papers containing elementary proofs of (1), many of which are referenced by Stark [1125], there are a fairly large number of books on complex analysis and advanced calculus in which (1) is proven by using Cauchy's residue calculus, Weierstrass's product theorem, Parseval's theorem, Fourier series expansions and so on. Some of these books were referred to by Choe [258], who seems to have rediscovered one of Euler's remarkably elementary proofs of (1)

Zeta and q-Zeta Functions and Associated Series and Integrals. DOI: 10.1016/B978-0-12-385218-2.00004-9

(*cf.*, e.g., Ayoub [81, p. 1079]), which is based on the integration of the following Taylor series expansion of $\arcsin x$ near $x = 0$:

$$\arcsin x = x + \sum_{k=1}^{\infty} \frac{1 \cdot 3 \cdot 5 \cdots (2k-1)}{2 \cdot 4 \cdot 6 \cdots (2k)} \frac{x^{2k+1}}{2k+1} \quad (-1 \leqq x \leqq 1) \tag{2}$$

with respect to t (where $x = \sin t$), by means of Wallis' integral formula in the form:

$$\int_0^{\pi/2} \sin^{2k+1} t \, dt = \frac{2 \cdot 4 \cdot 6 \cdots (2k)}{1 \cdot 3 \cdot 5 \cdots (2k+1)} \quad (k \in \mathbb{N}). \tag{3}$$

Motivated largely by the aforementioned (Euler's) proof of (1) detailed by Choe [258], Choi and Rathie [279] gave an essentially analogous derivation of (1), by appealing to the Gauss summation theorem 1.4(7). Subsequently, Choi, Rathie and Srivastava [281] showed how the derivation of (1) by Choi and Rathie [279] can be accomplished *without* using the Gauss summation theorem 1.4(7). They also presented several other evaluations of $\zeta(2)$ and related sums, including (for example) a fairly straightforward evaluation based on the theory of hypergeometric series.

Elementary and Hypergeometric Evaluations of $\zeta(2)$

Starting from the elementary integral:

$$\frac{1}{2}(\arcsin x)^2 = \int_0^x \frac{\arcsin t}{\sqrt{1-t^2}} \, dt \tag{4}$$

and replacing $\arcsin t$ by its Taylor series expansion given by (2), it is not difficult to obtain the representation:

$$\frac{1}{2}(\arcsin x)^2 = \sum_{k=0}^{\infty} \binom{2k}{k} \frac{2^{-2k}}{2k+1} \int_0^x \frac{t^{2k+1}}{\sqrt{1-t^2}} \, dt \quad (-1 \leqq x \leqq 1) \tag{5}$$

or, equivalently,

$$\frac{1}{2}(\arcsin x)^2 = \sum_{k=0}^{\infty} \frac{\left(\frac{1}{2}\right)_k}{k!} \frac{1}{2k+1} \int_0^x \frac{t^{2k+1}}{\sqrt{1-t^2}} \, dt \quad (-1 \leqq x \leqq 1). \tag{6}$$

Choi and Rathie [279, p. 396] express the second member of (6) as a hypergeometric ${}_2F_1$ series, by using the binomial expansion of $\left(1-t^2\right)^{-1/2}$ to evaluate the integral, set $x = 1$ and then apply the Gauss summation theorem 1.5(7).

The use of the Gauss summation theorem 1.5(7) in the above derivation of (1) by Choi and Rathie [279] can easily be avoided. Indeed, in the special case of (6) when $x = 1$, if we set $t = \sqrt{u}$ and apply the known integral formula 1.1(39) and 1.1(42), we readily find from (6) that

$$\sum_{k=0}^{\infty} \frac{1}{(2k+1)^2} = \frac{\pi^2}{8}, \tag{7}$$

which, in view of 2.3(1), is equivalent to (1).

By virtue of the relationships 2.5(26) and

$$\Phi\left(1, s, \frac{1}{2}\right) = \zeta\left(s, \frac{1}{2}\right) = (2^s - 1)\,\zeta(s), \tag{8}$$

2.5(17) immediately yields the hypergeometric representations:

$$\zeta(n,a) = a^{-n}\,{}_{n+1}F_n(1, a, \ldots, a;\ a+1, \ldots, a+1;\ 1) \quad (n \in \mathbb{N}\setminus\{1\}) \tag{9}$$

and

$$\zeta(n) = \left(\frac{2^n}{2^n - 1}\right) {}_{n+1}F_n\left(1, \frac{1}{2}, \ldots, \frac{1}{2};\ \tfrac{3}{2}, \ldots, \tfrac{3}{2};\ 1\right) \quad (n \in \mathbb{N}\setminus\{1\}). \tag{10}$$

In particular, by setting $n = 2$ in (10), we have

$$\zeta(2) = \frac{4}{3}\,{}_3F_2\left(1, \frac{1}{2}, \frac{1}{2};\ \tfrac{3}{2}, \tfrac{3}{2};\ 1\right), \tag{11}$$

which can be evaluated by applying Dixon's summation theorem 3.5(4) with

$$a = 1 \quad \text{and} \quad b = c = \frac{1}{2}$$

or by applying Whipple's summation theorem 3.5(6) with

$$a = b = \frac{1}{2}, \quad c = 1 \quad \text{and} \quad d = e = \frac{3}{2}.$$

We, thus, find from (11) and either one of the above applications that

$$\zeta(2) = \frac{4}{3}\left\{\Gamma\left(\tfrac{3}{2}\right)\right\}^3 \Gamma\left(\frac{1}{2}\right) = \frac{\pi^2}{6}. \tag{12}$$

Next, by appealing directly to the definition 2.3(1) or (9) with $a = 1$, we obtain the hypergeometric representation:

$$\zeta(n) = {}_{n+1}F_n(1, \ldots, 1;\ 2, \ldots, 2;\ 1) \qquad (n \in \mathbb{N}\setminus\{1\}). \tag{13}$$

The representations (10) and (13), together, yield the hypergeometric transformation:

$$
\begin{aligned}
&{}_{n+1}F_n\left(1, \frac{1}{2}, \ldots, \frac{1}{2};\ \tfrac{3}{2}, \ldots, \tfrac{3}{2};\ 1\right) \\
&\quad = (1-2^{-n})\,{}_{n+1}F_n(1, \ldots, 1;\ 2, \ldots, 2;\ 1) \qquad (n \in \mathbb{N} \setminus \{1\}),
\end{aligned} \tag{14}
$$

which, for $n = 2$, assumes the form:

$$
{}_3F_2\left(1, \frac{1}{2}, \frac{1}{2};\ \tfrac{3}{2}, \tfrac{3}{2};\ 1\right) = \frac{3}{4}\,{}_3F_2(1, 1, 1;\ 2, 2;\ 1). \tag{15}
$$

We remark in passing that, by making use of contiguous analogues of Dixon's summation theorem 3.5(4) and Whipple's summation theorem 3.5(6), which were given recently by Lavoie *et al.* [731, p. 268, Eq. (2); 732, p. 294, Eq. (4)] we can easily express the sums of many series [analogous to those occurring in (1) and (7)] in terms of $\zeta(2)$. The details are fairly straightforward.

Since

$$
\zeta(n) = \int_0^1 \cdots \int_0^1 \frac{dt_1 \ldots dt_n}{1 - t_1 \ldots t_n} \qquad (n \in \mathbb{N} \setminus \{1\}), \tag{16}
$$

it is not difficult to provide yet another solution of the Basler problem (1), by using the following change of variables:

$$
t_1 = \frac{v+u}{\sqrt{2}} \quad \text{and} \quad t_2 = \frac{v-u}{\sqrt{2}}
$$

in the double integral resulting from (16) when $n = 2$ (*cf.*, e.g., Ojha and Singh [874]).

The General Case of $\zeta(2n)$

Following Srivastava [1086], let us begin, by recalling the following summation formula for the Bernoulli polynomials (*cf.* Hansen [531, p. 342, Entry (50.11.10)]):

$$
\begin{aligned}
\sum_{k=0}^{[n/2]} \frac{(-n)_{2k}}{(2k)!} B_{2k}(x) B_{n-2k}(y) = \frac{1}{2}\big\{&(x+y-1)nB_{n-1}(x+y) \\
&+ (-1)^n (x-y)nB_{n-1}(x-y) \\
&- (n-1)\left[B_n(x+y) + (-1)^n B_n(x-y)\right]\big\},
\end{aligned} \tag{17}
$$

which, upon setting $x = 1$ and $y = 0$ (and replacing n by $2n$), yields

$$
\sum_{k=0}^{n} \binom{2n}{2k} B_{2k} B_{2n-2k} = -nB_{2n-1} - (2n-1)B_{2n} \qquad (n \in \mathbb{N}_0). \tag{18}
$$

Finally, by transposing the terms for $k = 0$ and $k = n$ in (18) to the right-hand side and making use of 1.7(7), we obtain

$$\sum_{k=1}^{n-1} \binom{2n}{2k} B_{2k} B_{2n-2k} = -(2n+1) B_{2n} \quad (n \in \mathbb{N} \setminus \{1\}), \tag{19}$$

which is, in view of the relationship 2.3(18), equivalent to 2.3(20), which (as noted there) can also be used to evaluate $\zeta(2n)$ $(n \in \mathbb{N} \setminus \{1\})$ with (1).

To give a *direct* proof of the summation formula 2.3(20), let us put (*cf.* Prudnikov [917])

$$\Omega(x) := \sum_{k=1}^{\infty} \exp\left(-\pi^2 k^2 x\right) \quad (x > 0), \tag{20}$$

so that

$$\int_0^x \Omega(x-t)\Omega(t)dt = x\Omega(x) + \frac{3}{2} \sum_{k=1}^{\infty} \frac{\exp\left(-\pi^2 k^2 x\right)}{\pi^2 k^2}, \tag{21}$$

which, by virtue of (1), readily yields

$$\lim_{x \to 0} \int_0^x \Omega(x-t)\Omega(t)dt = \frac{1}{4} \tag{22}$$

and

$$\int_0^x \Omega(x-t)\Omega(t)dt = O\left(xe^{-x}\right) \quad (x \to \infty). \tag{23}$$

Making use of (21) and (23), it is easily seen that

$$\begin{aligned} \mathcal{I}_s &:= \int_0^\infty \int_0^\infty (x+y)^{s-2} \Omega(x)\Omega(y) dx\, dy \\ &= \int_0^\infty x^{s-2} dx \int_0^x \Omega(x-t)\Omega(t)dt \\ &= \left(s + \tfrac{1}{2}\right) \frac{\Gamma(s-1)}{\pi^{2s}} \zeta(2s) \quad (\Re(s) > 1). \end{aligned} \tag{24}$$

Conversely, since

$$\int_0^\infty t^{s-1}\Omega(t)dt = \frac{\Gamma(s)}{\pi^{2s}}\zeta(2s) \quad \left(\Re(s) > 0;\ s \neq \tfrac{1}{2}\right), \tag{25}$$

which is an immediate consequence of the definition (20) and 2.3(1), an appeal to the Binomial theorem for $(x+y)^{n-2}$ in the form:

$$(x+y)^{n-2} = \sum_{k=1}^{n-1} \binom{n-2}{k-1} x^{k-1} y^{n-k-1} \quad (n \in \mathbb{N} \setminus \{1\}) \tag{26}$$

would lead us also to

$$\mathcal{I}_n = \frac{(n-2)!}{\pi^{2n}} \sum_{k=1}^{n-1} \zeta(2k)\zeta(2n-2k) \quad (n \in \mathbb{N} \setminus \{1\}). \tag{27}$$

Now, the summation formula 2.3(20) follows upon equating the two values of $\mathcal{I}_n$ given by (24) (*with* $s = n$) and (27).

Remark 1 The recursion formula (19) is a well-known (rather classical) result for Bernoulli numbers. It appears (for example) in Nielsen's book [861] and was also derived independently by Underwood [1177]. The equivalent result 2.3(20) was proven, by elementary methods by Williams [1229, p. 20, Theorem I] (see also Apostol [63, p. 427]).

Remark 2 In terms of the $\mathcal{L}$-function defined by [*cf.* Eq. 2.5(1)]

$$\mathcal{L}(s) = \sum_{k=0}^{\infty} \frac{(-1)^k}{(2k+1)^s} = 2^{-s}\,\Phi\left(-1, s, \tfrac{1}{2}\right) \quad (\Re(s) > 0), \tag{28}$$

Williams [1229, p. 22, Theorem II] gave an interesting companion of the result 2.3(20) in the form:

$$\sum_{k=1}^{n} \mathcal{L}(2k-1)\mathcal{L}(2n-2k+1) = \left(n - \tfrac{1}{2}\right)\left(1 - 2^{-2n}\right)\zeta(2n), \tag{29}$$

which appears erroneously in Hansen [531, p. 357, Entry (54.7.1)]. Since $\mathcal{L}(1)$ is the well-known Gregory series for $\dfrac{\pi}{4}$ (with $\mathcal{L}(2)$ being the familiar Catalan constant G), by setting $n = 1$ in (29), we immediately obtain

$$\zeta(2) = \frac{8}{3}\{\mathcal{L}(1)\}^2 = \frac{\pi^2}{6}. \tag{30}$$

Remark 3 In view of the constraint $\Re(s) > 0$ $\left(s \neq \frac{1}{2}\right)$ associated with the Mellin transform in (25), an earlier attempt by Kalla and Villalobos [623] to extend the summation formula 2.3(20), by expressing $\mathcal{I}_s$ as a non-terminating (infinite) sum analogous to (27), cannot be justified. Thus, the *main* result of Kalla and Villalobos [623, p. 17, Eq. (17)] holds true only in the finite form given already by the well-known result 2.3(20).

4.2 Rapidly Convergent Series for $\zeta(2n+1)$

On the subject of series representations for $\zeta(3)$, in addition to Euler's result 3.1(8), Chen and Srivastava [252] gave many series expressions for $\zeta(3)$, which are more rapidly convergent than that in 3.1(8), including

$$\zeta(3) = -\frac{8\pi^2}{5}\sum_{k=0}^{\infty}\frac{\zeta(2k)}{(2k+1)(2k+2)(2k+3)\,2^{2k}}. \tag{1}$$

And, as pointed out by (for example) Chen and Srivastava [252, pp. 180–181], another interesting series representation:

$$\zeta(3) = \frac{5}{2}\sum_{k=1}^{\infty}\frac{(-1)^{k-1}}{k^3\binom{2k}{k}}, \tag{2}$$

which played a key rôle in Apéry's proof of the irrationality of $\zeta(3)$ (see Apéry [56]), was proven independently by (among others) Hjortnaes [559], Gosper [496] and Apéry [56]. Furthermore, in their recent work on the Ray-Singer torsion and topological field theories, Nash and O'Connor [855, 856] obtained a number of remarkable integral expressions for $\zeta(3)$, including (for example) the following result [856, p. 1489 *et seq.*]:

$$\zeta(3) = \frac{2\pi^2}{7}\log 2 - \frac{8}{7}\int_0^{\pi/2} z^2 \cot z\, dz. \tag{3}$$

Since (Erdélyi *et al.* [421, p. 51, Eq. 1.20(3)])

$$z\cot z = -2\sum_{k=0}^{\infty}\zeta(2k)\left(\frac{z}{\pi}\right)^{2k} \qquad (|z| < \pi), \tag{4}$$

the result (3) is obviously equivalent to the series representation (*cf.* Dąbrowski [362, p. 202]; see also Chen and Srivastava [252, p. 191, Eq. (3.19)]):

$$\zeta(3) = \frac{2\pi^2}{7}\left(\log 2 + \sum_{k=0}^{\infty}\frac{\zeta(2k)}{(k+1)2^{2k}}\right). \tag{5}$$

Moreover, by integrating by parts, it is easily seen that

$$\int_0^{\pi/2} z^2 \cot z \, dz = -2 \int_0^{\pi/2} z \log \sin z \, dz, \tag{6}$$

so that the result (3) is equivalent also to the integral representation:

$$\zeta(3) = \frac{2\pi^2}{7} \log 2 + \frac{16}{7} \int_0^{\pi/2} z \log \sin z \, dz, \tag{7}$$

which was proven in the aforementioned 1772 paper by Euler (*cf.*, e.g., Ayoub [81, p. 1084]; see also Srivastava [1086]).

If follows from 3.2(7) that (*cf.*, e.g., Hansen [531, p. 359]; see also Srivastava [1084] where other references are also cited)

$$\sum_{k=0}^{\infty} \frac{(s)_{2k}}{(2k)!} \zeta(s+2k, a) t^{2k} = \frac{1}{2} \left[\zeta(s, a-t) + \zeta(s, a+t)\right] \quad (|t| < |a|). \tag{8}$$

In view of 2.3(3), the special case of the identity (8) when $a = 1$ and $t = \frac{1}{m}$ can be rewritten in the form:

$$\sum_{k=0}^{\infty} \frac{(s)_{2k}}{(2k)!} \frac{\zeta(s+2k)}{m^{2k}} = \begin{cases} (2^s - 1)\zeta(s) - 2^{s-1} & (m = 2) \\ \frac{1}{2}\left[(m^s - 1)\zeta(s) - m^s - \sum_{j=2}^{m-2} \zeta\left(s, \frac{j}{m}\right)\right] & (m \in \mathbb{N} \setminus \{1, 2\}), \end{cases} \tag{9}$$

where (as usual) an empty sum is interpreted as nil. In addition to the case $m = 2$, the formula (9) simplifies also in the cases when $m = 3$, 4 and 6, and we, thus, obtain the identities:

$$\sum_{k=0}^{\infty} \frac{(s)_{2k}}{(2k)!} \frac{\zeta(s+2k)}{3^{2k}} = \frac{1}{2}\left[(3^s - 1)\zeta(s) - 3^s\right], \tag{10}$$

$$\sum_{k=0}^{\infty} \frac{(s)_{2k}}{(2k)!} \frac{\zeta(s+2k)}{4^{2k}} = \frac{1}{2}\left[(4^s - 2^s)\zeta(s) - 4^s\right] \tag{11}$$

and

$$\sum_{k=0}^{\infty} \frac{(s)_{2k}}{(2k)!} \frac{\zeta(s+2k)}{6^{2k}} = \frac{1}{2}\left[(6^s - 3^s - 2^s + 1)\,\zeta(s) - 6^s\right], \tag{12}$$

respectively.

Identities of this kind seem to have first appeared in the work of Ramaswami [966], who actually proved the cases $m = 2$, 3 and 6 of the general result in (9). Each of these three identities of Ramaswami [966] can also be found in the work of Hansen [531, p. 357], who referred to Apostol [61] as his source for the identities (10) and (12) only. As a matter of fact, Apostol [61] reproduced the identities (10) and (12) from Ramaswami's work [966] and then proved an interesting *arithmetical* generalization of these identities (see also Klusch [673, p. 520]).

In its slightly variant form:

$$\sum_{k=1}^{\infty} \frac{(s+1)_{2k}}{(2k)!} \frac{\zeta(s+2k)}{2^{2k}} = (2^s - 2)\,\zeta(s), \tag{13}$$

the case $m = 2$ of the general result (9) was applied by Zhang and Williams [1251] (and, subsequently, by Cvijović and Klinowski [351]) with a view to finding two seemingly different series representations for $\zeta(2n+1)$ $(n \in \mathbb{N})$. Srivastava [1084] obtained much more rapidly converging series representations for $\zeta(2n+1)$ $(n \in \mathbb{N})$, chiefly by appealing appropriately to each of the aforementioned cases ($m = 2$, 3, 4 and 6) of the general result (9). Following his work (Srivastava [1084]), we begin, here, with the case $m = 2$ of the general result (9). Upon separating the first $n+1$ terms of the series occurring on the left-hand side, if we transpose the terms for $k = 0$ and $k = n$ to the right-hand side, we obtain

$$\begin{aligned} &\sum_{k=1}^{n-1} \frac{(s)_{2k}}{(2k)!} \frac{\zeta(s+2k)}{2^{2k}} + \sum_{k=n+1}^{\infty} \frac{(s)_{2k}}{(2k)!} \frac{\zeta(s+2k)}{2^{2k}} \\ &\quad = (2^s - 2)\zeta(s) - 2^{s-1} - \frac{(s)_{2n}}{(2n)!} \frac{\zeta(s+2n)}{2^{2n}}, \end{aligned} \tag{14}$$

which readily yields the identity:

$$\begin{aligned} &\sum_{k=1}^{n-1} \frac{(s)_{2k}}{(2k)!} 2^{2(n-k)} \zeta(s+2k) + \sum_{k=1}^{\infty} \frac{(s)_{2n+2k}}{(2n+2k)!} \frac{\zeta(s+2n+2k)}{2^{2k}} \\ &\quad = 2^{2n}(2^s - 2)\,\zeta(s) - 2^{s+2n-1} - \frac{(s)_{2n}}{(2n)!}\,\zeta(s+2n) \quad (n \in \mathbb{N}). \end{aligned} \tag{15}$$

Now, we apply the functional equation 2.3(12) in the first term on the right-hand side of (15) and divide both sides by $s+2n$. We, thus, find that

$$\sum_{k=1}^{n-1} \frac{(s)_{2k}}{(2k)!} 2^{2(n-k)} \left\{ \frac{\zeta(s+2k)}{s+2n} \right\} + \sum_{k=1}^{\infty} \frac{(s)_{2n}(s+2n+1)_{2k-1}}{(2n+2k)!} \frac{\zeta(s+2n+2k)}{2^{2k}}$$
$$= 2^{s+2n}(2^s-2)\pi^{s-1}\Gamma(1-s)\zeta(1-s) \left\{ \frac{\sin\left(\frac{1}{2}\pi s\right)}{s+2n} \right\} \tag{16}$$
$$- \left\{ \frac{2^{s+2n-1} + (s)_{2n}(2n)!\,\zeta(s+2n)}{s+2n} \right\} \quad (s \neq -2n;\ n \in \mathbb{N}).$$

It is easy to see from 1.6(10) that

$$\frac{d}{ds}\{(s)_{2n}\}\bigg|_{s=-2n} = -(2n)!\, H_{2n} \quad (n \in \mathbb{N}), \tag{17}$$

where H_n denotes the familiar harmonic numbers, defined by 3.2(36). We observe also that the following limit formula:

$$\lim_{s\to -2n} \left\{ \frac{\zeta(s+2k)}{s+2n} \right\} = \frac{(-1)^{n-k}}{2(2\pi)^{2(n-k)}} (2n-2k)!\,\zeta(2n-2k+1) \tag{18}$$
$$(k = 1, \ldots, n-1; \quad n \in \mathbb{N}\setminus\{1\})$$

is needed in the first sum on the left-hand side of (16) only when this sum is nonzero (that is, only when $n \in \mathbb{N}\setminus\{1\}$). Furthermore, by l'Hôpital's rule, we have

$$\lim_{s\to -2n} \left\{ \frac{2^{s+2n-1} + \frac{(s)_{2n}}{(2n)!}\zeta(s+2n)}{s+2n} \right\}$$
$$= \left[2^{s+2n-1}\log 2 + \frac{d}{ds}\{(s)_{2n}\} \cdot \frac{\zeta(s+2n)}{(2n)!} + \frac{(s)_{2n}}{(2n)!}\zeta'(s+2n) \right]\bigg|_{s=-2n} \tag{19}$$
$$= \frac{1}{2}(H_{2n} - \log\pi) \quad (n \in \mathbb{N}).$$

Finally, by letting $s \to -2n$ in (16) and making use of the limit relationships (18) and (19), we obtain Srivastava's first series representation for $\zeta(2n+1)$:

$$\zeta(2n+1) = (-1)^{n-1} \frac{(2\pi)^{2n}}{2^{2n+1}-1} \left[\frac{H_{2n} - \log\pi}{(2n)!} + \sum_{k=1}^{n-1} \frac{(-1)^k}{(2n-2k)!} \frac{\zeta(2k+1)}{\pi^{2k}} \right.$$
$$\left. + 2\sum_{k=1}^{\infty} \frac{(2k-1)!}{(2n+2k)!} \frac{\zeta(2k)}{2^{2k}} \right] \quad (n \in \mathbb{N}). \tag{20}$$

In precisely the same manner, we can apply the identities (10), (11) and (12) to prove the following *additional* series representations for $\zeta(2n+1)$ (Srivastava [1084, p. 389]):

$$\zeta(2n+1) = (-1)^{n-1} \frac{2(2\pi)^{2n}}{3^{2n+1}-1} \left[\frac{H_{2n} - \log\left(\frac{2}{3}\pi\right)}{(2n)!} + \sum_{k=1}^{n-1} \frac{(-1)^k}{(2n-2k)!} \frac{\zeta(2k+1)}{\left(\frac{2}{3}\pi\right)^{2k}} + 2\sum_{k=1}^{\infty} \frac{(2k-1)!}{(2n+2k)!} \frac{\zeta(2k)}{3^{2k}} \right] \quad (n \in \mathbb{N}); \tag{21}$$

$$\zeta(2n+1) = (-1)^{n-1} \frac{2(2\pi)^{2n}}{2^{4n+1}+2^{2n}-1} \left[\frac{H_{2n} - \log\left(\frac{1}{2}\pi\right)}{(2n)!} + \sum_{k=1}^{n-1} \frac{(-1)^k}{(2n-2k)!} \frac{\zeta(2k+1)}{\left(\frac{1}{2}\pi\right)^{2k}} + 2\sum_{k=1}^{\infty} \frac{(2k-1)!}{(2n+2k)!} \frac{\zeta(2k)}{4^{2k}} \right] \quad (n \in \mathbb{N}); \tag{22}$$

$$\zeta(2n+1) = (-1)^{n-1} \frac{2(2\pi)^{2n}}{3^{2n}(2^{2n}+1)+2^{2n}-1} \left[\frac{H_{2n} - \log\left(\frac{1}{3}\pi\right)}{(2n)!} + \sum_{k=1}^{n-1} \frac{(-1)^k}{(2n-2k)!} \frac{\zeta(2k+1)}{\left(\frac{1}{3}\pi\right)^{2k}} + 2\sum_{k=1}^{\infty} \frac{(2k-1)!}{(2n+2k)!} \frac{\zeta(2k)}{6^{2k}} \right] \quad (n \in \mathbb{N}). \tag{23}$$

Remarks and Observations

The series representation (20) is markedly different from each of the series representations for $\zeta(2n+1)$, which were given earlier by Zhang and Williams [1251, p. 1590, Eq. (3.13)] and (more recently) by Cvijović and Klinowski [351, p. 1265, Theorem A]. Since $\zeta(2k) \to 1$ as $k \to \infty$, the general term in the series representation (20) has the order estimate:

$$O\left(2^{-2k} \cdot k^{-2n-1}\right) \quad (k \to \infty;\ n \in \mathbb{N}),$$

whereas the general term in each of these earlier series representations has the order estimate:

$$O\left(2^{-2k} \cdot k^{-2n}\right) \quad (k \to \infty;\ n \in \mathbb{N}).$$

By suitably combining (20) and (22), it is fairly straightforward to obtain the series representation:

$$\zeta(2n+1) = (-1)^{n-1}\frac{2(2\pi)^{2n}}{(2^{2n}-1)(2^{2n+1}-1)}\left[\frac{\log 2}{(2n)!} + \sum_{k=1}^{n-1}\frac{(-1)^k(2^{2k}-1)}{(2n-2k)!}\frac{\zeta(2k+1)}{\pi^{2k}} - 2\sum_{k=1}^{\infty}\frac{(2k-1)!(2^{2k}-1)}{(2n+2k)!}\frac{\zeta(2k)}{2^{4k}}\right] \quad (n \in \mathbb{N}). \tag{24}$$

Making use of the relationship given in Problem 11 of Chapter 2, the series representation (24) can immediately be put in the form:

$$\zeta(2n+1) = (-1)^{n-1}\frac{2(2\pi)^{2n}}{(2^{2n}-1)(2^{2n+1}-1)}\left[\frac{\log 2}{(2n)!} + \sum_{k=1}^{n-1}\frac{(-1)^k(2^{2k}-1)}{(2n-2k)!}\frac{\zeta(2k+1)}{\pi^{2k}} + \frac{1}{2}\sum_{k=1}^{\infty}\frac{(-1)^{k-1}}{(2n+2k)!}\left(\frac{\pi}{2}\right)^{2k} E_{2k-1}(0)\right] \quad (n \in \mathbb{N}), \tag{25}$$

which is a slightly modified (and corrected) version of a result proven in a significantly different way by Tsumura [1169, p. 383, Theorem B].

Another interesting combination of the series representations (20) and (22) leads to the following variant of Tsumura's result (24) or (25):

$$\zeta(2n+1) = (-1)^{n-1}\frac{\pi^{2n}}{2^{2n+1}-1}\left[\frac{H_{2n}-\log\left(\frac{1}{4}\pi\right)}{(2n)!} + \sum_{k=1}^{n-1}\frac{(-1)^k(2^{2k+1}-1)}{(2n-2k)!}\frac{\zeta(2k+1)}{\pi^{2k}} - 4\sum_{k=1}^{\infty}\frac{(2k-1)!(2^{2k-1}-1)}{(2n+2k)!}\frac{\zeta(2k)}{2^{4k}}\right] \quad (n \in \mathbb{N}), \tag{26}$$

which is essentially the same as the *determinate* expression for $\zeta(2n+1)$, derived recently by Ewell [438, p. 1010, Corollary 3] by employing an entirely different technique from Srivastava's [1084].

Other similar combinations of the series representations (20) to (23) would yield the following companions of Ewell's result (26):

$$\zeta(2n+1) = (-1)^{n-1} \frac{2(2\pi)^{2n}}{(2^{2n+1}-1)(3^{2n}+1)} \left[\frac{H_{2n} - \log\left(\frac{1}{6}\pi\right)}{(2n)!} + \sum_{k=1}^{n-1} \frac{(-1)^k (2^{2k+1}-1)}{(2n-2k)!} \frac{\zeta(2k+1)}{\left(\frac{2}{3}\pi\right)^{2k}} - 4\sum_{k=1}^{\infty} \frac{(2k-1)!(2^{2k-1}-1)}{(2n+2k)!} \frac{\zeta(2k)}{6^{2k}} \right] \quad (n \in \mathbb{N}); \tag{27}$$

$$\zeta(2n+1) = (-1)^{n-1} \frac{2(2\pi)^{2n}}{(2^{2n}+1)(3^{2n+1}-1)} \left[\frac{2H_{2n} - \log\left(\frac{\pi^2}{27}\right)}{(2n)!} + \sum_{k=1}^{n-1} \frac{(-1)^k (3^{2k+1}-1)}{(2n-2k)!} \frac{\zeta(2k+1)}{\pi^{2k}} - 6\sum_{k=1}^{\infty} \frac{(2k-1)!\,(3^{2k-1}-1)}{(2n+2k)!} \frac{\zeta(2k)}{6^{2k}} \right] \quad (n \in \mathbb{N}); \tag{28}$$

$$\zeta(2n+1) = (-1)^{n-1} \frac{2(2\pi)^{2n}}{3^{2n+2}-2^{2n+3}+1} \left[\frac{H_{2n} - \log\left(\frac{8\pi}{27}\right)}{(2n)!} + \sum_{k=1}^{n-1} \frac{(-1)^k (3^{2k+1}-2^{2k+1})}{(2n-2k)!} \frac{\zeta(2k+1)}{(2\pi)^{2k}} - 12\sum_{k=1}^{\infty} \frac{(2k-1)!\,(3^{2k-1}-2^{2k-1})}{(2n+2k)!} \frac{\zeta(2k)}{6^{2k}} \right] \quad (n \in \mathbb{N}); \tag{29}$$

$$\zeta(2n+1) = (-1)^{n-1} \frac{2(2\pi)^{2n}}{2^{4n+3}+2^{2n+2}-3^{2n+2}-1} \left[\frac{H_{2n} - \log\left(\frac{27\pi}{128}\right)}{(2n)!} + \sum_{k=1}^{n-1} \frac{(-1)^k (4^{2k+1}-3^{2k+1})}{(2n-2k)!} \frac{\zeta(2k+1)}{(2\pi)^{2k}} - 24\sum_{k=1}^{\infty} \frac{(2k-1)!\,(4^{2k-1}-3^{2k-1})}{(2n+2k)!} \frac{\zeta(2k)}{12^{2k}} \right] \quad (n \in \mathbb{N}), \tag{30}$$

and

$$\zeta(2n+1) = (-1)^{n-1} \frac{2(2\pi)^{2n}}{3^{2n+1}(2^{2n}+1) - 2^{4n+2} + 2^{2n} - 1} \left[\frac{H_{2n} - \log\left(\frac{4\pi}{27}\right)}{(2n)!} + \sum_{k=1}^{n-1} \frac{(-1)^k (3^{2k+1} - 2^{2k+1})}{(2n-2k)!} \frac{\zeta(2k+1)}{\pi^{2k}} - 12 \sum_{k=1}^{\infty} \frac{(2k-1)!\,(3^{2k-1} - 2^{2k-1})}{(2n+2k)!} \frac{\zeta(2k)}{12^{2k}} \right] \quad (n \in \mathbb{N}). \tag{31}$$

Next, we recall another identity similar to (8) (see Srivastava [1084, p. 386, Eq. (1.7)]):

$$\sum_{k=0}^{\infty} \frac{(s)_{2k+1}}{(2k+1)!} \zeta(s+2k+1, a)\, t^{2k+1} = \frac{1}{2} [\zeta(s, a-t) - \zeta(s, a+t)] \quad (|t| < |a|). \tag{32}$$

By setting $t = 1/m$ and differentiating both sides with respect to s, we find from (32) that

$$\sum_{k=0}^{\infty} \frac{(s)_{2k+1}}{(2k+1)!\, m^{2k}} \left[\zeta'(s+2k+1, a) + \zeta(s+2k+1, a) \sum_{j=0}^{2k} \frac{1}{s+j} \right] = \frac{m}{2} \frac{d}{ds} \left\{ \zeta\left(s, a - \frac{1}{m}\right) - \zeta\left(s, a + \frac{1}{m}\right) \right\} \quad (m \in \mathbb{N} \setminus \{1\}). \tag{33}$$

In particular, when $m = 2$, (33) immediately yields

$$\sum_{k=0}^{\infty} \frac{(s)_{2k+1}}{(2k+1)!\, 2^{2k}} \left[\zeta'(s+2k+1, a) + \zeta(s+2k+1, a) \sum_{j=0}^{2k} \frac{1}{s+j} \right] = -\left(a - \frac{1}{2}\right)^{-s} \log\left(a - \frac{1}{2}\right). \tag{34}$$

By letting $s \to -2n-1$ ($n \in \mathbb{N}$) in the *further* special of this last identity (34) when $a = 1$, Wilton [1233, p. 92] obtained the following series representation for $\zeta(2n+1)$

(see also Hansen [531, p. 357, Entry (54.6.9)]):

$$\zeta(2n+1) = (-1)^{n-1}\pi^{2n}\left[\frac{H_{2n+1}-\log\pi}{(2n+1)!} + \sum_{k=1}^{n-1}\frac{(-1)^k}{(2n-2k+1)!}\,\frac{\zeta(2k+1)}{\pi^{2k}}\right.$$
$$\left. + 2\sum_{k=1}^{\infty}\frac{(2k-1)!}{(2n+2k+1)!}\,\frac{\zeta(2k)}{2^{2k}}\right] \quad (n \in \mathbb{N}), \tag{35}$$

which may be compared with the series representation (20). As a matter of fact, since

$$\frac{(2k)!}{(2n+2k)!} = \frac{(2k-1)!}{(2n+2k-1)!} - 2n\,\frac{(2k-1)!}{(2n+2k)!} \quad (n,k \in \mathbb{N}), \tag{36}$$

it is not difficult to deduce from (20) and (35) (with n replaced by $n-1$) that

$$\zeta(2n+1) = (-1)^n\,\frac{(2\pi)^{2n}}{n(2^{2n+1}-1)}\left[\sum_{k=1}^{n-1}\frac{(-1)^{k-1}k}{(2n-2k)!}\,\frac{\zeta(2k+1)}{\pi^{2k}}\right.$$
$$\left. + \sum_{k=0}^{\infty}\frac{(2k)!}{(2n+2k)!}\,\frac{\zeta(2k)}{2^{2k}}\right] \quad (n \in \mathbb{N}), \tag{37}$$

which is precisely the aforementioned *main* result of Cvijović and Klinowski [351, p. 1265, Theorem A]. As a matter of fact, in view of the derivative formula 2.3(22), the series representation (37) is essentially the same as a result given earlier by Zhang and Williams [1251, p. 1590, Eq. (3.13)] (see also Zhang and Williams [1251, p. 1591, Eq. (3.16)]), where an obviously more complicated (*asymptotic*) version of (37) was proven, by applying the same identity (13) above.

Observing also that

$$\frac{(2k)!}{(2n+2k+1)!} = \frac{(2k-1)!}{(2n+2k)!} - (2n+1)\,\frac{(2k-1)!}{(2n+2k+1)!} \quad (n,k \in \mathbb{N}), \tag{38}$$

we obtain yet another series representation for $\zeta(2n+1)$, by applying (20) and (35):

$$\zeta(2n+1) = (-1)^n\,\frac{2(2\pi)^{2n}}{r(2n-1)\,2^{2n}+1}\left[\sum_{k=1}^{n-1}\frac{(-1)^{k-1}k}{(2n-2k+1)!}\,\frac{\zeta(2k+1)}{\pi^{2k}}\right.$$
$$\left. + \sum_{k=0}^{\infty}\frac{(2k)!}{(2n+2k+1)!}\,\frac{\zeta(2k)}{2^{2k}}\right] \quad (n \in \mathbb{N}), \tag{39}$$

which provides a significantly simpler (and much more rapidly convergent) version of the other *main* result of Cvijović and Klinowski [351, p. 1265, Theorem B]:

$$\zeta(2n+1) = (-1)^n\,\frac{2(2\pi)^{2n}}{(2n)!}\sum_{k=0}^{\infty}\Omega_{n,k}\,\frac{\zeta(2k)}{2^{2k}} \quad (n \in \mathbb{N}), \tag{40}$$

where the coefficients $\Omega_{n,k}$ are given explicitly by

$$\Omega_{n,k} := \sum_{j=0}^{2n} \binom{2n}{j} \frac{B_{2n-j}}{(j+2k+1)(j+1)\,2^j} \qquad (n \in \mathbb{N};\ k \in \mathbb{N}_0), \tag{41}$$

in terms of the Bernoulli numbers (see Section 1.6). The definition (41) can be rewritten at once in the form:

$$\Omega_{n,k} = \sum_{j=0}^{n} \binom{2n}{2j} \frac{B_{2n-2j}}{(2j+2k+1)(2j+1)\,2^{2j}} - \frac{1}{(2n+2k)\,2^{2n}} \qquad (n \in \mathbb{N};\ k \in \mathbb{N}_0), \tag{42}$$

or, equivalently,

$$\begin{aligned} \Omega_{n,k} = (-1)^{n-1}\, &\frac{2(2n)!}{(2\pi)^{2n}} \sum_{j=0}^{n} \frac{(-\pi^2)^j\, \zeta(2n-2j)}{(2j+1)!(2j+2k+1)} \\ &- \frac{1}{(2n+2k)\,2^{2n}} \qquad (n \in \mathbb{N};\ k \in \mathbb{N}_0), \end{aligned} \tag{43}$$

by virtue of the relationship 2.3(18). Combining the partial fractions occurring in (42) or (43), it is easily seen that

$$\begin{aligned} \Omega_{n,k} = \frac{\prod_{\ell=0}^{n} \left\{(2k+2\ell+1)^{-1}\right\}}{(2n+2k)\,2^{2n}} &\left[\sum_{j=0}^{n} \binom{2n}{2j} \frac{2n+2k}{2j+1}\, 2^{2n-2j} B_{2n-2j} \right. \\ &\left. \cdot \prod_{\substack{\ell=0 \\ (\ell \neq j)}}^{n} (2k+2\ell+1) - \prod_{\ell=0}^{n} (2k+2\ell+1) \right] \qquad (n \in \mathbb{N};\ k \in \mathbb{N}_0). \end{aligned} \tag{44}$$

In view of the identity:

$$\sum_{j=0}^{n} \binom{2n}{2j} \frac{2^{2n-2j} B_{2n-2j}}{2j+1} = 1 = \sum_{j=0}^{n} \binom{2n}{2j} \frac{2^{2j} B_{2j}}{2n-2j+1}, \tag{45}$$

which is due essentially to Euler (*cf.*, e.g., Riordan [978, p. 123, Problem 12]), the expression inside brackets in (44) is a polynomial in k of degree n (not $n+1$), and, therefore,

$$\Omega_{n,k} = O(k^{-2}) \quad (k \to \infty;\ n \in \mathbb{N}). \tag{46}$$

It follows from (46) that the general term in (40) has the order estimate:

$$O(2^{-2k} \cdot k^{-2}) \quad (k \to \infty), \tag{47}$$

whereas the general term in our series representation (39) has precisely the same order estimate:

$$O(2^{-2k} \cdot k^{-2n-1}) \quad (k \to \infty;\ n \in \mathbb{N}) \tag{48}$$

as that in (20). Thus, even in the special case when $n = 1$, the series representing $\zeta(3)$ converges faster in (39) than in (40).

Various known series representations for $\zeta(2n+1)$ $(n \in \mathbb{N})$ of other types include those given (for example) by Ramanujan [965] (see also Berndt [121]), Glaisher [489] (see also Hansen [531, p. 359]), Koshliakov [695], Leshchiner [747], Grosswald [514, 515], Terras [1147], Cohen [334], Butzer *et al.* [197, 198], Dąbrowski [362] and others (see, e.g., Berndt [122, pp. 275 and 276]).

We conclude this section by remarking that a particular case of the series representation (23) when $n = 1$ was proven, by an entirely different method by Zhang and Williams [1251, p. 707, Theorem 9]. Furthermore, the following particular case of (37) when $n = 1$ (see 3.1(8)):

$$\zeta(3) = -\frac{4\pi^2}{7} \sum_{k=0}^{\infty} \frac{\zeta(2k)}{(2k+1)(2k+2)\,2^{2k}}, \tag{49}$$

which is contained in a 1772 paper entitled *Exercitationes Analyticae* by Euler (see, e.g., Ayoub [81, pp. 1084–1085]), was rediscovered by Ramaswami [966] and (more recently) by Ewell [436]. In fact, Euler's formula (49) was reproduced by Srivastava [1072, p. 7, Eq. (2.23)] from the work of Ramaswami [966]. In the *current* mathematical literature, however, Euler's formula (49) is being attributed to Ewell [436].

4.3 Further Series Representations

There are many further known series representations for $\zeta(2n+1)$ $(n \in \mathbb{N})$, which converge much more rapidly than those given by the defining series in 2.3(1). For example, we have the series representation (see also Equation 4.2(35)):

$$\begin{aligned}\zeta(2n+1) = (-1)^{n-1}\pi^{2n} \Bigg[&\frac{H_{2n+1} - \log \pi}{(2n+1)!} + \sum_{k=1}^{n-1} \frac{(-1)^k}{(2n-2k+1)!} \frac{\zeta(2k+1)}{\pi^{2k}} \\ &+ 2\sum_{k=1}^{\infty} \frac{(2k-1)!}{(2n+2k+1)!} \frac{\zeta(2k)}{2^{2k}} \Bigg] \quad (n \in \mathbb{N}),\end{aligned} \tag{1}$$

which was given over seven decades ago by Wilton [1233, p. 92] (see also Hansen [531, p. 357, Entry (54.6.9)]), and the following result given recently by Srivastava [1083, p. 10, Eq. (42)] (see also Srivastava [1082, p. 5, Eq. (3.4)]):

$$\zeta(2n+1) = (-1)^{n-1}\left(\frac{\pi}{2}\right)^{2n}\left[\frac{H_{2n+1}-\log\left(\frac{1}{2}\pi\right)}{(2n+1)!} + \frac{2(4^n-1)}{(2n+2)!}B_{2n+2}\log 2 - \frac{2^{2n+1}-1}{(2n+1)!}\zeta'(-2n-1) - \frac{2^{4n+3}}{(2n+1)!}\zeta'\left(-2n-1,\frac{1}{4}\right) + \sum_{k=1}^{n-1}\frac{(-1)^k}{(2n-2k+1)!}\frac{\zeta(2k+1)}{\left(\frac{1}{2}\pi\right)^{2k}} + 2\sum_{k=1}^{\infty}\frac{(2k-1)!}{(2n+2k+1)!}\frac{\zeta(2k)}{4^{2k}}\right] \quad (n\in\mathbb{N}), \tag{2}$$

where (*and in what follows*) a prime denotes the derivative of $\zeta(s)$ or $\zeta(s,a)$ with respect to s, an empty sum is to be interpreted as nil and H_n denotes the familiar harmonic numbers defined by 3.2(36).

Of the two seemingly analogous representations (1) and (2), the infinite series in (2) would obviously converge more rapidly with their general terms having the order estimates:

$$O\left(k^{-2n-2}\cdot m^{-2k}\right) \qquad (k\to\infty;\ n\in\mathbb{N};\ m=2 \text{ and } 4). \tag{3}$$

Srivastava and Tsumura [1111] derived three (presumably new) members of the class of the series representations (1) and (2). The general terms of the infinite series occurring in these three members [given in (29), (30) and (31) below] have the order estimates:

$$O\left(k^{-2n-2}\cdot m^{-2k}\right) \qquad (k\to\infty;\ n\in\mathbb{N};\ m=3,4,6), \tag{4}$$

which exhibit the fact that *each* of the three series representations derived here for $\zeta(2n+1)$ converges more rapidly than Wilton's result (1) and two of them (*cf.* Equations (30) and (31) below) at least as rapidly as Srivastava's result (2).

Srivastava and Tsumura [1111] begin, by defining the sequence $\{\gamma_n(x)\}_{n=0}^{\infty}$ by means of the generating function (*cf.* Eq. 1.6(2)):

$$F(x,t) := \frac{t-\log x}{e^t-x} = \sum_{n=0}^{\infty}\gamma_n(x)\frac{t^n}{n!} \quad \left(1\leqq x\leqq 1+c;\ c>0\right), \tag{5}$$

so that, clearly,

$$\gamma_n(1) = B_n \quad (n\in N_0). \tag{6}$$

Since the zeros of $e^t - x$ are given by

$$t = 2n\pi i + \log x \quad (n \in \mathbb{Z}), \tag{7}$$

the radius of convergence of the series in (5) is at least 2π. Hence, by the Cauchy-Hadamard theorem for absolute convergence (*cf.*, e.g., Whittaker and Watson [1225, p. 30]), we have

Lemma 4.1 (Srivastava and Tsumura [2000a]) *Let the sequence* $\{\gamma_n(x)\}_{n=0}^{\infty}$ *be defined by (5). Then there exists some nonnegative real number* κ*, such that*

$$\liminf_{n\to\infty} \left(\frac{|\gamma_n(x)|}{n!} \right)^{-1/n} = 2\pi + \kappa \quad (\kappa \geqq 0). \tag{8}$$

Following Srivastava and Tsumura [1111], we now consider the following Dirichlet series (cf. Eq. 2.3(1)):

$$\omega(s,x) := \sum_{n=1}^{\infty} \frac{x^{-n-1}}{n^s} + \frac{\log x}{s-1} \sum_{n=1}^{\infty} \frac{x^{-n-1}}{n^{s-1}} \quad \big(1 \leqq x \leqq 1+c;\ c > 0\big), \tag{9}$$

so that, obviously,

$$\omega(s,1) = \zeta(s). \tag{10}$$

In case $1 < x \leqq 1+c$ $(c > 0)$*, we can see that the function* $\omega(s,x)$ *is meromorphic, that is, holomorphic on the whole complex s-plane, except for a simple pole at* $s = 1$ *with residue*

$$\frac{\log x}{x-1}.$$

Lemma 4.2, below, provides an interesting generalization of the familiar relationship 2.3(10).

Lemma 4.2 (Srivastava and Tsumura [1111]) *Let* $\gamma_n(x)$ *and* $\omega(s,x)$ *be defined by (5) and (9), respectively. Then*

$$\omega(1-n,x) = -\frac{\gamma_n(x)}{n} \qquad (n \in \mathbb{N} \setminus \{1\}). \tag{11}$$

Making use of the generating function (5), we also have

Lemma 4.3 (Srivastava and Tsumura [1111]) *For* $n \in \mathbb{N}$ *and* $|\theta| < 2\pi$ $(\theta \in \mathbb{R})$,

$$
\begin{aligned}
&(2n+1)\sum_{k=1}^{\infty}\frac{x^{-k-1}}{k^{2n+2}}\sin(k\theta)+\sum_{k=1}^{\infty}\frac{x^{-k-1}}{k^{2n+1}}\left[\sin(k\theta)\log x-\theta\cos(k\theta)\right]\\
&=\frac{(-1)^n\theta^{2n+1}}{(2n+1)!}\,\frac{\log x}{x(x-1)}+\sum_{k=0}^{n-1}\frac{(-1)^k\theta^{2k+1}(2k+1)!}{(2n-2k)\,\omega(2n-2k+1,x)}\\
&\quad+\sum_{k=n+1}^{\infty}\frac{(-1)^k\theta^{2k+1}}{(2k+1)!}\,\gamma_{2k-2n}(x)\quad(1<x\leqq 1+c,\ c>0).
\end{aligned}
\tag{12}
$$

By applying Lemma 4.3, Srivastava and Tsumura [1111] first proved that

$$
\begin{aligned}
&(2n+1)\sum_{k=1}^{\infty}\frac{\sin(k\theta)}{k^{2n+2}}-\theta\sum_{k=1}^{\infty}\frac{\cos(k\theta)}{k^{2n+1}}-2n\theta\,\zeta(2n+1)\\
&=2(-1)^n\theta^{2n+1}\left(\sum_{k=1}^{n-1}\frac{(-1)^k\cdot k}{(2n-2k+1)!}\,\frac{\zeta(2k+1)}{\theta^{2k}}\right.\\
&\quad\left.-\sum_{k=0}^{\infty}\frac{(2k)!}{(2n+2k+1)!}\,\frac{\zeta(2k)}{(2\pi/\theta)^{2k}}\right)\quad(|\theta|<2\pi;\ \theta\in\mathbb{R};\ n\in\mathbb{N}).
\end{aligned}
\tag{13}
$$

Since each series in (13) is uniformly convergent with respect to θ on the open interval $(-2\pi, 2\pi)$, by executing termwise differentiation in (13) with respect to θ, they showed that (Srivastava and Tsumura [1111, p. 328, Theorem 2])

$$
\begin{aligned}
&2n\sum_{k=1}^{\infty}\frac{\cos(k\theta)}{k^{2n+1}}+\theta\sum_{k=1}^{\infty}\frac{\sin(k\theta)}{k^{2n}}-2n\,\zeta(2n+1)\\
&=2(-1)^n\theta^{2n}\left(\sum_{k=1}^{n-1}\frac{(-1)^k\cdot k}{(2n-2k)!}\,\frac{\zeta(2k+1)}{\theta^{2k}}-\sum_{k=0}^{\infty}\frac{(2k)!}{(2n+2k)!}\,\frac{\zeta(2k)}{(2\pi/\theta)^{2k}}\right)\\
&\qquad(|\theta|<2\pi;\ \theta\in\mathbb{R};\ n\in\mathbb{N}).
\end{aligned}
\tag{14}
$$

Formula (14) was proven *independently* by Katsurada [638, p. 81, Theorem 1], by using the Melin transform techniques.

In their special cases when $\theta = \pi$, Equations (13) and (14) readily yield a result given by Srivastava [1084, p. 393, Eq. (3.20)]:

$$\zeta(2n+1) = (-1)^n \frac{2(2\pi)^{2n}}{(2n-1)\,2^{2n}+1} \left[\sum_{k=1}^{n-1} \frac{(-1)^{k-1} \cdot k}{(2n-2k+1)!} \frac{\zeta(2k+1)}{\pi^{2k}} + \sum_{k=0}^{\infty} \frac{(2k)!}{(2n+2k+1)!} \frac{\zeta(2k)}{2^{2k}} \right] \quad (n \in \mathbb{N}) \tag{15}$$

and one of the two *main* results of Cvijović and Klinowski [351, p. 1265, Theorem A]:

$$\zeta(2n+1) = (-1)^n \frac{(2\pi)^{2n}}{n(2^{2n+1}-1)} \left[\sum_{k=1}^{n-1} \frac{(-1)^{k-1} \cdot k}{(2n-2k)!} \frac{\zeta(2k+1)}{\pi^{2k}} + \sum_{k=0}^{\infty} \frac{(2k)!}{(2n+2k)!} \frac{\zeta(2k)}{2^{2k}} \right] \quad (n \in \mathbb{N}). \tag{16}$$

Remark 1 The series representation (15) was given by Srivastava [1082, p. 4, Eq. (2.5)] (see also Srivastava [1084, p. 393, Eq. (3.20)]. Indeed, as already observed by Srivastava [*op. cit.*], (15) provides a significantly simpler (and much more rapidly convergent) version of one of the two *main* results of Cvijović and Klinowski [351, p. 1265, Theorem B].

Remark 2 The series representation (16) is the other *main* result of Cvijović and Klinowski [351, p. 1265, Theorem A], whose companion was referred to in Remark 1 above.

Remark 3 Since

$$\frac{(2k-1)!}{(2n+2k+1)!} = \frac{1}{2n+1} \left[\frac{(2k-1)!}{(2n+2k)!} - \frac{(2k)!}{(2n+2k+1)!} \right] \quad (n, k \in \mathbb{N}), \tag{17}$$

it is not difficult to obtain Wilton's result (1), by combining the series representation (15) with another result 4.2(20) of Srivastava [1082, p. 1, Eq. (1.3)] (see also Srivastava [1084, p. 389, Eq. (2.9)]).

With a view to applying (13) and (14) in their other special cases when

$$\theta = \frac{2}{3}\pi, \quad \frac{1}{2}\pi \quad \text{and} \quad \frac{1}{3}\pi, \tag{18}$$

we, now, recall several trigonometric sums given by **Lemma 4** (Srivastava and Tsumura [1111]). *For* $\Re(s) > 1$,

$$\sum_{n=1}^{\infty} \frac{\cos\left(\frac{2}{3}n\pi\right)}{n^s} = \frac{3^{1-s}-1}{2}\zeta(s), \tag{19}$$

$$\sum_{n=1}^{\infty} \frac{\sin\left(\frac{2}{3}n\pi\right)}{n^s} = \sqrt{3}\left\{\frac{3^{-s}-1}{2}\zeta(s) + 3^{-s}\zeta\left(s, \frac{1}{3}\right)\right\}, \tag{20}$$

$$\sum_{n=1}^{\infty} \frac{\cos\left(\frac{1}{2}n\pi\right)}{n^s} = 2^{-s}(2^{1-s}-1)\zeta(s), \tag{21}$$

$$\sum_{n=1}^{\infty} \frac{\sin\left(\frac{1}{2}n\pi\right)}{n^s} = (2^{-s}-1)\zeta(s) + 2^{1-2s}\zeta\left(s, \frac{1}{4}\right), \tag{22}$$

$$\sum_{n=1}^{\infty} \frac{\cos\left(\frac{1}{3}n\pi\right)}{n^s} = \frac{1}{2}\left(6^{1-s}-3^{1-s}-2^{1-s}+1\right)\zeta(s), \tag{23}$$

and

$$\sum_{n=1}^{\infty} \frac{\sin\left(\frac{1}{3}n\pi\right)}{n^s} = \sqrt{3}\left[\frac{3^{-s}-1}{2}\zeta(s) + 6^{-s}\left\{\zeta\left(s, \frac{1}{6}\right) + \zeta\left(s, \frac{1}{3}\right)\right\}\right]. \tag{24}$$

Srivastava and Tsumura [1111] first proved the following results:

$$\zeta(2n+1) = (-1)^n \frac{2(2\pi)^{2n}}{n(3^{2n+1}-1)}\left[\frac{(3^{2n}-1)\pi}{4\sqrt{3}(2n)!}B_{2n} + \frac{(-1)^n\pi}{\sqrt{3}(2\pi)^{2n}}\zeta\left(2n, \frac{1}{3}\right)\right.$$
$$\left. + \sum_{k=1}^{n-1}\frac{(-1)^{k-1}\cdot k}{(2n-2k)!}\frac{\zeta(2k+1)}{\left(\frac{2}{3}\pi\right)^{2k}} + \sum_{k=0}^{\infty}\frac{(2k)!}{(2n+2k)!}\frac{\zeta(2k)}{3^{2k}}\right] \quad (n \in \mathbb{N}) \tag{25}$$

$$\zeta(2n+1) = (-1)^n \frac{2(2\pi)^{2n}}{n(2^{4n+1}+2^{2n}-1)}\left[\frac{2^{2n-3}(2^{2n}-1)\pi}{(2n)!}B_{2n}\right.$$
$$+ \frac{(-1)^n\pi}{2(2\pi)^{2n}}\zeta\left(2n, \frac{1}{4}\right) + \sum_{k=1}^{n-1}\frac{(-1)^{k-1}\cdot k}{(2n-2k)!}\frac{\zeta(2k+1)}{\left(\frac{1}{2}\pi\right)^{2k}} \tag{26}$$
$$\left. + \sum_{k=0}^{\infty}\frac{(2k)!}{(2n+2k)!}\frac{\zeta(2k)}{4^{2k}}\right] \quad (n \in \mathbb{N}),$$

and

$$\zeta(2n+1) = (-1)^n \frac{2(2\pi)^{2n}}{n(6^{2n}+3^{2n}+2^{2n}-1)} \left[\frac{2^{2n-3}(3^{2n}-1)\pi}{\sqrt{3}\,(2n)!} B_{2n} + \frac{(-1)^n \pi}{2\sqrt{3}\,(2\pi)^{2n}} \left\{ \zeta\left(2n, \frac{1}{3}\right) + \zeta\left(2n, \frac{1}{6}\right) \right\} + \sum_{k=1}^{n-1} \frac{(-1)^{k-1} \cdot k}{(2n-2k)!} \frac{\zeta(2k+1)}{\left(\frac{1}{3}\pi\right)^{2k}} + \sum_{k=0}^{\infty} \frac{(2k)!}{(2n+2k)!} \frac{\zeta(2k)}{6^{2k}} \right] \quad (n \in \mathbb{N}), \tag{27}$$

in terms of the Bernoulli numbers B_n, defined by 1.7(2).

By comparing the series representation (26) with a known result, due to Srivastava [1083, p. 9, Eq. (41)] (see also Srivastava [1082, p. 5, Eq. (3.3)]), Srivastava and Tsumura [1111] derived an interesting identity involving the Zeta functions $\zeta(s)$ and $\zeta(s,a)$ (and their derivatives with respect to s), which is given below:

$$\frac{(-1)^n \pi}{2(2\pi)^{2n}} \zeta\left(2n, \frac{1}{4}\right) = \left\{ (2^{2n-2}-1)\log 2 - 2^{2n-3}(2^{2n}-1)\pi \right\} \frac{B_{2n}}{(2n)!} - \frac{2^{2n-1}-1}{2(2n-1)!} \zeta'(1-2n) - \frac{4^{2n-1}}{(2n-1)!} \zeta'\left(1-2n, \frac{1}{4}\right) \quad (n \in \mathbb{N}). \tag{28}$$

By applying (25), (26) and (27) and Srivastava's series representations 4.2(21), 4.2(22) and 4.2(23), Srivastava and Tsumura [1111] proved their *main* results given by

$$\zeta(2n+1) = (-1)^{n-1} \left(\frac{2\pi}{3}\right)^{2n} \left[\frac{H_{2n+1} - \log\left(\frac{2}{3}\pi\right)}{(2n+1)!} + \frac{(3^{2n+2}-1)\pi}{2\sqrt{3}\,(2n+2)!} B_{2n+2} + \frac{(-1)^{n-1}}{\sqrt{3}\,(2\pi)^{2n+1}} \zeta\left(2n+2, \frac{1}{3}\right) + \sum_{k=1}^{n-1} \frac{(-1)^k}{(2n-2k+1)!} \frac{\zeta(2k+1)}{\left(\frac{2}{3}\pi\right)^{2k}} + 2\sum_{k=1}^{\infty} \frac{(2k-1)!}{(2n+2k+1)!} \frac{\zeta(2k)}{3^{2k}} \right] \quad (n \in \mathbb{N}), \tag{29}$$

$$\zeta(2n+1) = (-1)^{n-1} \left(\frac{\pi}{2}\right)^{2n} \left[\frac{H_{2n+1} - \log\left(\frac{1}{2}\pi\right)}{(2n+1)!} + \frac{2^{2n}(2^{2n+2}-1)\pi}{(2n+2)!} B_{2n+2} + \frac{(-1)^{n-1}}{2(2\pi)^{2n+1}} \zeta\left(2n+2, \frac{1}{4}\right) + \sum_{k=1}^{n-1} \frac{(-1)^k}{(2n-2k+1)!} \frac{\zeta(2k+1)}{\left(\frac{1}{2}\pi\right)^{2k}} + 2\sum_{k=1}^{\infty} \frac{(2k-1)!}{(2n+2k+1)!} \frac{\zeta(2k)}{4^{2k}} \right] \quad (n \in \mathbb{N}), \tag{30}$$

and

$$\begin{aligned}\zeta(2n+1) = (-1)^{n-1}\left(\frac{\pi}{3}\right)^{2n}\Biggl[&\frac{H_{2n+1}-\log\left(\frac{1}{3}\pi\right)}{(2n+1)!}+\frac{2^{2n}(3^{2n+2}-1)\pi}{\sqrt{3}\,(2n+2)!}B_{2n+2}\\&+\frac{(-1)^{n-1}}{2\sqrt{3}\,(2\pi)^{2n+1}}\left\{\zeta\left(2n+2,\frac{1}{3}\right)+\zeta\left(2n+2,\frac{1}{6}\right)\right\}\\&+\sum_{k=1}^{n-1}\frac{(-1)^k}{(2n-2k+1)!}\frac{\zeta(2k+1)}{\left(\frac{1}{3}\pi\right)^{2k}}\\&+2\sum_{k=1}^{\infty}\frac{(2k-1)!(2n+2k+1)!}{}\frac{\zeta(2k)}{6^{2k}}\Biggr] \quad (n\in\mathbb{N}).\end{aligned} \tag{31}$$

Each of the main series representations (29), (30) and (31) belongs to the class containing Wilton's formula (1) and Srivastava's formula (2). If, for convenience, we denote the summands of the *infinite* series in the representations (1), (2), (29), (30) and (31), by $\mathcal{S}_k^{(j)}$ $(j=1,2,3,4,5)$, respectively, and apply Stirling's formula 1.1(34) and the fact that $\zeta(2k)\to 1$ as $k\to\infty$, we easily obtain the following order estimates (*cf.* Equations (3) and (4)):

$$\mathcal{S}_k^{(j)} = O\bigl(k^{-2n-2}\cdot m^{-2k}\bigr) \quad (k\to\infty;\ n\in\mathbb{N}) \tag{32}$$

($m=2$ when $j=1$; $m=3$ when $j=3$; $m=4$ when $j=2$ and $j=4$; $m=6$ when $j=5$).

Clearly, therefore, of these five series representations for $\zeta(2n+1)$, the result (31) involves the most rapidly convergent series. Conversely, the rate of convergence of the series involved in each of the Srivastava-Tsumura results (29), (30) and (31) is obviously much better than that of the series involved in Wilton's result (1). And, in particular, the rate of convergence of the series involved in the Srivastava-Tsumura result (30) is as good as that of the series involved in Srivastava's result (2).

4.4 Computational Results

Srivastava [1086] investigated rather systematically several interesting evaluations and representations of $\zeta(s)$ when $s\in\mathbb{N}\setminus\{1\}$. In one of many computationally useful special cases considered by him, it is observed that $\zeta(3)$ can be represented by means of a series which converges much more rapidly than that in Euler's celebrated formula 3.1(8), as well as the series 4.2(2) used recently by Apéry [56] in his proof of the irrationality of $\zeta(3)$. Symbolic and numeric computations using *Mathematica* (Version 4.0) for Linux show, among other things, that only 50 terms of the series (38) below are capable of producing an accuracy of seven decimal places.

Since

$$i\cot iz = \coth z = \frac{2}{e^{2z}-1}+1 \quad \left(i:=\sqrt{-1}\right), \tag{1}$$

by replacing z in the known expansion 4.2(4) by $\frac{1}{2}i\pi z$, it is easily seen that (*cf.*, e.g., Koblitz [681, p. 25]; see also Erdélyi *et al.* [421, p. 51, Eq. 1.20(1)])

$$\frac{\pi z}{e^{\pi z}-1}+\frac{\pi z}{2}=\sum_{k=0}^{\infty}\frac{(-1)^{k+1}\zeta(2k)}{2^{2k-1}}z^{2k} \qquad (|z|<2). \tag{2}$$

By setting $z = it$ in (2), multiplying both sides by t^{m-1} ($m \in \mathbb{N}$) and then integrating the resulting equation from $t = 0$ to $t = \tau$ ($0 < \tau < 2$), if we apply the (readily derivable) integral formula:

$$\int_0^{\tau} t^m e^{\lambda t}\,dt=\frac{(-1)^m m!e^{\lambda\tau}}{\lambda^{m+1}}\left(\sum_{j=0}^{m}\frac{(-\lambda\tau)^j}{j!}-e^{-\lambda\tau}\right) \qquad (\lambda\neq 0), \tag{3}$$

we obtain

$$\begin{aligned}
&m!\left(\frac{i}{\pi\tau}\right)^m\left\{\sum_{k=1}^{\infty}\frac{\cos(k\pi\tau)+i\sin(k\pi\tau)}{k^{m+1}}-\zeta(m+1)\right\}\\
&\quad+\sum_{k=1}^{\infty}\frac{\cos(k\pi\tau)}{k}+i\left(\frac{\pi\tau}{2(m+1)}+\sum_{k=1}^{\infty}\frac{\sin(k\pi\tau)}{k}\right)\\
&\quad+\sum_{j=1}^{m-1}\frac{m!}{(m-j)!}\left(\frac{i}{\pi\tau}\right)^j\sum_{k=1}^{\infty}\frac{\cos(k\pi\tau)+i\sin(k\pi\tau)}{k^{j+1}}\\
&=\sum_{k=0}^{\infty}\frac{\zeta(2k)}{k+\frac{1}{2}m}\left(\frac{\tau}{2}\right)^{2k} \qquad (m\in\mathbb{N};\ 0<\tau<2),
\end{aligned} \tag{4}$$

which, in the special case when $\tau = 1$, would immediately simplify to the form:

$$\begin{aligned}
&\frac{i\pi}{2(m+1)}-\left(2^{m+1}-1\right)\left(\frac{i}{2\pi}\right)^m m!\,\zeta(m+1)\\
&=\log 2+\sum_{j=1}^{m-1}\frac{m!}{(m-j)!}\left(2^j-1\right)\left(\frac{i}{2\pi}\right)^j\zeta(j+1)\\
&\quad+\sum_{k=0}^{\infty}\frac{\zeta(2k)}{\left(k+\frac{1}{2}m\right)2^{2k}} \qquad (m\in\mathbb{N}),
\end{aligned} \tag{5}$$

where we have also applied the last part of the definition 2.3(1). Furthermore, in (4), (5) and elsewhere in this section, an empty sum is interpreted (as usual) to be nil.

Now, for a given sequence $\{\Omega_n\}_{n=1}^{\infty}$, it is easily verified that

$$\sum_{j=1}^{m-1}\Omega_j=\sum_{j=1}^{[(m-1)/2]}\Omega_{2j}+\sum_{j=1}^{[m/2]}\Omega_{2j-1}, \tag{6}$$

where, just as in 4.1(17), $[\kappa]$ denotes the greatest integer in κ. Thus, we find from (5) that

$$\begin{aligned}\frac{i\pi}{2(m+1)} &- \left(2^{m+1}-1\right)\left(\frac{i}{2\pi}\right)^m m!\,\zeta(m+1)\\ &= \log 2 + \sum_{j=1}^{[(m-1)/2]} (-1)^j \binom{m}{2j} \frac{(2j)!\left(2^{2j}-1\right)}{(2\pi)^{2j}} \zeta(2j+1)\\ &+ i \sum_{j=1}^{[m/2]} (-1)^{j-1} \binom{m}{2j-1} \frac{(2j-1)!\left(2^{2j-1}-1\right)}{(2\pi)^{2j-1}} \zeta(2j)\\ &+ \sum_{k=0}^{\infty} \frac{\zeta(2k)}{\left(k+\frac{1}{2}m\right)2^{2k}} \quad (m \in \mathbb{N}).\end{aligned} \tag{7}$$

Setting $m = 2n \quad (n \in \mathbb{N})$ in (7) and then equating the real and imaginary parts in the resulting equation, we obtain (Srivastava [1086])

$$\begin{aligned}\zeta(2n+1) &= (-1)^{n-1} \frac{(2\pi)^{2n}}{(2n)!\left(2^{2n+1}-1\right)}\\ &\cdot \left[\log 2 + \sum_{j=1}^{n-1} (-1)^j \binom{2n}{2j} \frac{(2j)!\left(2^{2j}-1\right)}{(2\pi)^{2j}} \zeta(2j+1) + \sum_{k=0}^{\infty} \frac{\zeta(2k)}{(k+n)2^{2k}}\right] \quad (n \in \mathbb{N})\end{aligned} \tag{8}$$

and

$$\begin{aligned}\zeta(2n) &= (-1)^{n-1} \frac{(2\pi)^{2n-1}}{(2n)!(2^{2n-1}-1)}\\ &\cdot \left[\frac{\pi}{2(2n+1)} + \sum_{j=1}^{n-1} (-1)^j \binom{2n}{2j-1} \frac{(2j-1)!\left(2^{2j-1}-1\right)}{(2\pi)^{2j-1}} \zeta(2j)\right] \quad (n \in \mathbb{N}).\end{aligned} \tag{9}$$

Similarly, if we set $m = 2n+1$ $(n \in \mathbb{N}_0)$ in (7), we shall obtain the following results, which can provide series representations for $\zeta(2n+1)$ and $\zeta(2n)$ when $n \in \mathbb{N}$:

$$\begin{aligned}&\sum_{j=1}^{n} (-1)^{j-1} \binom{2n+1}{2j} \frac{(2j)!\left(2^{2j}-1\right)}{(2\pi)^{2j}} \zeta(2j+1)\\ &= \log 2 + \sum_{k=0}^{\infty} \frac{\zeta(2k)}{\left(k+n+\frac{1}{2}\right)2^{2k}} \quad (n \in \mathbb{N}_0)\end{aligned} \tag{10}$$

and

$$\zeta(2n) = (-1)^{n-1} \frac{(2\pi)^{2n-1}}{(2n-1)!\left(2^{2n}-1\right)} \cdot \left[\frac{\pi}{4n} + \sum_{j=1}^{n-1} (-1)^j \binom{2n-1}{2j-1} \frac{(2j-1)!\left(2^{2j-1}-1\right)}{(2\pi)^{2j-1}} \zeta(2j) \right] \quad (n \in \mathbb{N}). \tag{11}$$

Indeed, in its particular cases when $n = 0$ and $n \in \mathbb{N}$, the summation formula (10) immediately yields the known sum 3.4(493) and (Srivastava [1086])

$$\zeta(2n+1) = (-1)^{n-1} \frac{(2\pi)^{2n}}{(2n+1)!\left(2^{2n}-1\right)} \cdot \left[\log 2 + \sum_{j=1}^{n-1} (-1)^j \binom{2n+1}{2j} \frac{(2j)!\left(2^{2j}-1\right)}{(2\pi)^{2j}} \zeta(2j+1) + \sum_{k=0}^{\infty} \frac{\zeta(2k)}{\left(k+n+\frac{1}{2}\right) 2^{2k}} \right] \quad (n \in \mathbb{N}), \tag{12}$$

respectively.

Each of the recursion formulas (9) and (11) can be used to evaluate $\zeta(2n)$ $(n \in \mathbb{N})$. Formula (9) was proven, in a markedly different way, by (for example) Stark [1121, p. 199, Eq. (5)]. Formulas (9) and (11), together, yield the series identity (Srivastava [1086]):

$$\sum_{j=1}^{n-1} (-1)^{j-1} \binom{2n-1}{2j-1} \frac{(2j-1)!\left(2^{2j-1}-1\right)}{(2\pi)^{2j-1}} \zeta(2j) = \frac{\pi\left\{(2n-1)2^{2n-1}-2n\right\}}{4n(2n+1)\left(2^{2n-1}-1\right)} + \frac{2^{2n}-1}{2n\left(2^{2n-1}-1\right)} \sum_{j=1}^{n-1} (-1)^{j-1} \binom{2n}{2j-1} \frac{(2j-1)!\left(2^{2j-1}-1\right)}{(2\pi)^{2j-1}} \zeta(2j) \quad (n \in \mathbb{N}) \tag{13}$$

or, equivalently,

$$\sum_{j=1}^{n-1} (-1)^{j-1} \binom{2n-1}{2j-2} \left\{ 2(n-j)\left(2^{2n-1}-1\right) - 2^{2n-1} \right\} \frac{(2j-2)!\left(2^{2j-1}-1\right)}{(2\pi)^{2j-1}} \zeta(2j) = \frac{\pi\left\{(2n-1)2^{2n-1}-2n\right\}}{4n(2n+1)} \quad (n \in \mathbb{N}), \tag{14}$$

which can also be used as a recurrence relation for evaluating $\zeta(2n)$ $(n \in \mathbb{N})$.

An integral representation for $\zeta(2n+1)$, which is equivalent to the series representation (8), was given earlier by Dąbrowski [362, p. 203, Eq. (16); p. 206], who also mentioned the existence of (but did not fully state) the series representation (12). The series representation (8) is derived also in a forthcoming paper by Borwein *et al.* [153, Eq. (57)].

For $n=1$, (8) immediately yields the series representation 4.2(5) which, in conjunction with the known sum 3.4(493), would lead us readily to Euler's formula 3.1(8). Conversely, by setting $n=1$ in the series representation (12), we obtain

$$\zeta(3)=\frac{4\pi^2}{9}\left(\frac{1}{2}\log 2+\sum_{k=0}^{\infty}\frac{\zeta(2k)}{(2k+3)2^{2k}}\right), \tag{15}$$

which was derived *independently* by (for example) Glasser [491, p. 446, Eq. (12)], Zhang and Williams [1251, p. 1585, Eq. (2.13)] and Dąbrowski [362, p. 206] (see also Chen and Srivastava [252, p. 183, Eq. (2.15)]).

By suitably combining the series occurring in 4.2(5), 3.4(493) and (15), it is not difficult to derive several other series representations for $\zeta(3)$, which are analogous to Euler's formula 3.1(8). More generally, since

$$\begin{aligned}&\frac{\lambda k^2+\mu k+\nu}{(2k+2n-1)(2k+2n)(2k+2n+1)}\\&\quad=\frac{\mathcal{A}}{2k+2n-1}+\frac{\mathcal{B}}{2k+2n}+\frac{\mathcal{C}}{2k+2n+1},\end{aligned} \tag{16}$$

where, for convenience,

$$\mathcal{A}=\mathcal{A}_n(\lambda,\mu,\nu):=\frac{1}{2}\left[\lambda n^2-(\lambda+\mu)n+\frac{1}{4}(\lambda+2\mu+4\nu)\right], \tag{17}$$

$$\mathcal{B}=\mathcal{B}_n(\lambda,\mu,\nu):=-\left(\lambda n^2-\mu n+\nu\right), \tag{18}$$

and

$$\mathcal{C}=\mathcal{C}_n(\lambda,\mu,\nu):=\frac{1}{2}\left[\lambda n^2+(\lambda-\mu)n+\frac{1}{4}(\lambda-2\mu+4\nu)\right], \tag{19}$$

by appealing to (8), (10) *with* n replaced by $n-1$ and (12), we can derive the following unification of a large number of known (or new) series representations, including (for example) Euler's result 3.1(8) (Srivastava [1086]):

$$\zeta(2n+1) = \frac{(-1)^{n-1}(2\pi)^{2n}}{(2n)!\left\{\left(2^{2n+1}-1\right)\mathcal{B}+(2n+1)\left(2^{2n}-1\right)\mathcal{C}\right\}}$$
$$\cdot\left[\frac{1}{4}\lambda\log 2+\sum_{j=1}^{n-1}(-1)^j\binom{2n-1}{2j-2}\left\{2j(2j-1)\mathcal{A}+[\lambda(4n-1)-2\mu]nj\right.\right.$$
$$\left.+\lambda n\left(n+\frac{1}{2}\right)\right\}\frac{(2j-2)!\left(2^{2j}-1\right)}{(2\pi)^{2j}}\zeta(2j+1) \tag{20}$$
$$\left.+\sum_{k=0}^{\infty}\frac{\left(\lambda k^2+\mu k+\nu\right)\zeta(2k)}{(2k+2n-1)(k+n)(2k+2n+1)2^{2k}}\right] \quad (n\in\mathbb{N};\ \lambda,\mu,\nu\in\mathbb{C}),$$

where $\mathcal{A}$, $\mathcal{B}$ and $\mathcal{C}$ are given by (17), (18) and (19), respectively.

For $\lambda = 0$, the series representation (20) simplifies to the form:

$$\zeta(2n+1) = \frac{(-1)^{n-1}(2\pi)^{2n}}{(2n)!\left\{\left(2^{2n+1}-1\right)(\mu n-\nu)-\left(2^{2n}-1\right)\left(n+\frac{1}{2}\right)\left[\mu(n+\frac{1}{2})-\nu\right]\right\}}$$
$$\cdot\left[\sum_{j=1}^{n-1}(-1)^j\binom{2n-1}{2j-2}\left\{j(2j-1)\left[\nu-\mu\left(n-\frac{1}{2}\right)\right]-2\mu nj\right\}\right.$$
$$\left.\cdot\frac{(2j-2)!\left(2^{2j}-1\right)}{(2\pi)^{2j}}\zeta(2j+1)+\sum_{k=0}^{\infty}\frac{(\mu k+\nu)\zeta(2k)}{(2k+2n-1)(k+n)(2k+2n+1)2^{2k}}\right]$$
$$(n\in\mathbb{N};\ \mu,\nu\in\mathbb{C}). \tag{21}$$

Furthermore, by setting

$$\lambda=\mu=0 \quad\text{and}\quad \nu=1$$

in (20) or (alternatively) by setting

$$\mu=0 \quad\text{and}\quad \nu=1$$

in (21), we immediately obtain the series representation (Srivastava [1086]):

$$\zeta(2n+1)$$
$$= \frac{(-1)^{n-1}(2\pi)^{2n}}{(2n)!\left\{2^{2n}(2n-3)-2n+1\right\}}\left[\sum_{j=1}^{n-1}(-1)^j\binom{2n-1}{2j-2}\frac{(2j)!\left(2^{2j}-1\right)}{(2\pi)^{2j}}\zeta(2j+1)\right. \tag{22}$$
$$\left.+2\sum_{k=0}^{\infty}\frac{\zeta(2k)}{(2k+2n-1)(k+n)(2k+2n+1)2^{2k}}\right] \quad (n\in\mathbb{N}),$$

which, in the special case when $n = 1$, was given by Chen and Srivastava [252, p. 189, Eq. (2.45)].

Of the three representations (20), (21) and (22) for $\zeta(2n+1)$ $(n \in \mathbb{N})$, the infinite series in (22) converges most rapidly. Many other families of rapidly convergent series representations for $\zeta(2n+1)$ $(n \in \mathbb{N})$ can be found in the recent works of Srivastava [1083, 1084] (and, indeed, also in the numerous references already cited in each of these earlier works), including the results presented in previous sections (especially Section 4.2).

For various suitable special values of the parameters λ, μ, and ν, we can easily deduce from (20) and (21) several known (or new) series representations for $\zeta(2n+1)$ $(n \in \mathbb{N})$. For example, if we set

$$\mu = 2 \quad \text{and} \quad \nu = 2n+1$$

in our series representation (21), we shall obtain (Srivastava [1086])

$$\zeta(2n+1) = (-1)^{n-1} \frac{(2\pi)^{2n}}{(2n)!\left(2^{2n+1}-1\right)} \left[\sum_{j=1}^{n-1} (-1)^j \binom{2n-1}{2j-1} \cdot \frac{(2j)!\left(2^{2j}-1\right)}{(2\pi)^{2j}} \zeta(2j+1) - \sum_{k=0}^{\infty} \frac{\zeta(2k)}{(2k+2n-1)(k+n)2^{2k}}\right] \quad (n \in \mathbb{N}), \tag{23}$$

which, in the special case when $n = 1$, immediately yields Euler's formula 3.1(8).

The following additional series representations for $\zeta(2n+1)$ $(n \in \mathbb{N})$, which are analogous to (23), can also be deduced similarly from (21) (Srivastava [1086]):

$$\zeta(2n+1) = (-1)^{n-1} \frac{(2\pi)^{2n}}{(2n)!\left\{(2n-1)2^{2n}-2n\right\}} \left[2n \sum_{j=1}^{n-1} (-1)^j \binom{2n-1}{2j-2} \cdot \frac{(2j)!\left(2^{2j}-1\right)}{(2\pi)^{2j}} \zeta(2j+1) - \sum_{k=0}^{\infty} \frac{\zeta(2k)}{(k+n)(2k+2n+1)2^{2k}}\right] \quad (n \in \mathbb{N}) \tag{24}$$

and

$$\zeta(2n+1) = (-1)^{n-1} \frac{(2\pi)^{2n}}{(2n+1)!\left(2^{2n}-1\right)} \left[\sum_{j=1}^{n-1} (-1)^j \left(\frac{4nj-2j+1}{2j-1}\right) \binom{2n-1}{2j-2} \cdot \frac{(2j)!\left(2^{2j}-1\right)}{(2\pi)^{2j}} \zeta(2j+1) - 4 \sum_{k=0}^{\infty} \frac{\zeta(2k)}{(2k+2n-1)(2k+2n+1)2^{2k}}\right] \quad (n \in \mathbb{N}). \tag{25}$$

The special case of each of the last two series representations (24) and (25) when $n = 1$ was given by Zhang and Williams [1251, p. 1586]. For $n = 2$, Srivastava's formulas (22) to (25) would readily yield the following series representations for $\zeta(5)$:

$$\zeta(5) = \frac{\pi^2}{13}\zeta(3) - \frac{8\pi^4}{13}\sum_{k=0}^{\infty}\frac{\zeta(2k)}{(2k+3)(2k+4)(2k+5)2^{2k}}, \tag{26}$$

$$\zeta(5) = \frac{3\pi^2}{31}\zeta(3) + \frac{4\pi^4}{93}\sum_{k=0}^{\infty}\frac{\zeta(2k)}{(2k+3)(2k+4)2^{2k}}, \tag{27}$$

$$\zeta(5) = \frac{\pi^2}{11}\zeta(3) + \frac{\pi^4}{33}\sum_{k=0}^{\infty}\frac{\zeta(2k)}{(2k+4)(2k+5)2^{2k}}, \tag{28}$$

and

$$\zeta(5) = \frac{7\pi^2}{75}\zeta(3) + \frac{8\pi^4}{225}\sum_{k=0}^{\infty}\frac{\zeta(2k)}{(2k+3)(2k+5)2^{2k}}. \tag{29}$$

Next, with a view to *further* improving the rate of convergence in the most rapidly convergent series representations (22) and (26) considered in this work, we observe that

$$\begin{aligned}&\frac{1}{(2k+2n-1)(2k+2n)(2k+2n+1)(2k+2n+2)}\\ &\quad = \frac{1}{6}\left(\frac{1}{2k+2n-1} - \frac{1}{2k+2n+2}\right) - \frac{1}{2}\frac{1}{(2k+2n)(2k+2n+1)}.\end{aligned} \tag{30}$$

Thus, by applying the series representations (10) *with n* replaced by $n-1$, (8) *with n* replaced by $n+1$ and (24), we obtain (Srivastava [1086])

$$\begin{aligned}\zeta(2n+3) &= \frac{2\pi^2\left\{2^{2n+2} + n(2n-3)\left(2^{2n}-1\right) - 1\right\}}{(n+1)(2n+1)\left(2^{2n+3}-1\right)}\zeta(2n+1)\\ &+ (-1)^{n-1}\frac{(2\pi)^{2n+2}}{(2n+2)!\left(2^{2n+3}-1\right)}\left[\sum_{j=1}^{n-1}(-1)^j\left\{\binom{2n-1}{2j}\right.\right.\\ &\left. - \binom{2n+2}{2j} + 6n\binom{2n-1}{2j-2}\right\}\frac{(2j)!\left(2^{2j}-1\right)}{(2\pi)^{2j}}\zeta(2j+1)\\ &\left. + 12\sum_{k=0}^{\infty}\frac{\zeta(2k)}{(2k+2n-1)(2k+2n)(2k+2n+1)(2k+2n+2)2^{2k}}\right] \quad (n \in \mathbb{N}),\end{aligned} \tag{31}$$

where the series converges faster than that in (22).

In its special case when $n = 1$, (31) readily yields the following improved version of the series representation (26) above (*cf.* Zhang and Williams [1251, p. 1590, Eq. (3.14)]:

$$\zeta(5) = \frac{4\pi^2}{31}\zeta(3) + \frac{8\pi^4}{31}\sum_{k=0}^{\infty}\frac{\zeta(2k)}{(2k+1)(2k+2)(2k+3)(2k+4)2^{2k}} \tag{32}$$

in which $\zeta(3)$ can be replaced by its known value $-4\pi^2\zeta'(-2)$ given by 2.3(22) for $n = 1$.

Yet another rapidly convergent series representation for $\zeta(2n+3)$ $(n \in \mathbb{N})$, analogous to (31), can be derived by means of the identity:

$$\begin{aligned}\frac{1}{(2k+2n)(2k+2n+1)(2k+2n+2)(2k+2n+3)}\\ = \frac{1}{6}\left(\frac{1}{2k+2n} - \frac{1}{2k+2n+3}\right) - \frac{1}{2}\frac{1}{(2k+2n+1)(2k+2n+2)},\end{aligned} \tag{33}$$

together with the series representations (8), (12) *with* n replaced by $n+1$ and (23) *with* n replaced by $n+1$. We, thus, obtain the series representation (Srivastava [1086]):

$$\begin{aligned}\zeta(2n+3) &= \frac{2\pi^2\left\{\frac{1}{3}(2n+1)\left(2n^2-4n+3\right)\left(2^{2n}-1\right) - 2^{2n+1}+1\right\}}{(n+1)(2n+1)\left\{(2n-3)2^{2n+2}-2n\right\}}\zeta(2n+1)\\ &+ (-1)^{n-1}\frac{(2\pi)^{2n+2}}{(2n+2)!\left\{(2n-3)2^{2n+2}-2n\right\}}\left[\sum_{j=1}^{n-1}(-1)^j\left\{\binom{2n}{2j}\right.\right.\\ &\left.-\binom{2n+3}{2j} + 3\binom{2n+1}{2j-1}\right\}\frac{(2j)!\left(2^{2j}-1\right)}{(2\pi)^{2j}}\zeta(2j+1)\\ &\left.+ 12\sum_{k=0}^{\infty}\frac{\zeta(2k)}{(2k+2n)(2k+2n+1)(2k+2n+2)(2k+2n+3)2^{2k}}\right] \quad (n \in \mathbb{N}),\end{aligned} \tag{34}$$

which, in the special case when $n = 1$, yields

$$\zeta(5) = \frac{2\pi^2}{27}\zeta(3) - \frac{4\pi^4}{9}\sum_{k=0}^{\infty}\frac{\zeta(2k)}{(2k+2)(2k+3)(2k+4)(2k+5)2^{2k}}, \tag{35}$$

where the series obviously converges faster than that in (26).

Lastly, by applying the identity:

$$\begin{aligned}\frac{1}{2k(2k+2n-1)(2k+2n)(2k+2n+1)} &= \frac{1}{2n(2n-1)(2n+1)}\frac{1}{2k}\\ -\frac{1}{2(2n-1)}\frac{1}{2k+2n-1} + \frac{1}{2n}\frac{1}{2k+2n} &- \frac{1}{2(2n+1)}\frac{1}{2k+2n+1}\end{aligned} \tag{36}$$

in conjunction with the series representations (10) *with* n replaced by $n-1$, (8), (12) and the known result 3.4(26), we arrive at the following series representation for $\zeta(2n+1)$ $(n \in \mathbb{N})$ (Srivastava [1086]):

$$\zeta(2n+1) = (-1)^{n-1} \frac{(2\pi)^{2n}}{2\cdot(2n-1)!\left\{n-1-(n-2)2^{2n}\right\}} \left[\frac{12n^2-1}{2n^2\left(4n^2-1\right)^2} - \frac{\log\pi}{n\left(4n^2-1\right)} - \sum_{j=2}^{n-1}(-1)^j\binom{2n-2}{2j-3}\frac{(2j-1)!\left(2^{2j}-1\right)}{(2\pi)^{2j}}\zeta(2j+1) + \sum_{k=1}^{\infty}\frac{\zeta(2k)}{k(k+n)(2k+2n-1)(2k+2n+1)2^{2k}}\right] \quad (n \in \mathbb{N}), \tag{37}$$

where we have also applied the fact that $\zeta(0) = -\frac{1}{2}$.

For $n = 1$, (37) reduces immediately to Wilton's formula (*cf.* Hansen [531, p. 357, Entry (54.5.9)]; see also Chen and Srivastava [252, p. 181, Eq. (2.1)]):

$$\zeta(3) = \frac{\pi^2}{2}\left(\frac{11}{18} - \frac{1}{3}\log\pi + \sum_{k=1}^{\infty}\frac{\zeta(2k)}{k(k+1)(2k+1)(2k+3)2^{2k}}\right). \tag{38}$$

Furthermore, in its special case when $n = 2$, (37) would yield the following companion of the series representations (32) and (35):

$$\zeta(5) = \frac{2\pi^4}{45}\left(\log\pi - \frac{47}{60} - 30\sum_{k=1}^{\infty}\frac{\zeta(2k)}{k(k+2)(2k+3)(2k+5)2^{2k}}\right), \tag{39}$$

which does *not* contain a term involving $\zeta(3)$ on the right-hand side.

By eliminating $\zeta(2n+3)$ between the results (31) and (34), we can obtain a series representation for $\zeta(2n+1)$ $(n \in \mathbb{N})$, which would converge as rapidly as the series in (37). For $n \in \mathbb{N}$, we, thus, find that (Srivastava [1086])

$$\zeta(2n+1) = (-1)^{n-1}\frac{(2\pi)^{2n}}{(2n)!\Delta_n}\left[\sum_{j=1}^{n-1}(-1)^j\left(\left\{(2n-3)2^{2n+2}-2n\right\}\left\{\binom{2n-1}{2j} - \binom{2n+2}{2j} + 6n\binom{2n-1}{2j-2}\right\} - \left(2^{2n+3}-1\right)\cdot\left\{\binom{2n}{2j} - \binom{2n+3}{2j} + 3\binom{2n+1}{2j-1}\right\}\right)\frac{(2j)!\left(2^{2j}-1\right)}{(2\pi)^{2j}}\zeta(2j+1) + 12\sum_{k=0}^{\infty}\frac{(\xi_n k+\eta_n)\,\zeta(2k)}{(2k+2n-1)(2k+2n)(2k+2n+1)(2k+2n+2)(2k+2n+3)2^{2k}}\right], \tag{40}$$

where, for convenience,

$$\Delta_n := \left(2^{2n+3}-1\right)\left\{\frac{1}{3}(2n+1)\left(2n^2-4n+3\right)\left(2^{2n}-1\right)-2^{2n+1}+1\right\} \tag{41}$$
$$-\left\{(2n-3)2^{2n+2}-2n\right\}\left\{2^{2n+2}+n(2n-3)\left(2^{2n}-1\right)-1\right\},$$

$$\xi_n := 2\left\{(2n-5)2^{2n+2}-2n+1\right\}, \tag{42}$$

and

$$\eta_n := \left(4n^2-4n-7\right)2^{2n+2}-(2n+1)^2. \tag{43}$$

In its special case when $n=1$, (40) yields the following series representation (Srivastava [1086]):

$$\zeta(3) = -\frac{6\pi^2}{23}\sum_{k=0}^{\infty}\frac{(98k+121)\zeta(2k)}{(2k+1)(2k+2)(2k+3)(2k+4)(2k+5)2^{2k}}, \tag{44}$$

where the series obviously converges much more rapidly than that in *each* of the celebrated results 3.1(8) and 4.2(2).

We conclude this section by summarizing below the results of some symbolic and numeric computations with the series in (44) using *Mathematica* (Version 4.0) for Linux:

In[1] := (98k + 121)Zeta[2k] / ((2k + 1)(2k + 2)(2k + 3)(2k + 4)(2k + 5)2(2k))

$$\text{Out[1]} = \frac{(121+98k)\text{Zeta}[2k]}{2^{2k}(1+2k)(2+2k)(3+2k)(4+2k)(5+2k)}$$

In[2] := Sum [%, {k, 1, Infinity}] //Simplify

$$\text{Out[2]} = \frac{121}{240} - \frac{23\,\text{Zeta}[3]}{6\,\text{Pi}^2}$$

In[3] := N[%]

Out[3] = 0.0372903

In[4] := Sum [N[%1]//Evaluate, {k, 1, 50}]

Out[4] = 0.0372903

In[5] := N Sum [%1//Evaluate, {k, 1, Infinity}]

Out [5] = 0.0372903

Since $\zeta(0) = -\frac{1}{2}$, Out [2] evidently validates the series representation (44) symbolically. Furthermore, our numeric computations in Out [3], Out [4] and Out [5], together, exhibit the fact that only 50 terms ($k=1$ to $k=50$) of the series in (44) can produce an accuracy of seven decimal places (see, for details, Srivastava [1086]).

Problems

1. Prove that

$$\sum_{n=1}^{\infty} \frac{H_n}{n^2} = 2\sum_{n=1}^{\infty} \frac{1}{n^3} = 2\sum_{n=1}^{\infty} \frac{H_n}{(n+1)^2},$$

where H_n denotes the harmonic numbers, defined by 3.2(36).

(*Cf.* Equations 2.3(54) *and* 3.5(18); see also Klamkin [669, p. 195])

2. Prove that

$$\zeta(4) = \frac{36}{17}\sum_{k=1}^{\infty} \frac{1}{k^4\binom{2k}{k}}.$$

(Comtet [337, p. 89]; Cohen [334, p. 280])

3. Prove that

$$\zeta(3) = \frac{\pi^2}{6}\left\{\frac{3}{4} - \frac{1}{2}\log\frac{\pi}{3} + \sum_{k=1}^{\infty} \frac{\zeta(2k)}{k(2k+1)(2k+2)\,6^{2k}}\right\}.$$

(Zhang and Williams [1251, Section 2]; Chen and Srivastava [252, p. 184])

4. Prove that

$$\zeta(3) = \frac{\pi^2}{14}\left\{\frac{3}{2} - \log\frac{\pi}{4} - 4\sum_{k=1}^{\infty} \frac{(2^{2k-1}-1)\zeta(2k)}{k(2k+1)(2k+2)\,2^{4k}}\right\}.$$

(Ewell [438, p. 1011]; Chen and Srivastava [252, p. 187])

5. Prove the following series representations for $\zeta(3)$:

$$\zeta(3) = \frac{2\pi^2}{35}\left\{\frac{3}{2} - \log\frac{\pi}{6} - 4\sum_{k=1}^{\infty} \frac{(2^{2k-1}-1)\zeta(2k)}{k(2k+1)(2k+2)\,6^{2k}}\right\};$$

$$\zeta(3) = \frac{2\pi^2}{65}\left\{3 - \log\left(\frac{\pi^2}{27}\right) - 6\sum_{k=1}^{\infty} \frac{(3^{2k-1}-1)\zeta(2k)}{k(2k+1)(2k+2)\,6^{2k}}\right\};$$

$$\zeta(3) = \frac{4\pi^2}{25}\left\{\frac{3}{4} - \frac{1}{2}\log\left(\frac{8\pi}{27}\right) - 6\sum_{k=1}^{\infty} \frac{(3^{2k-1}-2^{2k-1})\zeta(2k)}{k(2k+1)(2k+2)\,6^{2k}}\right\};$$

$$\zeta(3) = \frac{4\pi^2}{31}\left\{\frac{3}{4} - \frac{1}{2}\log\left(\frac{27\pi}{128}\right) - 12\sum_{k=1}^{\infty} \frac{(4^{2k-1}-3^{2k-1})\zeta(2k)}{k(2k+1)(2k+2)\,12^{2k}}\right\};$$

and

$$\zeta(3) = \frac{4\pi^2}{37}\left\{\frac{3}{4} - \frac{1}{2}\log\left(\frac{4\pi}{27}\right) - 6\sum_{k=1}^{\infty} \frac{(3^{2k-1}-2^{2k-1})\zeta(2k)}{k(2k+1)(2k+2)\,12^{2k}}\right\}.$$

(Chen and Srivastava [252, p. 187])

6. For $n \in \mathbb{N} \setminus \{1\}$, show that

$$\zeta(2n+1) = \frac{(-1)^n (2\pi)^{2n}}{(2^{2n+1}-1)\,n} \left\{ \sum_{k=1}^{n-1} \frac{2^{2k}}{(2k-1)!\,(2n-2k)!\,(2k+1)^2} \right.$$
$$+ \sum_{k=1}^{n-1} \frac{2^{2k}}{(2k-1)!\,(2n-2k)!} \sum_{j=1}^{\infty} \frac{(-1)^j (2j-1)!}{2^{2j-1}\pi^{2j}}$$
$$\left. \cdot \sum_{i=0}^{\min(2j-2,2k)} \binom{2k}{i} \frac{(-1)^i \zeta(2j)}{2j-i-1} + \sum_{k=0}^{\infty} \frac{(2k)!\,\zeta(2k)}{(2k+4)!\,2^{2k}} \right\}.$$

(Zhang and Williams [1251, p. 1591])

7. Let

$$C_{p,q} := \sum_{j=2}^{\infty} \sum_{k=1}^{j-1} \frac{1}{j^p\, k^q} \qquad (p \in \mathbb{N} \setminus \{1\};\ q \in \mathbb{N})$$

and

$$C_{p,q,r} := \sum_{j=3}^{\infty} \sum_{k=2}^{j-1} \sum_{i=1}^{k-1} \frac{1}{j^p\, k^q\, i^r} \qquad (p \in \mathbb{N} \setminus \{1\};\ q,\ r \in \mathbb{N}).$$

Show that

$$C_{p,q,r} = (-1)^q \sum_{n=0}^{p-2} \left\{ \sum_{m=0}^{q+n-1} \frac{(q)_n\,(r)_m}{n!m!} C_{p-n,r+m,q+n-m} \right.$$
$$\left. + \sum_{m=0}^{r-1} \frac{(q)_{n+m}}{n!m!} C_{p-n,q+n+m,r-m} \right\}$$
$$+ \sum_{n=0}^{q-2} \frac{(p)_n}{n!} (-1)^n \zeta(p+n)\, C_{q-n,r} + (-1)^{q-1} \frac{(p+q-2)!}{(p-1)!\,(q-1)!}$$
$$\cdot \left\{ \sum_{m=0}^{r-2} (-1)^m \zeta(r-m) \left[C_{p+q-1,m+1} + \zeta(p+q+m) \right] \right.$$
$$\left. + (-1)^{r-1} \left[C_{p+q-1,r,1} + C_{p+q+r-1,1} + C_{p+q-1,r+1} + \zeta(p+q+r) \right] \right\},$$

where empty sums are interpreted (as usual) to be nil.

(*Cf.* Berndt [122, p. 253]; Markett [801, p. 122])

8. Prove that

$$\zeta(2n+1) = (-1)^n \frac{4(2\pi)^{2n}}{(2n+1)!} \sum_{k=0}^{\infty} \mathcal{R}_{2n+1,k}\, \zeta(2k) \qquad (n \in \mathbb{N}),$$

where the coefficients $\mathcal{R}_{2n+1,k}$ are rational and given by

$$\mathcal{R}_{2n+1,k} := \sum_{m=0}^{2n} \binom{2n}{m} \frac{(2n+1)\, B_{2n-m}}{2^{2k+m+1}\,(2k+m+1)\,(m+1)} \qquad (k \in \mathbb{N}_0)$$

and by the family of generating functions

$$\sum_{n=0}^{\infty} \mathcal{R}_{2n+1,k} \frac{t^{2n+1}}{(2n+1)!} = \frac{(2k-1)!}{t^{2k-1}\,(e^t-1)} \left(1 - e^{\frac{1}{2}t} \sum_{m=0}^{2k-1} \frac{(-1)^m}{m!} \left(\frac{t}{2}\right)^m\right) \qquad (k \in \mathbb{N};\ |t| < 2\pi).$$

(Cvijović and Klinowski [351, p. 1265]; see also 4.2(39) *et seq.*)

9. Prove that

$$\zeta(n) = \sum_{j=1}^{n-2} \sum_{p,q=1}^{\infty} \frac{1}{p^j\,(p+q)^{n-j}} \qquad (n \in \mathbb{N} \setminus \{1, 2\}).$$

(*cf.* Briggs [177]; see also Markett [801, p. 115])

10. Let a, b, c be real numbers greater than 1 and

$$P(a,b,c) = \sum_{r=1}^{\infty} r^{-a} \sum_{k=1}^{r} k^{-b} \sum_{l=1}^{k} l^{-c}.$$

Then show that

$$\begin{aligned} &P(a,b,c) + P(a,c,b) + P(b,c,a) + P(b,a,c) + P(c,a,b) + P(c,b,a) \\ &\quad = \zeta(a)\,\zeta(b)\,\zeta(c) + \zeta(a)\,\zeta(b+c) + \zeta(b)\,\zeta(c+a) + \zeta(c)\,\zeta(a+b) + 2\,\zeta(a+b+c), \end{aligned}$$

where ζ is the Riemann Zeta function, defined by 2.3(1).

(Sitaramachandrarao and Subbarao [1037, p. 471])

11. For each pair $(m,n) \in \mathbb{N}_0 \times \mathbb{N}$, one defines $A_{2m}(n)$ as follows:

(i) $A_{2m}(1) := B_{2m}$, and

(ii) for $n \in \mathbb{N} \setminus \{1\}$,

$$A_{2m}(n) := \sum \frac{\binom{2m}{2j_1,\,\ldots,\,2j_n}}{\{2j_1+1\}\,\{2\,(j_1+j_2)+1\} \cdots \{2\,(j_1+\cdots+j_{n-1})+1\}} \cdot B_{2j_1} \cdots B_{2j_n},$$

where the sum is taken over all $(j_1, \ldots, j_n) \in {\mathbb{N}_0}^n$, such that $j_1 + \cdots + j_n = m$ and $\binom{2m}{2j_1,\,\ldots,\,2j_n}$ is a multinomial coefficient; B_j are Bernoulli numbers, defined by 1.6(2). Then show that

$$\zeta(n) = \frac{2^{n-2}}{2^n - 1} \pi^2 \sum_{k=0}^{\infty} (-1)^k A_{2k}(n-2) \frac{\pi^{2k}}{(2k+2)!} \qquad (n \in \mathbb{N} \setminus \{1, 2\}).$$

(Ewell [437, p. 61])

12. Show that

$$\zeta(3) = \frac{2}{45}\pi^3 - 4\sum_{n=1}^{\infty} e^{-2\pi n}\,\sigma_{-3}(n)\left(2\pi^2 n^2 + \pi n + \frac{1}{2}\right)$$

and

$$\zeta(3) = \frac{7}{180}\pi^3 - 2\sum_{n=1}^{\infty} e^{-2\pi n}\,\sigma_{-3}(n),$$

where $\sigma_\nu(n)$ is given in Problem 20 of Chapter 2.

(*Cf.* Terras [1147, p. 181] and Grosswald [514]; see also Problem 21 of Chapter 2)

13. Consider the cosecant integral:

$$\mathcal{I}_s(\omega) := \int_0^{\pi/\omega} t^s \csc^2 t\, dt \quad (\Re(s) > 1; \quad \omega > 1).$$

Prove the following identities:

$$\mathcal{I}_s(\omega) = -\left(\frac{\pi}{\omega}\right)^s \cot\left(\frac{\pi}{\omega}\right) + s\int_0^{\pi/\omega} t^{s-1}\cot t\, dt \quad (\Re(s) > 1; \quad \omega > 1);$$

$$\mathcal{I}_s(\omega) = -\left(\frac{\pi}{\omega}\right)^s \cot\left(\frac{\pi}{\omega}\right) - 2s\left(\frac{\pi}{\omega}\right)^{s-1}\sum_{k=0}^{\infty}\frac{\zeta(2k)}{(s+2k-1)\omega^{2k}}$$

$$(\Re(s) > 1; \quad \omega > 1);$$

$$\int_0^{\pi/2} t^s \csc^2 t\, dt = -2s\left(\frac{\pi}{2}\right)^{s-1}\sum_{k=0}^{\infty}\frac{\zeta(2k)}{(s+2k-1)2^{2k}} \quad (\Re(s) > 1);$$

$$\int_0^{\pi/4} t^s \csc^2 t\, dt = -\left(\frac{\pi}{4}\right)^s - 2s\left(\frac{\pi}{4}\right)^{s-1}\sum_{k=0}^{\infty}\frac{\zeta(2k)}{(s+2k-1)4^{2k}} \quad (\Re(s) > 1);$$

$$\int_0^{\pi/3} t^s \csc^2 t\, dt = -\frac{1}{\sqrt{3}}\left(\frac{\pi}{3}\right)^s - 2s\left(\frac{\pi}{3}\right)^{s-1}\sum_{k=0}^{\infty}\frac{\zeta(2k)}{(s+2k-1)3^{2k}} \quad (\Re(s) > 1);$$

$$\int_0^{\pi/6} t^s \csc^2 t\, dt = -\sqrt{3}\left(\frac{\pi}{6}\right)^{s-1} - 2s\left(\frac{\pi}{6}\right)^{s-1}\sum_{k=0}^{\infty}\frac{\zeta(2k)}{(s+2k-1)6^{2k}} \quad (\Re(s) > 1);$$

$$\mathcal{I}_s(\omega) = -\left(\frac{\pi}{\omega}\right)^s\left[\cot\left(\frac{\pi}{\omega}\right) + i\right] - 2is\sum_{k=1}^{\infty}\frac{\gamma(s, -2k\pi i/\omega)}{(-2ki)^s} \quad (\Re(s) > 1;\ \omega > 1),$$

where $\gamma(z, \alpha)$ denotes the incomplete Gamma function, defined by 1.1(76);

$$\gamma(m+1, \alpha) = m!\left(1 - e^{-\alpha}\sum_{j=0}^{m}\frac{\alpha^j}{j!}\right) \quad (m \in \mathbb{N}_0);$$

$$\mathcal{I}_{2n}(\omega) = -\left(\frac{\pi}{\omega}\right)^{2n}\cot\left(\frac{\pi}{\omega}\right) - (2n)!\left(\frac{\pi}{\omega}\right)^{2n-1}\left[\frac{1}{(2n-1)!}\sum_{k=1}^{\infty}\frac{\cos(2k\pi/\omega)}{k}\right.$$
$$+\sum_{j=1}^{n-1}\frac{(-1)^j}{(2n-2j-1)!}\left(\frac{\omega}{2\pi}\right)^{2j}\sum_{k=1}^{\infty}\frac{\cos(2k\pi/\omega)}{k^{2j+1}}$$
$$\left.+\sum_{j=1}^{n}\frac{(-1)^j}{(2n-2j)!}\left(\frac{\omega}{2\pi}\right)^{2j-1}\sum_{k=1}^{\infty}\frac{\sin(2k\pi/\omega)}{k^{2j}}\right] \quad (n\in\mathbb{N};\ \omega>1);$$

$$\mathcal{I}_{2n+1}(\omega) = -\left(\frac{\pi}{\omega}\right)^{2n+1}\cot\left(\frac{\pi}{\omega}\right) + (-1)^n\frac{(2n+1)!}{2^{2n}}\zeta(2n+1)$$
$$-(2n+1)!\left(\frac{\pi}{\omega}\right)^{2n}\left[\frac{1}{(2n)!}\sum_{k=1}^{\infty}\frac{\cos(2k\pi/\omega))}{k} + \sum_{j=1}^{n}\frac{(-1)^j}{(2n-2j)!}\left(\frac{\omega}{2\pi}\right)^{2j}\right.$$
$$\left.\cdot\sum_{k=1}^{\infty}\frac{\cos(2k\pi/\omega)}{k^{2j+1}} + \sum_{j=1}^{n}\frac{(-1)^j}{(2n-2j+1)!}\left(\frac{\omega}{2\pi}\right)^{2j-1}\sum_{k=1}^{\infty}\frac{\sin(2k\pi/\omega)}{k^{2j}}\right]$$
$$(n\in\mathbb{N};\ \omega>1);$$

$$\sum_{k=0}^{\infty}\frac{\zeta(2k)}{(2k+2n-1)\omega^{2k}} = -\frac{1}{2}\log\left[2\sin\left(\frac{\pi}{\omega}\right)\right]$$
$$+\frac{(2n-1)!}{2}\left[\sum_{j=1}^{n-1}\frac{(-1)^j}{(2n-2j-1)!}\left(\frac{\omega}{2\pi}\right)^{2j}\mathrm{C}\ell_{2j+1}\left(\frac{2\pi}{\omega}\right)\right.$$
$$\left.+\sum_{j=1}^{n}\frac{(-1)^j}{(2n-2j)!}\left(\frac{\omega}{2\pi}\right)^{2j-1}\mathrm{C}\ell_{2j}\left(\frac{2\pi}{\omega}\right)\right]$$

and

$$\sum_{k=0}^{\infty}\frac{\zeta(2k)}{(2k+2n)\omega^{2k}} = (-1)^{n-1}\frac{(2n)!}{2^{2n+1}}\left(\frac{\omega}{\pi}\right)^{2n}\zeta(2n+1) - \frac{1}{2}\log\left[2\sin\left(\frac{\pi}{\omega}\right)\right]$$
$$+\frac{(2n)!}{2}\sum_{j=1}^{n}\frac{(-1)^j}{(2n-2j+1)!}\left(\frac{\omega}{2\pi}\right)^{2j}$$
$$\cdot\left[(2n-2j+1)\mathrm{C}\ell_{2j+1}\left(\frac{2\pi}{\omega}\right) + \frac{2\pi}{\omega}\mathrm{C}\ell_{2j}\left(\frac{2\pi}{\omega}\right)\right],$$

in terms of the generalized Clausen functions $\mathrm{C}\ell_{2n}$ and $\mathrm{C}\ell_{2n+1}$, defined by 2.4(80) and 2.4(81) and $n\in\mathbb{N}$ and $\omega>1$.

(Srivastava, Glasser, and Adamchik [1097])

14. Continuing Problem 13, show that

$$\zeta(2n+1) = (-1)^{n-1}\frac{(2\pi)^{2n}}{(2n+1)!\,(2^{2n}-1)}\left[\log 2 + 2\sum_{k=0}^{\infty}\frac{\zeta(2k)}{(2k+2n+1)2^{2k}}\right.$$
$$\left.+(2n+1)!\sum_{j=1}^{n-1}\frac{(-1)^j}{(2n-2j+1)!}\left(\frac{2^{2j}-1}{(2\pi)^{2j}}\right)\zeta(2j+1)\right]$$

and

$$\zeta(2n+1) = (-1)^{n-1} \frac{(2\pi)^{2n}}{(2n)!\left(2^{2n+1}-1\right)} \left[\log 2 + \sum_{k=0}^{\infty} \frac{\zeta 2k)}{(k+n)2^{2k}} \right.$$
$$\left. + (2n)! \sum_{j=1}^{n-1} \frac{(-1)^j}{(2n-2j)!} \left(\frac{2^{2j}-1}{(2\pi)^{2j}} \right) \zeta(2j+1) \right],$$

for $n \in \mathbb{N}$, respectively;

$$\sum_{k=1}^{\infty} \frac{\sin(kx+y)}{k^s} = \frac{(2\pi)^s}{2\Gamma(s)} \csc(\pi s) \left[\cos\left(y - \frac{1}{2}\pi s\right) \zeta\left(1-s,\ \frac{x}{2\pi}\right) \right.$$
$$\left. - \cos\left(y + \frac{1}{2}\pi s\right) \zeta\left(1-s,\ 1-\frac{x}{2\pi}\right) \right],$$

for $\Re(s) > 1$ and $0 < x < 2\pi$;

$$\sum_{k=1}^{\infty} \frac{\sin(2k\pi/\omega)}{k^s} = \frac{(2\pi)^s}{4\Gamma(s)} \csc\left(\frac{1}{2}\pi s\right) \left[\zeta\left(1-s,\ \frac{1}{\omega}\right) - \zeta\left(1-s,\ 1-\frac{1}{\omega}\right) \right],$$

for $\Re(s) > 1$ and $\omega > 1$;

$$\zeta(2n+1) = (-1)^{n-1} \frac{(2\pi)^{2n}}{(2n+1)!\left(3^{2n}-1\right)} \left[\log 3 + 4 \sum_{k=0}^{\infty} \frac{\zeta(2k)}{(2k+2n+1)3^{2k}} \right.$$
$$+ (2n+1)! \sum_{j=1}^{n-1} \frac{(-1)^j}{(2n-2j+1)!} \left(\frac{3^{2j}-1}{(2\pi)^{2j}} \right) \zeta(2j+1)$$
$$\left. - \frac{(2n+1)!}{\sqrt{3}} \sum_{j=1}^{n+1} \frac{(-1)^j}{(2n-2j+2)!} \frac{2\zeta\left(2j, \frac{1}{3}\right) - \left(3^{2j}-1\right)\zeta(2j)}{(2\pi)^{2j-1}} \right] \qquad (n \in \mathbb{N})$$

and

$$\zeta(2n+1) = (-1)^{n-1} \frac{(2\pi)^{2n}}{(2n)!\left(3^{2n+1}-1\right)} \left[\log 3 + 2 \sum_{k=0}^{\infty} \frac{\zeta(2k)}{(k+n)3^{2k}} \right.$$
$$+ (2n)! \sum_{j=1}^{n-1} \frac{(-1)^j}{(2n-2j)!} \left(\frac{3^{2j}-1}{(2\pi)^{2j}} \right) \zeta(2j+1)$$
$$\left. - \frac{(2n)!}{\sqrt{3}} \sum_{j=1}^{n} \frac{(-1)^j}{(2n-2j+1)!} \frac{2\zeta\left(2j, \frac{1}{3}\right) - \left(3^{2j}-1\right)\zeta(2j)}{(2\pi)^{2j-1}} \right] \qquad (n \in \mathbb{N});$$

$$\zeta(3) = \frac{\pi^2}{12} \left[\log 3 + 4 \sum_{k=0}^{\infty} \frac{\zeta(2k)}{(2k+3)3^{2k}} \right.$$
$$+ 2\sqrt{3} \sum_{j=1}^{2} \frac{(-1)^{j-1}}{(4-2j)!} \frac{2\zeta\left(2j, \frac{1}{3}\right) - \left(3^{2j}-1\right)\zeta(2j)}{(2\pi)^{2j-1}}$$

and

$$\zeta(3) = \frac{\pi^2}{13}\left[\log 3 + 2\sum_{k=0}^{\infty}\frac{\zeta(2k)}{(k+1)3^{2k}} + \frac{2}{\pi\sqrt{3}}\left\{\zeta\left(2,\frac{1}{3}\right) - 4\zeta(2)\right\}\right];$$

$$\begin{aligned}\zeta(2n+1) = (-1)^{n-1}\frac{(2\pi)^{2n}}{(2n+1)!\left(2^{2n}-1\right)}&\left[\log 2 + 4\sum_{k=0}^{\infty}\frac{\zeta(2k)}{(2k+2n+1)4^{2k}}\right.\\ &+ (2n+1)!\sum_{j=1}^{n-1}\frac{(-1)^j}{(2n-2j+1)!}\left(\frac{2^{2j}-1}{(2\pi)^{2j}}\right)\zeta(2j+1)\\ &\left.- (2n+1)!\sum_{j=1}^{n+1}\frac{(-1)^j}{(2n-2j+2)!}\frac{\zeta\left(2j,\frac{1}{4}\right) - 2^{2j-1}\left(2^{2j}-1\right)\zeta(2j)}{(2\pi)^{2j-1}}\right] \quad (n\in\mathbb{N})\end{aligned}$$

and

$$\begin{aligned}\zeta(2n+1) = (-1)^{n-1}\frac{(2\pi)^{2n}}{(2n)!\left(2^{4n+1}+2^{2n}-1\right)}&\left[\log 2 + 2\sum_{k=0}^{\infty}\frac{\zeta(2k)}{(k+n)4^{2k}}\right.\\ &+ (2n)!\sum_{j=1}^{n-1}\frac{(-1)^j}{(2n-2j)!}\left(\frac{2^{2j}-1}{(2\pi)^{2j}}\right)\zeta(2j+1)\\ &\left.- (2n)!\sum_{j=1}^{n}\frac{(-1)^j}{(2n-2j+1)!}\frac{\zeta\left(2j,\frac{1}{4}\right) - 2^{2j-1}\left(2^{2j}-1\right)\zeta(2j)}{(2\pi)^{2j-1}}\right] \quad (n\in\mathbb{N});\end{aligned}$$

$$\begin{aligned}\zeta(3) = \frac{2\pi^2}{9}&\left[\log 2 + 4\sum_{k=0}^{\infty}\frac{\zeta 2k)}{(2k+3)4^{2k}}\right.\\ &\left.+6\sum_{j=1}^{2}\frac{(-1)^{j-1}}{(4-2j)!}\frac{\zeta\left(2j,\frac{1}{4}\right) - 2^{2j-1}\left(2^{2j}-1\right)\zeta(2j)}{(2\pi)^{2j-1}}\right]\end{aligned}$$

and

$$\zeta(3) = \frac{2\pi^2}{35}\left[\log 2 + 2\sum_{k=0}^{\infty}\frac{\zeta(2k)}{(k+1)4^{2k}} + \frac{1}{\pi}\left\{\zeta\left(2,\frac{1}{4}\right) - 6\zeta(2)\right\}\right];$$

$$\begin{aligned}\zeta(2n+1) = (-1)^{n-1}\frac{(2\pi)^{2n}}{\left(2^{2n}-1\right)\left(3^{2n}-1\right)}&\left[-\frac{4}{(2n+1)!}\sum_{k=0}^{\infty}\frac{\zeta(2k)}{(2k+2n+1)6^{2k}}\right.\\ &+\sum_{j=1}^{n-1}\frac{(-1)^j}{(2n-2j+1)!}\left(\frac{\left(2^{2j}-1\right)\left(3^{2j}-1\right)}{(2\pi)^{2j}}\right)\zeta(2j+1)\\ &\left.+\frac{1}{\sqrt{3}}\sum_{j=1}^{n+1}\frac{(-1)^j}{(2n-2j+2)!}\frac{\zeta\left(2j,\frac{1}{3}\right) + \zeta\left(2j,\frac{1}{6}\right) - 2^{2j-1}\left(3^{2j}-1\right)\zeta(2j)}{(2\pi)^{2j-1}}\right] \quad (n\in\mathbb{N})\end{aligned}$$

and

$$\zeta(2n+1) = (-1)^{n-1}\frac{(2\pi)^{2n}}{2^{2n}+3^{2n}+6^{2n}-1}\left[\frac{2}{(2n)!}\sum_{k=0}^{\infty}\frac{\zeta(2k)}{(k+n)6^{2k}}\right.$$
$$-\sum_{j=1}^{n-1}\frac{(-1)^j}{(2n-2j)!}\left(\frac{(2^{2j}-1)(3^{2j}-1)}{(2\pi)^{2j}}\right)\zeta(2j+1)$$
$$\left.-\frac{1}{\sqrt{3}}\sum_{j=1}^{n}\frac{(-1)^j}{(2n-2j+1)!}\,\frac{\zeta\left(2j,\frac{1}{3}\right)+\zeta\left(2j,\frac{1}{6}\right)-2^{2j-1}\left(3^{2j}-1\right)\zeta(2j)}{(2\pi)^{2j-1}}\right] \quad (n\in\mathbb{N});$$

$$\zeta(3) = -\frac{\pi^2}{18}\left[2\sum_{k=0}^{\infty}\frac{\zeta(2k)}{(2k+3)6^{2k}}\right.$$
$$\left.+\sqrt{3}\sum_{j=1}^{2}\frac{(-1)^{j-1}}{(4-2j)!}\,\frac{\zeta\left(2j,\frac{1}{3}\right)+\left(2j,\frac{1}{6}\right)-2^{2j-1}\left(3^{2j}-1\right)\zeta(2j)}{(2\pi)^{2j-1}}\right]$$

and

$$\zeta(3) = \frac{\pi^2}{12}\left[\sum_{k=0}^{\infty}\frac{\zeta(2k)}{(k+1)6^{2k}}+\frac{1}{2\pi\sqrt{3}}\left\{\zeta\left(2,\frac{1}{3}\right)+\zeta\left(2,\frac{1}{6}\right)-16\zeta(2)\right\}\right].$$

(Srivastava, Glasser and Adamchik [1097])

15. Continuing Problems 13 and 14, show that

$$\sum_{k=0}^{\infty}\frac{\zeta(2k)}{(2k+2n+1)2^{2k}} = \frac{1}{2}\sum_{j=0}^{n}\frac{(-1)^{j-1}}{2j+1}\binom{2n+1}{2j}\left(\frac{2}{\pi}\right)^{2j}\int_0^{\infty}t^{2j+1}\,\mathrm{sech}^2 t\,dt,$$

for $n\in\mathbb{N}_0$;

$$\sum_{k=0}^{\infty}\frac{\zeta(2k)}{(2k+2n+1)2^{2k}}$$
$$=\frac{1}{2}\sum_{j=1}^{n}(-1)^{j-1}\binom{2n+1}{2j}\frac{(2j)!\left(2^{2j}-1\right)}{(2\pi)^{2j}}\zeta(2j+1)-\frac{1}{2}\log 2,$$

for $n\in\mathbb{N}_0$;

$$\zeta(2n+1) = (-1)^{n-1}\frac{(2\pi)^{2n}}{(2n+1)!\left(2^{2n}-1\right)}\left[\log 2+2\sum_{k=0}^{\infty}\frac{\zeta(2k)}{(2k+2n+1)2^{2k}}\right.$$
$$\left.+\sum_{j=1}^{n-1}(-1)^{j-1}\binom{2n+1}{2j}\frac{(2j)!\left(2^{2j}-1\right)}{(2\pi)^{2j}}\zeta(2j+1)\right],$$

for $n\in\mathbb{N}$.

(Srivastava, Glasser and Adamchik [1097, p. 842 *et seq.*])

16. Continuing Problems 13, 14 and 15, define

$$\mathcal{S}_p := \sum_{k=0}^{\infty} \frac{\zeta(2k)}{(2k+p)\omega^{2k}} = -\frac{\pi}{2\omega}\int_0^1 t^p \cot\left(\frac{\pi t}{\omega}\right) dt \quad (p \in \mathbb{N};\ |\omega| > 1).$$

Show that

$$\mathcal{S}_p = \frac{\pi i}{2(p+1)\omega} + \frac{1}{2}\left(-\frac{i\omega}{2\pi}\right)^p \int_1^{\Omega} \frac{(\log z)^p}{1-z} dz \quad (p \in \mathbb{N};\ \omega| > 1);$$

$$\mathcal{S}_p = \frac{\pi i}{2(p+1)\omega} - \frac{1}{2}\left(-\frac{i\omega}{2\pi}\right)^p \int_0^1 \{\log(1-(1-\Omega)t)\}^p \frac{dt}{t},$$

where, for convenience,

$$\Omega := \exp\left(\frac{2\pi i}{\omega}\right) \quad (|\omega| > 1);$$

or, equivalently,

$$\sum_{k=0}^{\infty} \frac{\zeta(2k)}{(2k+p)\omega^{2k}} = \frac{\pi i}{2(p+1)\omega} - \frac{p!}{2}\left(\frac{i\omega}{2\pi}\right)^p S_{1,p}\left(1 - e^{2\pi i/\omega}\right)$$

in terms of Nielsen's generalized Polylogarithmic function $S_{n,p}(z)$, defined by (*cf.*, e.g., Kölbig [686, p. 1233, Eq. (1.3)])

$$S_{n,p}(z) := \frac{(-1)^{n+p-1}}{(n-1)!p!}\int_0^1 (\log t)^{n-1}\{\log(1-zt)\}^p \frac{dt}{t} \quad (n, p \in \mathbb{N};\ z \in \mathbb{C});$$

$$S_{1,p}(z) = \zeta(p+1) + \sum_{k=0}^{p} \frac{(-1)^{k-1}}{k!}\{\log(1-z)\}^k \operatorname{Li}_{p-k+1}(1-z);$$

$$\sum_{k=0}^{\infty} \frac{\zeta(2k)}{(2k+p)\omega^{2k}} = \frac{\pi i}{2(p+1)\omega} - \frac{1}{2}\log\left(1 - e^{2\pi i/\omega}\right) - \frac{p!}{2}\left(\frac{i\omega}{2\pi}\right)^p \zeta(p+1) + \frac{1}{2}\sum_{k=1}^{p}\binom{p}{k}k!\left(\frac{i\omega}{2\pi}\right)^k \operatorname{Li}_k\left(e^{2\pi i/\omega}\right),$$

for $p \in \mathbb{N}$ and $|\omega| > 1$;

$$\int_0^1 t^p \cot(\nu t)dt = -\frac{i}{p+1} + \frac{1}{\nu}\log\left(1 - e^{2\nu i}\right) + \frac{p!}{\nu}\left(\frac{i}{2\nu}\right)^p \zeta(p+1) - \frac{1}{\nu}\sum_{k=1}^{p}\binom{p}{k}k!\left(\frac{i}{2\nu}\right)^k \operatorname{Li}_{k+1}\left(e^{2\nu i}\right),$$

for $p \in \mathbb{N}$ and $\nu \in \mathbb{C} \setminus \{0\}$.

(Srivastava, Glasser and Adamchik [1097, p. 843 *et seq.*])

17. Prove the following series representations for the values of $L(s, \chi)$:

(a) If $\chi(-1) = 1$ and $\chi \neq 1$, then

$$nL(2n+1, \chi) - n\sum_{l=1}^{\infty} \frac{\chi(l)}{l^{2n+1}} \cos\left(\frac{2l\pi x}{p}\right) - \frac{\pi x}{p}\sum_{l=1}^{\infty} \frac{\chi(l)}{l^{2n}} \sin\left(\frac{2l\pi x}{p}\right)$$

$$= (-1)^n \left(\frac{2\pi x}{p}\right)^{2n} \left[\sum_{k=1}^{n-1} \frac{(-1)^{k-1} \cdot k}{(2n-2k)!} \frac{L(2k+1, \chi)}{(2\pi x/p)^{2k}}\right.$$

$$\left. + \frac{\tau(\chi)}{p} \sum_{k=1}^{\infty} \frac{(2k)!}{(2n+2k)!} L(2k, \overline{\chi}) x^{2k}\right] \qquad (n \in \mathbb{N});$$

(b) If $\chi(-1) = -1$, then

$$L(2n, \chi) - \sum_{l=1}^{\infty} \frac{\chi(l)}{l^{2n}} \cos\left(\frac{2l\pi x}{p}\right)$$

$$= (-1)^n \left(\frac{2\pi x}{p}\right)^{2n-1} \left[\sum_{k=1}^{n-1} \frac{(-1)^{k-1}}{(2n-2k)!} \frac{L(2k, \chi)}{(2\pi x/p)^{2k-1}}\right.$$

$$\left. + \frac{2\tau(\chi)i}{p} \sum_{k=0}^{\infty} \frac{(2k)!}{(2n+2k)!} L(2k+1, \overline{\chi}) x^{2k+1}\right] \qquad (n \in \mathbb{N}),$$

where $L(s, \chi)$ is the Dirichlet L-function associated with a nontrivial Dirichlet character χ of modulus p and $\tau(\chi)$ denotes the Gauss sum defined by

$$\tau(\chi) := \sum_{k=1}^{p} \chi(k) \exp\left(\frac{2k\pi i}{p}\right) \quad \left(i := \sqrt{-1}\right).$$

(Katsurada [638, p. 82, Theorem 3])

18. Prove the following series representation:

$$\zeta(2n+1) + \frac{1}{2\pi x} \sum_{l=1}^{\infty} \frac{\sin(2\pi l x)}{l^{2n+2}}$$

$$= (-1)^{n-1} (2\pi x)^{2n} \left[\frac{1}{(2n+1)!} \left(\sum_{m=1}^{2n+1} \frac{1}{m} - \log(2\pi x)\right)\right.$$

$$\left. + \sum_{k=1}^{n-1} (-1)^k \frac{\zeta(2k+1)}{(2n-2k+1)!\,(2\pi x)^{2k}} + 2\sum_{k=1}^{\infty} \frac{(2k-1)!\,\zeta(2k)}{(2n+2k+1)!} x^{2k}\right]$$

$$(n \in \mathbb{N};\ x \in \mathbb{R};\ |x| \leq 1).$$

(Katsurada [638, p. 81, Theorem 2])

19. Prove the following series representations for the values of $L(s, \chi)$:

(a) If $\chi(-1) = 1$ and $\chi \neq 1$, then

$$L(2n+1, \chi) - \frac{p}{2\pi x} \sum_{l=1}^{\infty} \frac{\chi(l)}{l^{2n+2}} \sin\left(\frac{2l\pi x}{p}\right)$$

$$= (-1)^{n+1} \left(\frac{2\pi x}{p}\right)^{2n} \left[\sum_{k=0}^{n-1} \frac{(-1)^k}{(2n-2k+1)!} \frac{L(2k+1, \chi)}{(2\pi x/p)^{2k}}\right.$$

$$\left. + \frac{2\tau(\chi)}{p} \sum_{k=1}^{\infty} \frac{(2k-1)!}{(2n+2k+1)!} L(2k, \overline{\chi}) x^{2k}\right] \quad (n \in \mathbb{N});$$

(b) If $\chi(-1) = -1$, then

$$L(2n, \chi) - \frac{p}{2\pi x} \sum_{l=1}^{\infty} \frac{\chi(l)}{l^{2n+1}} \sin\left(\frac{2l\pi x}{p}\right)$$

$$= (-1)^{n+1} \left(\frac{2\pi x}{p}\right)^{2n-1} \left[\sum_{k=1}^{n-1} \frac{(-1)^k}{(2n+2k+1)!} \frac{L(2k, \chi)}{(2\pi x/p)^{2k-1}}\right.$$

$$\left. - \frac{2\tau(\chi) i}{p} \sum_{k=0}^{\infty} \frac{(2k)!}{(2n+2k+1)!} L(2k+1, \overline{\chi}) x^{2k+1}\right] \quad (n \in \mathbb{N}),$$

where $L(s, \chi)$ and $\tau(\chi)$ are as those given in Problem 17.

(Srivastava and Tsumura [1112])

20. Prove the following series representations for $\zeta(3)$:

$$\zeta(3) = -\frac{120}{1573}\pi^2$$

$$\cdot \sum_{k=0}^{\infty} \frac{8576k^2 + 24286k + 17283}{(2k+1)(2k+2)(2k+3)(2k+4)(2k+5)(2k+6)(2k+7)} \frac{\zeta(2k)}{2^{2k}};$$

$$\zeta(3) = -\frac{6}{17}\pi^2 \left[\frac{1}{120}(80H_4 - 31H_5 - 49\log\pi)\right.$$

$$\left. + 2\sum_{k=1}^{\infty} \frac{32k+49}{(2k)(2k+1)(2k+2)(2k+3)(2k+4)(2k+5)} \frac{\zeta(2k)}{2^{2k}}\right] (n \in \mathbb{N}),$$

where H_k denotes the harmonic numbers, defined by 3.2(36).

(Srivastava and Tsumura [1112, pp. 20 and 22])

5 Determinants of the Laplacians

During the last two decades, the problem of evaluation of the determinants of the Laplacians on Riemann manifolds has received considerable attention by many authors, including (among others) D'Hoker and Phong [377, 378], Sarnak [1004] and Voros [1201], who computed the determinants of the Laplacians on compact Riemann surfaces of constant curvature in terms of special values of the Selberg Zeta function. Although the first interest in the determinants of the Laplacians arose mainly for Riemann surfaces, it is also interesting and potentially useful to compute these determinants for classical Riemannian manifolds of higher dimensions, such as spheres. In this chapter, we are particularly concerned with the evaluation of the functional determinant for the n-dimensional sphere $\mathbf{S}^n$ with the standard metric.

In computations of the determinants of the Laplacians on manifolds of constant curvature, an important rôle is played by the closed-form evaluations of the series involving the Zeta function given in Chapter 3 (*cf.*, e.g., Choi and Srivastava [291, 292], Choi *et al.* [269]), as well as the theory of the multiple Gamma functions presented in Section 1.3 (*cf.*, e.g., Voros [1201], Vardi [1190], Choi [260] and Quine and Choi [954]).

5.1 The n-Dimensional Problem

Let $\{\lambda_n\}$ be a sequence, such that

$$0 = \lambda_0 < \lambda_1 \leqq \lambda_2 \leqq \cdots \leqq \lambda_n \leqq \cdots; \quad \lambda_n \uparrow \infty \quad (n \to \infty); \tag{1}$$

hence, we consider only such nonnegative increasing sequences. Then, we can show that

$$Z(s) := \sum_{n=1}^{\infty} \frac{1}{\lambda_n^s}, \tag{2}$$

which is known to converge absolutely in a half-plane $\Re(s) > \sigma$ for some $\sigma \in \mathbb{R}$.

Definition 5.1 (*cf.* Osgood *et al.* [881]). The determinant of the Laplacian Δ on the compact manifold M is defined to be

$$\det{}'\Delta := \prod_{\lambda_k \neq 0} \lambda_k, \tag{3}$$

Zeta and q-Zeta Functions and Associated Series and Integrals. DOI: 10.1016/B978-0-12-385218-2.00005-0

where $\{\lambda_k\}$ is the sequence of eigenvalues of the Laplacian Δ on M. The sequence $\{\lambda_k\}$ is known to satisfy the condition as in (1), but the product in (3) is always divergent; so, for the expression (3) to make sense, some sort of regularization procedure must be used. It is easily seen that, formally, $e^{-Z'(0)}$ is the product of nonzero eigenvalues of Δ. This product does not converge, but $Z(s)$ can be continued analytically to a neighborhood of $s = 0$. Therefore, we can give a *meaningful* definition:

$$\det{}'\Delta := e^{-Z'(0)}, \tag{4}$$

which is called the *Functional Determinant of the Laplacian* Δ on M.

Definition 5.2 The *order* μ of the sequence $\{\lambda_k\}$ is defined by

$$\mu := \inf\left\{\alpha > 0 \,\middle|\, \sum_{k=1}^{\infty} \frac{1}{\lambda_k^{\alpha}} < \infty\right\}. \tag{5}$$

The analogous and shifted analogous Weierstrass canonical products $E(\lambda)$ and $E(\lambda, a)$ of the sequence $\{\lambda_k\}$ are defined, respectively, by

$$E(\lambda) := \prod_{k=1}^{\infty}\left\{\left(1 - \frac{\lambda}{\lambda_k}\right)\exp\left(\frac{\lambda}{\lambda_k} + \frac{\lambda^2}{2\lambda_k^2} + \cdots + \frac{\lambda^{[\mu]}}{[\mu]\lambda_k^{[\mu]}}\right)\right\} \tag{6}$$

and

$$E(\lambda, a) := \prod_{k=1}^{\infty}\left\{\left(1 - \frac{\lambda}{\lambda_k + a}\right)\exp\left(\frac{\lambda}{\lambda_k + a} + \cdots + \frac{\lambda^{[\mu]}}{[\mu]\,(\lambda_k + a)^{[\mu]}}\right)\right\}, \tag{7}$$

where $[\mu]$ denotes the greatest integer part in the order μ of the sequence $\{\lambda_k\}$.

There exists the following relationship between $E(\lambda)$ and $E(\lambda, a)$ (see Voros [1201]):

$$E(\lambda, a) = \exp\left(\sum_{m=1}^{[\mu]} \mathcal{R}_{m-1}(-a)\frac{\lambda^m}{m!}\right)\frac{E(\lambda - a)}{E(-a)}, \tag{8}$$

where, for convenience,

$$\mathcal{R}_{[\mu]}(\lambda - a) := \frac{d^{[\mu]+1}}{d\lambda^{[\mu]+1}}\left\{-\log E(\lambda, a)\right\}. \tag{9}$$

The shifted series $Z(s, a)$ of $Z(s)$ in (2) by a is given by

$$Z(s, a) := \sum_{k=1}^{\infty} \frac{1}{(\lambda_k + a)^s}. \tag{10}$$

Formally, indeed, we have

$$Z'(0,-\lambda) = -\sum_{k=1}^{\infty} \log(\lambda_k - \lambda),$$

which, if we define

$$D(\lambda) := \exp\left[-Z'(0,-\lambda)\right], \tag{11}$$

immediately implies that

$$D(\lambda) = \prod_{k=1}^{\infty} (\lambda_k - \lambda).$$

In fact, Voros [1201] gave the relationship between $D(\lambda)$ and $E(\lambda)$ as follows:

$$\begin{aligned} D(\lambda) = \exp[-Z'(0)] \exp &\left[-\sum_{m=1}^{[\mu]} \mathrm{FP}Z(m) \frac{\lambda^m}{m} \right] \\ \cdot \exp &\left[-\sum_{m=2}^{[\mu]} C_{-m} \left(\sum_{k=1}^{m-1} \frac{1}{k} \right) \frac{\lambda^m}{m!} \right] E(\lambda), \end{aligned} \tag{12}$$

where an empty sum is interpreted to be nil and the *finite part* prescription is applied (as usual) as follows (*cf.* Voros [1201, p. 446]):

$$\mathrm{FP}f(s) := \begin{cases} f(s), & \text{if } s \text{ is not a pole}, \\ \lim\limits_{\epsilon \to 0} \left(f(s+\epsilon) - \frac{\text{Residue}}{\epsilon} \right), & \text{if } s \text{ is a simple pole}, \end{cases} \tag{13}$$

and

$$Z(-m) = (-1)^m \, m! \, C_{-m}. \tag{14}$$

Now consider the sequence of eigenvalues on the standard Laplacian Δ_n on $\mathbf{S}^n$. It is known from the work of Vardi [1190] (see also Terras [1148]) that the standard Laplacian Δ_n ($n \in \mathbb{N}$) has eigenvalues

$$\mu_k := k(k+n-1) \tag{15}$$

with multiplicity

$$\begin{aligned} \beta_k^n := \binom{k+n}{n} - \binom{k+n-2}{n} &= \frac{(2k+n-1)\,(k+n-2)!}{k!\,(n-1)!} \\ &= \frac{2k+n-1}{(n-1)!} \prod_{j=1}^{n-2} (k+j) \quad (k \in \mathbb{N}_0). \end{aligned} \tag{16}$$

From now on, we consider the shifted sequence $\{\lambda_k\}$ of $\{\mu_k\}$ in (15) by $\left(\frac{n-1}{2}\right)^2$ as a fundamental sequence. Then, the sequence $\{\lambda_k\}$ is written in the following simple and tractable form:

$$\lambda_k = \mu_k + \left(\frac{n-1}{2}\right)^2 = \left(k + \frac{n-1}{2}\right)^2 \tag{17}$$

with the same multiplicity as in (16).

We will exclude the zero mode, that is, start the sequence at $k = 1$ for later use. Furthermore, with a view to emphasizing n on $\mathbf{S}^n$, we choose the notations $Z_n(s)$, $Z_n(s,a)$, $E_n(\lambda)$, $E_n(\lambda,a)$ and $D_n(\lambda)$, whereas instead of $Z(s)$, $Z(s,a)$, $E(\lambda)$, $E(\lambda,a)$ and $D(\lambda)$, respectively.

We readily observe from (11) that

$$D_n\left(\left(\frac{n-1}{2}\right)^2\right) = \det{}'\Delta_n, \tag{18}$$

where $\det{}'\Delta_n$ denote the *determinants of the Laplacians* on $\mathbf{S}^n$ $(n \in \mathbb{N})$.

Choi [260] computed the determinants of the Laplacians on the n-dimensional unit sphere $\mathbf{S}^n$ $(n = 1, 2, 3)$ by factorizing the analogous Weierstrass canonical product of a shifted sequence $\{\lambda_k\}$ in (17) of eigenvalues of the Laplacians on $\mathbf{S}^n$ into multiple Gamma functions, whereas Choi and Srivastava [291, 292] and Choi *et al.* [269] made use of some closed-form evaluations of the series involving Zeta function given in Chapter 3 for the computation of the determinants of the Laplacians on $\mathbf{S}^n$ $(n = 2, 3, 4, 5, 6, 7)$. Quine and Choi [954] made use of zeta regularized products to compute $\det{}'\Delta_n$ and the determinant of the conformal Laplacian, $\det(\Delta_{\mathbf{S}^n} + n(n-2)/4)$.

In following three sections, we compute the determinants of the Laplacians on $\mathbf{S}^n$ $(n = 1, 2, 3, 4, 5, 6, 7)$ and $\det{}'(\Delta_{\mathbf{S}^n} + n(n-2)/4)$, by using the aforementioned methods.

5.2 Computations Using the Simple and Multiple Gamma Functions

Factorizations Into Simple and Multiple Gamma Functions

We begin by expressing $E_n(\lambda)$ $(n = 1, 2, 3)$ as the simple and multiple Gamma functions. Our results are summarized in the following proposition (see Choi [260]; see also Voros [1201])

Proposition 5.1 *The analogous Weierstrass canonical products $E_n(\lambda)$ $(n = 1, 2, 3)$ of the shifted sequence $\{\lambda_k\}$ in 5.1(17) can be expressed in terms of the simple and multiple Gamma functions as follows:*

$$E_1(\lambda) = \frac{1}{\left\{\Gamma(1-\sqrt{\lambda})\Gamma(1+\sqrt{\lambda})\right\}^2}, \tag{1}$$

$$E_2(\lambda) = \frac{\left\{\Gamma_2(\frac{1}{2})\right\}^4 e^{c_1\lambda}\Gamma\left(\frac{1}{2}-\sqrt{\lambda}\right)\Gamma\left(\frac{1}{2}+\sqrt{\lambda}\right)}{\pi\left(1-2\sqrt{\lambda}\right)\left(1+2\sqrt{\lambda}\right)\left\{\Gamma_2\left(\frac{1}{2}-\sqrt{\lambda}\right)\Gamma_2\left(\frac{1}{2}+\sqrt{\lambda}\right)\right\}^2}, \tag{2}$$

and

$$E_3(\lambda) = \frac{e^{c_2\lambda}}{1-\lambda}\,\frac{\Gamma_2(1-\sqrt{\lambda})\Gamma_2(1+\sqrt{\lambda})}{\{\Gamma_3(1-\sqrt{\lambda})\Gamma_3(1+\sqrt{\lambda})\}^2}, \tag{3}$$

where, for convenience,

$$c_1 := 2(\gamma - 1 + 2\log 2) \quad \textit{and} \quad c_2 := \log(2\pi) - \tfrac{3}{2},$$

and γ is the Euler-Mascheroni constant, defined by 1.1(3).

Proof. Here we will prove only the representation (2). First of all, it follows from 5.1(15) and 5.1(16) with $n = 2$ that the eigenvalues of Δ_2 on $\mathbf{S}^2$ have the sequence

$$\mu_k = k(k+1) \quad \text{with multiplicity} \quad 2k+1 \quad (k \in \mathbb{N}_0) \tag{4}$$

and the corresponding shifted sequence $\{\lambda_k\}$ of $\{\mu_k\}$ in (4) by $\frac{1}{4}$ [or 5.1(17) with $n = 2$] is

$$\lambda_k = \left(k + \frac{1}{2}\right)^2 \quad \text{with multiplicity} \quad 2k+1 \quad (k \in \mathbb{N}_0). \tag{5}$$

It is easily seen from the definition 5.1(5) that both of the sequences (4) and (5) are of order $\mu = 1$. ■

From 5.1(6) we find that the analogous Weierstrass canonical product of the shifted sequence $\{\lambda_k\}$ in (5) is

$$E_2(\lambda) = \prod_{k=1}^{\infty}\left[\left(1 - \frac{\lambda}{\left(k+\frac{1}{2}\right)^2}\right)\exp\left(\frac{\lambda}{\left(k+\frac{1}{2}\right)^2}\right)\right]^{2k+1}. \tag{6}$$

To express $E_2(\lambda)$ in terms of simple and multiple Gamma functions, we consider the associated product E_2^+, defined by

$$E_2^+(z) := \prod_{k=1}^{\infty}\left[\left(1-\frac{z}{k+\frac{1}{2}}\right)\exp\left(\frac{z}{k+\frac{1}{2}}+\frac{z^2}{2\left(k+\frac{1}{2}\right)^2}\right)\right]^{2k+1}. \tag{7}$$

Letting $\lambda = -w^2$ in (6) and using the definition (7), we obtain

$$E_2(\lambda) = E_2^+(iw)\,E_2^+(-iw). \tag{8}$$

It is seen from 1.4(3) that the G-function itself is related to the Barnes product $E^B(\lambda)$ given by

$$\begin{aligned} E^B(\lambda) :&= \prod_{k=1}^{\infty}\left[\left(1-\frac{\lambda}{k}\right)\exp\left(\frac{\lambda}{k}+\frac{\lambda^2}{2k^2}\right)\right]^k \\ &= (2\pi)^{\frac{\lambda}{2}}\,e^{\frac{1}{2}[(1+\gamma)\lambda^2-\lambda]}\,G(1-\lambda). \end{aligned} \tag{9}$$

Thus, $E_2^+(\lambda)$ can be further decomposed to involve the Barnes G-function, or rather the shifted Barnes product:

$$E^B\left(z,-\frac{1}{2}\right) = \prod_{k=1}^{\infty}\left[\left(1-\frac{z}{k-\frac{1}{2}}\right)\exp\left(\frac{z}{k-\frac{1}{2}}+\frac{z^2}{2\left(k-\frac{1}{2}\right)^2}\right)\right]^k. \tag{10}$$

It readily follows from (7) and (10) that

$$\begin{aligned} E_2^+(z) = &\frac{\left\{E^B\left(z,-\frac{1}{2}\right)\right\}^2}{(1-2z)^2\exp\left(4z+4z^2\right)} \\ &\cdot\left\{\prod_{k=1}^{\infty}\left[\left(1-\frac{z}{k+\frac{1}{2}}\right)\exp\left(\frac{z}{k+\frac{1}{2}}+\frac{z^2}{2\left(k+\frac{1}{2}\right)^2}\right)\right]\right\}^{-1}. \end{aligned} \tag{11}$$

Obviously, the sequence $\{\lambda_k\}$ with $\lambda_k = k$ is of order $\mu = 1$. It follows from 1.3(2) and 5.1(6) that the analogous Weierstrass canonical product $E(\lambda)$ of this sequence is

$$E(\lambda) = \prod_{k=1}^{\infty}\left(1-\frac{\lambda}{k}\right)e^{\frac{\lambda}{k}} = \frac{e^{\gamma\lambda}}{\Gamma(1-\lambda)}, \tag{12}$$

which, in view of 5.1(9), yields

$$\mathcal{R}_0(\lambda) = -\frac{d}{d\lambda}\{\log E(\lambda)\} = -\psi(1-\lambda) - \gamma, \tag{13}$$

where ψ is the Psi (or Digamma) function defined by 1.3(1).

Now, the shift equation (8) applied to (12) with $a = -\frac{1}{2}$ and $\mu = 1$ yields the classical result:

$$\prod_{k=0}^{\infty}\left(1 - \frac{z}{k+\frac{1}{2}}\right)\exp\left(\frac{z}{k+\frac{1}{2}}\right) = e^{(\gamma+2\log 2)z}\frac{\Gamma\left(\frac{1}{2}\right)}{\Gamma\left(\frac{1}{2}-z\right)}, \tag{14}$$

which, by virtue of 1.1(14), immediately becomes

$$\prod_{k=1}^{\infty}\left(1 - \frac{z}{k} + \frac{1}{2}\right)\exp\left(\frac{z}{k} + \frac{1}{2}\right) = \frac{\sqrt{\pi}\,e^{(\gamma+2\log 2)z}}{(1-2z)\,e^{2z}\,\Gamma\left(\frac{1}{2}-z\right)}. \tag{15}$$

If we apply 1.1(14), 1.3(53), 2.2(4), 2.3(2) and (15) to (11), we easily find that

$$E_2^{+}(z) = \frac{\Gamma\left(\frac{1}{2}-z\right)\left\{E^B\left(z, -\frac{1}{2}\right)\right\}^2}{\sqrt{\pi}\,(1-2z)\,2^{2z}\exp\left[(2+\gamma)z + \left(2+\frac{\pi^2}{4}\right)z^2\right]}. \tag{16}$$

Similarly, the shift equation 5.1(8) applied to (9) with $a = -\frac{1}{2}$ and $\mu = 2$ yields

$$\begin{aligned} E^B\left(z, -\tfrac{1}{2}\right) = (2\pi)^{\frac{1}{2}z}\exp\left[(\log 2)z + \left(\log 2 + \frac{\pi^2}{8}\right)z^2\right] \\ \cdot\exp\left[\tfrac{1}{2}\left\{(1+\gamma)z^2 + \gamma z\right\}\right]\frac{G\left(\frac{1}{2}-z\right)}{G\left(\frac{1}{2}\right)}. \end{aligned} \tag{17}$$

A combination of (16) and (17) gives

$$E_2^{+}(z) = \frac{(2\pi)^z\,\Gamma\left(\frac{1}{2}-z\right)}{\sqrt{\pi}\,(1-2z)}\left\{\frac{G\left(\frac{1}{2}-z\right)}{G\left(\frac{1}{2}\right)}\right\}^2\exp\left[-2z + (\gamma - 1 + 2\log 2)\,z^2\right]. \tag{18}$$

Finally, it follows from (8) and (18) that

$$\begin{aligned} E_2(\lambda) = \frac{\Gamma\left(\frac{1}{2}-iw\right)\Gamma\left(\frac{1}{2}+iw\right)\left\{G\left(\frac{1}{2}-iw\right)G\left(\frac{1}{2}+iw\right)\right\}^2}{\pi\left\{G\left(\frac{1}{2}\right)\right\}^2(1-2iw)(1+2iw)} \\ \cdot\exp\left[-2(\gamma - 1 + 2\log 2)\,w^2\right]. \end{aligned} \tag{19}$$

Since $\lambda = -w^2$, we have $w = \pm i\sqrt{\lambda}$. We see that the second member of (19) remains the same with w replaced by either $i\sqrt{\lambda}$ or $-i\sqrt{\lambda}$. In view of this observation and the relationship

$$G(z) = \frac{1}{\Gamma_2(z)},$$

(19) is seen to be equivalent to (2). This completes the proof of the proposition (2). Propositions (1) and (3) can be proven similarly.

Evaluations of $\det' \Delta_n$ $(n = 1, 2, 3)$

By making use of Proposition 5.1, we can compute the determinants of the Laplacians on $\mathbf{S}^n$ $(n = 1, 2, 3)$ explicitly. The results of these computations (as well as our computation of $\det' \Delta_4$) are summarized by

Proposition 5.2 *The following evaluations hold true*:

$$\det{}' \Delta_1 = 4\pi^2, \tag{20}$$

$$\det{}' \Delta_2 = \exp\left[\tfrac{1}{2} - 4\zeta'(-1)\right] = e^{\frac{1}{6}} A^4 = 3.195311496\ldots, \tag{21}$$

$$\det{}' \Delta_3 = \pi \exp\left[\frac{\zeta(3)}{2\pi^2}\right] = 3.338851215\ldots, \tag{22}$$

and

$$\begin{aligned} det' \Delta_4 &= \frac{1}{3} \exp\left(\frac{35,639,301}{2^{17}\cdot 5\cdot 7} - \frac{13}{3}\zeta'(-1) - \frac{2}{3}\zeta'(-3)\right) \\ &= \frac{1}{3} \cdot A^{\frac{13}{3}} \cdot C^{\frac{2}{3}} \cdot \exp\left(\frac{183,758,875}{2^{17}\cdot 3^3\cdot 7}\right), \end{aligned} \tag{23}$$

where A is the Glaisher-Kinkelin's constant, defined by 1.4(2), and C is the mathematical constant, defined by 1.4(70).

Proof. $\mathbf{det}'\Delta_1$: It follows from 5.1(15) and 5.1(16) with $n = 1$ that the sequence of eigenvalues of Δ_1 on $\mathbf{S}^1$ is

$$\mu_k = k^2 \quad \text{with multiplicity} \quad 2 \quad (k \in \mathbb{N}_0), \tag{24}$$

which corresponds to the shifted sequence $\{\lambda_k\}$ in 5.1(17). It is obvious from the definition 5.1(5) that the sequence $\{\mu_k\}$ in (24) (and also its shifted sequence $\{\lambda_k\}$) is of order $\mu = \frac{1}{2}$. Thus, its corresponding Zeta series $Z_1(s)$ is given by

$$Z_1(s) = \sum_{k=1}^{\infty} \frac{2}{k^{2s}} = 2\zeta(2s), \tag{25}$$

where $\zeta(s)$ is the Riemann Zeta function, defined by 2.3(1). It is seen from 5.1(4) and 2.2(20) that

$$\det{}'\Delta_1 = \exp[-4\zeta'(0)] = 4\pi^2,$$

which proves the evaluation (20). ■

Alternatively, even though it is not necessary to do so in the case of $\det{}'\Delta_1$, we can also prove (20), by making use of 5.1(12), 5.1(18) and (1). Indeed, since the shifted sequence considered above is of order $\mu = \frac{1}{2}$, it is found from 5.1(12), 5.1(18) and 1.1(13) that

$$\det{}'\Delta_1 = D_1(0) = \exp\left[-Z_1'(0)\right] \lim_{\lambda \to 0} E_1(\lambda) = \exp\left[-Z_1'(0)\right],$$

where $Z_1(s)$ is the same as in (25) and the evaluation (20) follows immediately.

det$'\Delta_2$: In this case, we saw that the shifted sequence $\{\lambda_k\}$ in (5) is of order $\mu = 1$. Therefore, we find from 5.1(12) and 5.1(18) that

$$\det{}'\Delta_2 = D_2\left(\tfrac{1}{4}\right) = \exp\left[-Z_2{}'(0) - \tfrac{1}{4}\mathrm{FP}Z_2(1)\right] \lim_{\lambda \to \frac{1}{4}} E_2(\lambda). \tag{26}$$

From (5) and 5.1(2), we have

$$Z_2(s) = \sum_{k=1}^{\infty} \frac{2k+1}{\left(k+\frac{1}{2}\right)^{2s}} = \left(2^{2s}-2\right)\zeta(2s-1) - 4^s, \tag{27}$$

where $\zeta(s)$ is the Riemann Zeta function, defined by 2.3(1). Since $Z_2(s)$ has a simple pole only at $s = 1$ with its residue 1 (see Section 2.3), it is seen from 5.1(13), 2.2(19) with $a = 1$ and (27) that

$$\begin{aligned}
\mathrm{FP}Z_2(1) &= \lim_{\epsilon \to 0}\left\{Z_2(1+\epsilon) - \frac{1}{\epsilon}\right\} \\
&= -4 + \lim_{\epsilon \to 0}\left\{\left(2^{2+2\epsilon}-2\right)\zeta(2\epsilon+1) - \frac{1}{\epsilon}\right\} \\
&= -4 + \lim_{\epsilon \to 0}\left[\left(2^{2+2\epsilon}-2\right)\left\{\zeta(2\epsilon+1) - \frac{1}{2\epsilon}\right\} + \frac{2^{1+2\epsilon}-2}{\epsilon}\right] \\
&= -4 + 2\gamma + \lim_{\epsilon \to 0}\left\{\frac{2^{1+2\epsilon}-2}{\epsilon}\right\},
\end{aligned}$$

which, upon employing l'Hôspital's rule, immediately yields

$$\mathrm{FP}Z_2(1) = 2\gamma + 4\log 2 - 4. \tag{28}$$

It is easy to see from 2.3(10), 1.7(7) and 2.1(31) that

$$Z_2{}'(0) = -\frac{13}{6}\log 2 - 2\zeta'(-1) = -\frac{1}{6} - \frac{13}{6}\log 2 + 2\log A. \tag{29}$$

It follows from (2) and 1.1(12) (with z replaced by $\frac{1}{2}+z$) that

$$\lim_{\lambda\to\frac{1}{4}} E_2(\lambda) = \frac{\left\{\Gamma_2\left(\frac{1}{2}\right)\right\}^4}{4} \exp\left[\frac{1}{2}(\gamma - 1 + 2\log 2)\right] \cdot \lim_{\lambda\to\frac{1}{4}} \frac{G\left(\frac{1}{2}-\sqrt{\lambda}\right) G\left(\frac{1}{2}-\sqrt{\lambda}\right)}{\left(\frac{1}{2}-\sqrt{\lambda}\right)\cosh\left(\pi i\sqrt{\lambda}\right)},$$

which, upon considering the following limit relationships:

$$\lim_{\lambda\to\frac{1}{4}} \frac{G\left(\frac{1}{2}-\sqrt{\lambda}\right)}{\cosh\left(\pi i\sqrt{\lambda}\right)} = \frac{1}{\pi} \quad \text{and} \quad \lim_{\lambda\to\frac{1}{4}} \frac{G\left(\frac{1}{2}-\sqrt{\lambda}\right)}{\frac{1}{2}-\sqrt{\lambda}} = 1, \tag{30}$$

yields

$$\lim_{\lambda\to\frac{1}{4}} E_2(\lambda) = \frac{\left\{\Gamma_2\left(\frac{1}{2}\right)\right\}^4}{4\pi} \exp\left[\frac{1}{2}(\gamma - 1 + 2\log 2)\right]. \tag{31}$$

Finally, if we substitute from (28), (29) and (31) into (26) and make use of 2.1(31), we are led at once to the evaluation (21).

det$'\Delta_3$: From 5.1(17) with $n=3$, we find that the shifted sequence $\{\lambda_k\}$ (by 1) of eigenvalues of the Laplacian Δ_3 on $\mathbf{S}^3$ is

$$\lambda_k = (k+1)^2 \quad \text{with multiplicity} \quad (k+1)^2 \quad (k \in \mathbb{N}_0), \tag{32}$$

which is easily seen to be of order $\mu = \frac{3}{2}$. Therefore, the analogous Weierstrass canonical product and its accompanying Zeta series for this sequence are

$$E_3(\lambda) = \prod_{k=1}^{\infty}\left[\left(1 - \frac{\lambda}{(k+1)^2}\right)\exp\left(\frac{\lambda}{(k+1)^2}\right)\right]^{(k+1)^2} \tag{33}$$

and

$$Z_3(s) = \sum_{k=1}^{\infty} \frac{(k+1)^2}{(k+1)^{2s}} = \zeta(2s-2) - 1. \tag{34}$$

It follows from 5.1(12) and 5.1(18) that

$$\det{}'\Delta_3 = D_3(1) = \exp\left[-Z_3{}'(0) - \mathrm{FP}Z_3(1)\right] \lim_{\lambda\to 1} E_3(\lambda). \tag{35}$$

It is also seen that $Z_3(s)$ has a simple pole only at $s = \frac{3}{2}$ with its residue $\frac{1}{2}$. Thus, we have from 5.1(13), 2.2(13) (with $a = 1$) and 2.3(22) (with $n = 1$) that

$$\mathrm{FP}Z_3(1) = Z_3(1) = \zeta(0) - 1 = -\tfrac{3}{2} \tag{36}$$

and

$$Z_3{}'(0) = 2\zeta'(-2) = -\frac{\zeta(3)}{2\pi^2}. \tag{37}$$

If we substitute from (36) and (37) into (35), we obtain

$$\det{}'\Delta_3 = \exp\left(\frac{3}{2} + \frac{\zeta(3)}{2\pi^2}\right) \lim_{\lambda\to 1} E_3(\lambda). \tag{38}$$

By taking the limit as $\lambda \to 1$ on both sides of (3) and applying 1.4(6) and Theorem 1.6(a), we obtain

$$\lim_{\lambda\to 1} E_3(\lambda) = 2\pi \exp\left(-\tfrac{3}{2}\right) \mathcal{A}, \tag{39}$$

where, for convenience,

$$\mathcal{A} = \lim_{\lambda\to 1} \left\{ \frac{\Gamma_2\left(1-\sqrt{\lambda}\right)\left\{\Gamma_3\left(1-\sqrt{\lambda}\right)\right\}^{-2}}{1-\lambda} \right\}. \tag{40}$$

It is readily seen from 2.1(27), 2.1(28), 2.1(24) and 2.2(17) that

$$\mathcal{A} = \exp\left[\zeta'(0) + 2\zeta'(-1) + \zeta'(-2)\right] \cdot \mathcal{B} \cdot \mathcal{C}, \tag{41}$$

where, for convenience, $\mathcal{B}$ and $\mathcal{C}$ are defined by

$$\mathcal{B} := \lim_{\lambda\to 1} \left\{ \frac{1}{1-\lambda} \left[\frac{(2\pi)^{\frac{1}{2}}}{\Gamma\left(1-\sqrt{\lambda}\right)} \right]^{\lambda} \right\}$$

and

$$\mathcal{C} := \lim_{\lambda\to 1} \left\{ \exp\left[-\zeta'(-2, 1-\sqrt{\lambda}) - 2\sqrt{\lambda}\,\zeta'(-1, 1-\sqrt{\lambda})\right] \right\}.$$

First, to determine the limit value $\mathcal{B}$, consider

$$\mathcal{B} = \lim_{\lambda \to 1} \left\{ \frac{\left[(2\pi)^{\frac{1}{2}}\right]^{\lambda}}{1+\sqrt{\lambda}} \cdot \frac{1}{\left(1-\sqrt{\lambda}\right)\left\{\Gamma\left(1-\sqrt{\lambda}\right)\right\}^{\lambda}} \right\},$$

which, upon setting $z = 1 - \sqrt{\lambda}$ and using 1.1(9) and 1.1(13), immediately becomes

$$\mathcal{B} = \left(\frac{\pi}{2}\right)^{\frac{1}{2}} \lim_{z \to 0} \left\{ \frac{1}{z\{\Gamma(z)\}^{1-2z+z^2}} \right\} = \left(\frac{\pi}{2}\right)^{\frac{1}{2}} \lim_{z \to 0} \left\{ \left(\frac{1}{\Gamma(z)}\right)^{z^2-2z} \right\}.$$

We, thus, find from 1.1(2) that

$$\mathcal{B} = \left(\frac{\pi}{2}\right)^{12} \lim_{z \to 0} z^{z^2-2z} = \left(\frac{\pi}{2}\right)^{12}. \tag{42}$$

Next, to evaluate the limit value $\mathcal{C}$, by setting $n = 1$ in 2.2(4), we have

$$\zeta(s,a) = \zeta(s,1+a) + a^{-s}, \tag{43}$$

which, upon differentiating with respect to s, yields

$$\zeta'(s,a) = \zeta'(s,1+a) - a^{-s}\log a. \tag{44}$$

Therefore, we see from (44) that

$$\zeta'\left(-1,1-\sqrt{\lambda}\right) = \zeta'\left(-1,2-\sqrt{\lambda}\right) - \left(1-\sqrt{\lambda}\right)\log\left(1-\sqrt{\lambda}\right)$$

and

$$\zeta'\left(-2,1-\sqrt{\lambda}\right) = \zeta'\left(-2,2-\sqrt{\lambda}\right) - \left(1-\sqrt{\lambda}\right)^2\log\left(1-\sqrt{\lambda}\right),$$

and so

$$\lim_{\lambda \to 1} \zeta'\left(-1,1-\sqrt{\lambda}\right) = \zeta'(-1) \quad \text{and} \quad \lim_{\lambda \to 1} \zeta'\left(-2,1-\sqrt{\lambda}\right) = \zeta'(-2),$$

which yield

$$\mathcal{C} = \exp\left[-2\zeta'(-1) - \zeta'(-2)\right]. \tag{45}$$

Finally, by substituting from (42) and (45) into (41) and using 2.2(20), we find from (39) that

$$\lim_{\lambda \to 1} E_3(\lambda) = \pi \exp\left(-\tfrac{3}{2}\right), \tag{46}$$

which, when substituted into (38), proves the evaluation (22).

A proof of the evaluation (23) will be presented in Section 5.3 in which each of the results (21), (22) and (23) is obtained by means of the summation formulas of Chapter 3 for series of Zeta functions (see also Problem 3 at the end of this chapter).

5.3 Computations Using Series of Zeta Functions

In this section, we compute $\det' \Delta_n$ ($n = 2, 3, 4, 5, 6, 7$) in a markedly different way from that detailed in Section 5.2 for $n = 1, 2, 3$. Our evaluations, here, are based largely on the summation formulas of Chapter 3 for series of Zeta functions.

det$'\Delta_2$: In view of 5.2(26), 5.2(28) and 5.2(29), it suffices to evaluate $E_2(\frac{1}{4})$. It follows from 5.2(6) that

$$E_2\left(\frac{1}{4}\right) = \prod_{k=1}^{\infty}\left[\left(1 - \frac{1}{(2k+1)^2}\right)^{2k+1} \exp\left(\frac{1}{2k+1}\right)\right], \tag{1}$$

which, upon taking logarithms and applying the Maclaurin series expansion of $\log(1+x)$, yields

$$\begin{aligned} \log E_2\left(\frac{1}{4}\right) &= -\sum_{n=2}^{\infty}\frac{1}{n}\left(\sum_{n=2}^{\infty}\frac{1}{(2k-1)^{2n-1}} - 1\right) \\ &= -\sum_{n=2}^{\infty}\frac{\zeta(2n-1)-1}{n} + \sum_{n=2}^{\infty}\frac{\zeta(2n-1)}{n\cdot 2^{2n-1}}. \end{aligned} \tag{2}$$

If we apply 3.4(520) and 3.4(529) on the right-hand side of (2), we obtain

$$\log E_2\left(\frac{1}{4}\right) = -1 + \frac{\gamma}{2} - \frac{7}{6}\log 2 + 6\log A, \tag{3}$$

which, upon combining 5.2(26) with 5.2(8), 5.2(29) and 2.1(31), proves the evaluation 5.2(21).

det$'\Delta_3$: By virtue of 5.2(38), it suffices to compute $E_3(1)$. Indeed, letting $\lambda = 1$ in 5.2(33) and taking logarithms of the resulting equation with the aid of the Maclaurin series expansion of $\log(1+x)$, we get

$$\log E_3(1) = -\sum_{k=2}^{\infty}\left(\sum_{n=1}^{\infty}\frac{1}{(n+1)k^{2n}}\right) = -\sum_{n=1}^{\infty}\frac{\zeta(2n)-1}{n+1}, \tag{4}$$

which, in view of 3.4(571), immediately yields

$$\log E_3(1) = \log\pi - \frac{3}{2}. \tag{5}$$

Now, if we make use of (5) in 5.2(38), we are led to the evaluation 5.2(22).

det$'\Delta_4$: Letting $n=4$ in the shifted sequence 5.1(17) of eigenvalues of Δ_4 on $\mathbf{S}^4$, we obtain the following sequence:

$$\lambda_k = \left(k+\tfrac{3}{2}\right)^2 \quad \text{with multiplicity} \quad \tfrac{1}{6}(k+1)(k+2)(2k+3) \quad (k \in \mathbb{N}_0), \tag{6}$$

which obviously has the order $\mu = 2$. Now, it follows from 5.1(12) and 5.1(18) that

$$\begin{aligned} \det{}'\Delta_4 &= D_4\left(\frac{9}{4}\right) \\ &= \exp\left(-Z_4'(0) - \frac{9}{4}\mathrm{FP}Z_4(1) - \frac{81}{32}\mathrm{FP}Z_4(2) - \frac{81}{32}C_{-2}\right)E_4\left(\frac{9}{4}\right). \end{aligned} \tag{7}$$

We can express $Z_4(s)$ for the sequence (6) in terms of the Riemann Zeta function as follows:

$$\begin{aligned} Z_4(s) &= \frac{1}{6}\sum_{k=1}^{\infty}\frac{(k+1)(k+2)(2k+3)}{\left(k+\frac{3}{2}\right)^{2s}} \\ &= \frac{2^{2s}}{6}\sum_{k=1}^{\infty}\frac{(k+1)(k+2)}{(2k+3)^{2s-1}} = \frac{2^{2s}}{6}\sum_{k=2}^{\infty}\frac{k(k+1)}{(2k+1)^{2s-1}} \\ &= \frac{2^{2s}}{24}\left(\sum_{k=2}^{\infty}\frac{1}{(2k+1)^{2s-3}} - \sum_{k=2}^{\infty}\frac{1}{(2k+1)^{2s-1}}\right) \\ &= \frac{2^{2s}}{24}\left(\sum_{k=1}^{\infty}\frac{1}{(2k-1)^{2s-3}} - \sum_{k=1}^{\infty}\frac{1}{(2k-1)^{2s-1}} - \frac{1}{3^{2s-3}} + \frac{1}{3^{2s-1}}\right), \end{aligned} \tag{8}$$

which, in view of 2.3(1), becomes

$$\begin{aligned} Z_4(s) &= \frac{1}{3}\left(2^{2s-3}-1\right)\zeta(2s-3) - \frac{1}{3}\left(2^{2s-3}-\frac{1}{4}\right)\zeta(2s-1) \\ &\quad - \frac{1}{3}\left(\frac{2}{3}\right)^{2s-3} + \frac{1}{8}\left(\frac{2}{3}\right)^{2s}. \end{aligned} \tag{9}$$

We observe from Section 2.3 that $Z_4(s)$ has simple poles at $s=1$ and $s=2$ with their residues $-\frac{1}{24}$ and $\frac{1}{6}$, respectively.

Using 2.3(10) and 5.1(14), we obtain

$$C_{-2} = \frac{1}{2}Z_4(-2) = -\frac{9,801,047}{2^{12}\cdot 3^3\cdot 5\cdot 7} \tag{10}$$

and

$$Z_4'(0) = \log\left(2^{-\frac{2869}{1440}}\cdot 3^2\right) + \frac{1}{12}\zeta'(-1) - \frac{7}{12}\zeta'(-3). \tag{11}$$

Now, we evaluate $\mathrm{FP}Z_4(1)$ and $\mathrm{FP}Z_4(2)$. Since $Z_4(s)$ has simple poles at $s = 1$ and $s = 2$, we have to use the second case of the definition of $\mathrm{FP}f(s)$ in 5.1(13) to compute the finite parts of $Z_4(s)$ for $s = 1$ and $s = 2$.

Using the expression in (9) for $Z_4(s)$ and 2.3(6) (or 2.2[19] with $a = 1$), we easily see that

$$\begin{aligned}\mathrm{FP}Z_4(1) &= \lim_{\epsilon\to 0}\left(Z_4(1+\epsilon)+\frac{1}{24\epsilon}\right)\\ &= -\frac{31}{72}-\frac{1}{3}\lim_{\epsilon\to 0}\left[\left(2^{2\epsilon-1}-\frac{1}{4}\right)\left\{\zeta(2\epsilon+1)-\frac{1}{2\epsilon}\right\}+\frac{2^{2\epsilon}-1}{4\epsilon}\right]\\ &= -\frac{31}{72}-\frac{\gamma}{12}-\frac{1}{6}\log 2.\end{aligned} \tag{12}$$

Similarly, we have

$$\begin{aligned}\mathrm{FP}Z_4(2) &= \lim_{\epsilon\to 0}\left(Z_4(2+\epsilon)-\frac{1}{6\epsilon}\right)\\ &= -\frac{16}{81}-\frac{7}{12}\zeta(3)+\frac{1}{3}\lim_{\epsilon\to 0}\left[\left(2^{2\epsilon+1}-1\right)\right.\\ &\quad\left.\cdot\left\{\zeta(2\epsilon+1)-\frac{1}{2\epsilon}\right\}+\frac{2^{2\epsilon+1}-2}{2\epsilon}\right]\\ &= -\frac{16}{81}+\frac{\gamma}{3}+\frac{2}{3}\log 2-\frac{7}{12}\zeta(3).\end{aligned} \tag{13}$$

Since the sequence in (6) is of order $\mu = 2$, its analogous Weierstrass canonical product $E_4(\lambda)$ is

$$\begin{aligned}E_4(\lambda) &= \prod_{k=1}^{\infty}\left(1-\frac{\lambda}{\left(k+\frac{3}{2}\right)^2}\right)^{\frac{1}{6}(k+1)(k+2)(2k+3)}\\ &\quad\cdot\exp\left[\frac{1}{6}(k+1)(k+2)(2k+3)\left(\frac{\lambda}{\left(k+\frac{3}{2}\right)^2}+\frac{\lambda^2}{2\left(k+\frac{3}{2}\right)^4}\right)\right].\end{aligned} \tag{14}$$

Upon setting $\lambda = \frac{9}{4}$ in (14) and taking the logarithms on both sides of the resulting equation, if we make use of 2.3(1) and the Maclaurin series expansion of $\log(1+x)$, we obtain

$$\log E_4\left(\frac{9}{4}\right) = -\frac{1}{6}\sum_{k=1}^{\infty}(k+1)(k+2)\left\{\sum_{n=3}^{\infty}\frac{3^{2n}}{n(2k+3)^{2n-1}}\right\}$$

$$= -\frac{1}{24}\sum_{n=3}^{\infty}\frac{3^{2n}}{n}\left[\sum_{k=1}^{\infty}\frac{1}{(2k-1)^{2n-3}} - \sum_{k=1}^{\infty}\frac{1}{(2k-1)^{2n-1}} - \frac{1}{3^{2n-3}} + \frac{1}{3^{2n-1}}\right]$$

$$= -\frac{1}{24}\sum_{n=3}^{\infty}\frac{1}{n}\left[3^{2n}\zeta(2n-3) - 8\left(\frac{3}{2}\right)^{2n}\zeta(2n-3) - 3^{2n}\zeta(2n-1)\right.$$

$$\left. +2\left(\frac{3}{2}\right)^{2n}\zeta(2n-1) - 24\right]. \tag{15}$$

Now, let $\alpha_1, \alpha_2, \alpha_3$, and α_4 denote the sums of the Zeta series occurring in 3.4(726), 3.4(729), 3.4(732) and 3.4(736), respectively. We then find from (15) that

$$\log E_4\left(\frac{9}{4}\right) = -\frac{1}{24}(2\alpha_1 - \alpha_2 + \alpha_3 - 8\alpha_4)$$
$$= -\frac{4}{9} + \frac{21}{32}\gamma - \frac{189}{128}\zeta(3) \tag{16}$$
$$+\frac{17}{4}\log A + \frac{5}{4}\log C + \log\left(2^{-\frac{979}{1440}}\cdot 3\right).$$

Finally, in view of 2.1(31) and 2.3(27), we are easily led from (7) and (9) through (16) to the evaluation 5.2(23).

det$'\Delta_5$: To evaluate $\det'\Delta_5$, we begin by setting $a = 3$, $n = 1$, $n = 2$ and $t = 2$ in 3.2(64), and then use 2.2(4), 2.2(20) and 2.3(22). We, thus, obtain

$$\sum_{n=1}^{\infty}\frac{\zeta(2n,3)}{n+1}2^{2n+2} = 10 + \log\left(3\cdot\pi^{-4}\right) \tag{17}$$

and

$$\sum_{n=1}^{\infty}\frac{\zeta(2n,3)}{n+2}2^{2n+4} = 20 - \frac{13\zeta(3)}{\pi^2} + \log\left(2^{288}\cdot 3\cdot\pi^{-16}\right). \tag{18}$$

By setting $n = 5$ in 5.1(17), we find that the shifted sequence of eigenvalues of Δ^5 of $\mathbf{S}^5$ is given as follows:

$$(k+2)^2 \quad \text{with multiplicity} \quad \frac{1}{12}(k+1)(k+2)^2(k+3) \quad (k\in\mathbb{N}). \tag{19}$$

It is seen that the sequence in (19) has the order $\mu = \frac{5}{2}$. We also have

$$\begin{aligned} Z_5(s) &= \frac{1}{12}\sum_{k=1}^{\infty}\frac{(k+1)(k+2)^2(k+3)}{(k+2)^{2s}} \\ &= \frac{1}{12}[\zeta(2s-4)-\zeta(2s-2)]+\frac{1}{3}\left(\frac{1}{2^{2s}}-\frac{1}{2^{2s-2}}\right). \end{aligned} \tag{20}$$

It is observed that $Z_5(s)$ has simple poles at $s = \frac{3}{2}$ and $s = \frac{5}{2}$. We, therefore, have

$$\text{FP}Z_5(1) = Z_5(1) = -\frac{5}{24} \quad \text{and} \quad \text{FP}Z_5(2) = Z_5(2) = -\frac{1}{12}\zeta(2) - \frac{5}{48}.$$

We also have

$$C_{-2} = \frac{1}{2}Z_5(-2) = -8 \quad \text{and} \quad Z_5'(0) = \frac{\zeta(5)}{8\pi^4} + \frac{\zeta(3)}{24\pi^2} + 2\log 2.$$

We, thus, find that

$$\begin{aligned} E_5(\lambda) = &\prod_{k=1}^{\infty}\left(1-\frac{\lambda}{(k+2)^2}\right)^{\frac{1}{12}(k+1)(k+2)^2(k+3)} \\ &\cdot\exp\left[\frac{1}{12}(k+1)(k+2)^2(k+3)\left(\frac{\lambda}{(k+2)^2}+\frac{\lambda^2}{2(k+2)^4}\right)\right], \end{aligned}$$

which, upon setting $\lambda = 4$ and taking logarithms on each side of the resulting equation and using (17) and (18), yields

$$\begin{aligned} \log E_5(4) &= -\frac{1}{12}\sum_{n=3}^{\infty}\frac{2^{2n}}{n}\left[\zeta(2n-4)-\zeta(2n-2)+\frac{1}{2^{2n-2}}-\frac{1}{2^{2n-4}}\right] \\ &= -\frac{1}{12}\left[\sum_{n=1}^{\infty}\frac{2^{2n+4}}{n+2}\zeta(2n,3)-\sum_{n=2}^{\infty}\frac{2^{2n+2}}{n+1}\zeta(2n,3)\right] \\ &= -\frac{\pi^2}{9}+\frac{13\zeta(3)}{12\pi^2}+\log\left(2^{-24}\cdot\pi\right). \end{aligned}$$

If we set $n = 5$ in 5.1(18) and use 5.1(12), we finally have

$$\begin{aligned} \det{}'\Delta_5 = D_5(4) &= \exp[-Z_5{}'(0) - 4\,\text{FP}Z_5(1) - 8\,\text{FP}Z_5(2) - 8\,C_{-2}]\,E_5(4) \\ &= \frac{\pi}{2^{26}}\exp\left[\frac{197}{3}-\frac{\zeta(3)}{24\pi^2}-\frac{\zeta(5)}{8\pi^4}\right]. \end{aligned} \tag{21}$$

det$'\Delta_6$: Next, by setting $n = 6$ in 5.1(17), we obtain the shifted sequence of eigenvalues of Δ_6 on $\mathbf{S}^6$ as follows:

$$\left(k+\frac{5}{2}\right)^2 \quad \text{with multiplicity} \quad \frac{1}{120}(2k+5)(k+4)(k+3)(k+2)(k+1). \tag{22}$$

We see that

$$\begin{aligned}Z_6(s) &= \frac{1}{120}\sum_{k=1}^{\infty}\frac{(2k+5)(k+4)(k+3)(k+2)(k+1)}{(k+\frac{5}{2})^{2s}} \\ &= \frac{1}{1920}\left[\left(2^{2s}-32\right)\zeta(2s-5) - 10\left(2^{2s}-8\right)\zeta(2s-3)\right. \\ &\qquad \left. +9\left(2^{2s}-2\right)\zeta(2s-1)\right] - (\tfrac{2}{5})^{2s}.\end{aligned} \tag{23}$$

It is observed that $Z_6(s)$ has simple poles at $s=1$, 2 and 3 with its residues $\frac{3}{640}$, $-\frac{1}{48}$ and $\frac{1}{120}$, respectively. It is also seen that the sequence in (22) has the order $\mu=3$. Now, we can find that

$$Z_6'(0) = -\frac{484051}{2^8\cdot 3^3\cdot 5\cdot 7}\log 2 + 2\log 5 - \frac{31}{960}\zeta'(-5) + \frac{7}{96}\zeta'(-3) - \frac{3}{320}\zeta'(-1),$$

$$\begin{aligned}\mathrm{FP}Z_6(1) &= \lim_{\epsilon\to 0}\left[Z_6(1+\epsilon) - \frac{3}{640\epsilon}\right] \\ &= -\frac{9323}{2^8\cdot 3^2\cdot 5^2} + \frac{3}{320}\gamma + \frac{3}{160}\log 2,\end{aligned}$$

$$\mathrm{FP}Z_6(2) = -\frac{4483}{2^5\cdot 3^2\cdot 5^4} + \frac{21}{320}\zeta(3) - \frac{1}{24}(\gamma + 2\log 2),$$

$$\mathrm{FP}Z_6(3) = -\frac{2^6}{5^6} + \frac{\gamma}{60} + \frac{1}{30}\log 2 - \frac{7}{24}\zeta(3) + \frac{93}{320}\zeta(5),$$

$$C_{-2} = \frac{1}{2}Z_6(-2) = -\frac{2217581021}{2^{15}\cdot 3^2\cdot 5\cdot 7\cdot 11},$$

and

$$C_{-3} = -\frac{1}{6}Z_6(-3) = \frac{62451523}{2^{18}\cdot 3^4\cdot 5^2\cdot 7\cdot 11\cdot 13} + \frac{5^6}{2^7\cdot 3}.$$

We also see that

$$\log E_6(\lambda) = -\frac{1}{60}\sum_{n=4}^{\infty}\frac{\lambda^n}{n}\left[\zeta\left(2n-5,\tfrac{7}{2}\right) - \tfrac{5}{2}\zeta\left(2n-3,\tfrac{7}{2}\right) + \frac{9}{16}\zeta\left(2n-1,\tfrac{7}{2}\right)\right], \tag{24}$$

which, for $\lambda = \frac{25}{4}$, yields

$$\begin{aligned}\log E_6\left(\frac{25}{4}\right) = -\frac{1}{60}&\left[\sum_{n=1}^{\infty}\frac{\zeta(2n+1,\frac{7}{2})}{n+3}\left(\frac{5}{2}\right)^{2n+6} - \frac{5}{2}\sum_{n=2}^{\infty}\frac{\zeta(2n+1,\frac{7}{2})}{n+2}\left(\frac{5}{2}\right)^{2n+4}\right. \\ &\left. + \frac{9}{16}\sum_{n=3}^{\infty}\frac{\zeta(2n+1,\frac{7}{2})}{n+1}\left(\frac{5}{2}\right)^{2n+2}\right].\end{aligned} \tag{25}$$

By setting $n = 0, 1, 2$, $a = \frac{7}{2}$ and $t = \frac{5}{2}$ in 3.2(67) and using some identities recorded in previous and present chapters, we have

$$\sum_{n=1}^{\infty} \frac{\zeta(2n+1, \frac{7}{2})}{n+1} \left(\frac{5}{2}\right)^{2n+2} = \frac{155}{12} - \frac{25}{4}\gamma - \frac{11}{12}\log 2 - \frac{5}{2}\log 3 - \frac{5}{2}\log 5 + 3\zeta'(-1); \tag{26}$$

$$\sum_{n=1}^{\infty} \frac{\zeta(2n+1, \frac{7}{2})}{n+2} \left(\frac{5}{2}\right)^{2n+4} = \frac{3325}{2^6} - \frac{625}{2^5}\gamma + \frac{3561553}{2^5.5.3831}\log 2 - \frac{53}{8}\log 3 - \frac{125}{8}\log 5 + 6\zeta'(-1) + \frac{15}{8}\zeta'(-3); \tag{27}$$

and

$$\sum_{n=1}^{\infty} \frac{\zeta(2n+1, \frac{7}{2})}{n+3} \left(\frac{5}{2}\right)^{2n+6} = \frac{623005}{2^8 \cdot 3^2} - \frac{15625}{2^6 \cdot 3}\gamma + \frac{253849}{2^6 \cdot 3^2 \cdot 7}\log 2 - \frac{485}{32}\log 3 - \frac{238627}{32}\log 5 + \frac{3133}{8}\zeta'(-1) + \frac{3}{2}\zeta'(-2) + \frac{507}{4}\zeta'(-3) + \frac{15}{8}\zeta'(-4) + \frac{63}{16}\zeta'(-5). \tag{28}$$

Applying (26), (27) and (28) to (25), we obtain

$$\log E_6\left(\frac{25}{4}\right) = -\frac{4639}{2^{10} \cdot 3^3} + \frac{1385}{2^8 \cdot 3^2}\gamma - \frac{246717677}{2^8 \cdot 3^2 \cdot 5 \cdot 7 \cdot 3831}\log 2 + \frac{118711}{2^6 \cdot 3 \cdot 5}\log 5 - \frac{6053}{2^6 \cdot 3 \cdot 5}\zeta'(-1) - \frac{1}{40}\zeta'(-2) - \frac{651}{2^6 \cdot 5}\zeta'(-3) - \frac{1}{2^5}\zeta'(-4) - \frac{21}{2^6 \cdot 5}\zeta'(-5) - \frac{413875}{2^{11} \cdot 3^2}\zeta(3) + \frac{96875}{2^{12}}\zeta(5). \tag{29}$$

If we set $n = 6$ in 5.1(18) and use 5.1(12), we get

$$\det{}'\Delta_6 = D_6\left(\frac{25}{4}\right) = \exp\left[-Z_6'(0) - \frac{25}{4}\mathrm{FP}Z_6(1) - \frac{625}{32}\mathrm{FP}Z_6(2) - \frac{15625}{192}\mathrm{FP}Z_6(3) - \frac{625}{32}\left(C_{-2} + \frac{25}{8}C_{-3}\right)\right] \cdot E_6\left(\frac{25}{4}\right),$$

which, upon using the above computations, yields

$$\det{}'\Delta_6 = 2^{-\frac{3990625}{735552}} \cdot 5^{\frac{116791}{960}} \cdot \exp\left[-\frac{38441354615245651}{5441253801984} - \frac{1511}{240}\zeta'(-1) - \frac{1}{40}\zeta'(-2) - \frac{2023}{960}\zeta'(-3) - \frac{1}{32}\zeta'(-4) - \frac{11}{480}\zeta'(-5)\right]. \tag{30}$$

det$'\Delta_7$: By setting $n = 7$ in 5.1(17), we see that the shifted sequence of eigenvalues of Δ_7 on $\mathbf{S}^7$ is given as follows:

$$\begin{aligned} \lambda_k = \mu_k + 9 = (k+3)^2 \quad &\text{with multiplicity} \\ &\frac{1}{360}(k+1)(k+2)(k+3)^2(k+4)(k+5). \end{aligned} \tag{31}$$

It is found that

$$\begin{aligned} Z_7(s) &= \frac{1}{360}\sum_{k=1}^{\infty}\frac{(k+1)(k+2)(k+3)^2(k+4)(k+5)}{(k+3)^{2s}} \\ &= \frac{1}{360}[\zeta(2s-6) - 5\zeta(2s-4) + 4\zeta(2s-2)] - 3^{-2s}, \end{aligned} \tag{32}$$

which shows us that $Z_7(s)$ has simple poles at $s = \frac{3}{2}$, $\frac{5}{2}$ and $\frac{7}{2}$ with their residues $\frac{1}{180}$, $-\frac{1}{144}$ and $\frac{1}{720}$, respectively. It is also observed that the sequence in (31) has the order $\mu = \frac{7}{2}$.

If we set $n = 7$ in 5.1(18) and use 5.1(12), we get

$$\begin{aligned} \det{}'\Delta_7 = D_7(9) = \exp\Bigg[&-Z_7{}'(0) - 9\mathrm{FP}Z_7(1) - \frac{81}{2}\mathrm{FP}\,Z_7(2) \\ &-243\,\mathrm{FP}\,Z_7(3) - \frac{81}{2}\left(C_{-2} + \frac{9}{2}C_{-3}\right)\Bigg]\cdot E_7(9). \end{aligned} \tag{33}$$

We can also easily verify each of the following evaluations:

$$C_{-2} = -\frac{81}{2}, \quad C_{-3} = \frac{243}{2}, \quad \mathrm{FP}\,Z_7(1) = -\frac{7}{60}, \quad \mathrm{FP}\,Z_7(2) = -\frac{7}{2^4\cdot 3^4} + \frac{\pi^2}{540},$$

$$\mathrm{FP}\,Z_7(3) = -\frac{161}{2^4\cdot 3^6\cdot 5} - \frac{\pi^2}{2^4\cdot 3^3} + \frac{\pi^4}{2^2\cdot 3^4\cdot 5^2},$$

and

$$Z_7{}'(0) = -\frac{\zeta(7)}{32\pi^6} - \frac{\zeta(5)}{48\pi^4} - \frac{\zeta(3)}{180\pi^2} + 2\log 3.$$

It is observed that

$$\begin{aligned} E_7(\lambda) = \prod_{k=1}^{\infty}\Bigg\{&\left(1 - \frac{\lambda}{(k+3)^2}\right)^{\frac{1}{360}(k+1)(k+2)(k+3)^2(k+4)(k+5)} \\ &\cdot\exp\Bigg[\frac{1}{360}(k+1)(k+2)(k+3)^2(k+4)(k+5) \\ &\cdot\left(\frac{\lambda}{(k+3)^2} + \frac{\lambda^2}{2(k+3)^4} + \frac{\lambda^3}{3(k+3)^6}\right)\Bigg]\Bigg\}, \end{aligned}$$

which, upon setting $\lambda = 9$ and taking logarithms on each side of the resulting equation and considering the Taylor-Maclaurin expansion of $\log(1-x)$, yields

$$\begin{aligned}\log E_7(9) = &-\frac{1}{360}\sum_{n=1}^{\infty}\frac{\zeta(2n,4)}{n+3}\cdot 3^{2n+6}+\frac{1}{72}\sum_{n=1}^{\infty}\frac{\zeta(2n,4)}{n+2}\cdot 3^{2n+4}\\ &-\frac{1}{90}\sum_{n=1}^{\infty}\frac{\zeta(2n,4)}{n+1}\cdot 3^{2n+2}-\frac{39}{80}\pi^2+\frac{3}{100}\pi^4+\frac{259}{240}.\end{aligned}\tag{34}$$

By setting $a = 4$, $n = 1, 2, 3$ and $t = 3$ in 3.2(64) and using some identities already recorded in this presentation, we have

$$\sum_{n=1}^{\infty}\frac{\zeta(2n,4)}{n+1}\cdot 3^{2n+2}=\frac{63}{2}-5\log 2+9\log 3+4\log 5-9\log\pi,\tag{35}$$

$$\begin{aligned}\sum_{n=1}^{\infty}\frac{\zeta(2n,4)}{n+2}\cdot 3^{2n+4}=&\frac{1071}{4}-29\log 2+243\log 3+16\log 5\\ &-81\log\pi-\frac{27\zeta(3)}{\pi^2},\end{aligned}\tag{36}$$

and

$$\begin{aligned}\sum_{n=1}^{\infty}\frac{\zeta(2n,4)}{n+3}\cdot 3^{2n+6}=&\frac{32211}{10}-125\log 2+3645\log 3-503936\log 5\\ &-729\log\pi-\frac{1215\zeta(3)}{2\pi^2}+\frac{405\zeta(5)}{2\pi^4}.\end{aligned}\tag{37}$$

If we apply (35), (36) and (37) to (34), we get

$$\begin{aligned}\log E_7(9) = &-\frac{789}{2^5\cdot 5}-\frac{137}{20}\log 3+1400\log 5\\ &+\log\pi+\frac{21\zeta(3)}{16\pi^2}-\frac{9\zeta(5)}{16\pi^4}-\frac{39}{80}\pi^2+\frac{3}{100}\pi^4.\end{aligned}\tag{38}$$

Finally, from (34) and (38), and other previously recorded results, we obtain

$$\det{}'\Delta_7 = 3^{-\frac{177}{20}}\cdot 5^{1400}\cdot\pi\cdot\exp\left[-\frac{1230367}{60}+\frac{949\zeta(3)}{720\pi^2}-\frac{13\zeta(5)}{24\pi^4}+\frac{\zeta(7)}{32\pi^6}\right].\tag{39}$$

5.4 Computations using Zeta Regularized Products

Quine and Choi [954] made use of the zeta regularized products to compute $\det'\Delta_n$ and the determinant of the conformal Laplacian, $\det(\Delta_{\mathbb{S}^n}+n(n-2)/4)$, which is introduced here. In general, the conformal Laplacian is defined to be

$$\Delta + \frac{(n-2)\,K}{4(n-1)},$$

where Δ is the Laplacian and K is the scalar curvature. For the sphere $\mathbf{S}^n$, $K = n(n-1)$. Computation of the above determinants is equivalent to computing derivatives at $s = 0$ of the zeta function

$$\sum_{k=1}^{\infty} \frac{\beta_k^n}{[k(k+n-1)]^s}$$

for the Laplacian and

$$\sum_{k=0}^{\infty} \frac{\beta_k^n}{[(k+n/2)(k+n/2-1)]^s}$$

for the conformal Laplacian and β_k^n being given in 5.1(16).

For simplicity, we restrict our discussion of the conformal Laplacian to the case when n is even. We consider the more general zeta function as in Section 5.1,

$$Z_n(s,\,a) = \sum_{k=1}^{\infty} \frac{\beta_k^n}{[(k+a)(k+n-1-a)]^s}$$

for integers a, $0 \leqq a \leqq n-1$, with $a = 0$ corresponding to the Laplacian and $a = n/2$ to the conformal Laplacian. If $a(n-1-a) \neq 0$, then

$$\det\left(\Delta_{\mathbf{S}^n} + a(n-1-a)\right) a\,(n-1-a)\,\exp\left(-Z_n'(0,\,a)\right),$$

and, if $a(n-1-a) = 0$, then

$$\det{}'\,\Delta_n = \exp\left(-Z_n'(0,\,a)\right).$$

We show below (Theorem 5.4) that for integers a, $\det(\Delta_{\mathbf{S}^n} + a(n-1-a))$, if $a(n-1-a) \neq 0$ and $(n-1)\det{}'\,\Delta_n$ are both of the form

$$\exp\left(\tau_n(a) + \sum_{k=1}^{n} \tau_k^n(a)\,\zeta'(-k+1)\right), \tag{1}$$

where ζ is the Riemann Zeta function defined by 2.3(1) and the numbers $\tau_n(a)$ and $\tau_k^n(a)$ are rational numbers for which we give explicit expressions in terms of coefficients of the Taylor expansion of β_k^n about $k = -a$ and $k = -(n-1)/2$. Using (1) and the functional equation for the Riemann Zeta function 2.3(11) or 2.3(12), it is easy to compute numerical values for the constants involved. Our computations show that $\tau_n(a) = 0$ for n odd and $\tau_k^n(a) = 0$, if n and k have opposite parity.

The technique that we introduce here for dealing with this computation, which is different from other approaches, is a lemma on zeta regularized products (Lemma 5.3), which may be useful in simplifying and understanding other computations of this type. This gives us, in addition, a way to factor our functional determinant $\det(\Delta_{S^n} + a(n-1-a))$ into multiple Gamma functions, as shown in Section 5.2, generalizing an equation of Voros [1201] and giving an alternate approach to computing it for integral a, $0 \leqq a \leqq n-1$ (Section 5.2).

We remark that choosing a to be an integer makes things considerably simpler because $\beta_k^n = 0$ for $k = -1, \ldots, -(n-1)$ and the computation reduces to computation of derivatives of the Riemann Zeta function. The same techniques could be used for noninteger values of a but would involve a more complicated expression involving derivatives of the Hurwitz zeta function.

A Lemma on Zeta Regularized Products and a Main Theorem

Let $\{\lambda_k\}$ be a sequence of nonzero complex numbers. If $Z(s)$ in 5.1(2) converges for $\Re(s) > s_0$ and has meromorphic continuation to a function meromorphic in $\Re(s) > \sigma$, for some $\sigma < 0$, with at most simple poles, then we say that the sequence is *zeta regularizable*. We define $\lambda_k^{-s} = \exp(-s \log \lambda_k)$, and the definition of Z depends on the choice of $\arg \lambda_k$.

For a zeta regularized sequence, we define the zeta regularized product (see 5.1(4))

$$\prod\nolimits_z \lambda_k = \exp\left(-Z'(0)\right).$$

We also define the product (see 5.1(11))

$$\prod\nolimits_z (\lambda_k - \lambda) = \exp\left(-Z'(0, -\lambda)\right),$$

where, for $\lambda \neq \lambda_k$, we adopt the convention that $\arg(\lambda_k - \lambda) \cong \arg \lambda_k$ for $|\lambda_k|$ large.

For λ_k, the sequence of nonzero eigenvalues of the Laplacian on a manifold, $\prod_z \lambda_k$ is called the *determinant of the Laplacian* and $\prod_z (\lambda_k + \lambda)$ the *functional determinant*, $\det'(\Delta + \lambda)$, as in Section 5.1.

Information on the formal properties of zeta regularized products can be found in [955] and [1201]. The use of zeta regularized products can be traced back as far as Barnes [97]. Our approach is to try to factor a zeta regularized product into simpler ones. One sees that the equality of $\prod_z (\lambda_k^2 - \lambda^2)$ and $\prod_z (\lambda_k - \lambda) \prod_z (\lambda_k + \lambda)$ is untrue without the introduction of an exponential factor. This factor can be computed from the relationship between the zeta regularized product and the Weierstrass product found in the above-cited references. We give a confirmation independent of this in the proof below.

Lemma 5.3 *Let λ_j be a zeta regularizable sequence, and let h be an integer, such that $\sum_{j=1}^{\infty} |\lambda_j|^{-h-1} < \infty$. For $\Re(s)$ sufficiently large, let $Z(s)$ be as in 5.1(2) and for $\lambda \neq \lambda_j$,* *let*

$$F(s)=\sum_{j=1}^{\infty}\left\{\left(\lambda_j^2-\lambda^2\right)^{-s}-\left(\lambda_j-\lambda\right)^{-s}-\left(\lambda_j+\lambda\right)^{-s}\right\}. \tag{2}$$

Then, we have

$$-F'(0)=\sum_{j=1}^{[h/2]}\operatorname*{Res}_{s=2j} Z(s)\left(1+\frac{1}{3}+\cdots+\frac{1}{2j-1}\right)\frac{\lambda^{2j}}{j}. \tag{3}$$

In terms of zeta regularized products, the result can be stated as

$$\prod\nolimits_z\left(\lambda_j^2-\lambda^2\right)=e^{F'(0)}\prod\nolimits_z\left(\lambda_j-\lambda\right)\prod\nolimits_z\left(\lambda_j+\lambda\right),$$

where $F'(0)$ is given by (3).

Proof. Define $G_k(s)$ by subtracting terms of the binomial expansion of the summand of (2),

$$G_k(s)=\left(\lambda_k^2-\lambda^2\right)^{-s}-\left(\lambda_k-\lambda\right)^{-s}-\left(\lambda_k+\lambda\right)^{-s}\\-\sum_{j=0}^{[h/2]}\left\{\binom{-s}{j}\lambda_k^{-2s-2j}-2\binom{-s}{2j}\lambda_k^{-s-2j}\right\}\lambda^{2j}. \tag{4}$$

Elementary estimates show that

$$G(s)=\sum_{k=1}^{\infty}G_k(s)$$

converges for $\Re(s)\geqq 0$. Since $G_k'(0)=0$, term by term differentiation gives

$$G'(0)=0. \tag{5}$$

Now, summing (4) over the index k gives

$$F(s)=\sum_{j=0}^{[h/2]}\left[\binom{-s}{j}Z(2s+2j)-2\binom{-s}{2j}Z(s+2j)\right]+G(s). \tag{6}$$

Differentiating (6) at $s=0$ and using (5) gives the result. ∎

Theorem 5.4 *For $n \in \mathbb{N} \setminus \{1\}$ and a an integer, such that $0 \leqq a \leqq n-1$ and $\Re(s) > n$, let*

$$Z_n(s, a) = \sum_{k=1}^{\infty} \beta_k^n \, [(k+a)(k+n-1-a)]^{-s}, \tag{7}$$

where β_k^n is given as in 5.1(16). For $k \in \mathbb{N}$ $(1 \leqq k \leqq n)$, define functions $t_j^n(a)$ by the expansion

$$\beta_{k-a}^n = \sum_{j=1}^{n} t_j^n(a) k^{j-1}. \tag{8}$$

The derivative at zero of (7) is given by

$$-Z_n'(0, a) = \tau_n(a) - \sum_{j=1}^{n} \left\{1 + (-1)^{n+j}\right\} t_j^n(a) \zeta'(1-j) - A, \tag{9}$$

where

$$\tau_n(a) = \left\{1 + (-1)^n\right\} \sum_{j=1}^{[n/2]} \frac{1}{2j} t_{2j}^n \left(\frac{n-1}{2}\right) \cdot \left(1 + \frac{1}{3} + \cdots + \frac{1}{2j-1}\right) \left(\frac{n-1}{2} - a\right)^{2j} \tag{10}$$

and $A = \log(n-1)$, if $a(n-1-a) = 0$ and $A = \log\{a(n-1-a)\}$ otherwise.

Proof. Let

$$F(s) = \sum_{k=1}^{\infty} \beta_k^n \left[\{(k+a)(k+n-1-a)\}^{-s} - (k+a)^{-s} - (k+n-1-a)^{-s}\right]. \tag{11}$$

Substituting $\lambda = (n-1)/2 - a$, $h = n$, gives us the situation described in Lemma 5.3. Hence, if for $\Re(s) > n$, we set

$$\widetilde{Z}_n(s, \lambda) = \sum_{k=1}^{\infty} \beta_k^n \left(k + \frac{n-1}{2} + \lambda\right)^{-s}, \tag{12}$$

we have

$$F'(0) = \sum_{j=1}^{[n/2]} \operatorname*{Res}_{s=2j} \widetilde{Z}_n(s, 0) \left(1 + \frac{1}{3} + \cdots + \frac{1}{2j-1}\right) \frac{\lambda^{2j}}{j}. \tag{13}$$

We, now, show that

$$\operatorname*{Res}_{s=k} \widetilde{Z}_n(s,\, 0) = t_k^n\left(\frac{n-1}{2}\right) \tag{14}$$

for $1 \leqq k \leqq n-1$ $(k \in \mathbb{N})$. Using (9), we may write

$$\begin{aligned}\widetilde{Z}_n(s,\, 0) &= \sum_{\ell=1}^{\infty}\sum_{j=1}^{n} t_j^n\left(\frac{n-1}{2}\right)\left(\ell + \frac{n-1}{2}\right)^{-s+j-1}\\ &= \sum_{j=1}^{n} t_j^n\left(\frac{n-1}{2}\right)\zeta\left(s-j+1,\, \frac{n+1}{2}\right),\end{aligned}$$

where $\zeta(s,\, a)$ is the Hurwitz Zeta function, defined by 2.2(1). Now, (14) easily follows. ■

Now, we find from (11) that

$$Z_n'(0,\, a) = F'(0) + \widetilde{Z}_n{}'(0,\, \lambda) + \widetilde{Z}_n{}'(0,\, -\lambda), \tag{15}$$

and we need to compute the last two terms on the right side of (15). First, note that $\beta_k^n = 0$ for $k = -1, -2, \cdots, -(n-1)$ and $\beta_0^n = 1$. Hence, if $a \neq 0$, we have

$$\begin{aligned}\widetilde{Z}_n'(0,\, -\lambda) &= \sum_{k=1}^{\infty} \beta_k^n\,(k+a)^{-s}\\ &= \sum_{k=1}^{\infty} \beta_{k-a}^n\, k^{-s} - a^{-s}\\ &= \sum_{k=1}^{\infty} t_k^n\,(a)\,\zeta(s-k+1) - a^{-s}.\end{aligned}$$

Thus, we have

$$\widetilde{Z}_n'(0,\, -\lambda) = \sum_{k=1}^{n} t_k^n\,(a)\,\zeta'(-k+1) + \log a \quad (a \neq 0). \tag{16}$$

If $a = 0$, we have

$$\widetilde{Z}_n'\left(0,\, -\frac{n-1}{2}\right) = \sum_{k=1}^{n} t_k^n\,(0)\,\zeta'(-k+1). \tag{17}$$

Similarly,

$$\widetilde{Z}_n'(0,\, \lambda) = \sum_{k=1}^{n} t_k^n\,(n-1-a)\,\zeta'(-k+1) + \log(n-1-a) \quad (a \neq n-1), \tag{18}$$

and, for $a = n - 1$,

$$\widetilde{Z}'_n\left(0, \frac{n-1}{2}\right) = \sum_{k=1}^{n} t_k^n(0)\,\zeta'(-k-1). \tag{19}$$

From $\beta_k^n = (-1)^{n+1}\,\beta_{-k+n+1}^n$, we deduce

$$t_k^n(n-1-a) = (-1)^{n+k}\,t_k^n(a). \tag{20}$$

Now, combining (13)–(20), we get (11).

Computations for small n

The following is a table of $\det'\Delta_n$, for $n = 2, \ldots, 6$, and the determinants of the conformal Laplacians $\det(\Delta_{\mathbb{S}^n} + n(n-2)/4)$, for $n = 4, 6, 8$. The first expression in each set of equations is obtained from Theorem 5.4. The second expression is obtained by replacing the values of ζ' at negative integers by its equivalent expression in terms of values of ζ and ζ' at positive integers obtained by differentiating the functional equation 2.3(14) or 2.3(22). Numeric evaluation of $\zeta'(k)$ ($k \in \mathbb{N}\setminus\{1\}$) is discussed in [1191]. The computations below were done using *Mathematica*.

$$\begin{aligned}
\det{}'\Delta_2 &= \exp\left(\frac{1}{2} - 4\zeta'(-1)\right)\\
&= \exp\left\{\frac{1}{6} + \frac{1}{3}(\gamma + \log(2\pi)) - \frac{2\zeta'(2)}{\pi^2}\right\}\\
&\cong 3.19531.
\end{aligned}$$

$$\begin{aligned}
\det{}'\Delta_3 &= \frac{1}{2}\exp\left(-2\zeta'(-2) - 2\zeta'(0)\right)\\
&= \frac{1}{2}\exp\left\{\log(2\pi) + \frac{\zeta(3)}{2\pi^2}\right\}\\
&\cong 3.33885.
\end{aligned}$$

$$\begin{aligned}
\det{}'\Delta_4 &= \frac{1}{3}\exp\left(\frac{15}{16} - \frac{2\zeta'(-3)}{3} - \frac{13\zeta'(-1)}{3}\right)\\
&= \frac{1}{3}\exp\left\{\frac{1267}{2160} + \frac{16}{45}(\gamma + \log(2\pi)) - \frac{13\zeta'(2)}{6\pi^2} + \frac{\zeta'(4)}{2\pi^4}\right\}\\
&\cong 1.73694.
\end{aligned}$$

$$\begin{aligned}\det{}'\Delta_5 &= \frac{1}{4}\exp\left(-\frac{\zeta'(-4)}{6}-\frac{23\,\zeta'(-2)}{6}-2\,\zeta'(0)\right)\\ &= \frac{1}{4}\exp\left\{\log(2\pi)+\frac{23\,\zeta(3)}{24\pi^2}-\frac{\zeta(5)}{8\pi^4}\right\}\\ &\cong 1.76292.\end{aligned}$$

$$\begin{aligned}\det{}'\Delta_6 &= \frac{1}{5}\exp\left(\frac{455}{432}-\frac{\zeta'(-5)}{30}-2\,\zeta'(-3)-\frac{149\,\zeta'(-1)}{30}\right)\\ &= \frac{1}{5}\exp\left\{\frac{303733}{453600}+\frac{751}{1890}(\gamma+\log(2\pi))\right.\\ &\qquad\left.-\frac{149\,\zeta'(2)}{60\pi^2}+\frac{3\,\zeta'(4)}{2\pi^4}-\frac{\zeta'(6)}{8\pi^6}\right\}\\ &\cong 1.29002.\end{aligned}$$

$$\begin{aligned}\det\left(\Delta_{\mathbb{S}^4}+2\right) &= \exp\left(-\frac{1}{144}-\frac{2\,\zeta'(-3)}{3}-\frac{\zeta'(-1)}{3}\right)\\ &= \exp\left\{-\frac{53}{2160}+\frac{1}{45}(\gamma+\log(2\pi))-\frac{\zeta'(2)}{6\pi^2}+\frac{\zeta'(4)}{2\pi^4}\right\}\\ &\cong 1.04562.\end{aligned}$$

$$\begin{aligned}\det\left(\Delta_{\mathbb{S}^6}+6\right) &= \exp\left(\frac{1}{1350}-\frac{\zeta'(-5)}{30}+\frac{\zeta'(-1)}{30}\right)\\ &= \exp\left\{\frac{1459}{453600}-\frac{1}{378}(\gamma+\log(2\pi))+\frac{\zeta'(2)}{60\pi^2}-\frac{\zeta'(6)}{8\pi^6}\right\}\\ &\cong 0.995257.\end{aligned}$$

$$\begin{aligned}\det\left(\Delta_{\mathbb{S}^8}+12\right) &= \exp\left(-\frac{5497}{50803200}-\frac{\zeta'(-7)}{1260}+\frac{\zeta'(-5)}{360}+\frac{\zeta'(-3)}{360}-\frac{\zeta'(-1)}{210}\right)\\ &= \exp\left\{-\frac{391481}{76204800}+\frac{23}{56700}(\gamma+\log(2\pi))\right.\\ &\qquad\left.-\frac{\zeta'(2)}{420\pi^2}-\frac{\zeta'(4)}{480\pi^4}+\frac{\zeta'(6)}{96\pi^6}+\frac{\zeta'(8)}{32\pi^8}\right\}\\ &\cong 1.00069.\end{aligned}$$

It should be *noted* that we need to compare the evaluation of $\det{}'\Delta_4$ in Sections 5.3 and 5.4 with those in other works (see Section 5.5).

5.5 Remarks and Observations

Even though each of the determinant expressions 5.2(20), 5.2(21) and 5.2(22) has been considered by other workers on this subject, our novel approach in Section 5.3 to the familiar problem of computation of the determinants of the Laplacians on the

n-sphere S^n $(n = 1, 2, 3)$ is relatively more natural than the earlier approaches to these and analogous problems.

The evaluations of $\det' \Delta_n$ $(n = 1, 2, 3)$ in Choi [260], Quine and Choi [954], Choi and Srivastava [291] and Kumagai [706] have the same results, whereas the evaluation of $\det' \Delta_4$ in Quine and Choi [954] is equal to that in Kumagai [706] and is not equal to that in Choi and Srivastava [292] (see also Section 5.3). It is the only difference that the number $35,639,301/(2^{17} \cdot 5 \cdot 7)$, appearing in the evaluation of $\det' \Delta_4$ in Choi and Srivastava [292] (see also Proposition 5.2), was evaluated as $15/16$ in Quine and Choi [954, p. 726] and Kumagai [706, p. 201, Corollary (iv)]. It seems to need to examine C_{-m} in 5.1(14) (see Voros [1201, p. 444, Eq. (3.3)]) again.

Vardi [1190] also computed $\det' \Delta_n$, by employing a direct analysis of the involved Zeta series in 5.1(2) with the original (or unshifted) sequence $\{\mu_k\}$ in 5.1(15). The results of $\det' \Delta_n$ $(n = 1, 2)$ in Proposition 5.2 agree with those of Vardi's (1988b). Yet Vardi's ingenious computation for $\det' \Delta_n$ should have some gaps (see Problem 5 below). Most recently, Kumagai (1999) evaluated $\det' \Delta_n$, by filling the gaps in Vardi's work (1988b), including a rather detailed discussion for the computations of the $\det' \Delta_n$ worked by several earlier authors (*cf.* Choi [260], Efrat [400], Quine and Choi [954], Vardi [1190]).

Quine and Choi [954] computed $\det' \Delta_n$ and $\det(\Delta_n + n(n-2)/4)$ the determinants of the Laplacians and the conformal Laplacians (respectively) on sphere, by using techniques due to Weisberger [1219] (see Section 5.4). Yet, their technique has the advantage of offering a unified approach. Branson and Ørsted [171] computed the conformal determinant $\det(\Delta_4 + 2)$, which agrees with the first expression as in Section 5.4 above (see also Quine and Choi [954, p. 727]).

Quine *et al.* [955] introduced the Zeta regularized product for the general sequences of nonzero complex numbers, instead of the increasing sequences of nonnegative real numbers given in 5.1(2), in which case their Zeta regularized product corresponds to the determinant of the Laplacian, defined by 5.1(4). They gave many examples for this extended concept of the determinant of the Laplacian.

Osgood *et al.* [881] noted that the modern quantum geometry of strings is concerned (to a large extent) with the set of all surfaces, varying metrics on those surfaces and determinants of the associated Laplacians. They investigated this determinant as a function of the metric on a given surface and, in particular, its extreme values when the metric is suitably restricted. They also observed an interesting fact related to the study of this chapter: *For all metrics σ on* $\mathbf{S}^2$ of area 4π,

$$\det' \Delta_\sigma \leqq \exp\left[\tfrac{1}{2} - 4\zeta'(-1)\right] = \det' \Delta_2, \tag{1}$$

where the equality holds true, if and only if σ is the standard round metric, Δ_σ being the Laplacian on $\mathbf{S}^2$ with the given metric σ.

Problems

1. Suppose that the series $\sum |f(n) n^{-s}|$ does not converge for all s or is divergent for all s. Show that there exists a real number σ_a, called the *abscissa of absolute convergence*, such that the series $\sum |f(n)\,n^{-s}|$ converges absolutely, if $\Re(s) = \sigma > \sigma_a$, but does not converge absolutely, if, $\sigma < \sigma_a$.

(*cf.* Apostol [65, p. 225])

2. Show that the analogous Weierstrass canonical product $E_n(\lambda)$ of the shifted sequence $\{\lambda_k\}$ in 5.1(17) for the standard Laplacian Δ_n on the n-dimensional unit sphere $\mathbf{S}^n$ can be expressed in terms of the multiple Gamma functions Γ_n given in Section 1.3 as follows:

$$E_n(\lambda) = \mathcal{R}_n(\lambda)\,\frac{\Gamma_{n-1}\left(\frac{n-1}{2} - \sqrt{\lambda}\right)\Gamma_{n-1}\left(\frac{n-1}{2} + \sqrt{\lambda}\right)}{\left\{\Gamma_n\left(\frac{n-1}{2} - \sqrt{\lambda}\right)\Gamma_n\left(\frac{n-1}{2} + \sqrt{\lambda}\right)\right\}^2},$$

where $\mathcal{R}_n(\lambda)$ is a meromorphic function with a simple pole only at $\lambda = \left(\frac{n-1}{2}\right)^2$ and $\Gamma_1(z) = \Gamma(z)$.

Show also that the above formula for $E_n(\lambda)$ holds true for all $n \in \mathbb{N}$, if we define $\Gamma_0(z) := \frac{1}{z}$.

(*cf.* Choi [259]; see also Theorem 1.4)

3. Let $\mathcal{R}_n(\lambda)$ be as in Problem 2 above. By using Proposition 5.1, show that

$$\mathcal{R}_1(\lambda) = -\frac{1}{\lambda},$$

$$\mathcal{R}_2(\lambda) = \frac{\left\{\Gamma_2\left(\frac{1}{2}\right)\right\}^4}{\pi\,(1-4\lambda)}\exp[2(\gamma - 1 + 2\log 2)\lambda],$$

and

$$\mathcal{R}_3(\lambda) = \frac{1}{1-\lambda}\exp\left[\left(\log(2\pi) - \frac{3}{2}\right)\lambda\right].$$

Also determine $\mathcal{R}_4(\lambda)$ explicitly and evaluate $\det'\Delta_4$ with the explicit form of $E_4(\lambda)$, by using the method presented in Section 5.2.

4. Show why Vardi's arguments leading to the evaluation:

$$\det'\Delta_3 = \exp\left[-2\zeta'(-2) + \zeta'(-1) - 3\zeta'(0)\right]$$

cannot be justified.

(*cf.* Vardi [1190])

5. Prove the inequality 5.5(1).

(*cf.* Osgood *et al.* [881])

6. Let $\mathbf{M}$ be a compact Riemann surface of genus $g \geq 2$. $\mathbf{M}$ can be identified with H/Γ, the action of a fuchsian group Γ on the upper complex half-plane $\mathrm{H} = \{z = x + iy \mid y > 0\}$ endowed with the Poincaré metric $\mathrm{d}s^2 = y^{-2}\left(\mathrm{d}x^2 + \mathrm{d}y^2\right)$. This is the classic model for hyperbolic geometry of constant negative curvature, $\mathcal{R} = -1$. Γ is a discrete subgroup of $\mathbf{PSL}(2,\mathbb{R}) = \mathbf{SL}(2,\mathbb{R})/\{\pm 1\}$, the group of Möbius transformations. From the Gauss-Bonnet theorem, one infers $\mathcal{R}\mathcal{A} = 2\pi\chi = 4\pi(1-g)$, where $\mathcal{A}$ denotes the area of $\mathbf{M}$ and χ its Euler characteristic, that is, $\mathcal{A} = 4\pi(g-1)$. In the Poincaré metric, the Laplacian on $\mathbf{M}$ (Laplace-Beltrami operator) is given by $\Delta = y^2(\partial^2/\partial x^2 + \partial^2/\partial y^2)$, and one is interested in the eigenvalue problem $-\Delta u = \lambda u$. The spectrum of Δ on $\mathbf{M}$ is discrete and real, $0 = \lambda_0 < \lambda_1 \leqq \lambda_2 \leqq \cdots$ with # (eigenvalues λ_n with $\lambda_n \leqq \lambda$)$\sim (\mathcal{A}/4\pi)\,\lambda$ asymptotically (Weyl's law).

Since the elements $\tau \in \binom{a\,b}{c\,d} \in \Gamma$ are hyperbolic, that is, $|\mathrm{Tr}\tau| = |a+d| > 2$, they are conjugate with Γ to a Möbius transformation of the form $z \to N(\tau)z$, $1 < N(\tau) < \infty$, where $N(\tau)$ is called the norm of τ. ($\tau_1, \tau_2 \in \Gamma$ are conjugate within Γ, if there exists a $\tau_3 \in \Gamma$, such that $\tau_1 = \tau_3 \tau_2 \tau_3^{-1}$.) The class of all elements in Γ that are conjugate to a given τ is called the conjugacy class of τ in Γ and denoted by $\{\tau\}$. The number $N(\tau)$ is, of course, the same within a conjugacy class and measure the "magnification." $N(\tau)$ has, however, another striking geometric interpretation, since there exists a unique relationship between the conjugacy classes in Γ and the homotopy classes of closed paths on the surface **M**. In each class, one defines a length $l(\tau)$ by the length of the shortest closed path measured by means of the Poincaré distance. One, then, obtains $N(\tau) = \exp[l(\tau)]$, $l(\tau) > 0$. Thus, the conjugacy classes in Γ can be uniquely parameterized by their length spectrum $\{l(\tau)\}$. Given any $\tau \in \Gamma$, there is a unique τ_0, such that $\tau = \tau_0^n$, $n \in \mathbb{N}$; τ_0 is called primitive element of Γ, since it cannot be expressed as a power of any other element of Γ. The corresponding closed orbit with length $l(\tau_0)$ is called a prime geodesic on **M**. Obviously, $l(\tau) = l(\tau_0^n) = n\,l(\tau_0)$, since in this case the prime geodesic is traversed n times. For the length spectrum of **M**, one has Huber's law $\nu(x) \sim e^x/x$ $(x \to \infty)$, where $\nu(x)$, is the number of inconjugate primitive τ's with $l(\tau) \leq x$.

(a) Prove the *Selberg trace formula*:

$$\sum_{n=0}^{\infty} h(\tau_n) = \frac{A}{2\pi} \int_{-\infty}^{\infty} t \tanh(\pi t)\, h(t)\, dt + \sum_{\{\tau\}_p} \sum_{n=1}^{\infty} \frac{l(\tau)}{\sinh\left(\frac{1}{2} n l(\tau)\right)} g(n l(\tau)),$$

where all series and the integral converges absolutely under the following conditions on the function $h(t)$: (i) $h(-t) = h(t)$, (ii) $h(t)$ is analytic in a strip $|\Im t| \leq \frac{1}{2} + \epsilon$ $(\epsilon > 0)$, (iii) $|h(t)| \leq a(1+|t|^2)^{-1-\epsilon}$ $(a > 0)$; the function $g(u)$ is the Fourier transform of $h(t)$:

$$g(u) = \frac{1}{2\pi} \int_{-\infty}^{\infty} \exp(-iut)\, h(t)\, dt;$$

the sum on the left-hand side runs over the eigenvalues of Δ parameterized in the form $\lambda_n = \frac{1}{4} + \tau_n^2$, that is, over the pairs $(\tau_n, -\tau_n)$, $\tau_n \in \mathbb{C}$ ($\tau = 0$ has to be counted twice, if $\frac{1}{4}$ happens to be an eigenvalue); the sum on the right-hand side is taken over all primitive conjugacy class in Γ, denoted by $\{\tau\}_p$.

By choosing

$$h(t) = \frac{1}{t^2 + \left(s - \frac{1}{2}\right)^2} - \frac{1}{t^2 + \left(z - \frac{1}{2}\right)^2} \qquad (\Re(s),\ \Re(z) > 1)$$

in the Selberg trace formula in (a),

(b) Prove that

$$\sum_{n=0}^{\infty} \left(\frac{1}{\lambda_n + s(s-1)} - \frac{1}{\lambda_n + z(z-1)} \right) = -2(g-1)[\psi(s) - \psi(z)] + \frac{1}{2s-1} \frac{Z'(s)}{Z(s)} - \frac{1}{2z-1} \frac{Z'(z)}{Z(z)},$$

where ψ is the Psi (or Digamma) function given in Section 1.2 and $Z(s)$ denotes the *Selberg Zeta function* on $\mathbf{M}$, defined by

$$Z(s) := \prod_{\{\tau\}_p} \prod_{n=0}^{\infty} \{1 - \exp[-(s+n)\, l(\tau)]\}.$$

(c) Prove that

$$\frac{1}{2s-1}\frac{Z'(s)}{Z(s)} = \mathcal{B} + \frac{1}{s(s-1)} + (g-1)\,[\psi(s)+\gamma] + \sum_{n=1}^{\infty}\left(\frac{1}{\lambda_n + s(s-1)} - \frac{1}{\lambda_n}\right),$$

where γ is the Euler-Mascheroni constant and

$$\mathcal{B} := \lim_{z\to 1+}\left(\frac{1}{2z-1}\frac{Z'(z)}{Z(z)} - \frac{1}{z(z-1)}\right) = \frac{1}{2}\frac{Z''(1)}{Z'(1)}.$$

(d) Deduce the following Laurent expansion about $s=1$:

$$\frac{1}{2s-1}\frac{Z'(s)}{Z(s)} = \frac{1}{s-1} + (\mathcal{B}-1) + \sum_{n=1}^{\infty} a_n\,(s-1)^n,$$

$$a_n = (-1)^{n+1}\left[1 + 2(g-1)\,\zeta(n+1) + \sum_{j=0}^{[n/2]}(-1)^{j+1}\binom{n-j}{j}\zeta_\Delta(n+1-j)\right] \quad (n\in\mathbb{N}),$$

where $\zeta(s)$ is the Riemann Zeta function and $\zeta_\Delta(s)$ denotes

$$\zeta_\Delta(s) := \sum_{n=1}^{\infty}\frac{1}{\lambda_n^s}.$$

(e) Prove that the finite part of $\zeta_\Delta(s)$ at $s=1$ is

$$\mathrm{FP}\,\zeta_\Delta(1) = 2(g-1)\,\gamma + \mathcal{B}\ (:= \gamma_\Delta).$$

(f) Prove that

$$\zeta_\Delta'(0) = 2(g-1)\,\mathcal{C} - \log(Z'(1)),$$

where

$$\mathcal{C} := -\frac{\pi}{2}\int_0^{\infty}\frac{t^2+\frac{1}{4}}{\cosh^2(\pi t)}\left[1 - \log\left(t^2+\frac{1}{4}\right)\right]dt$$

$$= \frac{1}{4} - \frac{1}{2}\log(2\pi) - 2\zeta'(-1).$$

Consider the functional determinant of Δ on $\mathbf{M}$

$$D_\Delta(z) := \det{}'(-\Delta + z) \quad (z \in \mathbb{C}),$$

where the prime indicates that the zero mode has been omitted and (by zeta regularization)

$$D_\Delta(0) := \exp[-\zeta'_\Delta(0)].$$

(g) Prove that

$$\begin{aligned} D_\Delta(z) &= D_\Delta(0)\exp(\gamma_\Delta z)\prod_{n=1}^{\infty}\left\{\left(1+\frac{z}{\lambda_n}\right)\exp\left(-\frac{z}{\lambda_n}\right)\right\} \\ &= Z'(1)\exp[\gamma_\Delta z - 2(g-1)C]\prod_{n=1}^{\infty}\left\{\left(1+\frac{z}{\lambda_n}\right)\exp\left(-\frac{z}{\lambda_n}\right)\right\}. \end{aligned}$$

(h) Prove the following fundamental relation for the Selberg Zeta function:

$$Z(s) = Z'(1)\, s(s-1)\, V(s),$$

where, for convenience,

$$\begin{aligned} V(s) :&= \exp[\gamma_\Delta s(s-1)]\left\{(2\pi)^{1-s}\exp[s(s-1)]\, G(s)\, G(s+1)\right]^{\frac{\mathcal{A}}{2\pi}} \\ &\cdot \prod_{n=1}^{\infty}\left\{\left[1+\frac{s(s-1)}{\lambda_n}\right]\exp\left[-\frac{s(s-1)}{\lambda_n}\right]\right\}, \end{aligned}$$

where $G(s)$ denotes the Barnes G-function given in Section 1.3.

(i) Prove that

$$Z(s) = s(s-1)\, D_\Delta(s(s-1))\left\{(2\pi)^{1-s}\exp[C+s(s-1)]\, G(s)\, G(s+1)\right]^{2(g-1)}.$$

(*Cf.* Steiner [1129]; see also Voros [1200])

7. Consider the function

$$z(s, a, b, c) = \sum_{n=1}^{\infty}\frac{P_d(cn)}{(cn+a)^s\,(cn+b)^s},$$

of the complex variable s for $\Re(s) > \frac{d+1}{2}$, where P_d is a polynomial of degree d and a, b, c real constants with $a, b > -1$, $c > 0$. Introduce the function of the complex variable z:

$$\phi(z, s, a, b, c) = \sum_{n=1}^{\infty}\frac{P_d(z)}{(cz+a)^s\,(cz+b)^s}.$$

Show that: For $\Re(s) > \frac{d+1}{2}$,

$$z(s, a, b, c) = \frac{1}{2}\phi(1, s, a, b, c) + \int_1^{\infty}\phi(z, s, a, b, c)\, dz + I(s),$$

where $I(s)$ is an entire function of s.

(See Spreafico [1063, Proposition 1])

6 q-Extensions of Some Special Functions and Polynomials

The theory of hypergeometric functions in one, two and more variables leads to a natural unification of much of the material of concern to the mathematical analysts from the seventeenth century to the present day. Functions of this type may be generalized along the lines of basic (or q-) number, resulting in the formation of q-extensions (or q-analogues). This was first systematically effected by E. Heine (1821–1881) in the middle of the nineteenth century, and the work was subsequently greatly extended by F. H. Jackson (1870–1960), W. N. Bailey (1893–1961), L. J. Slater, G. E. Andrews and many others up to the present day. In fact, in recent years, various families of basic (or q-) series and basic (or q-) polynomials have been investigated rather widely and extensively due mainly to their having been found to be potentially useful in such wide variety of fields as (for example) theory of partitions, number theory, combinatorial analysis, finite vector spaces, Lie theory, particle physics, nonlinear electric circuit theory, mechanical engineering, theory of heat conduction, quantum mechanics, cosmology and statistics (see, for details, [1099, pp. 350–351]). The books and monographs by (among others) W. N. Bailey [87], L. J. Slater [1040], H. Exton [439], H. M. Srivastava and P. W. Karlsson [1099, Chapter 9] and G. Gasper and M. Rahman [468] discussed extensively basic (or q-) hypergeometric functions in one, two and more variables. Here, in the present chapter, we choose to investigate some remarkable q-developments around (especially) the Zeta and related functions, which are reported in recent years. We also present the background material, involving (for example) Jackson's q-integral, the q-Gamma function and the q-Beta function, multiple q-Gamma functions, q-Bernoulli numbers and q-Bernoulli polynomials, q-Euler numbers and q-Euler polynomials, q-Zeta functions, multiple q-Zeta functions and so on.

6.1 q-Shifted Factorials and q-Binomial Coefficients

The *q-shifted factorial* $(a;q)_n$ is defined by

$$(a;q)_n := \begin{cases} 1 & (n=0) \\ \prod_{k=0}^{n-1}(1-aq^k) & (n \in \mathbb{N}), \end{cases} \tag{1}$$

Zeta and q-Zeta Functions and Associated Series and Integrals. DOI: 10.1016/B978-0-12-385218-2.00006-2

where $a, q \in \mathbb{C}$ and it is assumed that $a \neq q^{-m}$ $(m \in \mathbb{N}_0)$. It is noted that some other notations that have been used in the literature for the product $(a;q)_n$ in (1) are $(a)_{q,n}$, $[a]_n$, and even $(a)_n$, when the Pochhammer symbol 1.1(5) is not used and the base q is understood.

The *q-shifted factorial* for negative subscript is defined by

$$(a;q)_{-n} := \frac{1}{(1-aq^{-1})(1-aq^{-2})\cdots(1-aq^{-n})} \quad (n \in \mathbb{N}_0), \tag{2}$$

which yields

$$(a;q)_{-n} = \frac{1}{(aq^{-n};q)_n} = \frac{(-q/a)^n q^{\binom{n}{2}}}{(q/a;q)_n} \quad (n \in \mathbb{N}_0). \tag{3}$$

We also write

$$(a;q)_\infty := \prod_{k=0}^{\infty}(1-aq^k) \quad (a, q \in \mathbb{C};\ |q| < 1). \tag{4}$$

It is noted that, when $a \neq 0$ and $|q| \geqq 1$, the infinite product in (4) diverges. So, whenever $(a;q)_\infty$ is involved in a given formula, the constraint $|q| < 1$ will be tacitly assumed.

It follows from (1), (2) and (4) that

$$(a;q)_n = \frac{(a;q)_\infty}{(aq^n;q)_\infty} \quad (n \in \mathbb{Z}), \tag{5}$$

which can be extended to $n = \alpha \in \mathbb{C}$ as follows:

$$(a;q)_\alpha = \frac{(a;q)_\infty}{(aq^\alpha;q)_\infty} \quad (\alpha \in \mathbb{C};\ |q| < 1), \tag{6}$$

where the principal value of q^α is taken.

A list of easily-verified useful identities is given below:

$$(a;q)_n = \left(\frac{q^{1-n}}{a};q\right)_n (-a)^n q^{\binom{n}{2}} \quad (n \in \mathbb{Z}); \tag{7}$$

$$\left(\frac{a^{-1}q^{1-n}}{a};q\right)_n = (a;q)_n\left(-a^{-1}\right)^n q^{-\binom{n}{2}} \quad (n \in \mathbb{Z}); \tag{8}$$

$$(aq^{-n};q)_n = \left(\frac{q}{a};q\right)_n\left(-\frac{a}{q}\right)^n q^{-\binom{n}{2}} \quad (n \in \mathbb{Z}); \tag{9}$$

$$(a;q)_{n-k} = \frac{(a;q)_n}{(a^{-1}q^{1-n};q)_k}(-qa^{-1})^k q^{\binom{k}{2}-nk} \quad (n, k \in \mathbb{Z}); \tag{10}$$

$$(q^{-n};q)_k = \frac{(q;q)_n}{(q;q)_{n-k}}(-1)^k q^{\binom{k}{2}-nk} \quad (n,k\in\mathbb{Z}); \tag{11}$$

$$(aq^{-n};q)_k = \frac{(a;q)_k\,(qa^{-1};q)_n}{(a^{-1}q^{1-k};q)_n}q^{-nk} \quad (n,k\in\mathbb{Z}); \tag{12}$$

$$(aq^{-n};q)_{n-k} = \frac{(q/a;q)_n}{(q/a;q)_k}\left(-\frac{a}{q}\right)^{n-k} q^{\binom{k}{2}-\binom{n}{2}} \quad (n,k\in\mathbb{Z}); \tag{13}$$

$$(aq^k;q)_{n-k} = \frac{(a;q)_n}{(a;q)_k} \quad (n,k\in\mathbb{Z}); \tag{14}$$

$$(a;q)_{n+k} = (a;q)_n\,(aq^n;q)_k \quad (n,k\in\mathbb{Z}); \tag{15}$$

$$(aq^n;q)_k = \frac{(a;q)_k\,(aq^k;q)_n}{(a;q)_n} \quad (n,k\in\mathbb{Z}); \tag{16}$$

$$(aq^{2k};q)_{n-k} = \frac{(a;q)_n\,(aq^n;q)_k}{(a;q)_{2k}} \quad (n,k\in\mathbb{Z}); \tag{17}$$

$$(a;q)_{2n} = (a;q^2)_n\,(aq;q^2)_n \quad (n\in\mathbb{Z}); \tag{18}$$

$$(a^2;q^2)_n = (a;q)_n\,(-a;q)_n \quad (n\in\mathbb{Z}); \tag{19}$$

$$(aq^{-2n};q)_n = \frac{(q/a;q)_{2n}}{(q/a;q)_n}\left(-\frac{a}{q^2}\right)^n q^{-3\binom{n}{2}} \quad (n\in\mathbb{Z}); \tag{20}$$

$$(aq^{-kn};q)_n = \frac{(q/a;q)_{kn}}{(q/a;q)_{(k-1)n}}(-a)^n q^{\binom{n}{2}-kn^2} \quad (n,k\in\mathbb{Z}); \tag{21}$$

$$(aq^{kn};q)_n = \frac{(a;q)_{(k+1)n}}{(a;q)_{kn}} \quad (n,k\in\mathbb{Z}); \tag{22}$$

$$(aq^{jk};q)_{n-k} = \frac{(a;q)_n\,(aq^n;q)_{(j-1)k}}{(a;q)_{jk}} \quad (n,k,j\in\mathbb{Z}). \tag{23}$$

We, now, introduce some more q-notations, which would appear in this chapter quite frequently. The notation $[z]_q$ is defined by

$$[z]_q := \frac{1-q^z}{1-q} = \frac{q^z-1}{q-1} \quad (z\in\mathbb{C};\ q\in\mathbb{C}\setminus\{1\};\ q^z\neq 1). \tag{24}$$

A special case of (24) when $z\in\mathbb{N}$ is

$$[n]_q = \frac{q^n-1}{q-1} = 1+q+\cdots+q^{n-1} \quad (n\in\mathbb{N}), \tag{25}$$

which is called the *q-analogue* (or *q-extension*) of $n\in\mathbb{N}$, since

$$\lim_{q\to 1}[n]_q = \lim_{q\to 1}(1+q+\cdots+q^{n-1}) = n.$$

The *q-analogue* of $n!$ is then defined by

$$[n]_q! := \begin{cases} 1 & \text{if } n = 0, \\ [n]_q\,[n-1]_q \cdots [2]_q[1]_q & \text{if } n \in \mathbb{N}, \end{cases} \tag{26}$$

from which the *q-binomial coefficient* (or the *Gaussian polynomial* analogous to $\binom{n}{k}$) is defined by

$$\begin{bmatrix} n \\ k \end{bmatrix}_q := \frac{[n]_q!}{[n-k]_q!\,[k]_q!} \quad \left(n,\, k \in \mathbb{N}_0;\ 0 \leqq k \leqq n\right). \tag{27}$$

It is easily seen from (1) and (27) that

$$(q;q)_n = (1-q)^n\,[n]_q! \quad (n \in \mathbb{N}_0). \tag{28}$$

The q-binomial coefficient in (27) can be generalized in a similar way as in 1.1(20):

$$\begin{bmatrix} \alpha \\ k \end{bmatrix}_q := \frac{[\alpha]_{q;k}}{[k]_q!} \quad (\alpha \in \mathbb{C};\ k \in \mathbb{N}_0), \tag{29}$$

where $[\alpha]_{q;k}$ is defined by

$$[\alpha]_{q;k} := [\alpha]_q\,[\alpha-1]_q \cdots [\alpha-k+1]_q \quad (\alpha \in \mathbb{C};\ k \in \mathbb{N}_0). \tag{30}$$

The following notations are also frequently used:

$$(a_1, a_2, \ldots, a_m; q)_n := (a_1;q)_n\,(a_2;q)_n \cdots (a_m;q)_n \tag{31}$$

and

$$(a_1, a_2, \ldots, a_m; q)_\infty := (a_1;q)_\infty\,(a_2;q)_\infty \cdots (a_m;q)_\infty\,. \tag{32}$$

Now, the generalized binomial coefficient in (29) can be further generalized as follows:

$$\begin{bmatrix} \alpha \\ \beta \end{bmatrix}_q := \frac{\left(q^{\beta+1}, q^{\alpha-\beta+1}; q\right)_\infty}{\left(q, q^{\alpha+1}; q\right)_\infty} \quad (\alpha,\, \beta \in \mathbb{C};\ |q| < 1). \tag{33}$$

It is noted that, whenever there is no confusion, the notations $[z]_q$, $[n]_q!$ and $\begin{bmatrix} \alpha \\ \beta \end{bmatrix}_q$ are simply written as $[z]$, $[n]!$ and $\begin{bmatrix} \alpha \\ \beta \end{bmatrix}$, respectively. We record some known identities of $[z]_q$ and $\begin{bmatrix} \alpha \\ \beta \end{bmatrix}_q$ as follows:

$$[-n]_q = -\frac{1}{q^n}[n]_q \quad (n \in \mathbb{Z}); \tag{34}$$

$$[n]_{\frac{1}{q}} = \frac{1}{q^{n-1}}[n]_q \quad (n \in \mathbb{Z}); \tag{35}$$

$$[mn]_q = [m]_q [n]_{q^m} \quad (m, n \in \mathbb{Z}); \tag{36}$$

$$[n]_{\frac{1}{q}}! = \frac{1}{q^{n(n-1)/2}}[n]_q! \quad (n \in \mathbb{Z}); \tag{37}$$

$$\begin{bmatrix} n \\ k \end{bmatrix}_q = \begin{bmatrix} n \\ n-k \end{bmatrix}_q = \frac{(q;q)_n}{(q;q)_k (q;q)_{n-k}} \quad \left(n, k \in \mathbb{N}_0;\ 0 \leqq k \leqq n\right) \tag{38}$$

$$\begin{bmatrix} m \\ n \end{bmatrix}_q = \frac{[m]_q}{[n]_q}\begin{bmatrix} m-1 \\ n-1 \end{bmatrix}_q \quad (m \in \mathbb{Z};\ n \in \mathbb{N}); \tag{39}$$

$$\begin{bmatrix} m \\ n \end{bmatrix}_{\frac{1}{q}} = \frac{1}{q^{n(m-n)}}\begin{bmatrix} m \\ n \end{bmatrix}_q \quad (m \in \mathbb{Z};\ n \in \mathbb{N}_0); \tag{40}$$

$$\begin{bmatrix} m \\ k \end{bmatrix}_q \begin{bmatrix} k \\ j \end{bmatrix}_q = \begin{bmatrix} m \\ j \end{bmatrix}_q \begin{bmatrix} m-j \\ k-j \end{bmatrix}_q \quad (m \in \mathbb{Z};\ k, j \in \mathbb{N}_0;\ 0 \leqq j \leqq k); \tag{41}$$

$$\begin{bmatrix} \alpha \\ k \end{bmatrix}_q = \frac{\left(q^{-\alpha};q\right)_k}{(q;q)_k}\left(-q^{\alpha}\right)^k q^{-\binom{k}{2}} \quad (\alpha \in \mathbb{C};\ k \in \mathbb{N}_0); \tag{42}$$

$$\begin{bmatrix} k+\alpha \\ k \end{bmatrix}_q = \frac{\left(q^{\alpha+1};q\right)_k}{(q;q)_k} \quad (\alpha \in \mathbb{C};\ k \in \mathbb{N}_0); \tag{43}$$

$$\begin{bmatrix} -\alpha \\ k \end{bmatrix}_q = \begin{bmatrix} \alpha+k-1 \\ k \end{bmatrix}_q (-q^{-\alpha})^k q^{-\binom{k}{2}} \quad (\alpha \in \mathbb{C};\ k \in \mathbb{N}_0); \tag{44}$$

$$\begin{aligned} \begin{bmatrix} \alpha+1 \\ k \end{bmatrix}_q &= \begin{bmatrix} \alpha \\ k \end{bmatrix}_q q^k + \begin{bmatrix} \alpha \\ k-1 \end{bmatrix}_q \\ &= \begin{bmatrix} \alpha \\ k \end{bmatrix}_q + \begin{bmatrix} \alpha \\ k-1 \end{bmatrix}_q q^{\alpha+1-k} \quad (\alpha \in \mathbb{C};\ k \in \mathbb{N}_0); \end{aligned} \tag{45}$$

$$(z;q)_n = \sum_{k=0}^{n} \begin{bmatrix} n \\ k \end{bmatrix}_q (-z)^k q^{\binom{k}{2}} \quad \left(z \in \mathbb{C};\ n, k \in \mathbb{N}_0;\ 0 \leqq k \leqq n\right). \tag{46}$$

6.2 *q*-Derivative, *q*-Antiderivative and Jackson *q*-Integral

We begin by noting that F. J. Jackson was the first to develop q-calculus in a systematic way.

q-Derivative

The *q-derivative* of a function $f(t)$ is defined by

$$D_q\{f(t)\} := \frac{d_q}{d_q t}\{f(t)\} = \frac{f(qt)-f(t)}{(q-1)t}. \tag{1}$$

It is noted that

$$\lim_{q\to 1} D_q\{f(t)\} = \frac{d}{dt}\{f(t)\},$$

if $f(t)$ is differentiable.

We record some easily derivable q-derivative formulas:

$$D_q\{af(t)+b\,g(t)\} = a\,D_q\{f(t)\}+b\,D_q\{g(t)\}; \tag{2}$$

$$D_q\{f(t)\,g(t)\} = f(qt)\,D_q\{g(t)\}+g(t)\,D_q\{f(t)\}, \tag{3}$$

where the functions $f(t)$ and $g(t)$ are obviously interchangeable;

$$D_q\left\{\frac{f(t)}{g(t)}\right\} = \frac{g(qt)\,D_q\{f(t)\}-f(t)\,D_q\{g(t)\}}{g(t)\,g(qt)}, \tag{4}$$

so that, clearly,

$$\lim_{q\to 1}\left\{D_q\left\{\frac{f(t)}{g(t)}\right\}\right\} = \lim_{q\to 1}\left\{\frac{g(qt)\,D_q f(t)-f(qt)\,D_q\,g(t)}{g(t)\,g(qt)}\right\} = \frac{d}{dt}\left\{\frac{f(t)}{g(t)}\right\}. \tag{5}$$

It is known (see [618, pp. 3–4]) that there does not exist a general chain rule for q-derivatives. There is an exception given by

$$D_q\{f(u(x))\} = \left(D_{q^\beta}f\right)(u(x))\cdot D_q\{u(x)\}, \tag{6}$$

where $u = u(x) = \alpha\,x^\beta$, α and β being constants.

q-Antiderivative and Jackson *q*-Integral

The function $F(t)$ is a *q-antiderivative* of $f(t)$, if $D_q\{F(t)\} = f(t)$. It is denoted by

$$\int f(t)\,d_q t. \tag{7}$$

In ordinary calculus, an antiderivative is unique up to an additive constant. However, in the case of a q-antiderivative, it is known (see [618, p. 65, Proposition 18.1]) that, *for* $0<q<1$, *up to an additive constant, any function* $f(t)$ *has at most one* q*-antiderivative that is continuous at* $t=0$.

The *Jackson integral* of $f(t)$ is, thus, defined, *formally*, by

$$\int f(t)\, d_q t := (1-q)t \sum_{j=0}^{\infty} q^j f\left(q^j t\right), \tag{8}$$

which can be easily generalized as follows:

$$\int f(t)\, d_q g(t) = \sum_{j=0}^{\infty} f\left(q^j t\right)\left(g\left(q^j t\right) - g\left(q^{j+1} t\right)\right). \tag{9}$$

The following theorem gives a sufficient condition under which the formal series in (8) actually converges to a q-antiderivative.

Theorem 6.1 (618, p. 68, Theorem 19.1) *Suppose* $0 < q < 1$. *If* $|f(t)\, t^{\alpha}|$ *is bounded on the interval* $(0, A]$ *for some* $0 \leqq \alpha < 1$, *then the Jackson integral, defined by (7), converges to a function* $F(t)$ *on* $(0, A]$, *which is a q-antiderivative of* $f(t)$. *Moreover,* $F(t)$ *is continuous at* $t = 0$ *with* $F(0) = 0$.

Suppose that $0 < a < b$. The definite q-integral is defined as follows:

$$\int_0^b f(t)\, d_q t := (1-q)b \sum_{j=0}^{\infty} q^j f(q^j b) \tag{10}$$

and

$$\int_a^b f(t)\, d_q t = \int_0^b f(t)\, d_q t - \int_0^a f(t)\, d_q t. \tag{11}$$

A more general version of (10) is given by

$$\int_0^b f(t)\, d_q g(t) = \sum_{j=0}^{\infty} f\left(q^j b\right)\left(g\left(q^j b\right) - g\left(q^{j+1} b\right)\right). \tag{12}$$

The improper q-integral of $f(t)$ on $[0, \infty)$ is defined by

$$\begin{aligned} \int_0^{\infty} f(t)\, d_q t &:= \sum_{j=-\infty}^{\infty} \int_{q^{j+1}}^{q^j} f(t)\, d_q t \\ &= (1-q) \sum_{j=-\infty}^{\infty} f(q^j) q^j \quad (0 < q < 1) \end{aligned} \tag{13}$$

and

$$\begin{aligned}\int_0^\infty f(t)\,d_q t &= \sum_{j=-\infty}^{\infty} \int_{q^j}^{q^{j+1}} f(t)\,d_q t \\ &= \frac{q-1}{q} \sum_{j=-\infty}^{\infty} f(q^j)\,q^j \quad (q>1).\end{aligned} \tag{14}$$

As noted earlier (see [618, p. 71, Proposition 19.1]), *the improper q-integral, defined above, converges, if $t^\alpha f(t)$ is bounded in a neighborhood of $t=0$ with some $\alpha<1$ and for sufficiently large t with some $\alpha>1$.*

The formula for q-integration by parts, the fundamental theorem of q-calculus and q-Taylor formula are given as follows (see [618, Chapter 20]):

$$\int_a^b f(t)\,d_q g(t) = f(b)\,g(b) - f(a)\,g(a) - \int_a^b g(qt)\,d_q f(t) \quad (0 \leqq a < b \leqq \infty). \tag{15}$$

Theorem 6.2 (Fundamental Theorem of q-Calculus) *If $F(t)$ is an antiderivative of $f(t)$ and if $F(t)$ is continuous at $t=0$, then*

$$\int_a^b f(t)\,d_q t = F(b) - F(a) \quad (0 \leqq a < b \leqq \infty). \tag{16}$$

The q-analogue of $(t-a)^n$ is defined by the polynomial

$$(t-a)_q^n := \begin{cases} 1 & (n=0) \\ (t-a)\,(t-qa)\cdots(t-q^{n-1}a) & (n \in \mathbb{N}). \end{cases} \tag{17}$$

It is easy to see that

$$D_q\{(t-a)_q^n\} = [n]_q\,(t-a)_q^{n-1} \quad (n \in \mathbb{N}). \tag{18}$$

Theorem 6.3 (q-Taylor Formula with the Cauchy Remainder Term) *Suppose that $D_q^j\{f(t)\}$ is continuous at $t=0$ for any $j,\ n \in \mathbb{N}_0$ $(j \leqq n+1)$. Then a q-analogue of Taylor's formula with the Cauchy remainder is given as follows:*

$$f(b) = \sum_{j=0}^{n} \left(D_q^j f\right)(a)\,\frac{(b-a)_q^j}{[j]_q!} + \frac{1}{[n]_q!} \int_a^b D_q^{n+1}\{f(t)\}(b-qt)_q^n\,d_q t. \tag{19}$$

6.3 q-Binomial Theorem

We begin by recalling the well-known Ramanujan's ${}_1\Psi_1$-sum:

$$ {}_1\Psi_1(a;b;q,z) := \sum_{k=-\infty}^{\infty} \frac{(a;q)_k}{(b;q)_k} z^k = \frac{(az;q)_\infty \left(\frac{q}{az};q\right)_\infty (q;q)_\infty \left(\frac{b}{a};q\right)_\infty}{(z;q)_\infty \left(\frac{b}{az};q\right)_\infty (b;q)_\infty \left(\frac{q}{a};q\right)_\infty} \tag{1} $$

$$ (|q|<1;\ |a|>|q|;\ |b|<1;\ |b/a|<|z|<1). $$

A simple proof of (1) is given in [584] (see also [1080]). A special case of (1) when $b=q$ yields the *q-binomial theorem*:

$$ {}_1\Phi_0(a;-;q,z) := \sum_{k=0}^{\infty} \frac{(a;q)_k}{(q;q)_k} z^k = \frac{(az;q)_\infty}{(z;q)_\infty} \quad (|q|<1;\ |z|<1), \tag{2} $$

which were proven by several mathematicians, such as Cauchy [225] and Heine [549].

Two special cases of (2) when $a=0$ and when z is replaced by za^{-1} and $a\to\infty$ yield Euler's formulas:

$$ \sum_{k=0}^{\infty} \frac{z^k}{(q;q)_k} = \frac{1}{(z;q)_\infty} \quad (|q|<1;\ |z|<1) \tag{3} $$

and

$$ \sum_{k=0}^{\infty} \frac{(-1)^k q^{\binom{k}{2}}}{(q;q)_k} z^k = (z;q)_\infty \quad (|q|<1;\ |z|<1), \tag{4} $$

respectively.

It is observed that

$$ \lim_{q\downarrow 1} \frac{(q^a z;q)_\infty}{(z;q)_\infty} = \lim_{q\downarrow 1} {}_1\Phi_0\left(q^a;-;q,z\right) = {}_1F_0(a;-;z) = (1-z)^{-a} \tag{5} $$

$$ (|z|<1;\quad a\in\mathbb{C}), $$

which, by the principle of analytic continuation, holds true for $z\in\mathbb{C}$ cut along the positive real axis from 1 to ∞, with $(1-z)^{-a}$ positive when z is real and less than 1.

The special case of (2) when $a=q^{-n}$ $(n\in\mathbb{N}_0)$ gives

$$ {}_1\Phi_0(q^{-n};-;q,z) = (zq^{-n};q)_n = (-z)^n q^{-n(n+1)/2}\left(\frac{q}{z};q\right)_n \quad (n\in\mathbb{N}_0), \tag{6} $$

which, by the principle of analytic continuation, holds true for $z\in\mathbb{C}$ and is seen to be equivalent to 6.1(9).

A q-analogue of the classical exponential function e^z is defined by

$$e_q^z := \sum_{k=0}^{\infty} \frac{z^k}{[k]_q!}, \tag{7}$$

and another q-analogue of the classical exponential function e^z is defined by

$$E_q^z := \sum_{k=0}^{\infty} q^{k(k-1)/2} \frac{z^k}{[k]_q!} = (1+(1-q)z)_q^{\infty}. \tag{8}$$

It is easily seen, by applying (3) and (4), that

$$e_q^z E_q^{-z} = 1. \tag{9}$$

From 6.1(37), we also see that

$$e_{1/q}^z = E_q^z. \tag{10}$$

These two q-exponential functions under q-differentiation are given by

$$D_q\{e_q^z\} = e_q^z \quad \text{and} \quad D_q\{E_q^z\} = E_q^{qz}. \tag{11}$$

In general, we have

$$e_q^z e_q^w \neq e_q^{z+w}.$$

However, the following additive property does hold true:

$$e_q^z e_q^w = e_q^{z+w} \quad (wz = qzw). \tag{12}$$

Corresponding to the above-defined q-exponential functions e_q^z and E_q^z, the q-trigonometric functions are defined as follows:

$$\sin_q x := \frac{e_q^{ix} - e_q^{-ix}}{2i} \quad \text{and} \quad \mathrm{Sin}_q x := \frac{E_q^{ix} - E_q^{-ix}}{2i}; \tag{13}$$

$$\cos_q x := \frac{e_q^{ix} + e_q^{-ix}}{2} \quad \text{and} \quad \mathrm{Cos}_q x := \frac{E_q^{ix} + E_q^{-ix}}{2}. \tag{14}$$

It is easy to see from (10) that

$$\mathrm{Sin}_q x = \sin_{1/q} x \quad \text{and} \quad \mathrm{Cos}_q x = \cos_{1/q} x. \tag{15}$$

We can use (9) to obtain

$$\sin_q x \cdot \mathrm{Sin}_q x + \cos_q x \cdot \mathrm{Cos}_q x = 1, \tag{16}$$

which is the q-analogue of the elementary identity:

$$\sin^2 x + \cos^2 x = 1.$$

q-analogues of other trigonometric identities can also be obtained (see [468, p. 23]).

We apply the chain rule 6.2(6) and use (11) to find the following derivative formulas for the q-trigonometric functions:

$$D_q\{\sin_q x\} = \cos_q x \quad \text{and} \quad D_q\{\mathrm{Sin}_q x\} = \mathrm{Cos}_q qx \tag{17}$$

and

$$D_q\{\cos_q x\} = -\sin_q x \quad \text{and} \quad D_q\{\mathrm{Cos}_q x\} = -\mathrm{Sin}_q x. \tag{18}$$

Ramanujan's ${}_1\Psi_1$-sum (1) includes the Jacobi triple product identity [602] as a limiting case:

$$\begin{aligned}\sum_{k=-\infty}^{\infty} (-1)^k q^{k^2} z^k &= \lim_{c\to 0} {}_1\Psi_1(-1/c; 0; q^2, -qzc)\\ &= \left(zq; q^2\right)_\infty \left(\frac{q}{z}; q^2\right)_\infty \left(q^2; q^2\right)_\infty,\end{aligned} \tag{19}$$

which can be used to express the *theta functions* as follows (see, e.g., Whittaker and Watson [1225, Chapter 21], Gasper and Rahman [468, Section 1.6] and Bellman [113]):

$$\theta_1(x) = 2\sum_{k=0}^{\infty} (-1)^k q^{(k+1/2)^2} \sin(2k+1)x, \tag{20}$$

$$\theta_2(x) = 2\sum_{k=0}^{\infty} q^{(k+1/2)^2} \cos(2k+1)x, \tag{21}$$

$$\theta_3(x) = 1 + 2\sum_{k=1}^{\infty} q^{k^2} \cos 2kx \tag{22}$$

and

$$\theta_4(x) = 1 + 2\sum_{k=1}^{\infty} (-1)^k q^{k^2} \cos 2kx. \tag{23}$$

In terms of infinite products, upon replacing z in (19) by qe^{2ix}, $-qe^{2ix}$, $-e^{2ix}$ and e^{2ix}, respectively, we have

$$\theta_1(x) = 2q^{1/4} \sin x \prod_{k=1}^{\infty} \left(1 - q^{2k}\right)\left(1 - 2q^{2k} \cos 2x + q^{4k}\right), \tag{24}$$

$$\theta_2(x) = 2q^{1/4}\cos x \prod_{k=1}^{\infty}\left(1-q^{2k}\right)\left(1+2q^{2k}\cos 2x+q^{4k}\right), \tag{25}$$

$$\theta_3(x) = \prod_{k=1}^{\infty}\left(1-q^{2k}\right)\left(1+2q^{2k-1}\cos 2x+q^{4k-2}\right) \tag{26}$$

and

$$\theta_4(x) = \prod_{k=1}^{\infty}\left(1-q^{2k}\right)\left(1-2q^{2k-1}\cos 2x+q^{4k-2}\right). \tag{27}$$

We conclude this section by giving a widely-investigated generalization ${}_r\Phi_s$ of the function ${}_1\Phi_0(a;-;q,z)$ in (2), which is defined by

$$\begin{aligned} {}_r\Phi_s\left[\begin{matrix} a_1,\ldots,a_r; \\ & q,z \\ b_1,\ldots,b_s; \end{matrix}\right] &= {}_r\Phi_s(a_1,\ldots,a_r;b_1,\ldots,b_s;q,z) \\ &:= \sum_{k=0}^{\infty}(-1)^{(1-r+s)k}\, q^{(1-r+s)\binom{k}{2}} \\ &\quad \cdot \frac{(a_1;q)_k\cdots(a_r;q)_k}{(b_1;q)_k\cdots(b_s;q)_k}\,\frac{z^k}{(q;q)_k}, \end{aligned} \tag{28}$$

provided that the *generalized basic* (or q-) *hypergeometric series* in (28) converges.

6.4 q-Gamma Function and q-Beta Function

q-Gamma Function

The classic Gamma function $\Gamma(z)$ (see Section 1.1) was found by Euler, while he was trying to extend the factorial $n! = \Gamma(n+1)$ $(n \in \mathbb{N}_0)$ to real numbers. The q-factorial function $[n]_q!$ $(n \in \mathbb{N}_0)$, defined by 6.1(26), can be rewritten as follows:

$$(1-q)^{-n}\prod_{k=0}^{\infty}\frac{(1-q^{k+1})}{(1-q^{k+1+n})} = \frac{(q;q)_\infty}{(q^{n+1};q)_\infty}(1-q)^{-n} := \Gamma_q(n+1) \quad (0<q<1). \tag{1}$$

Replacing n by $x-1$ in (1), Jackson [592] defined the q-Gamma function $\Gamma_q(x)$ by

$$\Gamma_q(x) := \frac{(q;q)_\infty}{(q^x;q)_\infty}(1-q)^{1-x} \quad (0<q<1). \tag{2}$$

Note that it is not completely obvious that Jackson's q-Gamma function is the most natural extension of the above-defined $\Gamma_q(n+1)$ $(n \in \mathbb{N})$. Askey [73] has found analogues of many of the known facts about the Gamma function, and these strongly indicate that (2) is the natural extension of $\Gamma_q(n+1)$ $(n \in \mathbb{N})$, which reduces to $\Gamma(x)$ as $q \to 1$. Also note that, throughout this section, $0 < q < 1$ is assumed.

Setting $x = 1$ in (2) gives

$$\Gamma_q(1) = 1. \tag{3}$$

The q-Gamma function in (2) satisfies the fundamental functional relation $\Gamma(x+1) = x\Gamma(x)$ (see 1.1(9)):

$$\Gamma_q(x+1) = \frac{1-q^x}{1-q}\,\Gamma_q(x) = [x]_q\,\Gamma_q(x). \tag{4}$$

Indeed, by observing that

$$(q^{x+1};q)_\infty = \frac{(q^x;q)_\infty}{1-q^x}, \tag{5}$$

we find that

$$\begin{aligned}\Gamma_q(x+1) &= \frac{(q;q)_\infty}{\left(q^{x+1};q\right)_\infty}(1-q)^{-x}\\ &= \frac{1-q^x}{1-q}\,\frac{(q;q)_\infty}{(q^x;q)_\infty}(1-q)^{1-x} = \frac{1-q^x}{1-q}\,\Gamma_q(x).\end{aligned}$$

The q-Gamma function in (2) also satisfies the q-analogue of the Bohr-Mollerup theorem (see Theorem 1.1):

Theorem 6.4 (73, Theorem 3.1) *Let $f(x)$ be a function that satisfies each of the following properties:*

(a) $f(1) = 1$;
(b) *for some* $q \in (0,1)$,

$$f(x+1) = \frac{1-q^x}{1-q} f(x); \tag{6}$$

(c) $\log f(x)$ *is convex for* $x > 0$.

Then,

$$f(x) = \Gamma_q(x) = \frac{(q;q)_\infty}{(q^x;q)_\infty}(1-q)^{1-x} \quad (x > 0).$$

Proof. It is proven from (3) and (4) that $\Gamma_q(x)$ satisfies (a) and (b). Observe that, for $0 < q < 1$,

$$\begin{aligned}\frac{d}{dx}\log\Gamma_q(x) &= -\log(1-q) - \frac{d}{dx}\sum_{n=0}^{\infty}\log\left(1-q^{n+x}\right)\\ &= -\log(1-q) + \log q\sum_{n=0}^{\infty}\frac{q^{n+x}}{1-q^{n+x}}\end{aligned}$$

and

$$\frac{d^2}{dx^2} \log \Gamma_q(x) = (\log q)^2 \sum_{n=0}^{\infty} \frac{q^{n+x}}{\left(1 - q^{n+x}\right)^2}.$$

Thus, $\log \Gamma_q(x)$ is convex for $x > 0$. ■

Let $0 < x < 1$. The convexity of $\log f(x)$ implies

$$\log f(x+n) \leqq \log f(n) + x[\log f(n+1) - \log f(n)]$$

or

$$f(n+x) \leqq f(n) \left(\frac{1-q^n}{1-q}\right)^x.$$

The functional equation (6) then gives

$$\begin{aligned} f(x) &= \frac{(1-q)^n}{(q^x;q)_n} f(x+n) \leqq \frac{(1-q)^n}{(q^x;q)_n} f(n) \left(\frac{1-q^n}{1-q}\right)^x \\ &= \frac{(1-q)^n\,(q;q)_{n-1}\,(1-q^n)^x}{(q^x;q)_n\,(1-q)^{n-1}\,(1-q)^x} \end{aligned}$$

so

$$f(x) \leqq \frac{(q;q)_{n-1}}{(q^x;q)_n} (1-q)^{1-x} \left(1-q^n\right)^x \quad (0 < q < 1,\ 0 < x < 1). \tag{7}$$

To find a lower bound for $f(x)$, apply the convexity of $\log f(x)$ at $n+x$, $n+1$ and $n+x+1$ to obtain

$$\log f(n+1) \leqq \log f(n+x) + (1-x)\,[\log f(n+x+1) - \log f(n+x)]$$

or

$$f(n+1) \leqq f(n+x) \left(\frac{1-q^{n+x}}{1-q}\right)^{1-x}.$$

As above, the functional equation (6) gives

$$\begin{aligned} f(x) &= \frac{(1-q)^n}{(q^x;q)_n} f(x+n) \geqq \frac{(1-q)^n}{(q^x;q)_n} \left(\frac{1-q}{1-q^{n+x}}\right)^{1-x} f(n+1) \\ &= \frac{(1-q)^{1-x}\,(q;q)_n}{\left(1-q^{n+x}\right)^{1-x}\,(q^x;q)_n}, \end{aligned}$$

so

$$f(x) \geqq \frac{(q;q)_n\,(1-q)^{1-x}}{(q^x;q)_n\left(1-q^{n+x}\right)^{1-x}} \quad (0<q<1,\ 0<x<1). \tag{8}$$

Setting $n \to \infty$ in (7) and (8) gives

$$f(x) = \frac{(q;q)_\infty}{(q^x;q)_\infty}\,(1-q)^{1-x} = \Gamma_q(x).$$

The functional equation (6) then gives $f(x) = \Gamma_q(x)$ for $x > 0$.

Remark 1 The q-gamma function has simple poles at $x = 0,\ -1,\ -2,\ \ldots$ and the residues

$$\operatorname*{Res}_{x=-n} \Gamma_q(x) = \frac{(1-q)^{n+1}}{\left(q^{-n};q\right)_n \log q^{-1}} \quad (n \in \mathbb{N}_0). \tag{9}$$

Indeed,

$$\begin{aligned}
\lim_{x\to -n} (x+n)\,\Gamma_q(x) &= \lim_{x\to -n} \frac{(x+n)\,(q;q)_\infty}{(1-q^x)\left(1-q^{x+1}\right)} \cdots (1-q)^{1-x} \\
&= \frac{(1-q)^{n+1}}{\left(1-q^{-n}\right)\cdots\left(1-q^{-1}\right)} \lim_{x\to -n} \frac{x+n}{1-q^{x+n}} \\
&= \frac{(1-q)^{n+1}}{\left(q^{-n};q\right)_n \log q^{-1}}.
\end{aligned}$$

The q-gamma function has no zeros, so its reciprocal is an entire function with zeros at $x = -n$ $(n \in \mathbb{N}_0)$.

$$\frac{1}{\Gamma_q(x)} = (1-q)^{x-1} \prod_{n=0}^{\infty} \frac{1-q^{n+x}}{1-q^{n+1}}, \tag{10}$$

which also has zeros at $x = -n + (2\pi i k/\log q)$ $(k \in \mathbb{Z};\ n \in \mathbb{N}_0)$.

Applying 6.3(3) and 6.3(4) to the definition of the q-Gamma function (2) yields

$$\Gamma_q(x) = (q;q)_\infty\,(1-q)^{1-x} \sum_{k=0}^{\infty} \frac{q^{kx}}{(q;q)_k} \tag{11}$$

and

$$\frac{1}{\Gamma_q(x)} = \frac{(1-q)^{x-1}}{(q;q)_\infty} \sum_{k=0}^{\infty} \frac{(-1)^k\, q^{\binom{k}{2}}\, q^{kx}}{(q;q)_k}. \tag{12}$$

Now, the generalized binomial coefficient $\begin{bmatrix}\alpha\\ \beta\end{bmatrix}_q$, defined by 6.1(33), can, in terms of q-Gamma function, be rewritten like 1.1(41) as follows:

$$\begin{bmatrix}\alpha\\ \beta\end{bmatrix}_q = \frac{\Gamma_q(\alpha+1)}{\Gamma_q(\beta+1)\,\Gamma_q(\alpha-\beta+1)}. \tag{13}$$

The q-factorial $[n]_q!$, defined by 6.1(26), is a monotone increasing function of q for $q>0$, so it is natural to ask if something happens for the q-Gamma function. Since $\Gamma_q(1)=\Gamma_q(2)=1$, something different may happen for $1<x<2$ than for other $x>0$.

Theorem 6.5 (73, Theorem 4.1) *The following inequalities hold:*

$$\Gamma_r(x) \leqq \Gamma_q(x) \leqq \Gamma(x) \quad (0<x\leqq 1 \text{ or } x\geqq 2,\ 0<r<q<1), \tag{14}$$

$$\Gamma(x) \leqq \Gamma_q(x) \leqq \Gamma_r(x) \quad (1\leqq x\leqq 2,\ 0<r<q<1). \tag{15}$$

Also

$$\lim_{q\to 1-} \Gamma_q(x) = \Gamma(x). \tag{16}$$

Proof. Only the proof of (16) is provided. By (14) and (15), $\Gamma_q(x)$ is a monotone function of q, which is bounded, and so has a limit. By Theorem 6.4, the limit satisfies the assumptions of the Bohr-Mollerup theorem (see Theorem 1.1) and so is $\Gamma(x)$. ■

A q-analogue of Legendre duplication formula for the Gamma function 1.1(29) is given as follows:

$$\Gamma_q(2x)\,\Gamma_{q^2}\left(\frac{1}{2}\right) = \Gamma_{q^2}(x)\,\Gamma_{q^2}\left(x+\frac{1}{2}\right)(1+q)^{2x-1}. \tag{17}$$

Indeed, we have

$$\begin{aligned}\frac{\Gamma_{q^2}(x)\,\Gamma_{q^2}\left(x+\frac{1}{2}\right)}{\Gamma_{q^2}\left(\frac{1}{2}\right)} &= \frac{(q^2;q^2)_\infty\,(q;q^2)_\infty\,\left(1-q^2\right)^{1-x+1/2-x}}{(q^{2x};q^2)_\infty\,(q^{2x+1};q^2)_\infty\,\left(1-q^2\right)^{1/2}}\\ &= \frac{(q;q)_\infty}{(q^{2x};q)_\infty}\left(1-q^2\right)^{1-2x} = \Gamma_q(2x)\,(1+q)^{1-2x}.\end{aligned}$$

Also, a q-analogue of Gauss multiplication formula for the Gamma function 1.1(51) is similarly proven:

$$\begin{aligned}&\Gamma_q(nx)\,\Gamma_{q^n}\left(\frac{1}{n}\right)\Gamma_{q^n}\left(\frac{2}{n}\right)\cdots\Gamma_{q^n}\left(\frac{n-1}{n}\right)\\ &= \Gamma_{q^n}(x)\,\Gamma_{q^n}\left(x+\frac{1}{n}\right)\cdots\Gamma_{q^n}\left(x+\frac{n-1}{n}\right)[n]_q^{nx-1} \quad (n\in\mathbb{N}).\end{aligned} \tag{18}$$

q-Beta Function

A special case of 6.2(10) when $b=1$ yields

$$\int_0^1 f(t)\,d_q t = (1-q)\sum_{n=0}^{\infty} q^n f(q^n). \tag{19}$$

Now, try to get a q-analogue of the Beta function, defined by 1.1(59), and keep the integrand $f(t)=t^{\alpha-1}(1-t)^{\beta-1}$ of 1.1(59) in mind. Since q^n already appears in (19), it is natural to use $t^{\alpha-1}$ in itself. Even though there is no hope of evaluating a sum that contains a factor that $(1-q^n)^{\beta-1}$, in view of 6.3(2) and 6.3(5), there seems to be an alternative of a q-analogue of $(1-t)^{\beta-1}$ as follows:

$$\sum_{n=0}^{\infty} \frac{(q^{1-\beta};q)_n}{(q;q)_n} t^n = \frac{(tq^{1-\beta};q)_\infty}{(t;q)_\infty}. \tag{20}$$

(20) is almost the right q-analogue of $(1-t)^{\beta-1}$. The one thing wrong is that the function being integrated in 1.1(39) is not $(1-t)^{\beta-1}$ but $(1-t)_+^{\beta-1}$, since the range of integration in 1.1(39) stops at $t=1$. The first point in the sequence $t=q^n$, which lies to the right of $t=1$, is $t=q^{-1}$. So, by shifting the variable t by tq^β, Askey [73, Eq. (5.7)] chose a natural candidate for the q-Beta function:

$$\begin{aligned} B_q(\alpha,\beta) &:= \int_0^1 t^{\alpha-1}\frac{(tq;q)_\infty}{(tq^\beta;q)_\infty}\,d_q t \quad (\Re(\alpha)>0;\ \beta\in\mathbb{C}\setminus\mathbb{Z}_0^-) \\ &= (1-q)\sum_{n=0}^{\infty} q^{n\alpha}\frac{(q^{n+1};q)_\infty}{(q^{n+y};q)_\infty}. \end{aligned} \tag{21}$$

We use 6.1(5) and 6.3(2) to prove a relationship between the q-Gamma function and the q-Beta function (see 1.1(42)), which shows a natural choice in (21).

Theorem 6.6 *We have*

$$B_q(\alpha,\beta) = \frac{\Gamma_q(\alpha)\,\Gamma_q(\beta)}{\Gamma_q(\alpha+\beta)}. \tag{22}$$

Proof.

$$\begin{aligned} B_q(\alpha,\beta) &= (1-q)\frac{(q;q)_\infty}{(q^\beta;q)_\infty}\sum_{n=0}^{\infty}\frac{(q^\beta;q)_n}{(q;q)_n}q^{n\alpha} \\ &= \frac{(1-q)\,(q;q)_\infty\,(q^{\alpha+\beta};q)_\infty}{(q^\beta;q)_\infty\,(q^\alpha;q)_\infty} \\ &= \frac{\frac{(q;q)_\infty}{(q^\alpha;q)_\infty}(1-q)^{1-\alpha}\,\frac{(q;q)_\infty}{(q^\beta;q)_\infty}(1-q)^{1-\beta}}{\frac{(q;q)_\infty}{(q^{\alpha+\beta};q)_\infty}(1-q)^{1-\alpha-\beta}} = \frac{\Gamma_q(\alpha)\,\Gamma_q(\beta)}{\Gamma_q(\alpha+\beta)}. \end{aligned}$$

■

Although it is not possible to change variables in a sum (and so in a q-integral), there are times when a change of variables in an ordinary integral will lead to another integral that can be approximated by a q-integral. For example, setting $u = ct$ in 1.1(41) yields

$$\frac{B(x,y)}{c^x} = \int_0^\infty \frac{t^{x-1}\,dt}{(1+ct)^{x+y}} \quad (\Re(x) > 0;\ \Re(y) > 0). \tag{23}$$

Askey [73] observed that Ramanujan's sum 6.3(1) gives a q-integral extension of (23). Indeed, we use 6.1(5) to rewrite 6.3(1) as follows:

$$\sum_{n=-\infty}^{\infty} \frac{(bq^n;q)_\infty}{(aq^n;q)_\infty} x^n = \frac{(ax;q)_\infty \left(\frac{q}{ax};q\right)_\infty (q;q)_\infty \left(\frac{b}{a};q\right)_\infty}{(x;q)_\infty \left(\frac{b}{ax};q\right)_\infty (a;q)_\infty \left(\frac{q}{a};q\right)_\infty},$$

which, upon setting $x = q^\alpha$, $a = -c$, $b = -cq^{\alpha+\beta}$ and using the q-binomial theorem 6.3(2) to replace $\left(-cq^{\alpha+\beta+n};q\right)_\infty / (-cq^n;q)_\infty$, gives

$$\begin{aligned}&\sum_{n=-\infty}^{\infty} \left[\sum_{k=0}^{\infty} \frac{(q^{\alpha+\beta};q)_k}{(q;q)_k} (-cq^n)^k\right] q^{\alpha n} \\ &\quad = \frac{(-cq^\alpha;q)_\infty (-c^{-1}q^{1-\alpha};q)_\infty (q;q)_\infty (q^{\alpha+\beta};q)_\infty}{(-c;q)_\infty (-c^{-1}q;q)_\infty (q^\alpha;q)_\infty (q^\beta;q)_\infty}.\end{aligned} \tag{24}$$

If we use 6.2(13) and 6.3(2), (24) can be rewritten as

$$\begin{aligned}&\int_0^\infty \sum_{k=0}^{\infty} \frac{(q^{\alpha+\beta};q)_k}{(q;q)_k} (-cx)^k x^{\alpha-1}\,d_q x \\ &\quad = \frac{(-cq^\alpha;q)_\infty (-c^{-1}q^{1-\alpha};q)_\infty \Gamma_q(\alpha)\Gamma_q(\beta)}{(-c;q)_\infty (-c^{-1}q;q)_\infty \Gamma_q(\alpha+\beta)} \\ &\quad = \int_0^\infty \frac{(-cq^{\alpha+\beta}x;q)_\infty}{(-cx;q)_\infty} x^{\alpha-1}\,d_q x.\end{aligned} \tag{25}$$

In view of 6.3(5), it is seen that

$$\lim_{q\to 1} \frac{(-cq^\alpha;q)_\infty \left(-c^{-1}q^{1-\alpha};q\right)_\infty}{(-c;q)_\infty \left(-c^{-1}q;q\right)_\infty} = (1+c)^{-\alpha}\left(1+\frac{1}{c}\right)^\alpha = c^{-\alpha}$$

and

$$\lim_{q\to 1} \frac{\left(-cq^{\alpha+\beta}x;q\right)_\infty}{(-cx;q)_\infty} = (1+cx)^{-\alpha-\beta}.$$

6.5 A q-Extension of the Multiple Gamma Functions

Motivated by Theorem 6.4 and Theorem 1.5, Nishizawa [865, 866] and Ueno and Nishizawa [1176] presented a q-analogue of the multiple Gamma functions (see Section 1.4) and its properties. Here, we introduce their works without proof. We assume $0 < q < 1$ throughout this section.

We begin by giving q-analogues of Gauss's and Euler's product forms of the Gamma function 1.1(4) and 1.1(7) (see [865, Equations (3.2) and (3.3)]):

$$\Gamma_q(z+1) = \lim_{N\to\infty} \frac{[1][2]\cdots[N]}{[z+1][z+2]\cdots[z+N]} [N+1]^z \tag{1}$$

and

$$\Gamma_q(z+1) = \prod_{n=1}^{\infty} \left\{ \left(\frac{[n+1]}{[n]} \right)^z \left(\frac{[z+n]}{[n]} \right)^{-1} \right\}, \tag{2}$$

where $[z] = [z]_q$ throughout this section.

Nishizawa [865, Definition 4.1] defines the q-analogue of Vignéras's G_r-function (see Theorems 1.4 and 1.5) as follows:

Definition 6.7 For $z \in \mathbb{C}$ with $\Re(z) > 0$,

$$G_0(z+1;q) := [z+1],$$

$$G_r(z+1;q) := (1-q)^{-\frac{z}{r}} \prod_{n=1}^{\infty} \left\{ \left(\frac{1-q^{z+n}}{1-q^n} \right)^{(-1)^r \binom{n+r-2}{r-1}} (1-q^n)^{g_r(z,n)} \right\} \quad (r \in \mathbb{N}),$$

where

$$g_1(z,n) := 0,$$

$$g_r(z,n) := \sum_{k=1}^{r-1} (-1)^{k-1} \binom{z}{r-k} \binom{n+k-2}{k-1} \quad (r \in \mathbb{N} \setminus \{1\}).$$

We note that $G_1(z;q) = \Gamma_q(z)$. The infinite products of these functions are absolutely convergent. Nishizawa [865, Theorem 4.2] (see also [866, Theorem 3.1]) proves a q-analogue of Theorem 1.5:

Theorem 6.8 *A unique hierarchy of functions exists, which satisies*

(i) $G_r(z+1;q) = G_{r-1}(z;q)\, G_r(z;q)$,
(ii) $G_r(1;q) = 1$,
(iii) $\frac{d^{r+1}}{dz^{r+1}} \log G_{r+1}(z+1;q) \geqq 0 \quad (z \geqq 0)$,
(iv) $G_0(z;q) = [z]$,

where

$$g_r(z,u) = \binom{z-u}{r-1} - \binom{-u}{r-1}.$$

$G_r(z+1;q)$ is expressed as the following infinite product representation:

$$G_r(z+1;q) := (1-q)^{-\binom{z}{r}} \prod_{k=1}^{\infty} \left\{ \left(\frac{1-q^{z+k}}{1-q^k} \right)^{\binom{-k}{r-1}} \left(1-q^k\right)^{g_r(z,k)} \right\} \quad (r \in \mathbb{N}). \quad (3)$$

The case of Theorem 6.8 when $r = 1$ corresponds to Askey's theorem (see Theorem 6.4). So, the sequence $\{G_r(z;q)\}$ includes a q-Gamma function. We call an element of the sequence a *multiple q-Gamma function*. The expression (3) can be regarded as a q-analogue of the Weierstrass product form for the function $G_r(z)$ given in Theorem 1.4.

Nishizawa [865, Proposition 4.4] (see also [866, Proposition 3.2]) derives the following counterpart of (1) and (2) for the function $G_r(z;q)$:

Theorem 6.9 *If $\Re(z) > 0$, then*

$$G_r(z+1;q) = \lim_{N\to\infty} \left\{ \frac{G_{r-1}(1;q)\cdots G_{r-1}(N;q)}{G_{r-1}(z+1;q)\cdots G_{r-1}(z+N;q)} \cdot \prod_{m=1}^{r} G_{r-m}(N+1;q)^{\binom{z}{m}} \right\}$$

and

$$G_r(z+1;q) = \prod_{n=1}^{\infty} \left\{ \frac{G_{r-1}(n;q)}{G_{r-1}(z+n;q)} \prod_{m=1}^{r} \left(\frac{G_{r-m}(n+1;q)}{G_{r-m}(n;q)} \right)^{\binom{z}{m}} \right\}.$$

Koornwinder [694, Theorem B.2] proved rigorously the following result:

$$\lim_{q\uparrow 1} \Gamma_q(z+1) = \Gamma(z+1) \quad (z \in \mathbb{C} \setminus \{-1, -2, \ldots\}). \quad (4)$$

By making use of the Euler-Maclaurin summation formula (see 2.7(21)), Ueno and Nishizawa [1176, Proposition 4.1] (see also Nishizawa [866, Proposition 3.3]) obtain an expansion formula of $\log G_n(z+1;q)$ as follows:

Theorem 6.10 *Suppose that $\Re(z) > -1$ and $m > n$. They have*

$$\log G_n(z+1;q) = \left\{\binom{z+1}{n} + \sum_{r=1}^{n} \frac{B_r}{r!}\left(-\frac{d}{dz}\right)^{r-1}\binom{z}{n-1}\right\} \log\frac{1-q^{z+1}}{1-q}$$
$$+\sum_{r=1}^{n}\left\{\left(-\frac{d}{dz}\right)^{r-1}\binom{z}{n-1}\right\}\int_1^{z+1}\frac{\xi^r}{r!}\frac{q^{\xi}\log q}{1-q^{\xi}}\,d\xi \tag{5}$$
$$+\sum_{j=0}^{n-1} G_{n,j}(z)\,C_j(q) + \sum_{r=1}^{m}\frac{B_r}{r!}F_{n,r-1}(z;q) - R_{n,m}(z;q),$$

where

$$F_{n,r-1}(z;q) := \left[\frac{d^{r-1}}{dt^{r-1}}\left\{\binom{-t}{n-1}\log\frac{1-q^{z+t}}{1-q^{z+1}}\right\}\right]_{t=1},$$
$$C_j(q) := -\sum_{r=1}^{n+1}\frac{B_r}{r!}f_{j+1,r-1}(1;q) + \int_1^{\infty}\frac{(-1)^n\widetilde{B}_{n+1}(t)}{(n+1)!}f_{j+1,n+1}(t;q)\,dt,$$
$$f_{j+1,r-1}(t;q) := \frac{d^{r-1}}{dt^{r-1}}\left(t^j\log\frac{1-q^t}{1-q}\right),$$
$$R_{n,m}(z;q) := \frac{(-1)^{m-1}}{m!}\int_1^{\infty}\widetilde{B}_m(t)\frac{d^m}{dt^m}\left\{\binom{-t}{n-1}\log\frac{1-q^{z+t}}{1-q^{z+1}}\right\}dt.$$

The formula (5) is a generalization of Moak's [838, Theorem 2]. As remarked by Daalhuis [361], the formula (5) is not an asymptotic expansion. Yet, Ueno and Nishizawa [1176] (see also Nishizawa [866]) observes that each term of (5) converges uniformly as $q \uparrow 1$. Thus, they obtain the following theorem (see Nishizawa [866, Theorem 3.4]):

Theorem 6.11 *As $q \to 1-0$, $G_n(z+1;q)$ converges to $G_n(z+1)$ uniformly on any compact set in the domain $\mathbb{C} \setminus \{-1, -2, \ldots\}$.*

6.6 *q*-Bernoulli Numbers and *q*-Bernoulli Polynomials

Carlitz ([215] and [217]) introduced q-extensions of the classic Bernoulli numbers and polynomials. Since then, many authors have studied this and related subjects (see, e.g., [227, 228, 255, 363, 493, 626, 649, 652, 653, 659, 660, 682, 971, 992, 1006, 1100, 1168, 1170] and [1174]). In the frequently cited [215], Carlitz introduced the relation

$$(q\eta+1)^k = \eta_k \quad (k \in \mathbb{N}\setminus\{1\}), \quad \eta_0 = 1, \quad \eta_1 = 0 \tag{1}$$

as an inductive definition for a certain sequence $\{\eta_k\}_{k=0}^{\infty}$ of functions $\eta_k = \eta_{k;q}$, depending rationally on the parameter q. This definition is interpreted according to the umbral calculus convention that, after expanding the left side into monomials, one replaces each power η^j by the corresponding sequence element η_j. The same paper exhibits polynomials $\eta_k(x) = \eta_{k;q}(x)$ in q^x, determined by a difference equation analogous to that governing the Bernoulli polynomials (see 1.7(4)), namely

$$\eta_{k;q}(x+1) - \eta_{k;q}(x) = k q^x [x]_q^{k-1}, \quad \eta_k(0) = \eta_k, \tag{2}$$

where $[x]_q$ is defined by 6.1(24). In what follows, we require, for convenience, that $0 < q < 1$.

Because η_k and $\eta_k(x)$ do not remain finite at $q = 1$, Carlitz [215] used η_k and $\eta_k(x)$ in (1) and (2) to define a set of numbers $\beta_k = \beta_{k;q}$, by

$$\beta_k = \eta_k + (q-1)\,\eta_{k+1}, \tag{3}$$

and a set of polynomials $\beta_k(x) = \beta_{k;q}(x)$, by the recurrence relation

$$q^x \beta_k(x) = \eta_k(x) + (q-1)\,\eta_{k+1}(x), \quad \beta_k := \beta_k(0) \quad (k \in \mathbb{N}_0). \tag{4}$$

In the limit $q \to 1$, β_k reduces to the Bernoulli number B_k. These numbers $\beta_{k;q}$ and polynomials $\beta_{k;q}(x)$ are q-analogues of the ordinary Bernoulli numbers B_k and polynomials $B_k(x)$ (see Section 1.6).

Recently, a variety of works on q-Bernoulli numbers and polynomials and q-Euler numbers and polynomials have appeared, for example, Cenkci and Can [227], Ryoo [992] and Kim [654]. Kim, in particular, has published 50 papers on this and related subjects; a few are cited here. Although many authors, including those whose works are mentioned above, have treated more general generating functions of q-eta and q-epsilon polynomials than those discussed here, we, first, revisit the seminal work of Carlitz [215], substituting formal generating series for the difference equations employed by him (see (1.2)). In fact, Carlitz himself noted [215, p. 988, lines 4–6] that it is easy to define numbers and polynomials, as well as generating functions, of higher order. We do propose certain minor corrections to [215]; corrected statements appear here, for example, as Equations (74) and (80).

Here, we will recover the identities for η_k and $\eta_k(x)$ given by Carlitz [215]. In place of the difference relations (1) and (2), defining η_k and $\eta_k(x)$, we apply the (now-standard) formal generating series

$$G_q(x,t) := -t \sum_{k=0}^{\infty} q^{k+x} e^{[k+x]_q t} \equiv \sum_{k=1}^{\infty} \eta_{k;q}(x) \frac{t^k}{k!} \tag{5}$$

and its specialization

$$G_q(t) := G_q(0,t) = -t \sum_{k=0}^{\infty} q^k e^{[k]_q t} \equiv \sum_{k=1}^{\infty} \eta_{k;q} \frac{t^k}{k!}. \tag{6}$$

It is evident from (5) and (6) that

$$\eta_{k;q} = \eta_{k;q}(0) \quad (k \in \mathbb{N}). \tag{7}$$

To retain as close as possible a correspondence with Carlitz [215], we observe that the absence of any term involving $\eta_{0;q}$ in the generating series permits us to define $\eta_{0;q} = 1$ as in (1).

It follows from 6.1(24) that

$$[x+y]_q = [x]_q + q^x [y]_q. \tag{8}$$

Using (8), we find that

$$\eta_{k;q}(x) = \sum_{j=0}^{k} \binom{k}{j} [x]_q^{k-j} q^{xj} \eta_{j;q} \quad (k \in \mathbb{N}_0). \tag{9}$$

Indeed, by (5), (6) and (8), we have

$$\begin{aligned}
\sum_{k=0}^{\infty} \eta_{k;q}(x) \frac{t^k}{k!} &= -t \sum_{k=0}^{\infty} q^{k+x} e^{([x]_q + [k]_q q^x)\, t} \\
&= e^{[x]_q t} \left(-q^x t \sum_{k=0}^{\infty} q^k e^{[k]_q q^x t} \right) \\
&= e^{[x]_q t} G_q\left(q^x t\right) = \left(\sum_{j=0}^{\infty} [x]_q^j \frac{t^j}{j!} \right) \left(\sum_{j=0}^{\infty} \eta_{j;q} q^{xj} \frac{t^j}{j!} \right) \\
&= \sum_{k=0}^{\infty} \left(\sum_{j=0}^{k} [x]_q^{k-j} \eta_{j;q} q^{xj} \frac{1}{j!(k-j)!} \right) t^k,
\end{aligned}$$

which, upon equating the coefficients of t^k, yields (9).

From (5), we get q-difference equation for $G_{x;q}(t)$:

$$e^t G_{x;q}(qt) = t q^x e^{[x]_q t} + G_{x;q}(t), \tag{10}$$

which easily yields

$$\sum_{j=0}^{k+1} \binom{k+1}{j} q^j \eta_{j;q}(x) = (k+1) q^x [x]_q^k + \eta_{k+1;q}(x) \quad (k \in \mathbb{N}_0). \tag{11}$$

The special cases of (10) and (11) when $x = 0$, respectively, give the q-difference equation for $G_q(t)$:

$$e^t G_q(qt) = G_q(t) + t \tag{12}$$

and

$$\eta_{k;q} = \begin{cases} \frac{1}{q-1} & \text{if } k = 1, \\ \sum_{j=0}^{k} \binom{k}{j} \eta_{j;q} q^j & \text{if } k \in \mathbb{N} \setminus \{1\}. \end{cases} \tag{13}$$

We find from (9) and (13) that

$$\eta_{k;q}(1) = \eta_{k;q}(0) = \eta_{k;q} \quad (k \in \mathbb{N} \setminus \{1\}). \tag{14}$$

It is seen from (5) that

$$G_{x+1;q}(t) - G_{x;q}(t) = t\, q^x\, e^{[x]_q t}, \tag{15}$$

which gives

$$\eta_{k+1;q}(x+1) - \eta_{k+1;q}(x) = (k+1)\, q^x\, [x]_q^k \quad (k \in \mathbb{N}_0). \tag{16}$$

Considering

$$e^{[j+x]_q t} = e^{\frac{t}{1-q}}\, e^{\frac{q^{j+x}}{q-1} t} \tag{17}$$

in (5), we find that

$$\begin{aligned} \sum_{k=0}^{\infty} \eta_{k;q}(x) \frac{t^k}{k!} &= -t \sum_{\ell=0}^{\infty} \frac{q^{(\ell+1)x} t^\ell}{(q-1)^\ell\, \ell!} \sum_{j=0}^{\infty} q^{(\ell+1)j} \sum_{\ell=0}^{\infty} \frac{(-1)^\ell\, t^\ell}{(q-1)^\ell\, \ell!} \\ &= -t \sum_{\ell=0}^{\infty} \frac{q^{(\ell+1)x} t^\ell}{\left(1 - q^{\ell+1}\right)(q-1)^\ell\, \ell!} \sum_{\ell=0}^{\infty} \frac{(-1)^\ell\, t^\ell}{(q-1)^\ell\, \ell!} \\ &= -t \sum_{k=0}^{\infty} \left\{ \sum_{j=0}^{k} (-1)^{k-j} \frac{q^{(j+1)x}}{j!(k-j)!\left(1 - q^{j+1}\right)} \right\} \frac{t^k}{(q-1)^k}, \end{aligned}$$

which, upon equating the coefficients of t^{k+1} and assuming $0/[0]_q = 1$, after a little simplification, yields an explicit expression of $\eta_{k;q}(x)$:

$$(q-1)^k\, \eta_{k;q}(x) = \sum_{j=0}^{k} (-1)^{k-j} \binom{k}{j} \frac{j}{[j]_q} q^{jx} \quad (k \in \mathbb{N}), \tag{18}$$

which can be equivalently written as:

$$\sum_{j=0}^{k} \binom{k}{j} (q-1)^j\, \eta_{j;q}(x) = \frac{k}{[k]_q} q^{kx} \quad (k \in \mathbb{N}), \tag{19}$$

by using the inverse relation technique in combinatorial analysis. Indeed, using a well-known series manipulation

$$\sum_{\ell=0}^{k}\sum_{j=0}^{\ell} A_{\ell,j} = \sum_{j=0}^{k}\sum_{\ell=j}^{k} A_{\ell,j} \tag{20}$$

and a binomial identity

$$\binom{k}{\ell}\binom{\ell}{j} = \binom{k}{j}\binom{k-j}{\ell-j}, \tag{21}$$

we have

$$\begin{aligned}\sum_{\ell=0}^{k}\binom{k}{\ell}(q-1)^{\ell}\,\eta_{\ell;q}(x) &= \sum_{j=0}^{k}\binom{k}{j}\frac{j}{[j]_q}q^{jx}\sum_{\ell=j}^{k}(-1)^{\ell-j}\binom{k-j}{\ell-j}\\ &= \sum_{j=0}^{k}\binom{k}{j}\frac{j}{[j]_q}q^{jx}(1-1)^{k-j} = \frac{k}{[k]_q}q^{kx}.\end{aligned}$$

The special cases of (18) and (19) when $x=0$, respectively, gives

$$(q-1)^k\,\eta_{k;q} = \sum_{j=0}^{k}(-1)^{k-j}\binom{k}{j}\frac{j}{[j]_q} \quad (k \in \mathbb{N}), \tag{22}$$

or, equivalently,

$$\sum_{j=0}^{k}\binom{k}{j}(q-1)^j\,\eta_{j;q} = \frac{k}{[k]_q} \quad (k \in \mathbb{N}). \tag{23}$$

We give multiplication formula for $\eta_{k;q}(x)$:

$$[m]_q^{k-1}\sum_{j=0}^{m-1}\eta_{k;q^m}\left(\frac{x+j}{m}\right) = \eta_{k;q}(x) \quad (k \in \mathbb{N}_0;\ m \in \mathbb{N}), \tag{24}$$

or, equivalently,

$$[m]_q^{k-1}\sum_{j=0}^{m-1}\eta_{k;q^m}\left(x+\frac{j}{m}\right) = \eta_{k;q}(mx) \quad (k \in \mathbb{N}_0;\ m \in \mathbb{N}). \tag{25}$$

Indeed, we have

$$\sum_{k=0}^{\infty}[m]_q^{k-1}\sum_{j=0}^{m-1}\eta_{k;q^m}\left(\frac{x+j}{m}\right)\frac{t^k}{k!}=\frac{1}{[m]_q}\sum_{j=0}^{m-1}\sum_{k=0}^{\infty}\eta_{k;q^m}\left(\frac{x+j}{m}\right)\frac{\left([m]_q t\right)^k}{k!}$$

$$=-t\sum_{j=0}^{m-1}\sum_{s=0}^{\infty}q^{ms+j+x}\,e^{[ms+j+x]_q t}$$

$$=-t\sum_{\ell=0}^{\infty}q^{\ell+x}\,e^{[\ell+x]_q t}=\sum_{k=0}^{\infty}\eta_{k;q}(x)\,\frac{t^k}{k!}.$$

It follows from (5) and (8) that

$$G_{x+y;q}(t)=e^{[y]_q t}\,G_{x+y;q}\left(q^y t\right), \tag{26}$$

which yields

$$\eta_{k;q}(x+y)=\sum_{j=0}^{k}\binom{k}{j}\eta_{j;q}(x)\,q^{yj}\,[y]_q^{k-j}\quad (k\in\mathbb{N}_0). \tag{27}$$

If we replace q and x by $\frac{1}{q}$ and $1-x$, respectively, in (18) and use 6.1(35), we get

$$\eta_{k;q^{-1}}(1-x)=(-1)^k\,q^{k-1}\,\eta_{k;q}(x)\quad (k\in\mathbb{N}), \tag{28}$$

the special case $x=1$ of which, in view of (12), yields

$$\eta_{k;q^{-1}}(0)=(-1)^k\,q^{k-1}\,\eta_{k;q}(1)=(-1)^k\,q^{k-1}\,\eta_{k;q}\quad (k\in\mathbb{N}\setminus\{1\}). \tag{29}$$

q-Stirling Numbers of the Second Kind

From the initial condition $\Delta^0 f(x)=f(x)$ and the recurrence relation

$$\Delta^{n+1}f(x)=\Delta^n f(x+1)-q^n\,\Delta^n f(x)\quad (n\in\mathbb{N}_0), \tag{30}$$

one may define (see [593], [869, Chapter 1] and [1209]) a sequence $\{\Delta^n\}_{n=0}^{\infty}$ of q-difference operators in which the index identifies the position of the pertinent operator within the sequence but is not directly interpreted as an operator exponent. It is easy to prove that

$$\Delta^n f(x)=\sum_{j=0}^{n}(-1)^j\,q^{\frac{1}{2}j(j-1)}\begin{bmatrix}n\\ j\end{bmatrix}_q f(x+n-j), \tag{31}$$

where $\begin{bmatrix}n\\ j\end{bmatrix}_q$ is defined by 6.1(29).

If $f(x)$ is a polynomial in q^x of degree $\leqq n$, it is obvious that we may put

$$f(x) = \sum_{j=0}^{n} \alpha_j \, [x]_{q;j}, \tag{32}$$

where $[x]_{q;j}$ is defined by 6.1(30) and α_j is independent of x. To determine the coefficients α_j, by making use of the easily proved formula

$$\Delta^n [x]_{q;j} = [j]_{q;n} \, [x]_{q;j-n} \, q^{n(x-j+n)}, \tag{33}$$

apply Δ^n to both members of (32) and put $x = 0$ to finally get

$$\alpha_j = \frac{\Delta^j f(0)}{[j]_q!}. \tag{34}$$

q-Stirling numbers of the second kind $S_q(n,j)$ are defined by taking $f(x) = [x]_q^n$ in (32) and writing, for convenience, as

$$[x]_q^n = \sum_{j=0}^{n} q^{\frac{1}{2}j(j-1)} \, S_q(n,j) \, [x]_{q;j}, \tag{35}$$

where $S_q(n,j)$ is independent of x. It follows from (35) that

$$\begin{aligned} [x]_q^{n+1} &= \sum q^{\frac{1}{2}j(j-1)} \, S_q(n,j) \, [x]_{q;j} \left([x-j]_q \, q^j + [j]_q \right) \\ &= \sum q^{\frac{1}{2}j(j-1)} \left(S_q(n,j-1) + [j]_q \, S_q(n,j) \right) [x]_{q;j}, \end{aligned}$$

which yields a recursion formula

$$S_q(n+1,j) = S_q(n,j-1) + [j]_q \, S_q(n,j). \tag{36}$$

It follows immediately from (36) that $S_q(n,j)$ is a polynomial in q with integral coefficients.

We use (32) and (31) to express $S_q(n,j)$ explicitly:

$$S_q(n,j) = \frac{q^{-\frac{1}{2}j(j-1)}}{[j]_q!} \sum_{\ell=0}^{j} (-1)^\ell \, q^{\frac{1}{2}\ell(\ell-1)} \begin{bmatrix} j \\ \ell \end{bmatrix}_q [j-\ell]_q^n \quad (n, j \in \mathbb{N}_0), \tag{37}$$

the right side of which vanishes for $n < j$ and is equivalently written as

$$S_q(n,j) = \frac{1}{[j]_q!} \sum_{\ell=0}^{j} (-1)^{j-\ell} \, q^{\frac{1}{2}\ell(\ell+1-2j)} \begin{bmatrix} j \\ \ell \end{bmatrix}_q [\ell]_q^n \quad (n, j \in \mathbb{N}_0). \tag{38}$$

Note that, upon taking the limit $q \to 1$ on (38), $S_q(n,j)$ reduces to the familiar Stirling numbers of the second kind (see 1.6(21)).

A formula of a different sort may also be introduced:

$$\begin{bmatrix} n \\ j \end{bmatrix}_q = \sum_{\ell=j}^{n} \binom{n}{\ell} (q-1)^{\ell-j} S_q(\ell,j), \tag{39}$$

or, equivalently,

$$(q-1)^{n-j} S_q(n,j) = \sum_{\ell=j}^{n} (-1)^{n-\ell} \binom{n}{\ell} \begin{bmatrix} \ell \\ j \end{bmatrix}_q. \tag{40}$$

Note that (38) and (40) are, in fact, equivalent.

We define a slightly-generalized form of $S_q(n,j)$ by means of

$$[x+y]_q^n = \sum_{j=0}^{n} q^{\frac{1}{2}(y+j)(y+j-1)} S_q(n,j)(y)\,[x]_{q;j}, \tag{41}$$

so that $S_q(n,j)(0) = S_q(n,j)$. As in getting (36) and (37), we obtain a recursion formula

$$S_q(n+1,j)(y) = S_q(n,j-1)(y) + [y+j]_q\, S_q(n,j)(y), \tag{42}$$

which shows that $S_q(n,j)(y)$ is a polynomial in q^y; an explicit formula

$$q^{\frac{1}{2}(y+j)(y+j-1)} S_q(n,j)(y) = \frac{1}{[j]_q!} \sum_{\ell=0}^{j} (-1)^{\ell} q^{\frac{1}{2}\ell(\ell-1)} \begin{bmatrix} j \\ \ell \end{bmatrix}_q [y+j-\ell]_q^n. \tag{43}$$

It is easily seen that

$$q^{\frac{1}{2}y(y-1+2j)} S_q(n,j)(y) = \sum_{\ell=j}^{n} \binom{n}{\ell} [y]_q^{n-\ell} q^{\ell y} S_q(\ell,j). \tag{44}$$

It follows from either (43) or (44) that

$$q^{\frac{1}{2}y(y-1)} \sum_{\ell=j}^{n} \binom{n}{\ell} (q-1)^{\ell-j} S_q(\ell,j)(y) = q^{(n-j)y} \begin{bmatrix} n \\ j \end{bmatrix}_q. \tag{45}$$

The Polynomial $\beta_k(x) = \beta_{k;q}(x)$

Let $\{a_j\}_{j=0}^{\infty}$ be an arbitrary sequence of numbers and define

$$f_n(x) = \sum_{j=0}^{n} \binom{n}{j} a_j\,[x]_q^{n-j} q^{jx}. \tag{46}$$

Then, $f_n(0) = a_n$ and

$$f_n(x+y) = \sum_{j=0}^{n} \binom{n}{j} f_j(y)\, [x]_q^{n-j}\, q^{jx}, \tag{47}$$

the special case $y = 0$ of which reduces to (46).

The polynomial $\eta_n(x) = \eta_{n;q}(x)$ and the number $\eta_n = \eta_{n;q}(0)$ do not remain finite when $q = 1$. Carlitz [215], therefore, introduces a polynomial $\beta_n(x) = \beta_{n;q}(x)$ and number $\beta_n = \beta_{n;q}$ that will approach a finite limit for $q = 1$. We, first, define

$$\beta_n = \eta_n + (q-1)\,\eta_{n+1} \quad (n \in \mathbb{N}_0). \tag{48}$$

Repeated application of (48) leads to

$$\sum_{j=0}^{k} \binom{k}{j} (q-1)^j\, \eta_{n-k+j} = \sum_{j=0}^{k-1} \binom{k-1}{j} (q-1)^j\, \beta_{n-k+j},$$

which, for $n = k$, becomes

$$\sum_{j=0}^{n} \binom{n}{j} (q-1)^j\, \eta_j = \sum_{j=0}^{n-1} \binom{n-1}{j} (q-1)^j\, \beta_j,$$

which, upon comparing with (23) and (22), leads at once to

$$\sum_{j=0}^{n} \binom{n}{j} (q-1)^j\, \beta_j = \frac{n+1}{[n+1]_q} \tag{49}$$

and

$$(q-1)^n\, \beta_n = \sum_{j=0}^{n} (-1)^{n-j} \binom{n}{j} \frac{j+1}{[j+1]_q}. \tag{50}$$

Carlitz [215] defines

$$(q-1)^n\, \beta_n(x) = \sum_{j=0}^{n} (-1)^{n-j} \binom{n}{j} \frac{j+1}{[j+1]_q}\, q^{jx}, \tag{51}$$

which implies (4). It follows from (50) and (51) that

$$\beta_n(x) = \sum_{j=0}^{n} \binom{n}{j} \beta_j\, q^{jx}\, [x]_q^{n-j} := \left(q^x \beta + [x]_q\right)^n, \tag{52}$$

which, in view of (47), generalizes the following result:

$$\beta_n(x+y) = \sum_{j=0}^{n} \binom{n}{j} \beta_j(y)\, q^{jx}\, [x]_q^{n-j} := \left(q^x \beta(y) + [x]_q\right)^n. \tag{53}$$

Analogous to (16), we find that

$$q^{x+1}\, \beta_n(x+1) - q^x \beta_n(x) = n\, q^x\, [x]_q^{n-1} + (q-1)\,(n+1)\, q^x\, [x]_q^n,$$

so that

$$q\, \beta_n(x+1) - \beta_n(x) = n\, q^x\, [x]_q^{n-1} + (q-1)\, [x]_q^n, \tag{54}$$

which, for $x = 0$, becomes

$$q(q\beta+1)^n - \beta_n = \begin{cases} 1 & (n=1), \\ 0 & (n \in \mathbb{N} \setminus \{1\}). \end{cases} \tag{55}$$

Analogous to (25) and (28), we have

$$[k]_q^{n-1} \sum_{j=0}^{k-1} q^j\, \beta_{n;q^k}\left(x + \frac{j}{k}\right) = \beta_{n;q}(kx) \quad (k \in \mathbb{N}) \tag{56}$$

and

$$\beta_{n;q^{-1}}(1-x) = (-q)^n\, \beta_{n;q}(x), \tag{57}$$

which, for $x = 1$, implies

$$\beta_{n;q^{-1}}(0) = (-q)^n\, \beta_{n;q}(1) = (-1)^n\, q^{n-1}\, \beta_n \quad (n \in \mathbb{N} \setminus \{1\}). \tag{58}$$

It follows from (35) that

$$\sum_{\ell=0}^{k-1} q^\ell\, [\ell]_q^n = \sum_{j=0}^{n} q^{\frac{1}{2}j(j+1)}\, S_q(n,j)\, \frac{[k]_{q;j+1}}{[j+1]_q}. \tag{59}$$

Conversely, in view of (16), the left member of (59) is

$$\begin{aligned} \frac{1}{n+1}\left(\eta_{n+1}(k) - \eta_{n+1}\right) &= \frac{1}{n+1} \sum_{j=1}^{n+1} \binom{n+1}{j} [k]_q^j\, q^{(n+1-j)k}\, \eta_{n+1-j} \\ &\quad + \left(q^{(n+1)k} - 1\right) \frac{\eta_{n+1}}{n+1} \\ &= \sum_{j=0}^{n} \binom{n}{j} [k]_q^{j+1}\, q^{(n-j)k}\, \frac{\eta_{n-j}}{j+1} + \left(q^{(n+1)k} - 1\right) \frac{\eta_{n+1}}{n+1}. \end{aligned}$$

Comparison with (59) yields an identity in k. Considering

$$[k]_{q;j+1} = [k]_q\,[k-1]_{q;j} \quad \text{and} \quad \lim_{k\to 0} \frac{q^{(n+1)k}-1}{[k]_q} = (n+1)(q-1),$$

to divide both members of this identity by $[k]_q$ and then let $k \to 0$, we obtain

$$\eta_n + (q-1)\eta_{n+1} = \sum_{j=0}^{n} q^{\frac{1}{2}j(j+1)}\, S_q(n,j)\, \frac{[-1]_{q;j}}{[j+1]_q},$$

which, in view of (48), yields

$$\beta_n = \sum_{j=0}^{n} (-1)^j\, S_q(n,j)\, \frac{[j]_q!}{[j+1]_q}. \tag{60}$$

It is obvious from (60) and (37) that β_n remains finite for $q = 1$. Indeed,

$$\lim_{q\to 1} \beta_n = \sum_{j=0}^{n} \frac{1}{j+1} \sum_{\ell=0}^{j} (-1)^{j-\ell} \binom{j}{\ell} (j-\ell)^n = B_n, \tag{61}$$

where B_n is the n-th Bernoulli number in Nörlund's notation (see 1.5(30)).

Substitution from (37) in (60) leads to the explicit formula

$$\beta_n = \sum_{j=0}^{n} \frac{1}{[j+1]_q} \sum_{\ell=0}^{j} (-1)^{\ell}\, q^{\frac{1}{2}\ell(\ell+1-2j)} \begin{bmatrix} j \\ \ell \end{bmatrix}_q [\ell]_q^n, \tag{62}$$

which can, upon using (52), be more generally expressed as

$$\beta_n(x) = \sum_{j=0}^{n} \frac{1}{[j+1]_q} \sum_{\ell=0}^{j} (-1)^{\ell}\, q^{\frac{1}{2}\ell(\ell+1-2j)} \begin{bmatrix} j \\ \ell \end{bmatrix}_q [x+\ell]_q^n. \tag{63}$$

Comparison with (63) and (43) yields a generalized formula of (60):

$$\beta_n(x) = \sum_{j=0}^{n} (-1)^j\, q^{\frac{1}{2}x(x-1+2j)}\, S_q(n,j)(x)\, \frac{[j]_q!}{[j+1]_q}. \tag{64}$$

6.7 *q*-Euler Numbers and *q*-Euler Polynomials

We begin by defining a set polynomials $\epsilon_{k;q}(x) = \epsilon_k(x)$ and a set of numbers $\epsilon_{k;q}(0) = \epsilon_{k;q} = \epsilon_k$ by means of the following generating functions given by formal series:

$$H_{x;q}(t) \equiv [2]_q \sum_{j=0}^{\infty} (-1)^j q^j e^{[j+x]_q t} := \sum_{k=0}^{\infty} \epsilon_{k;q}(x)\, \frac{t^k}{k!} \tag{1}$$

and

$$H_q(t) \equiv [2]_q \sum_{j=0}^{\infty} (-1)^j q^j e^{[j]_q t} := \sum_{k=0}^{\infty} \epsilon_{k;q} \frac{t^k}{k!}. \tag{2}$$

It is easy to see from (1) that

$$q H_{x+1;q}(t) + H_{x;q}(t) = [2]_q e^{[x]_q t}, \tag{3}$$

which immediately yields

$$q \epsilon_k(x+1) + \epsilon_k(x) = [2]_q [x]_q^k \quad (k \in \mathbb{N}_0). \tag{4}$$

It is observed that $\epsilon_k(x)$ is uniquely determined by the functional equation (4). From (4) follows a summation formula

$$(-1)^{k-1} q^k \epsilon_n(x+k) + \epsilon_n(x) = [2]_q \sum_{j=0}^{k-1} (-1)^j q^j [x+j]_q^n. \tag{5}$$

Considering 6.6(17) in (1), analogous to 6.6(18), we obtain an explicit formula for $\epsilon_{k;q}(x)$:

$$(q-1)^k \epsilon_{k;q}(x) = (q+1) \sum_{j=0}^{k} (-1)^{k-j} \binom{k}{j} \frac{q^{xj}}{1+q^{j+1}}, \tag{6}$$

or, equivalently,

$$\sum_{j=0}^{k} \binom{k}{j} (q-1)^j \epsilon_{j;q}(x) = \frac{(q+1) q^{kx}}{1+q^{k+1}}. \tag{7}$$

The special cases of (6) and (7) when $x = 0$, respectively, give

$$(q-1)^k \epsilon_{k;q} = \sum_{j=0}^{k} (-1)^{k-j} \binom{k}{j} \frac{q+1}{1+q^{j+1}} \tag{8}$$

and

$$\sum_{j=0}^{k} \binom{k}{j} (q-1)^j \epsilon_{j;q} = \frac{q+1}{1+q^{k+1}}. \tag{9}$$

Replacing x and q, respectively, by $1-x$ and q^{-1} in (6) leads to

$$\epsilon_{k;q^{-1}}(1-x) = (-1)^k q^k \epsilon_{k;q}(x). \tag{10}$$

From 6.6(8) we find that

$$H_{x+y;q}(t) = e^{[x]_q t} H_{y;q}\left(q^x t\right), \tag{11}$$

which gives

$$\epsilon_{k;q}(x+y) = \sum_{j=0}^{k} \binom{k}{j} [x]_q^{k-j} q^{jx} \epsilon_{j;q}(y) := \left(q^x \epsilon(y) + [x]_q\right)^k. \tag{12}$$

The special case of (12) when $y = 0$ becomes

$$\epsilon_{k;q}(x) = \sum_{j=0}^{k} \binom{k}{j} [x]_q^{k-j} q^{jx} \epsilon_{j;q} := \left(q^x \epsilon + [x]_q\right)^k. \tag{13}$$

Rearranging the series in (1) as even and odd terms, we see that

$$t q^x H_{x;q}(t) = G_{\frac{x+1}{2};q^2}[(q+1)t] - G_{\frac{x}{2};q^2}[(q+1)t], \tag{14}$$

which yields

$$(k+1)\, q^x \epsilon_{k;q}(x) = (q+1)^{k+1}\left\{\eta_{k+1;q^2}\left(\frac{x+1}{2}\right) - \eta_{k+1;q^2}\left(\frac{x}{2}\right)\right\} \quad (k \in \mathbb{N}_0), \tag{15}$$

serves to express $q^x \epsilon_{k;q}(x)$ in terms of η-polynomials.

Corresponding to 6.6(25), we get the multiplication formulas

$$[m]_q^k \sum_{j=0}^{m-1} (-q)^j \epsilon_{k;q^m}\left(x + \frac{j}{m}\right) = \frac{q^m+1}{q+1} \epsilon_{k;q}(mx) \quad (m \text{ is odd}) \tag{16}$$

and

$$[2]_q [m]_q^k \sum_{j=0}^{m-1} (-1)^{j+1} \eta_{k+1;q^m}\left(x + \frac{j}{m}\right) = (k+1) q^{mx} \epsilon_{k;q}(mx) \quad (m \text{ is even}). \tag{17}$$

Note that, for $m = 2$, (17) reduces to (15).

To get explicit formulas of a simpler kind, Carlitz [215] starts by noting the function (*cf.* [861, Chapter VII])

$$Y_{n;q}(x) := \sum_{k=0}^{n} (-1)^{n-k} \frac{q^{n-k}}{\prod_{j=k}^{n}\left(q^{j+1}+1\right)} \begin{bmatrix} x \\ k \end{bmatrix}_q, \tag{18}$$

which is straightforwardly proven to satisfy

$$q\, Y_{n;q}(x+1) + Y_{n;q}(x) = \begin{bmatrix} x \\ n \end{bmatrix}_q. \tag{19}$$

In view of (4) and 6.6(35), we have

$$q\,\epsilon_{n;q}(x+1)+\epsilon_{n;q}(x)=[2]_q\sum_{k=0}^{n} q^{\frac{1}{2}k(k-1)}\,S_q(n,k)\,[k]_q!\,\big(q\,Y_{k;q}(x+1)+Y_{k;q}(x)\big),$$

which yields

$$\epsilon_{n;q}(x)=[2]_q\sum_{k=0}^{n} q^{\frac{1}{2}k(k-1)}\,S_q(n,k)\,[k]_q!\,Y_{k;q}(x). \tag{20}$$

The special case of (20) when $x=0$ implies

$$\epsilon_{n;q}=\sum_{k=0}^{n}(-1)^k q^{\frac{1}{2}k(k+1)}\,\frac{S_q(n,k)\,[k]_q!}{\prod_{j=1}^{k}\left(q^{j+1}+1\right)}, \tag{21}$$

which, using 6.6(37), becomes

$$\epsilon_{n;q}=\sum_{k=0}^{n}\frac{(-1)^k q^k}{\prod_{j=1}^{k}\left(q^{j+1}+1\right)}\sum_{\ell=0}^{k}(-1)^{\ell} q^{\frac{1}{2}\ell(\ell-1)}\begin{bmatrix}k\\ \ell\end{bmatrix}_q [k-\ell]_q^n. \tag{22}$$

Conversely, if we use 6.6(41), we obtain

$$\begin{aligned}&q\,\epsilon_{n;q}(x+y+1)+\epsilon_{n;q}(x+y)\\&\quad=[2]_q\sum_{k=0}^{n} q^{\frac{1}{2}(y+k)(y+k-1)}\,S_q(n,k)(y)\,[k]_q!\,\big(q\,Y_{k;q}(x+1)+Y_{k;q}(x)\big),\end{aligned}$$

which gives

$$\epsilon_{n;q}(x+y)=[2]_q\sum_{k=0}^{n} q^{\frac{1}{2}(y+k)(y+k-1)}\,S_q(n,k)(y)\,[k]_q!\,Y_{k;q}(x). \tag{23}$$

The special case of (23) when $x=0$ becomes

$$\epsilon_{n;q}(y)=\sum_{k=0}^{n}(-1)^k q^{\frac{1}{2}(y+k)(y+k-1)+k}\,\frac{S_q(n,k)(y)\,[k]_q!}{\prod_{j=1}^{k}\left(q^{j+1}+1\right)}, \tag{24}$$

which, upon using 6.6(43), gives

$$\epsilon_{n;q}(y)=\sum_{k=0}^{n}\frac{(-1)^k q^k}{\prod_{j=1}^{k}\left(q^{j+1}+1\right)}\sum_{\ell=0}^{k}(-1)^{\ell} q^{\frac{1}{2}\ell(\ell-1)}\begin{bmatrix}k\\ \ell\end{bmatrix}_q [y+k-\ell]_q^n. \tag{25}$$

Remark 2 It is, in view of (22), observed that the product

$$\prod_{j=1}^{n}\left(q^{j+1}+1\right)\epsilon_{n;q}$$

is a polynomial in q, which, for $q=1$, reduces to the number C_n (see [869, p. 27]). Likewise, (25) indicates that the product

$$2^{-n}(q+1)^n\prod_{j=1}^{n}\left(q^{2(j+1)}+1\right)\epsilon_{n;q^2}\left(\frac{1}{2}\right)$$

is a polynomial in q, which, for $q=1$, reduces to the Euler number E_n.

6.8 The q-Apostol-Bernoulli Polynomials $\mathcal{B}_k^{(n)}(x;\lambda)$ of Order n

We begin this section by setting $a=q^n$ $(n\in\mathbb{N})$ in the q-binomial theorem 6.3(2) to obtain

$$\frac{1}{(z;q)_n}=\frac{1}{\prod_{k=0}^{n-1}\left(1-zq^k\right)}=\sum_{k=0}^{\infty}\frac{[n]_{q;k}}{[k]_q\,!}z^k. \tag{1}$$

A q-extension of the Apostol-Bernoulli polynomials $\mathcal{B}_k^{(n)}(x;\lambda;q)$ of order $n\in\mathbb{N}$ (see Section 1.7) is defined, here, by means of the following generating function:

$$\begin{aligned}G_{x;\lambda;q}^{(n)}(t)&\equiv(-t)^n\sum_{k=0}^{\infty}\frac{[n]_{q;k}}{[k]_q\,!}\lambda^k q^{k+x}e^{[k+x]_q t}\\&:=\sum_{k=0}^{\infty}\mathcal{B}_k^{(n)}(x;\lambda;q)\frac{t^k}{k!},\end{aligned} \tag{2}$$

and a q-extension of the Apostol-Bernoulli numbers $\mathcal{B}_k^{(n)}(\lambda;q)$ of order $n\in\mathbb{N}$ is defined, here, by means of the following generating function:

$$\begin{aligned}G_{\lambda;q}^{(n)}(t)&\equiv(-t)^n\sum_{k=0}^{\infty}\frac{[n]_{q;k}}{[k]_q\,!}\lambda^k q^k e^{[k]_q t}\\&:=\sum_{k=0}^{\infty}\mathcal{B}_k^{(n)}(\lambda;q)\frac{t^k}{k!}.\end{aligned} \tag{3}$$

Remark 3 Here, for convenience, only the case of order $\alpha = n \in \mathbb{N}$ is considered. It is easy to see that, when $q \to 1$, the generating functions of $\mathcal{B}_k^{(n)}(x;\lambda;q)$ and $\mathcal{B}_k^{(n)}(\lambda;q)$ in (2) and (3) would tend, respectively, to those of $\mathcal{B}_k^{(n)}(x;\lambda)$ and $\mathcal{B}_k^{(n)}(\lambda)$ given by 1.8(13) and 1.8(14). It is also observed that

$$\mathcal{B}_k^{(n)}(0;\lambda;q) = \mathcal{B}_k^{(n)}(\lambda;q) \quad (k \in \mathbb{N}_0) \tag{4}$$

and

$$\mathcal{B}_k^{(1)}(x;\lambda;q) = \mathcal{B}_k(x;\lambda;q) \quad \text{and} \quad \mathcal{B}_k^{(1)}(\lambda;q) = \mathcal{B}_k(\lambda;q), \tag{5}$$

which were, very recently, introduced and investigated by Cenki and Can [227, Eqs. (4) and (6)]. We note that

$$\mathcal{B}_k^{(1)}(1;q) = \mathcal{B}_k(q)$$

are the familiar q-Bernoulli numbers that were considered by many authors (see, e.g., [215], [228], [649], [659], [660], [682] and [1100]).

A q-extension of the Apostol-Euler polynomials $\mathcal{E}_k^{(n)}(x;\lambda;q)$ of order $n \in \mathbb{N}$ is defined, here, by means of the following generating function:

$$\begin{aligned} H_{x;\lambda;q}^{(n)}(t) &\equiv 2^n \sum_{k=0}^{\infty} \frac{[n]_{q;k}}{[k]_q!} (-\lambda)^k q^{k+x} e^{[k+x]_q t} \\ &:= \sum_{k=0}^{\infty} \mathcal{E}_k^{(n)}(x;\lambda;q) \frac{t^k}{k!}, \end{aligned} \tag{6}$$

and a q-extension of the Apostol-Euler numbers $\mathcal{E}_k^{(n)}(\lambda;q)$ of order n is defined here by means of the following generating function:

$$\begin{aligned} H_{\lambda;q}^{(n)}(t) &\equiv 2^n \sum_{k=0}^{\infty} \frac{[n]_{q;k}}{[k]_q!} (-\lambda)^k q^k e^{[k]_q t} \\ &:= \sum_{k=0}^{\infty} \mathcal{E}_k^{(n)}(\lambda;q) \frac{t^k}{k!}. \end{aligned} \tag{7}$$

Remark 4 It is easy to see that, when $q \to 1$, the generating functions of $\mathcal{E}_k^{(n)}(x;\lambda;q)$ and $\mathcal{E}_k^{(n)}(\lambda;q)$ in (6) and (7) would tend, respectively, to those of $\mathcal{E}_k^{(n)}(x;\lambda)$ and $\mathcal{E}_k^{(n)}(\lambda)$ given by 1.8(15) and 1.8(16). It is also observed that

$$\mathcal{E}_k^{(n)}(0;\lambda;q) = \mathcal{E}_k^{(n)}(\lambda;q) \quad (k \in \mathbb{N}_0). \tag{8}$$

We, first, examine (2) and (3) to get the following relationship between these two generating functions:

$$G^{(n)}_{x;\lambda;q}(t) = G^{(n)}_{\lambda;q}\left(q^{x}\,t\right)\,e^{[x]_q t}\,q^{(1-n)x}, \tag{9}$$

which, upon considering (2) and (3) again, yields

$$\mathcal{B}^{(n)}_{k}(x;\lambda;q) = \sum_{j=0}^{k}\binom{k}{j}\mathcal{B}^{(n)}_{j}(\lambda;q)\,[x]_q^{k-j}\,q^{(j+1-n)x} \quad (k \in \mathbb{N}_0). \tag{10}$$

Next, by considering the following identity:

$$e^{[k]_q t} = e^{\frac{t}{1-q}}\,e^{-\frac{q^k}{1-q}t},$$

we obtain

$$\begin{aligned} G^{(n)}_{\lambda;q}(t) &= (-t)^n\,e^{\frac{t}{1-q}}\sum_{k=0}^{\infty}\frac{[n]_{q;k}\,\lambda^k\,q^k}{[k]_q\,!}\sum_{j=0}^{\infty}\frac{(-1)^j\,q^{jk}}{j!\,(1-q)^j}\,t^j \\ &= (-t)^n\,e^{\frac{t}{1-q}}\sum_{j=0}^{\infty}\frac{(-1)^j\,t^j}{j!\,(1-q)^j}\sum_{k=0}^{\infty}\frac{[n]_{q;k}}{[k]_q\,!}\left(\lambda\,q^{j+1}\right)^k. \end{aligned}$$

In view of (1), the last sum is equal to $1/\left(\lambda\,q^{j+1};q\right)_n$, and we, thus, find that

$$\sum_{k=0}^{\infty}\mathcal{B}^{(n)}_{k}(\lambda;q)\,\frac{t^k}{k!} = (-t)^n\sum_{k=0}^{\infty}\frac{1}{(1-q)^k}\sum_{j=0}^{k}\frac{(-1)^j}{\left(\lambda\,q^{j+1};q\right)_n}\,\frac{t^k}{j!\,(k-j)!},$$

which, upon equating the coefficients of t^{n+k} on both sides, yields the following explicit expression for $\mathcal{B}^{(n)}_{k}(\lambda;q)$:

$$\mathcal{B}^{(n)}_{n+k}(\lambda;q) = (-1)^n\,\frac{(k+1)_n}{(1-q)^k}\sum_{j=0}^{k}\binom{k}{j}\frac{(-1)^j}{\left(\lambda\,q^{j+1};q\right)_n} \quad (k \in \mathbb{N}_0;\ n \in \mathbb{N}). \tag{11}$$

In light of (5), a special case of (11) when $n = 1$ reduces to the following result:

$$\mathcal{B}_k(\lambda;q) = \frac{1}{(1-q)^{k-1}}\sum_{j=0}^{k}\binom{k}{j}\frac{(-1)^j\,j}{1-\lambda\,q^j} \quad (k \in \mathbb{N}_0), \tag{12}$$

which provides a corrected version of the corresponding formula given in [227, p. 215].

We, now, give the following q-difference equation for $G_{\lambda;q}^{(n)}(t)$:

$$\lambda\, e^t\, q^{1-n}\, G_{\lambda;q}^{(n)}(qt) = G_{\lambda;q}^{(n)}(t) + t\, G_{q\lambda;q}^{(n-1)}(t) \quad (n \in \mathbb{N}), \tag{13}$$

where

$$G_{q\lambda;q}^{(0)}(t) := 1.$$

Indeed, we have

$$\lambda\, e^t\, q^{1-n}\, G_{\lambda;q}^{(n)}(qt) = (-t)^n \sum_{k=0}^{\infty} \frac{[k]_q}{[n+k-1]_q} \frac{[n]_{q;k}}{[k]_q\,!} \lambda^k\, q^k\, e^{[k]_q t},$$

which, upon considering the following relationship:

$$\frac{[k]_q}{[n+k-1]_q} = 1 - \frac{[n-1]_q}{[n+k-1]_q} q^k,$$

leads us to (13).

The special case of (13) when $n = 1$ reduces immediately to a known result [227, p. 216, Eq. (5)].

From (3) and (13) it is easy to derive the following q-difference equation for $\mathcal{B}_k^{(n)}(\lambda; q)$:

$$\lambda\, q^{1-n} \sum_{j=0}^{k+1} \binom{k+1}{j} q^j\, \mathcal{B}_j^{(n)}(\lambda; q) = \mathcal{B}_{k+1}^{(n)}(\lambda; q) + (k+1)\, \mathcal{B}_k^{(n-1)}(q\lambda; q) \tag{14}$$

$$(k \in \mathbb{N}_0;\ n \in \mathbb{N}).$$

Remark 5 Since the q-difference equation for $G_{\lambda;q}^{(n)}(t)$ in (13) holds true, if and only if the q-difference equation for $\mathcal{B}_k^{(n)}(\lambda; q)$ in (14) holds true and also since the coefficients of a power series are uniquely determined, we conclude that the generating function $G_{\lambda;q}^{(n)}(t)$ of the q-extension of the Apostol-Bernoulli numbers $\mathcal{B}_k^{(n)}(\lambda; q)$ of order n can be determined as a solution of the q-difference equation for $G_{\lambda;q}^{(n)}(t)$ in (13).

Remark 6 The following q-distribution relation holds true for $\mathcal{B}_k^{(n)}(x; \lambda; q)$:

$$[m]_q^{k-1} \sum_{j=0}^{m-1} \lambda^j\, \mathcal{B}_k^{(n)}\left(\frac{x+j}{m}; \lambda^m; q^m\right) = \mathcal{B}_k^{(n)}(x; \lambda; q) \quad (k \in \mathbb{N}_0;\ m, n \in \mathbb{N}). \tag{15}$$

The special case of (15) when $n = 1$ was proven by Cenki and Can [227, p. 216, Lemma 5]. Here we give an equivalence statement that (15) holds true for $n \in \mathbb{N} \setminus \{1\}$. Indeed, we have

$$
\begin{aligned}
I_{x;\lambda;q}^{(n;m)}(t) &\equiv \sum_{k=0}^{\infty} [m]_q^{k-1} \sum_{j=0}^{m-1} \lambda^j \mathcal{B}_k^{(n)} \left(\frac{x+j}{m}; \lambda^m; q^m \right) \frac{t^k}{k!} \\
&= \frac{1}{[m]_q} \sum_{j=0}^{m-1} \lambda^j \sum_{k=0}^{\infty} \mathcal{B}_k^{(n)} \left(\frac{x+j}{m}; \lambda^m; q^m \right) \frac{([m]_q t)^k}{k!} \\
&= \frac{1}{[m]_q} \sum_{j=0}^{m-1} \lambda^j G_{\frac{x+j}{m};\lambda^m;q^m}^{(n)} ([m]_q t).
\end{aligned}
$$

By applying (9) and the identity 6.5(8) to the last equation, we get

$$
I_{x;\lambda;q}^{(n;m)}(t) = \frac{q^{(1-n)x} e^{[x]_q t}}{[m]_q} \sum_{j=0}^{m-1} \lambda^j q^{(1-n)j} G_{\lambda^m;q^m}^{(n)} \left(q^{x+j} [m]_q t \right) e^{[j]_q q^x t}. \tag{16}
$$

Moreover, in view of (2) and (9), we have

$$
\sum_{k=0}^{\infty} \mathcal{B}_k^{(n)} (x; \lambda; q) \frac{t^k}{k!} = G_{\lambda;q}^{(n)} \left(q^x t \right) e^{[x]_q t} q^{(1-n)x}. \tag{17}
$$

Now, it is easily seen from (16) and (17) that (15) holds true, if and only if the following result holds true:

$$
G_{\lambda;q}^{(n)}(t) = \frac{1}{[m]_q} \sum_{j=0}^{m-1} \lambda^j q^{(1-n)j} G_{\lambda^m;q^m}^{(n)} \left(q^j [m]_q t \right) e^{[j]_q t} \quad (m, n \in \mathbb{N}). \tag{18}
$$

Let $P_{\lambda;q}^{(n)}(t)$ be the right-hand side of (18). In view of Remark 5, it suffices to show that $P_{\lambda;q}^{(n)}(t)$ satisfies the following q-difference equation:

$$
\lambda e^t q^{1-n} P_{\lambda;q}^{(n)}(qt) = P_{\lambda;q}^{(n)}(t) + t P_{q\lambda;q}^{(n-1)}(t) \quad (n \in \mathbb{N} \setminus \{1\}). \tag{19}
$$

Indeed, if we begin with

$$
\begin{aligned}
\lambda e^t q^{1-n} P_{\lambda;q}^{(n)}(qt) &= \frac{1}{[m]_q} \sum_{j=1}^{m} \lambda^j q^{(1-n)j} G_{\lambda^m;q^m}^{(n)} \left(q^j [m]_q t \right) e^{[j]_q t} \\
&= P_{\lambda;q}^{(n)}(t) + \frac{1}{[m]_q} \Big(\lambda^m q^{(1-n)m} G_{\lambda^m;q^m}^{(n)} \left(q^m [m]_q t \right) e^{[m]_q t} \\
&\qquad - G_{\lambda^m;q^m}^{(n)} \left([m]_q t \right) \Big)
\end{aligned}
$$

and apply (13) to the last equation, we obtain

$$\lambda e^t q^{1-n} P_{\lambda;q}^{(n)}(qt) = P_{\lambda;q}^{(n)}(t) + t\, G_{(q\lambda)^m;q^m}^{(n-1)}\left([m]_q t\right). \tag{20}$$

Now, it is seen from (19) and (20) that an equivalence condition for the validity of the q-distribution relation for $\mathcal{B}_k^{(n)}(x;\lambda;q)$ in (15) can be written as in Remark 7 below.

Remark 7 The q-difference equation (15) holds true, if and only if

$$P_{q\lambda;q}^{(n)}(t) = G_{(q\lambda)^m;q^m}^{(n)}\left([m]_q t\right) \quad (m, n \in \mathbb{N}),$$

that is, if and only if the following result holds true for $n, m \in \mathbb{N}$:

$$\left([m]_q - 1\right) G_{(q\lambda)^m;q^m}^{(n)}\left([m]_q t\right) = \sum_{j=1}^{m-1} (q\lambda)^j q^{(1-n)j} G_{(q\lambda)^m;q^m}^{(n)}\left(q^j [m]_q t\right) e^{[j]_q t}, \tag{21}$$

where an empty sum is understood (as usual) to be nil.

6.9 The q-Apostol-Euler Polynomials $\mathcal{E}_k^{(n)}(x;\lambda)$ of Order n

By applying the methodology and techniques used above in getting some identities for the generating functions of the q-extensions of the Apostol-Bernoulli polynomials and numbers, we can derive the following corresponding identities involving the generating functions of the q-extensions of the Apostol-Euler polynomials and numbers:

$$H_{x;\lambda;q}^{(n)}(t) = H_{\lambda;q}^{(n)}\left(q^x t\right) e^{[x]_q t} q^x. \tag{1}$$

$$\mathcal{E}_k^{(n)}(x;\lambda;q) = \sum_{j=0}^{k} \binom{k}{j} \mathcal{E}_j^{(n)}(\lambda;q)\, [x]_q^{k-j} q^{(j+1)x} \quad (k \in \mathbb{N}_0;\ n \in \mathbb{N}). \tag{2}$$

$$\mathcal{E}_k^{(n)}(\lambda;q) = \frac{1}{(1-q)^k} \sum_{j=0}^{k} \binom{k}{j} \frac{(-1)^j}{\left(-\lambda q^{j+1};q\right)_n} \quad (k \in \mathbb{N}_0;\ n \in \mathbb{N}). \tag{3}$$

The q-difference equation for $H_{\lambda;q}^{(n)}(t)$ is given by

$$H_{\lambda;q}^{(n)}(t) - 2H_{q\lambda;q}^{(n-1)}(t) = -\lambda e^t q H_{\lambda;q}^{(n)}(qt) \quad (n \in \mathbb{N}), \tag{4}$$

where

$$H_{q\lambda;q}^{(0)}(t) := 1.$$

The q-difference equation for $\mathcal{E}_k^{(n)}(\lambda; q)$ is given by

$$\mathcal{E}_k^{(n)}(\lambda; q) - 2\,\mathcal{E}_k^{(n-1)}(q\lambda; q) = -\lambda \sum_{j=0}^{k} \binom{k}{j} \mathcal{E}_j^{(n)}(\lambda; q)\, q^{j+1} \quad (k \in \mathbb{N}_0;\ n \in \mathbb{N}). \tag{5}$$

The following relationship holds true between $\mathcal{B}_k^{(n)}(x; \lambda; q)$ and $\mathcal{E}_k^{(n)}(x; \lambda; q)$:

$$(-2)^n\, \mathcal{B}_{n+k}^{(n)}(x; -\lambda; q) = (k+1)_n\, \mathcal{E}_k^{(n)}(x; \lambda; q) \quad (k \in \mathbb{N}_0;\ n \in \mathbb{N}), \tag{6}$$

which follows immediately, by using the following generating-function relationship:

$$G_{x;-\lambda;q}^{(n)}(t) = \left(-\frac{t}{2}\right)^n H_{x;\lambda;q}^{(n)}(t) \quad (n \in \mathbb{N}). \tag{7}$$

Remark 8 We *conjecture* that the following q-distribution relation holds true for $\mathcal{E}_k^{(n)}(x; \lambda; q)$:

$$[m]_q^{k-1} \sum_{j=0}^{m-1} (-\lambda)^j\, \mathcal{E}_k^{(n)}\left(\frac{x+j}{m}; \lambda^m; q^m\right) = \mathcal{E}_k^{(n)}(x; \lambda; q) \quad (k \in \mathbb{N}_0;\ m, n \in \mathbb{N}). \tag{8}$$

When $n = 1$, (8) is easily verified. The other cases of our *conjectured* relationship (8) when $n \in \mathbb{N} \setminus \{1\}$ remain to be proven.

6.10 A Generalized q-Zeta Function

Many authors have tried to give q-analogues of the Riemann Zeta function $\zeta(s)$, defined by 2.3(1), and its related functions (see, e.g., [227, 255, 363, 493, 626, 652, 1006, 1100, 1168–1170] and [1174]). Here, we give a q-analogue of the generalized zeta function $\zeta(s, a)$, defined by 2.2(1), among other things, by just following the method of Kaneko *et al.* [626], who mainly used Euler-Maclaurin summation formula (see 2.7(21) or 2.7(22)) to present and investigate a q-analogue of the Riemann zeta function $\zeta(s)$, defined by 2.3(1), and gave a good and reasonable explanation that their q-analogue may be a best choice. They [626] also commented that q-analogue of $\zeta(s, a)$ can be achieved by modifying their method.

An Auxiliary Function Defining Generalized q-Zeta Function

We begin this section by presenting a function $g_q(a; s, t)$, defined by

$$g_q(a; s, t) := \sum_{n=0}^{\infty} \frac{q^{(n+a)t}}{[n+a]_q^s} \quad (0 < q < 1;\ 0 < a \leqq 1). \tag{1}$$

Remark 1 It is easy to see that, by the ratio test, the series in (1) converges absolutely for $\Re(t) > 0$ and any $s \in \mathbb{C}$. The series in (1) also converges to the generalized Zeta function $\zeta(s,a)$, defined by 2.2(1), as $q \uparrow 1$, if $\Re(s) > 1$ and $\Re(t) > 0$. So, we can define $g_q(a; s, t)$ in (1) as a q-analogue of the generalized Zeta function $\zeta(s,a)$. However, Kaneko *et al.* [626] took several examples to contend that $t = s-1$ in $g_q(1; s, t)$ seems the best possible choice to define a q-analogue of the Riemann Zeta function $\zeta(s)$, defined by 2.3(1).

Theorem 6.12 *For* $0<q<1$, *as a function of* $(s,t) \in \mathbb{C}^2$, $g_q(a; s, t)$ *in* (1) *is continued meromorphically, by means of the series expansion:*

$$g_q(a; s, t) = (1-q)^s \sum_{r=0}^{\infty} \binom{s+r-1}{r} \frac{q^{a(t+r)}}{1-q^{t+r}} \tag{2}$$

with simple poles at

$$t = -r + \frac{2\ell\pi}{\log q} \quad (r \in \mathbb{N}_0;\ \ell \in \mathbb{Z}). \tag{3}$$

Proof. It follows immediately from the binomial theorem that

$$\begin{aligned} g_q(a; s, t) &= (1-q)^s \sum_{n=0}^{\infty} q^{(n+a)t} \left(1 - q^{n+a}\right)^{-s} \\ &= (1-q)^s \sum_{n=0}^{\infty} q^{(n+a)t} \sum_{r=0}^{\infty} \binom{s+r-1}{r} q^{(n+a)r} \\ &= (1-q)^s \sum_{r=0}^{\infty} \binom{s+r-1}{r} \sum_{n=0}^{\infty} q^{(n+a)(t+r)} \\ &= (1-q)^s \sum_{r=0}^{\infty} \binom{s+r-1}{r} \frac{q^{a(t+r)}}{1-q^{t+r}}. \end{aligned}$$

■

In view of Remark 1, we also choose to define a q-analogue of the generalized Zeta function $\zeta(s,a)$ in 2.2(1) by

$$\zeta_q(s,a) := g_q(a; s, s-1) := \sum_{n=0}^{\infty} \frac{q^{(n+a)(s-1)}}{[n+a]_q^s} \quad (0<q<1;\ 0<a \leqq 1). \tag{4}$$

We list, here, some properties that easily follow from Theorem 6.12 as

Theorem 6.13

(a) *The function* $\zeta_q(s,a)$ *has simple poles at points in* $1 + 2\pi i\mathbb{Z}/\log q$ *and in the set*

$$\left\{ \ell + \frac{2\pi i m}{\log q} \,\middle|\, \ell \in \mathbb{Z}_{\leqq 0},\ m \in \mathbb{Z} \setminus \{0\} \right\},$$

where $\mathbb{Z}_{\leqq 0} := \mathbb{Z} \setminus \mathbb{N}$. *In particular,* $s = 1$ *is a simple pole of* $\zeta_q(s,a)$ *with its residue* $(q-1)/\log q$.

(b) *For* $m \in \mathbb{N}_0$, *the limiting value*

$$\lim_{s \to -m} \zeta_q(s,a) =: \zeta_q(-m,a)$$

exists and is given explicitly by

$$\zeta_q(-m,a) = (1-q)^{-m} \left\{ \sum_{r=0}^{m} (-1)^r \binom{m}{r} \frac{q^{(1-a)(m+1-r)}}{q^{m+1-r}-1} + \frac{(-1)^{m+1}}{(m+1)\log q} \right\}. \tag{5}$$

Now, we have

Theorem 6.14

$$\lim_{q \uparrow 1} \zeta_q(-m,a) = -\frac{B_{m+1}(a)}{m+1} \quad (m \in \mathbb{N}_0), \tag{6}$$

where $B_{m+1}(a)$ *denotes the Bernoulli polynomials (see Section* 1.7*).*

Proof. In view of (5), it is equivalent to show that

$$\lim_{q \uparrow 1} (1-q)^{-m} \left\{ \sum_{r=0}^{m} (-1)^r \binom{m}{r} \frac{q^{(1-a)(m+1-r)}}{q^{m+1-r}-1} + \frac{(-1)^{m+1}}{(m+1)\log q} \right\} = -\frac{B_{m+1}(a)}{m+1},$$

which, upon multiplying $(-1)^{m+1}(m+1)$ and setting $m+1-r = r'$ and then dropping the prime on r in the resulting equation, yields

$$\lim_{q \uparrow 1} (1-q)^{-m} \left\{ (m+1) \sum_{r=1}^{m+1} (-1)^r \binom{m}{r-1} \frac{q^{(1-a)r}}{q^r-1} + \frac{1}{\log q} \right\} = (-1)^m B_{m+1}(a). \tag{7}$$

Using 1.6(1)

$$\begin{aligned} \frac{q^{(1-a)r}}{q^r-1} &= \frac{e^{(1-a)r\log q}}{e^{r\log q}-1} = \frac{1}{r} \frac{r\log q\, e^{(1-a)r\log q}}{e^{r\log q}-1} \frac{1}{\log q} \\ &= \frac{1}{r} \sum_{k=0}^{\infty} B_k(1-a) \frac{(r\log q)^k}{k!} \frac{1}{\log q}, \end{aligned}$$

we have

$$\begin{aligned}
&(m+1)\sum_{r=1}^{m+1}(-1)^r\binom{m}{r-1}\frac{q^{(1-a)r}}{q^r-1}\\
&=(m+1)\sum_{r=1}^{m+1}(-1)^r\binom{m}{r-1}\frac{1}{r}\sum_{k=0}^{\infty}B_k(1-a)\frac{(r\log q)^k}{k!}\frac{1}{\log q}\\
&=\sum_{k=0}^{\infty}\left(\sum_{r=1}^{m+1}(-1)^r\binom{m+1}{r}r^k\right)B_k(1-a)\frac{(\log q)^{k-1}}{k!}.
\end{aligned}$$

The inner sum of the last expression can be evaluated as

$$\begin{aligned}
\sum_{r=1}^{m+1}(-1)^r\binom{m+1}{r}r^k&=\left(x\frac{d}{dx}\right)^k\left\{(1-x)^{m+1}-1\right\}\Big|_{x=1}\\
&=\begin{cases}-1 & (k=0),\\ 0 & (0<k<m+1),\\ (-1)^{m+1}(m+1)! & (k=m+1),\end{cases}
\end{aligned}$$

and we find

$$\begin{aligned}
&(m+1)\sum_{r=1}^{m+1}(-1)^r\binom{m}{r-1}\frac{q^{(1-a)r}}{q^r-1}\\
&=-\frac{1}{\log q}+(-1)^{m+1}B_{m+1}(1-a)(\log q)^m+O\left((\log q)^{m+1}\right)\qquad(q\to 1).
\end{aligned}$$

From this observation, the expansion

$$\log q=q-1+O\left((q-1)^2\right)\qquad(q\to 1)$$

and the known relation 1.6(10), we prove the desired result (7):

$$\begin{aligned}
&\lim_{q\to 1}(1-q)^{-m}\left\{(m+1)\sum_{r=1}^{m+1}(-1)^r\binom{m}{r-1}\frac{q^{(1-a)r}}{q^r-1}+\frac{1}{\log q}\right\}\\
&=(-1)^{m+1}B_{m+1}(1-a)\lim_{q\to 1}\frac{(\log q)^m}{(1-q)^m}=-B_{m+1}(1-a)=(-1)^m B_{m+1}(a).
\end{aligned}$$

■

Definition 6.8 In view of Theorem 6.14, it is natural to define the q-Bernoulli polynomials $\mathcal{B}_{n;q}(a)$ by

$$\mathcal{B}_{n;q}(a):=-n\,\zeta_q(1-n,a)\quad(n\in\mathbb{N}).\tag{8}$$

If we use (8), then we get some interesting properties and relations, which are summarized as in the following theorem:

Theorem 6.15

(a) *$\mathcal{B}_{n;q}(a)$ can be expressed in an explicit form:*

$$(q-1)^n\,\mathcal{B}_{n;q}(a) = \sum_{r=0}^{n} (-1)^r \binom{n}{r} \frac{r}{[r]_q}\, q^{(1-a)r} \quad (n \in \mathbb{N}), \tag{9}$$

where the term with $r = 0$ is understood to be

$$\lim_{r\to 0} \frac{r}{q^r - 1} = \frac{1}{\log q}. \tag{10}$$

So $\mathcal{B}_{0;q}(a)$ may be defined as

$$\mathcal{B}_{0;q}(a) := \frac{q-1}{\log q}. \tag{11}$$

(9) can be written equivalently as follows:

$$\sum_{r=0}^{n} (-1)^r \binom{n}{r} (q-1)^r\, \mathcal{B}_{r;q}(a) = \frac{n\,q^{(1-a)n}}{[n]_q}. \tag{12}$$

(b) *We have a relationship between the Carlitz's q-Bernoulli polynomials $\beta_n(a)$ in 6.6(51) and $\mathcal{B}_{n;q}(a)$ here:*

$$q^a\,\beta_n(a) = (-1)^n \left\{\mathcal{B}_{n;q}(1-a) + (1-q)\,\mathcal{B}_{n+1;q}(1-a)\right\} \quad (n \in \mathbb{N}). \tag{13}$$

(c) *We have a relationship between $\beta_k(a)$ and $\zeta_q(-k, a)$: For $n \in \mathbb{N}$,*

$$q^a\,\beta_n(a) = (-1)^n \left\{(q-1)\,(n+1)\,\zeta_q(-n, 1-a) - n\,\zeta_q(1-n, 1-a)\right\}. \tag{14}$$

(d) *We have a relationship between $\epsilon_{n;q}(2a)$ in Section 6.7 and $\zeta_{q^2}(-n, a)$:*

$$q^{2a}\,\epsilon_{n;q}(2a) = (-1)^n\,(q+1)^{n+1} \left\{\zeta_{q^2}\left(-n,\, \frac{1}{2} - a\right) - \zeta_{q^2}\,(-n,\, 1-a)\right\} \quad (n \in \mathbb{N}_0). \tag{15}$$

Proof. In view of (5), $\mathcal{B}_{n;q}(a)$ can be expressed as

$$\begin{aligned}\mathcal{B}_{n;q}(a) &= (q-1)^{-n+1} \left\{\sum_{r=1}^{n} (-1)^r \binom{n}{r} \frac{r}{q^r - 1}\, q^{(1-a)r} + \frac{1}{\log q}\right\} \\ &= (q-1)^{-n+1} \left\{\sum_{r=0}^{n} (-1)^r \binom{n}{r} \frac{r\,q^{(1-a)r}}{q^r - 1}\right\} \quad (n \in \mathbb{N}),\end{aligned}$$

which is seen equal to (9). As in getting 6.6(19), (9) can be written as (12). ∎

Using the familiar binomial identity

$$\binom{n+1}{r} = \binom{n}{r} + \binom{n}{r-1},$$

we find from 6.6(51) that

$$(q-1)^n \beta_n(a) = (-1)^n q^{-a} \left\{ \sum_{r=1}^{n} (-1)^r \binom{n}{r} \frac{r}{[r]_q} q^{ra} - \sum_{r=1}^{n+1} (-1)^r \binom{n+1}{r} \frac{r}{[r]_q} q^{ra} \right\},$$

which, in view of (9), yields the relationship (13).

(14) follows easily from (8) and (13).

Note that, if we also assume the convention of (10) in the term with $j = 0$ of Equation 6.6(18), it is easy to see from 6.6(18) and (9) the following relation:

$$\eta_{n;q}(a) = (-1)^n \mathcal{B}_{n;q}(1-a) \quad (n \in \mathbb{N}_0). \tag{16}$$

Now, an application of (8) and (16) to 6.7(15) yields (15).

Remark 2 Substitution of (16) for (13) is seen to come back to the defining relation 6.6(4). Taking the limit on each side of (14) as $q \uparrow 1$, together with (6), reduces immediately to a familiar identity 1.7(10). Replacing q and a by q^2 and $\frac{1}{4}$, respectively, in (15) and letting $q \uparrow 1$ in the resulting equation, together with (6), in view of Remark 1, we obtain

$$E_n = 2^n E_n\left(\frac{1}{2}\right) = \frac{2^{2n+1}}{n+1} \left\{ B_{n+1}\left(\frac{3}{4}\right) - B_{n+1}\left(\frac{1}{4}\right) \right\} \quad (n \in \mathbb{N}_0), \tag{17}$$

which is a special case of the well-known relationship between Euler polynomials $E_n(x)$ and Bernoulli polynomials $B_n(x)$ (see [1094, p. 65, Eq. (60)]; see also [869, p. 25, Eq. (25)]).

Application of Euler-Maclaurin Summation Formula

Here, we mainly show

Theorem 6.16

$$\lim_{q \uparrow 1} \zeta_q(s,a) = \zeta(s,a). \tag{18}$$

We present, for convenience, a special case of the Euler-Maclaurin summation formula 2.7(21) *when* $a = 0$, $b = N \in \mathbb{N}$, $K = M + 1 \in \mathbb{N}$:

$$\sum_{n=0}^{N} f(n) = \int_{0}^{N} f(x)\,dx + \frac{1}{2}\left(f(0) + f(N)\right) + \sum_{k=1}^{M} \frac{B_{k+1}}{(k+1)!}\left(f^{(k)}(N) - f^{(k)}(0)\right)$$
$$- \frac{(-1)^{M+1}}{(M+1)!} \int_{0}^{N} \widetilde{B}_{M+1}(x) f^{(M+1)}(x)\,dx, \tag{19}$$

where $f \in C^{\infty}[0, \infty)$, $N \in \mathbb{N}$ *and* $M \in \mathbb{N}_0$.

The Hurwitz (or generalized) Zeta function $\zeta(s, a)$, *defined by* 2.2(1), *is restricted as*

$$\zeta(s, a) := \sum_{k=0}^{\infty} (k + a)^{-s} \quad (\Re(s) > 1;\ 0 < a \leqq 1). \tag{20}$$

Applying $f(x) = (x + a)^{-s}$ $(\Re(s) > 1)$ to (19) and taking the limit of each side of the resulting equation as $N \to \infty$ gives

$$\zeta(s, a) = \frac{a^{-s+1}}{s-1} + \frac{1}{2a^s} + \sum_{k=1}^{M} \frac{B_{k+1}}{(k+1)!} (s)_k\, a^{-s-k}$$
$$- \frac{(s)_{M+1}}{(M+1)!} \int_{0}^{\infty} \widetilde{B}_{M+1}(x)\,(x + a)^{-s-M-1}\,dx. \tag{21}$$

Remark 3 It is observed that the integral involved in (21) converges for $\Re(s) > -M$. So, $\zeta(s, a)$ can be continued analytically to $\Re(s) > -M$. Since M can be arbitrarily large, $\zeta(s, a)$ can be continued analytically to the whole s-plane, except for an obvious simple pole at $s = 1$.

The special case of (19) when $M = 1$ is given as

$$\sum_{n=0}^{N} f(n) = \int_{0}^{N} f(x)\,dx + \frac{1}{2}\left(f(0) + f(N)\right) + \frac{1}{12}\left(f'(N) - f'(0)\right)$$
$$- \frac{1}{2} \int_{0}^{N} \widetilde{B}_2(x) f^{(2)}(x)\,dx. \tag{22}$$

Consider a function

$$f_q(x; s, a) := \frac{q^{(x+a)(s-1)}}{\left(1 - q^{x+a}\right)^s} \quad (\Re(s) > 1;\ 0 < a \leqq 1) \tag{23}$$

ready for application to (22) to compute the following:

$$\frac{\partial}{\partial x} f_q(x; s, a) = (\log q)\, q^{(x+a)(s-1)} \frac{s-1+q^{x+a}}{\left(1-q^{x+a}\right)^{s+1}};$$

$$\frac{\partial^2}{\partial x^2} f_q(x; s, a) = (\log q)^2\, q^{(x+a)(s-1)} \frac{s(s+1) - 3s\left(1-q^{x+a}\right) + \left(1-q^{x+a}\right)^2}{\left(1-q^{x+a}\right)^{s+2}};$$

and, in general,

$$\frac{\partial^k}{\partial x^k} f_q(x; s, a) = (\log q)^k\, q^{(x+a)(s-1)} \left(1-q^{x+a}\right)^{-s-2} p_q(x; s, a),$$

where $p_q(x; s, a)$ is a polynomial in s and q^{x+a}.

Applying (23) to (22), taking the limit of the resulting equation as $N \to \infty$ and considering

$$\begin{aligned}\int_0^\infty \frac{q^{(x+a)(s-1)}}{\left(1-q^{x+a}\right)^s}\, dx &= \int_0^\infty \frac{q^{-x-a}}{\left(q^{-x-a}-1\right)^s}\, dx \\ &= \left[\frac{\left(q^{-x-a}-1\right)^{1-s}}{(s-1)\log q}\right]_0^\infty = -\frac{q^{a(s-1)}\left(1-q^a\right)^{1-s}}{(s-1)\log q},\end{aligned} \tag{24}$$

we readily find that, for $\Re(s) > 1$,

$$\begin{aligned}\zeta_q(s, a) = (1-q)^s \sum_{n=0}^\infty \frac{q^{(n+a)(s-1)}}{\left(1-q^{n+a}\right)^s} &= -\frac{q^{a(s-1)}(1-q)^s\left(1-q^a\right)^{1-s}}{(s-1)\log q} \\ &+ \frac{(1-q)^s q^{a(s-1)}}{2(1-q^a)^s} - \frac{1}{12}(\log q) q^{a(s-1)}(1-q)^s \frac{s-1+q^a}{(1-q^a)^{s+1}} - (1-q)^s \\ &\cdot \frac{(\log q)^2}{2} \int_0^\infty \widetilde{B}_2(x) q^{(x+a)(s-1)} \frac{s(s+1) - 3s\left(1-q^{x+a}\right) + \left(1-q^{x+a}\right)^2}{\left(1-q^{x+a}\right)^{s+2}} dx.\end{aligned} \tag{25}$$

Remark 4 Unlike in the classic case represented by (21), the integral in (25) cannot be made to converge by simply choosing M sufficiently large, instead of $M = 1$, because the presence of the factor $q^{(x+a)(s-1)}$ in $f_q^{(M+1)}(x; s, a)$ implies that necessary $\Re(s) > 1$.

Therefore, in this case, we use the known Fourier expansion of the periodic Bernoulli polynomials (see [1225, Chapter IX, Miscellaneous Exercise 12]):

$$\widetilde{B}_k(x) = -k! \sum_{n \in \mathbb{Z} \setminus \{0\}} \frac{e^{2\pi i n x}}{(2\pi i n)^k}. \tag{26}$$

The equality in (26) is valid for all real numbers x when $k \geqq 2$, in which case the sum is absolutely and uniformly convergent.

Substituting (26) with $k = 2$ into (25) and interchanging the summation and integration, we find that

$$\begin{aligned}\zeta_q(s,a) = &-\frac{q^{a(s-1)}(1-q)^s(1-q^a)^{1-s}}{(s-1)\log q} + \frac{(1-q)^s q^{a(s-1)}}{2(1-q^a)^s} \\ &-\frac{1}{12}(\log q)\, q^{a(s-1)}(1-q)^s \frac{s-1+q^a}{(1-q^a)^{s+1}} \\ &+(1-q)^s(\log q)^2 \sum_{n\in\mathbb{Z}\setminus\{0\}} \frac{1}{(2\pi i n)^2} I_n(x;s,a,q),\end{aligned} \tag{27}$$

where

$$I_n(x;s,a,q) := \int_0^\infty e^{2\pi i n x} q^{(x+a)(s-1)} \frac{s(s+1)-3s\left(1-q^{x+a}\right)+\left(1-q^{x+a}\right)^2}{\left(1-q^{x+a}\right)^{s+2}}\,dx.$$

By letting $q^{x+a} = u$ in $I_n(x;s,a,q)$, we obtain

$$\begin{aligned}I_n(x;s,a,q) = &-\frac{e^{-2\pi i n a}}{\log q}\big[s(s+1)B_{q^a}(s+\delta n-1,-s-1) \\ &-3sB_{q^a}(s+\delta n-1,-s)+B_{q^a}(s+\delta n-1,-s+1)\big],\end{aligned} \tag{28}$$

where

$$\delta := \frac{2\pi i}{\log q} \quad \text{and} \quad B_t(\alpha,\beta) = \int_0^t u^{\alpha-1}(1-u)^{\beta-1}\,du \quad (\Re(\alpha) > 0) \tag{29}$$

is the incomplete Beta function given in 1.1(61).

It follows from (27) and (28) that

$$\begin{aligned}\zeta_q(s,a) = &-\frac{q^{a(s-1)}(1-q)^s(1-q^a)^{1-s}}{(s-1)\log q} + \frac{(1-q)^s q^{a(s-1)}}{2(1-q^a)^s} \\ &-\frac{1}{12}(\log q)q^{a(s-1)}(1-q)^s \frac{s-1+q^a}{(1-q^a)^{s+1}} \\ &-(1-q)^s\log q \sum_{n\in\mathbb{Z}\setminus\{0\}} \frac{e^{-2\pi i n a}}{(2\pi i n)^2}\big[s(s+1)B_{q^a}(s+\delta n-1,-s-1) \\ &-3sB_{q^a}(s+\delta n-1,-s)+B_{q^a}(s+\delta n-1,-s+1)\big].\end{aligned} \tag{30}$$

Note that each of the incomplete Beta functions in (30) converges absolutely for $\Re(s) > 1$ and is uniformly bounded with respect to n. Indeed, we see that

$$\begin{aligned} \left|B_{q^a}(s-1+\delta n, -s+\nu)\right| &\leqq \int_0^{q^a} |u|^{\Re(s-2)} \left|u^{\delta n}\right| |1-u|^{\Re(-s+\nu-1)}\, du \\ &= \int_0^{q^a} u^{\sigma-2}(1-u)^{-\sigma+\nu-1}\, du \quad (n \in \mathbb{Z}\setminus\{0\};\ \sigma = \Re(s);\ \nu = -1, 0, 1). \end{aligned} \tag{31}$$

Hence, the sum in (30) converges absolutely.

Repeated use of integration by parts yields the formula

$$\begin{aligned} B_t(\alpha,\beta) &= \sum_{k=1}^{M-1} (-1)^{k-1} \frac{(1-\beta)_{k-1}}{(\alpha)_k} t^{\alpha+k-1}(1-t)^{\beta-k} \\ &\quad + (-1)^{M-1} \frac{(1-\beta)_{M-1}}{(\alpha)_M} B_t(\alpha+M-1, \beta-M+1) \quad (M \in \mathbb{N}\setminus\{1\}). \end{aligned} \tag{32}$$

Applying (32) to $B_{q^a}(s-1+\delta n, -s-1)$, we have

$$\begin{aligned} B_{q^a}(s-1+\delta n, -s-1) &= \sum_{k=1}^{M-1} (-1)^{k-1} \frac{(2+s)_{k-1}}{(s+\delta n-1)_k} q^{a(s+\delta n+k-2)} \left(1-q^a\right)^{-s-1-k} \\ &\quad + (-1)^{M-1} \frac{(2+s)_{M-1}}{(s+\delta n-1)_{M-1}} \\ &\quad \cdot B_{q^a}(s-2+M+\delta n, -s-M), \end{aligned}$$

which, in view of (31), allows us to carry out the analytic continuation of $B_{q^a}(s-1+\delta n, -s-1)$ as a function of s into the region $\Re(s) > 2-M$. From this, we have

$$\begin{aligned} &\sum_{n\in\mathbb{Z}\setminus\{0\}} \frac{e^{-2\pi i n a}}{(2\pi i n)^2} s(s+1) B_{q^a}(s+\delta n-1, -s-1) \\ &= \sum_{k=1}^{M-1} (-1)^{k-1} \sum_{n\in\mathbb{Z}\setminus\{0\}} \frac{e^{-2\pi i n a}}{(2\pi i n)^2} \frac{(s)_{k+1}}{(s-1+\delta n)_k} q^{a(s+\delta n+k-2)} \left(1-q^a\right)^{-s-1-k} \\ &\quad - (-1)^{M-1} \log q \sum_{n\in\mathbb{Z}\setminus\{0\}} \frac{(s)_{M+1}}{(2\pi i n)^2 (s+\delta n-1)_{M-1}} \\ &\quad \cdot \int_0^\infty e^{2\pi i n x} q^{(x+a)(s-2+M)} \left(1-q^{x+a}\right)^{-s-M-1} dx \end{aligned}$$

from which we obtain

$$\begin{aligned}
&\lim_{q\uparrow 1}(1-q)^s \log q \sum_{n\in\mathbb{Z}\setminus\{0\}} \frac{e^{-2\pi ina}}{(2\pi in)^2}\, s(s+1)\, B_{q^a}(s+\delta n-1,-s-1) \\
&=\sum_{k=1}^{M-1}\sum_{n\in\mathbb{Z}\setminus\{0\}} \frac{(s)_{k+1}}{(2\pi in)^2}\,\frac{1}{a^{k+s+1}\,(2\pi in)^k} \\
&\quad -\sum_{n\in\mathbb{Z}\setminus\{0\}} \frac{(s)_{M+1}}{(2\pi in)^2}\int_0^\infty \frac{e^{2\pi inx}}{(2\pi in)^{M-1}\,(x+a)^{M+s+1}}\,dx \qquad (33)\\
&=-\sum_{k=1}^{M-1}\frac{B_{k+2}}{(k+2)!}\,(s)_{k+1}\,a^{-k-s-1}+(s)_{M+1}\int_0^\infty \frac{\widetilde{B}_{M+1}(x)}{(M+1)!}\,(x+a)^{-M-s-1}\,dx.
\end{aligned}$$

We also have

$$\lim_{q\uparrow 1}\frac{(1-q)^s\,(1-q^a)^{1-s}}{\log q}=-a^{1-s} \quad \text{and} \quad \lim_{q\uparrow 1}\frac{(\log q)\,(1-q)^s}{(1-q^a)^{s+1}}=-\frac{1}{a^{s+1}}. \qquad (34)$$

Taking the limit of each side of (30) as $q\uparrow 1$ and considering (33) and (34), we obtain

$$\begin{aligned}
\lim_{q\uparrow 1}\zeta_q(s,a) &= \frac{a^{-s+1}}{s-1}+\frac{1}{2a^s}+\sum_{k=1}^{M}\frac{B_{k+1}}{(k+1)!}\,(s)_k\,a^{-s-k} \\
&\quad -\frac{(s)_{M+1}}{(M+1)!}\int_0^\infty \widetilde{B}_{M+1}(x)\,(x+a)^{-s-M-1}\,dx, \qquad (35)
\end{aligned}$$

where we have used the following results:

$$\lim_{q\uparrow 1}(1-q)^s \log q \sum_{n\in\mathbb{Z}\setminus\{0\}} \frac{e^{-2\pi ina}}{(2\pi in)^2}\, s\, B_{q^a}(s+\delta n-1,-s)=0$$

and

$$\lim_{q\uparrow 1}(1-q)^s \log q \sum_{n\in\mathbb{Z}\setminus\{0\}} \frac{e^{-2\pi ina}}{(2\pi in)^2}\, s\, B_{q^a}(s+\delta n-1,-s+1)=0.$$

Hence, it follows from (21) and (35) that (18) is proven.

We conclude this section by noting that all identities and properties in this section, obtained by setting $a=1$, are seen to recover those in Kaneko *et al.* [626].

6.11 Multiple q-Zeta Functions

Zhao [1261] defined and investigated a q analogue of the multiple Zeta functions defined by 2.7(1) based on the work of Kaneko *et al.* [626], which was studied in Section 6.8. Here, we introduce Zhao's work [1261].

It is known (see Section 2.7) that the multiple Zeta function $\zeta(s_1, \ldots, s_d)$ can be extended to a meromorphic function on $\mathbb{C}^d \setminus \mathcal{D}_d$, where

$$\mathcal{D}_d := \left\{ (s_1, \ldots, s_d) \in \mathbb{C}^d \;\middle|\; \begin{array}{l} s_d = 1,\ s_{d-1} + s_d = 2,\ 1,\ -2m\,(m \in \mathbb{N}_0), \\ s_j + \cdots + s_d \in \mathbb{Z}_{\leqq d-j+1}\,(j \leqq d-2) \end{array} \right\}, \tag{1}$$

$\mathbb{Z}_{\mathcal{P}}$ denoting the set of integers satisfying any property $\mathcal{P}$. To find a q-analogue of the multiple Zeta functions $\zeta(s_1, \ldots, s_d)$, defined by 2.7(1), Zhao [1261] defines an auxiliary function of $2d$ complex variables $s_1, \ldots, s_d, t_1, \ldots, t_d \in \mathbb{C}$

$$g_q(s_1, \ldots, s_d; t_1, \ldots, t_d) := \sum_{0<k_1<\cdots<k_d} \frac{q^{k_1 t_1 + \cdots + k_d t_d}}{[k_1]^{s_1} \cdots [k_d]^{s_d}}, \tag{2}$$

which converges, if $\Re\left(t_j + \cdots + t_d\right) > 0$, for all $j = 1, \ldots, d$. Zhao [1261] also defines a multiple q-Zeta function by specialization of g_q:

$$\begin{aligned} \zeta_q(s_1, \ldots, s_d) &:= g_q(s_1, \ldots, s_d; s_1 - 1, \ldots, s_d - 1) \\ &= \sum_{0<k_1<\cdots<k_d} \frac{q^{k_1(s_1-1) + \cdots + k_d(s_d-1)}}{[k_1]^{s_1} \cdots [k_d]^{s_d}}. \end{aligned} \tag{3}$$

Note that the special case of (3) when $d = 1$ is the same as the q-analogue of the Riemann Zeta function $\zeta(s)$ given by Kaneko *et al.* [626] (see also Section 6.8). We give a notation $\mathcal{D}_d(q)$ by

$$\mathcal{D}_d(q) := \left\{ (s_1, \ldots, s_d) \in \mathbb{C}^d \;\middle|\; \begin{array}{l} s_d = 1 + \frac{2\pi i}{\log q}\mathbb{Z},\ s_d \in \mathbb{Z}_{\leqq 0} + \frac{2\pi i}{\log q}\mathbb{Z}_{\neq 0}, \\ s_j + \cdots + s_d \in \mathbb{Z}_{\leqq d-j+1} + \frac{2\pi i}{\log q}\mathbb{Z}\,(j < d) \end{array} \right\}, \tag{4}$$

the last condition of which is vacuous, if $d = 1$.

Analytic Continuation of g_q and ζ_q

We start with a simple lemma:

Lemma 6.17 *For all $\frac{1}{2} < q < 1$, we have*

$$1 \leqq \frac{1 - q^k}{1 - q} = [k]_q < 2k \quad (k \in \mathbb{N}).$$

Proof. Since $0 < q^k \leqq q < 1$, the first inequality is obvious. ■

Let $f(q) = 2k(1-q) - \left(1-q^k\right)$. We find that $f(q) \geqq 0$ for all $0 < q < 1$. Indeed, for all such q,

$$f'(q) = -2k + kq^{k-1} \leqq -2k + k = -k < 0.$$

So, f is strictly decreasing, and the positivity of f follows from $f(1) = 0$. This implies that $[k]_q < 2k$ as desired.

Theorem 6.18 *The function $g_q(s_1, \ldots, s_d; t_1, \ldots, t_d)$ in (2) converges, if $\Re(t_j + \cdots + t_d) > 0$, for all $j = 1, \ldots, d$. It can be analytically continued to a meromorphic function over $\mathbb{C}^{2d}$, by means of the series expansion*

$$g_q(s_1, \ldots, s_d; t_1, \ldots, t_d) = (1-q)^{wt(\vec{s})} \sum_{r_1, \ldots, r_d = 0}^{\infty} \prod_{j=1}^{d} \left[\binom{s_j + r_j - 1}{r_j} \frac{q^{j(r_j + t_j)}}{1 - q^{r_j + t_j + \cdots + r_d + t_d}} \right], \tag{5}$$

where $wt(\vec{s}) := s_1 + \cdots + s_d$. It has simple poles at

$$t_j + \cdots + t_d \in \mathbb{Z}_{\leqq 0} + \frac{2\pi i}{\log q} \mathbb{Z}.$$

Proof. Assume $|\Re(s_j)| < N_j \in \mathbb{N}$ and let $\tau_j := \Re(t_j)$ for all $j = 1, \ldots, d$. By Lemma 6.16, we have

$$|g_q(s_1, \ldots, s_d; t_1, \ldots, t_d)| < \sum_{0 < k_1 < \cdots < k_d} \prod_{j=1}^{d} (2k_j)^{N_j} q^{k_j \tau_j}. \tag{6}$$

For simplicity, let $k = k_{d-1}$ (when $d = 1$, take $k_0 = 0$), $n = k_d$, $N = N_d$ and $\tau = \tau_d$. Then, by root test, $\sum_{n>k} n^N q^{n\tau}$ converges, and moreover,

$$\sum_{n>k} n^N q^{n\tau} = \frac{1}{\tau^N} \left(q \frac{d}{dq} \right)^N \sum_{n>k} q^{n\tau} = \left(1 - q^\tau\right)^{-N} f_N(k; q, \tau) q^{k\tau}, \tag{7}$$

where $f_N(x; q, \tau) = \sum_{\ell=0}^{N} c_\ell x^\ell$ is a polynomial of degree N, whose coefficients depend only on the constants N, q and τ. This proves the first part of the theorem when $d = 1$. ■

In the general case, it follows from (6) and (7) that

$$|g_q(s_1, \ldots, s_d; t_1, \ldots, t_d)| < \frac{2^{N_1 + \cdots + N_d}}{(1 - q^\tau)^N} \sum_{\ell=0}^{N} c_\ell \cdot \sum_{0 < k_1 < \cdots < k_{d-1} < k} \left(\prod_{j=1}^{d-2} (k_j)^{N_j} q^{k_j \tau_j} \right) k^{N_{d-1} + \ell} q^{k(\tau_{d-1} + \tau_d)}.$$

Hence, the first part of the theorem follows from an easy induction on d. By using the binomial expansion for $(1-x)^{-s}$, we get

$$\begin{aligned} &g_q(s_1,\ldots,s_d;t_1,\ldots,t_d) \\ &= (1-q)^{wt(\vec{s})} \sum_{0<k_1<\cdots<k_d} \sum_{r_1,\ldots,r_d=0}^{\infty} \left(\prod_{j=1}^{d} \binom{s_j+r_j-1}{r_j} q^{k_j(r_j+t_j)} \right). \end{aligned}$$

As $0<q<1$, the series converges absolutely by Stirling's formula. So, we can exchange the summations. The theorem follows immediately from the next lemma, by taking $z_j = q^{r_j+t_j}$ for $j=1,\ldots,d$.

Lemma 6.19 *Let $z_j \in \mathbb{C}$, such that $|z_j|<1$ for $j=1,\ldots,d$. Then, we find that*

$$\sum_{0<k_1<\cdots<k_d} \prod_{j=1}^{d} z_j^{k_j} = \prod_{j=1}^{d} \frac{z_j\cdots z_d}{1-z_j\cdots z_d}. \tag{8}$$

Proof. When $d=1$, this is clear. For $d \geqq 2$, we denote the right hand side of (8), by $F_d(z_1,\ldots,z_d)$. Then, we have

$$\begin{aligned} \sum_{0<k_1<\cdots<k_d} \prod_{j=1}^{d} z_j^{k_j} &= \sum_{0<k_1<\cdots<k_{d-1}} \prod_{j=1}^{d-1} z_j^{k_j} \sum_{k_{d-1}<k_d} z_d^{k_d} \\ &= \sum_{0<k_1<\cdots<k_{d-1}} \prod_{j=1}^{d-1} z_j^{k_j} \frac{z_d^{k_{d-1}+1}}{1-z_d} = \frac{z_d}{1-z_d} F_{d-1}(z_1,\ldots,z_{d-2},z_{d-1}z_d). \end{aligned}$$

The lemma follows immediately by induction. ■

Recalling $\zeta_q(s_1,\ldots,s_d) = g_q(s_1,\ldots,s_d;s_1-1,\ldots,s_d-1)$, we have the following immediate consequence:

Theorem 6.20 *The q-multiple Zeta function $\zeta_q(s_1,\ldots,s_d)$, defined by (3), can be extended to a meromorphic function with simple poles lying along $\mathcal{D}_d(q)$:*

$$\begin{aligned} &\zeta_q(s_1,\ldots,s_d) \\ &= (1-q)^{wt(\vec{s})} \sum_{r_1,\ldots,r_d=0}^{\infty} \prod_{j=1}^{d} \left[\binom{s_j+r_j-1}{r_j} \frac{q^{j(r_j+s_j-1)}}{1-q^{r_j+s_j+\cdots+r_d+s_d-d+j-1}} \right]. \end{aligned} \tag{9}$$

To see the effect of taking different specializations of t_j in g_q, we define the shifting operators $\mathbf{S}_j$ $(1 \leqq j \leqq d)$ on the multiple Zeta function as follows:

$$\mathbf{S}_j \zeta_q(s_1,\ldots,s_d) = \zeta_q(s_1,\ldots,s_d) + (1-q)\zeta_q(s_1,\ldots,s_{j-1},\ldots,s_d). \tag{10}$$

It is obvious that these operators are commutative.

Theorem 6.21 *Let $n_1, \dots, n_d$ be nonnegative integers. Then, we have*

$$g_q(s_1,\dots,s_d; s_1-1-n_1,\dots,s_d-1-n_d) = \mathbf{S}_1^{n_1}\circ\cdots\circ\mathbf{S}_d^{n_d}\,\zeta_q(s_1,\dots,s_d)$$
$$=\sum_{r_1=0}^{n_1}\cdots\sum_{r_d=0}^{n_d}\left[\prod_{j=1}^{d}\binom{n_j}{r_j}(1-q)^{r_j}\right]\zeta_q(s_1-r_1,\dots,s_d-r_d).$$

Proof. We sketch the proof only in the case $n_1=\cdots=n_{d-1}=0$. The general case is completely similar. In the rest of this section, we always let $\mathbf{S}$ be the shifting operator on the last variable. Suppose $n_d=n=1$. Then, we find

$$g_q(s_1,\dots,s_d; s_1-1,\dots,s_{d-1}-1,s_d-2)$$
$$=\sum_{0<k_1<\cdots<k_d}\frac{q^{k_1(s_1-1)+\cdots+k_{d-1}(s_{d-1}-1)+k_d(s_d-2)}}{[k_1]^{s_1}\cdots[k_d]^{s_d}}$$
$$=\sum_{0<k_1<\cdots<k_d}\frac{q^{k_1(s_1-1)+\cdots+k_{d-1}(s_{d-1}-1)}}{[k_1]^{s_1}\cdots[k_{d-1}]^{s_{d-1}}}\;\frac{q^{k_d(s_d-2)}\left(1-q^{k_d}\right)+q^{k_d(s_d-1)}}{[k_d]^{s_d}}$$
$$=(1-q)\,\zeta_q(s_1,\dots,s_{d-1},s_d-1)+\zeta_q(s_1,\dots,s_d)=\mathbf{S}\zeta_q(s_1,\dots,s_d).$$

The rest follows easily by induction. ■

The next corollary answers an implicit question in Kaneko *et al.* [626].

Corollary 6.22 *Let $n\in\mathbb{N}$. The specialization of t of $g_q(s;t):=g_q(1;s,t)$ in* 6.8(1) *to $s-1-n$ is*

$$g_q(s;s-1-n)=\mathbf{S}^n\zeta_q(s)=\sum_{r=0}^{n}\binom{n}{r}(1-q)^r\,\zeta_q(s-r), \tag{11}$$

where $\zeta_q(s-r):=\zeta_q(s-r,1)$ in 6.10(4).

We observe that one effect of the shifting operator is to bring more poles. Essentially, $\mathbf{S}^n$ shifts all the poles of $\zeta_q(s)$, by n to the right on the complex plane.

Analytic Continuation of Multiple Zeta Functions

We begin this subsection, by giving a simple observation.

Lemma 6.23 *For every real number x, we have*

$$\left|\widetilde{B}_M(x)\right|\leqq\frac{4M!}{(2\pi)^M}\quad(M\in\mathbb{N}\setminus\{1\}), \tag{12}$$

where $\widetilde{B}_k(x)$ denote the periodic Bernoulli polynomials given in 6.10(26).

Proof. It follows from 6.10(26) and the fact that $\zeta(M)\leqq\zeta(2)=\frac{\pi^2}{6}<2$ for $M\geqq 2$. ■

In Section 2.7, by two different methods, $\zeta(s_1,\dots,s_d)$ can be continued to a meromorphic function on $\mathbb{C}^d$ with simple poles along $\mathcal{D}_d$, defined by (1). However, neither approach is suitable for the present purpose. So, we follow the idea in Section 6.8 to prove a third approach in the rest of this subsection. The same idea will also be used to deal with the q-multiple Zeta functions.

To simplify our notation, in the definition of the multiple Zeta function $\zeta(s_1,\ldots,s_d)$, we replace s_d, n_{d-1} and n_d by s, k and n, respectively. Taking $f(x)=1/x^s$ and $m=k$ and ∞ in a Euler-Maclaurin summation formula 2.7(22), we have, for $\Re(s)>1$,

$$\begin{aligned}\sum_{n>k}\frac{1}{n^s}=\sum_{n=1}^{\infty}\frac{1}{n^s}-\sum_{n=1}^{k}\frac{1}{n^s}&=\int_k^\infty f(x)\,dx-\frac{1}{2}f(k)\\&-\sum_{r=1}^{M}\frac{B_{r+1}}{(r+1)!}f^{(r)}(k)-\frac{(-1)^{M+1}}{(M+1)!}\int_k^\infty \widetilde{B}_{M+1}(x)f^{(M+1)}(x)\,dx\\&=\frac{1}{(s-1)\,k^{s-1}}-\frac{1}{2k^s}+\sum_{r=1}^{M}\frac{B_{r+1}}{(r+1)!}\frac{(s)_r}{k^{s+r}}-\frac{(s)_{M+1}}{(M+1)!}\int_k^\infty\frac{\widetilde{B}_{M+1}(x)}{x^{s+M+1}}\,dx,\end{aligned}$$

where we have used the fact that $B_{2k+1}=0$ $(k\in\mathbb{N})$. In view of the observation

$$\zeta(s_1,\ldots,s_d)=\sum_{0<k_1<k_2<\cdots<k_{d-1}}\frac{1}{k_1{}^{s_1}\,k_2{}^{s_2}\cdots k_{d-1}{}^{s_{d-1}}}\sum_{n>k}\frac{1}{n^s},$$

we have

Theorem 6.24 *For all $(s_1,\ldots,s_d)\in\mathbb{C}^d\setminus\mathcal{D}_d$ and $M>1+|\sigma_d|+|\sigma_{d-1}|$, we have*

$$\begin{aligned}\zeta(s_1,\ldots,s_d)&=\sum_{r=0}^{M+1}\frac{B_r}{r!}(s_d)_{r-1}\,\zeta(s_1,\ldots,s_{d-1}+s_d+r-1)\\&-\frac{(s_d)_{M+1}}{(M+1)!}\sum_{0<k_1<k_2<\cdots<k_{d-1}}\frac{1}{k_1{}^{s_1}\,k_2{}^{s_2}\cdots k_{d-1}{}^{s_{d-1}}}\int_{k_{d-1}}^\infty\frac{\widetilde{B}_{M+1}(x)}{x^{s_d+M+1}}\,dx,\end{aligned}\tag{13}$$

where $(s)_{-1}:=1/(s-1)$. This provides an analytic continuation of $\zeta(s_1,\ldots,s_d)$ to $\mathbb{C}^d\setminus\mathcal{D}_d$.

Proof. We only need to show that the series in (13) converges. Lemma 6.23 implies (if $d=2$, then take $k_0=1$)

$$\sum_{k_{d-1}=k_{d-2}}^{\infty}\left|\frac{1}{k_{d-1}^{s_{d-1}}}\int_{k_{d-1}}^\infty\frac{\widetilde{B}_{M+1}(x)}{x^{s_d+M+1}}\,dx\right|\leqq\frac{4\,(M+1)!}{(2\pi)^{M+1}\,(M-|\sigma_d|)}\sum_{k_{d-1}=k_{d-2}}^{\infty}\frac{1}{k_{d-1}^{M-|\sigma_{d-1}|-|\sigma_d|}},$$

which converges (absolutely) whenever $M>1+|\sigma_d|+|\sigma_{d-1}|$.

Now, we are ready to prove the main theorem of this section. ■

Theorem 6.25 *The multiple q-Zeta function $\zeta_q(s_1,\ldots,s_d)$, defined by (3), can be extended to a meromorphic function on $\mathbb{C}^d$ with simple poles along $\mathcal{D}_d(q)$ given in (4). Further, for all $(s_1,\ldots,s_d)\in\mathbb{C}^d\setminus\mathcal{D}_d$,*

$$\lim_{q\uparrow 1}\zeta_q(s_1,\ldots,s_d)=\zeta(s_1,\ldots,s_d). \tag{14}$$

Proof. Fix $(s_1,\ldots,s_d)\in\mathbb{C}^d$, such that $\sigma_j+\cdots+\sigma_d<d-j+1$ for all $j=1,\ldots,d$. When $d=1$, this is Theorem 6.16 when $a=1$. So, we assume $d\in\mathbb{N}\setminus\{1\}$ and proceed by induction. The key is a recursive formula for $\zeta_q(s_1,\ldots,s_d)$, similar to (13) for $\zeta(s_1,\ldots,s_d)$. To derive this formula, we appeal to the Euler-Maclaurin summation formula 2.7(22) again. Hence, we set

$$F(x):=\frac{q^{x(s-1)}}{(1-q^x)^s}$$

as in 6.8(17). Then, we have

$$F'(x)=(\log q)\,q^{x(s-1)}\,\frac{s-1+q^x}{(1-q^x)^{s+1}}$$

and

$$F''(x)=(\log q)^2\,q^{x(s-1)}\,\frac{s(s+1)-3s(1-q^x)+(1-q^x)^2}{(1-q^x)^{s+2}}.$$

In definition (3), for simplicity, we replace s_d, k_{d-1} and k_d by s, k and n, respectively. We, now, take $M=1, f(x)=F(x+k-1)$, let $m\to\infty$ in 2.7(22) and get

$$\begin{aligned}
\sum_{n>k}\frac{q^{n(s-1)}}{[n]_q^s}&=(1-q)^s\left(-F(k)+\sum_{n=1}^{\infty}f(n)\right)\\
&=(1-q)^s\left(\int_k^{\infty}F(x)\,dx-\frac{1}{2}F(k)-\frac{1}{12}F'(k)-\frac{1}{2}\int_1^{\infty}\widetilde{B}_2(x)f''(x)\,dx\right)\\
&=\frac{q-1}{(s-1)\log q}\,\frac{q^{k(s-1)}}{[k]_q^{s-1}}-\frac{1}{2}\,\frac{q^{k(s-1)}}{[k]_q^s}+\frac{1}{12}\,\frac{\log q}{q-1}\,\frac{q^{k(s-1)}\left(s+q^k-1\right)}{[k]_q^{s+1}}\\
&\quad-\frac{(1-q)^s(\log q)^2}{2}\int_k^{\infty}\widetilde{B}_2(x)\,q^{x(s-1)}\,\frac{s(s+1)-3s(1-q^x)+(1-q^x)^2}{(1-q^x)^{s+2}}\,dx,
\end{aligned} \tag{15}$$

because $\widetilde{B}_2(x+k-1)=\widetilde{B}_2(x)$ by periodicity.

By the same argument as in Section 6.10, using the incomplete Beta integral given in 6.10(29), we obtain from 6.10(26) the following expression for the last line in (15):

$$= -\sum_{n\in\mathbb{Z}\setminus\{0\}} \frac{(1-q)^s \log q}{(2\pi i n)^2} \sum_{\nu=\pm 1,0} a_\nu(s) B_{q^k}(s-1+\delta n,\ -s+\nu) \tag{16}$$

where $\delta = 2\pi i/\log q$, $a_{-1}(s) = s(s+1)$, $a_0(s) = -3s$ and $a_1(s) = 1$. ∎

Repeatedly applying integration by parts on these incomplete Beta integrals, we get for $\nu = \pm 1, 0$ and positive integer $M \geqq 2$

$$\begin{aligned} B_{q^k}(s-1+\delta n,\ -s+\nu) &= \sum_{r=1}^{M-1} (-1)^{r-1} \frac{(s+1-\nu)_{r-1}}{(s-1+\delta n)_r} \frac{q^{k(s+r-2)}}{\left(1-q^k\right)^{s+r-\nu}} \\ &+ (-1)^{M-1} \frac{(s+1-\nu)_{M-1}}{(s-1+\delta n)_{M-1}} B_{q^k}(s-2+M+\delta n,\ -s-M+1+\nu). \end{aligned} \tag{17}$$

Set $\vec{s}\,' = (s_1, \ldots, s_{d-2})$, if $d \geqq 3$ and $\vec{s}\,' = \emptyset$, if $d = 2$. Putting (3), (15), (16) and (17) together and applying Theorem 6.21, we get

$$\begin{aligned} \zeta_q\left(\vec{s}\,',\ s_{d-1},\ s_d\right) &= \frac{q-1}{(s-1)\log q} \zeta_q\left(\vec{s}\,',\ s_{d-1}+s_d-1\right) - \frac{1}{2}\mathbf{S}\zeta_q\left(\vec{s}\,',\ s_{d-1}+s_d\right) \\ &+ \frac{s}{12}\frac{\log q}{q-1}\mathbf{S}^2\zeta_q\left(\vec{s}\,',\ s_{d-1}+s_d+1\right) \\ &+ \frac{\log q}{12}\mathbf{S}\zeta_q\left(\vec{s}\,',\ s_{d-1}+s_d\right) - \sum_{\nu=\pm 1,0}(C_\nu + D_\nu), \end{aligned} \tag{18}$$

where C_ν and D_ν are contributions from the sum involving $B_{q^k}(\cdots,\ -s+\nu)$. We write

$$T(q,s,n,r) := \prod_{j=0}^{r-1} (2\pi i n + (s-1+j)\log q)^{-1} \tag{19}$$

to express C_ν and D_ν explicitly as follows:

$$C_{-1} = \sum_{r=1}^{M-1} \sum_{n\in\mathbb{Z}\setminus\{0\}} \frac{T(q,s,n,r)}{(2\pi i n)^2} \left(\frac{\log q}{q-1}\right)^{r+1} (s)_{r+1}\, \mathbf{S}^3 \zeta_q\left(\vec{s}\,',\ s_{d-1}+s_d+r+1\right)$$

and

$$D_{-1} = \sum_{0<k_1<\cdots<k_{d-2}} \frac{q^{k_1(s_1-1)+\cdots+k_{d-2}(s_{d-2}-1)}}{[k_1]_q^{s_1}\cdots[k_{d-2}]_q^{s_{d-2}}} \sum_{k=k_{d-2}+1}^{\infty} R(M,q,k,s_{d-1},s),$$

where we replace the index k_{d-1} by k and

$$R(M,q,k,s_{d-1},s) := -\frac{q^{k_{d-1}(s_{d-1}-1)}}{[k_{d-1}]_q^{s_{d-1}}} \sum_{n\in\mathbb{Z}\setminus\{0\}} \frac{T(q,s,n,M-1)}{(2\pi in)^2}\left(\frac{\log q}{q-1}\right)^{M+1} \cdot (s)_{M+1}\int_{k_{d-1}}^{\infty} e^{2\pi inx} q^{x(s-2+M)}[x]_q^{-s-M-1}\,dx; \tag{20}$$

$$C_0 = 3\log q \sum_{r=1}^{M-1}\sum_{n\in\mathbb{Z}\setminus\{0\}} \frac{T(q,s,n,r)}{(2\pi in)^2}\left(\frac{\log q}{q-1}\right)^{r} \cdot (s)_r \mathbf{S}^2\zeta_q\left(\vec{s}\,', s_{d-1}+s_d+r\right)$$

and

$$D_0 = -3\log q \sum_{0<k_1<\cdots<k_{d-2}<k_{d-1}} \frac{q^{k_1(s_1-1)+\cdots+k_{d-1}(s_{d-1}-1)}}{[k_1]_q^{s_1}\cdots[k_{d-1}]_q^{s_{d-1}}} \sum_{n\in\mathbb{Z}\setminus\{0\}} \frac{T(q,s,n,M-1)}{(2\pi in)^2} \cdot \left(\frac{\log q}{q-1}\right)^{M}(s)_M \int_{k_{d-1}}^{\infty} e^{2\pi inx} q^{x(s-2+M)}[x]_q^{-s-M}\,dx;$$

$$C_1 = (\log q)^2 \sum_{r=1}^{M-1}\sum_{n\in\mathbb{Z}\setminus\{0\}} \frac{T(q,s,n,r)}{(2\pi in)^2}\left(\frac{\log q}{q-1}\right)^{r-1}(s)_{r-1}\mathbf{S}\zeta_q\left(\vec{s}\,', s_{d-1}+s_d+r-1\right)$$

and

$$D_1 = -(\log q)^2 \sum_{0<k_1<\cdots<k_{d-2}<k_{d-1}} \frac{q^{k_1(s_1-1)+\cdots+k_{d-1}(s_{d-1}-1)}}{[k_1]_q^{s_1}\cdots[k_{d-1}]_q^{s_{d-1}}} \sum_{n\in\mathbb{Z}\setminus\{0\}} \frac{T(q,s,n,M-1)}{(2\pi in)^2} \cdot \left(\frac{\log q}{q-1}\right)^{M-1}(s)_{M-1} \int_{k_{d-1}}^{\infty} e^{2\pi inx} q^{x(s-2+M)}[x]_q^{-s-M+1}\,dx.$$

The next crucial step is to control the summation over k_{d-1} and show that they converge uniformly with respect to q. When $0 < q \leqq 1/2$, this is clear. The only nontrivial part is when $q \uparrow 1$. So, we assume $1/2 < q < 1$. Note that

$$\lim_{q\uparrow 1} T(q,s,n,r) = \frac{1}{(2\pi in)^r} \quad \text{and} \quad \lim_{q\uparrow 1}\frac{\log q}{q-1} = 1.$$

Lemma 6.26 *Let* $s_d = s = \sigma + i\tau$ *and*

$$q_0 := \begin{cases} \max\left\{1/2,\ e^{(6-2\pi)/\tau}\right\} & (\tau > 0), \\ 1/2 & (\tau \leqq 0). \end{cases}$$

Then, for all $q_0 < q < 1$ *and* $k \in \mathbb{N}$, *we have*

$$\left|\frac{\log q}{q-1}\right| < 2 \quad \textit{and} \quad |T(q,s,n,r)| < \frac{1}{(6n)^r}.$$

Proof. Let $f(q) = 2(1-q) + \log q$. Then, $f'(q) = -2 + 1/q < 0$ whenever $q > q_0$. So, f is strictly decreasing on $q_0 < q$. Then, we have $f(q) > f(1) = 0$ whenever $q_0 < q < 1$. This implies that $2(1-q) > -\log q$ from which $\log q/(q-1) < 2$. ■

To bound $T(q,s,n,r)$, we consider each of its factors in definition (19). For each $0 \leqq j < r$, we have

$$\begin{aligned}|2\pi in + (s-1+j)\log q|^2 &= \{(\sigma - 1 + j)\log q\}^2 + (2\pi n + \tau \log q)^2 \\ &\geqq (2\pi n + \tau \log q)^2,\end{aligned}$$

which is independent of j $(j = 0, 1, \ldots, r-1)$. If $\tau \leqq 0$, then clearly $|2\pi in + (s-1+j)\log q| > 6n$. If $\tau > 0$, then it follows from $q > e^{(6-2\pi)/\tau}$ that

$$2\pi n + \tau \log q > 2\pi n + 6 - 2\pi \geqq 6(n-1) + 6 = 6n$$

as desired.

Next, we want to bound the integral terms in D_{-1}. Let $|\sigma_d| < N$ and $|\sigma_{d-1}| < N'$ for some positive integers N and N'. Fix an arbitrary $x > k = k_{d-1}$ and a positive integer

$$M > 16 + 2N + 6\sum_{j=1}^{d-1}(N_j + 1),$$

where we recall $|\Re(s_j)| < N_j \in \mathbb{N}$ $(j = 1, 2, \ldots, d)$ and so $N = N_d$ and $N' = N_{d-1}$.

Then, we have

$$q^{-k(M/6-N'-1)}\left|q^{k(s'-1)}\,q^{x(s-2+M)}\,[x]_q^{-s-M-1}\right| < g(q),$$

where $s' = s_{d-1}$, and, for convenience, $g(q)$ is given by

$$g(q) = q^{x(M-N-2)-\frac{kM}{6}}\,[x]_q^{-M+N-1}. \tag{21}$$

Lemma 6.27 *Let* $1/2 < q < 1$. *Then,* $g(q)$ *is increasing as a function of* q, *so that*

$$g(q) \leqq \lim_{q\uparrow 1} g(q) = \frac{1}{x^{M-N+1}}.$$

Proof. Taking the logarithmic derivative of $g(q)$, we have

$$\frac{g'(q)}{g(q)} = \frac{h(q)}{q(1-q)(1-q^x)},$$

where

$$h(q) := (1-q)\left(1-q^x\right)\left(x(M-N-2)-\frac{kM}{6}\right)$$
$$+(M-N+1)\left\{xq^x(1-q)-q+q^{x+1}\right\}.$$

Then, we get

$$h'(q) = \left\{(x+1)q^x - xq^{x-1} - 1\right\}\left\{x(M-N-2)-\frac{kM}{6}\right\}$$
$$+(M-N+1)\left\{x^2q^{x-1} - x(x+1)q^x - 1 + (x+1)q^x\right\}.$$

Clearly, we have $h'(1) = 0$. Moreover, it is easily seen that

$$\begin{aligned} q^{2-x}h''(q) &= (kM/6+3x)\left\{x^2(1-q)-x(1+q)\right\}+qx(x+1)(M-N+1) \\ &> qx(x+1)(M-N+1) - x(1+q)(kM/6+3x) \\ &\geqq qx^2\left\{M-N+1-\frac{1+q}{q}(3+M/6)\right\} \\ &> qx^2(M/2-N-8) > 0, \end{aligned}$$

where we used the fact that if $q > 1/2$, then $(1+q)/q < 3$. This implies that $h'(q)$ is increasing, by recalling $h'(1) = 0$, so that $h'(q) < 0$, for all $1/2 < q < 1$. It follows that $h(q)$ is decreasing. Since $h(1) = 0$, we find that $h(q) > 0$ for all such q. Therefore, $g'(q) > 0$, and so $g(q)$ is increasing for all $1/2 < q < 1$. This completes the proof of the lemma. ∎

We, now, can bound the innermost sum of D_{-1}. It follows from Lemma 6.17, Lemma 6.26 and Lemma 6.27 that (if $d = 2$, then take $k_0 = 1$)

$$\begin{aligned} \sum_{k=1+k_{d-2}}^{\infty} \left|R(M,q,k,s',s)\right| &< \sum_{k=1+k_{d-2}}^{\infty} (2k)^{N'} \frac{2\zeta(M+1)}{4\pi^2 6^{M-1}} 2^{M+1} \\ &\quad \cdot (N)_{M+1} q^{k(M/6-N'-1)} \int_k^{\infty} \frac{dx}{x^{M-N+1}} \\ &< \frac{(M+1)!}{M-N}\binom{M+N}{M+1} \sum_{k=1+k_{d-2}}^{\infty} \frac{q^{k(M/6-N'-1)}}{k^{M-N-N'}} \\ &< (M+1)!\,(2M)^{M+1} \sum_{k=1+k_{d-2}}^{\infty} q^{k(M/6-N'-1)}, \end{aligned}$$

since $2^{N'+M+2}\zeta(M+1) < 4\pi^2 6^{M-1}$ and $M-N > 2$.

Therefore, by Lemma 6.17, we have

$$|D_{-1}| < (2M)^{2M+2} \sum_{0<k_1<\cdots<k_{d-2}} \left(\prod_{j=1}^{d-2} k_j^{N_j} q^{k_j(-N_j-1)} \right) q^{k_{d-2}(M/6-N_{d-1}-1)},$$

which converges as proven in Theorem 6.18.

Exactly the same argument can apply to the integral terms in D_0 and D_1. These convergence results imply two things. First, we can show by induction on d that (18) gives rise to an analytic continuation of $\zeta_q(s_1, \ldots, s_d)$ as a meromorphic function on $\mathbb{C}^d \setminus \mathcal{D}_d(q)$. Second, also by induction on d, we, now, can conclude that it is legitimate to take the limit $q \uparrow 1$ inside the sums of C_ν and D_ν to get (note that, for any $n \in \mathbb{N}$,

$$\lim_{q\uparrow 1} \mathbf{S}^n \zeta_q(\vec{s}) = \zeta(\vec{s}) \left(\vec{s} \in \mathbb{C}^{d-1} \setminus \mathcal{D}_{d-1}(q)\right))$$

$$\begin{aligned}
\lim_{q\uparrow 1} \zeta_q(\vec{s}', s_{d-1}, s_d) &= \frac{1}{s-1} \zeta(\vec{s}', s_{d-1}+s_d-1) - \frac{1}{2} \zeta(\vec{s}', s_{d-1}+s_d) \\
&\quad + \frac{s}{12} \zeta(\vec{s}', s_{d-1}+s_d+1) - \sum_{r=1}^{M-1} \sum_{n\in\mathbb{Z}\setminus\{0\}} \frac{1}{(2\pi i n)^{r+2}} (s)_{r+1} \zeta(\vec{s}', s_{d-1}+s_d+r+1) \\
&\quad + \sum_{0<k_1<\cdots<k_{d-1}} \frac{1}{k_1^{s_1}\cdots k_{d-1}^{s_{d-1}}} \sum_{n\in\mathbb{Z}\setminus\{0\}} \frac{(s)_{M+1}}{(2\pi i n)^{M+1}} \int_{k_{d-1}}^{\infty} e^{2\pi i n x} x^{-s-M-1}\, dx \\
&= \sum_{r=0}^{M+1} \frac{B_r}{r!} (s_d)_{r-1} \zeta(s_1, \ldots, s_{d-1}+s_d+r-1) \\
&\quad - \frac{1}{(M+1)!} \sum_{0<k_1<\cdots<k_{d-1}} \frac{1}{k_1^{s_1}\cdots k_{d-1}^{s_{d-1}}} \int_{k_{d-1}}^{\infty} \widetilde{B}_{M+1}(x) \frac{(s)_{M+1}}{x^{s+M+1}}\, dx,
\end{aligned}$$

by 6.10(26) and its specialization with $x=1$:

$$\sum_{n\in\mathbb{Z}\setminus\{0\}} \frac{1}{(2\pi i n)^{r+2}} = -\frac{\widetilde{B}_{r+2}(1)}{(r+2)!} = -\frac{B_{r+2}}{(r+2)!}.$$

The main theorem, Theorem 6.25, now mostly follows from Theorem 6.24. The poles at $s_d = m - \frac{2\pi i}{\log q} n$ are given, by the first term in the formula (18) when $m = 1$ and $n = 0$ and by the terms $T(q, s_d, n, r)$, as defined in (19), if $m \leqq 1$ and $n \neq 0$. The locations of the other poles are obtained by induction, using those poles of the q-generalized Zeta function presented in Theorem 6.13. This completes the proof of the main theorem.

Special Values of $\zeta_q(s_1, s_2)$

For integers $n_1, \ldots, n_d$, we set

$$\zeta_q(n_1, \ldots, n_d) = \lim_{s_1 \to n_1} \cdots \lim_{s_d \to n_d} \zeta_q(s_1, \ldots, s_d)$$

and

$$\zeta_q^R(n_1, \ldots, n_d) = \lim_{s_d \to n_d} \cdots \lim_{s_1 \to n_1} \zeta_q(s_1, \ldots, s_d)$$

if the limits exist. Interesting phenomena occur already in the case $d = 2$, and these should be generalized to arbitrary depth. By Theorem 6.20, we get

$$\zeta_q(s_1, s_2) = (1-q)^{s_1+s_2} \sum_{r_1, r_2 = 0}^{\infty} \binom{s_1 + r_1 - 1}{r_1} \binom{s_2 + r_2 - 1}{r_2} \cdot \frac{q^{2s_2+2r_2+s_1+r_1-3}}{\left(1-q^{s_2+r_2-1}\right)\left(1-q^{s_2+r_2+s_1+r_1-2}\right)}$$

$$= (1-q)^{s_1+s_2} \left\{ \frac{q^{2s_2+s_1-3}}{\left(1-q^{s_2-1}\right)\left(1-q^{s_2+s_1-2}\right)} + \frac{s_1 q^{2s_2+s_1-2}}{\left(1-q^{s_2-1}\right)\left(1-q^{s_2+s_1-1}\right)} \right.$$

$$+ \frac{s_2 q^{2s_2+s_1-1}}{\left(1-q^{s_2}\right)\left(1-q^{s_2+s_1-1}\right)} + \frac{s_1 s_2 q^{2s_2+s_1}}{\left(1-q^{s_2}\right)\left(1-q^{s_2+s_1}\right)}$$

$$\left. + \frac{s_1(s_1+1) q^{2s_2+s_1-1}}{2\left(1-q^{s_2-1}\right)\left(1-q^{s_2+s_1}\right)} + \frac{s_2(s_2+1) q^{2s_2+s_1+1}}{2\left(1-q^{s_2+1}\right)\left(1-q^{s_2+s_1}\right)} + \cdots \right\}.$$

Clearly, we have

$$\zeta_q(0,0) = \lim_{s_1 \to 0} \lim_{s_2 \to 0} \zeta_q(s_1, s_2) = \frac{1}{\left(q^2-1\right)(q-1)} - \frac{3}{2(q-1)\log q} + \frac{1}{\log^2 q};$$

$$\zeta_q^R(0,0) = \lim_{s_2 \to 0} \lim_{s_1 \to 0} \zeta_q(s_1, s_2)$$

$$= \frac{1}{\left(q^2-1\right)(q-1)} - \frac{1}{(q-1)\log q} + \frac{q}{2(q-1)\log q}.$$

We readily find that

$$\lim_{q \uparrow 1} \zeta_q(0,0) = \frac{1}{3} \quad \text{and} \quad \lim_{q \uparrow 1} \zeta_q^R(0,0) = \frac{5}{12}, \tag{22}$$

which is consistent with those limiting values in 2.7(18).

Problems

1. Show that

$$\lim_{q\uparrow 1}(1-q)^s \log q \sum_{n\in\mathbb{Z}\setminus\{0\}} \frac{e^{-2\pi ina}}{(2\pi in)^2}\, sB_{q^a}(s+\delta n-1,-s)=0.$$

(See the last part of Section 6.10)

2. Show the limiting values in 6.11(22):

$$\lim_{q\uparrow 1}\zeta_q(0,0)=\frac{1}{3} \quad\text{and}\quad \lim_{q\uparrow 1}\zeta_q^R(0,0)=\frac{5}{12}.$$

We may use the following asymptotic formulas:

$$\frac{1}{t+2}=\frac{1}{2}-\frac{t}{4}+\frac{t^2}{8}+O\left(t^3\right) \quad (t\to 0);$$

$$\frac{1}{\log(1+t)}=\frac{1}{t}+\frac{1}{2}-\frac{t}{12}+\frac{t^2}{24}+O\left(t^3\right) \quad (t\to 0);$$

$$\frac{1}{\log^2(1+t)}=\frac{1}{t^2}+\frac{1}{t}+\frac{1}{12}+0\cdot t+O\left(t^2\right) \quad (t\to 0).$$

3. For notations, here, see Problem 48 in Chapter 2. The q-analogues of MZVs, MZSVs and the multiple polylogarithm with equality are defined, respectively, as follows:

$$\zeta_q[k_1,\ldots,k_r]:=\sum_{n_1>\cdots>n_r>0}\frac{q^{n_1(k_1-1)+\cdots+n_r(k_r-1)}}{[n_1]^{k_1}\cdots[n_r]^{k_r}},$$

$$\zeta_q^*[k_1,\ldots,k_r]:=\sum_{n_1\geqq\cdots\geqq n_r\geqq 1}\frac{q^{n_1(k_1-1)+\cdots+n_r(k_r-1)}}{[n_1]^{k_1}\cdots[n_r]^{k_r}}$$

and

$$\mathrm{Li}^*_{k_1,\ldots,k_r}[t]:=\sum_{n_1\geqq\cdots\geqq n_r\geqq 1}\frac{t^{n_1}}{[n_1]^{k_1}\cdots[n_r]^{k_r}}.$$

Prove the following results:

(a) Sum Formula. For $r<k$ $(r,\,k\in\mathbb{N})$,

$$\sum_{\mathbf{k}\in I_0(k,r)}\zeta^*[\mathbf{k}]=\frac{1}{k-1}\binom{k-1}{r-1}\sum_{\ell=0}^{r-1}\binom{r-1}{\ell}(k-1-\ell)(1-q)^{\ell}\,\zeta_q[k-\ell].$$

(b) Cyclic Sum Formula. For $(k_1,\,\ldots,\,k_r)\in I_0(k,\,r)$,

$$\sum_{i=1}^{r}\sum_{j=0}^{k_i-2}\zeta_q^*[k_i-j,\,k_{i+1},\,\ldots,\,k_r,\,k_1,\,\ldots,\,k_{i-1},\,j+1]$$

$$=\sum_{\ell=0}^{r}(k-\ell)\binom{r}{\ell}(1-q)^{\ell}\,\zeta_q[k-\ell+1],$$

where the empty sum is to be interpreted as nil.

(c) Generating Function of Multiple Polylogarithms.

$$\sum_{k>r>0} \left\{ \sum_{\mathbf{k} \in I_0(k,r)} \mathrm{Li}^*_{\mathbf{k}}[t] \right\} u^{k-r-1} v^{r-1}$$
$$= \frac{1}{1-u-v} \int_0^t \frac{(s;q)_\infty}{(bs;q)_\infty} {}_2\Phi_1(a,b;aq;q,s)\, d_q s,$$

where ${}_2\Phi_1$ is the q-hypergeometric series (see 6.3(28)), the integral is the Jackson q-integral given in 6.2(10),

$$q^{-a-1} = \frac{1}{1-(1-q)(u+v)} \quad \text{and} \quad b = \frac{1-(1-q)u}{1-(1-q)(u+v)}.$$

(see [873, Theorems 1 and 2 and Corollary 3])

a. For $m \in \mathbb{N}_0$,

$$\sum_{n=1}^{\infty} \frac{q^{n(k-m)}}{[n]^{k+1}} = \sum_{\ell=0}^{m} \binom{m}{\ell} (1-q)^\ell \zeta_q[k-\ell+1].$$

(See [873, p. 3033])

4. Continuing Problem 4, denote the q-analogue of MZVs by MqZVs, for short. Define a generating function of MqZVs as follows:

$$\Phi_0[x,y,z] := \sum_{k,r,s=0}^{\infty} \left\{ \sum_{\substack{\mathbf{k} \\ |\mathbf{k}|=k,\, \mathrm{dep}\,\mathbf{k}=r,\, \mathrm{ht}\,\mathbf{k}=s}} \zeta_q[\mathbf{k}] \right\} x^{k-r-s} y^{r-s} z^{s-1}.$$

Then,

$$1 + (z-xy)\,\Phi_0 = \exp\left\{ \sum_{n=2}^{\infty} \zeta_q[n] \sum_{m=0}^{\infty} \frac{(q-1)^m}{m+n} \left(x^{m+n} + y^{m+n} - \alpha^{m+n} - \beta^{m+n} \right) \right\},$$

where $\alpha^{m+n} + \beta^{m+n}$ is a polynomial in x, y and z, determined by

$$\alpha + \beta = x + y + (q-1)(z-xy) \quad \text{and} \quad \alpha\beta = z.$$

(See [875, Theorem 1])

5. If $r \in \mathbb{N}$ and $z \in \mathbb{C} \setminus \{q^{-m}[m]_q \mid m \in \mathbb{N}\}$, then show that

$$\sum_{k_1 > \cdots > k_r > 0} \frac{q^{k_1}}{[k_1]_q} \sum_{j=1}^{r} \frac{1}{[k_j]_q - z q^{k_j}} = \sum_{m=1}^{\infty} \frac{q^{rm}}{[m]_q^r \left([m]_q - z q^m\right)}.$$

(See [169, Theorem 2])

6. Let n, m and $s_1, s_2, \ldots, s_m$ be nonnegative integers. Define the multiply-nested sums

$$Z_n[s_1, \ldots, s_m] := \sum_{n \geqq k_1 \geqq \cdots \geqq k_m \geqq 1} \prod_{j=1}^{m} q^{k_j} [k_j]_q^{-s_j},$$

$$A_n[s_1, \ldots, s_m] := \sum_{n \geqq k_1 \geqq \cdots \geqq k_m \geqq 1} (-1)^{k_1+1} q^{k_1(k_1+1)/2} \begin{bmatrix} n \\ k_1 \end{bmatrix}_q \prod_{j=1}^{m} q^{(s_j-1)k_j} [k_j]_q^{-s_j},$$

with the understanding that $Z_0[s_1, \ldots, s_m] = A_0[s_1, \ldots, s_m] = 0$ and, with empty argument lists, $Z_n[\,] = A_n[\,] = 1$, if $n \in \mathbb{N}$ and $m = 0$. The sums are over all integers k_j, satisfying the indicated inequalities.

For convenience, $\mathbf{Cat}_{j=1}^{m}\{s_j\}$ denotes the concatenated argument sequence $s_1, \ldots, s_m$ and $\{s\}^m = \mathbf{Cat}_{j=1}^{m}\{s\}$ $(m \in \mathbb{N}_0)$ consecutive copies of the argument s.

For n, r and $a_1, b_1, \ldots, a_r, b_r \in \mathbb{N}$, show that

$$Z_n\left[\underset{j=1}{\overset{r-1}{\mathbf{Cat}}}\left\{\{1\}^{a_j-1}, b_j+1\right\}, \{1\}^{a_r-1}, b_r\right] = A_n\left[a_1, \{1\}^{b_1-1}, \underset{j=2}{\overset{r}{\mathbf{Cat}}}\left\{a_j+1, \{1\}^{b_j-1}\right\}\right].$$

(See [166, Theorem 1])

7. If $n \in \mathbb{N}$ and $s > 1$, then show that

$$n\zeta\left[\{s\}^n\right] = \sum_{k=1}^{n} (-1)^{k+1} \zeta\left[\{s\}^{n-k}\right] \sum_{j=0}^{k-1} \binom{k-1}{j} (1-q)^j \zeta[ks-j].$$

(See [167, Theorem 1])

8. For $s, t \in \mathbb{N} \setminus \{1\}$, show that

$$\begin{aligned}\zeta[s]\zeta[t] = &\sum_{a=0}^{s-1}\sum_{b=0}^{s-1-a} \binom{a+t-1}{t-1}\binom{t-1}{b}(1-q)^b \zeta[t+a, s-a-b] \\ &+ \sum_{a=0}^{t-1}\sum_{b=0}^{t-1-a} \binom{a+s-1}{s-1}\binom{s-1}{b}(1-q)^b \zeta[s+a, t-a-b] \\ &- \sum_{j=1}^{\min(s,t)} \frac{(s+t-j-1)!}{(s-j)!\,(t-j)!} \frac{(1-q)^j}{(j-1)!} \varphi[s+t-j],\end{aligned}$$

where, for convenience, the sum $\varphi[s]$ is given by

$$\varphi[s] := \sum_{n=1}^{\infty} \frac{(n-1)\,q^{(s-1)n}}{[n]_q^s} = \sum_{n=1}^{\infty} \frac{n\,q^{(s-1)n}}{[n]_q^s} - \zeta[s].$$

(See [168, Theorem 2.1])

9. Let $\sigma_j(n)$ denote the sum of the jth powers of the divisors of n, given by

$$\sigma_j(n) := \sum_{d|n} d^j \quad \text{and} \quad S_j(q) := \sum_{n=1}^{\infty} \sigma_j(n)\,q^n \quad (q \in \mathbb{C};\ |q| < 1).$$

These generating functions $S_j(q)$ are known to have the following expansions as Lambert series:

$$S_j(q) = \sum_{n=1}^{\infty} \frac{n^j q^n}{1-q^n}.$$

The simplest case is $j = 0$ when $\sigma_0(n) = d(n)$, the divisor function counting the numbers of divisors of n. It was shown by Uchimura [1171] that

$$\sum_{n=1}^{\infty} d(n)\, q^n = \sum_{n=1}^{\infty} n q^n \prod_{j=n+1}^{\infty} \left(1 - q^j\right).$$

Uchimura [1173] studied the following more general expressions:

$$U_k(q) := \sum_{n=1}^{\infty} n^k q^n \prod_{j=n+1}^{\infty} \left(1 - q^j\right) \qquad (k \in \mathbb{N}).$$

Show that

$$U_k(q) = \sum \frac{k!}{n_1! \cdots n_k!} \left(\frac{S_0(q)}{1!}\right)^{n_1} \left(\frac{S_1(q)}{2!}\right)^{n_2} \cdots \left(\frac{S_{k-1}(q)}{k!}\right)^{n_k},$$

where the summation extends over all integers

$$n_1, \ldots, n_k \geqq 0 \quad \text{with} \quad n_1 + 2n_2 + \cdots + kn_k = k.$$

(See [1173])

10. Continued from Problem 9, show that
(a) for $k \in \mathbb{N}$, we have

$$S_{k-1}(q) = \sum \frac{(-1)^{n_1+\cdots+n_k-1}}{n_1! \cdots n_k!} (n_1 + \cdots + n_k - 1)! \left(\frac{U_1(q)}{1!}\right)^{n_1} \cdots \left(\frac{U_k(q)}{k!}\right)^{n_k},$$

where the summation extends over all integers

$$n_1, \ldots, n_k \geqq 0 \quad \text{with} \quad n_1 + 2n_2 + \cdots + kn_k = k;$$

(See [383, Theorem 1])

(b) for $k \in \mathbb{N}$, we have

$$\sum_{m=1}^{\infty} \frac{(-1)^{m-1} q^{\binom{m+1}{2}}}{(q;q)_m (1-q^m)^k} = \sum_{i=1}^{k} \left\{ \sum_{j=0}^{k-i} \binom{k-1}{j+i-1} \frac{s(j+i,i)}{(j+i)!} \right\} U_i(q),$$

where the $s(m,n)$ are the Stirling numbers of the first kind (see Section 1.5).

(See [383, Theorem 3])

11. Let $m, n \in \mathbb{N}$, $\tau = m - n + 1$, a, b, z, q be variables and

$$\mathbb{A} = \left\{ \frac{a - bq}{c - zq}, \frac{a - bq^2}{c - zq^2}, \ldots, \frac{a - bq^n}{c - zq^n} \right\}.$$

Show that

$$\sum_{1\leqq i_1\leqq i_2\leqq\cdots\leqq i_r\leqq n} \frac{a-bq^{i_1}}{c-zq^{i_1}}\,\frac{a-bq^{i_2}}{c-zq^{i_2}}\cdots\frac{a-bq^{i_r}}{c-zq^{i_r}}$$

$$=\frac{c^n\,(zq/c;q)_n}{(q;q)_n\,(az-bc)^{n-1}}\sum_{i=1}^{n}\binom{n}{i}_q \frac{(-1)^{i-1}\,q^{\binom{i+1}{2}-ni}\left(1-q^i\right)\left(a-b\,q^i\right)^m}{\left(c-z\,q^i\right)^{\tau+1}}.$$

(See [464, Proposition 2.1])

12. Show that

$$\sum_{k=1}^{n}\binom{n}{k}(-1)^k f(k)=\frac{(-1)^n}{2\pi i}\int_C \frac{n!}{z(z-1)\cdots(z-n)}f(z)\,dz,$$

where C is a positively-oriented (counter-clockwise) closed contour in the complex z-plane, which encloses the poles at $z=1,2,\ldots,n$ and no other poles of the integrand. By extending the contour of integration and taking the extra negative residues into account, one can get a number of identities, provided that the integral goes to zero, which, incidentally, it does for a large class of functions $f(z)$.

(See [916, p. 552])

13. Let

$$\psi_q(z):=\frac{d}{dz}\left\{\frac{\Gamma_q(z)}{\Gamma_q(z)}\right\}.$$

Show that

$$\psi_q(z)\sim\ln\left(\frac{1-q^z}{1-q}\right)+\frac{\ln q}{2\left(q^{-z}-1\right)}-\sum_{k=1}^{\infty}\frac{B_{2k}}{(2k)!}\left\{\left(\frac{d}{dt}\right)^{2k-1}\left(\frac{q^{z+t}\ln q}{1-q^{z+t}}\right)\right\}\Bigg|_{t=0}$$

$$=\ln\left(\frac{1-q^z}{1-q}\right)+\frac{\ln q}{2\left(q^{-z}-1\right)}-\sum_{k=1}^{\infty}\frac{B_{2k}}{(2k)!}\left\{\frac{\ln q}{(1-q)^z}\right\}^{2k} q^z P_{2k-2}\left(q^z\right),$$

where $P_n(z)$ is a polynomial of degree n, satisfying the following recurrence relation:

$$P_n(z)=\left(z-z^2\right)P'_{n-1}(z)+(nz+1)P_{n-1}(z)\quad(n\in\mathbb{N}),\qquad P_0=1,$$

where the *principal* branch of the logarithm is taken.

(See [838, Theorem 1])

14. Moak [836, Equation (1.1)] defines Γ_q $(q>1)$ as follows:

$$\Gamma_q(x)=\frac{q^{\binom{x}{2}}\left(q^{-1};q^{-1}\right)_\infty(q-1)^{1-x}}{\left(q^{-x};q^{-1}\right)_\infty}.$$

Show that this Γ_q $(q>1)$ also satisfies the functional equation

$$\Gamma_q(x+1)=[x]_q\,\Gamma_q(x).$$

15. One of the most interesting q-analogs of harmonic numbers H_n seems to be $H_n(q)$, given by

$$H_n(q) = \sum_{k=1}^{n} \frac{q^k}{1-q^k}.$$

Show that

$$\sum_{k=1}^{n} \frac{(-1)^{k-1}}{1-q^k}\, q^{k(k+1)/2} \binom{n}{k}_q = H_n(q).$$

(See [1185]; see also [52])

16. Let p be an odd prime and let $n \in \mathbb{N}$ and $r \in \mathbb{Z}$. If $r \in \mathbb{N}$, then show that

$$\sum_{\substack{j=1 \\ (j,p)=1}}^{np} \frac{1}{j^r} = -\sum_{k=1}^{\infty} \binom{-r}{k} L_p\left(r+k,\, \omega^{1-k-r}\right) (pn)^k$$

and

$$\sum_{\substack{j=1 \\ (j,2)=1}}^{2n} \frac{1}{j^r} = -\sum_{k=1}^{\infty} \binom{-r}{k} L_2\,(r+k,\, 1)\;(2n)^k,$$

where $L_p(s, \chi)$ (p a prime) denotes the p-adic L-function attached to a Dirichlet character χ and ω the Teichmüller character.

(See [1210, Theorem 1])

17. The $\underline{q}$-logarithm function is defined by

$$l_{\underline{q}}(x) = \sum_{n=1}^{\infty} \frac{(-1)^{n-1}\,(x-1)^n}{[n]_{\underline{q}}} \qquad (|x-1| < 1),$$

where

$$\underline{q} = (q_1, \ldots, q_r) \quad (0 < q_1, \ldots, q_r < 1) \quad \text{and} \quad [n]_{\underline{q}} = [n]_{q_1} \cdots [n]_{q_r}.$$

Let $0 < q_j < 1$ $(j = 1, \ldots, r)$. The $\underline{q}$-Mahler measure $m_{\underline{q}}(f)$ of a rational function $f(x_1, \ldots, x_n) \in \mathbb{C}(x_1, \ldots, x_n)$ is defined as follows:

$$m_{\underline{q}}(f) = \Re\left\{ \int_0^1 \cdots \int_0^1 l_{\underline{q}}\left(f\left(e^{2\pi i\theta_1}, \ldots, e^{2\pi i\theta_n}\right)\right) d\theta_1 \cdots d\theta_n \right\}.$$

Show that

$$m_{\underline{q}}\left(az\frac{x+1}{y+1}+1\right) = \frac{4}{\pi^2} \sum_{\substack{n \geqq 1 \\ (n \text{ odd})}} \frac{a^n}{[n]_{\underline{q}}\, n^2} \qquad (0 < a \leqq 1).$$

(See Gon *et al.* [493, Theorem 1])

18. Subject to suitable convergence conditions, if we, suppose that

$$c_n = \sum_{m=0}^{\infty} a_{m+n} b_m,$$

then show that

$$\sum_{m=0}^{\infty} b_m \sum_{n=-\infty}^{\infty} a_n = \sum_{n=-\infty}^{\infty} c_n.$$

(See Andrews [46, Transformation Lemma])

19. If $h(z)$ is given by

$$h(z) = \left(-zq^2;\, q^4\right)_\infty \left(\sum_{n=0}^{\infty} \frac{q^{4n^2} z^{2n}}{\left(-zq^2;\, q^4\right)_n} + \sum_{n=1}^{\infty} q^{2n^2} z^{-n} \left(-z^{-1} q^2;\, q^4\right)_n\right),$$

then show that

$$h(z) + zq\, h\left(zq^2\right) = \frac{\left(q^2;\, q^2\right)_\infty \left(-zq;\, q^2\right)_\infty \left(-z^{-1} q;\, q^2\right)_\infty}{\left(-q^2;\, q^2\right)_\infty \left(q;\, q^5\right)_\infty \left(q^4;\, q^5\right)_\infty}.$$

(See Andrews [46, Theorem 2])

20. The coefficients $B_n(\alpha, \theta, q)$ are defined by

$$\sum_{n=-\infty}^{\infty} B_n(\alpha, \theta, q)\, x^n = \frac{\prod_{n=1}^{\infty} \left(1 + 2xq^n \cos\theta + x^2 q^{2n}\right)}{\prod_{n=1}^{\infty} \left(1 + \alpha\, q^n\, x\, e^{i\theta}\right)}.$$

Show that

$$\sum_{n=-\infty}^{\infty} q^{-n} B_{n+m}(\alpha, \theta, q)\, B_n(\beta, \theta, q) = \frac{(\alpha q)_\infty\, (\beta q)_\infty\, (1/\alpha)_m \left(-\alpha\, q\, e^{i\theta}\right)^m}{\{(q)_\infty\}^2}$$

$$\cdot \sum_{n=-\infty}^{\infty} \left(\frac{q^m}{\alpha}\right)_n (1/\beta)_n \left(\alpha\, \beta\, q\, e^{2i\theta}\right)^n.$$

(See Bhargava *et al.* [132, Theorem])

21. A q-analogue of the usual logarithm is defined by

$$\ell_q(x) = \sum_{n=1}^{\infty} \frac{(-1)^{n-1} (x-1)^n}{[n]_q}.$$

Recall the quantum dilogarithm given by (see Kirillov [667]):

$$\mathrm{Li}_{2,q}(x) = \sum_{n=1}^{\infty} \frac{x^n}{n\, [n]_q}.$$

Kurokawa and Wakayama [719] defined the q-function $F_q(x)$ by

$$F_q(x) := \prod_{n=1}^{\infty} \left(1 - q^{-n} e^{2\pi i x}\right)^{q-1}.$$

Show that

$$\int_0^{\frac{\pi}{2}} \ell_q\left(1 - a e^{2ix}\right) dx = -\frac{1}{2i} \left\{\mathrm{Li}_{2,q}(-a) - \mathrm{Li}_{2,q}(a)\right\}$$

$$= \frac{1}{2i} \log\left(\frac{F_q\left(\frac{\log a}{2\pi i} + \frac{1}{2}\right)}{F_q\left(\frac{\log a}{2\pi i}\right)}\right) \qquad (0 < a \leqq 1).$$

(See Kurokawa and Wakayama [719, Theorem 1.2])

22. Show that

$$\int_0^1 \zeta(s,x)\, d_q x = \frac{1}{[1-s]_q} + \sum_{k=0}^{\infty} \binom{-s}{k} \frac{\zeta(s+k)}{[k+1]_q} \qquad (\Re(s) < 1)$$

and

$$\int_0^1 \zeta(s,1-x)\, d_q x = \sum_{k=0}^{\infty} (-1)^k \binom{-s}{k} \frac{\zeta(s+k)}{[k+1]_q} \qquad (s \in \mathbb{C} \setminus \{1\}),$$

where $\zeta(s,x)$ and $\zeta(s)$ denote the generalized (or Hurwitz) zeta function and the Riemann zeta function, and the Jackson integral of a suitable function f is defined by

$$\int_0^1 f(x)\, d_q x = (q-1) \sum_{n=1}^{\infty} f\left(q^{-n}\right) q^{-n} \qquad (q > 1).$$

(See Kurokawa *et al.* [715, Theorem 2])

23. For $n \in \mathbb{N}$, define the Apéry number $A(n)$ by

$$A(n) := \sum_{j=0}^{n} \binom{n+j}{j}^2 \binom{n}{j}^2$$

and suppose that the integers $a(n)$ are given by the following generating function:

$$\sum_{n=1}^{\infty} a(n)\, q^n = q \prod_{n=1}^{\infty} \left(1 - q^{2n}\right)^4 \left(1 - q^{4n}\right)^4.$$

Show that, for every odd prime p,

$$A\left(\frac{p-1}{2}\right) \equiv a(p) \pmod{p}.$$

(Beukers [127]; see also Ahlgren and Ono [14, Eq. (1.6)])

24. Derive the q-Pfaff-Saalschütz formula:

$$ {}_3\Phi_2\left[\begin{array}{c} a,\, b,\, q^{-n}; \\ \\ c,\, abc^{-1}q^{1-n}; \end{array} q,\, q\right] = \frac{\left(\frac{c}{a};\, q\right)_n \left(\frac{c}{b};\, q\right)_n}{(c;\, q)_n \left(\frac{c}{ab};\, q\right)_n}.$$

(See, e.g., Wang [1206, Theorem 4.1]; see also Srivastava ([1070] *and* [1077]) for more general results)

25. Show that

$$\sum_{k=1}^{n} q^{4(n-k)} \frac{\left(1-q^k\right)^2}{(1-q)^2} \frac{\left(1+q^2-2q^{k+1}\right)}{\left(1-q^2\right)} = \begin{bmatrix} n+1 \\ 2 \end{bmatrix}_q^2.$$

(Zhao and Feng [1259, Eq. (2.1)])

26. The q-analogue of the Psi function $\psi(x)$ (Section 1.3) is defined as the logarithmic derivative of the q-Gamma function, that is,

$$\psi_q(x) = \frac{\Gamma_q'(x)}{\Gamma_q(x)},$$

so that we find from 6.2(2) that

$$\begin{aligned}\psi_q(x) &= -\operatorname{Log}(1-q) + \operatorname{Log} q \sum_{n=0}^{\infty} \frac{q^{n+x}}{1-q^{n+x}} \\ &= -\operatorname{Log}(1-q) + \frac{\operatorname{Log} q}{1-q} \sum_{n=1}^{\infty} \frac{q^{nx}}{1-q^n} \qquad (x>0),\end{aligned}$$

which implies that

$$\psi_q(x) = -\operatorname{Log}(1-q) + + \frac{\operatorname{Log} q}{1-q} \int_0^q \frac{t^{x-1}}{1-t}\, d_q t.$$

For $q \in (0, 1)$, show that

$$\psi_{q,m}(x)\, \psi_{q,n}(x) \geqq \left(\psi_{q,\frac{m+n}{2}}(x)\right)^2 \quad \left(x>0;\ m,\ n,\ \frac{m+n}{2} \in \mathbb{N}\right),$$

where $\psi_{q,n}(x) := \psi_q^{(n)}(x)$.

(Brahim [170]; see also [699])

27. For a suitably bounded sequence $\{\Omega_n\}_{n\in\mathbb{N}_0}$ of essentially arbitrary complex numbers, show that the following multiple q-summation formula holds true:

$$\sum_{k=0}^{\infty}\sum_{k_1,\ldots,k_r=0}^{n}\Omega(k)\frac{(q^{-n};q)_{k_1}\cdots(q^{-n};q)_{k_r}(\gamma q^n;q)_{k+k_1+\cdots+k_r}}{(q;q)_{k_1}\cdots(q;q)_{k_r}(\gamma;q)_{k+k_1+\cdots+k_r}}\frac{(\beta_1;q)_{k_1}\cdots(\beta_r;q)_{k_r}}{(\beta_1 q;q)_{k_1}\cdots(\beta_r q;q)_{k_r}}$$

$$=(\beta_1\cdots\beta_r)^n\frac{[(q;q)_n]^r(\gamma/(\beta_1\cdots\beta_r);q)_n}{(\gamma;q)_n(\beta_1 q;q)_n\cdots(\beta_r q;q)_n}\sum_{k=0}^{\infty}\Omega(k)\frac{(\gamma q^n/(\beta_1\cdots\beta_r);q)_k}{(\gamma/(\beta_1\cdots\beta_r);q)_k},$$

provided that the series involved are absolutely convergent.

(Lin and Srivastava [766, p. 319]; see also Problem 62 of Chapter 1 for the limit case when $q\to 1$)

28. For the basic (or q-) Genocchi polynomials $G_{n;q}^{(\alpha)}(x)$ and the basic (or q-) Genocchi numbers $\widetilde{G}_{n;q}^{(\alpha)}$ and $G_{n;q}^{(\alpha)}$ (see, e.g., Section 1.7), defined for $q,\alpha\in\mathbb{C}$ $(|q|<1)$ by

$$(2z)^{\alpha}\sum_{n=0}^{\infty}\frac{([\alpha]_q)_n}{[n]_q!}(-1)^n q^{n+x}e^{z[n+x]_q}=\sum_{n=0}^{\infty}G_{n;q}^{(\alpha)}(x)\frac{z^n}{n!}$$

and

$$\widetilde{G}_{n;q}^{(\alpha)}:=2^n\,G_{n;q}^{(\alpha)}\left(\frac{\alpha}{2}\right),$$

respectively, so that

$$G_{n;q}(x):=G_{n;q}^{(1)}(x)\quad\text{and}\quad\widetilde{G}_{n;q}:=2^n\,G_{n;q}\left(\frac{1}{2}\right)\qquad(n\in\mathbb{N}_0),$$

derive the following limit formulas:

$$\lim_{q\to 1}\left\{G_{n;q}^{(\alpha)}(x)\right\}=G_n^{(\alpha)}(x)\quad\text{and}\quad\lim_{q\to 1}\left\{G_{n;q}^{(\alpha)}\right\}=G_n^{(\alpha)}.$$

(*cf.*, e.g., Srivastava [1091]; see also the many references relevant to this subject, which are cited therein, as well as in this book)

Remark 5 For basic (or q-) extensions of the Apostol-Genocchi polynomials $G_n(x;\lambda)$ and the Apostol-Genocchi polynomials $G_n^{(\alpha)}(x;\lambda)$ of (real or complex) order α, defined by 1.8(56), the interested reader may refer to a number of recent works by (among others) Luo [786].

29. For suitably bounded sequences $\{\Omega_n\}_{n\in\mathbb{N}_0}$ and $\{\Lambda_{m,n}\}_{m,n\in\mathbb{N}_0}$, show that

$$\sum_{n=0}^{\infty} \frac{\prod_{j=1}^{\ell} (c_j;q)_n}{\prod_{j=1}^{m} (d_j;q)_n} \frac{\Omega_n t^n}{(q;q)_n} \sum_{k=0}^{[n/N]} \frac{(q^{-n};q)_{Nk} \prod_{j=1}^{m} (q^{1-n}/d_j;q)_{Nk}}{\prod_{j=1}^{\ell} (q^{1-n}/c_j;q)_{Nk}}$$

$$\cdot q^{\left(1-(\ell-m-1)n\right)Nk} \frac{\Lambda_{n,k} z^k}{(q;q)_k}$$

$$= \sum_{n,k=0}^{\infty} q^{-\frac{1}{2}(\ell-m-1)Nk(Nk+1)} \frac{\prod_{j=1}^{\ell} \left\{c_j^{Nk} \cdot (c_j;q)_n\right\}}{\prod_{j=1}^{m} \left\{d_j^{Nk} \cdot (d_j;q)_n\right\}} \Omega_{n+Nk} \Lambda_{n+Nk,k} \frac{t^n}{(q;q)_n}$$

$$\cdot \frac{\left(z\left\{(-1)^{\ell-m+1} t\right\}^N\right)^k}{(q;q)_k} \qquad (N \in \mathbb{N}).$$

Also deduce its special case, involving the basic (or q-) hypergeometric function ${}_r\Phi_s$, defined by 6.3(28).

(See Srivastava [1066, p. 334, Eq. (3.9)])

30. For $q, \alpha \in \mathbb{C}$ ($|q| < 1$), let the q-Bernoulli polynomials $B_{n;q}^{(\alpha)}(x)$ of order α in q^x and the q-Euler polynomials $E_{n;q}^{(\alpha)}(x)$ of order α in q^x be defined by means of the following generating functions:

$$(-t)^\alpha \sum_{n=0}^{\infty} \frac{([\alpha]_q)_n}{[n]_q!} q^{n+x} e^{[n+x]_q t} = \sum_{n=0}^{\infty} B_{n;q}^{(\alpha)}(x) \frac{t^n}{n!}$$

and

$$2^\alpha \sum_{n=0}^{\infty} \frac{([\alpha]_q)_n}{[n]_q!} (-1)^n q^{n+x} e^{[n+x]_q t} = \sum_{n=0}^{\infty} E_{n;q}^{(\alpha)}(x) \frac{t^n}{n!},$$

respectively. Show that each of the following relationships holds true:

$$B_{n;q}^{(\alpha)}(x+y) = \sum_{k=0}^{n} \binom{n}{k} \left[\frac{1}{2}\left(q^{(k-\alpha)x} B_{k;q}^{(\alpha)}(y) + q^{n-k-x-\alpha+2} \mathfrak{B}_{k;q;x}^{(\alpha)}(y)\right)\right.$$

$$\left. + \frac{1}{2} k q^{n-k-x-\alpha+2} \mathfrak{B}_{k-1;q;x}^{(\alpha-1)}(y)\right] E_{n-k;q}(x) \qquad (n \in \mathbb{N}_0;\ \alpha \in \mathbb{C}),$$

$$E_{n;q}^{(\alpha)}(x+y) = \sum_{k=0}^{n} \frac{1}{k+1} \binom{n}{k} \left[q^{n-k-x-\alpha+1}\left(2\mathfrak{E}_{k+1;q;x}^{(\alpha-1)}(y) - \mathfrak{E}_{k+1;q;x}^{(\alpha)}(y)\right)\right.$$

$$\left. - q^{(k+1)x} E_{k+1;q}^{(\alpha)}(y)\right] B_{n-k;q}(x) + \frac{2^\alpha q^y (q^{n+1}-1)}{(n+1)(-q;q)_\alpha} B_{n+1;q}(x)$$

$$(n \in \mathbb{N}_0;\ \alpha \in \mathbb{C})$$

and

$$B_{n;q}(x) := B^{(1)}_{n;q}(x) = \sum_{k=0}^{n} \binom{n}{k} \left[\frac{1}{2} \left(q^{(k-1)x} B_{k;q} + q^{n-k-x+1} B_{k;q;x}(0) \right) \right.$$
$$\left. + \frac{1}{2}\, k q^{n-k} (1 - q^{x-1})^{k-1} \right] E_{n-k;q}(x),$$

where the polynomials $\mathfrak{B}^{(\alpha)}_{n;q;y}(x)$ and $\mathfrak{E}^{(\alpha)}_{n;q;y}(x)$ in q^x are as follows:

$$\mathfrak{B}^{(\alpha)}_{n;q;y}(x+1) = \sum_{k=0}^{n} \binom{n}{k} q^{(k-\alpha+1)y} B^{(\alpha)}_{k;q}(x)$$

and

$$\mathfrak{E}^{(\alpha)}_{n;q;y}(x+1) = \sum_{k=0}^{n} \binom{n}{k} q^{(k+1)y} E^{(\alpha)}_{k;q}(x),$$

which yield the following obvious relationships:

$$\mathfrak{B}^{(\alpha)}_{n;q;1}(x) = B^{(\alpha)}_{n;q}(x) \quad \text{and} \quad \mathfrak{E}^{(\alpha)}_{n;q;1}(x) = E^{(\alpha)}_{n;q}(x), \tag{2.15}$$

respectively. We also assume (as usual) that

$$\mathfrak{B}_{n;q;y}(x) := B^{1}_{n;q;y}(x) \quad \text{and} \quad \mathfrak{E}_{n;q;y}(x) := E^{1}_{n;q;y}(x).$$

(See Luo and Srivastava [790, p. 249, Theorem 1; p. 251, Theorem 2]; see also Srivastava and Pintér [1105, p. 379, Theorem 1; p. 380, Theorem 2], Cheon [254, p. 368, Theorem 3] and Sections 1.7 and 6.6 to 6.9)

31. Making use of the definitions and notations in Problem 30 above, show that

$$q^{\alpha-1} B^{(\alpha)}_{n;q}(x+1) - B^{(\alpha)}_{n;q}(x) = n B^{(\alpha-1)}_{n-1;q}(x) \quad (n \in \mathbb{N} \setminus \{1\}),$$
$$B^{(\alpha)}_{n;q}(x+y) = \sum_{k=0}^{n} \binom{n}{k} B^{(\alpha)}_{k;q}(x) q^{(k-\alpha+1)y} [y]_q^{n-k},$$
$$q^{\alpha-1} E^{(\alpha)}_{n;q}(x+1) + E^{(\alpha)}_{n;q}(x) = 2 E^{(\alpha-1)}_{n;q}(x) \quad (n \in \mathbb{N} \setminus \{1\})$$

and

$$E^{(\alpha)}_{n;q}(x+y) = \sum_{k=0}^{n} \binom{n}{k} E^{(\alpha)}_{k;q}(x) q^{(k+1)y} [y]_q^{n-k},$$
$$q^{\alpha-1} \mathfrak{B}^{(\alpha)}_{n;q;y}(x+1) - \mathfrak{B}^{(\alpha)}_{n;q;y}(x) = n \mathfrak{B}^{(\alpha-1)}_{n-1;q;y}(x) \quad (n \in \mathbb{N} \setminus \{1\})$$

and

$$q^{\alpha-1} \mathfrak{E}^{(\alpha)}_{n;q;y}(x+1) + \mathfrak{E}^{(\alpha)}_{n;q;y}(x) = 2 \mathfrak{E}^{(\alpha-1)}_{n;q;y}(x) \quad (n \in \mathbb{N}_0).$$

(See Luo and Srivastava [790, pp. 245–248, Lemmas 1 to 3])

7 Miscellaneous Results

Over one and a half centuries ago, von Staudt and Clausen proved independently (and published almost simultaneously) a useful theorem for the calculation of the higher-order Bernoulli numbers, which has recently been generalized and proven in many different ways in various investigations dealing with evaluations of the Bernoulli and Euler polynomials at rational arguments. Closed-form summations of several trigonometric series have also been of recent concern by various authors. In addition to presenting some of these developments in this chapter, we establish several integrals involving a certain periodic function and various other families of definite integrals involving logarithms of the Gamma function, by introducing some not-too-familiar mathematical constants that stem from the use of the Euler-Maclaurin summation formula 1.3(68).

7.1 A Set of Useful Mathematical Constants

There are some classes of mathematical constants involved naturally in the Gamma and multiple Gamma functions. Here, we introduce those known mathematical constants associated with the Gamma and multiple Gamma functions and show how they are involved, if possible (see Choi [265]).

Euler-Mascheroni Constant γ

In addition to the brief history of γ and its various integral representations in Chapter 1, it is remarked that the Euler-Mascheroni constant γ is the third important mathematical constant next to π and e, whose transcendence were shown by Ferdinand Lindemann in 1882 and Charles Hermite in 1873, respectively. The mathematical constants π, e and γ are often referred to as the *holy trinity*. Very recently, in a tantalizing blend of history and mathematics, Havil takes us on a journey through logarithms and the harmonic series, the two defining elements of γ as in 1.1(3), toward the first account of γ's place in mathematics in his book [543]. Finch [449, pp. 35–40] recorded a fairly comprehensive bibliography of as many as 136 references on the γ.

We can show that the infinite product

$$\prod_{n=1}^{\infty}\left(1+\frac{z}{n}\right)e^{-\frac{z}{n}}$$

Zeta and q-Zeta Functions and Associated Series and Integrals. DOI: 10.1016/B978-0-12-385218-2.00007-4

converges in the finite complex plane $\mathbb{C}$ to an entire function that has simple zeros at $z=-1,-2,-3,\ldots$, this argument yields that

$$\prod_{n=1}^{\infty}\left(1+\frac{z}{n}\right)^{-1} e^{\frac{z}{n}}$$

converges on every compact subset in $\mathbb{C}\setminus\{-1,-2,-3,\ldots\}$ to a function with simple poles at $z=-1,-2,\ldots$. Using this fact, Weierstrass defined the Gamma function, $\Gamma(z)$, is a meromorphic function on $\mathbb{C}$ with simple poles at $z=0,-1,-2,\ldots$, given by (see 1.1(2))

$$\Gamma(z)=\frac{e^{-cz}}{z}\prod_{n=1}^{\infty}\left(1+\frac{z}{n}\right)^{-1} e^{\frac{z}{n}}, \tag{1}$$

where c *is a constant chosen so that* $\Gamma(1)=1$. Indeed, it is easy to show that the chosen constant c in (1) corresponds to the Euler-Mascheroni constant γ in 1.1(3).

Series Representations for γ

Here, we summarize some known series representations for γ (some of which are seen already recorded in Chapter 3, yet, for completeness, rewritten here) and point out that one of those series representations seems *incorrectly* recorded (see [264, 284, 543]).

We start by recalling a well-known series representation for γ:

$$\gamma=\sum_{k=2}^{\infty}(-1)^k\frac{\zeta(k)}{k}, \tag{2}$$

where the Riemann Zeta function $\zeta(s)$ is defined by 2.3(1).

$$\gamma=1-\sum_{k=2}^{\infty}\frac{\zeta(k)-1}{k}. \tag{3}$$

$$\gamma=1-\log 2+\sum_{k=2}^{\infty}(-1)^k\frac{\zeta(k)-1}{k}. \tag{4}$$

$$\gamma=1-\frac{1}{2}\log 2-\sum_{k=1}^{\infty}\frac{\zeta(2k+1)-1}{2k+1}. \tag{5}$$

$$\gamma=\log 2-\sum_{k=1}^{\infty}\frac{\zeta(2k+1)-1}{k+1}. \tag{6}$$

$$\gamma=\log 2-\sum_{k=1}^{\infty}\frac{\zeta(2k+1)}{(2k+1)\,2^{2k}}. \tag{7}$$

$$\gamma=1-\log\frac{3}{r2}-\sum_{k=1}^{\infty}\frac{\zeta(2k+1)-1}{(2k+1)\,2^{2k}}. \tag{8}$$

$$\gamma = 2 - 2\log 2 - \sum_{k=1}^{\infty} \frac{\zeta(2k+1)-1}{(k+1)(2k+1)}. \tag{9}$$

$$\gamma = 1 + \frac{1}{3}\log\frac{8}{15} - \sum_{k=1}^{\infty} \frac{\zeta(2k+1)-1}{2k+1}\left(\frac{3}{2}\right)^{2k}. \tag{10}$$

$$\gamma = 1 - \frac{1}{2}\log 6 + \sum_{k=2}^{\infty} (-1)^k \frac{\zeta(k)-1}{k}\, 2^{k-1}. \tag{11}$$

$$\gamma = 1 - \sum_{k=1}^{\infty} \frac{\zeta(2k+1)}{(k+1)(2k+1)}, \tag{12}$$

which is commented on in the work of Ramanujan [964], who gave a general formula containing (11) as a very special case.

$$\gamma = \log 2 - 2\sum_{k=1}^{\infty} k \sum_{j=A_{k-1}+1}^{A_k} \frac{1}{(3j)^3 - 3j}, \tag{13}$$

where

$$A_k := \frac{1}{2}(3^k - 1) \quad (k = 0,\, 1,\, 2,\, \ldots).$$

$$\gamma = \frac{1}{2} - \frac{1}{3} + \sum_{k=2}^{\infty} k \sum_{i=0}^{2^k-1} \frac{(-1)^i}{2^k + i}. \tag{14}$$

$$\gamma = \frac{1}{2} + \sum_{k=1}^{\infty} k \sum_{i=2^{k-1}}^{2^k-1} 12i(2i+1)(2i+2). \tag{15}$$

$$\gamma = 1 - \sum_{k=1}^{\infty} k \sum_{i=2^{k-1}+1}^{2^k} \frac{1}{(2i-1)(2i)}. \tag{16}$$

Campbell [208, p. 200] recorded an interesting series representation for γ that can easily be written in the form:

$$\gamma = 1 - \log\frac{3}{2} - \sum_{k=1}^{\infty} \frac{\zeta(2k+1)-1}{(2k+1)^{2k}}. \tag{17}$$

We note that the expression of γ in (17) *seems to be the most rapidly convergent series* among the ever-known series representations of γ, by observing $1 < \zeta(2k+1) < 2$ for each positive integer k and the following rough estimations:

$$0 < \sum_{k=21}^{\infty} \frac{\zeta(2k+1)-1}{(2k+1)^{2k}} < 10^{-65} \quad \text{and} \quad 0 < \sum_{k=41}^{\infty} \frac{\zeta(2k+1)-1}{(2k+1)^{2k}} < 10^{-153}.$$

However, when (17) is compared with (8), it is carefully concluded that the expression (17) is *incorrectly* recorded.

Choi and Nash [277] obtained a class of integral representations of the Glaisher-Kinkelin constant A, defined by 1.4(2):

$$\log A = \frac{1}{8} + \frac{1}{24}\left(3 + \frac{2}{p^2-1}\right)\log p - \frac{1}{p^2-1}\sum_{j=1}^{(p-1)/2}\left[j\log j + p\int_0^{\left[\frac{j}{p}\right]}\left(t-\frac{j}{p}\right)\{\psi(1+t)+\psi(1-t)\}\,dt\right], \tag{18}$$

where p is an odd positive integer greater than 1.

By using the theory of the double Gamma function Γ_2, the following log-Gamma integral was evaluated by Choi and Srivastava [289, p. 100, Eq. (2.29)] (see also 3.4(496)):

$$\int_0^{\frac{1}{2}} \log\Gamma(t+1)\,dt = -\frac{1}{2} - \frac{7}{24}\log 2 + \frac{1}{4}\log\pi + \frac{3}{2}\log A, \tag{19}$$

which was equivalent to 2.2(37). Gosper [498, p. 71] made use of the Polylogarithm function (*cf.* [752]) to evaluate the log-Gamma integral (19) in the form:

$$\int_0^{\frac{1}{2}} \log\Gamma(t+1)\,dt = -\frac{1}{2} + \frac{\gamma}{8} - \frac{1}{6}\log 2 + \frac{3}{8}\log\pi - \frac{3\zeta'(2)}{4\pi^2}, \tag{20}$$

so that, by comparing (19) and (20), we obtain the following relationship among π, γ, A and $\zeta'(2)$:

$$\zeta'(2) = \pi^2\left(\frac{\gamma}{6} + \frac{1}{6}\log(2\pi) - 2\log A\right), \tag{21}$$

or, equivalently,

$$\gamma = \frac{6\zeta'(2)}{\pi^2} + \log\frac{A^{12}}{2\pi}. \tag{22}$$

Conversely, if we apply the known relationship 2.1(31) in (21), we find that (see [294, p. 240, Eq. (18)])

$$\zeta'(-1) = \frac{1}{12}[1-\gamma-\log(2\pi)] + \frac{\zeta'(2)}{2\pi^2}, \tag{23}$$

which can also be obtained by appealing appropriately to Riemann's functional equation for $\zeta(z)$ (see 2.3(11)). The relationship (22) is recorded *erroneously* in the work of Voros [1201, p. 453, Eq. (6.25)].

Bendersky [114] (see also [6, p. 199]) presented a set of constants, including B and C, defined, respectively, by 1.3(69) and 1.3(70): there exists constants D_k, defined by

$$\log D_k := \lim_{n\to\infty}\left(\sum_{m=1}^{n} m^k \log m - p(n,k)\right) \quad (k \in \mathbb{N}_0), \tag{24}$$

where the definition of $p(n,k)$ in Adamchik [6, p. 198, Eq. (20)] is *corrected* here as follows:

$$\begin{aligned} p(n,k) :=& \frac{nk}{2}\log n + \frac{n^{k+1}}{k+1}\left(\log n - \frac{1}{k+1}\right) \\ &+ k!\sum_{j=1}^{k} \frac{n^{k-j}B_{j+1}}{(j+1)!\,(k-j)!}\left[\log n + (1-\delta_{kj})\sum_{\ell=1}^{j}\frac{1}{k-\ell+1}\right] \end{aligned}$$

and δ_{kj} is the Kronecker symbol, defined by $\delta_{kj} = 0$ $(k \neq j)$ and $\delta_{kj} = 1$ $(k = j)$.

For the constants D_k $(k \in \mathbb{N}_0)$, defined in (24), we can show that

$$D_0 = (2\pi)^{\frac{1}{2}}, \quad D_1 = A, \quad D_2 = B \quad \text{and} \quad D_3 = C$$

and

$$\log D_k = \frac{B_{k+1}H_k}{k+1} - \zeta'(-k) \quad (k \in \mathbb{N}_0), \tag{25}$$

where B_n are the Bernoulli numbers and H_n are the harmonic numbers.

The constants introduced in this section can be seen to be involved in the theory of multiple Gamma functions. For example, the following log-multiple Gamma integral (see [269, p. 523, Eq. (2.50)])

$$\begin{aligned} \int_0^{\frac{3}{2}} \log\Gamma_3(t+2)\,dt =& \int_0^1 \log G(t+1)\,dt + \int_0^1 \log\Gamma_3(t+1)\,dt \\ &+ \int_0^{\frac{1}{2}} \log\Gamma(t+1)\,dt + 2\int_0^{\frac{1}{2}} \log G(t+1)\,dt + \int_0^{\frac{1}{2}} \log\Gamma_3(t+1)\,dt \\ =& -\frac{259}{768} - \frac{29}{1920}\log 2 + \frac{9}{16}\log\pi - \frac{15}{16}\log A - \frac{5}{4}\log B + \frac{15}{16}\log C. \end{aligned} \tag{26}$$

A Class of Constants Analogous to $\{D_k\}$

Setting $f(x) = \frac{1}{x}$, $a = 1$, $b = N$, $K = 1$ in the Euler-Maclaurin summation formula 2.7(21) and taking the limit of the resulting equation as $N \to \infty$, we obtain the *Euler-Mascheroni constant* γ given in 1.1(3).

By appealing to the Leibniz's rule for differentiation, we find that the mth order derivative of $f(x) := x^{k-1} \log x$ $(k \in \mathbb{N})$ is

$$f^{(m)}(x) = \alpha_{k,m} x^{k-m-1} \log x + \beta_{k,m} x^{k-m-1} \quad \left(0 \leqq m \leqq k-1;\ m \in \mathbb{N}_0\right), \tag{27}$$

where the constants $\alpha_{k,m}$ and $\beta_{k,m}$ are given as follows:

$$\alpha_{k,m} = \frac{(k-1)!}{(k-m-1)!} \quad \text{and} \quad \beta_{k,m} = \frac{(k-1)!}{(k-m-1)!} \sum_{j=1}^{m} \frac{1}{k-m-1+j} \tag{28}$$
$$\left(0 \leqq m \leqq k-1;\ m \in \mathbb{N}_0\right).$$

Note also that

$$f^{(k)}(x) = \frac{\alpha_{k,k-1}}{x} \quad \text{and} \quad f^{(k+1)}(x) = -\frac{\alpha_{k,k-1}}{x^2}. \tag{29}$$

We prove the following identity:

$$\sum_{m=1}^{k-1} \frac{(-1)^m}{m!} B_m \beta_{k,m-1} = \frac{(-1)^k}{k} \sum_{m=1}^{k} \binom{k}{m} \frac{1}{m} B_{k-m} + \frac{1}{k^2} \quad (k \in \mathbb{N}), \tag{30}$$

the empty sum being interpreted (as usual, in what follows) to be nil.

Indeed, since (30) holds trivially for $k = 1$, we assume $k \in \mathbb{N} \setminus \{1\}$. Let

$$\mathcal{S}_k := (-1)^k k \sum_{m=1}^{k-1} \frac{(-1)^m}{m!} B_m \beta_{k,m-1} \quad (k \in \mathbb{N} \setminus \{1\}),$$

which, upon setting $n = k - m$ and using (28), yields

$$\mathcal{S}_k = \sum_{n=1}^{k-1} (-1)^n \binom{k}{n} B_{k-n} \sum_{j=1}^{k-n-1} \frac{1}{n+j}. \tag{31}$$

Define a function $\mathcal{H}(x)$ given by

$$\mathcal{H}(x) := \sum_{n=1}^{k-1} (-1)^n \binom{k}{n} B_{k-n} \sum_{j=1}^{k-n-1} x^{n-1+j}. \tag{32}$$

Since the innermost sum is a finite geometric series and

$$\mathcal{S}_k = \int_0^1 \mathcal{H}(x)\,dx,$$

we find that

$$\mathcal{H}(x) = \frac{h(x) - x^{k-1}h(1)}{1-x}, \tag{33}$$

where, for convenience,

$$h(x) := \sum_{n=1}^{k-1} \binom{k}{n} B_{k-n}\,(-x)^n.$$

It follows from Equations 1.6(3) and 1.6(4) that

$$\begin{aligned} h(x) &= B_k(-x) - B_k - (-x)^k \\ &= B_k(1-x) - B_k - k(-x)^{k-1} - (-x)^k, \end{aligned} \tag{34}$$

which, upon setting $x = 1$, yields

$$h(1) = (-1)^k(k-1). \tag{35}$$

Putting (34) and (35) into (33), we get

$$\mathcal{H}(x) = \frac{B_k(1-x) - B_k}{1-x} + (-1)^k x^{k-1},$$

which, upon integrating from 0 to 1 and changing the variable $t = 1 - x$, gives

$$\mathcal{S}_k = \int_0^1 \frac{B_k(t) - B_k}{t}\,dt + \frac{(-1)^k}{k}. \tag{36}$$

Applying Equation 1.7(3) to (36), we obtain

$$\mathcal{S}_k = \sum_{m=1}^{k} \binom{k}{m} \frac{1}{m} B_{k-m} + (-1)^k \frac{1}{k}, \tag{37}$$

which, upon dividing by $(-1)^k k$, leads to the desired identity (30).

Setting $f(x) = x^{k-1}\log x$ $(k \in \mathbb{N})$, $a = 1$, $b = N$ $(N \in \mathbb{N} \setminus \{1\})$ and $K = k+1$ in 2.7(21) and using (30), we can obtain a class of mathematical constants $\{C_k\}_{k\in\mathbb{N}}$, defined for $k \in \mathbb{N}$ by

$$\begin{aligned} C_k := \lim_{N\to\infty} \Bigg[& \sum_{n=1}^{N} n^{k-1}\log n - \frac{N^k}{k}\log N + \frac{N^k}{k^2} \\ & - \sum_{m=1}^{k} \frac{(-1)^m}{m!} B_m\, \alpha_{k,m-1} N^{k-m}\log N - \sum_{m=1}^{k-1} \frac{(-1)^m}{m!} B_m\, \beta_{k,m-1} N^{k-m} \Bigg], \end{aligned} \tag{38}$$

where $\alpha_{k,m}$ and $\beta_{k,m}$ are defined by (28) (see [274]).

Remark 1 In the process of getting the constants C_k in (38), we find that

$$\begin{aligned} C_k = & -\frac{(-1)^k}{k} \sum_{m=1}^{k} \binom{k}{m} \frac{1}{m} B_{k-m} + \frac{(-1)^k}{k(k+1)} B_{k+1} \\ & - \frac{(-1)^k}{k(k+1)} \int_1^{\infty} \frac{B_{k+1}(\{x\})}{x^2}\, dx \quad (k \in \mathbb{N}). \end{aligned} \tag{39}$$

Remark 2 It follows from the Stirling's formula 1.1(32) for $n!$ that $C_1 = \log\sqrt{2\pi}$. Note also that $C_2 = \log A$, where A is the *Glaisher-Kinkelin constant* given in 1.4(2), $C_3 = \log B$ and $C_4 = \log C$, where the constants B and C are given in 1.4(69) and 1.4(70), respectively.

Remark 3 We find that the constants $\{C_k\}_{k\in\mathbb{N}}$ are connected with Bendersky-Adamchik constants $\{D_k\}_{k\in\mathbb{N}}$ in (24) as follows:

$$\log D_{k-1} = C_k \ (k \in \mathbb{N}). \tag{40}$$

Remark 4 As a matter of fact, Choi and Lee [274] introduced the constants $\{C_k\}_{k\in\mathbb{N}}$ in (38) to evaluate the following family of series associated with the Riemann zeta function:

$$\sum_{m=2}^{\infty} \frac{(-1)^m}{m+k}\zeta(m) = \frac{1}{k} + \frac{\gamma}{k+1} + \sum_{\ell=1}^{k} (-1)^{\ell} \binom{k}{\ell-1} C_{\ell} \quad (k \in \mathbb{N}), \tag{41}$$

which may be evaluated as a special case of a generalized theorem very recently developed [297, Theorem 7.2, p. 403].

Other Classes of Mathematical Constants

Choi *et al.* [310, p. 112, Eq. (36)] introduced a class of mathematical constants A_p $(p > 0)$, defined by

$$\begin{aligned}\log A_p := \lim_{n\to\infty}\Bigg[\sum_{k=1}^{n}\left(k+\frac{1}{p}\right)\log\left(k+\frac{1}{p}\right)\\ -\left\{\frac{n^2}{2}+\left(\frac{1}{2}+\frac{1}{p}\right)n+\frac{1}{2p^2}+\frac{1}{2p}+\frac{1}{12}\right\}\log\left(n+\frac{1}{p}\right)+\frac{n^2}{4}+\frac{n}{2p}\Bigg],\end{aligned} \tag{42}$$

by making use of the Euler-Maclaurin summation formula 1.4(68). It is easy to see that (*cf.* [310, p. 114])

$$\log A_1 = \log A - \frac{1}{4}, \tag{43}$$

where A denotes the Glaisher-Kinkelin constant. Choi *et al.* [310, p. 112, Eq. (37)] also presented another class of mathematical constants C_p $(p > 1)$:

$$\begin{aligned}\log C_p := \lim_{n\to\infty}\Bigg[\sum_{k=1}^{n}\left(k-\frac{1}{p}\right)\log\left(k-\frac{1}{p}\right)\\ -\left\{\frac{n^2}{2}+\left(\frac{1}{2}-\frac{1}{p}\right)n+\frac{1}{2p^2}-\frac{1}{2p}+\frac{1}{12}\right\}\log\left(n-\frac{1}{p}\right)+\frac{n^2}{4}-\frac{n}{2p}\Bigg],\end{aligned} \tag{44}$$

where (obviously)

$$C_p = A_{-p}. \tag{45}$$

In terms of the mathematical constants A_p and C_p, the following log-Gamma integrals were evaluated earlier by Choi *et al.* [310, p. 114, Eqs. (39) and (40)]:

$$\int_0^{1/p}\log\Gamma(1+t)\,dt = \log A - \log A_p + \frac{1}{2p}\log(2\pi) - \frac{1}{2p} - \frac{3}{4p^2} \quad (p > 0) \tag{46}$$

and

$$\int_0^{1/p}\log\Gamma(1-t)\,dt = \log C_p - \log A + \frac{1}{2p}\log(2\pi) - \frac{1}{2p} + \frac{3}{4p^2} \quad (p > 1). \tag{47}$$

With a view to deriving a relationship between

$$A_p \quad \text{and} \quad \zeta'\left(-1, \frac{1}{p}\right) \quad (p > 0)$$

analogous to that in 2.1(31), we set

$$f(x) := \left(x + \frac{1}{p}\right)^{-s} \qquad (p > 0)$$

in the Euler-Maclaurin summation formula 1.4(68) with $a = 0$. We, then, obtain

$$\begin{aligned}\sum_{k=1}^{n}\left(k+\frac{1}{p}\right)^{-s} \sim C(s,p) + \frac{\left(n+\frac{1}{p}\right)^{1-s} - p^{s-1}}{1-s} + \frac{1}{2}\left(n+\frac{1}{p}\right)^{-s} \\ - \sum_{r=1}^{m}\frac{B_{2r}}{(2r)!}(s)_{2r-1}\left(n+\frac{1}{p}\right)^{-s-2r+1} + \mathcal{R}(s,p;n;m) \\ (\Re(s) > -2m-1;\ m \in \mathbb{N}_0),\end{aligned} \tag{48}$$

where $C(s,p)$ is a constant depending on s and p, $(s)_n$ denotes the Pochhammer symbol defined by 1.1(5) and the remainder part $\mathcal{R}(s,p;n;m)$ satisfies the following limit relationship:

$$\lim_{n\to\infty} \mathcal{R}(s,p;n;m) = 0 \quad (\Re(s) > -2m-1;\ m \in \mathbb{N}_0).$$

In fact, $C(s,p)$ can be expressed explicitly as follows, by applying the same $f(x)$ to 1.4(68) and reducing the domain to $\Re(s) > 1$:

$$\begin{aligned}C(s,p) &= \lim_{n\to\infty}\left[\sum_{k=1}^{n}\left(k+\frac{1}{p}\right)^{-s} - \frac{\left(n+\frac{1}{p}\right)^{1-s} - p^{s-1}}{1-s}\right] \\ &= \zeta\left(s,\frac{1}{p}\right) - p^s - \frac{p^{s-1}}{s-1} \qquad (\Re(s) > 1).\end{aligned} \tag{49}$$

Now, it follows from (48) and (49) that

$$\begin{aligned}\zeta\left(s,\frac{1}{p}\right) = \lim_{n\to\infty}\Bigg[\sum_{k=1}^{n}\left(k+\frac{1}{p}\right)^{-s} - \frac{\left(n+\frac{1}{p}\right)^{1-s}}{1-s} - \frac{1}{2}\left(n+\frac{1}{p}\right)^{-s} \\ + \sum_{r=1}^{m}\frac{B_{2r}}{(2r)!}(s)_{2r-1}\left(n+\frac{1}{p}\right)^{-s-2r+1}\Bigg] + p^s \\ (m \in \mathbb{N},\ \Re(s) > -2m-1,\ \text{and } s \neq 1;\ m = 0 \text{ and } \Re(s) > 1).\end{aligned} \tag{50}$$

We note that the right-hand side of (50) is analytic for

$$\Re(s) > -2m-1 \quad (m \in \mathbb{N}_0) \quad \text{and} \quad s \neq 1.$$

Thus, we can differentiate (50) with respect to s under the limit sign and set $s = -1$ in the resulting equation. We, therefore, obtain (see [296, Eq. (3.24)])

$$\begin{aligned}\zeta'\left(s,\frac{1}{p}\right) &= \lim_{n\to\infty}\left[-\sum_{k=1}^{n}\left(k+\frac{1}{p}\right)^{-s}\log\left(k+\frac{1}{p}\right)-\frac{\left(n+\frac{1}{p}\right)^{1-s}}{(s-1)^2}\right.\\
&+\frac{\left(n+\frac{1}{p}\right)^{1-s}}{1-s}\log\left(n+\frac{1}{p}\right)+\frac{1}{2}\left(n+\frac{1}{p}\right)^{-s}\log\left(n+\frac{1}{p}\right)\\
&\left.+\sum_{r=1}^{m}\frac{B_{2r}}{(2r)!}(s)_{2r-1}\left(n+\frac{1}{p}\right)^{-s-2r+1}\left\{\sum_{j=1}^{2r-1}\frac{1}{s+j-1}-\log\left(n+\frac{1}{p}\right)\right\}\right]\\
&+p^s\log p \quad (m\in\mathbb{N},\ \Re(s)>-2m-1,\ \text{and } s\neq 1;\ m=0 \text{ and } \Re(s)>1),\end{aligned} \tag{51}$$

where we have also used the following elementary identity:

$$\frac{d}{ds}(s)_n = (s)_n\left(\sum_{k=1}^{n}\frac{1}{s+k-1}\right) \quad (n\in\mathbb{N}_0). \tag{52}$$

By setting $s = -1$ and $s = -2$ in (51), we find that

$$\begin{aligned}\zeta'\left(-1,\frac{1}{p}\right) &= -\lim_{n\to\infty}\left[\sum_{k=1}^{n}\left(k+\frac{1}{p}\right)\log\left(k+\frac{1}{p}\right)\right.\\
&-\left\{\frac{n^2}{2}+\left(\frac{1}{p}+\frac{1}{2}\right)n+\frac{1}{2p^2}+\frac{1}{2p}+\frac{1}{12}\right\}\log\left(n+\frac{1}{p}\right)\\
&\left.+\frac{n^2}{4}+\frac{n}{2p}\right]-\frac{1}{4p^2}+\frac{1}{12}+\frac{1}{p}\log p \quad (p>0)\end{aligned} \tag{53}$$

and

$$\begin{aligned}\zeta'\left(-2,\frac{1}{p}\right) &= -\lim_{n\to\infty}\left[\sum_{k=1}^{n}\left(k+\frac{1}{p}\right)^2\log\left(k+\frac{1}{p}\right)-\left\{\frac{n^3}{3}+\left(\frac{1}{p}+\frac{1}{2}\right)n^2\right.\right.\\
&\left.+\left(\frac{1}{p^2}+\frac{1}{p}+\frac{1}{6}\right)n+\frac{1}{3p^3}+\frac{1}{2p^2}+\frac{1}{6p}\right\}\log\left(n+\frac{1}{p}\right)+\frac{n^3}{9}\\
&\left.+\frac{n^2}{3p}+\left(\frac{1}{3p^2}-\frac{1}{12}\right)n\right]-\frac{1}{9p^3}+\frac{1}{12p}+\frac{\log p}{p^2} \quad (p>0).\end{aligned} \tag{54}$$

Comparing (42) and (53), we finally obtain the desired relationship:

$$\log A_p = -\zeta'\left(-1,\frac{1}{p}\right)+\frac{1}{12}-\frac{1}{4p^2}+\frac{1}{p}\log p \quad (p>0), \tag{55}$$

which, in view of (43), immediately yields 2.1(31) when $p = 1$.

Next, in view of the well-known reflection formula (see 1.1(12)):

$$\Gamma(1+z)\Gamma(1-z) = \frac{\pi z}{\sin(\pi z)},$$

we can evaluate the following integral (*cf.* [310, p. 114, Eq. (43)]):

$$\int_0^{\pi/p} \log\sin t\, dt = \pi\left(\log A_p - \log C_p - \frac{1}{p}\log(2p)\right) \quad (p > 1), \tag{56}$$

which, by virtue of another known result [289, p. 95, Eq. (2.2)]:

$$\int_0^{\pi/p} \log\sin t\, dt = \frac{\pi}{p}\log\left(\frac{\sin\left(\frac{\pi}{p}\right)}{2\pi}\right) + \pi\log\left[\frac{G\left(1+\frac{1}{p}\right)}{G\left(1-\frac{1}{p}\right)}\right] \tag{57}$$

yields the following relationship between the mathematical constants A_p and C_p:

$$A_p = \left[\frac{p}{\pi}\sin\left(\frac{\pi}{p}\right)\right]^{1/p}\frac{G\left(1+\frac{1}{p}\right)}{G\left(1-\frac{1}{p}\right)}C_p \quad (p > 1). \tag{58}$$

To get a more general class of mathematical constants (see [296]) than those given by (42), we begin by differentiating the function

$$f(x) := \left(x+\frac{1}{p}\right)^q \log\left(x+\frac{1}{p}\right) \quad (p > 0;\ q \in \mathbb{N}) \tag{59}$$

l times ($l \in \mathbb{N}$). We, thus, obtain

$$f^{(l)}(x) = \left(x+\frac{1}{p}\right)^{q-l}\left\{\left(\prod_{j=1}^{l}(q-j+1)\right)\log\left(x+\frac{1}{p}\right) + P_l(q)\right\} \quad (l \in \mathbb{N}), \tag{60}$$

where $P_l(q)$ is a polynomial of degree $l-1$ in q satisfying the following recurrence relation:

$$P_l(q) := \begin{cases} (q-l+1)P_{l-1}(q) + \prod_{j=1}^{l-1}(q-j+1) & (l \in \mathbb{N}\setminus\{1\}), \\ 1 & (l = 1). \end{cases} \tag{61}$$

In fact, by mathematical induction on $l \in \mathbb{N}$, we can give an explicit expression for $P_l(q)$ as follows:

$$P_l(q) = \left(\prod_{j=1}^{l} (q-j+1)\right)\left(\sum_{j=1}^{l} \frac{1}{q-j+1}\right) \quad (l \in \mathbb{N}). \tag{62}$$

By substituting from (59) and (60) into the Euler-Maclaurin summation formula 1.4(68) with $a = 0$, we get a class of mathematical constants $C_{p,q}$, defined by

$$\begin{aligned}\log C_{p,q} := \lim_{n\to\infty} \Bigg[& \sum_{k=1}^{n} \left(k+\frac{1}{p}\right)^q \log\left(k+\frac{1}{p}\right) + \frac{1}{(q+1)^2}\left(n+\frac{1}{p}\right)^{q+1} \\ & - \frac{1}{q+1}\left(n+\frac{1}{p}\right)^{q+1} \log\left(n+\frac{1}{p}\right) - \frac{1}{2}\left(n+\frac{1}{p}\right)^q \log\left(n+\frac{1}{p}\right) \\ & - \sum_{r=1}^{[(q+1)/2]} \frac{B_{2r}}{(2r)!}\left(n+\frac{1}{p}\right)^{q-2r+1} \\ & \cdot \left\{\left(\prod_{j=1}^{2r-1}(q-j+1)\right)\log\left(n+\frac{1}{p}\right) + P_{2r-1}(q)\right\}\Bigg] \quad (p > 0;\ q \in \mathbb{N}),\end{aligned} \tag{63}$$

where $[x]$ denotes (as usual) the greatest integer less than or equal to x.

Setting $q = 1$ in (63) and comparing the resulting equation with (42), it is easy to get a relationship between A_p and $C_{p,1}$ as follows:

$$C_{p,1} = \exp\left(\frac{1}{4p^2} - \frac{1}{12}\right) A_p \quad (p > 0). \tag{64}$$

If we set $s = -q$ ($q \in \mathbb{N}$) in (51), we obtain

$$\begin{aligned}-\zeta'\left(-q, \frac{1}{p}\right) = \lim_{n\to\infty} \Bigg[& \sum_{k=1}^{n} \left(k+\frac{1}{p}\right)^q \log\left(k+\frac{1}{p}\right) + \frac{1}{(q+1)^2}\left(n+\frac{1}{p}\right)^{q+1} \\ & - \frac{1}{q+1}\left(n+\frac{1}{p}\right)^{q+1} \log\left(n+\frac{1}{p}\right) - \frac{1}{2}\left(n+\frac{1}{p}\right)^q \log\left(n+\frac{1}{p}\right) \\ & - \sum_{r=1}^{m} \frac{B_{2r}}{(2r)!}\left(n+\frac{1}{p}\right)^{q-2r+1} \left\{\left(\prod_{j=1}^{2r-1}(q-j+1)\right)\log\left(n+\frac{1}{p}\right)\right. \\ & \left. + \left(\prod_{j=1}^{2r-1}(q-j+1)\right)\left(\sum_{j=1}^{2r-1}\frac{1}{q-j+1}\right)\right\}\Bigg] - p^{-q}\log p \\ & (m, q \in \mathbb{N};\ q < 2m+1).\end{aligned} \tag{65}$$

Thus, by comparing (63) and (65) and applying (62), we get the following relationship between $C_{p,q}$ and $\zeta'\left(-q, \frac{1}{p}\right)$:

$$\log C_{p,q} = -\zeta'\left(-q, \frac{1}{p}\right) + p^{-q}\log p \quad (p > 0;\ q \in \mathbb{N}). \tag{66}$$

We conclude this section by remarking that, in a mild sense, the constants $C_{p,q}$ are generalizations of the Bendersky-Adamchik constants D_k in (24), because, in view of (25) and (66), there is a relationship between D_k and $C_{1,k}$:

$$\log C_{1,k} = \log D_k - \frac{B_{k+1} H_k}{k+1} \quad (k \in \mathbb{N}). \tag{67}$$

7.2 Log-Sine Integrals Involving Series Associated with the Zeta Function and Polylogarithms

Motivated essentially by their potential for applications in a wide range of mathematic and physical problems, the Log-Sine integrals have been evaluated, in the existing literature on the subject, in many different ways. Very recently, Choi *et al.* [270] showed how nicely some general formulas analogous to the generalized Log-Sine integral $\mathrm{Ls}_n^{(m)}\left(\frac{\pi}{3}\right)$ can be obtained using the theory of Polylogarithms. Relevant connections of the results presented here with those obtained in earlier works are also indicated precisely.

For the Log-Sine integrals $\mathrm{Ls}_n(\theta)$ of order n, defined by (see 2.4(82))

$$\mathrm{Ls}_n(\theta) := -\int_0^\theta \left(\log\left|2\sin\frac{x}{2}\right|\right)^{n-1} dx \quad (n \in \mathbb{N} \setminus \{1\}), \tag{1}$$

the recurrence relation 2.4(111) holds true when $\theta = \pi$ (see, e.g., [752, p. 218, Eq. (7.112)]).

By using an idea analogous to that of Shen [1024], Beumer [128] presented a recursion formula for

$$D(n) := \frac{(-1)^{n-1}}{2\cdot(n-1)!} \int_0^\pi \left[\log\left(\sin\frac{x}{2}\right)\right]^{n-1} dx \quad (n \in \mathbb{N})$$

in the following form:

$$\sum_{k=1}^{2n-1} (-1)^{k-1} D(k) D(2n-k) = (-1)^{n+1} \frac{2^{2n}-1}{(2n)!} \pi^{2n} B_{2n} \quad (n \in \mathbb{N}), \tag{2}$$

where B_n are the Bernoulli numbers (see Section 1.7), and

$$D(1) = \frac{\pi}{2} \quad \text{and} \quad D(2) = \frac{\pi}{2} \log 2.$$

In fact, by mainly analyzing the generalized binomial theorem and the familiar Weierstrass canonical product form of the Gamma function $\Gamma(z)$ (see 1.1(2)), Shen [1024, p. 1396, Eq. (19)] evaluated the Log-Sine integral $\mathrm{Ls}_{k+1}(2\pi)$ as follows:

$$\frac{1}{2\pi} \int_0^{2\pi} \left(\log \left|2 \sin \frac{x}{2}\right|\right)^k dx = (-1)^k \frac{k!}{2^k} \sum_{n \geqq \frac{k}{2}}^{\infty} \sigma_k^n \quad (k, n \in \mathbb{N}), \tag{3}$$

where σ_k^n are given in terms of the Stirling numbers $s(n, k)$ of the first kind (see, for details, Section 1.5; see also [969]), by

$$\sigma_k^n = \sum_{m=1}^{k-1} \frac{s(n, k-m)}{n!} \frac{s(n, m)}{n!}.$$

More recently, Batir [102] presented integral representations, involving Log-Sine terms, for some series associated with

$$\binom{2k}{k}^{-1} k^{-n} \quad \text{and} \quad \binom{2k}{k}^{-2} k^{-n},$$

and for some closely-related series, by using a number of elementary properties of Polylogarithms. Lewin [752, pp. 102–103; p. 164] presented the following integral formulas:

$$\int_0^{\frac{\pi}{2}} \log \left(2 \sin \frac{x}{2}\right) dx = -G \tag{4}$$

and

$$\int_0^{\frac{\pi}{2}} x \log \left(2 \sin \frac{x}{2}\right) dx = \frac{35}{32} \zeta(3) - \frac{1}{2} \pi G, \tag{5}$$

where G denotes the Catalan constant, defined by 1.4(16).

Several other authors have concentrated on the problem of evaluation of the Log-Sine integral $\mathrm{Ls}_n(\theta)$ and the generalized Log-Sine integral $\mathrm{Ls}_n^{(m)}(\theta)$ of order n and index m, defined by

$$\mathrm{Ls}_n^{(m)}(\theta) := -\int_0^{\theta} x^m \left(\log \left|2 \sin \frac{x}{2}\right|\right)^{n-m-1} dx \tag{6}$$

with the *argument* θ given by $\theta = \frac{\pi}{3}$. (Throughout this section, we choose the *principal branch* of the logarithm function $\log z$ in case z is a complex variable.) In particular, van der Poorten [1181] proved that

$$\int_0^{\frac{\pi}{3}} \log^2\left(2\sin\frac{x}{2}\right) dx = \frac{7}{108}\pi^3 \tag{7}$$

and

$$\int_0^{\frac{\pi}{3}} x\log^2\left(2\sin\frac{x}{2}\right) dx = \frac{17}{6480}\pi^4. \tag{8}$$

Zucker [1267] established the following two integral formulas:

$$\int_0^{\frac{\pi}{3}} \left\{\log^4\left(2\sin\frac{x}{2}\right) - \frac{3}{2}x^2\log^2\left(2\sin\frac{x}{2}\right)\right\} dx = \frac{253}{3240}\pi^5 \tag{9}$$

and

$$\int_0^{\frac{\pi}{3}} \left\{x\log^4\left(2\sin\frac{x}{2}\right) - \frac{x^3}{2}\log^2\left(2\sin\frac{x}{2}\right)\right\} dx = \frac{313}{408240}\pi^6. \tag{10}$$

Zhang and Williams [1253] extensively investigated $\mathrm{Ls}_n\left(\frac{\pi}{3}\right)$ and $\mathrm{Ls}_n^{(m)}\left(\frac{\pi}{3}\right)$ along with other integrals to present two general formulas (see [1253, p. 272, Eqs. (1.6) and (1.7)]), which include the integral formulas (7) to (10) as special cases. We choose to recall here one more explicit special case of the Zhang-Williams integral formulas as follows:

$$\begin{aligned}\int_0^{\frac{\pi}{3}} &\left\{\log^6\left(2\sin\frac{x}{2}\right) - \frac{15}{4}x^2\log^4\left(2\sin\frac{x}{2}\right)\right.\\ &\left.+\frac{15}{16}x^4\log^2\left(2\sin\frac{x}{2}\right)\right\} dx = \frac{77821}{2^6\cdot 3^6\cdot 7}\pi^7.\end{aligned} \tag{11}$$

The following well-known formula is recorded (see [531, p. 334, Entry (50.5.16)]):

$$x\cdot\cot x = 1 + \sum_{n=1}^{\infty}(-1)^n\frac{2^{2n}B_{2n}}{(2n)!}x^{2n} \quad (|x| < \pi). \tag{12}$$

Analogous Log-Sine Integrals

Choi *et al.* [270] showed how nicely some general formulas analogous to the generalized Log-Sine integral $\mathrm{Ls}_n^{(m)}\left(\frac{\pi}{3}\right)$ can be obtained, by using the theory of Polylogarithms. Indeed, by carrying out repeated integration by parts in 2.4(71) in conjunction with 2.4(3), we obtain

$$\begin{aligned}\mathrm{Li}_n(z) - \mathrm{Li}_n(w) &= \int_w^z \mathrm{Li}_{n-1}(t)\,\frac{dt}{t}\\ &= \left(\sum_{k=1}^{n-2} \frac{(-1)^{k-1}}{k!}\,(\log t)^k\,\mathrm{Li}_{n-k}(t) + \frac{(-1)^{n-1}}{(n-1)!}\,(\log t)^{n-1}\log(1-t)\right)\Bigg|_{t=w}^{z}\\ &\quad + \frac{(-1)^{n-1}}{(n-1)!}\int_w^z (\log t)^{n-1}\,\frac{dt}{1-t} \qquad (n \in \mathbb{N}\setminus\{1\}),\end{aligned} \tag{13}$$

where (and elsewhere in this section) an empty sum is understood to be nil; in particular, we have

$$\begin{aligned}\zeta(n) = \mathrm{Li}_n(1) &= \frac{(-1)^{n-1}}{(n-1)!}\int_0^1 (\log t)^{n-1}\,\frac{dt}{1-t}\\ &= \frac{(-1)^{n-1}}{(n-1)!}\int_0^1 [\log(1-t)]^{n-1}\,\frac{dt}{t} \qquad (n \in \mathbb{N}\setminus\{1\}),\end{aligned} \tag{14}$$

which, by substituting $t = u^{-1}$, yields

$$(-1)^{n-1}\int_1^{e^{-i\theta}} (\log t)^{n-1}\,\frac{dt}{1-t} = \int_1^{e^{i\theta}} (\log t)^{n-1}\,\frac{dt}{1-t} + \frac{1}{n}\,(i\theta)^n \qquad (0 \leqq \theta \leqq \pi). \tag{15}$$

Furthermore, it is easily observed, by setting $t = 1 - e^{ix}$, that

$$\int_0^{1-e^{i\theta}} (\log t)^{n-1}\,\frac{dt}{1-t} = -i\int_0^{\theta}\left[\frac{1}{2}\,i\,(x-\pi) + \log\left|2\sin\frac{x}{2}\right|\right]^{n-1} dx, \tag{16}$$

or, equivalently, that

$$\int_1^{1-e^{i\theta}} (\log t)^{n-1} \frac{dt}{1-t} = -i \int_0^{\theta} \left[\frac{1}{2} i(x-\pi) + \log\left|2\sin\frac{x}{2}\right|\right]^{n-1} dx + (-1)^n (n-1)!\,\zeta(n) \quad (n \in \mathbb{N}\setminus\{1\}) \tag{17}$$

in view of (14). In its special case when $n = 2m+1$ ($m \in \mathbb{N}$) and $w = 1$, (13) yields

$$\begin{aligned}\int_1^{z} (\log t)^{2m} \frac{dt}{1-t} &= (2m)!\,\mathrm{Li}_{2m+1}(z) - (2m)!\,\zeta(2m+1)\\ &+ (2m)! \sum_{k=1}^{2m-1} \frac{(-1)^k}{k!} (\log z)^k\, \mathrm{Li}_{2m+1-k}(z)\\ &- (\log z)^{2m} \log(1-z) \quad (m \in \mathbb{N}).\end{aligned} \tag{18}$$

Putting $z = e^{i\frac{\pi}{3}}$ in (18) and using the following elementary identity:

$$1 - e^{i\frac{\pi}{3}} = e^{-i\frac{\pi}{3}}, \tag{19}$$

we get

$$\begin{aligned}\int_1^{e^{i\frac{\pi}{3}}} (\log t)^{2m} \frac{dt}{1-t} &= i(-1)^m \left(\frac{\pi}{3}\right)^{2m+1} - (2m)!\,\zeta(2m+1)\\ &+ (2m)! \sum_{k=0}^{2m-1} \frac{(-1)^k}{k!} \left(i\frac{\pi}{3}\right)^k \mathrm{Li}_{2m+1-k}\left(e^{i\frac{\pi}{3}}\right) \quad (m \in \mathbb{N}).\end{aligned} \tag{20}$$

We, now, separate the even and odd parts of the sum occurring in (20) and make use of 2.4(79) and 2.4(85). We, thus, obtain

$$\begin{aligned}\sum_{k=0}^{2m-1} \frac{(-1)^k}{k!} \left(i\frac{\pi}{3}\right)^k \mathrm{Li}_{2m+1-k}\left(e^{i\frac{\pi}{3}}\right) &= \sum_{k=0}^{2m-1} \frac{(-1)^{[k/2]}}{k!} \left(\frac{\pi}{3}\right)^k \mathrm{Cl}_{2m+1-k}\left(\frac{\pi}{3}\right)\\ &+ i \sum_{k=0}^{2m-1} \frac{(-1)^{\left[\frac{k+1}{2}\right]}}{k!} \left(\frac{\pi}{3}\right)^k \mathrm{Gl}_{2m+1-k}\left(\frac{\pi}{3}\right).\end{aligned} \tag{21}$$

Upon substituting from (21) into (20), and equating the real and imaginary parts on each side of the resulting equation, we obtain

$$\Re\left(\int_1^{e^{i\frac{\pi}{3}}} (\log t)^{2m} \frac{dt}{1-t}\right) = -(2m)!\,\zeta(2m+1)$$
$$+ (2m)! \sum_{k=0}^{2m-1} \frac{(-1)^{[k/2]}}{k!} \left(\frac{\pi}{3}\right)^k \mathrm{Cl}_{2m+1-k}\left(\frac{\pi}{3}\right) \quad (m \in \mathbb{N}) \tag{22}$$

and

$$\Im\left(\int_1^{e^{i\frac{\pi}{3}}} (\log t)^{2m} \frac{dt}{1-t}\right) = (-1)^m \left(\frac{\pi}{3}\right)^{2m+1}$$
$$+ (2m)! \sum_{k=0}^{2m-1} \frac{(-1)^{\left[\frac{k+1}{2}\right]}}{k!} \left(\frac{\pi}{3}\right)^k \mathrm{Gl}_{2m+1-k}\left(\frac{\pi}{3}\right) \quad (m \in \mathbb{N}). \tag{23}$$

Setting $n = 2m+1$ $(m \in \mathbb{N})$ and $\theta = \frac{\pi}{3}$ in (15), we have

$$\int_1^{e^{-i\frac{\pi}{3}}} (\log t)^{2m} \frac{dt}{1-t} = \int_1^{e^{i\frac{\pi}{3}}} (\log t)^{2m} \frac{dt}{1-t}$$
$$+ \frac{(-1)^m}{2m+1} \left(\frac{\pi}{3}\right)^{2m+1} i \quad (m \in \mathbb{N}). \tag{24}$$

Also, by putting $n = 2m+1$ $(m \in \mathbb{N})$ and $\theta = \frac{\pi}{3}$ in (17), and using (19), we obtain

$$\int_1^{e^{-i\frac{\pi}{3}}} (\log t)^{2m} \frac{dt}{1-t} = -(2m)!\,\zeta(2m+1)$$
$$- i \int_0^{\frac{\pi}{3}} \left[\frac{1}{2} i(x-\pi) + \log\left(2\sin\frac{x}{2}\right)\right]^{2m} dx \quad (m \in \mathbb{N}). \tag{25}$$

Now, by using the binomial theorem, we find that

$$\left[\frac{1}{2} i(x-\pi) + \log\left(2\sin\frac{x}{2}\right)\right]^{2m}$$
$$= \sum_{k=0}^{m} (-1)^k \binom{2m}{2k} \left(\frac{x-\pi}{2}\right)^{2k} \log^{2m-2k}\left(2\sin\frac{x}{2}\right) \tag{26}$$
$$+ i \sum_{k=1}^{m} (-1)^{k+1} \binom{2m}{2k-1} \left(\frac{x-\pi}{2}\right)^{2k-1} \log^{2m+1-2k}\left(2\sin\frac{x}{2}\right) \quad (m \in \mathbb{N}).$$

Substituting (26) into the integrand on the right-hand side of (25) and equating the real and imaginary parts of each side of the resulting equation, we obtain

$$\begin{aligned} \Re\left(\int_1^{e^{-i\frac{\pi}{3}}} (\log t)^{2m} \frac{dt}{1-t}\right) &= -(2m)!\,\zeta(2m+1) \\ &+ \sum_{k=1}^{m} \frac{(-1)^{k+1}}{2^{2k-1}} \binom{2m}{2k-1} \int_0^{\frac{\pi}{3}} (x-\pi)^{2k-1} \log^{2m+1-2k}\left(2\sin\frac{x}{2}\right) dx \quad (m \in \mathbb{N}) \end{aligned} \tag{27}$$

and

$$\begin{aligned} \Im\left(\int_1^{e^{-i\frac{\pi}{3}}} (\log t)^{2m} \frac{dt}{1-t}\right) \\ = \sum_{k=0}^{m} \frac{(-1)^{k+1}}{2^{2k}} \binom{2m}{2k} \int_0^{\frac{\pi}{3}} (x-\pi)^{2k} \log^{2m-2k}\left(2\sin\frac{x}{2}\right) dx \quad (m \in \mathbb{N}). \end{aligned} \tag{28}$$

Finally, from (22), (23), (24), (27) and (28), we can deduce the following analogous Log-Sine integral formulas:

$$\begin{aligned} \sum_{k=1}^{m} \frac{(-1)^{k+1}}{2^{2k-1}} \binom{2m}{2k-1} \int_0^{\pi/3} (x-\pi)^{2k-1} \log^{2m+1-2k}\left(2\sin\frac{x}{2}\right) dx \\ = (2m)! \sum_{k=0}^{2m-1} \frac{(-1)^{[k/2]}}{k!} \left(\frac{\pi}{3}\right)^k \mathrm{Cl}_{2m+1-k}\left(\frac{\pi}{3}\right) \quad (m \in \mathbb{N}) \end{aligned} \tag{29}$$

and

$$\begin{aligned} \sum_{k=0}^{m} \frac{(-1)^{k+1}}{2^{2k}} \binom{2m}{2k} \int_0^{\frac{\pi}{3}} (x-\pi)^{2k} \log^{2m-2k}\left(2\sin\frac{x}{2}\right) dx \\ = (-1)^m \left(1 + \frac{1}{2m+1}\right) \left(\frac{\pi}{3}\right)^{2m+1} \\ + (2m)! \sum_{k=0}^{2m-1} \frac{(-1)^{\left[\frac{k+1}{2}\right]}}{k!} \left(\frac{\pi}{3}\right)^k \mathrm{Gl}_{2m+1-k}\left(\frac{\pi}{3}\right) \quad (m \in \mathbb{N}). \end{aligned} \tag{30}$$

Just as in our derivations of (29) and (30), if we set

$$n = 2m \quad (m \in \mathbb{N}),\ z = e^{i\frac{\pi}{3}} \quad \text{and} \quad w = 1$$

in (13) and apply (15) and (17) with $\theta = \frac{\pi}{3}$, we obtain some other analogous Log-Sine integral formulas as follows:

$$\begin{aligned}\sum_{k=0}^{m-1} \frac{(-1)^k}{2^{2k+1}} \binom{2m-1}{2k+1} \int_0^{\frac{\pi}{3}} (x-\pi)^{2k+1} \log^{2m-2-2k}\left(2\sin\frac{x}{2}\right) dx \\ = (-1)^{m+1}\left(1+\frac{1}{2m}\right)\left(\frac{\pi}{3}\right)^{2m} - 2(2m-1)!\,\zeta(2m) \\ + (2m-1)! \sum_{k=0}^{2m-2} \frac{(-1)^{\left[\frac{k}{2}\right]}}{k!}\left(\frac{\pi}{3}\right)^k \mathrm{Gl}_{2m-k}\left(\frac{\pi}{3}\right) \quad (m \in \mathbb{N})\end{aligned} \tag{31}$$

and

$$\begin{aligned}\sum_{k=0}^{m-1} \frac{(-1)^k}{2^{2k}} \binom{2m-1}{2k} \int_0^{\frac{\pi}{3}} (x-\pi)^{2k} \log^{2m-1-2k}\left(2\sin\frac{x}{2}\right) dx \\ = (2m-1)! \sum_{k=0}^{2m-2} \frac{(-1)^{\left[\frac{k-1}{2}\right]}}{k!}\left(\frac{\pi}{3}\right)^k \mathrm{Cl}_{2m-k}\left(\frac{\pi}{3}\right) \quad (m \in \mathbb{N}).\end{aligned} \tag{32}$$

Remarks on $\mathrm{Cl}_n(\theta)$ and $\mathrm{Gl}_n(\theta)$

In view of Equations (29) to (32), we need to express the generalized Clausen function $\mathrm{Cl}_n(\theta)$ and its associated Clausen function $\mathrm{Gl}_n(\theta)$ as explicitly as possible, at least at the argument $\theta = \pi/3$. To do this, we begin by rewriting a known formula

$$\begin{aligned}\sum_{k=1}^{\infty} \frac{\cos(2\pi kx)}{k^n} + i \sum_{k=1}^{\infty} \frac{\sin(2\pi kx)}{k^n} \\ = \frac{(2\pi i)^{n-1}}{(n-1)!}\left\{\zeta'(1-n,\, x) + (-1)^{n-1}\zeta'(1-n,\, 1-x) - \pi i \frac{B_n(x)}{n}\right\} \\ (0 < x < 1;\ n \in \mathbb{N}\setminus\{1\}),\end{aligned} \tag{33}$$

which was proven by Adamchik [6, Eq. (9)] (see also [829, Eq. (21)]), who used Lerch's transform formula [745] and where $B_n(x)$ are the Bernoulli polynomials of degree n in x (see, for details, Section 1.7). Now, replacing n by $2n$ and $2n+1$ in (33) and equating the real and imaginary parts in each case, we obtain the following formulas 2.4(109) and

$$\mathrm{Cl}_n(2\pi x) = (-1)^{1+\left[\frac{1}{2}(n+1)\right]} \frac{(2\pi)^{n-1}}{(n-1)!} \left\{ \zeta'(1-n, x) + (-1)^{n+1} \zeta'(1-n, 1-x) \right\}$$
$$(0 < x < 1;\; n \in \mathbb{N} \setminus \{1\}). \qquad (34)$$

Srivastava et al. [1097, Eq. (3.8) and Eq. (3.17)] presented the following formulas:

$$\mathrm{Cl}_{2n+1}\left(\frac{\pi}{3}\right) = \frac{1}{2}\left(1-2^{-2n}\right)\left(1-3^{-2n}\right)\zeta(2n+1) \quad (n \in \mathbb{N}) \qquad (35)$$

and

$$\mathrm{Cl}_{2n}\left(\frac{\pi}{3}\right) =, \frac{\sqrt{3}}{6^{2n}} \left\{ \zeta\left(2n, \frac{1}{3}\right) + \zeta\left(2n, \frac{1}{6}\right) - 2^{2n-1}\left(3^{2n}-1\right)\zeta(2n) \right\} \quad (n \in \mathbb{N}). \qquad (36)$$

Cvijović and Klinowski [349, Eq. (10a)] proved formulas that can be specialized in the following form:

$$\mathrm{Cl}_n\left(\frac{2\pi p}{q}\right) = \frac{1}{q^n} \sum_{k=1}^{q} \zeta\left(n, \frac{k}{q}\right)$$
$$\left\{ \frac{1+(-1)^n}{2} \sin\left(\frac{2k\pi p}{q}\right) + \frac{1-(-1)^n}{2} \cos\left(\frac{2k\pi p}{q}\right) \right\} \qquad (37)$$
$$(n,\; q+1 \in \mathbb{N} \setminus \{1\}; \quad p \in \mathbb{Z}),$$

which, upon replacing n by $2n+1$ and $2n$ with $p = 1$ and $q = 6$, leads to (35) and (36), respectively.

We are ready to consider some explicit expressions of (29) to (32). Upon setting $m = 1, 2, 3$ in (30) and $m = 2, 3$ in (31), if we apply 2.4(109), we obtain the following explicit analogous Log-Sine integral formulas:

$$\int_0^{\frac{\pi}{3}} \log^2\left(2 \sin \frac{x}{2}\right) dx = \frac{7}{108}\pi^3, \qquad (38)$$

$$\int_0^{\frac{\pi}{3}} (x-\pi) \log^2\left(2 \sin \frac{x}{2}\right) dx = -\frac{403}{2^4 \cdot 3^4 \cdot 5}\pi^4, \qquad (39)$$

$$\int_0^{\frac{\pi}{3}} \left[\frac{3}{2}(x-\pi)^2 \log^2\left(2 \sin \frac{x}{2}\right) - \log^4\left(2 \sin \frac{x}{2}\right)\right] dx = \frac{73}{2^4 \cdot 3^4 \cdot 5}\pi^5, \qquad (40)$$

$$\int_0^{\frac{\pi}{3}} \left[2\,(x-\pi)\log^4\left(2\sin\frac{x}{2}\right) - (x-\pi)^3\log^2\left(2\sin\frac{x}{2}\right)\right] dx = -\frac{39883}{2^4\cdot 3^6\cdot 5\cdot 7}\,\pi^6, \tag{41}$$

and

$$\int_0^{\frac{\pi}{3}} \left[\log^6\left(2\sin\frac{x}{2}\right) - \frac{15}{4}\,(x-\pi)^2\log^4\left(2\sin\frac{x}{2}\right) + \frac{15}{16}\,(x-\pi)^4\log^2\left(2\sin\frac{x}{2}\right)\right] dx = \frac{697}{2^6\cdot 3^6\cdot 7}\,\pi^7. \tag{42}$$

It should be noted that the integral formulas (7) to (11) can easily be deduced from these last integral formulas (38) to (42).

Upon setting $m = 1, 2$ in (29) and (30), if we apply (35) and (36), we obtain

$$\int_0^{\frac{\pi}{3}} (x-\pi)\log\left(2\sin\frac{x}{2}\right) dx = \frac{2}{3}\,\zeta(3) - \frac{4\sqrt{3}}{81}\,\pi^3 + \frac{\sqrt{3}\pi}{54}\left\{\zeta\left(2,\frac{1}{3}\right) + \zeta\left(2,\frac{1}{6}\right)\right\}. \tag{43}$$

$$\begin{aligned} &\int_0^{\frac{\pi}{3}} \left\{2(x-\pi)\log^3\left(2\sin\frac{x}{2}\right) - \frac{1}{2}\,(x-\pi)^3\log\left(2\sin\frac{x}{2}\right)\right\} dx \\ &= \frac{100}{9}\,\zeta(5) - \frac{4\pi^2}{9}\,\zeta(3) - \frac{8\sqrt{3}}{243}\,\pi^5 + \frac{\sqrt{3}\pi}{162}\left\{\zeta\left(4,\frac{1}{3}\right) + \zeta\left(4,\frac{1}{6}\right)\right\} \\ &\quad - \frac{\sqrt{3}\pi^3}{243}\left\{\zeta\left(2,\frac{1}{3}\right) + \zeta\left(2,\frac{1}{6}\right)\right\}. \end{aligned} \tag{44}$$

$$\int_0^{\frac{\pi}{3}} \log\left(2\sin\frac{x}{2}\right) dx = \frac{2\sqrt{3}}{27}\,\pi^2 - \frac{\sqrt{3}}{36}\left\{\zeta\left(2,\frac{1}{3}\right) + \zeta\left(2,\frac{1}{6}\right)\right\}. \tag{45}$$

$$\begin{aligned} &\int_0^{\frac{\pi}{3}} \left\{\log^3\left(2\sin\frac{x}{2}\right) - \frac{3}{4}\,(x-\pi)^2\log\left(2\sin\frac{x}{2}\right)\right\} dx \\ &= \frac{2\sqrt{3}}{243}\,\pi^4 + \frac{2\pi}{3}\,\zeta(3) + \frac{\sqrt{3}\pi^2}{108}\left\{\zeta\left(2,\frac{1}{3}\right) + \zeta\left(2,\frac{1}{6}\right)\right\} \\ &\quad - \frac{\sqrt{3}}{216}\left\{\zeta\left(4,\frac{1}{3}\right) + \zeta\left(4,\frac{1}{6}\right)\right\}. \end{aligned} \tag{46}$$

Further Remarks and Observations

We consider the following general integral:

$$\mathrm{Ls}_{m+2}^{(m)}(z)=\int_0^z x^m \log\left(2\sin\frac{x}{2}\right)dx \quad (m\in\mathbb{N}_0). \tag{47}$$

Applying integration by parts in (47) and using (12) and 2.3(18), we obtain

$$\begin{aligned}\int_0^z x^m \log\left(2\sin\frac{x}{2}\right)dx &= \frac{z^{m+1}}{m+1}\log\left(2\sin\frac{z}{2}\right)-\frac{z^{m+1}}{(m+1)^2}\\ &+\frac{2z^{m+1}}{m+1}\sum_{k=1}^{\infty}\frac{\zeta(2k)}{m+2k+1}\left(\frac{z}{2\pi}\right)^{2k} \quad (|z|<2\pi;\ m\in\mathbb{N}).\end{aligned} \tag{48}$$

If the known formulas Eq. (2.16) in [269, p. 515] and Eq. (2.13) in [269, p. 514] are used, the infinite series in (48) can be expressed as finite series as follows:

$$\begin{aligned}\int_0^z x^{2m} \log\left(2\sin\frac{x}{2}\right)dx &= \frac{z^{2m+1}}{2m+1}\log\left(2\sin\frac{z}{2}\right)+\frac{(2\pi)^{2m+1}}{2m+1}\sum_{k=0}^{2m+1}\binom{2m+1}{k}\\ &\cdot\left[\zeta'\left(-k,1-\frac{z}{2\pi}\right)+(-1)^k\zeta'\left(-k,1+\frac{z}{2\pi}\right)\right]\left(\frac{z}{2\pi}\right)^{2m+1-k}\\ &\qquad (|z|<2\pi;\ m\in\mathbb{N}_0)\end{aligned} \tag{49}$$

and

$$\begin{aligned}\int_0^z x^{2m-1} \log\left(2\sin\frac{x}{2}\right)dx &= \frac{z^{2m}}{2m}\log\left(2\sin\frac{z}{2}\right)+(-1)^{m+1}(2m-1)!\,\zeta(2m+1)\\ &+\frac{(2\pi)^{2m}}{2m}\sum_{k=0}^{2m}\binom{2m}{k}\left[\zeta'\left(-k,1-\frac{z}{2\pi}\right)+(-1)^k\zeta'\left(-k,1+\frac{z}{2\pi}\right)\right]\left(\frac{z}{2\pi}\right)^{2m-k}\\ &\qquad (|z|<2\pi;\ m\in\mathbb{N}).\end{aligned} \tag{50}$$

The special cases of (49) when $m=0$ and (50) when $m=1$ would, in light of some of the identities in Chapter 2, readily yield, respectively, the following formulas:

$$\begin{aligned}\int_0^{\frac{\pi}{3}} \log\left(2\sin\frac{x}{2}\right)dx &= 2\pi\left[\zeta'\left(-1,\frac{5}{6}\right)-\zeta'\left(-1,\frac{1}{6}\right)\right]=-\mathrm{Cl}_2\left(\frac{\pi}{3}\right)\\ &=-\frac{\pi}{3}\log(2\pi)+2\pi\log\left(\frac{\Gamma_2\left(\frac{5}{6}\right)}{\Gamma_2\left(\frac{7}{6}\right)}\right)\end{aligned} \tag{51}$$

and

$$\int_0^{\frac{\pi}{3}} x \log\left(2 \sin\frac{x}{2}\right) dx = \frac{2\zeta(3)}{3} + \frac{2\pi^2}{3}\left[\zeta'\left(-1, \frac{5}{6}\right) - \zeta'\left(-1, \frac{1}{6}\right)\right]$$
$$= \frac{2\zeta(3)}{3} - \frac{\pi^2}{9}\log(2\pi) + \frac{2\pi^2}{3}\log\left(\frac{\Gamma_2\left(\frac{5}{6}\right)}{\Gamma_2\left(\frac{7}{6}\right)}\right), \tag{52}$$

where Γ_2 denotes the double Gamma function (see [94, 96]; see also Section 1.4).

Here, it should be remarked in passing that, in view of Equation (34) and Equations (49) to (52), closed-form expressions for derivatives of $\zeta(s, a)$ at the negative integer values of s and rational values of a are needed, some of which were given by Adamchik [6] and by Miller and Adamchik [829].

We note also that Srivastava *et al.* [1097] studied extensively and systematically some definite integrals in conjunction with series involving the Zeta function, such as in (48).

If (34) is used, (49) and (50) are readily rewritten as follows:

$$\int_0^{z} x^{2m} \log\left(2 \sin\frac{x}{2}\right) dx = (2m)! \sum_{k=2}^{2m+2} \frac{(-1)^{k+[\frac{1}{2}(k+1)]}}{(2m+2-k)!}\mathrm{Cl}_k(z)\, z^{2m+2-k} \tag{53}$$
$$(|z| < 2\pi;\ m \in \mathbb{N}_0)$$

and

$$\int_0^{z} x^{2m-1} \log\left(2 \sin\frac{x}{2}\right) dx = (-1)^{m+1}(2m-1)!\,\zeta(2m+1)$$
$$+ (2m-1)! \sum_{k=2}^{2m+1} \frac{(-1)^{k+[\frac{1}{2}(k+1)]}}{(2m+1-k)!}\mathrm{Cl}_k(z)\, z^{2m+1-k} \tag{54}$$
$$(|z| < 2\pi;\ m \in \mathbb{N}).$$

By setting $m = 1$ in (53) and $m = 2$ in (54) with $z = \frac{\pi}{3}$, we obtain

$$\int_0^{\frac{\pi}{3}} x^2 \log\left(2 \sin\frac{x}{2}\right) dx = -\frac{2\sqrt{3}}{729}\pi^4 - \frac{2\pi}{9}\zeta(3)$$
$$- \frac{\sqrt{3}\pi^2}{324}\left\{\zeta\left(2, \frac{1}{3}\right) + \zeta\left(2, \frac{1}{6}\right)\right\} + \frac{\sqrt{3}}{648}\left\{\zeta\left(4, \frac{1}{3}\right) + \zeta\left(4, \frac{1}{6}\right)\right\} \tag{55}$$

and

$$\int_0^{\frac{\pi}{3}} x^3 \log\left(2\sin\frac{x}{2}\right) dx = -\frac{2\sqrt{3}}{243}\pi^5 - \frac{\pi^2}{9}\zeta(3) - \frac{29}{9}\zeta(5) - \frac{\sqrt{3}\pi^3}{972}\left\{\zeta\left(2,\frac{1}{3}\right) + \zeta\left(2,\frac{1}{6}\right)\right\} + \frac{\sqrt{3}\pi}{648}\left\{\zeta\left(4,\frac{1}{3}\right) + \zeta\left(4,\frac{1}{6}\right)\right\}. \tag{56}$$

By using 1.3(53), Shen's process (see [1024, pp. 1393–1394]) can be shortened considerably and it may also be easier to apply to other situations of a similar nature (see Section 3.5). For example, the formula (3) can be stated as follows:

$$\frac{1}{2\pi}\int_0^{2\pi} \left[\log\left(2\sin\frac{x}{2}\right)\right]^n dx = \frac{(-1)^n n!}{2^n} a_n \quad (n \in \mathbb{N}_0), \tag{57}$$

where the coefficients a_n are defined by

$$\frac{2^{-2z}\,\Gamma\left(\frac{1}{2}-z\right)}{\sqrt{\pi}\,\Gamma(1-z)} := \sum_{n=0}^{\infty} a_n z^n$$

and given by the following simple recursion formula:

$$a_1 = 0 \quad \text{and} \quad (n+1)\,a_{n+1} = 2\sum_{k=1}^{n} a_{n-k}\left(2^k - 1\right)\zeta(k+1) \quad (n \in \mathbb{N}).$$

From the equation (7.160) in [752, p. 230], the following formulas can be obtained:

$$(-1)^m \int_0^{\frac{\pi}{3}} \left(x - \frac{\pi}{3}\right)^{2m+1} \log\left(2\sin\frac{x}{2}\right) dx = -\frac{1}{2}(2m+1)!\left(1 - 2^{-2m-2}\right)\left(1 - 3^{-2m-2}\right)\zeta(2m+3) + (2m+1)!\sum_{k=0}^{m}(-1)^k\left(\frac{\pi}{3}\right)^{2k}\frac{\zeta(2m+3-2k)}{(2k)!} \quad (m \in \mathbb{N}_0); \tag{58}$$

$$\int_0^{\frac{\pi}{3}} \left(\log\frac{\sin x}{\sin(x+\pi/3)}\right)^{2m+1} dx = (-1)^{m+1}\left(\frac{3m+4}{2m+3}\right)\left(\frac{\pi}{3}\right)^{2m+3} + (2m+2)!\sum_{k=0}^{m}(-1)^k\left(\frac{\pi}{3}\right)^{2k+1}\frac{\zeta(2m+2-2k)}{(2k+1)!} \quad (m \in \mathbb{N}_0); \tag{59}$$

$$
\begin{aligned}
&(-1)^{m+1}\int_0^{\frac{\pi}{3}}\left(x-\frac{\pi}{3}\right)^{2m}\log\left(2\sin\frac{x}{2}\right)dx+\frac{1}{2m+1}\int_0^{\frac{\pi}{3}}\left[\log\left(\frac{\sin x}{\sin\left(x+\frac{\pi}{3}\right)}\right)\right]^{2m}dx\\
&\quad=(2m)!\sum_{k=1}^{m}(-1)^k\left(\frac{\pi}{3}\right)^{2k-1}\frac{\zeta(2m+3-2k)}{(2k-1)!}\qquad(m\in\mathbb{N});
\end{aligned}
\tag{60}
$$

and

$$
\begin{aligned}
&\sum_{k=0}^{m}(-1)^k\left(\frac{\pi}{3}\right)^{2k}\frac{\zeta(2m+2-2k)}{(2k)!}=\frac{1}{2}(-1)^m\left(\frac{6m+5}{(2m+2)!}\right)\left(\frac{\pi}{3}\right)^{2m+2}\\
&\quad+\frac{1}{2}\left(1-2^{-2m-1}\right)\left(1-3^{-2m-1}\right)\zeta(2m+2)\quad(m\in\mathbb{N}).
\end{aligned}
\tag{61}
$$

The special cases of (58) and (59) when $m=0$ yield, respectively,

$$
\int_0^{\frac{\pi}{3}}\left(x-\frac{\pi}{3}\right)\log\left(2\sin\frac{x}{2}\right)dx=\frac{2}{3}\zeta(3)
\tag{62}
$$

and

$$
\int_0^{\frac{\pi}{3}}\log\left(\frac{\sin x}{\sin\left(\left(x+\frac{\pi}{3}\right)\right)}\right)dx=\frac{5\pi^3}{81},
\tag{63}
$$

which is a corrected version of the second equation in [752, p. 230, Eq. (7.161)].

The formula (62) is recorded in [752, p. 230] and can be obtained from (51) and (52).

7.3 Applications of the Gamma and Polygamma Functions Involving Convolutions of the Rayleigh Functions

The Gamma function and its nth logarithmic derivatives (i.e., the Polygamma or the Psi functions), which were introduced in Chapter 1, have found many interesting and useful applications in a variety of subjects in pure and applied mathematics. Choi and Srivastava [299] mainly applied these functions to treat convolutions of the Rayleigh functions by recalling a general identity expressing a certain class of series as Psi-functions and to evaluate a class of log-sine integrals in an algorithmic way. They [299] also evaluated some Euler sums and gave much simpler Psi-function expressions for some known parameterized multiple sums. Here, we choose to discuss only the application to convolutions of the Rayleigh functions in [299].

The Gamma function with its several equivalent forms (see Section 1.1) and the polygamma functions given in 1.3(52) have been used, in the vast mathematical literature, in many different ways. For example, the generalized Goldbach-Euler series has been shown to be connected with the ψ-function as follows (*cf.* [821, p. 88, Eq. (5)]):

$$\sum_{n=2}^{\infty}\sum_{k=0}^{\infty}\frac{1}{(pk+r)^n-1}=\frac{1}{p}\left[\psi\left(\frac{r}{p}\right)-\psi\left(\frac{r-1}{p}\right)\right], \tag{1}$$

where $p\in\mathbb{N}$ and $r=p$ or $r=p+1$; we exclude the case $r=p=1$. For other applications, see the works by (for example) Doelder [391] and Shen [1024], as well as others.

Series Expressible in Terms of the ψ-Function

We recall one of the main results of Al-Saqabi *et al.* [21] as Theorem 7.1 below.

Theorem 7.1. (Al-Saqabi *et al.* [21]) *Suppose that $P(x)$ is a polynomial in x of degree $\leqq r-2$ $(r\in\mathbb{N}\setminus\{1\})$. Then,*

$$\begin{aligned}\sum_{n=1}^{\infty}&\frac{P(n)}{(\lambda_1+\mu_1 n)\cdots(\lambda_r+\mu_r n)}\\&=-\sum_{k=1}^{r}\mu_k^{r-2}P\left(-\frac{\lambda_k}{\mu_k}\right)\psi\left(1+\frac{\lambda_k}{\mu_k}\right)\prod_{\substack{j=1\\(j\neq k)}}^{r}(\lambda_j\mu_k-\lambda_k\mu_j)^{-1},\end{aligned} \tag{2}$$

provided that the parameters $\lambda_1,\ldots,\lambda_r$ and $\mu_1,\ldots,\mu_r$ are constrained by

$$\lambda_j\mu_k\neq\lambda_k\mu_j\quad(j\neq k;\quad j,k=1,\ldots,r).$$

An interesting generalization of Theorem 7.1 was derived by Wu *et al.* [1237]. Here, in our present investigation, we write a special case of (2) when $P(x)\equiv 1$ as follows:

$$\sum_{k=0}^{\infty}\frac{1}{(\mu_1 k+\lambda_1)(\mu_2 k+\lambda_2)\cdots(\mu_r k+\lambda_r)}=-\left(\prod_{j=1}^{r}\mu_j\right)^{-1}\sum_{j=1}^{r}u_j\,\psi\left(\frac{\lambda_j}{\mu_j}\right), \tag{3}$$

where $r\in\mathbb{N}\setminus\{1\}$, $\mu_j\neq 0$ $(j=1,2,\ldots,r)$, λ_j/μ_j's are distinct complex numbers and

$$u_j=\left(\prod_{\substack{1\leqq i\leqq r\\ i\neq j}}\left(\frac{\lambda_i}{\mu_i}-\frac{\lambda_j}{\mu_j}\right)\right)^{-1}\qquad(j=1,\ldots,r).$$

The following further special case of (3) when $r = 3$ is recorded by Hansen [531, p. 106, Entry (6.1.65)]:

$$\sum_{k=0}^{\infty} \frac{1}{(\mu_1 k+\lambda_1)(\mu_2 k+\lambda_2)(\mu_3 k+\lambda_3)} = \frac{\mu_1\,\psi(\lambda_1/\mu_1)}{(\mu_2\lambda_1-\mu_1\lambda_2)(\mu_1\lambda_3-\mu_3\lambda_1)} + \frac{\mu_2\,\psi(\lambda_2/\mu_2)}{(\mu_1\lambda_2-\mu_2\lambda_1)(\mu_2\lambda_3-\mu_3\lambda_2)} + \frac{\mu_3\,\psi(\lambda_3/\mu_3)}{(\mu_1\lambda_3-\mu_3\lambda_1)(\mu_3\lambda_2-\mu_2\lambda_3)}. \tag{4}$$

Upon setting $\mu_j = 1$ $(j = 1, 2, 3)$, $\lambda_1 = a$, $\lambda_2 = b$ and $\lambda_3 = c$, (4) yields

$$\sum_{k=1}^{\infty} \frac{1}{(k+a)(k+b)(k+c)} = \frac{\psi(a)}{(a-b)(c-a)} + \frac{\psi(b)}{(b-a)(c-b)} + \frac{\psi(c)}{(c-a)(b-c)} - \frac{1}{abc}. \tag{5}$$

Now, taking partial derivatives in (5) with respect to the variables a, b and c, we can express the following sum:

$$\sum_{k=1}^{\infty} \frac{1}{(k+a)^i(k+b)^j(k+c)^k} \qquad (i, j, k \in \mathbb{N}) \tag{6}$$

in terms of the Psi-function and the polygamma functions. Only the case $i = j = 3$ and $k = 1$ of (6) is recorded here for our use in the next subsection:

$$\begin{aligned}\sum_{k=1}^{\infty} \frac{1}{(k+a)^3(k+b)^3(k+c)} &= \frac{\psi''(a)}{2(a-b)^3(c-a)} + \frac{\psi''(b)}{2(b-a)^3(c-b)} \\ &+ \frac{4a-b-3c}{(a-b)^4(a-c)^2}\,\psi'(a) + \frac{4b-a-3c}{(b-a)^4(b-c)^2}\,\psi'(b) \\ &+ \frac{6(c-a)^2-3(a-b)(c-a)+(a-b)^2}{(a-b)^5(c-a)^3}\,\psi(a) \\ &+ \frac{6(c-b)^2-3(b-a)(c-b)+(b-a)^2}{(b-a)^5(c-b)^3}\,\psi(b) \\ &+ \frac{\psi(c)}{(c-a)^3(b-c)^3} - \frac{1}{a^3b^3c}.\end{aligned} \tag{7}$$

It may be remarked in passing that all of the interesting known or new special cases of the results, including (2) of Al-Saqabi *et al.* [21] (especially those that can be found in Hansen [531], Chrystal [313], Jolley [613] and Prudnikov *et al.* [918]), were mentioned explicitly by Al-Saqabi *et al.* [21]. Of course, Eq. (7) may be derived from a very specialized case of Eq. (25) in Wu *et al.* [1237, p. 6].

Convolutions of the Rayleigh Functions

Varlamov [1192, 1193] systematically investigated convolutions of the Rayleigh functions with respect to the Bessel index and attempted to exhibit their usefulness for constructing global-in-time solutions of semilinear evolution equations in circular domains.

Convolutions of the Rayleigh functions with respect to the Bessel index are defined as follows:

$$R_l(m) = \sum_{k=-\infty}^{\infty} \sigma_l(m-k)\,\sigma_l(k) \quad (l \in \mathbb{N};\ m \in \mathbb{Z}),$$

where Rayleigh functions $\sigma_l(\nu)$ are defined as (see [1215, p. 502])

$$\sigma_l(\nu) = \sum_{n=1}^{\infty} \frac{1}{\lambda_{\nu,n}^{2l}} \quad (l,\ n \in \mathbb{N}),$$

ν being the index of the Bessel function $J_\nu(x)$, whose zeros are $\lambda_{\nu,n}$. Varlamov [1192, 1193] succeeded in treating the Rayleigh functions, $\sigma_l(\nu)$ as special functions, by presenting some references for functions of the Rayleigh type, as well as in their own right (see the references cited by Varlamov [1192, 1193]).

Varlamov [1192] expressed $R_1(m)$ in terms of the ψ-function and made a simple use of the expression of the ψ-function to give the following asymptotic formula for $R_1(m)$:

$$R_1(m) \sim \frac{\ln|m|}{4\,|m|} \quad (|m| \to \infty).$$

Varlamov [1193] also presented a general formula for $R_l(m)$ in terms of $\sigma_l(\nu)$ and expressed $R_2(m)$ in terms of the ψ-function, together with its asymptotic formula as $|m| \to \infty$, by making use of the ψ-function, as follows:

$$R_2(m) \sim \frac{1}{2^8}\left[\left(\frac{2\pi^2}{3}-5\right)\frac{1}{|m|^3} - \left(\frac{2\pi^2}{3}-5\right)\frac{4}{|m|^4} + \frac{34\pi^2/3 - 165 + 24(\gamma + \ln|m|)}{|m|^5} + O\left(\frac{\ln|m|}{|m|^6}\right)\right] \quad (|m| \to \infty).$$

Furthermore, Varlamov [1193] provided an example of an initial-boundary-value problem leading to an explicit appearance of $R_1(m)$ and $R_2(m)$. To show a usefulness of the polygamma functions, we express $R_3(m)$ in terms of the polygamma functions and give its asymptotic formula as $|m| \to \infty$.

It is easy to get the following formula of $R_3(m)$ from one of Varlamov's results [1193, Eq. (4.1)]:

$$R_3(m) = \frac{1}{2^{10}}\{\mathcal{A}_3(m) + 2\mathcal{B}_3(m)\} \quad (m \in \mathbb{Z}), \tag{8}$$

where

$$\mathcal{A}_3(m) = \sum_{k=0}^{|m|} \frac{1}{(k+1)^3\,(k+2)\,(k+3)\,(|m|-k+1)^3\,(|m|-k+2)\,(|m|-k+3)}$$

and

$$\mathcal{B}_3(m) = \sum_{k=1}^{\infty} \frac{1}{(k+1)^3\,(k+2)\,(k+3)\,(|m|+k+1)^3\,(|m|+k+2)\,(|m|+k+3)}.$$

We can, now, express $\mathcal{A}_3(m)$ and $\mathcal{B}_3(m)$ in terms of the polygamma functions. We, first, decompose $\mathcal{A}_3(m)$ into partial fractions as follows:

$$\begin{aligned}
\mathcal{A}_3(m) = {} & \frac{1}{4(|m|+2)^5\,(|m|+3)^3\,(|m|+4)^3} \Big\{(3|m|^3+17|m|^2+14|m|-28)^2 \\
& - 2(|m|+3)\,(|m|+4)\left(|m|^4-4|m|^3-62|m|^2-150|m|-88\right)\Big\} \sum_{k=0}^{|m|} \frac{1}{k+1} \\
& - \frac{3|m|^3+17|m|^2+14|m|-28}{2(|m|+2)^4\,(|m|+3)^2\,(|m|+4)^2} \sum_{k=0}^{|m|} \frac{1}{(k+1)^2} \\
& + \frac{1}{(|m|+2)^3\,(|m|+3)\,(|m|+4)} \sum_{k=0}^{|m|} \frac{1}{(k+1)^3} \\
& - \frac{2}{(|m|+3)^3\,(|m|+4)\,(|m|+5)} \sum_{k=0}^{|m|} \frac{1}{k+2} \\
& + \frac{1}{4(|m|+4)^3\,(|m|+5)\,(|m|+6)} \sum_{k=0}^{|m|} \frac{1}{k+3},
\end{aligned}$$

so that, by using 1.2(7), we have

$$\begin{aligned}
\mathcal{A}_3(m) = {} & \frac{1}{4(|m|+2)^5\,(|m|+3)^3\,(|m|+4)^3} \Big\{(3|m|^3+17|m|^2+14|m|-28)^2 \\
& -2(|m|+3)\,(|m|+4)\left(|m|^4-4|m|^3-62|m|^2-150|m|-88\right)\Big\} \left(\psi(|m|+2)+\gamma\right) \\
& - \frac{3|m|^3+17|m|^2+14|m|-28}{2(|m|+2)^4\,(|m|+3)^2\,(|m|+4)^2} \left(\psi'(|m|+2)-\psi'(1)\right) \\
& + \frac{1}{2(|m|+2)^3\,(|m|+3)\,(|m|+4)} \left(\psi''(|m|+2)-\psi''(1)\right)
\end{aligned}$$

$$- \frac{2}{(|m|+3)^3\,(|m|+4)\,(|m|+5)}\,(\psi(|m|+3)-\psi(2))$$
$$+ \frac{1}{4(|m|+4)^3\,(|m|+5)\,(|m|+6)}\,(\psi(|m|+4)-\psi(3)),$$

where γ denotes the Euler-Mascheroni constant, defined by 1.1(3).

Next, we decompose $\mathcal{B}_3(m)$ into the following partial fractions:

$$\mathcal{B}_3(m) = \frac{\mathcal{S}_1(m)}{|m|(|m|+1)} - \frac{\mathcal{S}_2(m)}{|m|(|m|-1)} + \frac{\mathcal{S}_3(m)}{|m|(|m|-1)} - \frac{\mathcal{S}_4(m)}{|m|(|m|+1)},$$

where

$$\mathcal{S}_1(m) = \sum_{k=1}^{\infty} \frac{1}{(k+1)^3\,(k+|m|+1)^3\,(k+2)},$$
$$\mathcal{S}_2(m) = \sum_{k=1}^{\infty} \frac{1}{(k+1)^3\,(k+|m|+1)^3\,(k+3)},$$
$$\mathcal{S}_3(m) = \sum_{k=1}^{\infty} \frac{1}{(k+1)^3\,(k+|m|+1)^3\,(k+|m|+2)},$$

and

$$\mathcal{S}_4(m) = \sum_{k=1}^{\infty} \frac{1}{(k+1)^3\,(k+|m|+1)^3\,(k+|m|+3)}.$$

Using (7), 1.3(53) and other needed identities listed in Chapters 1 and 2, we can express $\mathcal{S}_i(m)$ $(i=1,\,2,\,3,\,4)$ as follows:

$$\mathcal{S}_1(m) = \frac{\zeta(3,\,|m|+1)}{|m|^3\,(|m|-1)} - \frac{\pi^2}{6}\,\frac{|m|+3}{|m|^4} + \frac{4|m|-3}{|m|^4\,(|m|-1)^2}\,\zeta(2,\,|m|+1)$$
$$+ \frac{\gamma(|m|^2+3|m|+6)}{|m|^5} - \frac{10\,|m|^2-15\,|m|+6}{|m|^5\,(|m|-1)^3}\,\psi(|m|+1)$$
$$+ \frac{\zeta(3)}{|m|^3} + \frac{1-\gamma}{(|m|-1)^3} - \frac{1}{2\,(|m|+1)^3}.$$

$$\mathcal{S}_2(m) = \frac{\zeta(3,\,|m|+1)}{|m|^3\,(|m|-2)} - \frac{\pi^2}{24}\,\frac{|m|+6}{|m|^4} + \frac{2(2|m|-3)}{|m|^4\,(|m|-2)^2}\,\zeta(2,\,|m|+1)$$
$$+ \frac{\gamma}{8}\,\frac{|m|^2+6|m|+24}{|m|^5} - \frac{2(2|m|^2-15|m|+12)}{|m|^5\,(|m|-2)^3}\,\psi(|m|+1)$$
$$+ \frac{\zeta(3)}{2\,|m|^3} + \frac{3-2\gamma}{16(|m|-2)^3} - \frac{1}{3\,(|m|+1)^3}.$$

$$\begin{aligned} \mathcal{S}_3(m) &= \frac{\zeta(3)}{|m|^3\,(|m|+1)} - \frac{\zeta(3,\,|m|+1)}{|m|^3} - \frac{\pi^2}{6}\,\frac{4|m|+3}{|m|^4\,(|m|+1)^2} \\ &\quad + \frac{|m|-3}{|m|^4}\,\zeta(2,\,|m|+1) + \frac{\gamma\,(10\,|m|^2+15\,|m|+6)}{|m|^5\,(|m|+1)^3} \\ &\quad + \frac{|m|^2-3|m|+6}{|m|^5}\,\psi(|m|+1) - \frac{\psi(|m|+2)}{(|m|+1)^3} - \frac{1}{(|m|+1)^3\,(|m|+2)}. \end{aligned}$$

$$\begin{aligned} \mathcal{S}_4(m) &= -\frac{\zeta(3,\,|m|+1)}{2\,|m|^3} - \frac{\pi^2}{3}\,\frac{2|m|+3}{|m|^4\,(|m|+2)^2} + \frac{|m|-6}{4\,|m|^4}\,\zeta(2,\,|m|+1) \\ &\quad + \frac{2\gamma\,(5\,|m|^2+15\,|m|+12)}{|m|^5\,(|m|+2)^3} + \frac{|m|^2-6|m|+24}{8\,|m|^5}\,\psi(|m|+1) \\ &\quad + \frac{\zeta(3)}{|m|^3\,(|m|+2)} - \frac{\psi(|m|+3)}{8\,(|m|+2)^3} - \frac{1}{(|m|+1)^3\,(|m|+3)}. \end{aligned}$$

Finally, by applying 1.3(65) to each of the above expressions, we get the following asymptotic formulas:

$$\mathcal{A}_3(m) = \left(\frac{13}{8} + \frac{\pi^2}{4} + \zeta(3)\right)\frac{1}{|m|^5} + O\left(\frac{1}{|m|^6}\right) \qquad (|m| \to \infty)$$

and

$$\mathcal{B}_3(m) = \left(\frac{31}{48} - \frac{\pi^2}{8} + \frac{\zeta(3)}{2}\right)\frac{1}{|m|^5} + O\left(\frac{1}{|m|^6}\right) \qquad (|m| \to \infty),$$

which, upon using (8), yields a simple asymptotic formula for $R_3(m)$ as follows:

$$R_3(m) = \frac{1}{2^{10}}\left(\frac{35}{12} + 2\zeta(3)\right)\frac{1}{|m|^5} + O\left(\frac{1}{|m|^6}\right) \qquad (|m| \to \infty).$$

7.4 Bernoulli and Euler Polynomials at Rational Arguments

We begin by recalling a theorem involving the Bernoulli numbers discovered in 1840 (and proven independently and published almost simultaneously) by C. von Staudt (1798–1867) and T. Clausen (1801–1885).

Theorem 7.2. [von Staudt-Clausen Theorem] *Let B_n denote the Bernoulli numbers, defined by 1.7(2). Then,*

$$B_{2n} = I_n - \sum_{p-1|2n} \frac{1}{p} \qquad (n \in \mathbb{N}), \tag{1}$$

where I_n is an integer and the sum is taken over all primes p, such that $p-1$ divides $2n$.

As noted by Carathéodory [210, p. 283], the von Staudt-Clausen theorem (1) is very useful for the calculation of higher-order Bernoulli numbers, since the value of I_n follows even from a preliminary estimate of B_{2n} that one might obtain in one way or another (see also Problem 6 of Chapter 1).

Bartz and Rutkowski [99] gave a simple proof of a generalization of the von Staudt-Clausen theorem (1) to hold true for the Bernoulli polynomials, which also reduces to another interesting and inspiring result (Almkvist and Meurman [19, p. 104, Theorem 2]):

$$k^n\left[B_n\left(\tfrac{h}{k}\right)-B_n\right]\in\mathbb{Z}\quad(h,\,k\in\mathbb{N}),\tag{2}$$

by Almkvist and Meurman [19, p. 104, Theorem 2]), who proved (2) in an elementary (yet rather complicated) way. Subsequently, simpler proofs of the Almkvist-Meurman theorem (2) have been given by Sury [1137] and Clarke and Slavutsk II [1041]. An analogous result involving the Euler polynomials, instead, was proven by Fox [459]. All the aforecited works are concerned with evaluations of the Bernoulli and Euler polynomials at rational arguments.

The remainder of this section is based largely upon the work of Srivastava [1087], who gave several remarkably shorter proofs of each of the Cvijović-Klinowski summation formulas and also considered their various generalizations and analogues.

The Cvijović-Klinowski Summation Formulas

By employing a number of properties and characteristics of certain Dirichlet and trigonometric series, Cvijović and Klinowski [349] evaluated the Bernoulli polynomials $B_n(x)$ with $n\in\mathbb{N}\setminus\{1\}$ and the Euler polynomials $E_n(x)$ with $n\in\mathbb{N}$ for $0\leqq x\leqq 1$ when x is a rational number. For the sake of ready reference, we recall, here, the *main* results of Cvijović and Klinowski [349] as Theorem 7.3 and Theorem 7.4 below:

Theorem 7.3 *In terms of the Hurwitz Zeta function* $\zeta(s,a)$, *defined by* 2.2(1), *the Bernoulli polynomials* $B_n(x)$ *at rational arguments are given by*

$$B_{2n-1}\left(\frac{p}{q}\right)=(-1)^n\frac{2(2n-1)!}{(2q\pi)^{2n-1}}\sum_{j=1}^{q}\zeta\left(2n-1,\frac{j}{q}\right)\sin\left(\frac{2jp\pi}{q}\right)\tag{3}$$

$$(n\in\mathbb{N}\setminus\{1\};\ p\in\mathbb{N}_0;\ q\in\mathbb{N};\ 0\leq p\leq q)$$

and

$$B_{2n}\left(\frac{p}{q}\right)=(-1)^{n-1}\frac{2(2n)!}{(2q\pi)^{2n}}\sum_{j=1}^{q}\zeta\left(2n,\frac{j}{q}\right)\cos\left(\frac{2jp\pi}{q}\right)\tag{4}$$

$$\left(n\in\mathbb{N};\ p\in\mathbb{N}_0;\ q\in\mathbb{N};\ 0\leqq p\leqq q\right).$$

Theorem 7.4 *In terms of the Hurwitz Zeta function $\zeta(s,a)$, defined by 2.2(1), the Euler polynomials $E_n(x)$ at rational arguments are given by*

$$E_{2n-1}\left(\frac{p}{q}\right) = (-1)^n \frac{4(2n-1)!}{(2q\pi)^{2n}} \sum_{j=1}^{q} \zeta\left(2n, \frac{2j-1}{q}\right) \cos\left(\frac{(2j-1)p\pi}{q}\right) \tag{5}$$

$$\left(n \in \mathbb{N};\ p \in \mathbb{N}_0;\ q \in \mathbb{N};\ 0 \leqq p \leqq q\right)$$

and

$$E_{2n}\left(\frac{p}{q}\right) = (-1)^n \frac{4(2n)!}{(2q\pi)^{2n+1}} \sum_{j=1}^{q} \zeta\left(2n+1, \frac{2j-1}{2q}\right) \sin\left(\frac{(2j-1)p\pi}{q}\right) \tag{6}$$

$$\left(n \in \mathbb{N};\ p \in \mathbb{N}_0;\ q \in \mathbb{N};\ 0 \leqq p \leqq q\right).$$

As pointed out by Cvijović and Klinowski [349, p. 1535], the formula (3) was derived earlier, in a completely different way, by Almkvist and Meurman [19, p. 107, Proposition 10], who applied this formula to show *eventually* that their assertion holds true.

Srivastava's Shorter Proofs of Theorem 7.3 and Theorem 7.4

First Proof of Theorem 7.3. We begin by recalling the following known result (*cf.*, e.g., Magnus *et al.* [795, p. 27]):

$$B_n(x) = -\frac{2 \cdot n!}{(2\pi)^n} \sum_{k=1}^{\infty} \frac{1}{k^n} \cos\left(2k\pi x - \frac{n\pi}{2}\right) \tag{7}$$

$$\left(n \in \mathbb{N} \setminus \{1\} \quad \text{and} \quad 0 \leqq x \leqq 1;\ n = 1 \quad \text{and} \quad 0 < x < 1\right),$$

which, in view of the definition in 2.2(1) *and* the elementary series identity:

$$\sum_{k=1}^{\infty} f(k) = \sum_{j=1}^{q} \sum_{k=0}^{\infty} f(qk+j) \quad (q \in \mathbb{N}), \tag{8}$$

immediately yields

$$B_n\left(\frac{p}{q}\right) = -\frac{2 \cdot n!}{(2q\pi)^n} \sum_{j=1}^{q} \zeta\left(n, \frac{j}{q}\right) \cos\left(\frac{2jp\pi}{q} - \frac{n\pi}{2}\right) \tag{9}$$

$$\left(n \in \mathbb{N} \setminus \{1\};\ p \in \mathbb{N}_0;\ q \in \mathbb{N};\ 0 \leqq p \leqq q\right),$$

where we have also set

$$x = \frac{p}{q} \quad \left(p \in \mathbb{N}_0;\ q \in \mathbb{N};\ 0 \leqq p \leqq q\right).$$

Upon replacing n in (9) by $2n-1$ and $2n$, respectively, we obtain the assertions (3) and (4) of Theorem 7.3. ■

Second Proof of Theorem 7.3. Upon setting $s=n$ $(n \in \mathbb{N} \setminus \{1\})$ in 2.2(8) and applying 2.1(17) (with $\ell = n-1$) to the left member of the resulting identity, we arrive once again at the formula (9), which unifies the assertions (3) and (4) of Theorem 7.3, the missing case $p=0$ for the Bernoulli numbers B_n being easily verified *directly* by means of the well-known identities 2.3(18) and

$$\zeta(s) = q^{-s} \sum_{j=1}^{q} \zeta\left(s, \frac{j}{q}\right) \quad (q \in \mathbb{N}). \tag{10}$$

■

First Proof of Theorem 7.4. We, now, show that Theorem 7.4 is actually a simple consequence of Theorem 7.3. Indeed, by appealing to the known relationship 1.7(60), we find from (9) (with n replaced by $n+1$) that

$$\begin{aligned} E_n\left(\frac{p}{q}\right) &= \frac{2}{n+1}\left[-\frac{2(n+1)!}{(2q\pi)^{n+1}} \sum_{j=1}^{q} \zeta\left(n+1, \frac{j}{q}\right) \sin\left(\frac{2jp\pi}{q} - \frac{n\pi}{2}\right)\right. \\ &\quad \left. + 2^{n+1} \frac{2(n+1)!}{(4q\pi)^{n+1}} \sum_{j=1}^{2q} \zeta\left(n+1, \frac{j}{2q}\right) \sin\left(\frac{jp\pi}{q} - \frac{n\pi}{2}\right)\right] \\ &= \frac{4 \cdot n!}{(2q\pi)^{n+1}} \left[\sum_{j=1}^{2q} \zeta\left(n+1, \frac{j}{2q}\right) \sin\left(\frac{jp\pi}{q} - \frac{n\pi}{2}\right)\right. \\ &\quad \left. - \sum_{j=1}^{q} \zeta\left(n+1, \frac{j}{p}\right) \sin\left(\frac{2jp\pi}{q} - \frac{n\pi}{2}\right)\right] \\ &\qquad \left(n \in \mathbb{N};\ p \in \mathbb{N}_0;\ q \in \mathbb{N};\ 0 \leqq p \leqq q\right). \end{aligned} \tag{11}$$

Obviously, since

$$\begin{aligned} &\sum_{j=1}^{2q} \zeta\left(n+1, \frac{j}{2q}\right) \sin\left(\frac{jp\pi}{q} - \frac{n\pi}{2}\right) \\ &\quad = \sum_{j=1}^{q} \zeta\left(n+1, \frac{2j-1}{2q}\right) \sin\left(\frac{(2j-1)p\pi}{q} - \frac{n\pi}{2}\right) \\ &\quad + \sum_{j=1}^{q} \zeta\left(n+1, \frac{j}{q}\right) \sin\left(\frac{2jp\pi}{q} - \frac{n\pi}{2}\right), \end{aligned} \tag{12}$$

upon separating the even and odd terms, (11) leads us immediately to the formula:

$$E_n\left(\frac{p}{q}\right)=\frac{4\cdot n!}{(2q\pi)^{n+1}}\sum_{j=1}^{q}\zeta\left(n+1,\frac{2j-1}{2q}\right)\sin\left(\frac{(2j-1)p\pi}{q}-\frac{n\pi}{2}\right) \tag{13}$$

$$\left(n\in\mathbb{N};\ p\in\mathbb{N}_0;\ q\in\mathbb{N};\ 0\leqq p\leqq q\right),$$

which yields the assertions (5) and (6) of Theorem 7.3 when n is replaced by $2n-1$ and $2n$, respectively. ■

Second Proof of Theorem 7.4. Just as in our first proof of Theorem 7.3, the unification (13) of the assertions (5) and (6) of Theorem 7.4 can be proven *directly* by merely applying the series identity (8) in the following known result (*cf.*, e.g., Magnus *et al.* [795, p. 30]; see also 1.7(56) and 1.7(57)):

$$E_n(x)=\frac{4\cdot n!}{\pi^{n+1}}\sum_{k=1}^{\infty}\frac{1}{(2k-1)^{n+1}}\sin\left((2k-1)\pi x-\frac{n\pi}{2}\right) \tag{14}$$

$$(n\in\mathbb{N}\quad\text{and}\quad 0\leqq x\leqq 1;\ n=0\quad\text{and}\quad 0<x<1),$$

with, of course,

$$x=\frac{p}{q}\quad\left(p\in\mathbb{N};\ q\in\mathbb{N};\ 0\leqq p\leqq q\right).$$

■

Formulas Involving the Hurwitz-Lerch Zeta Function

By virtue of 1.3(53), each of the formulas (9) and (13) and their even and odd integer versions given by Theorem 7.3 and Theorem 7.4 can easily be restated in terms of the Polygamma function.

For the general Hurwitz-Lerch Zeta function $\Phi(z,s,a)$, defined by 2.5(1), it is easily seen by using the series identity (8) that

$$\Phi(z,s,a)=q^{-s}\sum_{j=1}^{q}\Phi\left(z^q,s,\frac{a+j-1}{q}\right)z^{j-1}, \tag{15}$$

which, in the special case when

$$z=\exp\left(\frac{2p\pi i}{q}\right)\quad(p\in\mathbb{Z};\ q\in\mathbb{N}),$$

yields the summation formula:

$$L\left(\frac{p}{q},s,a\right)=q^{-s}\sum_{j=1}^{q}\zeta\left(s,\frac{a+j-1}{q}\right)\exp\left(\frac{2(j-1)p\pi i}{q}\right) \tag{16}$$

for the generalized Zeta function $L(x, s, a)$, defined by 2.5(11), in terms of the Hurwitz Zeta function $\zeta(s, a)$.

For $z = 1$, (15) reduces at once to the identity [Hansen (1975, p. 360, Entry (54.13.1))]:

$$\zeta(s, a) = q^{-s} \sum_{j=1}^{q} \zeta\left(s, \frac{a+j-1}{q}\right), \tag{17}$$

which, for $a = 1$, yields the well-known result (10). Conversely, by setting $a = \frac{1}{2}$ in (15) and (16), we have

$$\sum_{n=1}^{\infty} \frac{z^n}{(2n-1)^s} = (2q)^{-s} \sum_{j=1}^{q} \Phi\left(z^q, s, \frac{2j-1}{2q}\right) z^{j-1} \tag{18}$$

and

$$\sum_{n=0}^{\infty} \frac{e^{(2n+1)p\pi i/q}}{(2n+1)^s} = (2q)^{-s} \sum_{j=1}^{q} \zeta\left(s, \frac{2j-1}{2q}\right) \exp\left(\frac{(2j-1)p\pi i}{q}\right), \tag{19}$$

respectively.

Lastly, in their special cases when $a = 1$, (15) and (16) yield the following companions of the summation formulas (18) and (19), respectively:

$$\sum_{n=1}^{\infty} \frac{z^n}{n^s} =: \mathrm{Li}_s(z) = q^{-s} \sum_{j=1}^{q} \Phi\left(z^q, s, \frac{j}{q}\right) z^j \tag{20}$$

and

$$\sum_{n=1}^{\infty} \frac{e^{2np\pi i/q}}{n^s} =: L\left(\frac{p}{q}, s\right) = q^{-s} \sum_{j=1}^{q} \zeta\left(s, \frac{j}{q}\right) \exp\left(\frac{2jp\pi i}{q}\right). \tag{21}$$

In particular, when $s = \nu$ ($\nu > 1$), by simply equating the real and imaginary parts in (19) and (21), we immediately obtain the following summation formulas involving trigonometric series:

$$\sum_{n=0}^{\infty} \frac{\cos[(2n+1)p\pi/q]}{(2n+1)^\nu} = (2q)^{-\nu} \sum_{j=1}^{q} \zeta\left(\nu, \frac{2j-1}{2q}\right) \cos\left(\frac{(2j-1)p\pi}{q}\right); \tag{22}$$

$$\sum_{n=0}^{\infty} \frac{\sin[(2n+1)p\pi/q]}{(2n+1)^\nu} = (2q)^{-\nu} \sum_{j=1}^{q} \zeta\left(\nu, \frac{2j-1}{2q}\right) \sin\left(\frac{(2j-1)p\pi}{q}\right); \tag{23}$$

$$\sum_{n=1}^{\infty} \frac{\cos(2np\pi/q)}{n^{\nu}} = q^{-\nu} \sum_{j=1}^{q} \zeta\left(\nu, \frac{j}{q}\right) \cos\left(\frac{2jp\pi}{q}\right); \tag{24}$$

$$\sum_{n=1}^{\infty} \frac{\sin(2np\pi/q)}{n^{\nu}} = q^{-\nu} \sum_{j=1}^{q} \zeta\left(\nu, \frac{j}{q}\right) \sin\left(\frac{2jp\pi}{q}\right). \tag{25}$$

The assertions made by the Lemma of Cvijović and Klinowski [349, p. 1530] are *precisely* the special summation formulas (19) and (21) *with* $s = \nu$ $(\nu > 1)$. Formula (21) with $s = \nu$ $(\nu > 1)$ is, in fact, also one of the three *main* results in another paper by Cvijović and Klinowski [350, p. 207, Eq. (7)]; the other two *main* results of Cvijović and Klinowski [350, p. 207, Eqs. (8a) and (8b)] are essentially the same as (23) and (22), respectively, *with* p replaced by $2p$. Cvijović and Klinowski [350, p. 208, Eqs. (9a) and (9b)] also gave the special summation formulas (23) and (22), respectively, *with* $q = Q$.

As already pointed out by Srivastava [1087], each of the trigonometric sums (22) to (25), and indeed also various *further* special cases of many of the summation formulas considered here, can be found derived in the work of Cvijović and Klinowski [349].

An Application of Lerch's Functional Equation 2.5(29)

In view of the summation formula (16), Lerch's functional equation 2.5(29) with

$$a = \frac{p}{q} \quad (p \in \mathbb{Z};\ q \in \mathbb{N})$$

yields

$$L\left(x, 1-s, \frac{p}{q}\right) = \frac{\Gamma(s)}{(2p\pi)^s} \left\{ \sum_{j=1}^{q} \zeta\left(s, \frac{x+j-1}{q}\right) \exp\left[\left(\frac{s}{2} - \frac{2(x+j-1)p}{q}\right)\pi i\right] \right.$$
$$\left. + \sum_{j=1}^{q} \zeta\left(s, \frac{j-x}{q}\right) \exp\left[\left(-\frac{s}{2} + \frac{2(j-x)p}{q}\right)\pi i\right] \right\}. \tag{26}$$

In terms of a *mild* generalization $\mathcal{B}_n(a, \alpha)$ of the classical Bernoulli polynomials $B_n(x)$, defined by 2.5(41), by setting $s = n$ $(n \in \mathbb{N})$ in (26) and using 2.5(40), Srivastava [1087] obtained the following novel analogue of (9) for the generalized Bernoulli polynomials $\mathcal{B}_n(a, \alpha)$:

$$\mathcal{B}_n\left(\frac{p}{q}, e^{2\pi i x}\right) = -\frac{n!}{(2q\pi)^n} \left\{ \sum_{j=1}^{q} \zeta\left(n, \frac{x+j-1}{q}\right) \exp\left[\left(\frac{n}{2} - \frac{2(x+j-1)p}{q}\right)\pi i\right] \right.$$
$$\left. + \sum_{j=1}^{q} \zeta\left(n, \frac{j-x}{q}\right) \exp\left[\left(-\frac{n}{2} + \frac{2(j-x)p}{q}\right)\pi i\right] \right\}, \tag{27}$$

which holds true whenever each side exists. Indeed, in its *special* case when $x \in \mathbb{Z}$, Srivastava's formula (27) can easily be shown to reduce to the known result (9).

7.5 Closed-Form Summation of Trigonometric Series

By Weierstrass's M-test, it is easily seen that each of the classical trigonometric series:

$$C_\nu(\alpha) := \sum_{k=0}^{\infty} \frac{\cos(2k+1)\alpha}{(2k+1)^\nu} \quad \text{and} \quad S_\nu(\alpha) := \sum_{k=0}^{\infty} \frac{\sin(2k+1)\alpha}{(2k+1)^\nu} \tag{1}$$

converges uniformly for all real values of α when $\nu > 1$. Furthermore, since

$$C_\nu(-\alpha) = C_\nu(\alpha) = -C_\nu(\pi - \alpha) \tag{2}$$

and

$$S_\nu(-\alpha) = -S_\nu(\alpha) = -S_\nu(\pi - \alpha), \tag{3}$$

it is sufficient to evaluate $C_\nu(\alpha)$ and $S_\nu(\alpha)$ only over the range $0 \leqq \alpha \leqq \frac{1}{2}\pi$.

The series $C_{2n+1}(\alpha)$ and $S_{2n}(\alpha)$ $(n \in \mathbb{N})$ converge extremely slowly, and the problem of their numerical evaluation was addressed by (among others) Dempsey *et al.* [373], Boersma and Dempsey [139] and Gautschi [473]. For the particular case when $n = 1$, Dempsey *et al.* [373] developed a procedure based on Plana's summation formula, together with Romberg's method of integration, which significantly improves the convergence and accuracy when compared with those resulting from *direct* summation. Boersma and Dempsey [139], conversely, transformed each of $C_{2n+1}(\alpha)$ and $S_{2n}(\alpha)$ into a rapidly convergent series that is well-suited for their computation (when, e.g., $n = 1$).

Recently, Cvijović and Klinowski [350] considered the general series in (1) when $\nu > 1$ and α is a rational multiple of 2π. In terms of the Riemann Zeta function $\zeta(s)$ and the Hurwitz Zeta function $\zeta(s, a)$, defined by 2.3(1) and 2.2(1), respectively, they first showed that (Cvijović and Klinowski [350, p. 207, Eqs. (8a) and (8b)])

$$\begin{aligned} C_\nu\left(\frac{2p\pi}{q}\right) = \frac{1}{q^\nu}\Bigg\{&\zeta(\nu)\left(1 - \frac{1}{2^\nu}\right) \\ &+ \sum_{j=1}^{q-1} \zeta\left(\nu, \frac{j}{q}\right)\left[\cos\left(\frac{2jp\pi}{q}\right) - \frac{1}{2^\nu}\cos\left(\frac{4jp\pi}{q}\right)\right]\Bigg\} \end{aligned} \tag{4}$$

and

$$S_\nu\left(\frac{2p\pi}{q}\right) = \frac{1}{q^\nu}\sum_{j=1}^{q-1} \zeta\left(\nu, \frac{j}{q}\right)\left[\sin\left(\frac{2jp\pi}{q}\right) - \frac{1}{2^\nu}\sin\left(\frac{4jp\pi}{q}\right)\right] \tag{5}$$

$$(p, q \in \mathbb{N};\ \nu > 1).$$

Following the work of Srivastava [1085]), we now show that the summation formulas (4) and (5), as well as their equivalent variants when α is a rational multiple of π (*cf.* Cvijović and Klinowski [350, p. 208, Eqs. (9a) and (9b)]), would follow easily from some general results involving the family of Dirichlet series in 2.5(1), which is usually called Lerch's function or Lerch's transcendent.

For convenience, we recall, here, the summation formulas 6.1(24), 6.1(25), 6.1(22) and 6.1(23) in the following (slightly modified) form:

$$\sum_{k=1}^{\infty} \frac{\cos(2kp\pi/q)}{(2k)^{\nu}} = (2q)^{-\nu} \sum_{j=1}^{q} \zeta\left(\nu, \frac{j}{q}\right) \cos\left(\frac{2jp\pi}{q}\right), \tag{6}$$

$$\sum_{k=1}^{\infty} \frac{\sin(2kp\pi/q)}{(2k)^{\nu}} = (2q)^{-\nu} \sum_{j=1}^{q} \zeta\left(\nu, \frac{j}{q}\right) \sin\left(\frac{2jp\pi}{q}\right), \tag{7}$$

$$\begin{aligned} C_{\nu}\left(\frac{p\pi}{q}\right) &:= \sum_{k=0}^{\infty} \frac{\cos\left[(2k+1)p\pi/q\right]}{(2k+1)^{\nu}} \\ &= (2q)^{-\nu} \sum_{j=1}^{q} \zeta\left(\nu, \frac{2j-1}{2q}\right) \cos\left(\frac{(2j-1)p\pi}{q}\right), \end{aligned} \tag{8}$$

and

$$\begin{aligned} S_{\nu}\left(\frac{p\pi}{q}\right) &:= \sum_{k=0}^{\infty} \frac{\sin\left[(2k+1)p\pi/q\right]}{(2k+1)^{\nu}} \\ &= (2q)^{-\nu} \sum_{j=1}^{q} \zeta\left(\nu, \frac{2j-1}{2q}\right) \sin\left(\frac{(2j-1)p\pi}{q}\right), \end{aligned} \tag{9}$$

respectively. Since (*cf.* Equation 6.1(17) with $q = 2$)

$$\zeta(s,a) = 2^{-s}\left[\zeta\left(s, \frac{1}{2}a\right) + \zeta\left(s, \frac{1}{2}a + \frac{1}{2}\right)\right], \tag{10}$$

the second member of the summation formula (4) can be rewritten in the form:

$$\begin{aligned} &q^{-\nu} \sum_{j=1}^{q} \zeta\left(\nu, \frac{j}{q}\right)\left[\cos\left(\frac{2jp\pi}{q}\right) - 2^{-\nu}\cos\left(\frac{4jp\pi}{q}\right)\right] \\ &= (2q)^{-\nu}\left\{\sum_{j=1}^{q}\left[\zeta\left(\nu, \frac{j}{2q}\right) + \zeta\left(\nu, \frac{j+q}{2q}\right)\right]\cos\left(\frac{2jp\pi}{q}\right)\right. \\ &\qquad \left. - \sum_{j=1}^{q} \zeta\left(\nu, \frac{j}{q}\right)\cos\left(\frac{4jp\pi}{q}\right)\right\} \end{aligned}$$

$$= (2q)^{-\nu}\left\{\sum_{j=1}^{q}\zeta\left(\nu,\frac{2j-1}{2q}\right)\cos\left(\frac{2(2j-1)p\pi}{q}\right)\right. \tag{11}$$

$$+\sum_{j=1}^{q}\zeta\left(\nu,\frac{2j}{2q}\right)\cos\left(\frac{4jp\pi}{q}\right)$$

$$\left.-\sum_{j=1}^{q}\zeta\left(\nu,\frac{j}{q}\right)\cos\left(\frac{4jp\pi}{q}\right)\right\}$$

$$= (2q)^{-\nu}\sum_{j=1}^{q}\zeta\left(\nu,\frac{2j-1}{2q}\right)\cos\left(\frac{2(2j-1)p\pi}{q}\right),$$

where we have also applied the trigonometric identity:

$$\cos\left(\frac{2jp\pi}{q}\right) = \cos\left(\frac{2(j+q)p\pi}{q}\right) \quad \left(1 \leqq j \leqq q;\ j,\, q \in \mathbb{N};\ p \in \mathbb{Z}\right).$$

Thus, the summation formula (4) is reduced to its *equivalent* form:

$$C_\nu\left(\frac{2p\pi}{q}\right) = (2q)^{-\nu}\sum_{j=1}^{q}\zeta\left(\nu,\frac{2j-1}{2q}\right)\cos\left(\frac{2(2j-1)p\pi}{q}\right), \tag{12}$$

which is precisely (8) with p replaced rather trivially by $2p$.

Next, since the $j=q$ term in (5) is obviously nil, it is not difficult to reduce the summation formula (5) similarly to its *equivalent* form:

$$S_\nu\left(\frac{2p\pi}{q}\right) = (2q)^{-\nu}\sum_{j=1}^{q}\zeta\left(\nu,\frac{2j-1}{2q}\right)\sin\left(\frac{2(2j-1)p\pi}{q}\right), \tag{13}$$

which is precisely (9) with p replaced rather trivially by $2p$.

Each of the summation formulas (6) to (9) can, indeed, be proven *directly* by means of the series identity 6.1(8). The derivation of (4) and (5), by Cvijović and Klinowski [350], made use of the relationship:

$$\chi_\nu(z) = \mathrm{Li}_\nu(z) - 2^{-\nu}\,\mathrm{Li}_\nu(z^2), \tag{14}$$

where

$$\chi_\nu(z) := \sum_{k=0}^{\infty}\frac{z^{2k+1}}{(2k+1)^\nu} \quad \text{and} \quad \mathrm{Li}_\nu(z) := \sum_{k=1}^{\infty}\frac{z^k}{k^\nu} \quad \left(|z| \leqq 1;\ \Re(\nu) > 1\right) \tag{15}$$

denote Legendre's Chi function and the Polylogarithm function of order ν (*cf.*, e.g., Lewin [752]). In fact, as observed by Cvijović and Klinowski [350, p. 208], the (equivalent) summation formulas (8) and (9) are obtained when we make use of the relationship:

$$\chi_\nu(z) = \frac{1}{2}\left[\mathrm{Li}_\nu(z) - \mathrm{Li}_\nu(-z)\right], \tag{16}$$

instead of (14).

Problems

1. Borwein *et al.* [145] presented some very interesting parameterized classes of multiple sums whose many special cases reduce to known Euler and related sums, by making a considerable use of computer algebra systems (as they noted). Their main formulas are recalled here as follows:

$$\sum_{n=1}^{\infty} \frac{1}{n(n-x)} \sum_{m=1}^{n-1} \frac{1}{m-x} = \sum_{n=1}^{\infty} \frac{1}{n^2(n-x)} \qquad (x \in \mathbb{C} \setminus \mathbb{N}); \tag{a}$$

$$\begin{aligned}\sum_{n=1}^{\infty} \frac{1}{n^2-x^2} \sum_{m=1}^{n-1} \frac{my}{m^2-y^2} &= \sum_{s>0,\, t\geqq 0} \zeta(2s,\, 2t+1)\, x^{2s-2} y^{2t+1} \\ &= \frac{\pi}{2}\left[x\cot(\pi x) + 2y\cot(\pi y)\right] S_1(x,y) \\ &\quad + \frac{\pi}{2} x\cot(\pi x)\, S_2(x,y) - \frac{1}{2} S_3(x,y) \\ &\qquad \left(x,\, y \in \mathbb{C} \setminus (\mathbb{Z} \setminus \{0\})\right);\end{aligned} \tag{b}$$

$$\zeta(s,t) := \sum_{n>m\geqq 1} \frac{1}{n^s m^t} = \sum_{n=1}^{\infty} \frac{1}{n^s} \sum_{m=1}^{n-1} \frac{1}{m^t};$$

$$S_1(x,y) := \sum_{n=1}^{\infty} \frac{ny}{\left[(n+x)^2-y^2\right]\left[(n-x)^2-y^2\right]};$$

$$S_2(x,y) := \sum_{n=1}^{\infty} \frac{ny\left(4y^2-x^2\right)}{\left(n^2-y^2\right)\left[(n+x)^2-y^2\right]\left[(n-x)^2-y^2\right]};$$

$$S_3(x,y) := \sum_{n=1}^{\infty} \frac{ny}{\left(n^2-y^2\right)\left(n^2-x^2\right)}.$$

In fact, Borwein *et al.* [145] expressed $S_j(x,y)\ (j = 1, 2, 3)$ in terms of the ψ-function and $\cot(x)$. Show that $S_j(x,y)\ (j = 1, 2, 3)$ can be expressed in terms of the *psi*-function, only in a much simpler way as follows:

$$S_1(x,y) = -\frac{1}{8x}\left[\psi(1+x+y) - \psi(1+x-y) - \psi(1-x+y) + \psi(1-x-y)\right];$$

$$\begin{aligned} S_2(x,y) &= \frac{x+2y}{8x^2}\,[\psi(1+x-y)-\psi(1-x+y)+\psi(1-y)-\psi(1+y)] \\ &\quad + \frac{x-2y}{8x^2}\,[\psi(1-x-y)-\psi(1+x+y)+\psi(1-y)-\psi(1+y)]; \\ S_3(x,y) &= \frac{y}{2\,(y^2-x^2)}\,[\psi(1+x)+\psi(1-x)-\psi(1+y)-\psi(1-y)]. \end{aligned}$$

(Choi and Srivastava [299])

2. Upon replacing x by t in (a) and integrating the resulting identity with respect to t from 0 to x, Borwein *et al.* [145] found that

$$\sum_{n=1}^{\infty}\frac{1}{n^2}\log\left(1-\frac{x}{n}\right)=\sum_{n=1}^{\infty}\frac{1}{n}\sum_{m=1}^{n-1}\frac{1}{n-m}\log\left(\frac{1-x/m}{1-x/n}\right) \qquad (x\in\mathbb{C}\setminus\mathbb{N}). \tag{c}$$

Use (c) and prove the following formulas:

$$\begin{aligned} \sum_{n=1}^{\infty}\frac{1}{n}\sum_{m=1}^{n-1}\frac{1}{m}\arctan\left(\frac{m}{n^2-mn+1}\right) &= \sum_{n=1}^{\infty}\frac{\arctan(1/n)}{n^2} \\ &= \sum_{k=0}^{\infty}\frac{(-1)^k}{2k+1}\,\zeta(2k+3) \end{aligned}$$

and

$$\sum_{n=1}^{\infty}\frac{1}{n}\sum_{m=1}^{n-1}\frac{1}{m}\log\frac{n^2\left\{(n-m)^2+1\right\}}{(n-m)^2\,(n^2+1)}=\sum_{k=1}^{\infty}\frac{(-1)^{k-1}}{k}\,\zeta(2k+2).$$

(Choi and Srivastava [299]; see also Borwein *et al.* [145, p. 331])

3. Show that

$$\int_0^{\infty}\frac{P(x)}{z+x}\,dx=\left(z-\frac{1}{2}\right)\log z-z+\frac{1}{2}\log(2\pi)-\log\Gamma(z),$$

where $P(x)$ is given as in (1) and $|\arg(z)|\leqq\pi-\epsilon\ (0<\epsilon<\pi)$.

(Rainville [959, p. 30, Theorem 12])

4. Show that the constant A_4 in 7.1(42) with $p=4$ can be expressed in the form:

$$\log A_4=\frac{5}{64}+\frac{1}{2}\log 2-\frac{1}{8}\log A-\frac{G}{2\pi},$$

where A is the Glaisher-Kinkelin constant, defined by 1.3(2), and G is the Catalan constant, defined by 1.3(16).

(Choi and Srivastava [289, p. 100, Eq. (2.30)])

5. Evaluate the following integral:

$$\int_1^{\infty}P(x)\log\frac{2x-1}{2x+1}\,dx,$$

where $P(x)=x-[x]-\frac{1}{2}$.

(Choi *et al.* [310, pp. 108–109, Eq. (3.7)]; Zhang [1249, p. 69, Eq. (14)])

6. Assume that $n \in \mathbb{N}$ and set $w = \exp(i\pi/n)$. Let $\zeta(s,a)$ and $\chi_\nu(z)$ be the Hurwitz Zeta and Legendre Chi functions, defined by 2.2(1) and 6.2(15), respectively. Then show that

$$\zeta\left(\nu, \frac{2j-1}{2n}\right) = \frac{1}{n}\sum_{k=1}^{n}(2n)^\nu\,\chi_\nu\left(w^k\right) w^{-k(2j-1)} \quad (j = 1, \ldots, n),$$

and

$$\chi_\nu\left(w^k\right) = \frac{1}{(2n)^\nu}\sum_{j=1}^{n}\zeta\left(\nu, \frac{2j-1}{2n}\right) w^{k(2j-1)} \quad (j = 1, \ldots, n).$$

(Cvijović and Klinowski [352, p. 1625, Eqs. (7a) and (7b)])

7. Prove that

$$\zeta\left(2n, \frac{2p-1}{2q}\right) + \zeta\left(2n, 1 - \frac{2p-1}{2q}\right)$$
$$= (-1)^n \pi^{2n}\frac{(2q)^{2n}}{(2n-1)!}\sum_{r=1}^{q} E_{2n-1}\left(\frac{r}{q}\right)\cos\left(\frac{(2p-1)r\pi}{q}\right),$$

and

$$\zeta\left(2n+1, \frac{2p-1}{2q}\right) - \zeta\left(2n+1, 1 - \frac{2p-1}{2q}\right)$$
$$= (-1)^n \pi^{2n+1}\frac{(2q)^{2n}}{(2n)!}\sum_{r=1}^{q} E_{2n}\left(\frac{r}{q}\right)\sin\left(\frac{(2p-1)r\pi}{q}\right) \quad (n \in \mathbb{N}).$$

(Cvijović and Klinowski [352, p. 1628, Eqs. (15a) and (15b)])

8. Prove that, if $f(x) \in \mathbb{Q}[x]$ is integer-valued on $\mathbb{N}_0$, that is, if $f(k) \in \mathbb{Z}$ for all $k \in \mathbb{N}_0$, and p is prime, then $f(a) \in \mathbb{Z}_{(p)}$ for all $a \in \mathbb{Z}_{(p)}$, where $\mathbb{Z}_{(p)}$ denotes the ring of rational numbers whose denominator is not divisible by the prime p.

(Clarke and Slavutsk II [325, p. 22, Proposition 1])

9. Show that, for $\alpha > -1$,

$$\int_0^1 \frac{\rho^\alpha}{\rho^2 - 2\rho\cos\theta + 1}\,d\rho = \frac{1}{2q}\sum_{k=1}^{q-1}\left(\frac{\sin(k\theta)}{\sin\theta}\right)\psi\left(1 + \frac{\alpha - k}{2q}\right),$$

where

$$\theta = \frac{p\pi}{q} \quad (p, q \in \mathbb{N};\ 0 < p < 2q;\ (p,q) = 1),$$

and (p,q) denotes that p and q are relatively prime.

(de Doelder [391, p. 327])

10. Prove that

$$\log \frac{G\left(1+\frac{p}{2q}\right)}{G\left(1-\frac{p}{2q}\right)} = -\frac{p}{2q}\log\left(\frac{1}{\pi}\sin\left(\frac{p\pi}{2q}\right)\right)$$
$$+\frac{1}{8\pi q^2}\sum_{r=1}^{q-1}\left[\psi'\left(1-\frac{r}{2q}\right)-\psi'\left(\frac{r}{2q}\right)\right]\sin\left(\frac{pr\pi}{q}\right)$$
$$(p\in\mathbb{N};\ q\in\mathbb{N}\setminus\{1\};\ 1\leqq p\leqq 2q-1;\ (p,q)=1),$$

where G denotes the Barnes G-function considered in Section 1.3 and (p,q) is the same notation as in Problem 8.

(Choi and Srivastava [289, p. 95, Eq. (2.2)])

11. Evaluate the following integral:

$$\int_1^\infty \frac{P(x)}{2x+1}dx = -\frac{3}{4}+\frac{1}{4}\log 2+\frac{1}{2}\log 3,$$

where $P(x)=x-[x]-\frac{1}{2}$.

(Zhang [1249, p. 68, Eq. (11)]; see also [1094, Section 6.3])

12. Evaluate the following integrals:

$$\int_0^{\frac{\pi}{p}} \frac{x^{2m}}{\sin x}dx = 2(-1)^m(2m)!\left(1-2^{-2m-1}\right)\zeta(2m+1)$$
$$-\frac{(2m)!}{p}\left(\frac{\pi}{p}\right)^{2m}\sum_{k=1}^{m}\frac{(-1)^k(2\pi)^{-2k}}{(2m-2k)!}\sum_{\ell=0}^{p-1}\cos\left(\frac{(2\ell+1)\pi}{p}\right)\zeta\left(2k+1,\frac{2\ell+1}{2p}\right)$$
$$-2(2m)!\left(\frac{\pi}{p}\right)^{2m+1}\sum_{k=1}^{m}\frac{(-1)^k(2\pi)^{-2k}}{(2m-2k+1)!}\sum_{\ell=0}^{p-1}\sin\left(\frac{(2\ell+1)\pi}{p}\right)\zeta\left(2k,\frac{2\ell+1}{2p}\right)$$
$$-2\left(\frac{\pi}{p}\right)^{2m}\gamma(p)\quad(m\in\mathbb{N};\ p\in\mathbb{N}\setminus\{1\})$$

and

$$\int_0^{\frac{\pi}{p}} \frac{x^{2m+1}}{\sin x}dx$$
$$=\frac{(2m+1)!}{2p^2}\left(\frac{\pi}{p}\right)^{2m}\sum_{k=0}^{m}\frac{(-1)^k(2\pi)^{-2k}}{(2m-2k)!}\sum_{\ell=0}^{p-1}\sin\left(\frac{(2\ell+1)\pi}{p}\right)\zeta\left(2k+2,\frac{2\ell+1}{2p}\right)$$
$$-\frac{(2m+1)!}{p}\left(\frac{\pi}{p}\right)^{2m+1}\sum_{k=1}^{m}\frac{(-1)^k(2\pi)^{-2k}}{(2m-2k+1)!}\sum_{\ell=0}^{p-1}\cos\left(\frac{(2\ell+1)\pi}{p}\right)\zeta\left(2k+1,\frac{2\ell+1}{2p}\right)$$
$$-2\left(\frac{\pi}{p}\right)^{2m+1}\gamma(p)\quad(m\in\mathbb{N}_0;\ p\in\mathbb{N}\setminus\{1\}),$$

where $\gamma(p)$ is given by

$$\gamma(p) = \sum_{j=0}^{\infty} \sum_{\ell=0}^{p-1} \frac{1}{2pj + 2\ell + 1} \cos\left(\frac{(2\ell+1)\pi}{p}\right)$$

and (*as usual*) an empty sum is understood to be nil.

(Cho *et al.* [257, Equations (20) and (22)])

13. Show that

$$\sum_{n=1}^{\infty} H_n \frac{x^n}{n} = \frac{1}{2} [\log(1-x)]^2 + \sum_{n=1}^{\infty} \frac{x^n}{n^2} \qquad (|x| < 1).$$

(Alzer *et al.* [31, Eq. (3.47)]; see also [531, p. 63, Entry (5.13.15)])

14. Show that

$$\zeta(s) = \frac{1}{1-2^{1-s}} \sum_{n=0}^{\infty} \frac{1}{2^{n+1}} \sum_{k=0}^{n} (-1)^k \binom{n}{k} \frac{1}{(k+1)^s} \qquad (s \in \mathbb{C} \setminus \{1\}).$$

(Coffey [330, Eq. (11)]; see also [542])

Bibliography

[1] M. Abouzahra and L. Lewin, The polylogarithm in algebraic number fields, *J. Number Theory* **21** (1985), 214–244.

[2] M. Abouzahra, L. Lewin and X. Hongnian, Polylogarithms in the field of Omega (a root of a given cubic): Functional equations and ladders, *Aequationes Math.* **33** (1987), 23–45.

[3] M. Abramowitz and I. A. Stegun (Editors), *Handbook of Mathematical Functions with Formulas, Graphs, and Mathematical Tables*, Tenth Printing, National Bureau of Standards, Applied Mathematics Series **55**, National Bureau of Standards, Washington, D.C., 1972; Reprinted by Dover Publications, New York, 1965 (see also [880]).

[4] J. Aczél, *Lectures on Functional Equations and Their Applications*, Academic Press, New York and London, 1966.

[5] V. S. Adamchik, On Stirling numbers and Euler sums *J. Comput. Appl. Math.* **79** (1997), 119–130.

[6] V. S. Adamchik, Polygamma functions of negative order, *J. Comput. Appl. Math.* **100** (1998), 191–199.

[7] V. S. Adamchik, On the Barnes function, in *Proceedings of the* 2001 *International Symposium on Symbolic and Algebraic Computation* (London, Ontario; July 22–25, 2001), pp. 15–20 (electronic), ACM Press, New York, 2001.

[8] V. S. Adamchik, The multiple Gamma function and its application to computation of series, *Ramanujan J.* **9** (2005), 271–288.

[9] V. S. Adamchik, On the Hurwitz function for rational arguments, *Appl. Math. Comput.* **187** (2007), 3–12.

[10] V. S. Adamchik and K. S. Kölbig, A definite integral of a product of two polylogarithms, *SIAM J. Math. Anal.* **19** (1988), 926–938.

[11] V. S. Adamchik and H. M. Srivastava, Some series of the Zeta and related functions, *Analysis* **18** (1998), 131–144.

[12] A. W. Addison, A series representation for Euler's constant, *Amer. Math. Monthly* **74** (1967), 823–824.

[13] L. V. Ahlfors, *Complex Analysis*, Third edition, McGraw-Hill Book Company, New York, Toronto and London, 1979.

[14] S. Ahlgren and K. Ono, A Gaussian hypergeometric series evaluation and Apéry number congruences, *J. Reine Angew. Math.* **518** (2000), 187–212.

[15] J. R. Airey and H. A. Webb, The practical importance of the confluent hypergeometric function, *Philos. Mag.* **36** (1918), 129–141.

[16] S. Akiyama, S. Egami and Y. Tanigawa, Analyitic continuation of multiple Zeta functions and their values at non-postive integers, *Acta Arith.* **98** (2001), 107–116.

[17] S. Akiyama and Y. Tanigawa, Multiple Zeta values at non-postive integers, *Ramanujan J.* **5** (2001), 327–351.

[18] W. P. Alexeiewsky, Über eine Classe von Funktionen, die der Gammafunktion analog sind, *Leipzig Weidmannsche Buchhandlung* **46** (1894), 268–275.

[19] G. Almkvist and A. Meurman, Values of Bernoulli polynomials and Hurwitz's Zeta function at rational arguments, *C. R. Math. Rep. Acad. Sci. Canada* **13** (1991), 104–108.

[20] W. A. Al-Salam and A. Verma, Some remarks on q-Beta integral, *Proc. Amer. Math. Soc.* **85** (1982), 360–362.

[21] B. N. Al-Saqabi, S. L. Kalla and H. M. Srivastava, A certain family of infinite series associated with Digamma functions, *J. Math. Anal. Appl.* **159** (1991), 361–372.

[22] H. Alzer, Some Gamma function inequalities, *Math. Comput.* **60** (1993), 337–346.

[23] H. Alzer, On some inequalities for the Gamma and Psi functions, *Math. Comput.* **66** (1997), 373–389.

[24] H. Alzer, Inequalities for the Gamma function, *Proc. Amer. Math. Soc.* **128** (1999), 141–147.

[25] H. Alzer, Inequalities for the volume of the unit ball in $\mathbb{R}^n$, *J. Math. Anal. Appl.* **252** (2000), 353–363.

[26] H. Alzer, Inequalities involving $\Gamma(x)$ and $\Gamma(1/x)$, J. Comput. Appl. Math. **192** (2006), 460–480.

[27] H. Alzer, Sub-and superadditive property of Euler's Gamma function, *Proc. Amer. Math. Soc.* **135** (2007), 3641–3648.

[28] H. Alzer, A superadditive property of Hadamard's Gamma function, *Abh. Math. Semin. Univ. Hamburg* **79** (2009), 11–23.

[29] H. Alzer and N. Batir, Monotonicity properties of the Gamma function, *Appl. Math. Lett.* **20** (2007), 778–781.

[30] H. Alzer, J. L. Brenner and O. G. Ruehr, On Mathieu's inequality, *J. Math. Anal. Appl.* **218** (1998), 607–610.

[31] H. Alzer, D. Karayannakis and H. M. Srivastava, Series representations for some mathematical constants, *J. Math. Anal. Appl.* **320** (2006), 145–162.

[32] H. Alzer and S. Ruscheweyh, A subadditive property of the Gamma function, *J. Math. Anal. Appl.* **285** (2003), 564–577.

[33] H. Alzer and J. Wells, Inequalities for the Polygamma functions, *SIAM J. Math. Anal.* **29** (1998), 1459–1466.

[34] T. Amdeberhan, Faster and faster convergent series for $\zeta(3)$, *Electronic J. Combin.* **3** (1) (1996), 1–2 (electronic).

[35] M. J. Anastassiadis, Fonctions semi-monotones et semi-convexes et solutions d'une équation fonctionnelle, *Bull. Sci. Math.* (*Sér.* 2) **76** (1952), 148–160.

[36] M. J. Anastassiadis, Sur les solutions logarithmiquement convexes ou concaves d'une équation fonctionnelle, *Bull. Sci. Math.* (*Sér.* 2) **81** (1957), 78–87.

[37] M. J. Anastassiadis, Une propriété de la fonction Gamma, *Bull. Sci. Math.* (*Sér.* 2) **81** (1957), 116–118.

[38] M. J. Anastassiadis, Définitions fonctionnelles de la fonction $B(x, y)$, *Bull. Sci. Math.* (*Sér.* 2) **83** (1959), 24–32.

[39] M. J. Anastassiadis, *Définition des Fonctions Eulériennes par des Équations Fonctionnelles*, Gauthier-Villars, Paris, 1964.

[40] G. D. Anderson, R. W. Barnard, K. C. Richards, M. K. Vamanamurthy and M. Vuorinen, Inequalities for zero-balanced hypergeometric functions, *Trans. Amer. Math. Soc.* **347** (1995), 1713–1723.

[41] G. D. Anderson, M. K. Vamanamurthy and M. Vuorinen, Functional inequalities for complete elliptic integrals and their ratios, *SIAM J. Math. Anal.* **21** (1990), 536–549.

[42] G. D. Anderson, M. K. Vamanamurthy and M. Vuorinen, Functional inequalities for hypergeometric functions and complete elliptic integrals, *SIAM J. Math. Anal.* **23** (1992), 512–524.

[43] G. D. Anderson and S.-L. Qiu, A monotoneity property of the Gamma function, *Proc. Amer. Math. Soc.* **125** (1997), 3355–3362.

[44] G. E. Andrews, A simple proof of Jacobi's triple product identity, *Proc. Amer. Math. Soc.* **16** (1965), 333–334.

[45] G. E. Andrews, On Ramanujan's summation of ${}_1\psi_1(a; b; z)$, *Proc. Amer. Math. Soc.* **22** (1969), 552–553.

[46] G. E. Andrews, On a transformation of bilateral series with applications, *Proc. Amer. Math. Soc.* **25** (1970), 554–558.

[47] G. E. Andrews, *Number Theory*, W. B. Saunders Company, Philadelphia, London and Toronto, 1971.

[48] G. E. Andrews, An introduction to Ramanujan's "Lost" Notebook, *Amer. Math. Monthly* **86** (1979), 89–108.

[49] L. C. Andrews, *Special Functions of Mathematics for Engineers*, Second edition, McGraw-Hill Book Company, New York, Toronto and London, 1992.

[50] G. E. Andrews and R. Askey, Another q-extension of the Beta function, *Proc. Amer. Math. Soc.* **81** (1981), 97–100.

[51] G. E. Andrews, R. Askey and R. Roy, *Special Functions*, Encyclopedia of Mathematics and Its Applications, Vol. 71, Cambridge University Press, Cambridge, London and New York, 1999.

[52] G. E. Andrews and K. Uchimura, Identities in combinatorics. IV: Differentiation and harmonic numbers, *Utilitas Math.* **28** (1985), 265–269.

[53] J. Anglesio, Problems and solutions, *Amer. Math. Monthly* **103** (1996), 427.

[54] J. Anglesio, Problems and solutions, *Amer. Math. Monthly* **103** (1996), 903.

[55] T. Aoki and Y. Ohno, Sum relations for multiple Zeta values and connection formulas for the Gauss hypergeometric functions, *Publ. RIMS Kyoto Univ.* **41** (2005), 329–337.

[56] R. Apéry, Irrationalité de $\zeta(2)$ et $\zeta(3)$, in *Journées Arithmétiques de Luminy* (Colloq. Internat. CNRS, Centre Univ. Luminy, Luminy, 1978), pp. 11–13, *Astérisque* **61** (1979), Soc. Math. France, Paris.

[57] T. M. Apostol, Generalized Dedekind sums and transformation formulae of certain Lambert series, *Duke Math. J.* **17** (1950), 147–157.

[58] T. M. Apostol, On the Lerch Zeta-function, *Pacific J. Math.* **1** (1951), 161–167.

[59] T. M. Apostol, Remark on the Hurwitz Zeta function, *Proc. Amer. Math. Soc.* **2** (1951), 690–693.

[60] T. M. Apostol, Addendum to "On the Lerch Zeta-function," *Pacific J. Math.* **2** (1952), 10.

[61] T. M. Apostol, Some series involving the Riemann Zeta function, *Proc. Amer. Math. Soc.* **5** (1954), 239–243.

[62] T. M. Apostol, Dirichlet L-functions and character power sums, *J. Number Theory* **2** (1970), 223–234.

[63] T. M. Apostol, Another elementary proof of Euler's formula for $\zeta(2n)$, *Amer. Math. Monthly* **80** (1973), 425–431.

[64] T. M. Apostol, *Mathematical Analysis*, Second edition, Addison-Wesley Publishing Company, Menlo Park (California), London and Don Mills (Ontario), 1974.

[65] T. M. Apostol, *Introduction to Analytic Number Theory*, Springer-Verlag, New York, Heidelberg and Berlin, 1976.

[66] T. M. Apostol, *Modular Functions and Dirichlet Series in Number Theory*, Springer-Verlag, New York, Heidelberg and Berlin, 1976.

[67] T. M. Apostol, An elementary view of Euler's summation formula, *Amer. Math. Monthly* **106** (1999), 409–418.

[68] T. M. Apostol and T. H. Vu, Dirichlet series related to the Riemann Zeta function, *J. Number Theory* **19** (1984), 85–102.

[69] P. Appell, Sur la nature arithmétique de la constanted'Euler, *C. R. Acad. Sci. Paris Sér. I Math.* **15** (1926), 897–899.

[70] T. Arakawa and M. Kaneko, Multiple Zeta values, poly-Bernoulli numbers, and related Zeta-functions, *Nagoya Math. J.* **153** (1999), 189–209.

[71] G. B. Arfken and H. Weber, *Mathematical Methods for Physicists*, Fifth edition, Academic Press, New York, 2000.

[72] E. Artin, *The Gamma Function*, Holt, Rinehart and Winston, New York, Chicago, San Francisco, Toronto and London, 1964.

[73] R. Askey, The q-Gamma and q-Beta functions, *Applicable Anal.* **8** (1978), 125–141.

[74] R. Askey, Ramanujan's extensions of the Gamma and Beta functions, *Amer. Math. Monthly* **87** (1980), 346–359.

[75] R. Askey, A q-extension of Cauchy's Beta integral, *SIAM Rev.* **26** (1984), 270.

[76] R. D. Atanassov and U. V. Tsoukrovski, Some properties of a class of logarithmically completely monotonic functions, *C. R. Acad. Bulgare Sci.* **41** (2) (1988), 21–23.

[77] M. Atiyah, R. Bott and V. K. Patodi, On the heat equation and the index theorem, *Invent. Math.* **19** (1973), 279–330.

[78] F. V. Atkinson, The mean value of the Riemann Zeta-function, *Acta Math.* **81** (1949), 353–376.

[79] J. Aular de Durán, S. L. Kalla and H. M. Srivastava, Fractional calculus and the sums of certain families of infinite series, *J. Math. Anal. Appl.* **190** (1995), 738–754.

[80] J.-C. Aval, Multivariate Fuss-Catalan numbers, *Discrete Math.* **308** (2008), 4660–4669.

[81] R. G. Ayoub, Euler and the Zeta function, *Amer. Math. Monthly* **81** (1974), 1067–1086.

[82] R. G. Ayoub, Partial triumph or total failure? *Math. Intelligencer* **7** (2) (1985), 55–58.

[83] R. L. Bagby, A simple proof that $\Gamma'(1) = -\gamma$, *Amer. Math. Monthly* **117** (2010), 83–85.

[84] D. H. Bailey, Numerical results on the transcendence of constants involving π, e, and Euler's constant, *Math. Comput.* **50** (1988), 275–281.

[85] D. H. Bailey, J. M. Borwein and R. E. Crandall, On the Khintchine constant, *Math. Comput.* **66** (1997), 417–431.

[86] D. H. Bailey, J. M. Borwein and R. Girgenshon, Experimental evaluation of Euler sums, *Experiment. Math.* **3** (1994), 17–30.

[87] W. N. Bailey, *Generalized Hypergeometric Series*, Cambridge Tracts in Mathematics and Mathematical Physics, Vol. 32, Cambridge University Press, Cambridge, London and New York, 1935; Reprinted by Stechert-Hafner Service Agency, New York and London, 1964.

[88] W. N. Bailey, Series of hypergeometric type which are infinite in both directions, *Quart. J. Math. Oxford Ser.* **7** (1936), 105–115.

[89] W. N. Bailey, On the basic bilateral hypergeometric series ${}_2\Psi_2$, *Quart. J. Math. Oxford Ser.* **1** (2) (1950), 194–198.

[90] W. N. Bailey, On the simplification of some identities of the Rogers-Ramanujan type, *Proc. London Math. Soc.* (*Ser.* 3) **1** (1951), 217–221.

[91] U. Balakrishnan, A series for $\zeta(s)$, *Proc. Edinburgh Math. Soc.* (*Ser.* 2) **31** (1988), 205–210.

[92] E. R. Barnes and W. E. Kaufman, The Euler-Mascheroni constant, *Amer. Math. Monthly* **72** (1965), 1023.
[93] E. W. Barnes, The theory of the Gamma function, *Messenger Math.* **29** (1899), 64–128.
[94] E. W. Barnes, The theory of the G-function, *Quart. J. Math.* **31** (1899), 264–314.
[95] E. W. Barnes, Genesis of the double Gamma function, *Proc. London Math. Soc.* **31** (*Ser.* 1) (1900), 358–381.
[96] E. W. Barnes, The theory of the double Gamma function, *Philos. Trans. Roy. Soc. London Ser. A* **196** (1901), 265–388.
[97] E. W. Barnes, On the theory of the multiple Gamma functions, *Trans. Cambridge Philos. Soc.* **19** (1904), 374–439.
[98] R. G. Bartle, *The Elements of Real Analysis*, Second edition, John Wiley and Sons, New York, Chichester, Brisbane and Toronto, 1976.
[99] K. Bartz and J. Rutkowski, On the von Staudt-Clausen theorem, *C. R. Math. Rep. Acad. Sci. Canada* **15** (1993), 46–48.
[100] A. Basu and T. M. Apostol, A new method for investigating Euler sums, *Ramanujan J.* **4** (2000), 397–419.
[101] P. T. Bateman and E. Grosswald, On Epstein's Zeta function, *Acta Arith.* **9** (1964), 365–373.
[102] N. Batir, Integral representations of some series involving $\binom{2k}{k}^{-1} k^{-n}$ and some related series, *Appl. Math. Comput.* **147** (2004), 645–667.
[103] N. Batir, On some properties of Digamma and Polygamma functions, *J. Math. Anal. Appl.* **328** (2007), 452–465.
[104] N. Batir, Inequalities for the Gamma function, *Arch. Math. (Basel)* **91** (2008), 554–563.
[105] N. Batir, Inequalities for the double Gamma function, *J. Math. Anal. Appl.* **351** (2009), 182–185.
[106] N. Batir and M. Cancan, A double inequality for the double Gamma function, *Internat. J. Math. Anal. (Ruse)* **2** (2008), 329–335.
[107] N. Batir and M. Cancan, New complete monotonicity properties of the Gamma function, *Adv. Stud. Contemp. Math.* **19** (2009), 165–170.
[108] A. Byad and T. Kim, Identities for the Bernoulli, the Euler and the Genocchi numbers and polynomials, *Adv. Stud. Contemp. Math.* **20** (2010), 247–253.
[109] R. W. Beals and R. S. C. Wong, *Special Functions. A Graduate Text*, Cambridge Studies in Advanced Mathematics, Vol. 126, Cambridge University Press, Cambridge, London and New York, 2010.
[110] E. F. Beckenbach, Convex functions, *Bull. Amer. Math. Soc.* **54** (1948), 439–460.
[111] E. F. Beckenbach, Superadditivity inequalities, *Pacific J. Math.* **14** (1964), 421–438.
[112] P. Beckmann, *A History of* π (Pi), Second edition, Golem Press, Boulder, Colorado, 1971.
[113] R. Bellman, *A Brief Introduction to Theta Functions*, Holt, Rinehart and Winston, New York, 1961.
[114] L. Bendersky, Sur la fonction Gamma généralisée, *Acta Math.* **61** (1933), 263–322.
[115] C. Berg, Integral representation of some functions related to the Gamma function, *Mediterr. J. Math.* **1** (2004), 433–439.
[116] C. Berg and H. L. Pedersen, A completely monotone function related to the Gamma function, *J. Comput. Appl. Math.* **133** (2001), 219–230.
[117] B. C. Berndt, Two new proofs of Lerch's functional equation, *Proc. Amer. Math. Soc.* **32** (1972), 403–408.
[118] B. C. Berndt, On the Hurwitz Zeta-function, *Rocky Mountain J. Math.* **2** (1972), 151–157.

[119] B. C. Berndt, Generalized Dedekind Eta-functions and generalized Dedekind sums, *Trans. Amer. Math. Soc.* **178** (1973), 495–508.

[120] B. C. Berndt, Elementary evaluation of $\zeta(2n)$, *Math. Mag.* **48** (1975), 148–153.

[121] B. C. Berndt, Modular transformations and generalizations of several formulae of Ramanujan, *Rocky Mountain J. Math.* **7** (1977), 147–189.

[122] B. C. Berndt, *Ramanujan's Notebooks*, Part II, Springer-Verlag, Berlin, Heidelberg and New York, 1989.

[123] B. C. Berndt and W.-C. Chu, Two entries on bilateral hypergeometric series in Ramanujan's lost notebook, *Proc. Amer. Math. Soc.* **135** (2007), 129–134.

[124] J. Berstel and D. Perrin, The origins of combinatorics on words, *European J. Combin.* **28** (2007), 996–1022.

[125] N. Bettaibi and J. E. Kamel, Asymptotic expansions for a class of q-integral tranforms, *Appl. Math. Comput.* **198** (2008), 433–444.

[126] F. Beukers, A note on the irrationality of $\zeta(2)$ and $\zeta(3)$, *Bull. London Math. Soc.* **11** (1979), 268–272.

[127] F. Beukers, Another congruence for the Apéry numbers, *J. Number Theory* **25** (1987), 201–210.

[128] M. G. Beumer, Some special integrals, *Amer. Math. Monthly* **68** (1961), 645–647.

[129] W. A. Beyer and M. S. Waterman, Error analysis of a computation of Euler's constant, *Math. Comput.* **28** (1974), 599–604.

[130] W. A. Beyer and M. S. Waterman, Decimals and partial quotients of Euler's constant and $\ln 2$, *Math. Comput.* **28** (1974), 667.

[131] W. A. Beyer and M. S. Waterman, Decimals and partial quotients of Euler's constant and $\ln 2$, *Math. Comput.* **32** (1978), 317–318.

[132] S. Bhargava, D. D. Somashekara and D. Mamta, A new convolution identity deducible from the remarkable formula of Ramanujan, *Taiwanese J. Math.* **11** (2007), 399–406.

[133] H. Bigler, Über Gammafunktionen mit beliebigem parameter, *J. Reine Angew. Math.* **102** (1888), 237–254.

[134] G. Bleimann and E. L. Stark, Kernels of finite oscillations and convolution integrals, *Acta Math. Acad. Sci. Hungar.* **35** (1980), 419–429.

[135] C. R. Blyth and P. K. Pathak, A note on easy proofs of Stirling's theorem, *Amer. Math. Monthly* **93** (1986), 376–379.

[136] S. Bochner, On Riemann's functional equation with multiple Gamma factors, *Ann. of Math.* **67** (2) (1958), 29–41.

[137] S. Bochner and K. Chandrasekharan, On Riemann's functional equation, *Ann. of Math.* **63** (2) (1956), 336–360.

[138] R. P. Boas Jr. and J. W. Wrench Jr., Partial sums of the harmonic series, *Amer. Math. Monthly* **78** (1971), 864–870.

[139] J. Boersma and J. P. Dempsey, On the numerical evaluation of Legendre's Chi-function, *Math. Comput.* **59** (1992), 157–163.

[140] P. E. Böhmer, *Differenzengleichungen und bestimmte Integrale*, Koehler, Leipzig, 1939.

[141] G. Boros, O. Espinosa and V. H. Moll, On some families of integrals solvable in terms of Polygamma and Negapolygamma functions, *Integral Transforms Spec. Funct.* **14** (2003), 187–203.

[142] H. Bohr and I. Mollerup, *Loerebog i Matematisk Analyse*, Vol. III, Kopenhagen, 1922.

[143] G. Boole, *A Treatise on the Calculus of Finite Differences*, Macmillan and Company, London, 1872; Second edition (Reprinted), Dover Publications, New York, 1960.

[144] D. Borwein and J. M. Borwein, On an intriguing integral and some series related to $\zeta(4)$, *Proc. Amer. Math. Soc.* **123** (1995), 1191–1198.

[145] D. Borwein, J. M. Borwein and D. M. Bradley, Parametric Euler sum idenities, *J. Math. Anal. Appl.* **316** (2006), 328–338.

[146] D. Borwein, J. M. Borwein and R. Girgensohn, Explicit evaluation of Euler sums, *Proc. Edinburgh Math. Soc.* **38** (2) (1995), 277–294.

[147] J. M. Borwein, Some modular identities of Ramanujan useful in approximating π, *Proc. Amer. Math. Soc.* **95** (1985), 365–371.

[148] J. M. Borwein and P. B. Borwein, *Pi and the AGM*, John Wiley and Sons, New York, Chichester, Brisbane, Toronto and Singapore, 1987.

[149] J. M. Borwein and D. M. Bradley, Empirically determined Apéry-like formulae for $\zeta(4n+3)$, *Experimental Math.* **6** (1997), 181–194.

[150] J. M. Borwein, D. M. Bradley and D. J. Broadhurst, Evaluations of k-fold Euler/Zagier sums: A compendium of results for arbitrary k, *Electronic J. Combin.* **4** (2) (1997), 1–21 (electronic).

[151] J. M. Borwein, D. M. Bradley, D. J. Broadhurst and P. Lisoněk, Combinatorial aspects of multiple Zeta values, *Electronic J. Combin.* **5** (Research Paper No. 38) (1998), 1–12 (electronic).

[152] J. M. Borwein, D. M. Bradley, D. J. Broadhurst and P. Lisoněk, Special values of multiple polylogarithms, *Trans. Amer. Math. Soc.* **353** (2001), 907–941.

[153] J. M. Borwein, D. M. Bradley and R. E. Crandall, Computational strategies for the Riemann Zeta function, *J. Comput. Appl. Math.* **121** (2000), 247–296.

[154] J. M. Borwein and R. Girgensohn, Evaluation of triple Euler sums, *Electronic J. Combin.* **3** (1) (1996), 1–27 (electronic).

[155] L. Bourguet, Note sur les intégrales Eulériennes, *Acta Math.* **1** (1882), 295–296.

[156] L. Bourguet, Sur quelques intégrales définies, *Acta Math.* **1** (1882), 363–367.

[157] L. Bourguet, Sur les intégrales Eulériennes et quelques autres fonctions uniformes, *Acta Math.* **2** (1883), 231–232.

[158] L. Bourguet, Sur la fonction Eulérienne, *Acta Math.* **2** (1883), 296–298.

[159] F. Bowman, Note on the integral $\int_0^{\frac{1}{2}\pi}$ $(\log\sin\theta)^n\,d\theta$, *J. London Math. Soc.* **22** (1947), 172–173.

[160] D. Bowman and D. M. Bradley, *Multiple Polylogarithms: A Brief survey* in *q-Series with Applications to Combinatorics, Number Theory, and Physics* (Papers from the Conference held at the University of Illinois, Urbana, Illinois; October 26–28, 2000) (B.C. Berndt and K. Ono, Editors), pp. 71–92, Contemporary Mathematics **291**, American Mathematical Society, Providence, Rhode Island, 2001.

[161] K. N. Boyadzhiev, A series transformation formula and related polynomials, *Internat. J. Math. Math. Sci.* **23** (2005), 3849–3866.

[162] K. N. Boyadzhiev, Apostol-Bernoulli functions, derivative polynomials and Eulerian polynomials, *Adv. Appl. Discrete Math.* **1** (2) (2008), 109–122.

[163] A. V. Boyd, Gurland's inequality for the Gamma function, *Skand. Aktuarietidskr.* **1960** (1960), 134–135.

[164] A. V. Boyd, Note on a paper by Uppuluri, *Pacific J. Math.* **22** (1967), 9–10.

[165] D. M. Bradley, A class of series acceleration formulae for Catalan's constant, *Ramanujan J.* **3** (1999), 159–173.

[166] D. M. Bradley, Duality for finite multiple harmonic q-series, *Discrete Math.* **300** (2005), 44–56.

[167] D. M. Bradley, Multiple q-Zeta values, *J. Algebra* **283** (2005), 752–798.

[168] D. M. Bradley, A q-analog of Euler's decomposition formula for the double Zeta function, *Internat. J. Math. Math. Sci.* **21** (2005), 3453–3458.

[169] D. M. Bradley, On the sum formula for multiple q-Zeta values, *Rocky Mountain J. Math.* **37** (2007), 1427–1434.

[170] K. Brahim, Turán-type inequalties for some q-special functions, *J. Inequal. Pure Appl. Math.* **10** (2) (2009), 1–5 (electronic).

[171] T. P. Branson and B. Ørsted, Explicit functional determinants in four dimensions, *Proc. Amer. Math. Soc.* **113** (1991), 669–682.

[172] R. P. Brent, Computation of the regular continued fraction for Euler's constant, *Math. Comput.* **31** (1977), 771–777.

[173] R. P. Brent, γ and e^{γ} to 20700D and their regular continued fractions to 20000 partial quotients, *Math. Comput.* **32** (1978), 311.

[174] R. P. Brent and E. M. McMillan, Some new algorithms for high-precision computation of Euler's constant, *Math. Comput.* **34** (1980), 305–312.

[175] E. H. M. Brietzke, An identity of Andrews and a new method for the Riordan array proof of combinatorial identities, *Discrete Math.* **308** (2008), 4246–4262.

[176] W. E. Briggs, Some constants associated with the Riemann Zeta-function, *Michigan Math. J.* **3** (1955–1956), 117–121.

[177] W. E. Briggs, Zeta function identity, Problem 1302, *Math. Mag.* **62** (1989), 275–276.

[178] W. E. Briggs, S. Chowla, A. J. Kempner and W. E. Mientka, On some infinite series, *Scripta Math.* **21** (1955), 28–30.

[179] W. E. Briggs and S. Chowla, The power series coefficients of $\zeta(s)$, *Amer. Math. Monthly* **62** (1955), 323–325.

[180] K. Bringmann and K. Ono, An arithmetic formula for the partition function, *Proc. Amer. Math. Soc.* **135** (2007), 3507–3514.

[181] K. Bringmann and H. Swisher, On a conjecture of Koike on identities between Thompson series and Rogers-Ramanujan functions, *Proc. Amer. Math. Soc.* **135** (2007), 2317–2326.

[182] D. J. Broadhurst and D. Kreimer, Association of multiple Zeta values with positive knots via Feynman diagrams, up to 9 loops, *Phys. Lett. B* **393** (1997), 403–412.

[183] T. J. I'A. Bromwich, *An Introduction to the Theory of Infinite Series*, Second edition, Macmillan and Company, London, 1926.

[184] J. W. Brown and R. V. Churchill, *Complex Variables and Applications*, Seventh edition, McGraw-Hill International Editions, McGraw-Hill Book Company, New York, Toronto and London, 2003.

[185] P. S. Bruckman and C. Georghiou, An Euler sum, *Fibonacci Quart.* **20** (1982), 186–187.

[186] A. M. Bruckner, Minimal superadditive extensions of superadditive functions, *Pacific J. Math.* **10** (1960), 1155–1162.

[187] A. M. Bruckner, Tests for the superadditivity of functions, *Proc. Amer. Math. Soc.* **13** (1962), 126–130.

[188] A. M. Bruckner and E. Ostrow, Some function classes related to the class of convex functions, *Pacific J. Math.* **12** (1962), 1203–1215.

[189] V. Brun, On the problem of partitioning the circle so as to visualize Leibniz' formula for π, *Nordisk Mat. Tidskr.* **3** (1955), 159–166.

[190] V. Brun, Leibniz' formula for π deduced by a "mapping" of the circular disc, *Nordisk Mat. Tidskr.* **18** (1970), 73–81.

[191] J. Brüning, Heat equation asymptotics for singular Sturm-Liouville operators, *Math. Ann.* **268** (1984), 173–196.

[192] Yu. A Brychkov, *Handbook of Special Functions: Derivatives, Integrals, Series and Other Formulas*, Capman and Hall (CRC Press), Boca Raton, London and New York, 2008.

[193] Yu. A Brychkov, On multiple sums of special functions, *Integral Transforms Spec. Funct.* **21** (2010), 245–253.

[194] H. M. Bui and J. P. Keating, On the mean values of Dirichlet L-functions, *Proc. London Math. Soc.* **95** (3) (2007), 273–298.

[195] R. G. Buschman and H. M. Srivastava, Asymptotic behavior of some power series with ζ-functions in the coefficients, *Monatsh. Math.* **115** (1993), 291–298.

[196] J. Bustoz and M. E.-H. Ismail, On Gamma function inequalities, *Math. Comput.* **47** (1986), 659–667.

[197] P. L. Butzer and M. Hauss, Riemann Zeta function: Rapidly converging series and integral representations, *Appl. Math. Lett.* **5** (2) (1992), 83–88.

[198] P. L. Butzer and M. Hauss, Integral and rapidly converging series representations of the Dirichlet L-functions $L_1(s)$ and $L_{-4}(s)$, *Atti Sem. Mat. Fis. Univ. Modena* **40** (1992), 329–359.

[199] P. L. Butzer and F. Jongmans, Eugène Catalan and the rise of Russian science, *Acad. Roy. Belg. Bull. Cl. Sci. (Ser.* 6) **2** (1991), 59–90.

[200] P. L. Butzer, C. Markett and M. Schmidt, Stirling numbers, central factorial numbers, and representations of Riemann Zeta function, *Resultate Math.* **19** (1991), 257–274.

[201] P. L. Butzer, T. K. Pogány and H. M. Srivastava, A linear ODE for the Omega function associated with the Euler function $E_\alpha(z)$ and the Bernoulli function $B_\alpha(z)$, *Appl. Math. Lett.* **19** (2006), 1073–1077.

[202] P. L. Butzer and M. Schmidt, Central factorial numbers and their role in finite difference calculus and approximation, *Colloq. Math. Soc. János Bolyai* **58** (1990), 128–150.

[203] P. L. Butzer, M. Schmidt and E. L. Stark, Observations on the history of central B-splines, *Arch. History Exact Sci.* **39** (1988), 137–156.

[204] P. L. Butzer, M. Schmidt, E. L. Stark and L. Vogt, Central factorial numbers; their main properties and some applications, *Numer. Funct. Anal. Optim.* **10** (1989), 419–488.

[205] P. L. Butzer and E. L. Stark, On a trigonometric convolution operator with kernel having two zeros of simple multiplicity, *Acta Math. Acad. Sci. Hungar.* **20** (1969), 451–461.

[206] D. K. Callebaut, Generalization of the Cauchy-Schwarz inequality, *J. Math. Anal. Appl.* **12** (1965), 491–494.

[207] C. Callias, The heat equation with singular coefficients, *Comm. Math. Phys.* **88** (1983), 357–385.

[208] R. Campbell, *Les Intégrales Eulériennes et Leurs Applications*, Dunod, Paris, 1966.

[209] I. N. Cangül, V. Kurt, Y. Simsek, H. K. Pak and S.-H. Rim, An invariant p-adic q-integral associated with q-Euler numbers and polynomials, *J. Nonlinear Math. Phys.* **14** (2007), 8–14.

[210] C. Carathéodory, *Theory of Functions of a Complex Variable*, Vol. 1 (Second English edition translated by F. Steinhardt), Chelsea Publishing Company, New York, 1958.

[211] E. Carletti and G. M. Bragadin, On Dirichlet series associated with polynomials, *Proc. Amer. Math. Soc.* **121** (1994), 33–37.

[212] E. Carletti and G. M. Bragadin, On Minakshisundaram-Pleijel zeta functions on spheres, *Proc. Amer. Math. Soc.* **122** (1994), 993–1001.

[213] L. Carlitz, On abelian fields, *Trans. Amer. Math. Soc.* **35** (1933), 122–136.

[214] L. Carlitz, Generalized Bernoulli and Euler numbers, *Duke Math. J.* **8** (1941), 585–589.

[215] L. Carlitz, q-Bernoulli numbers and polynomials, *Duke Math. J.* **15** (1948), 987–1000.

[216] L. Carlitz, Some congruences for Bernoulli numbers of higher order, *Quart. J. Math. Oxford (Ser.* 2) **4** (1953), 112–116.

[217] L. Carlitz, q-Bernoulli and Eulerian numbers, *Trans. Amer. Math. Soc.* **76** (1954), 332–350.

[218] L. Carlitz, Multiplication formulas for products of Bernoulli and Euler polynomials, *Pacific J. Math.* **9** (1959), 661–666.

[219] L. Carlitz, A multiple sum, *Rev. Mat. Hisp.-Amer.* (*Ser.* 4) **22** (1962), 167–170.

[220] G. F. Carrier, M. Krook and C. E. Pearson, *Functions of a Complex Variable: Theory and Technique*, Hod Books, Ithaca, New York, 1983.

[221] P. Cassou-Noguès, Analogues p-adiques des fonctions Γ-multiples, in *Journées Arithmétiques de Luminy* (Colloq. Internat. CNRS, Centre Univ. Luminy, Luminy, 1978), pp. 43–55, *Astérisque* **61**, Soc. Math. France, Paris, 1979.

[222] E. Catalan, Sur une transformation de série, *Mém. Soc. Roy. Sci. Liège Sér.* 2 **12** (1864), 206–215.

[223] E. Catalan, Mémoire sur la transformation des séries et sur quelques intégrales défines, *Mém. Acad. Roy. Belg.* **33** (1865), 1–50.

[224] E. Catalan, Recherches sur la constante G, et sur les inégrales Eulériennes, *Mémo. Acad. Imp. Sci. St.-Pétersbourg Sér.* 7 **31** (3) (1883), 1–51.

[225] A. L. Cauchy, Mémoire sur les fonctions dont plusieurs valeurs sont liées entre elles par une équation linéaire, et sur diverses transformations de produits composés d'un nombre indéfini de facteurs, *C. R. Acad. Sci. Paris* **17** (1843), p. 523, *Oeuvres de Cauchy*, 1re série, Vol. 8, Gauthier-Villars, Paris, 1893, 42–50.

[226] A. Cayley, The numerical value of $\Pi i = \Gamma(1+i)$, *Messenger Math.* **23** (1893/1894), 36–38.

[227] M. Cenkci and M. Can, Some results on q-analogue of the Lerch Zeta function, *Adv. Stud. Contemp. Math.* **12** (2006), 213–223.

[228] M. Cenkci, M. Can and V. Kurt, p-adic interpolation functions and Kummer-type congruences for q-twisted and q-generalized twisted Euler numbers, *Adv. Stud. Contemp. Math.* **9** (2004), 203–216.

[229] M. Cenkci, M. Can and V. Kurt, q-Extensions of Genocchi numbers, *J. Korean Math. Soc.* **43** (2006), 183–198.

[230] P. Cerone and C. T. Lenard, On integral forms of generalized Mathieu series, *J. Inequal. Pure Appl. Math.* **4** (5) (2003), 1–11 (electronic).

[231] K. Chakraborty, S. Kanemitsu and H.-L. Li, On the values of a class of Dirichlet series at rational arguments, *Proc. Amer. Math. Soc.* **138** (2010), 1223–1230.

[232] K. Chakraborty, S. Kanemitsu and H. Tsukada, *Vistas of Special Functions*, Vol. II, World Scientific Publishing Company, Singapore, Hackensack (New Jersey), London and Hong Kong, 2010 (see [631] for Vol. I).

[233] H. H. Chan, Triple product identity, quintuple product identity and Ramanujan's differential equations for the classical Eisenstein series, *Proc. Amer. Math. Soc.* **135** (2007), 1987–1992.

[234] K. Chandrasekharan and R. Narasimhan, Sur l'ordre moyen de quelques fonctions arithmétiques, *C. R. Acad. Sci. Paris* **251** (1960), 1333–1335.

[235] C.-H. Chang and H. M. Srivastava, A note on Bernoulli identities associated with the Weierstrass $\wp$-function, *Integral Transforms Spec. Funct.* **18** (2007), 245–253.

[236] C.-H. Chang, H. M. Srivastava and T.-C. Wu, Some families of Weierstrass-type functions and their applications, *Integral Transforms Spec. Funct.* **19** (2008), 621–632.

[237] J. B. Chassan, A statistical derivation of a pair of trigonometric inequalities, *Amer. Math. Monthly* **62** (1955), 353–356.

[238] M. A. Chaudhry, A. Qadir, H. M. Srivastava and R. B. Paris, Extended hypergeometric and confluent hypergeometric functions, *Appl. Math. Comput.* **159** (2004), 589–602.

[239] J. Chaudhuri, Some special integrals, *Amer. Math. Monthly* **74** (1967), 545–548.

[240] C.-P. Chen, The Riemann hypothesis and Gamma conditions, *J. Math. Anal. Appl.* **173** (1993), 258–275.

[241] C.-P. Chen, Monotonicity and convexity for the Gamma function, *J. Inequal. Pure Appl. Math.* **6** (4) (2005), Article 100.

[242] C.-P. Chen, Complete monotonicity properties for a ratio of Gamma functions, *Univ. Beograd. Publ. Elektrotehn. Fak. Ser. Mat.* **16** (2005), 26–28.

[243] C.-P. Chen, Inequalities associated with Barnes G-function, *Exposition. Math.* **29** (2011), 119–125.

[244] C.-P. Chen, A remark on the inequality of Batir for the double Gamma function, Preprint 2010.

[245] C.-P. Chen and F. Qi, Logarithmically completely monotonic functions relating to the Gamma function, *J. Math. Anal. Appl.* **321** (2006), 405–411.

[246] C.-P. Chen, F. Qi and H. M. Srivastava, Some properties of functions related to the Gamma and Psi functions, *Integral Transforms Spec. Funct.* **21** (2010), 153–164.

[247] C.-P. Chen and H. M. Srivastava, A class of two-sided inequalities involving the Psi and Polygamma functions, *Integral Transforms Spec. Funct.* **21** (2010), 523–528.

[248] C.-P. Chen and H. M. Srivastava, Some inequalities and monotonicity properties associated with the Gamma and Psi functions and the Barnes G-function, *Integral Transforms Spec. Funct.* **22** (2011), 1–15.

[249] C.-P. Chen and H. M. Srivastava, New representations for the Lugo and Euler-Mascheroni constants, *Appl. Math. Lett.* **24** (2011), 1239–1244.

[250] C.-P. Chen, H. M. Srivastava, L. Li and S. Manyama, Inequalities and monotonicity properties for the Psi (or Digamma) function and estimates for the Euler-Mascheroni constant, *Integral Transforms Spec. Funct.* **22** (2011), 681–693.

[251] K.-J. Chen and H. M. Srivastava, A generalization of two q-identities of Andrews, *J. Combin. Theory Ser. A* **95** (2001), 381–386.

[252] M.-P. Chen and H. M. Srivastava, Some families of series representations for the Riemann $\zeta(3)$, *Resultate Math.* **33** (1998), 179–197.

[253] X.-J. Chen and W.-C. Chu, Dixon's ${}_3F_2(1)$-series and identities involving harmonic numbers and the Riemann Zeta function, *Discrete Math.* **310** (2010), 83–91.

[254] G.-S. Cheon, A note on the Bernoulli and Euler polynomials, *Appl. Math. Lett.* **16** (3) (2003), 365–368.

[255] I. Cherednik, On q-analogues of Riemann's Zeta function, *Selecta Math.* **7** (2001), 447–491.

[256] Y. J. Cho, J. Choi, and M. Jung, Note on an evaluation of $\zeta(p)$, *Indian J. Pure Appl. Math.* **37** (2006), 259–263.

[257] Y. J. Cho, M. Jung, J. Choi and H. M. Srivastava, Closed-form evaluations of definite integrals and associated infinite series involving the Riemann Zeta function, *Internat. J. Comput. Math.* **83** (2006), 461–472.

[258] B. R. Choe, An elementary proof of $\Sigma_{n=1}^{\infty} 1/n^2 = \pi^2/6$, *Amer. Math. Monthly* **94** (1987), 662–663.

[259] J. Choi, On a generalization of the Hurwitz Zeta function $\zeta(s, a)$, *Indian J. Pure Appl. Math.* **23** (1992), 83–91.

[260] J. Choi, Determinant of Laplacian on S^3, *Math. Japon.* **40** (1994), 155–166.

[261] J. Choi, A proof of Stirling's formula, *Comm. Korean Math. Soc.* **9** (1994), 853–855.

[262] J. Choi, Explicit formulas for the Bernolli polynomials of order n, *Indian J. Pure Appl. Math.* **27** (1996), 667–674.

[263] J. Choi, A duplication formula for the double Gamma function Γ_2, *Bull. Korean Math. Soc.* **33** (1996), 289–294.

[264] J. Choi, Integral and series representations for the Euler's constant, in *Proceedings of the Seventh Conference on Real and Complex Analysis* (Hiroshima, Japan; October 15–16), pp. 43–55, Hiroshima University, Hiroshima, Japan, 2003.
[265] J. Choi, Some mathematical constants, *Appl. Math. Comput.* **187** (2007), 122–140.
[266] J. Choi, Certain summation formulas involving harmonic numbers and generalized harmonic numbers, *Appl. Math. Comput.* **218** (2011), 734–740.
[267] J. Choi, P. J. Anderson and H. M. Srivastava, Some q-extensions of the Apostol-Bernoulli and the Apostol-Euler polynomials of order n, and the multiple Hurwitz Zeta function, *Appl. Math. Comput.* **199** (2008), 723–737.
[268] J. Choi, P. J. Anderson and H. M. Srivastava, Carlitz's q-Bernoulli and q-Euler numbers and polynomials and a class of generalized q-Hurwitz Zeta functions, *Appl. Math. Comput.* **215** (2009), 1185–1208.
[269] J. Choi, Y. J. Cho and H. M. Srivastava, Series involving the Zeta function and multiple Gamma functions, *Appl. Math. Comput.* **159** (2004), 509–537.
[270] J. Choi, Y. J. Cho and H. M. Srivastava, Log-sine integrals involving series associated with the Zeta function and Polylogarithms, *Math. Scand.* **105** (2009), 199–217.
[271] J. Choi and D. Cvijović, Values of the Polygamma functions at rational arguments, *J. Phys. A: Math. Theoret.* **40** (2007), 15019–15028 (see also [272] below).
[272] J. Choi and D. Cvijović, Corrigendum: Values of the Polygamma functions at rational arguments [*J. Phys. A: Math. Theoret.* **40** (2007), 15019–15028], *J. Phys. A: Math. Theoret.* **43** (2010), 239801 (see also [271] above).
[273] J. Choi, D. S. Jang and H. M. Srivastava, A generalization of the Hurwitz-Lerch Zeta function, *Integral Transforms Spec. Funct.* **19** (2008), 65–79.
[274] J. Choi and J. Lee, Closed-form evaluation of a class of series associated with the Riemann zeta function, in *Proceedings of the Eleventh International Conference on Finite or Infinite Dimensional Complex Analysis and Applications* (Chiang Mai, Thailand; July 27–31), pp. 36–53, Chiang Mai University, Chiang Mai, Thailand, 2003.
[275] J. Choi, J. Lee and H. M. Srivastava, A generalization of Wilf's formula, *Kodai Math. J.* **26** (2003), 44–48.
[276] J. Choi and Y. M. Nam, The first Eulerian integral, *Kyushu J. Math.* **49** (1995), 421–427.
[277] J. Choi and C. Nash, Integral representation of the Kinkelin's constant A, *Math. Japon.* **45** (1997), 223–230.
[278] J. Choi and J. R. Quine, E. W. Barnes' approach of the multiple Gamma functions, *J. Korean Math. Soc.* **29** (1992), 127–140.
[279] J. Choi and A. K. Rathie, An evaluation of $\zeta(2)$, *Far East J. Math. Sci.* **5** (1997), 393–398.
[280] J. Choi, A. K. Rathie and S. Malani, Kummer's theorem and its contiguous identities, *Taiwanese J. Math.* **11** (2007), 1521–1527.
[281] J. Choi, A. K. Rathie and H. M. Srivastava, Some hypergeometric and other evaluations of $\zeta(2)$ and allied series, *Appl. Math. Comput.* **104** (1999), 101–108.
[282] J. Choi and T. Y. Seo, Evaluation of some infinite series, *Indian J. Pure Appl. Math.* **28** (1997), 791–796.
[283] J. Choi and T. Y. Seo, The double Gamma function, *East Asian Math. J.* **13** (1997), 159–174.
[284] J. Choi and T. Y. Seo, Integral formulas for Euler's constant, *Comm. Korean Math. Soc.* **13** (1998), 683–689.
[285] J. Choi and T. Y. Seo, Identities involving series of the Riemann Zeta function, *Indian J. Pure Appl. Math.* **30** (1999), 649–652.

[286] J. Choi and T. Y. Seo, Series representations for the Euler-Mascheroni constant γ, *East Asian Math. J.* **18** (2002), 75–84.

[287] J. Choi and H. M. Srivastava, A reducible case of double hypergeometric series involving the Riemann Zeta function, *Bull. Korean Math. Soc.* **33** (1996), 107–110.

[288] J. Choi and H. M. Srivastava, Sums associated with the Zeta function, *J. Math. Anal. Appl.* **206** (1997), 103–120.

[289] J. Choi and H. M. Srivastava, Certain classes of series involving the Zeta function, *J. Math. Anal. Appl.* **231** (1999), 91–117.

[290] J. Choi and H. M. Srivastava, Certain classes of infinite series, *Monatsh. Math.* **127** (1999), 15–25.

[291] J. Choi and H. M. Srivastava, An application of the theory of the double Gamma function *Kyushu J. Math.* **53** (1999), 209–222.

[292] J. Choi and H. M. Srivastava, Certain classes of series associated with the Zeta function and multiple Gamma functions *J. Comput. Appl. Math.* **118** (2000), 87–109.

[293] J. Choi and H. M. Srivastava, Evaluation of higher-order derivatives of the Gamma function, *Univ. Beograd. Publ. Elektrotehn. Fak. Ser. Mat.* **11** (2000), 9–18.

[294] J. Choi and H. M. Srivastava, A certain family of series associated with the Zeta and related functions, *Integral Transforms Spec. Funct.* **12** (2001), 237–250.

[295] J. Choi and H. M. Srivastava, A certain family of series associated with the Zeta and related functions, *Hiroshima Math. J.* **32** (2002), 417–429.

[296] J. Choi and H. M. Srivastava, A family of log-Gamma integrals and associated results, *J. Math. Anal. Appl.* **303** (2005), 436–449.

[297] J. Choi and H. M. Srivastava, Certain families of series associated with the Hurwitz-Lerch Zeta function, *Appl. Math. Comput.* **170** (2005), 399–409.

[298] J. Choi and H. M. Srivastava, Explicit evaluation of Euler and related sums, *Ramanujan J.* **10** (2005), 51–70.

[299] J. Choi and H. M. Srivastava, Some applications of the Gamma and Polygamma functions involving convolutions of the Rayleigh functions, multiple Euler sums and log-sine integrals, *Math. Nachr.* **282** (2009), 1709–1723.

[300] J. Choi and H. M. Srivastava, A note on a multiplication formula for the multiple Gamma function Γ_n, *Italian J. Pure Appl. Math.* **23** (2008), 179–188.

[301] J. Choi and H. M. Srivastava, Integral representations for the Gamma function, the Beta function, and the double Gamma function, *Integral Transforms Spec. Funct.* **20** (2009), 859–869.

[302] J. Choi and H. M. Srivastava, Integral representations for the Euler-Mascheroni constant γ, *Integral Transforms Spec. Funct.* **21** (2010), 675–690.

[303] J. Choi and H. M. Srivastava, Mathieu series and associated sums involving the Zeta functions, *Comput. Math. Appl.* **59** (2010), 861–867.

[304] J. Choi and H. M. Srivastava, The multiple Hurwitz Zeta function and the multiple Hurwitz-Euler Eta function, *Taiwanese J. Math.* **15** (2011), 501–522.

[305] J. Choi and H. M. Srivastava, Inequalities for the multiple Gamma functions, Preprint 2010.

[306] J. Choi, H. M. Srivastava and V. S. Adamchik, Multiple Gamma and related functions, *Appl. Math. Comput.* **134** (2003), 515–533.

[307] J. Choi, H. M. Srivastava and Y. Kim, Applications of a certain family of hypergeometric summation formulas associated with Psi and Zeta functions, *Comm. Korean Math. Soc.* **16** (2001), 319–332.

[308] J. Choi, H. M. Srivastava and J. R. Quine, Some series involving the Zeta function, *Bull. Austral. Math. Soc.* **51** (1995), 383–393.

[309] J. Choi, H. M. Srivastava, T. Y. Seo and A. K. Rathie, Some families of infinite series, *Soochow J. Math.* **25** (1999), 209–219.

[310] J. Choi, H. M. Srivastava and N.-Y. Zhang, Integrals involving a function associated with the Euler-Maclaurin summation formula, *Appl. Math. Comput.* **93** (1998), 101–116.

[311] S. Chowla, Note on Dirichlet's L-functions, *Acta Arith.* **1** (1935), 113–114.

[312] L. M. Christophe Jr., M. V. Finn and J. A. Crow, Two series involving the Zeta function values, *Math. Mag.* **59** (1986), 176–178.

[313] G. Chrystal, *Algebra: An Elementary Text-Book*, Part II, Seventh edition, Chelsea Publishing Company, New York, 1964.

[314] J.-T. Chu, A modified Wallis product and some applications, *Amer. Math. Monthly* **69** (1962), 402–404.

[315] W.-C. Chu, Hypergeometric series and the Riemann Zeta function, *Acta Arith.* **82** (1997), 103–118.

[316] W.-C. Chu and L. de Donno, Hypergeometric series and harmonic number identities, *Adv. Appl. Math.* **34** (2005), 123–137.

[317] W.-C. Chu and P. Magli, Summation formulae on reciprocal sequences, *European J. Combin.* **28** (2007), 921–930.

[318] W.-C. Chu and H. M. Srivastava, Ordinary and basic bivariate hypergeometric transformations associated with the Appell and Kampé de Fériet functions, *J. Comput. Appl. Math.* **156** (2003), 355–370.

[319] W.-C. Chu and C.-Y. Wang, Convolution formulae for Bernoulli numbers, *Integral Transforms Spec. Funct.* **21** (2010), 437–457.

[320] W.-C. Chu and X.-X. Wang, Abel's method on summation by parts and hypergeometric contiguous relations, *Integral Transforms Spec. Funct.* **18** (2007), 771–807.

[321] W.-C. Chu and C.-A. Wei, Legendre inversions and balanced hypergeometric series identities, *Discrete Math.* **308** (2008), 541–549.

[322] W.-C. Chu and W.-L. Zhang, Transformations of Kummer-type for ${}_2F_2$-series and their q-analogues, *J. Comput. Appl. Math.* **216** (2008), 467–473.

[323] W.-C. Chu and D.-Y. Zheng, Infinite series with harmonic numbers and central binomial coefficients, *Internat. J. Number Theory* **5** (2009), 429–448.

[324] G. V. Chudnovsky, Padé approximations to the generalized hypergeometric functions. I, *J. Math. Pures Appl.* **58** (1979), 445–476.

[325] F. Clarke and I. Sh. Slavutskii, The integrality of the values of Bernoulli polynomials and of generalized Bernoulli numbers, *Bull. London Math. Soc.* **29** (1997), 22–24.

[326] M. W. Coffey, On some log-cosine integrals related to $\zeta(3)$, $\zeta(4)$, and $\zeta(6)$, *J. Comput. Appl. Math.* **159** (2003), 205–215.

[327] M. W. Coffey, Relations and positivity results for the derivatives of the Riemann ξ function, *J. Comput. Appl. Math.* **166** (2004), 525–534.

[328] M. W. Coffey, On one-dimensional digamma and polygamma series related to the evaluation of Feynman diagrams, *J. Comput. Appl. Math.* **183** (2005), 84–100.

[329] M. W. Coffey, One integral in three ways: moments of a quantum distribution, *J. Phys. A: Math. Gen.* **39** (6) (2006), 1425–1431.

[330] M. W. Coffey, New results on the Stieltjes constants: Asymptotic and exact evaluation, *J. Math. Anal. Appl.* **317** (2006), 603–612.

[331] M. W. Coffey, On some series representations of the Hurwitz zeta function, *J. Comput. Appl. Math.* **216** (2008), 297–305.

[332] M. W. Coffey, New results concerning power series expansions of the Riemann xi function and the Li/Keiper constants, *Proc. Roy. Soc. London Ser. A Math. Phys. Engrg. Sci.* **464** (2008), 711–731.

[333] H. Cohen, Sums involving the values at negative integers of *LL*-functions of quadratic characters, *Math. Ann.* **217** (1975), 271–285.

[334] H. Cohen, Généralisation d'une construction de R. Apéry, *Bull. Soc. Math. France* **109** (1981), 269–281.

[335] H. Cohen, *Number Theory*, Vol. I: *Tools and Diophantine Equations*; Vol. II: *Analytic and Modern Tools*, Graduate Texts in Mathematics, **239** and **240**, Springer-Verlag, Berlin, Heidelberg and New York, 2007.

[336] H. Cohen, L. Lewin and D. Zagier, A sixteenth-order polylogarithm ladder, *Experiment. Math.* **1** (1992), 25–34.

[337] L. Comtet, *Advanced Combinatorics: The Art of Finite and Infinite Expansions* (Translated from the French by J. W. Nienhuys), Reidel, Dordrecht and Boston, 1974.

[338] K. Conrad, A q-analogue of Mahler expansions. I, *Adv. Math.* **153** (2000), 185–230.

[339] J. B. Conway, *Functions of One Complex Variable*, Second edition, Springer-Verlag, New York, Heidelberg and Berlin 1978.

[340] R. Cooper, On a duplication formula, *Math. Gazette* **41** (1957), 217–218.

[341] E. T. Copson, *Theory of Functions of a Complex Variable*, Clarendon (Oxford University) Press, Oxford, London and New York 1935.

[342] H. S. M. Coxeter, The functions of Schläfli and Lobatschefsky, *Quart. J. Math. Oxford Ser.* **6** (1935), 13–29.

[343] H. Cramér, Contribution to the theory of statistical estimation, *Skand. Aktuar.* **29** (1946), 85–94.

[344] R. E. Crandall and J. P. Buhler, On the evaluation of Euler sums, *Experiment. Math.* **3** (1994), 275–285.

[345] D. Cvijović, Integral representations of the Legendre Chi-function, *J. Math. Anal. Appl.* **332** (2007), 1056–1062.

[346] D. Cvijović, Closed-form evaluations of some families of definite tangent and secant integrals, *Integral Transforms Spec. Funct.* **18** (2007), 569–579.

[347] D. Cvijović, Derivative polynomials and closed-form higher derivative formulae, *Appl. Math. Comput.* **215** (2009), 3002–3006.

[348] D. Cvijović, The Dattoli-Srivastava conjectures concerning generating functions involving the harmonic numbers, *Appl. Math. Comput.* **215** (2010), 4040–4043.

[349] D. Cvijović and J. Klinowski, New formulae for the Bernoulli and Euler polynomials at rational arguments, *Proc. Amer. Math. Soc.* **123** (1995), 1527–1535.

[350] D. Cvijović and J. Klinowski, Closed-form summation of some trigonometric series, *Math. Comput.* **64** (1995), 205–210.

[351] D. Cvijović and J. Klinowski, New rapidly convergent series representations for $\zeta(2n+1)$, *Proc. Amer. Math. Soc.* **125** (1997), 1263–1271.

[352] D. Cvijović and J. Klinowski, Values of the Legendre Chi and Hurwitz Zeta functions at rational arguments, *Math. Comput.* **68** (1999), 1623–1630.

[353] D. Cvijović and H. M. Srivastava, Closed-form summation of the Dowker and related sums, *J. Math. Phys.* **48** (2007), 043507: 1–12.

[354] D. Cvijović and H. M. Srivastava, Summation of a family of finite secant sums, *Appl. Math. Comput.* **190** (2007), 590–598.

[355] D. Cvijović and H. M. Srivastava, Closed-form evaluations of certain cosecant integrals and associated series involving the Riemann zeta function, *Adv. Stud. Contemp. Math.* **15** (2007), 1–12.

[356] D. Cvijović and H. M. Srivastava, Closed-form evaluations of certain definite integrals by employing the Cauchy integral theorem, *Numer. Algorithms* **49** (2008), 129–141.

[357] D. Cvijović and H. M. Srivastava, Evaluations of some classes of the trigonometric moment integrals, *J. Math. Anal. Appl.* **351** (2009), 244–256.

[358] D. Cvijović and H. M. Srivastava, Some discrete Fourier transform pairs associated with the Lipschitz-Lerch Zeta function, *Appl. Math. Lett.* **22** (2009), 1081–1084.

[359] D. Cvijović) and H. M. Srivastava, Asymptotics of the Landau constants and their relationship with hypergeometric functions, *Taiwanese J. Math.* **13** (2009), 855–870.

[360] D. Cvijović, J. Klinowski and H. M. Srivastava, Some polynomials associated with Williams' limit formula for $\zeta(2n)$, *Math. Proc. Cambridge Philos. Soc.* **135** (2003), 199–209.

[361] A. B. O. Daalhuis, Asymptotics expansions for q-Gamma function, q-exponential and q-Bessel functions, *J. Math. Anal. Appl.* **186** (1994), 896–913.

[362] A. Dąbrowski, A note on the values of the Riemann Zeta function at positive odd integers, *Nieuw Arch. Wisk.* **14** (4) (1996), 199–207.

[363] A. Dąbrowski, A note on p-adic q-ζ-functions, *J. Number Theory* **64** (1997), 100–103.

[364] G. Damamme and Y. Hellegouarch, Transcendence of the values of the Carlitz Zeta function by Wade's method, *J. Number Theory* **39** (1991), 257–278.

[365] D. A. Darling, An integral giving Euler's constant, *Amer. Math. Monthly* **104** (1997), 881.

[366] G. Dattoli and H. M. Srivastava, A note on harmonic numbers, umbral calculus and generating functions, *Appl. Math. Lett.* **21** (2008), 686–693.

[367] H. Daudé, P. Flajolet and B. Vallée, An average-case analysis of the Gaussian algorithm for lattice reduction, *Combin. Probab. Comput.* **6** (1997), 397–433.

[368] H. T. Davis, An extension to Polygamma functions of a theorem of Gauss, *Bull. Amer. Math. Soc.* **41** (1935), 243–247.

[369] H. T. Davis, *The Summation of Series*, The Principia Press of Trinity University, San Antonio, Texas, 1962.

[370] W. A. Day, On monotonicity of the relaxation functions of viscoelastic materials, *Proc. Cambridge Philos. Soc.* **67** (1970), 503–508.

[371] N. G. de Bruijn, Integralen voor de ζ-functie van Riemann, *Mathematica (Zutphen)* **B5** (1937), 170–180.

[372] N. J. De Lillo, *Advanced Calculus with Applications*, Macmillan and Company, New York, 1982.

[373] K. M. Dempsey, L. Dajin and J. P. Dempsey, Plana's summation formula for $\Sigma_{m=1,3,\ldots}^{\infty}$ $m^{-2}\sin(m\alpha)$, $m^{-3}\cos(m\alpha)$, $m^{-2}A^m$, $m^{-3}A^m$, *Math. Comput.* **55** (1992), 693–703.

[374] A. Denjoy, Une démonstration de l'identité fondamentale de la fonction $\zeta(s)$ de Riemann, *J. Analyse Math.* **3** (1954), 197–206.

[375] K. Devlin, *The Millennium Problems*, Basic Books, New York, 2002.

[376] P. H. Diananda, Some inequalities related to an inequality of Mathieu, *Math. Ann.* **250** (1980), 95–98.

[377] E. D'Hoker and D. H. Phong, On determinant of Laplacians on Riemann surface, *Comm. Math. Phys.* **104** (1986), 537–545.

[378] E. D'Hoker and D. H. Phong, Multiloop amplitudes for the bosonic polyakov string, *Nucl. Phys. B* **269** (1986), 204–234.

[379] E. D'Hoker and D. H. Phong, Functional determinants on Mandelstam diagrams, *Comm. Math. Phys.* **124** (1989), 629–645.

[380] P. Diaconis and D. Freeman, An elementary proof of Stirling's formula, *Amer. Math. Monthly* **93** (1986), 123–125.

[381] L. A. Dikiĭ, Trace formulas for Strum-Liouville differential operators, *Amer. Math. Soc. Transl.* **18** (1961), 81–115.

[382] K. Dilcher, On generalized gamma functions related to the Laurent coefficients of the Riemann Zeta function, *Aequationes Math.* **48** (1994), 55–85.

[383] K. Dilcher, Some q-series identities related to divisor functions, *Discrete Math.* **145** (1995), 83–93.

[384] A. Dinghas, Zur throrie der Gammafunktion, *Math.-Phys. Semester.* **6** (1958/1959), 245–252.

[385] A. Dinghas, Zur Charakterisierung der Riemannschen Zeta-Funktion durch Hadamardsche Funktionalgleichungen, *Rev. Roumaine Math. Pures Appl.* **19** (1974), 995–1007.

[386] W. Dittrich and M. Reuter, Effective QCD-Lagrangian with ξ-function regularization, *Phys. Lett. B* **128** (1983), 321–326.

[387] A. L. Dixon and W. L. Ferrar, A class of discontinuous integrals, *Quart. J. Math. Oxford Ser.* **7** (1936), 81–96.

[388] G. B. Djordjević and H. M. Srivastava, Some generalizations of the incomplete Fibonacci and the incomplete Lucas polynomials, *Adv. Stud. Contemp. Math.* **11** (2005), 11–32.

[389] G. B. Djordjević and H. M. Srivastava, Incomplete generalized Jacobsthal and Jacobsthal-Lucas numbers, *Math. Comput. Modelling* **42** (2005), 1049–1056.

[390] G. B. Djordjević and H. M. Srivastava, Some generalizations of certain sequences assoociated with the Fibonacci numbers, *J. Indonesian Math. Soc.* **12** (2006), 99–112.

[391] P. J. De Doelder, On some series containing $\psi(x) - \psi(y)$ and $\left(\psi(x) - \psi(y)\right)^2$ for certain values of x and y, *J. Comput. Appl. Math.* **37** (1991), 125–141.

[392] J. Dougall, On Vandermonde's theorem and some more general expansions, *Proc. Edinburgh Math. Soc.* **25** (1907), 114–132.

[393] J. S. Dowker, Effective action in spherical domains, *Comm. Math. Phys.* **162** (1994), 633–647.

[394] J. S. Dowker and K. Kirsten, The Barnes ζ-function, sphere determinants and Glaisher-Kinkelin-Bendersky constants, *Anal. Appl.* **3** (2005), 45–68.

[395] J. Dufresnoy et Ch. Pisot, Sur la relation fonctionnelle $f(x+1) - f(x) = \varphi(x)$, *Bull. Soc. Math. Belg.* **15** (1963), 259–270.

[396] J. Dutka, On some Gamma function inequalities, *SIAM J. Math. Anal.* **16** (1985), 180–185.

[397] A. Eagle, A simple theory of the Gamma function, *Math. Gazette* **14** (1928/1929), 118–127.

[398] H. M. Edwards, *Riemann's Zeta Function*, Academic Press, New York and London, 1974; Dover Publications, Mineola, New York, 2001.

[399] J. Edwards, *A Treatise on the Integral Calculus with Applications*: *Examples and Problems*, Vols. 1 and 2, Chelsea Publishing Company, New York 1954.

[400] I. Efrat, Determinants of Laplacians on surfaces of finite volume, *Comm. Math. Phys.* **119** (1988), 443–451; Erratum, *ibid.* **138** (1991), 607.

[401] S. Egami, Reciprocity laws of multiple Zeta functions and generalized Dedekind sums, in *Analytic Number Theory and Related Topics* (Tokyo, 1991), pp. 17–27, World Scientific Publishing Company, Singapore, London and River Edge (New Jersey), 1993.

[402] S. Egami and K. Matsumoto, Asymptotic expansions of multiple Zeta functions and power mean values of Hurwitz Zeta functions, *J. London Math. Soc.* (*Ser.* 2) **66** (2002), 41–60.

[403] Á. Elbert, Asymptotic expansion and continued fraction for Mathieu's series, *Period. Math. Hungar.* **13** (1982), 1–8.

[404] Á. Elbert and A. Laforgia, On some properties of the Gamma function, *Proc. Amer. Math. Soc.* **128** (2000), 2667–2673.

[405] Á. Elbert and A. Laforgia, An inequality for the product of two integrals relating to the incomplete Gamma functions, *J. Inequal. Appl.* **5** (2000), 39–51.

[406] N. Elezović, C. Giordano and J. Pečarić, The best bounds in Gautschi's inequality, *Math. Inequal. Appl.* **3** (2000), 239–252.

[407] N. Elezović, H. M. Srivastava and Ž. Tomovski, Integral representations and integral transforms of some families of Mathieu type series, *Integral Transforms Spec. Funct.* **19** (2008), 481–495.

[408] C. J. Eliezer and D. E. Daykin, Generalizations and applications of Cauchy-Schwarz inequalities, *Quart. J. Math. Oxford* (*Ser.* 2) **18** (1967), 357–360.

[409] E. Elizalde, Derivative of the generalized Riemann Zeta function $\zeta(z,q)$ at $z=-1$, *J. Phys. A: Math. Gen.* **18** (1985), 1637–1640.

[410] E. Elizalde, An asymptotic expansion for the first derivative of the generalized Riemann Zeta function, *Math. Comput.* **47** (1986), 347–350.

[411] E. Elizalde, *Ten Physical Applications of Spectral Zeta Functions*, Lecture Notes in Physics, Springer-Verlag, Berlin, Heidelberg and New York, 1995.

[412] E. Elizalde, S. D. Odintsov, A. Romeo, A. A. Bytsenko and S. Zerbini, *Zeta Regularization Techniques with Applications* World Scientific Publishing Company, Singapore, New Jersey, London and Hong Kong, 1994.

[413] E. Elizalde and A. Romeo, An integral involving the generalized Zeta function *Internat. J. Math. Math. Sci.* **13** (1990), 453–460.

[414] E. Elizalde and J. Soto, ζ-regularized Lagrangians for massive quarks in constant background mean-fields, *Ann. Phys.* **162** (1985), 192–211.

[415] D. Elliott and P. R. Johnston, Gauss-Legendre quadrature for the evaluation of integrals involving the Hankel function, *J. Comput. Appl. Math.* **211** (2008), 23–35.

[416] C. Elsner, On a sequence transformation with integral coefficients for Euler's constant, *Proc. Amer. Math. Soc.* **123** (1995), 1537–1541.

[417] J. Elstrodt, Note on the Selberg trace formula for the Picard group, *Abh. Math. Sem. Univ. Hamburg* **55** (1985), 207–209.

[418] O. Emersleben, Über die Reihe $\sum_{k=1}^{\infty} k/(k^2+c^2)^2$, *Math. Ann.* **125** (1952), 165–171.

[419] B. J. English and G. Rousseau, Bounds for certain harmonic sums, *J. Math. Anal. Appl.* **206** (1997), 428–441.

[420] T. Erber, The Gamma function inequalities of Gurland and Gautschi, *Skand. Aktuarietidskr.* **44** (1961), 27–28.

[421] A. Erdélyi, W. Magnus, F. Oberhettinger and F. G. Tricomi, *Tables of Integral Transforms*, Vol. I, McGraw-Hill Book Company, New York, Toronto and London, 1954.

[422] A. Erdélyi, W. Magnus, F. Oberhettinger and F. G. Tricomi, *Higher Transcendental Functions*, Vol. III, McGraw-Hill Book Company, New York, Toronto and London, 1955.

[423] O. Espinosa and V. H. Moll, A generalized polygamma function, *Integral Trans. Special Funct.* **15** (2004), 101–115.

[424] T. Estermann, Elementary evaluation of $\zeta(2k)$, *J. London Math. Soc.* **22** (1947), 10–13.

[425] A. Eswarathasan and E. Levine, p-Integral harmonic sums, *Discrete Math.* **91** (1991), 249–257.

[426] L. Euler, Methods gereralis summandi progressiones, *Comment. Acad. Sci. Petropolit.* **6** (1732/1733), 68–97 (see also *Opera Omnia (Collected Works)*, Series Prima **XIV**, 42–72).

[427] L. Euler, De progressionibus harmonicus observationes, *Comment. Acad. Sci. Imper. Petropolit.* **7** (1734/1735), 150–161.

[428] L. Euler, Methodvs vniversalis seriervm convergentivm svmmas, *Comment. Acad. Sci. Imper. Petropolit.* **8** (1736), 3–9.

[429] L. Euler, Methodvs vniversalis series svmmandi, *Comment. Acad. Sci. Imper. Petropolit.* **8** (1736), 147–158.

[430] L. Euler, Inventio summae cuiusgue seriei ex dato termino generali, *Comment. Acad. Sci. Petropolit.* **8** (1736), 9–22 (see also *Opera Omnia (Collected Works)*, Series Prima **XIV**, 108–123).

[431] L. Euler, De seriebus quibusdam considerationes. *Comment. Acad. Sci. Petropolit.* **12** (1740), 53–96 (see also *Opera Omnia (Collected Works)*, Series Prima **XIV**, 407–461).

[432] L. Euler, De seriebus divergentibus, *Novi Comment. Acad. Sci. Petropolit.* **5** (1754/1755), 205–237 (see also *Opera Omnia (Collected Works)*, Series Prima **XIV**, 585–617).

[433] L. Euler, Remarques sur un beau rapport entre les series des puissances tant directes que reciproques. *Mém. Acad. Sci. Berlin* **17** (1761), 83–106 (see also *Opera Omnia (Collected Works)*, Series Prima **XV**, 70–90).

[434] L. Euler, De summis serierum numeros Bernoullianos involventium, *Novi Comment. Acad. Sci. Imper. Petropolit.* **14** (1769), 129–167.

[435] L. Euler, Meditationes circa singulare serierum genus, *Novi Comment. Acad. Sci. Imper. Petropolit.* **20** (1775), 140–186.

[436] J. A. Ewell, A new series representation for $\zeta(3)$, *Amer. Math. Monthly* **97** (1990), 219–220.

[437] J. A. Ewell, On values of the Riemann Zeta function at integral arguments, *Canad. Math. Bull.* **34** (1991), 60–66.

[438] J. A. Ewell, On the Zeta function values $\zeta(2k+1)$, $k = 1, 2, \ldots$, *Rocky Mountain J. Math.* **25** (1995), 1003–1012.

[439] H. Exton, *q-Hypergeometric Functions and Applications*, Halsted Press (Ellis Horwood Limited, Chichester), John Wiley and Sons, New York, Chichester, Brisbane and Toronto, 1983.

[440] P. Eymard and J.-P. Lafon, *The Number* π (Translated from the 1999 French original by S. S. Wilson), American Mathematical Society, Providence, Rhode Island, 2004.

[441] J.-P. Fang, Extensions of q-Chu-Vandermonde's identity, *J. Math. Anal. Appl.* **339** (2008), 845–852.

[442] J.-P. Fang, q-differential operator identities and applications, *J. Math. Anal. Appl.* **332** (2007), 1393–1407.

[443] W. Feller, Completely monotone functions and sequences, *Duke Math. J.* **5** (1939), 661–674.

[444] W. L. Ferrar, Summation formulae and their relation to Dirichlet's series, *Compositio Math.* **1** (1935), 344–360.

[445] W. L. Ferrar, Summation formulae and their relation to Dirichlet's series. II, *Compositio Math.* **4** (1936/1937), 394–405.

[446] C. Ferreira and J. L. López, An asymtotic expansion of the double Gamma function, *J. Approx. Theory* **111** (2001), 298–314.

[447] C. Ferreira and J. L. López, Asymptotic expansions of the Hurwitz-Lerch Zeta function, *J. Math. Anal. Appl.* **298** (2004), 210–224.

[448] E. M. Ferreira and J. Sesma, Zeros of the Macdonald function of complex order, *J. Comput. Appl. Math.* **211** (2008) 223–231.

[449] S. R. Finch, *Mathematical Constants*, Encyclopedia of Mathematics and Its Applications, Vol. 94, Cambridge University Press, Cambridge, New York, Port Melbourne, Madrid and Cape Town, 2003.

[450] N. J. Fine, Note on the Hurwitz Zeta-function, *Proc. Amer. Math. Soc.* **2** (1951), 361–364.

[451] A. M. Fink, Kolmogorov-Landau inequalities for monotone functions, *J. Math. Anal. Appl.* **90** (1982), 251–258.

[452] R. A. Fisher, Theory of statistical estimation, *Proc. Cambridge Philos. Soc.* **22** (1924/1925), 700–725.

[453] P. Flajolet and H. Prodinger, On Stirling numbers for complex arguments and Hankel contours, *SIAM J. Discrete Math.* **12** (1999), 155–159.

[454] P. Flajolet and B. Salvy, Euler sums and contour integral representations, *Experiment. Math.* **7** (1998), 15–35.

[455] P. Flajolet and L. Vepstas, On differences of zeta values, *J. Comput. Appl. Math.* **216** (2008), 297–305.

[456] H. G. Forder, Duplication formulae, *Math. Gazette* **41** (1957), 215–217.

[457] A. R. Forsyth, *Theory of Functions of a Complex Variable*, Cambridge University Press, Cambridge, London and New York, 1900.

[458] B. L. Foster and R. R. Phelps, Problem 4712 proposed by J. V. Whittaker, *Amer. Math. Monthly* **63** (1956), 669; *ibid.* **64** (1957), 677–678.

[459] G. J. Fox, Euler polynomials at rational numbers, *C. R. Math. Rep. Acad. Sci. Canada* **21** (1999), 87–90.

[460] G. J. Fox, A p-adic L-function of two variables, *Enseign. Math.* **46** (2000), 225–278.

[461] R. D. Fray, Congruence properties of ordinary and q-Bernoulli coefficients, *Duke Math. J.* **34** (1967), 467–480.

[462] C. L. Frenzen, Error bounds for the asymptotic expansion of the ratio of two Gamma functions with complex argument, *SIAM J. Math. Anal.* **23** (1992), 505–511.

[463] E. Friedman and S. Ruijsenaars, Shintani-Barnes Zeta and Gamma functions, *Adv. Math.* **187** (2004), 362–395.

[464] A. M. Fu and A. Lascoux, q-identities from Lagrange and Newton interpolation, *Adv. Appl. Math.* **31** (2003), 527–531.

[465] J. M. Gandhi, A conjectured representation of Genocchi numbers, *Amer. Math. Monthly* **77** (1970), 505–506.

[466] M. Garg, K. Jain and S. L. Kalla, A further study of general Hurwitz-Lerch zeta function, *Algebras Groups Geom.* **25** (2008), 311–319.

[467] M. Garg, K. Jain and H. M. Srivastava, Some relationships between the generalized Apostol-Bernoulli polynomials and Hurwitz-Lerch Zeta functions, *Integral Transforms Spec. Funct.* **17** (2006), 803–815.

[468] G. Gasper and M. Rahman, *Basic Hypergeometric Series* (with a Foreword by Richard Askey), Encyclopedia of Mathematics and Its Applications, Vol. 35, Cambridge University Press, Cambridge, New York, Port Chester, Melbourne and Sydney, 1990; Second edition, Encyclopedia of Mathematics and Its Applications, Vol. 96, Cambridge University Press, Cambridge, London and New York, 2004.

[469] C. F. Gauss, Disquisitiones Generales Circa Seriem Infinitam $\left[\frac{\alpha\beta}{1\cdot\gamma}\right]x + \left[\frac{\alpha(\alpha+1)\beta(\beta+1)}{1\cdot 2\cdot\gamma(\gamma+1)}\right]x^2 + \left[\frac{\alpha(\alpha+1)(\alpha+2)\beta(\beta+1)(\beta+2)}{1\cdot 2\cdot\gamma(\gamma+1)(\gamma+2)}\right]x^3 +$ etc. Pars Prior. *Comm. Soc. Regia Sci. Göttingen Rec.* **2** (1813), 3–46. Reprinted in Gesammelte Werke (Göttingen), Bd. 3 (1866), 123–163.

[470] W. Gautschi, Some elementary inequalities relating to the Gamma and incomplete Gamma function, *J. Math. Phys.* **38** (1959), 77–81.

[471] W. Gautschi, A harmonic mean inequality for the Gamma function, *SIAM J. Math. Anal.* **5** (1974), 278–281.

[472] W. Gautschi, Some mean value inequalities for the Gamma function, *SIAM J. Math. Anal.* **5** (1974), 282–292.

[473] W. Gautschi, On certain slowly convergent series occurring in plate contact problems, *Math. Comput.* **57** (1991), 325–338.

[474] W. Gautschi, The incomplete gamma function since Tricomi, in *Tricomi's Ideas ans Contemporary Applied Mathematics*, Atti Convegni Lincei, Vol. 147, pp. 203–237, Accademia Nazionale dei Lincei, Rome, 1998.

[475] W. Gautschi, The Hardy-Littlewood function: an exercise in slowly convergent series, *J. Comput. Appl. Math.* **178** (2005), 249–254.

[476] W. Gautschi, On Euler's attempt to compute logarithms by interpolation: A commentary to his letter of February 16, 1734 to Daniel Bernoulli, *J. Comput. Appl. Math.* **219** (2007), 408–415.

[477] W. Gautschi, F. E. Harris and N. M. Temme, Expansions of the exponential integral in incomplete Gamma functions, *Appl. Math. Lett.* **16** (2003), 1095–1099.

[478] I. M. Gel'fand and G. E. Shilov, *Generalized Functions*, Vol. I, Academic Press, New York and London, 1964.

[479] C. Georghiou and A. N. Philippou, Harmonic sums and the Zeta function, *Fibonacci Quart.* **21** (1983), 29–36.

[480] I. Gerst, Some series for Euler's constant, *Amer. Math. Monthly* **76** (1969), 273–275.

[481] D. P. Giesy, Still another elementary proof that $\Sigma\, 1/k^2 = \pi^2/6$, *Math. Mag.* **45** (1972), 148–149.

[482] J. W. L. Glaisher, On the history of Euler's constant, *Messenger Math.* **1** (1872), 25–30.

[483] J. W. L. Glaisher, On the integrals $\int_0^\infty \sin(x^n)\,dx$ and $\int_0^\infty \cos(x^n)\,dx$, *Messenger Math.* **1** (1872), 106–111.

[484] J. W. L. Glaisher, On the product $1^1\,2^2\cdots n^n$, *Messenger Math.* **7** (1877), 43–47.

[485] J. W. L. Glaisher, On products and series involving prime numbers only, *Quart. J. Math.* **26** (1893), 1–74.

[486] J. W. L. Glaisher, Note on a relation connecting constants analogous to Euler's constant, *Messenger Math.* **24** (1894/1895), 24–27.

[487] J. W. L. Glaisher, Expressions for Gamma functions in terms of complete elliptic integrals, *Messenger Math.* **24** (1894/1895), 27–47.

[488] J. W. L. Glaisher, The Bernoullian function, *Quart. J. Math.* **29** (1898), 1–168.

[489] J. W. L. Glaisher, Summations of certain numerical series, *Messenger Math.* **42** (1912/1913), 19–34.

[490] J. W. L. Glaisher, Relations connecting quantities of the form $1 + 2^{-n} + 3^{-n} + 4^{-n}$&c., *Messenger Math.* **44** (1914/1915), 1–10.

[491] M. L. Glasser, Some integrals of the arctangent function, *Math. Comput.* **22** (1968), 445–447.

[492] D. V. Gokhale, On an inequality for Gamma function, *Skand. Aktuarietidskr.* **46** (1963), 213–215.

[493] Y. Gon, N. Kurokawa and H. Oyanagi, Multiple q-Mahler measures and Zeta functions, *J. Number Theory* **124** (2007), 328–345.

[494] S. M. Gonek, Mean values of the Riemann Zeta-function and its derivatives, *Invent. Math.* **75** (1984), 123–141.

[495] B. González, J. Matera and H. M. Srivastava, Some q-generating functions and associated families of generalized hypergeometric polynomials, *Math. Comput. Modelling* **34** (1–2) (2001), 133–175.

[496] R. W. Gosper Jr., A calculus of series rearrangements, in *Algorithms and Complexity: New Directions and Recent Results* (J. F. Traub, Editor), pp. 121–151, Academic Press, New York, London and Toronto, 1976.

[497] R. W. Gosper Jr., Decision procedure for indefinite hypergeometric summation, *Proc. Nat. Acad. Sci. U. S. A.* **75** (1978), 40–42.

[498] R. W. Gosper Jr., $\int_{n/4}^{m/6} \ln \Gamma(z)\,dz$, *Fields Institute Comm.* **14** (1997), 71–76.

[499] H. W. Gould, Explicit formulas for Bernoulli numbers, *Amer. Math. Monthly* **79** (1972), 44–51.

[500] H. W. Gould, *Cominatorial Identities: A Standardized Set of Tables Listing* 500 *Binomial Coefficient Summations*, Morgantown, West Virginia, 1972.

[501] H. W. Gould, Euler's formula for nth differences of powers, *Amer. Math. Monthly* **85** (1978), 450–467.

[502] H. W. Gould and H. M. Srivastava, Some combinatorial identities associated with the Vandermonde convolution, *Appl. Math. Comput.* **84** (1997), 97–102.

[503] X. Gourdon and P. Sebah, *Constants and Records of Computation* [http://numbers.computation.free.fr/Constants/Miscellaneous/Records.html.]

[504] S. P. Goyal and R. K. Laddha, On the generalized Riemann Zeta functions and the generalized Lambert transform, *Gaṇita Sandesh* **11** (1997), 99–108.

[505] I. S. Gradshteyn and I. M. Ryzhik, *Tables of Integrals, Series, and Products* (Corrected and Enlarged edition prepared by A. Jeffrey), Academic Press, New York, 1980; Sixth edition, 2000.

[506] J. H. Graf, Berechnung von $\Gamma(a)\Gamma\left(a+\frac{1}{n}\right)\Gamma\left(a+\frac{2}{n}\right)\cdots\Gamma\left(a+\frac{n-1}{n}\right):\Gamma(na)$, *Arch. Math. Phys.* **11** (1907), 206–209.

[507] R. L. Graham, D. E. Knuth and O. Patashnik, *Concrete Mathematics*, Addison-Wesley Publishing Company, Reading, Massachusetts, 1989.

[508] J.-P. Gram, Note sur le calcul de la fonction $\zeta(s)$ de Riemann, *Overs. Over. D. K. D. Vidensk. Selsk. Forh.* **1895** (1895), 303–308.

[509] R. Greenberg, D. C. B. Marsh and A. E. Danese, A Zeta-function summation, *Amer. Math. Monthly* **74** (1967), 80–81.

[510] D. H. Greene and D. E. Knuth, *Mathematics for the Analysis of Algorithms*, Third edition, Birkhäuser, Boston, Basel and Berlin, 1990.

[511] J. Greene, Hypergeometric series over finite fields, *Trans. Amer. Math. Soc.* **301** (1987), 77–101.

[512] A. Z. Grinshpan and M. E.-H. Ismail, Completely monotonic functions involving the Gamma and q-Gamma functions, *Proc. Amer. Math. Soc.* **134** (2006), 1153–1160.

[513] B. H. Gross and N. Koblitz, Gauss sums and the p-adic Γ-function, *Ann. of Math.* **109** (1979), 569–581.

[514] E. Grosswald, Die Werte der Riemannschen Zetafunktion an ungeraden Argumenatellen, *Nachr. Wiss. Göttingen Math.-Phys. Kl.* II **1970** (1970), 9–13.

[515] E. Grosswald, Comments on some formulae of Ramanujan, *Acta. Arith.* **21** (1972), 25–34.

[516] J. Guillera and J. Sondow, Double integrals and infinite products for some classical constants via analytic continuations of Lerch's transcendent, *Ramanujan J.* **16** (2008), 247–270.

[517] A. P. Guinand, Some formulae for the Riemann Zeta-function, *J. London Math. Soc.* **22** (1947), 14–18.

[518] B.-N. Guo and F. Qi, Inequalities for generalized weighted mean values of convex function, *Math. Inequal. Appl.* **4** (2001), 195–202.

[519] B.-N. Guo and F. Qi, Generalization of Bernoulli polynomials, *Internat. J. Math. Ed. Sci. Tech.* **33** (2002), 428–431.

[520] B.-N. Guo and F. Qi, Properties and applications of a function involving exponential functions, *Comm. Pure Appl. Anal.* **8** (2009), 1231–1249 (see also [927]).

[521] S. Guo, F. Qi and H. M. Srivastava, Necessary and sufficient conditions for two classes of functions to be logarithmically completely monotonic, *Integral Transforms Spec. Funct.* **18** (2007), 819–826.

[522] S. Guo, F. Qi and H. M. Srivastava, Supplements to a class of logarithmically completely monotonic functions associated with the Gamma function, *Appl. Math. Comput.* **197** (2008), 768–774.

[523] V. J. W. Guo and M. J. Schlosser, Curious extensions of Ramanujan's ${}_1\Psi_1$ summation formula, *J. Math. Anal. Appl.* **334** (2007), 393–403.

[524] V. J. W. Guo, Bijective proofs of Gould's and Rothe's identities, *Discrete Math.* **308** (2008), 1756–1759.

[525] P. L. Gupta, R. C. Gupta, S.-H. Ong and H. M. Srivastava, A class of Hurwitz-Lerch Zeta distributions and their applications in reliability, *Appl. Math. Comput.* **196** (2008), 521–531.

[526] J. Gurland, On Wallis' formula, *Amer. Math. Monthly* **63** (1956), 643–645.

[527] J. Gurland, An inequality satisfied by the Gamma function, *Skand. Aktuarietidskr.* **39** (1956), 171–172.

[528] J. M. Gutiérrez, M. A. Hernández, P. J. Miana and N. Romero, New identities in the Catalan triangle, *J. Math. Anal. Appl.* **341** (2008), 52–61.

[529] J. Hadamard, Sur une propriété de la fonction $\zeta(s)$, *Bull. Soc. Math. France* **39** (1909), 59–60.

[530] R. L. Hall and N. Saad, Perturbation expansions for the spiked harmonic oscillator and related series involving the Gamma function *J. Phys. A: Math. Gen.* **33** (2000), 5531–5537.

[531] E. R. Hansen, *A Table of Series and Products*, Prentice-Hall, Englewood Cliffs, New Jersey, 1975.

[532] G. H. Hardy, A new proof of Kummer's series for $\log\Gamma(a)$, *Messenger Math.* **31** (1901/1902), 31–33.

[533] G. H. Hardy, On the expression of the double Zeta-function and double Gamma function in terms of elliptic functions, *Trans. Cambridge Philos. Soc.* **20** (1905), 395–427.

[534] G. H. Hardy, On the roots of the equation $\frac{1}{\Gamma(x+1)} = c$, *Proc. London Math. Soc.* **2** (2) (1904), 1–7.

[535] G. H. Hardy, On double Fourier series and especially these which represent the double Zeta-function and incommensurable parameters, *Quart. J. Math.* **37** (1906), 53–79.

[536] G. H. Hardy, Note on Dr. Vacca's series for γ, *Quart. J. Pure Appl. Math.* **43** (1912), 215–216.

[537] G. H. Hardy, *Divergent Series*, Clarendon (Oxford University) Press, Oxford, London and New York, 1949; Second (Textually Unaltered) edition, Chelsea Publishing Company, New York, 1991.

[538] G. H. Hardy and E. M. Wright, *An Introduction to the Theory of Numbers*, Oxford University (Clarendon) Press, Oxford, London and New York, 1954.

[539] F. E. Harris, Incomplete Bessel, generalized incomplete gamma, or leaky aquifer functions, *J. Comput. Appl. Math.* **215** (2008), 260–269.

[540] H. Haruki, A new characterization of Euler's Gamma function by a functional equation, *Aequationes Math.* **31** (1986), 173–183.

[541] M. Hashimoto, S. Kanemitsu and M. Toda, On Gauss' formula for ψ and finite expressions for the L-series at 1, *J. Math. Soc. Japan* **60** (2008), 219–236.

[542] H. Hasse, Ein summierungsverfahren für die Riemannsche ζ-Reihe, *Math. Zeitschr.* **32** (1930), 458–464.

[543] J. Havil, *Gamma* (*Exploring Euler's Constant*), Princeton University Press, Princeton and Oxford, 2003.

[544] Y.-A. He and W.-P. Zhang, Some sum relations involving Bernoulli and Euler polynomials, *Integral Transforms Spec. Funct.* **22** (2011), 207–215.

[545] E. Hecke, Über die Bestimmung Dirichletscher Reihen durch ihre Funktionalgleichung, *Math. Ann.* **112** (1936), 664–699 (see also *Mathematische Werke*, Vol. 33, 591–626).

[546] H. Heilbronn, On the class-number in imaginary quadratic fields, *Quart. J. Math. Oxford Ser.* **5** (1934), 150–160.

[547] H. Heilbronn, On real characters, *Acta Arith.* **2** (1936), 212–213.

[548] H. Heilbronn and E. H. Linfoot, On the imaginary quadratic corpora of class-number one, *Quart. J. Math. Oxford Ser.* **5** (1934), 752–755.

[549] E. Heine, Untersuchungen über die Reihe $\cdots$, *J. Reine Angew. Math.* **34** (1847), 285–328.

[550] E. Heine, *Handbuch der Kugelfunktionen*; *Theorie und Anwendungen*, Reimer, Berlin, 1878.

[551] D. A. Hejhal, The Selberg trace formula for congruence subgroups, *Bull. Amer. Math. Soc.* **81** (1975), 752–755.

[552] D. A. Hejhal, *The Selberg Trace Formula for* $\mathbf{PSL}(2,\mathbb{R})$, Vol. 1, Lecture Notes in Mathematics **548**, Springer-Verlag, Berlin, Heidelberg and New York, 1976.

[553] D. A. Hejhal, *The Selberg Trace Formula for* $\mathbf{PSL}(2,\mathbb{R})$, Vol. 2, Lecture Notes in Mathematics **1001**, Springer-Verlag, Berlin, Heidelberg, New York and Tokyo, 1983.

[554] P. Henrici, *Applied and Computational Complex Analysis*, John Wiley and Sons, New York, Chichester, Brisbane, Toronto and Singapore, 1986.

[555] S. Herrmann, Genocchi numbers and f-vectors of simplicial balls, *European J. Combin.* **29** (2008), 1087–1091.

[556] M. Hervé, Itération des transformations analytiques dans le bicercle-unité *Ann. Sci. École Norm. Sup.* **71** (3) (1954), 1–28.

[557] D. Hilbert, *Mathematical Problems* (Translated into English by M. W. Newson with the approval of D. Hilbert), *Bull. Amer. Math. Soc.* **8** (1902), 437–445 and 478–479.

[558] J. M. Hill, Laplace transforms and the Riemann Zeta function, *Integral Transforms Spec. Funct.* **18** (2007), 193–205.

[559] M. M. Hjortnaes, Overfϕring av rekken $\Sigma_{k=1}^{\infty}(1/k^3)$ til et bestemt integral, in *Proceedings of the Twelfth Scandanavian Mathematical Congress* (Lund; August 10–15, 1953), pp. 211–213, Lund 1954.

[560] M. E. Hoffman, Multiple harmonic series, *Pacific J. Math.* **152** (1992), 275–290.

[561] M. E. Hoffman, The algebra of multiple harmonic series, *J. Algebra* **194** (1997), 477–495.

[562] M. E. Hoffman, Derivative polynomials, Euler polynomials, and associated integer sequences, *Electronic J. Combin.* **6** (1999), 1–13 (electronic).

[563] O. Hölder, Über eine Transcendente Funktion, *Göttingen, Dieterichsche Verlags-Buchhandlung* **1886** (1886), 514–522.

[564] O. Hölder, Über eine von Abel untersuchte Transzendente und eine merkwürdige Funktionalbeziehung, *Ber. Verhandlungen Saechsischen Akad. Wissenschaften Leipzig. Math.-Phys. Kl.* **80** (1928), 312–325.

[565] F. Holme, En enkel beregning av $\Sigma_{k=1}^{\infty} \frac{1}{k^2}$, *Nordisk Mat. Tidskr.* **18** (1970), 91–92.

[566] A. F. Horadam, Genocchi polynomials, in *Proceedings of the Fourth International Conference on Fibonacci Numbers and Their Applications*, Kluwer Academic Publishers, Dordrecht, Boston and London, 1991, 145–166.

[567] A. F. Horadam, Negative order Genocchi polynomials, *Fibonacci Quart.* **30** (1992), 21–34.

[568] A. F. Horadam, Generation of Genocchi polynomials of first order by recurrence relations, *Fibonacci Quart.* **30** (1992), 239–243.

[569] R. A. Horn, On infinitely divisible matrices, kernels and functions, *Zeitschr. Wahrscheinlichkeitstheorie Verw. Geb.* **8** (1967), 219–230.

[570] S. J. X. Hou and J. Zeng, A q-analog of dual sequences with applications, *European J. Combin.* **28** (2007), 214–227.

[571] R. M. Hovstad, The series $\Sigma_{k=1}^{\infty} \frac{1}{k^{2p}}$, the area of the unit circle and Leibniz' formula, *Nordisk Mat. Tidskr.* **20** (1972), 92–98.

[572] Yi Huang, Some new integral formulas for Euler's constant, *Comm. Korean Math. Soc.* **26** (2011) (in press).

[573] P. Humbert, Sur les polynômes hypergéométriques, *C. R. Acad. Sci. Paris* **172** (1921), 1282–1283.

[574] P. Humbert, Formule de multiplication pour la fonction de Kummer, $\Phi(\alpha, \gamma, x)$, *C. R. Acad. Sci. Paris* **173** (1921), 217–219.

[575] A. Hurwitz, Einige eigenschaften der Dirichlet'schen funktionen $F(s) = \Sigma\,(D/n)\cdot 1/n^s$, die bei der bestimmung der classenanzahlen binärer quadratischer formen auftreten, *Zeitschr. Math. Phys.* **27** (1882), 86–101.

[576] A. Hurwitz, Über die entwickelungscoefficienten der lemniscatischen functionen, *Math. Ann.* **51** (1899), 196–226.

[577] J. I. Hutchinson, On the roots of the Riemann Zeta function, *Trans. Amer. Math. Soc.* **27** (1925), 49–60.

[578] J. I. Hutchinson, Properties of functions represented by the Dirichlet series $\Sigma\,(a\nu + b)^{-s}$, or by linear combinations of such series, *Trans. Amer. Math. Soc.* **31** (1929), 322–344.

[579] K.-W. Hwang, Y.-H. Kim and T. Kim, Interpolation functions of q-extensions of Apostol's type Euler polynomials, *J. Inequal. Appl.* **2009** (2009), 451217.

[580] A. E. Ingham, *The Distribution of Prime Numbers*, Cambridge University Press, Cambridge, London and New York, 1932.

[581] K. Ireland and M. Rosen, *A Classical Introduction to Modern Number Theory*, Springer-Verlag, Berlin, Heidelberg and New York, 1982.

[582] M. Ishibashi, Multiple cotangent and generalized Eta functions, *Ramanujan J.* **4** (2000), 221–229.

[583] H. Ishikawa and K. Matsumoto, An explicit formula of Atkinson type for the product of the Riemann Zeta-function and a Dirichlet polynomial, *Central European J. Math.* **9** (2011), 102–106.

[584] M. E.-H. Ismail, A simple proof of Ramanujan's ${}_1\Psi_1$ sum, *Proc. Amer. Math. Soc.* **63** (1977), 185–186.

[585] M. E.-H. Ismail, L. Lorch and M. Muldoon, Completely monotonic functions associated with the Gamma function and its q-analogues, *J. Math. Anal. Appl.* **116** (1986), 1–9.

[586] A. Ivić, *The Riemann Zeta-Function*, John Wiley and Sons, New York, Chichester, Brisbane and Toronto, 1985.

[587] F. H. Jackson, A generalization of the functions $\Gamma(n)$ and x^n, *Proc. Roy. Soc. London* **74** (1904), 64–72.

[588] F. H. Jackson, The basic gamma function and the elliptic functions, *Proc. Roy. Soc. London* **76** (1905), 127–144.

[589] F. H. Jackson, On q-functions and a certain difference operator, *Trans. Roy. Soc. Edinburgh* **46** (1908), 253–281.

[590] F. H. Jackson, Generalization of the differential operative symbol with an extended form of Boole's equation $\Theta(\Theta-1)(\Theta-2)\cdots(\Theta-n+1)=x^n\frac{d^n}{dx^n}$, *Messenger Math.* **38** (1908/1909), 57–61.

[591] F. H. Jackson, q-form of Taylor's theorem, *Messenger Math.* **38** (1908/1909), 62–64.

[592] F. H. Jackson, On q-definite integrals, *Quart. J. Pure Appl. Math.* **41** (1910), 193–203.

[593] F. H. Jackson, q-difference equations, *Amer. J. Math.* **32** (1910), 305–314.

[594] F. H. Jackson, The q-series corresponding to Taylor's series, *Messenger Math.* **39** (1910), 26–28.

[595] F. H. Jackson, Transformations of q-series, *Messenger Math.* **39** (1910), 145–153.

[596] F. H. Jackson, The product of q-hypergeometric functions, *Messenger Math.* **40** (1911), 92–100.

[597] F. H. Jackson, The q-integral analogous to Borel's integral, *Messenger Math.* **47** (1917), 57–64.

[598] F. H. Jackson, Summation of q-hypergeometric series, *Messenger Math.* **50** (1921), 101–112.

[599] F. H. Jackson, Examples of a generalization of Euler's transformation for power series, *Messenger Math.* **57** (1928), 169–187.

[600] F. H. Jackson, Basic integration, *Quart. J. Math. Oxford* (*Ser.* 2) **2** 1951, 1–16.

[601] M. Jackson, On Lerch's transcendant and the basic bilateral hypergeometric series ${}_2\Psi_2$, *J. London Math. Soc.* **25** (1950), 189–196.

[602] C. G. J. Jacobi, Fundamenta Nova Theoriae Functionum Ellipticarum, Regiomonti. Sumptibus fratrum Bornträger; Reprinted in *Gesammelte Werke* **1** (1881), 49–239, Reimer, Berlin; Reprinted by Chelsea Publishing Company, New York, 1969.

[603] C. G. J. Jacobi, De usu legitimo formulae summatoriae Maclaurinianae, *J. Reine Angew. Math.* **12** (1834), 263–272.

[604] L.-C. Jang and T. Kim, On the distribution of the q-Euler polynomials and the q-Genocchi polynomials of higher order, *J. Inequal. Appl.* **2008** (2008), 723615.

[605] V. K. Jain and H. M. Srivastava, Some families of multilinear q-generating functions and combinatorial q-series identities, *J. Math. Anal. Appl.* **192** (1995), 413–418.

[606] W. Janos and H. M. Srivastava, Solution to Problem 2.98 *Univ. Beograd Publ. Elektroteh. Fak. Ser. Mat.* **10** (1999), 121–122.

[607] J. L. W. V. Jensen, Sur la fonction $\zeta(s)$ de Riemann, *C. R. Acad. Sci. Paris* **104** (1887), 1156–1159.

[608] J. L. W. V. Jensen, Sur les fonctions convexes et les inégalités entre les valeurs moyennes, *Acta Math.* **30** (1906), 175–193.

[609] J. L. W. V. Jensen, Gammafunctionernes Theori i elementaer Fremstilling, *Nyd. Tidds. Math.* **II B** (1891), 33–56, 57–72 and 83–85; English translation: An elementary exposition of the theory of the Gamma function, *Ann. Math.* **17** (1915), 124–166.

[610] C.-G. Ji and W.-P. Li, Values of coefficients of cyclotomic polynomials, *Discrete Math.* **308** (2008), 5860–5863.

[611] F. John, Special solutions of certain difference equations, *Acta Math.* **71** (1939), 175–189.

[612] W. W. Johnson, Note on the numerical transcendents S_n and $s_n = S_n - 1$, *Bull. Amer. Math. Soc.* **12** (1905/1906), 477–482.

[613] L. B. W. Jolley, *Summation of Series*, Second edition, Dover Publications, New York, 1961.

[614] C. Jordan, *Calculus of Finite Differences*, Third edition, Chelsea Publishing Company, New York, 1965.

[615] P. F. Jordan, Infinite sums of Psi functions, *Bull. Amer. Math. Soc.* **79** (1973), 681–683.

[616] M. Jung, Y. J. Cho and J. Choi, Euler sums evaluatable from integrals, *Comm. Korean Math. Soc.* **19** (2004), 545–555.

[617] M. Kac, Can one hear the shape of a drum?, *Amer. Math. Monthly* **73** (1966), 1–23.

[618] V. Kac and P. Cheung, *Quantum Calculus*, Springer-Verlag, Berlin, Heidelberg and New York, 2002.

[619] J.-P. Kahane and S. Mandelbrojt, L'équation fonctionnelle de Riemann et la formule sommatoire de Poisson, *Ann. Sci. École Norm. Sup.* **75** (3) (1958), 57–80.

[620] H.-H. Kairies, Zur axiomatischen charakterisierung der Gammafunktion, *J. Reine Angew. Math.* **236** (1969), 103–111.

[621] H.-H. Kairies, Über die logarithmische der Gammafunktion, *Math. Ann.* **184** (1970), 157–162.

[622] H.-H. Kairies and M. E. Muldoon, Some characterizations of q-factorial functions, *Aequationes Math.* **25** (1982), 67–76.

[623] S. Kalla and A. Villalobos, Transformadas de Laplace y Mellin en teoria de numeros, *Math. Notae* **35** (1991), 13–23.

[624] K. Kamano, The multiple Hurwitz Zeta function and a generalization of Lerch's formula, *Tokyo J. Math.* **29** (2006), 61–73.

[625] M. Kaneko, A recurrence formula for the Bernoulli numbers, *Proc. Japan Acad. Ser. A Math. Sci.* **71** (1995), 192–193.

[626] M. Kaneko, N. Kurokawa and M. Wakayama, A variation of Euler's approach to the Riemann Zeta function, *Kyushu J. Math.* **57** (2003), 175–192.

[627] S. Kanemitsu, On some sums involving Farey fractions, *Math. J. Okayama Uni.* **20** (1978), 101–113.

[628] S. Kanemitsu, H. Kumagai and M. Yoshimoto, Sums involving the Hurwitz Zeta function, *Ramanujan J.* **5** (2001), 5–19.

[629] S. Kanemitsu, M. Katsurada and M. Yoshimoto, On the Hurwitz-Lerch Zeta-function, *Aequationes Math.* **59** (2000), 1–19.

[630] S. Kanemitsu, Y. Tanigawa, H. Tsukada and M.Yoshimoto, Contributions to the theory of the Hurwitz Zeta-function, *Hardy-Ramanujan J.* **30** (2007), 31–55.

[631] S. Kanemitsu and H. Tsukada, *Vistas of Special Functions*, Vol. I, World Scientific Publishing Company, Singapore, Hackensack (New Jersey), London and Hong Kong, 2007 (see [232] for Vol. II).

[632] B. K. Karande and N. K. Thakare, On the unification of Bernoulli and Euler polynomials, *Indian J. Pure Appl. Math.* **6** (1975), 98–107.

[633] P. W. Karlsson, E. D. Krupnikov and H. M. Srivastava, Some hypergeometric transformation and reduction formulas involving the Kampé de Fériet functions, *Internat. J. Math. Statist. Sci.* **9** (2000), 211–226.

[634] P. W. Karlsson and H. M. Srivastava, A note on Henrici's triple product theorem, *Proc. Amer. Math. Soc.* **110** (1990), 85–88.

[635] P. W. Karlsson and H. M. Srivastava, Transformations of multiple q-series with quasi-arbitrary terms, *J. Math. Anal. Appl.* **231** (1999), 241–254.

[636] K. Katayama, On Ramanujan's formula for values of Riemann Zeta-function at positive odd integers, *Acta Arith.* **22** (1973), 149–155.

[637] M. Katsurada, Power series with the Riemann Zeta-function in the coefficients, *Proc. Japan Acad. Ser. A Math. Sci.* **72** (1996), 61–63.

[638] M. Katsurada, Rapidly convergent series representations for $\zeta(2n+1)$ and their χ-analogue, *Acta Arith.* **40** (1999), 79–89.

[639] M. Katsurada and K. Matsumoto, Asymptotic expansions of the mean values of Dirichlet L-functions, *Math. Zeischr.* **208** (1991), 23–39.

[640] D. K. Kazarinoff, On Wallis' formula, *Edinburgh Math. Notes* **40** (1956), 19–21.

[641] T. Kazumaki, Asymptotic expansions for a class of zeta-functions, *Ramanujan J.* **24** (2011), 331–343.

[642] J. D. Kečkić and P. M. Vasić, Some inequalities for the Gamma function, *Publ. Inst. Math. (Beograd) (N. S.)* **11** (25) (1971), 107–114.

[643] W. Kemp, On Gamma function inequalities, *Skand. Aktuarietidskr.* **56** (1973), 65–69.

[644] F. K. Kenter, A matrix representation for Euler's constant, γ, *Amer. Math. Monthly* **106** (1999), 452–454.

[645] D. Kershaw, Some extensions of W. Gautschi's inequalties for the Gamma function, *Math. Comput.* **41** (1983), 607–611.

[646] A. Khintchine, Metrische Kettenbruchprobleme, *Compositio Math.* **1** (1934), 361–382.

[647] A. Khintchine, *Continued Fractions*, Third edition, University of Chicago Press, Chicago, 1964.

[648] A. A. Kilbas, H. M. Srivastava and J. J. Trujillo, *Theory and Applications of Fractional Differential Equations*, North-Holland Mathematical Studies, Vol. 204, Elsevier (North-Holland) Science Publishers, Amsterdam, London and New York, 2006.

[649] T. Kim, On a q-analogue of the p-adic log gamma functions and related integrals, *J. Number Theory* **76** (1999), 320–329.

[650] T. Kim, On the analogs of Euler numbers and polynomials associated with p-adic q-integral on $\mathbb{Z}_p$ at $q=-1$, *J. Math. Anal. Appl.* **331** (2007), 779–792.

[651] T. Kim, On p-adic interpolating function for q-Euler numbers and its derivatives, *J. Math. Anal. Appl.* **339** (2008), 598–608.

[652] T. Kim, q-Riemann Zeta function, *Internat. J. Math. Math. Sci.* **2004** (2004), 599–605.

[653] T. Kim, A note on the q-multiple Zeta function, *Adv. Stud. Contemp. Math.* **8** (2004), 111–113.

[654] T. Kim, q-Euler numbers and polynomials associated with p-adic q-integrals, *J. Nonlinear Math. Phys.* **14** (1) (2007), 15–27.

[655] T. Kim, On the q-extension of Euler numbers and Genocchi numbers, *J. Math. Anal. Appl.* **326** (2007), 1458–1465.

[656] T. Kim, On the multiple q-Genocchi and Euler numbers, *Russian J. Math. Phys.* **15** (2008), 481–486.

[657] T. Kim, L.-C. Jang and H. K. Pak, A note on q-Euler numbers and Genocchi numbers, *Proc. Japan Acad. Ser. A Math. Sci.* **77** (2001), 139–141.

[658] T. Kim, Y.-H. Kim and K.-W. Hwang, On the q-extensions of the Bernoulli and Euler numbers. related identities and Lerch zeta function, *Proc. Jangjeon Math. Soc.* **12** (2009), 77–92.

[659] T. Kim and S.-H. Rim, A note on p-adic Carlitz q-Bernoulli numbers, *Bull. Austral. Math. Soc.* **62** (2000), 227–234.

[660] T. Kim and S.-H. Rim, Generalized Carlitz's q-Bernoulli numbers in the p-adic number field, *Adv. Stud. Contemp. Math.* **2** (7) (2000), 9–11.

[661] T. Kim and S.-H. Rim, On the twisted q-Euler numbers and polynomials associated with basic q-l-functions, *J. Math. Anal. Appl.* **336** (2007), 738–744.

[662] Y. Kim, Infinite series associated with Psi and Zeta functions, *Honam Math. J.* **22** (2000), 53–60.

[663] Y.-H. Kim, W.-J. Kim and L.-C. Jang, On the q-extension of Apostol-Euler numbers and polynomials, *Abstr. Appl. Anal.* **2008** (2008), 296159.

[664] C. H. Kimberling, Two-dimensional complete monotonicity with diagonalization, *Amer. Math. Monthly* **80** (1973), 789–791.

[665] C. H. Kimberling, A probabilistic interpretation of complete monotonicity, *Aequationes Math.* **10** (1974), 152–164.

[666] V. H. Kinkelin, Über eine mit der Gamma Funktion verwandte Transcendente und deren Anwendung auf die Integralrechnung, *J. Reine Angew. Math.* **57** (1860), 122–158.

[667] A. N. Kirillov, Dilogarithm identities, *Progr. Theoret. Phys. Suppl.* **118** (1995), 61–142.

[668] G. Klambauer, *Mathematical Analysis*, Marcel Dekker, New York, 1975.

[669] M. S. Klamkin, A summation problem; Advanced Problem, 4431 *Amer. Math. Monthly* **58** (1951), 195; *ibid.* **59** (1952), 471–472.

[670] M. S. Klamkin, Another summation, *Amer. Math. Monthly* **62** (1955), 129–130.

[671] D. Klusch, Asymptotic equalities for the Lipschitz-Lerch Zeta-function, *Arch. Math. (Basel)* **49** (1987), 38–43.

[672] D. Klusch, A hybrid version of a theorem of Atkinson, *Rev. Roumaine Math. Pures Appl.* **34** (1989), 721–728.

[673] D. Klusch, On the Taylor expansion of the Lerch Zeta-Function, *J. Math. Anal. Appl.* **170** (1992), 513–523.

[674] J. C. Kluyver, On certain series of Mr. Hardy, *Quart. J. Pure Appl. Math.* **50** (1927), 185–192.

[675] K. Knopp, *Funktionentheorie*, Vol. II, Fifth edition, Sammlung Göschen 703, Walter de Gruyter, Berlin, 1941; English translation by F. Bagemihl: *Theory of Functions*, Dover Publications, New York, 1947.

[676] K. Knopp, *Theory and Application of Infinite Series*, Second English edition (Translated from the Second German edition and revised in accordance with the Fourth German edition by R. C. H. Young), Hafner Publishing Company, New York, 1951.

[677] M. I. Knopp, *Modular Functions in Analytic Number Theory*, Markham Publishing Company, Chicago, 1970.

[678] D. E. Knuth, Euler's constant to 1271 places *Math. Comput.* **16** (1962), 275–281.

[679] D. E. Knuth, Evaluation of Porter's constant, *Comput. Math. Appl.* **2** (2) (1976), 137–139.

[680] D. E. Knuth, Two notes on notation, *Amer. Math. Monthly* **99** (1992), 403–422.

[681] N. Koblitz, *p-Adic Numbers, p-Adic Analysis, and Zeta-Functions*, Graduate Texts in Mathematics, Vol. 58, Springer-Verlag, New York, Berlin and Heidelberg, 1977.

[682] N. Koblitz, On Carlitz's q-Bernoulli numbers, *J. Number Theory* **14** (1982), 332–339.

[683] W. Koepf, *Hypergeometric Summation*, Vieweg ünd Sohn Verlagsgesell-schaft mbH, Braunschweig and Wiesbaden, 1998.

[684] K. S. Kölbig, Closed expressions for $\int_0^1 t^{-1}\log^{n-1} t \log^p(1-t)\,dt$, *Math. Comput.* **39** (1982), 647–654.

[685] K. S. Kölbig, On the integral $\int_0^{\frac{\pi}{2}} \log^n \cos x \log^p \sin x\,dx$, *Math. Comput.* **40** (1983), 565–570.

[686] K. S. Kölbig, Nielsen's generalized Polylogarithms, *SIAM J. Math. Anal.* **17** (1986), 1232–1258.

[687] K. S. Kölbig, The Polygamma function and the derivatives of the cotangent function for rational arguments, CERN-IT-Reports, CERN-CN-96-005, 1996.

[688] K. S. Kölbig, The Polygamma function $\psi^{(k)}(x)$ for $x = 1/4$ and $x = 3/4$, *J. Comput. Appl. Math.* **75** (1996), 43–46.

[689] Y. Komatu, Elementary inequalities for Mills' ratio, *Res. Statist. Appl. Res. Un. Japan. Sci. Engrs.* **4** (1955), 69–70.

[690] Y. Komori, K. Matsumoto and H. Tsumura, Zeta and L-functions and Bernoulli polynomials of root systems, *Proc. Japan Acad. Ser. A Math. Sci.* **84** (2008), 57–62.

[691] Y. Komori, K. Matsumoto and H. Tsumura, On multiple Bernoulli polynomials and multiple L-functions of root systems, *Proc. London Math. Soc.* **100** (3) (2010), 303–347.

[692] Y. Komori, K. Matsumoto and H. Tsumura, Functional equations and functional relations for the Euler double Zeta-function and its generalization of Eisenstein type, *Publ. Math. Debrecen* **77** (2010), 15–31.

[693] S. Kondo, *Value of Euler Constant* [http://ja0hxv.calico.jp/pai/egamma.html.]

[694] T. H. Koornwinder, Jacobi functions as limit cases of q-ultraspherical polynomials, *J. Math. Anal. Appl.* **148** (1990), 44–54.

[695] N. S. Koshlliakov, Application of the theory of sum-formulae to the investigation of a class of one-valued analytical functions in the theory of numbers, *Messenger Math.* **58** (1928/1929), 1–23.

[696] S. Koumandos, On Ruijsenaars' asymptotic expansion of the logarithm of the double Gamma function, *J. Math. Anal. Appl.* **341** (2008), 1125–1132.

[697] S.-Y. Koyama and N. Kurokawa, Values of the double sine function, *J. Number Theory* **123** (2007), 204–223.

[698] S.-Y. Koyama and N. Kurokawa, Multiple Eisenstein series and multiple cotangent functions, *J. Number Theory* **128** (2008), 1769–1774.

[699] C. Krattenthaler and H. M. Srivastava, Summations for basic hypergeometric series involving a q-analogue of the Digamma function, *Comput. Math. Appl.* **32** (3) (1996), 73–91.

[700] L. Kronecker, Bemerkungen über die darstellung von reihen durch integrale, J. Reine Angew. Math. **105** (1889), 157–159.

[701] W. Krull, Bemerkungen zur differenzengleichung $g(x+1) - g(x) = \varphi(x)$, *Math. Nachr.* **1** (1948), 365–376.

[702] W. Krull, Bemerkungen zur differenzengleichung $g(x+1) - g(x) = \varphi(x)$. II *Math. Nachr.* **2** (1949), 251–262.

[703] T. Kubota, *Elementary Theory of Eisenstein Series*, Kodansha Limited, Tokyo; John Wiley and Sons, New York, London, Sydney and Toronto, 1973.

[704] T. Kubota and H. W. Leopoldt, Eine p-adische theorie der zetawerte, *J. Reine Angew. Math.* **214/215** (1964), 328–339.

[705] M. Kuczma, *Functional Equations in a Single Variable*, PWN-Polish Scientific Publishers, Warszawa, 1968.

[706] H. Kumagai, The determinant of the Laplacian on the n-sphere, *Acta Arith.* **91** (1999), 199–208.

[707] E. E. Kummer, Über die hypergeometrische Reihe, *J. Reine Angew. Math.* **15** (1836), 39–83 and 127–172.

[708] E. E. Kummer, Über die transcendenten, welche aus wiederholten integrationen rationaler formeln entstehen, *J. Reine Angew. Math.* **21** (1840), 74–90.

[709] E. E. Kummer, Beitrag zur theorie der funktion $\Gamma(x) = \int_0^\infty e^{-v} v^{x-1}\, dv$, *J. Reine Angew. Math.* **35** (1847), 1–4.

[710] N. Kurokawa, Limit values of Eisenstein series and multiple cotangent functions, *J. Number Theory* **128** (2008), 1775–1783.

[711] N. Kurokawa, Multiple sine functions and Selberg Zeta functions, *Proc. Japan Acad. Ser. A Math. Sci.* **67** (1991), 61–64.

[712] N. Kurokawa, Gamma factors and Plancherel measures, *Proc. Japan Acad. Ser. A Math. Sci.* **68** (1992), 256–260.

[713] N. Kurokawa, Multiple zeta functions: An example, in *Zeta Functions in Geometry* (Tokyo, 1990), pp. 219–226, *Adv. Stud. Pure Math.* **21**, Kinokuniya, Tokyo, 1992.

[714] N. Kurokawa and S.-Y. Koyama, Multiple sine functions, *Forum Math.* **15** (2003), 839–876.

[715] N. Kurokawa, K. Mimachi and M. Wakayama, Jackson's integral of the Hurwitz Zeta function, *Rend. Circ. Mat. Palermo* **56** (2007), 43–56.

[716] N. Kurokawa, H. Ochiai and M. Wakayama, Multiple trigonometry and Zeta functions, *J. Ramanujan Math. Soc.* **17** (2002), 101–113.

[717] N. Kurokawa and M. Wakayama, On $\zeta(3)$, *J. Ramanujan Math. Soc.* **16** (2001), 205–214.

[718] N. Kurokawa and M. Wakayama, Absolute tensor products, *Internat. Math. Res. Notices* **2005** (2004), 249–260.

[719] N. Kurokawa and M. Wakayama, A q-logarithmic analogue of Euler's sine integral, *Rend. Sem. Mat. Univ. Padova* **114** (2005), 51–62.

[720] B. Kurt, A further symmetric relation on the analogue of the Apostol-Bernoulli and the analogue of the Apostol-Genocchi polynomials, *Appl. Math. Sci. (Ruse)* **3** (2009), 2757–2764.

[721] V. Kurt, The multiplication formulae for the Genocchi polynomials, *Proc. Jangjeon Math. Soc.* **13** (2010), 89–96.

[722] V. Kurt and M. Cenkci, A new approach to q-Genocchi numbers and polynomials, *Bull. Korean Math. Soc.* **47** (2010), 575–583.

[723] G. Labelle and L. Laforest, Combinatorial variations on multidimensional quadtrees, *J. Combin. Theory Ser. A* **69** (1995), 1–16.

[724] A. Laforgia, Further inequalities for the Gamma function, *Math. Comput.* **42** (1984), 597–600.

[725] A. Laforgia and P. Natalini, Turán-type inequalities for some special functions, *J. Inequal. Pure Appl. Math.* **7** (1) (2006) http://jipam.vu.edu.au/article.php?sid=638].

[726] B. Lahcene, On extended generalized Gamma function, in *First Conference on Mathematical Sciences*, 406–417, *Proceedings of the conference held at Zarqu Private University* (Zarka; April 18–20), 2006.

[727] E. Lammel, Ein Beweis, dass die Riemannsche Zetafinktion $\zeta(s)$ in $|s-1| \leqq 1$ keine Nullstelle besitzt, *Univ. Nac. Tucumán Rev. Ser. A* **16** (1966), 209–217.

[728] Y. Lan, A limit formula for $\zeta(2k+1)$, *J. Number Theory* **78** (1999), 271–286.

[729] E. Landau, *Handbuch der Lehre von der Verteilung der Primzahlen*, Second edition, Chelsea Publishing Company, New York, 1953.

[730] L. J. Lange, An elegant continued fraction for π, *Amer. Math. Monthly* **106** (1999), 456–458.

[731] J.-L. Lavoie, F. Grondin, A. K. Rathie and K. Arora, Generalizations of Dixon's theorem on the sum of a ${}_3F_2$, *Math. Comput.* **62** (1994), 267–276.

[732] J.-L. Lavoie, F. Grondin and A. K. Rathie, Generalizations of Whipple's theorem on the sum of a ${}_3F_2$, *J. Comput. Appl. Math.* **72** (1996), 293–300.

[733] I. B. Lazarević and A. Lupaş, Functional equations for Wallis and Gamma functions, *Univ. Beograd. Publ. Elektrotehn. Fak. Ser. Mat. Fiz.* **461–497** (1974), 245–251.

[734] T. Q. T. Le and J. Murakami, Kontsevich's integral for the Homfly polynomial and relations between values of multiple Zeta functions, *Topology Appl.* **62** (1995), 193–206.

[735] H. Lee, B. M. Ok and J. Choi, Notes on some identities involving the Riemann Zeta function, *Comm. Korean Math. Soc.* **17** (2002), 165–173.

[736] J. Lee, Integrals of Bernoulli polynomials and series of Zeta function, *Comm. Korean Math. Soc.* **14** (1999), 707–716.

[737] A. M. Legendre, *Traité des Fonctions Elliptiques et des Intégrales Euleŕiennes*, Vol. 2, Huzard-Courcier, Paris, 1826.

[738] D. H. Lehmer, Factorization of certain cyclotomic functions, *Ann. of Math.* **34** (2) (1933), 461–479.

[739] D. H. Lehmer, Note on an absolute constant of Khintchine, *Amer. Math. Monthly* **46** (1939), 148–152.

[740] D. H. Lehmer, On the maxima and minima of Bernoulli polynomials, *Amer. Math. Monthly* **47** (1940), 533–538.

[741] D. H. Lehmer, Euler constants for arithmetical progressions, *Acta Arith.* **27** (1975), 125–142.

[742] D. H. Lehmer, Interesting series involving the central binomial coefficient, *Amer. Math. Monthly* **92** (1985), 449–457.

[743] J. Lehner and M. Newman, Sums involving Farey fractions, *Acta Arith.* **15** (1969), 181–187.

[744] A. K. Lenstra, H. W. Lenstra Jr. and L. Lovász, Factoring polynomials with rational coefficients, *Math. Ann.* **261** (1982), 515–534.

[745] M. Lerch, Note sur la fonction $\mathcal{R}(w,x,s) = \Sigma_{n=0}^{\infty}(e^{2\pi inx}/(n+w)^s)$, *Acta Math.* **11** (1887), 19–24.

[746] M. Lerch, Démonstration élémentaire de la formule, *Enseign. Math.* **5** (1903), 450–453.

[747] D. Leshchiner, Some new identities for $\zeta(k)$, *J. Number Theory* **13** (1981), 355–362.

[748] W. J. LeVeque, *Topics in Number Theory*, Vol. II Addison-Wesley Publishing Company, Reading (Massachusetts) and London, 1956.

[749] S. Levit and U. Smilansky, A theorem on infinite products of eigenvalues of Sturm-Liouville type operators, *Proc. Amer. Math. Soc.* **65** (1977), 299–302.

[750] P. Lévy, Sur le développement en fraction continue d'un nombre choisi au hasard, *Compositio Math.* **3** (1936), 286–303.

[751] L. Lewin, *Dilogarithms and Associated Functions*, Macdonald, London, 1958.

[752] L. Lewin, *Polylogarithms and Associated Functions*, Elsevier (North-Holland), New York, London and Amsterdam, 1981.

[753] L. Lewin, The dilogarithm in algebraic fields, *J. Austral. Math. Soc. Ser. A* **33** (1982), 302–330.

[754] L. Lewin, The inner structure of the dilogarithm in algebraic fields, *J. Number Theory* **19** (1984), 345–373.
[755] L. Lewin, The order-independence of the polylogarithmic ladder structure-implications for a new category of functional equations, *Aequationes Math.* **30** (1986), 1–20.
[756] L. Lewin (Editor), *Structural Properties of Polylogarithms*, Mathematical Surveys and Monographs, Vol. 37, American Mathematical Society, Providence, Rhode Island, 1991.
[757] L. Lewin and E. Rost, Polylogarithmic functional equations: A new category of results developed with the help of computer algebra (MACSYMA), *Aequationes Math.* **31** (1986), 223–242.
[758] H.-L. Li, M. Hashimoto and S. Kanemitsu, Examples of the Hurwitz transform, *J. Math. Soc. Japan* **61** (2009), 651–660.
[759] H.-L. Li, S. Kanemitsu and H. Tsukada, Modular relation interpretation of the series involving the Riemann Zeta values, *Proc. Japan Acad. Ser. A Math. Sci.* **84** (2008), 154–158.
[760] N. Y. Li and T. Mansour, An identity involving Narayana numbers, *European J. Combin.* **29** (2008), 672–675.
[761] J. J. Y. Liang and J. Todd, The Stieltjes constants, *J. Res. Nat. Bur. Standards Sect. B Math. Sci.* **76** (1972), 161–178.
[762] P. Lichtenbaum, Über die $\chi(s) = \Sigma_{k=0}^{\infty}(-1)^k(2k+1)^{-s}$, *Math. Zeitschr.* **33** (1931), 641–647.
[763] S. Lichtenbaum, Values of zeta-functions at non-negative integers, in Number Theory, *Lecture Notes in Math.* **1068**, Springer-Verlag, Berlin, Heidelberg and New York, 1984.
[764] S.-D. Lin, Y.-S. Chao and H. M. Srivastava, Some expansions of the exponential integral in series of the incomplete Gamma function, *Appl. Math. Lett.* **18** (2005), 513–520.
[765] S.-D. Lin and H. M. Srivastava, Some families of the Hurwitz-Lerch Zeta functions and associated fractional derivative and other integral representations, *Appl. Math. Comput.* **154** (2004), 725–733.
[766] S.-D. Lin and H. M. Srivastava, Some closed-form evaluations of multiple hypergeometric and q-hypergeometric series, *Acta Appl. Math.* **86** (2005), 309–327.
[767] S.-D. Lin, H. M. Srivastava and P.-Y. Wang, Some expansion formulas for a class of generalized Hurwitz-Lerch Zeta functions, *Integral Transforms Spec. Funct.* **17** (2006), 817–827.
[768] S.-D. Lin, H. M. Srivastava and S.-T. Tu, Basic (or q-) extensions of some multiple infinite sums derived by means of fractional calculus, *J. Fract. Calc.* **27** (2005), 43–57.
[769] E. Lindelöf, *Le Calcul des Résidus*, Gauthier-Villars, Paris, 1905.
[770] R. Lipschitz, Untersuchung einer aus vier Elementen gebildeten Reihe, *J. Reine Angew. Math.* **54** (1857), 313–328.
[771] R. Lipschitz, Untersuchung der Eigenschaften einer Gattung von unendlichen Reihen, *J. Reine Angew. Math.* **105** (1889), 127–156.
[772] G.-D. Liu and H. M. Srivastava, Explicit formulas for the Nörlund polynomials $B_n^{(x)}$ and $b_n^{(x)}$, *Comput. Math. Appl.* **51** (2006), 1377–1384.
[773] H.-M. Liu and W.-P. Wang, Some identities on the Bernoulli, Euler and Genocchi polynomials via power sums and alternate power sums, *Discrete Math.* **309** (2009), 3346–3363.
[774] L. Lorch, Inequalities for ultraspherical polynomials and the Gamma function, *J. Approx. Theory* **40** (1984), 115–120.
[775] L. Lorch, M. E. Muldoon and P. Szegö, Higher monotonicity properties of certain Sturm-Liouville functions, III *Canad. J. Math.* **22** (1970), 1238–1265.

[776] J. H. Loxton, Special values of the Dilogarithm function, *Acta Arith.* **43** (1984), 155–166.

[777] D.-Q. Lu, Some properties of Bernoulli polynomials and their generalizations, *Appl. Math. Lett.* **24** (2011), 746–751.

[778] L. G. Lucht, Mittelwertungleichungen für Lösungen gewisser differenzengleichungen, *Aequationes Math.* **39** (1990), 204–209.

[779] Y. L. Luke, *The Special Functions and Their Approximations*, Vols. I and II, Academic Press, New York and London, 1969.

[780] Q.-M. Luo, On the Apostol-Bernoulli polynomials, *Central European J. Math.* **2** (2004), 509–515.

[781] Q.-M. Luo, Apostol-Euler polynomials of higher order and Gaussian hypergeometric functions, *Taiwanese J. Math.* **10** (2006), 917–925.

[782] Q.-M. Luo, Fourier expansions and integral representations for the Apostol-Bernoulli and Apostol-Euler polynomials, *Math. Comput.* **78** (2009), 2193–2208.

[783] Q.-M. Luo, The multiplication formulas for the Apostol-Bernoulli and Apostol-Euler polynomials of higher order, *Integral Transforms Spec. Funct.* **20** (2009), 377–391.

[784] Q.-M. Luo, Fourier expansions and integral representations for the Genocchi polynomials, *J. Integer Seq.* **12** (2009), 09.1.4.

[785] Q.-M. Luo, q-Extensions for the Apostol-Genocchi polynomials, *Gen. Math.* **17** (2009), 113–125.

[786] Q.-M. Luo, Extensions of the Genocchi polynomials and their Fourier expansions and integral representations, *Osaka J. Math.* **48** (2011), 291–309.

[787] Q.-M. Luo, B.-N. Guo, F. Qi and L. Debnath, Generalizations of Bernoulli numbers and polynomials, *Internat. J. Math. Math. Sci.* **2003** (59) (2003), 3769–3776.

[788] Q.-M. Luo and H. M. Srivastava, Some generalizations of the Apostol-Bernoulli and Apostol-Euler polynomials, *J. Math. Anal. Appl.* **308** (2005), 290–302.

[789] Q.-M. Luo and H. M. Srivastava, Some relationships between the the Apostol-Bernoulli and Apostol-Euler polynomials, *Comput. Math. Appl.* **51** (2006), 631–642.

[790] Q.-M. Luo and H. M. Srivastava, q-Extensions of some relationships between the Bernoulli and Euler polynomials, *Taiwanese J. Math.* **15** (2011), 241–257.

[791] Q.-M. Luo and H. M. Srivastava, Some generalizations of the Apostol-Genocchi polynomials and the Stirling numbers of the second kind, *Appl. Math. Comput.* **217** (2011), 5702–5728.

[792] P. Luschny, Is the Gamma function misdefined? Or: Hadamard versus Euler-who found the better gamma function? [http://www.luschny.de/math/factorial/hadamard/HadamardsGammaFunction.html].

[793] I. G. Macdonald, *Symmetric Functions and Hall Polynomials*, Clarendon (Oxford University) Press, Oxford, London and New York, 1979.

[794] P. A. MacMahon, *Combinatorial Analysis*, Vol. I, Chelsea Publishing Company, New York, 1915.

[795] W. Magnus, F. Oberhettinger and R. P. Soni, *Formulas and Theorems for the Special Functions of Mathematical Physics*, Third Enlarged edition, Die Grundlehren der Mathematischen Wissenschaften in Einzeldarstellungen mit besonderer Berücksichtingung der Anwendungsgebiete, Bd. **52**, Springer-Verlag, Berlin, Heidelberg and New York, 1966.

[796] K. Mahler, An application of Jensen's formula to polynomials, *Mathematika* **7** (1960), 98–100.

[797] L. Maligranda, J. E. Pečarić and L. E. Persson, Stolarsky's inequality with general weights, *Proc. Amer. Math. Soc.* **123** (1995), 2113–2118.

[798] C. J. Malmstén, Sur la formule $hu_x = \Delta u_x - \frac{h}{2}\Delta u_x + \frac{B_1 h^2}{1\cdot 2}\cdot\Delta u_x'' - \frac{B_2 h^4}{1\cdots 4}\cdot\Delta u_x^{IV} +$ etc., *J. Reine Angew. Math.* **35** (1847), 55–82.

[799] Y. Manin, Lectures on Zeta functions and motives, *Astérisque* **228** (1995), 121–163.

[800] H. B. Mann, An inequality suggested by the theory of statistical inference, *Illinois J. Math.* **6** (1962), 131–136.

[801] C. Markett, Triple sums and the Riemann Zeta function, *J. Number Theory* **48** (1994), 113–132.

[802] L. Mascheroni, *Adnotationes ad calculum integralem Euleri*, Vols. 1 and 2, Ticino, Italy, 1790 and 1792. Reprinted in L. Euler, *Leonhardi Euleri Opera Omnia*, Ser. 1, Vol. 12, pp. 415–542, B. G. Teubner, Leipzig, 1915.

[803] É. L. Mathieu, *Traité de Physique Mathématique.* VI–VII: *Theory de l'Elasticité des Corps Solides* (Part 2), Gauthier-Villars, Paris, 1890.

[804] K. Matsumoto, Asymptotic series for double Zeta, double Gamma, and Hecke L-functions, *Math. Proc. Cambridge Philos. Soc.* **123** (1998), 385–405 (see also [805], [807] and [1005]).

[805] K. Matsumoto, Corrigendum and addendum to "Asymptotic series for double Zeta, double Gamma, and Hecke L-functions," *Math. Proc. Cambridge Philos. Soc.* **132** (2002), 377–384 (see also [804], [807] and [1005]).

[806] K. Matsumoto, The analytic continuation and the asymptotic behaviour of certain multiple zeta functions. I, *J. Number Theory* **101** (2003), 223–243.

[807] K. Matsumoto, Asymptotic expansions of double Gamma-functions and related remarks, in *Analytic Number Theory* (C. Jia and K. Matsumoto, Editors), pp. 243–268, Series on Developments in Mathematics, Vol. 6, Kluwer Academic Publishers, Dordrecht, Boston and London, 2002 (see also [804], [805] and [1005]).

[808] K. Matsumoto, Functional equations for double Zeta-functions, *Math. Proc. Cambridge Philos. Soc.* **136** (2004), 1–7.

[809] K. Matsumoto, The analytic continuation and the asymptotic behaviour of certain multiple Zeta-functions. III, *Comment. Math. Univ. St. Paul.* **54** (2005), 163–186.

[810] K. Matsumoto, The analytic theory of multiple Zeta-functions and its applications, *Sugaku Expositions* **23** (2010), 143–167.

[811] K. Matsumoto and L. Weng, Zeta-functions defined by two polynomials, in *Number Theoretic Methods: Future Trends* (S. Kanemitsu and C. Jia, Editors), pp. 233–262, Series on Developments in Mathematics, Vol. 8, Kluwer Academic Publishers, Dordrecht, Boston and London, 2002.

[812] Y. Matsuoka, Asymptotic formula for Vallée Poussin's singular integrals, *Sci. Rep. Kagoshima Univ.* **9** (1960), 25–34.

[813] Y. Matsuoka, On the values of a certain Dirichlet series at rational integers, *Tokyo J. Math.* **5** (1982), 399–403.

[814] A. E. Mayer, Konvexe lösung der funktionalgleichung, *Acta Math.* **70** (1939), 57–62.

[815] H. P. McKean, Selberg's trace formula as applied to a compact Riemann surface, *Comm. Pure Appl. Math.* **25** (1972), 225–246.

[816] H. P. McKean and I. M. Singer, Curvature and the Eigenvalues of the Laplacian, *J. Differential Geom.* **1** (1967), 43–69.

[817] Hj. Mellin, Über die transcendente function $Q(x) = \Gamma(x) - P(x)$, *Acta Math.* **2** (1883), 261–295.

[818] Hj. Mellin, Eine verallgemeinerung der gleichung $\Gamma(1+x)\Gamma(1-x) = \pi x/\sin\pi x$, *Acta. Math.* **3** (1883/1984), 102–104.

[819] Hj. Mellin, Zur theorie der Gammafunktion, *Acta Math.* **8** (1886), 37–80.

[820] Hj. Mellin, Om en ny klass af transcendenta funktioner, hvilka äro nära beslägtade med Gammafunktionen, *Acta Soc. Sci. Fenn.* **14** (1886), 355–385.

[821] Z. A. Melzak, *Companion to Concrete Mathematics*, Vol. I: *Mathematical Techniques and Various Applications*, John Wiley and Sons, New York, London, Sydney and Toronto, 1973.

[822] Z. A. Melzak, *Companion to Concrete Mathematics*, Vol. II: *Mathematical Ideas, Modelling and Applications*, John Wiley and Sons, New York, London, Sydney and Toronto, 1976.

[823] P. K. Menon, Some series involving the Zeta function, *Math. Student* **29** (1961), 77–80.

[824] M. Merkle, On log-convexity of a ratio of Gamma functions, *Univ. Beograd. Publ. Elektrotehn. Fak. Ser. Mat.* **8** (1997), 114–119.

[825] M. Mihoubi, Bell polynomials and binomial type sequences, *Discrete Math.* **308** (2008), 2450–2459.

[826] M. Mikolás, Farey series and their connection with the prime number problem, I *Acta Sci. Math. (Szeged)* **13** (1950), 93–117.

[827] M. Mikolás, Integral formulae of arithmetical characteristics relating to the Zeta-function of Hurwitz, *Publ. Math. Debrecen* **5** (1957), 44–53.

[828] A. R. Miller, Summations for certain series containing the Digamma function, *J. Phys. A: Math. Gen.* **39** (2006), 3011–3020.

[829] J. Miller and V. S. Adamchik, Derivatives of the Hurwitz Zeta function for rational arguments, *J. Comput. Appl. Math.* **100** (1998), 201–206.

[830] J. Milnor, On polylogarithms, Hurwitz zeta functions, and the Kubert identities, *Enseign. Math.* **29** (1983), 281–322.

[831] S. Minakshisundaram, A generalization of Epstein Zeta functions. *Canad. J. Math.* **1** (1949), 320–327.

[832] S. Minakshisundaram and A. Pleijel, Some properties of the eigenfunctions of the Laplace-operator on Riemann manifolds, *Canad. J. Math.* **1** (1949), 242–256.

[833] H. Minc, Upper bounds for permanents of (0, 1)-matrices, *Bull. Amer. Math. Soc.* **69** (1963), 789–791.

[834] H. Minc and L. Sathre, Some inequalities involving $(r!)^{1/r}$, *Proc. Edinburgh Math. Soc. (Ser. 2)* **14** (1964/1965), 41–46.

[835] K. Mitchell, Tables of the function $\int_0^z \frac{-\log|1-y|}{y}\,dy$, with an account of some properties of this and related functions, *Philos. Mag.* **40** (1949), 351–368.

[836] D. S. Moak, The q-Gamma function for $q > 1$, *Aequationes Math.* **20** (1980), 278–285.

[837] D. S. Moak, The q-Gamma function for $x < 0$, *Aequationes Math.* **21** (1980), 179–191.

[838] D. S. Moak, The q-analogue of Stirling formula, *Rocky Mountain J. Math.* **14** (1984), 403–414.

[839] G. Monegato and A. Strozzi, The numerical evaluation of two integral transforms, *J. Comput. Appl. Math.* **211** (2008), 173–180.

[840] G. F. Montgomery, On the transmission error function for Meteor-Burst communication, *Proc. Inst. Radio Engrs.* **46** (1958), 1423–1424.

[841] L. J. Mordell, Some applications of Fourier series in the analytic theory of numbers, *Proc. Cambridge Philos. Soc.* **24** (1928), 585–596.

[842] L. J. Mordell, Poisson's summation formula and the Riemann zeta function, *J. London Math. Soc.* **4** (1929), 285–291.

[843] L. J. Mordell, On the evaluation of some multiple series, *J. London Math. Soc.* **33** (1958), 368–371.

[844] L. J. Mordell, On the evaluation of some multiple series, *J. London Math. Soc.* **33** (1958), 371–375.

[845] Y. Morita, A p-adic analogue of the Γ-function, *J. Fac. Sci. Univ. Tokyo Sect. 1 A* **22** (1975), 255–256.

[846] C. Mortici, Improved convergence towards generalized Euler-Mascheroni constant, *Appl. Math. Comput.* **215** (2010), 3443–3448.

[847] C. Mortici, Best estimates of the generalized Stirling formula, *Appl. Math. Comput.* **215** (2010), 4044–4048.

[848] C. Mortici, Estimating the Digamma and Trigamma functions by completely monotonicity arguments, *Appl. Math. Comput.* **217** (2010), 4081–4085.

[849] Y. Motohashi, A note on the mean value of the zeta and L-functions. I, *Proc. Japan Acad. Ser. A Math. Sci.* **61** (1985), 222–224.

[850] Y. Motohashi, *Spectral Theory of the Riemann Zeta-Function*, Cambridge University Press, Cambridge, New York and Melbourne, 1997.

[851] M. E. Muldoon, Some characterizations of the Gamma function involving the notion of complete monotonicity, *Aequationes Math.* **8** (1972), 212–215.

[852] M. E. Muldoon, Some monotonicity properties and characterizations of the Gamma function, *Aequationes Math.* **18** (1978), 54–63.

[853] M. R. Murty and N. Saradha, Transcendental values of the digamma function, *J. Number Theory* **125** (2007), 298–318.

[854] T. S. Nanjundiah, On a formula of A. C. Dixon, *Proc. Amer. Math. Soc.* **9** (1958), 308–311.

[855] C. Nash and D. O'Connor, Ray-Singer torsion, topological field theories and the Riemann Zeta function at $s = 3$, in *Low-Dimensional Topology and Quantum Field Theory, Proceedings of a NATO Advanced Research Workshop* held at the Isaac Newton Institute at Cambridge, U.K.; September 6–12, 1992) (H. Osborn, Editor), pp. 279–288, Plenum Press, New York and London, 1993.

[856] C. Nash and D. O'Connor, Determinants of Laplacians, the Ray-Singer torsion on lens spaces and the Riemann Zeta function, *J. Math. Phys.* **36** (1995), 1462–1505.

[857] I. P. Natanson, *Constructive Function Theory*. Vol. 1: *Uniform Approximation* (Translated from the Russian by Alexis N. Obolensky), Frederick Ungar Publishing Company, New York, 1964.

[858] E. H. Neville, A trigonometrical inequality, *Proc. Cambridge Philos. Soc.* **47** (1951), 629–632.

[859] T. A. Newton, Derivation of a factorial function by method of analogy, *Amer. Math. Monthly* **68** (1961), 917–920.

[860] N. Nielsen, *Handbuch der Theorie der Gammafunktion*, Druck und Verlag von B. G. Teubner, Leipzig, 1906.

[861] N. Nielsen, *Traité Élémentaire des Nombres de Bernoulli*, Gauthier-Villars, Paris, 1923.

[862] N. Nielsen, *Die Gammafunktion*, Chelsea Publishing Company, Bronx, New York, 1965.

[863] K. Nishimoto and H. M. Srivastava, Certain classes of infinite series summable by means of fractional calculus, *J. College Engrg. Nihon Univ. Ser. B* **30** (1989), 97–106.

[864] K. Nishimoto, C.-E Yen and M.-L. Lin, Some integral forms for a generalized Zeta function, *J. Fract. Calc.* **22** (2002), 91–97.

[865] M. Nishizawa, On a q-analogue of the multiple Gamma functions, *Lett. Math. Phys.* **37** (1996), 201–209.

[866] M. Nishizawa, Multiple Gamma function, ite q- and elliptic analogue, *Rocky Mountain J. Math.* **32** (2002), 793–811.

[867] M. Nishizawa, Infinite product representations for multiple Gamma function, arXiv:math/0404077v1 [math.CA], Preprint 2004.

[868] N. E. Nörlund, Mémoire sur les ploynômes de Bernoulli, *Acta Math.* **43** (1920), 121–196.

[869] N. E. Nörlund, *Vorlesungen über Differentzenrechnung*, Springer-Verlag, Berlin, 1924; Reprinted by Chelsea Publishing Company, Bronx, New York, 1954.

[870] M. A. Nyblom, On the evaluation of a definite integral involving nested square root functions, *Rocky Mountain J. Math.* **37** (2007), 1301–1304.

[871] F. Oberhettinger, Note on the Lerch Zeta function, *Pacific J. Math.* **6** (1956), 117–120.

[872] Y. Ohno, A generalization of the duality and sum formulas on the multiple zeta values, *J. Number Theory* **74** (1999), 39–43.

[873] Y. Ohno and J.-I. Okuda, On the sum formula for the q-analogue of non-strict multiple zeta values, *Proc. Amer. Math. Soc.* **135** (2007), 3029–3037.

[874] P. K. Ojha and I. N. Singh, A discussion on two methods for finding the value of the Riemann Zeta function $\zeta(s)$ when $s = 2$, *Math. Ed. (Siwan)* **33** (1999), 24–28.

[875] J.-I. Okuda and Y. Takeyama, On relations for the multiple q-zeta values, *Ramanujan J.* **14** (2007), 379–387.

[876] J.-I. Okuda and K. Ueno, Relations for multiple zeta values and Mellin transforms of multiple polylogarithms, *Publ. RIMS Kyoto Univ.* **40** (2004), 537–564.

[877] K. B. Oldham, J. Myland and J. Spanier, *An Atlas of Functions. With Equator, the Atlas Function Calculator*, Second edition [with 1 CD-ROM (Windows)], Springer, New York, 2009.

[878] I. Olkin, An inequality satisfied by the Gamma function, *Skand. Aktuarietidskr.* **42** (1959), 37–39.

[879] F. W. J. Olver, *Asymptotics and Special Functions*, Series on Computer Science and Applied Mathematics, Academic Press, New York and London, 1974.

[880] F. W. J. Olver, D. W. Lozier, R. F. Boisvert and C. W. Clark (Editors), *NIST Handbook of Mathematical Functions* [With 1 CD-ROM (Windows, Macintosh and UNIX)], U.S. Department of Commerce, National Institute of Standards and Technology, Washington, D. C., 2010; Cambridge University Press, Cambridge, London and New York, 2010 (see also [3]).

[881] B. Osgood, R. Phillips and P. Sarnak, Extremals of determinants of Laplacians, *J. Funct. Anal.* **80** (1988), 148–211.

[882] S. Ostrovska, q-Bernstein polynomials of the Cauchy kernel, *Appl. Math. Comput.* **198** (2008), 261–270.

[883] A. M. Ostrowski, On the zeros of Bernoulli polynomials of even order, *Enseign. Math.* **6** (1960), 27–47.

[884] A. Ostrowski, Note on Poisson's treatment of the Euler-Maclaurin formula, *Comment. Math. Helv.* **44** (1969), 202–206.

[885] A. Ostrowski, On the remainder term of the Euler-Maclaurin formula, *J. Reine Angew. Math.* **239/240** (1969), 268–286.

[886] H. Ozden, Y. Simsek and H. M. Srivastava, A unified presentation of the generating functions of the generalized Bernoulli, Euler and Genocchi polynomials, *Comput. Math. Appl.* **60** (2010), 2779–2787.

[887] H. Pan and H.-Q. Cao, A congruence involving products of q-binomial coefficients, *J. Number Theory* **121** (2006), 224–233.

[888] I. Papadimitriou, A simple proof of the formula $\Sigma_{k=1}^{\infty} k^{-2} = \pi^2/6$, *Amer. Math. Monthly* **80** (1973), 424–425.

[889] J. M. Patin, A very short proof of Stirling's formula, *Amer. Math. Monthly* **96** (1989), 41–42.

[890] P. Paule and C. Schneider, Computer proofs of a new family of harmonic number identities, *Adv. Appl. Math.* **31** (2003), 359–378.

[891] A. Petojević and H. M. Srivastava, Computation of the Mordell-Tornheim zeta values, *Proc. Amer. Math. Soc.* **136** (2008), 2719–2728.

[892] A. Petojević and H. M. Srivastava, Computation of Euler's type sums of the products of Bernoulli numbers, *Appl. Math. Lett.* **22** (2009), 796–801.

[893] F. Pillichshammer, Euler's constant and averages of fractional parts, *Amer. Math. Monthly* **117** (2010), 78–82.

[894] M. L. V. Pitteway and C. M. A. Castle, On the subtractive version of Euclid's algorithm, *Bull. Inst. Math. Appl.* **24** (1988), 17–21.

[895] L. Pochhammer, Bemerkungen über das integral $\Gamma(a)$, *Math. Ann.* **41** (1892), 157–166.

[896] I. Podlubny, *Fractional Differential Equations: An Introduction to Fractional Derivatives, Fractional Differential Equations, to Methods of Their Solution and Some of Their Applications*, Mathematics in Science and Engineering, Vol. 198, Academic Press, New York, London, Tokyo and Toronto, 1999.

[897] T. K. Pogány, Integral representation of a series which includes the Mathieu **a**-series, *J. Math. Anal. Appl.* **296** (2004), 309–313.

[898] T. K. Pogány, Testing Alzer's inequality for Mathieu series $S(r)$, *Math. Maced.* **2** (2004), 1–4.

[899] T. K. Pogány, Integral representation of Mathieu $(\mathbf{a}, \lambda)$-series, *Integral Transforms Spec. Funct.* **16** (2005), 685–689.

[900] T. K. Pogány, Integral expressions for Mathieu-type series whose terms contain Fox's H-function, *Appl. Math. Lett.* **20** (2007), 764–769.

[901] T. K. Pogány and H. M. Srivastava, Some Mathieu-type series associated with the Fox-Wright function, *Comput. Math. Appl.* **57** (2009), 127–140.

[902] T. K. Pogány, H. M. Srivastava and Ž. Tomovski, Some families of Mathieu **a**-series and alternating Mathieu **a**-series, *Appl. Math. Comput.* **173** (2006), 69–108.

[903] T. K. Pogány and Ž. Tomovski, On multiple generalized Mathieu series, *Integral Transforms Spec. Funct.* **17** (2006), 285–293.

[904] T. K. Pogány and Ž. Tomovski, On Mathieu-type series whose terms contain generalized hypergeometric function ${}_pF_q$ and Meijer's G-function, *Math. Comput. Modelling* **47** (2008), 952–969.

[905] H. O. Pollak, A Remark on "Elementary inequalities for Mills' ratio", *Res. Statist. Appl. Res. Un. Japan. Sci. Engrs.* **4** (1956), 40.

[906] G. Pólya and G. Szegö, *Problems and Theorems in Analysis* (Translated from the German by D. Aeppli), Vol. I, Springer-Verlag, New York, Heidelberg and Berlin, 1972.

[907] B. S. Popov, Expressions of Laguerre polynomials through Bernoulli polynomials, *Mat. Bilten* **22** (48) (1998), 15–18.

[908] E. L. Post, The generalized Gamma functions, *Ann. Math.* **20** (2) (1918/1919), 202–217.

[909] K. Postelmans and W. N. Assche, Irrationality of $\zeta_q(1)$ and $\zeta_q(2)$, *J. Number Theory* **126** (2007), 119–154.

[910] E. O. Powell, An integral related to the radiation integrals, *Philos. Mag.* **34** (1943), 600–607.

[911] E. O. Powell, A table of the generalized Riemann Zeta function in a particular case, *Quart. J. Mech. Appl. Math.* **5** (1952), 116–123.

[912] A. Prabhu and H. M. Srivastava, Some limit formulas for the Gamma and Psi (or Digamma) functions at their singularities, *Integral Transforms Spec. Funct.* **22** (2011), 587–592.

[913] M. Prévost, A new proof of the irrationality of $\zeta(2)$ and $\zeta(3)$ using Padeé approximants, *J. Comput. Appl. Math.* **67** (1996), 219–235.

[914] M. Prévost, Legendre modified moments for Euler's constant, *J. Comput. Appl. Math.* **219** (2007), 484–492.

[915] M. Prévost, Padeé approximation and Apostol-Bernoulli and Apostol-Euler polynomials, *J. Comput. Appl. Math.* **233** (2010), 3005–3017.

[916] H. Prodinger, Some applications of the q-Rice formula, *Random Structures Algorithms* **19** (2001), 552–557.

[917] A. P. Prudnikov, On the Euler problem of summation of harmonic series, *Integral Transforms Spec. Funct. Inform. Bull.* **1** (1) (1996), 5–7 (in Russian).

[918] A. P. Prudnikov, Yu. A. Bryčkov and O. I. Maričev, *Integrals and Series: Supplementary Chapters*, Nauka, Moscow, 1986 (in Russian).

[919] A. P. Prudnikov, Yu. A. Brychkov and O. I. Marichev, *Integrals and Series*, Vol. 1: *Elementary Functions*, Gordon and Breach Science Publishers, New York, London and Tokyo, 1986.

[920] F. E. Prym, Zur theorie der Gammafunktion, *J. Reine Angew. Math.* **82** (1877), 165–172.

[921] F. Qi, Generalized weighted mean values with two parameters, *Proc. Roy. Soc. London Ser. A Math. Phys. Engrg. Sci.* **454** (1998), 2723–2732.

[922] F. Qi, Generalized abstract mean values, *J. Inequal. Pure Appl. Math.* **1** (1) (2000), Article 4.

[923] F. Qi, Monotonicity results and inequalities for the Gamma and incomplete Gamma functions, *Math. Inequal. Appl.* **5** (2002), 61–67.

[924] F. Qi, The extended mean values: Definition, properties, monotonicities, comparison, convexities, generalizations, applications, *Cubo Mat. Ed.* **5** (3) (2003), 63–90.

[925] F. Qi, An integral expression and some inequalities of Mathieu series, *Rostock. Math. Kolloq.* **58** (2004), 37–46.

[926] F. Qi, Certain logarithmically N-alternating monotonic functions involving Gamma and q-Gamma functions, *RGMIA Res. Rep. Coll.* **8** (3) (2005), 413–422 (see also *Nonlinear Funct. Anal. Appl.* **12** (2007), 675–685).

[927] F. Qi, Monotonicity and logarithmic convexity for a class of elementary functions involving the exponential function, *RGMIA Res. Rep. Coll.* **9** (3) (2006), Article 3 (see also [520]).

[928] F. Qi, Three-log-convexity for a class of elementary functions involving exponential function, *J. Math. Anal. Approx. Theory* **1** (2006), 100–103.

[929] F. Qi, The best bounds in Kershaw's inequality and two completely monotonic functions, *RGMIA Res. Rep. Coll.* **9** (4) (2006), Article 2 (see also [940]).

[930] F. Qi, A completely monotonic function involving divided difference of Psi and an equivalent inequality involving sum, *RGMIA Res. Rep. Coll.* **9** (4) (2006), Article 5 (see also [934]).

[931] F. Qi, A completely monotonic function involving divided differences of Psi and Polygamma functions and an application, *RGMIA Res. Rep. Coll.* **9** (4) (2006), Article 8 (see also [942]).

[932] F. Qi, A class of logarithmically completely monotonic functions and the best bounds in the first Kershaw's double inequality, *J. Comput. Appl. Math.* **206** (2007), 1007–1014.

[933] F. Qi, Three classes of logarithmically completely monotonic functions involving Gamma and Psi functions, *Integral Transforms Spec. Funct.* **18** (2007), 503–509.

[934] F. Qi, A completely monotonic function involving the divided difference of the Psi function and an equivalent inequality involving sums, *ANZIAM J.* **48** (2007), 523–532 (see also [930]).

[935] F. Qi and C.-P. Chen, A complete monotonicy property of the Gamma function, *J. Math. Anal. Appl.* **296** (2004), 603–607.

[936] F. Qi and C.-P. Chen, A complete monotonicity property of the Gamma function, *J. Math. Anal. Appl.* **296** (2004), 603–607 (see also [939]).

[937] F. Qi, S.-X. Chen and W.-S. Cheung, Logarithmically completely monotonic functions concerning Gamma and Digamma functions, *Integral Transforms Spec. Funct.* **18** (2007), 435–443.

[938] F. Qi, R.-Q. Cui, C.-P. Chen and B.-N. Guo, Some completely monotonic functions involving Polygamma functions and an application, *J. Math. Anal. Appl.* **310** (2005), 303–308.

[939] F. Qi and B.-N. Guo, Complete monotonicities of functions involving the Gamma and Digamma functions, *RGMIA Res. Rep. Coll.* **7** (1) (2004), 63–72 (see also [936] *and* [941]).

[940] F. Qi and B.-N. Guo, Completely monotonic functions involving divided differences of the Di- and Tri-Gamma functions and some applications, *Comm. Pure Appl. Anal.* **8** (2009), 1975–1989 (see also [929]).

[941] F. Qi and B.-N. Guo, Some logarithmically completely monotonic functions related to the Gamma function, *J. Korean Math. Soc.* **47** (2010), 1283–1297 (see also [939]).

[942] F. Qi and B.-N. Guo, A class of completely monotonic functions involving divided differences of the Psi and Tri-Gamma functions and some applications, *J. Korean Math. Soc.* **48** (2011), 655–667 (see also [930]).

[943] F. Qi, B.-N. Guo and C.-P. Chen, Some completely monotonic functions involving the Gamma and Polygamma functions, *RGMIA Res. Rep. Coll.* **7** (1) (2004), 31–36 (see also [944]).

[944] F. Qi, B.-N. Guo and C.-P. Chen, Some completely monotonic functions involving the Gamma and Polygamma functions, *J. Austral. Math. Soc.* **80** (2006), 81–88 (see also [943]).

[945] F. Qi, B.-N. Guo and C.-P. Chen, The best bounds in Gautschi-Kershaw inequalities, *Math. Inequal. Appl.* **9** (2006), 427–436.

[946] F. Qi, B.-N. Guo, S. Guo and S.-X. Chen, A function involving Gamma function and having logarithmically absolute convexity, *Integral Transforms Spec. Funct.* **18** (2007), 837–843.

[947] F. Qi, S. Guo and S.-X. Chen, A new upper bound in the second Kershaw's double inequality and its generalizations, *J. Comput. Appl. Math.* **220** (2008), 111–118.

[948] F. Qi, D.-W. Niu and J. Cao, Logarithmically completely monotonic functions involving the Gamma and Polygamma functions, *J. Math. Anal. Approx. Theory* **1** (1) (2006), 66–74.

[949] F. Qi and S.-L. Xu, The function $(b^x - a^x)/x$: Inequalities and properties, *Proc. Amer. Math. Soc.* **126** (1998), 3355–3359.

[950] F. Qi, Q. Yang and W. Li, Two logarithmically completely monotonic functions connected with Gamma function, *Integral Transforms Spec. Funct.* **17** (2006), 539–542.

[951] F. Qi and S.-Q. Zhang, Note on monotonicity of generalized weighted mean values, *Proc. Roy. Soc. London Ser. A Math. Phys. Engrg. Sci.* **455** (1999), 3259–3260.

[952] S.-L. Qiu and M. Vuorinen, Some properties of the Gamma and Psi functions, with applications, *Math. Comput.* **74** (2004), 723–742.

[953] M. H. Quenouille, The evaluation of probabilities in a normal multivariate distribution, with special reference to the correlation ratio, *Proc. Edinburgh Math. Soc.* (*Ser.* 2) **8** (1947/1950), 95–100.

[954] J. R. Quine and J. Choi, Zeta regularized products and functional determinants on spheres, *Rocky Mountain J. Math.* **26** (1996), 719–729.

[955] J. R. Quine, S. H. Heydari and R. Y. Song, Zeta regularized products, *Trans. Amer. Math. Soc.* **338** (1993), 213–231.

[956] H. Rademacher, *Topics in Analytic Number Theory*, Springer-Verlag, Berlin, Heidelberg and New York, 1973.

[957] D. Răducanu and H. M. Srivastava, A new class of analytic functions defined by means of a convolution operator involving the Hurwitz-Lerch Zeta function, *Integral Transforms Spec. Funct.* **18** (2007), 933–943.

[958] M. Rahman, A simple evaluation of Askey and Wilson's q-Beta integral, *Proc. Amer. Math. Soc.* **92** (1984), 413–417.

[959] E. D. Rainville, *Special Functions*, Macmillan Company, New York, 1960; Reprinted by Chelsea Publishing Company, Bronx, New York, 1971.

[960] W. Raji, A new proof of the transformation law of Jacobi's theta function $\theta(w, \tau)$, *Proc. Amer. Math. Soc.* **135** (2007), 3127–3132.

[961] P. M. Rajikovič, M. D. Petkovič and P. Barry, The Hankel transform of the sum of consecutive generalized Catalan numbers, *Integral Transform. Spec. Funct.* **18** (2007), 285–296.

[962] K. Ramachandra, Some applications of Kronecker's limit formulas, *Ann. of Math.* **80** (1964), 104–148.

[963] S. Ramanujan, Question 327, *J. Indian Math. Soc.* **3** (1911), 209.

[964] S. Ramanujan, A series for Euler's constant γ, *Messenger Math.* **46** (1916/1917), 73–80.

[965] S. Ramanujan, *Notebooks of Srinivasa Ramanujan*, Vols. I and II, Tata Institute of Fundamental Research, Bombay, 1957.

[966] V. Ramaswami, Notes on Riemann's ζ-function, *J. London Math. Soc.* **9** (1934), 165–169.

[967] G. Rasch, Notes on the Gamma-function, *Ann. of Math.* **32** (1931), 591–599.

[968] Th. M. Rassias, S. N. Singh and H. M. Srivastava, Some q-generating functions associated with basic multiple hypergeometric series, *Comput. Math. Appl.* **27** (1) (1994), 33–39.

[969] Th. M. Rassias and H. M. Srivastava, Some classes of infinie series associated with the Riemann Zeta and Polygamma functions and generalized harmonic numbers, *Appl. Math. Comput.* **131** (2002), 593–605.

[970] Th. M. Rassias and T. Trif, Log-convex solutions of the second order to the functional equation $f(x+1) = g(x)f(x)$, *J. Math. Anal. Appl.* **331** (2007), 1440–1451.

[971] D. Rawlings, Bernoulli trials and number theory, *Amer. Math. Monthly* **101** (1994), 948–952.

[972] R. Remmert, *Theory of Complex Functions*, Springer-Verlag, Berlin, Heidelberg and New York, 1989.

[973] R. Remmert, Wielandt's theorem about the Γ-function, *Amer. Math. Monthly* **103** (1996), 214–220.

[974] R. Remmert, *Classical Topics in Complex Function Theory*, Springer-Verlag, Berlin, Heidelberg and New York, 1998.

[975] B. Richmond and D. Merlini, Srirling numbers for complex arguments, *SIAM J. Discrete Math.* **10** (1997), 73–82.

[976] B. Richmond and G. Szekeres, Some formulas related to Dilogarithms, the Zeta function and the Andrews-Gordon identities, *J. Austral. Math. Soc. Ser. A* **31** (1981), 362–373.

[977] B. Riemann, Über die Anzahl der Primzahlen unter einer gegebenen Grösse, *Monatsber. Akad. Berlin* **1859** (1859), 671–680.
[978] J. Riordan, *Combinatorial Identities*, Wiley Series in Probability and Mathematical Statistics, John Wiley and Sons, New York, London and Sydney, 1968.
[979] J. Riordan and P. R. Stein, Proof of a conjecture on Genocchi numbers, *Discrete Math.* **5** (1973), 381–388.
[980] H. Robbins, A remark on Stirling's formula, *Amer. Math. Monthly* **62** (1955), 26–29.
[981] L. J. Rogers, On function sum theorems connected with the series $\Sigma_{n=1}^{\infty} \frac{x^n}{x^2}$, *Proc. London Math. Soc.* **4** (2) (1907), 169–189.
[982] S. Roman, *The Umbral Calculus*, Academic Press, New York and London, 1984.
[983] R. A. Rosenbaum, Sub-additive functions, *Duke Math. J.* **17** (1950), 227–247.
[984] J. B. Rosser, Real roots of real Dirichlet L-series, *J. Res. Nat. Bur. Standards* **45** (1950), 505–514.
[985] J. B. Rosser, Transformations to speed the convergence of series, *J. Res. Nat. Bur. Standards* **46** (1951), 56–64.
[986] C. H. Rowe, A proof of the asymptotic series for $\log \Gamma(z)$ and $\log \Gamma(z+a)$, *Ann. of Math.* **32** (1931), 10–16.
[987] H. Ruben, Variance bounds and orthogonal expansions in Hilbert space with an application to inequalities for Gamma functions and π, *J. Reine Angew. Math.* **225** (1967), 147–153.
[988] W. Rudin, *Principles of Mathematical Analysis*, Third edition, McGraw-Hill Book Company, New York, Toronto and London, 1976.
[989] S. Ruijsenaars, First order analytic difference equations and integrable quantum systems, *J. Math. Phys.* **38** (1997), 1069–1146.
[990] S. Ruijsenaars, On Barnes' multiple Zeta and Gamma functions, *Adv. Math.* **156** (2000), 107–132.
[991] G. Rutledge and R. D. Douglass, $\int_0^1 \frac{\log u}{u} \log^2(1+u)\,du$ and related definite integrals, *Amer. Math. Monthly* **41** (1934), 29–36.
[992] C. S. Ryoo, A note on q-Bernoulli numbers and polynomials, *Appl. Math. Lett.* **20** (2007), 524–531.
[993] C. S. Ryoo, A numerical computation on the structure of the roots of q-extension of Genocchi polynomials, *Appl. Math. Lett.* **21** (2008), 348–354.
[994] C. S. Ryoo, On the q-extension of Genocchi polynomials, *J. Concr. Appl. Math.* **8** (2010), 616–622.
[995] C. S. Ryoo, A numerical computation of the roots of q-Euler polynomials, *J. Comput. Anal. Appl.* **12** (2010), 148–156.
[996] C. S. Ryoo and T. Kim, On the real roots of the q-Hurwitz Zeta-function, *Proc. Jangjeon Math. Soc.* **11** (2008), 205–214.
[997] F. Saidak, On the logarithmic derivative of the Euler product, *Tatra Mountain Math. Publ.* **29** (2004), 113–122.
[998] B. E. Sagan, Congruence properties of q-analogs, *Adv. Math.* **95** (1992), 127–143.
[999] M. Saigo and H. M. Srivastava, The behavior of the zero-balanced hypergeometric series ${}_pF_{p-1}$ near the boundary of its convergence region, *Proc. Amer. Math. Soc.* **110** (1990), 71–76.
[1000] A. A. Samoletov, A sum containing factorials, *J. Comput. Appl. Math.* **131** (2001), 503–504.
[1001] H. F. Sandham, Advanced Problem 4353, *Amer. Math. Monthly* **56** (1949), 414.
[1002] H. F. Sandham, A logarithmic transcendent, *J. London Math. Soc.* **24** (1949), 83–91.

[1003] J. Sándor, A bibliography on gamma functions: Inequalities and applications. [http://www.math.ubbcluj.ro/ jsandor/letolt/art42.pdf].

[1004] P. Sarnak, Determinants of Laplacians, *Comm. Math. Phys.* **110** (1987), 113–120.

[1005] Y. Sasaki, The first derivative multiple Zeta values at non-positive integers, *Ramanujan J.* **21** (2010), 267–284.

[1006] J. Satoh, q-Analogue of Riemann's ζ-function and q-Euler numbers, *J. Number Theory* **31** (1989), 346–362.

[1007] R. K. Saxena, C. Ram, N. Dudi and S. L. Kalla, Generalized Gamma-type functions involving Kummer's confluent hypergeometric function and associated probability distributions, *Integral Transforms Spec. Funct.* **18** (2007), 679–687 (see also [1108]).

[1008] H. Schmidt, Eine bemerkung zum aufbau der lehre von der Γ-funktion, *Arch. Math. (Basel)* **9** (1958), 297–299.

[1009] W. O. Schumann, *Elektrische Durchbruchfeldstärke von Gasen*, Springer-Verlag, Berlin and Heidelberg, 1923.

[1010] R. Schuster, A generalization of the Barnes G-function, *Zeitschr. Anal. Anwendungen* **11** (1992), 229–236.

[1011] I. J. Schwatt, Note on the expansion of a function, *Philos. Mag.* **29** (1915), 65–67.

[1012] I. J. Schwatt, The expansion of $\tan^n x$ by Maclaurin theorem, *Tôhoku Math. J.* **33** (1) (1931), 150–152.

[1013] A. Selberg, Bemerkninger om et multipelt integral, *Nordisk Mat. Tidskr.* **26** (1944), 71–78.

[1014] A. Selberg, Harmonic analysis and discontinuous groups in weakly symmetric Riemannian spaces with applications to Dirichlet series, *J. Indian Math. Soc. (N. S.)* **20** (1956), 47–87.

[1015] A. Selberg, *Collected Papers*, Vol. I, Springer-Verlag, Berlin, Heidelberg, New York, London, Paris and Tokyo, 1989.

[1016] A. Selberg, *Collected Papers*, Vol. II, Springer-Verlag, Berlin, Heidelberg, New York, London, Paris, Tokyo, Hong Kong, Barcelona and Budapest, 1991.

[1017] A. Selberg and S. Chowla, On Epstein's Zeta-function, *J. Reine Angew. Math.* **227** (1967), 86–110.

[1018] J. B. Sellliah, An inequality satisfied by the Gamma function, *Canad. Math. Bull.* **19** (1976), 85–87.

[1019] T. Y. Seo, J. Choi, J. S. Gang and B. M. Ok, Note on the analytic continuation of the multiple Hurwitz Zeta function $\zeta_n(s, a)$, *Kyungpook Math. J.* **33** (1993), 127–131.

[1020] R. E. Shafer and O. P. Lossers, Euler's constant, *Amer. Math. Monthly* **76** (1969), 1077–1079.

[1021] R. Shail, A class of infinite sums and integrals, *Math. Comput.* **70** (2001), 789–799.

[1022] J. D. Shallit and K. Zikan, A theorem of Goldbach, *Amer. Math. Monthly* **93** (1986), 402–403.

[1023] H. J. Sharpe, On the Gamma functions of a complex variable, *Messenger Math.* **13** (1883/1884), 111–112.

[1024] L.-C. Shen, Remarks on some integrals and series involving the Stirling numbers and $\zeta(n)$, *Trans. Amer. Math. Soc.* **347** (1995), 1391–1399.

[1025] R. A. Silverman, *Complex Analysis with Applications*, Dover Publications, New York, 1984.

[1026] T. Shintani, A proof of the classical Kronecker limit formula, *Tokyo J. Math.* **3** (1980), 191–199.

[1027] K. Shiratani, Kummer's congruence for generalized Bernoulli numbers and its application, *Mem. Fac. Sci. Kyushu Univ. Ser. A* **26** (1971), 119–138.

[1028] S. T. Shovelton, A formula in finite differences and its application to mechanical quadrature, *Messenger Math.* **38** (1908/1909), 49–57.

[1029] R. Simion, Combinatorial statistics on type-B analogues of noncrossing partitions and restricted permutations, *Electronic J. Combin.* **7** (2000), Article R9.

[1030] Y. Simsek, V. Kurt and D. Kim, New approach to the complete sum of products of the twisted (h,q)-Bernoulli numbers and polynomials, *J. Nonlinear Math. Phys.* **14** (1) (2007), 44–56.

[1031] Y. Simsek and H. M. Srivastava, A family of p-adic twisted interpolation functions associated with the modified Bernoulli numbers, *Appl. Math. Comput.* **216** (2010), 2976–2987.

[1032] Y. Simsek and S. Yang, Transformation of four Titchmarsh-type infinite integrals and generalized Dedekind sums associated with Lambert series, *Adv. Stud. Contemp. Math.* **9** (2004), 195–202.

[1033] R. J. Singh and D. P. Verma, Some series involving Riemann Zeta function, *Yokohama Math. J.* **31** (1983), 1–4.

[1034] R. Sitaramachandrarao, A formula of S. Ramanujan, *J. Number Theory* **25** (1987), 1–19.

[1035] R. Sitaramachandrarao and A. Sivaramsarma, Some identities involving the Riemann Zeta function, *Indian J. Pure Appl. Math.* **10** (1979), 602–607.

[1036] R. Sitaramachandrarao and A. Sivaramsarma, Two identities due to Ramanujan, *Indian J. Pure Appl. Math.* **11** (1980), 1139–1140.

[1037] R. Sitaramachandrarao and M. V. Subbarao, Transformation formulae for multiple series, *Pacific J. Math.* **113** (1984), 471–479.

[1038] I. S. Skau and E. S. Selmer, Noen anvendelser av finn holmes methode for beregning av $\Sigma_{k=1}^{\infty}\frac{1}{k^2}$, *Nordisk Mat. Tidskr.* **19** (1971), 120–124.

[1039] L. J. Slater, Further identities of the Rogers-Ramanujan type, *Proc. London Math. Soc.* **54** (2) (1951/1952), 147–167.

[1040] L. J. Slater, *Generalized Hypergeometric Functions*, Cambridge University Press, Cambridge, London and New York, 1966.

[1041] D. V. Slavić, On inequalities for $\Gamma(x+1)/\Gamma(x+1/2)$, *Univ. Beograd. Publ. Elektrotehn. Fak. Ser. Mat. Fiz.* **498–541** (1975), 17–20.

[1042] J. R. Smart, On the values of the Epstein zeta function, *Glasgow Math. J.* **14** (1973), 1–12.

[1043] A. Sofo, Series associated with Polygamma functions, *Scientia Ser. A: Math. Sci.* **17** (2009), 95–104.

[1044] A. Sofo and H. M. Srivastava, A family of sums containing factorials, *Integral Transforms Spec. Funct.* **20** (2009), 393–399.

[1045] A. Sofo and H. M. Srivastava, Identities for the harmonic numbers and binomialcoefficients, *Ramanujan J.* **25** (2011), 93–113.

[1046] J.-W. Son, A remark on the q-analogue of some special functions, *Comm. Korean Math. Soc.* **20** (2005), 623–640 (Korean).

[1047] S. H. Son, Basic functional equations of the Rogers-Ramanujan functions, *Rocky Mountain J. Math.* **37** (2007), 652–662.

[1048] J. Sondow, Analytic continuation of Riemann's Zeta function and values at negative integers via Euler's transformation of series, *Proc. Amer. Math. Soc.* **120** (1994), 421–424.

[1049] J. Sondow, The Riemann hypothesis, simple zeros and the asymptotic convergence degree of improper Riemann sums, *Proc. Amer. Math. Soc.* **126** (1998), 1311–1314.

[1050] J. Sondow, Criteria for irrationality of Euler's constant, *Proc. Amer. Math. Soc.* **131** (2003), 3335–3344.

[1051] J. Sondow, Double integrals for Euler's constant and $\ln(4/\pi)$ and an analog of Hadjicostas's formula, *Amer. Math. Monthly* **112** (2005), 61–65.

[1052] J. Sondow, A hypergeometric approach, via linear forms involving logarithms, to criteria for irrationality of Euler's constant. With an appendix by Sergey Zlobin, *Math. Slovaca* **59** (2009), 307–314.

[1053] J. Sondow and P. Hadjicostas, The generalized-Euler-constant function $\gamma(z)$ and a generalization of Somos's quadratic recurrence constant, *J. Math. Anal. Appl.* **332** (2007), 292–314.

[1054] J. Sondow and W. Zudilin, Euler's constant, q-logarithms, and formulas of Ramanujan and Gosper, *Ramanujan J.* **12** (2006), 225–244.

[1055] M. N. Sonin, Sur les termes complémentaires de la Formule sommatoire d'Euler et de celle de Stirling, *Ann. Sci. École Norm. Sup.* **3** (1889), 257–262.

[1056] J. Spanier and K. B. Oldham, *An Atlas of Functions*, Hemisphere Publishing Corporation (Harper and Row Publishers), Washington D. C., New York and London, 1987.

[1057] W. Spence, *An Essay on the Theory of the Various Orders of Logarithmic Transcendents; with an Inquiry into Their Applications to the Integral Calculus and the Summation of Series*, John Murray and Archibald Constable and Company, Edinburgh and London, 1809.

[1058] M. R. Spiegel, *Mathematical Handbook*, McGraw-Hill Book Company, New York, 1968.

[1059] M. R. Spiegel, *Fourier Analysis*, Schaum's Outline Series, McGraw-Hill Book Company, New York, 1974.

[1060] M. R. Spiegel and J. Liu, *Mathematical Handbook of Formulas and Tables*, Second edition, Schaum's Outline Series, McGraw-Hill Book Company, New York, 1999.

[1061] O. Spiess, *Die Summe der reziproken Quadratzahlen*, in *Festschrift zum 60 Geburtstag von Prof. Dr. Andreas Speiser*, (L. V. Ahlfors *et al.*, Editors), pp. 66–86, Füssli, Zürich, 1955.

[1062] M. Z. Spivey, Combinatorial sums and finite differences, *Discrete Math.* **307** (2007), 3130–3146.

[1063] M. Spreafico, Zeta functions and regularized determinants on projective spaces, *Rocky Mountain J. Math.* **35** (2003), 1499–1511.

[1064] R. Sprugnoli, Negation of binomial coefficients, *Discrete Math.* **308** (2008), 5070–5077.

[1065] G. K. Srinivasan, The Gamma function: An eclectic tour, *Amer. Math. Monthly* **114** (2007), 297–315.

[1066] H. M. Srivastava, A family of q-generating functions, *Bull. Inst. Math. Acad. Sinica* **12** (1984), 327–336.

[1067] H. M. Srivastava, A certain class of q-series transformations, *J. Math. Anal. Appl.* **107** (1985), 498–508.

[1068] H. M. Srivastava, Summation of a class of series involving the Riemann Zeta function, *Rev. Técn. Fac. Ingr. Univ. Zulia* **9** (1986), 79–82.

[1069] H. M. Srivastava, Some infinite series associated with the Riemann Zeta function, *Yokohama Math. J.* **35** (1987), 47–50.

[1070] H. M. Srivastava, A transformation for an n-balanced ${}_3\Phi_2$, *Proc. Amer. Math. Soc.* **101** (1987), 108–112.

[1071] H. M. Srivastava, A note on a generalization of a q-series transformation of Ramanujan, *Proc. Japan Acad. Ser. A Math. Sci.* **63** (1987), 143–145.

[1072] H. M. Srivastava, A unified presentation of certain classes of series of the Riemann Zeta function, *Riv. Mat. Univ. Parma (Ser.* 4) **14** (1988), 1–23.

[1073] H. M. Srivastava, Sums of certain series of the Riemann Zeta function, *J. Math. Anal. Appl.* **134** (1988), 129–140.

[1074] H. M. Srivastava, Some explicit formulas for the Bernoulli and Euler numbers and polynomials, *Internat. J. Math. Ed. Sci. Tech.* **19** (1988), 79–82.

[1075] H. M. Srivastava, Littlewood's teaser, *Internat. J. Math. Ed. Sci. Tech.* **19** (1988), 201–203.

[1076] H. M. Srivastava, Sums of a certain class of q-series, *Proc. Japan Acad. Ser. A Math. Sci.* **65** (1989), 8–11.

[1077] H. M. Srivastava, An extension of the q-Saalschützian theorem, *Acta Math. Hungar.* **53** (1989), 115–118.

[1078] H. M. Srivastava, A class of infinite series involving binomial and factorial coefficients, *Internat. J. Math. Ed. Sci. Tech.* **21** (1990), 75–79.

[1079] H. M. Srivastava, A simple algorithm for the evaluation of a class of generalized hypergeometric series, *Stud. Appl. Math.* **86** (1992), 79–86.

[1080] H. M. Srivastava, Some convolution identities based upon Ramanujan's bilateral sum, *Bull. Austral. Math. Soc.* **49** (1994), 433–437.

[1081] H. M. Srivastava, A certain family of sub-exponential series, *Internat. J. Math. Ed. Sci. Tech.* **25** (1994), 211–216.

[1082] H. M. Srivastava, Certain families of rapidly convergent series representations for $\zeta(2n+1)$, *Math. Sci. Res. Hot-Line* **1** (6) (1997), 1–6 (Research Announcement).

[1083] H. M. Srivastava, Further series representations for $\zeta(2n+1)$, *Appl. Math. Comput.* **97** (1998), 1–15.

[1084] H. M. Srivastava, Some rapidly converging series for $\zeta(2n+1)$, *Proc. Amer. Math. Soc.* **127** (1999), 385–396.

[1085] H. M. Srivastava, A note on the closed-form summation of some trigonometric series, *Kobe J. Math.* **16** (1999), 177–182.

[1086] H. M. Srivastava, Some simple algorithms for the evaluations and representations of the Riemann Zeta function at positive integer arguments, *J. Math. Anal. Appl.* **246** (2000), 331–351.

[1087] H. M. Srivastava, Some formulas for the Bernoulli and Euler polynomials at rational arguments, *Math. Proc. Cambridge Philos. Soc.* **129** (2000), 77–84.

[1088] H. M. Srivastava, Some families of rapidly convergent series representations for the Zeta functions, *Taiwanese J. Math.* **4** (2000), 569–598.

[1089] H. M. Srivastava, Remarks on a sum containing factorials, *J. Comput. Appl. Math.* **142** (2002), 441–444.

[1090] H. M. Srivastava, Leonard Euler (1707–1783) and the computational aspects of some Zeta-function series, in *Proceedings of the Global KMS Day International Conference celebrating the* 60th *Anniversary of the Korean Mathematical Society* (Seoul, Republic of Korea; October 27–29, 2006); *J. Korean Math. Soc.* **44** (2007), 1163–1184.

[1091] H. M. Srivastava, Some generalizations and basic (or q-) extensions of the Bernoulli, Euler and Genocchi polynomials, *Appl. Math. Inform. Sci.* **5** (2011), 390–444.

[1092] H. M. Srivastava and A. A. Attiya, An integral operator associated with the Hurwitz-Lerch Zeta function and differential subordination, *Integral Transforms Spec. Funct.* **18** (2007), 207–216.

[1093] H. M. Srivastava, M. A. Chaudhry, A. Qadir and A. Tassaddiq, Some extensions of the Fermi-Dirac and Bose-Einstein functions with applications to the family of the Zeta and related functions, *Russian J. Math. Phys.* **18** (2011), 107–121 (see also [1145] below).

[1094] H. M. Srivastava and J. Choi, *Series Associated with the Zeta and Related Functions*, Kluwer Academic Publishers, Dordrecht, Boston and London, 2001.

[1095] H. M. Srivastava, M. Garg and S. Choudhary, A new generalization of the Bernoulli and related polynomials, *Russian J. Math. Phys.* **17** (2010), 251–261.

[1096] H. M. Srivastava, M. Garg and S. Choudhary, Some new families of generalized Euler and Genocchi polynomials, *Taiwanese J. Math.* **15** (2011), 283–305.

[1097] H. M. Srivastava, M. L. Glasser and V. S. Adamchik, Some definite integrals associated with the Riemann Zeta function, *Zeitschr. Anal. Anwendungen* **19** (2000), 831–846.

[1098] H. M. Srivastava and V. K. Jain, q-Series identities and reducibility of basic double hypergeometric functions, *Canad. J. Math.* **38** (1986), 215–231.

[1099] H. M. Srivastava and P. W. Karlsson, *Multiple Gaussian Hypergeometric Series*, Halsted Press (Ellis Horwood Limited, Chichester), John Wiley and Sons, New York, Chichester, Brisbane and Toronto, 1985.

[1100] H. M. Srivastava, T. Kim and Y. Simsek, q-Bernoulli numbers and polynomials associated with multiple q-Zeta functions and basic L-series, *Russian J. Math. Phys.* **12** (2005), 201–228.

[1101] H. M. Srivastava, J.-L. Lavoie and R. Tremblay, A class of addition theorems, *Canad. Math. Bull.* **26** (1983), 438–445.

[1102] H. M. Srivastava and H. L. Manocha, *A Treatise on Generating Functions*, Halsted Press (Ellis Horwood Limited, Chichester), John Wiley and Sons, New York, Chichester, Brisbane and Toronto, 1984.

[1103] H. M. Srivastava and E. A. Miller, A simple reducible case of double hypergeometric series involving Catalan's constant and Riemann's ζ-function, *Internat. J. Math. Ed. Sci. Tech.* **21** (1990), 375–377.

[1104] H. M. Srivastava and E. A. Miller, A recurrence relation for the Bernoulli numbers, *Internat. J. Math. Sci.* **1** (2002), 69–73.

[1105] H. M. Srivastava and Á. Pintér, Remarks on some relationships between the Bernoulli and Euler polynomials, *Appl. Math. Lett.* **17** (2004), 375–380.

[1106] H. M. Srivastava, D. Jankov, T. K. Pogány and R. K. Saxena, Two-sided inequalities for the extended Hurwitz-Lerch Zeta function, *Comput. Math. Appl.* **62** (2011), 516–522.

[1107] H. M. Srivastava, R. K. Saxena, T. K. Pogány and R. Saxena, Integral and computational representations of the extended Hurwitz-Lerch Zeta function, *Integral Transforms Spec. Funct.* **22** (2011), 487–506.

[1108] H. M. Srivastava, R. K. Saxena and C. Ram, A unified presentation of the Gamma-type functions occurring in diffraction theory and associated probability distributions, *Appl. Math. Comput.* **162** (2005), 921–947.

[1109] H. M. Srivastava and Ž. Tomovski, Some problems and solutions involving Mathieu's series and its generalizations, *J. Inequal. Pure Appl. Math.* **5** (2) (2004), 1–13 (electronic).

[1110] H. M. Srivastava and P. G. Todorov, An explicit formula for the generalized Bernoulli polynomials, *J. Math. Anal. Appl.* **130** (1988), 509–513.

[1111] H. M. Srivastava and H. Tsumura, A certain class of rapidly convergent series representations for $\zeta(2n+1)$, *J. Comput. Appl. Math.* **118** (2000), 323–335.

[1112] H. M. Srivastava and H. Tsumura, New rapidly convergent series representations for $\zeta(2n+1)$, $L(2n,\chi)$, and $L(2n+1,\chi)$, *Math. Sci. Res. Hot-Line* **4** (7) (2000), 17–24 (Research Announcement).

[1113] H. M. Srivastava and H. Tsumura, Certain classes of rapidly convergent series representations for $L(2n,\chi)$ and $L(2n+1,\chi)$, *Acta Arith.* **100** (2001), 195–201.

[1114] H. M. Srivastava and H. Tsumura, Inductive construction of rapidly convergent series representations for $\zeta(2n+1)$, *Internat. J. Comput. Math.* **80** (2003), 1161–1173.

[1115] R. Srivastava, A simplified overview of certain relations among infinite series that arose in the context of fractional calculus, *J. Math. Anal. Appl.* **162** (1991), 152–158.

[1116] R. Srivastava, Classes of series identities and associated hypergeometric reduction formulas, *Appl. Math. Comput.* **215** (2009), 235–242.

[1117] R. Srivastava, Some combinatorial series identities and rational sums, *Integral Transforms Spec. Funct.* **20** (2009), 83–91.

[1118] R. Srivastava, Some families of combinatorial and other series identities and their applications, *Appl. Math. Comput.* **218** (2011), 1077–1083.

[1119] E. L. Stark, $1-1/4+1/9+-\ldots=\pi^2/12$, *Praxis Math.* **12** (1970), 1-3.

[1120] E. L. Stark, A new method of evaluating the sums of $\sum_{k=1}^{\infty}(-1)^{k+1}k^{-2p}$, $p=1,2,3,\ldots$ and related series, *Elem. Math.* **27** (1972), 32–34.

[1121] E. L. Stark, The series $\sum_{k=1}^{\infty}k^{-s}$, $s=2,3,4,\cdots$, once more, *Math. Mag.* **47** (1974), 197–202.

[1122] E. L. Stark, Another proof of the formula $\Sigma_{k=1}^{\infty}1/k^2=\pi^2/6$, *Amer. Math. Monthly* **76** (1969), 552–553.

[1123] E. L. Stark, $1-1/4+1/9+-\cdots=\pi^2/12$, *Praxis Math.* **12** (1970), 1-3.

[1124] E. L. Stark, A new method of evaluating the sums of $\sum_{k=1}^{\infty}(-1)^{k+1}k^{-2p}$, $p=1,2,3,\ldots$ and related series, *Elem. Math.* **27** (1972), 32–34.

[1125] E. L. Stark, The series $\Sigma_{k=1}^{\infty}k^{-s}$, $s=2,3,4,\cdots$, once more, *Math. Mag.* **47** (1974), 197–202.

[1126] J. F. Steffensen, *Interpolation*, Second edition, Chelsea Publishing Company, New York, 1950.

[1127] F. Steiner, On Selberg's Zeta function for compact Riemann surfaces, *Phys. Lett. B* **188** (1987), 447–454.

[1128] T. J. Stieltjes, Een en ander over de integraal $\int_0^1 l\Gamma(x+u)\,du$, *Nieuw Arch. Wisk.* **2** (1878), 100–104.

[1129] T. J. Stieltjes, Table des valeurs des sommes $S_k=\Sigma_1^{\infty}n^{-k}$, *Acta Math.* **10** (1887), 299–302.

[1130] T. J. Stieltjes, Sur le dе́veloppement de log $\Gamma(a)$, *J. Math. Pures Appl. (Sér. 4)* **5** (1889), 425–444.

[1131] G. Stoica, O Relaţie de recurenţă ın studiul funcţiei Zeta a lui Riemann (A recurrence formula in the study of the Riemann Zeta function), *Stud. Cerc. Mat.* **39** (1987), 261–264.

[1132] K. B. Stolarsky, From Wythoff's Nim to Chebyshev's inequality, *Amer. Math. Monthly* **98** (1991), 889–900.

[1133] M. V. Subbarao and R. Sitaramachandrarao, On some infinite series of L. J. Mordell and their analogues, *Pacific J. Math.* **119** (1985), 245–255.

[1134] Z.-H. Sun, Congruences involving Bernoulli and Euler numbers, *J. Number Theory* **128** (2008), 280–312.

[1135] Z.-H. Sun, Congruences involving Bernoulli polynomials, *Discrete Math.* **308** (2008), 71–112.

[1136] Z.-W. Sun, On sums of binomial coefficients and their applications, *Discrete Math.* **308** (2008), 4231–4245.

[1137] B. Sury, The value of Bernoulli polynomials at rational arguments, *Bull. London Math. Soc.* **25** (1993), 327–329.

[1138] D. Suryanarayana, Sums of the Riemann Zeta function, *Math. Student* **42** (1974), 141–143.

[1139] A. Swaminathan, Convexity of the incomplete beta functions, *Integral Transforms Spec. Funct.* **18** (2007), 503–509.

[1140] D. W. Sweeney, On the computation of Euler's constant, *Math. Comput.* **17** (1963), 170–178.

[1141] G. Szegö, *Orthogonal Polynomials*, Fourth edition, Amererican Mathematical Society Colloquium Publications, Vol. 23, American Mathematical Society, Providence, Rhode Island, 1975.

[1142] G. Szekeres, Some asymptotic formulae in the theory of partitions. II, *Quart. J. Math. Oxford* (*Ser.* 2) **4** (1953), 96–111.

[1143] J. Szmidt, The Selberg trace formula for the Picard group $\mathbf{SL}(2, \mathbb{Z}[i])$, *Acta Arith.* **42** (1983), 391–424.

[1144] V. E. Tarasov, Fractional derivative as fractional power of derivative, *Internat. J. Math.* **18** (2007), 281–299.

[1145] A. Tassaddiq, Fourier transform representation of the extended Fermi-Dirac and Bose-Einstein functions with applications to the family of the Zeta and related functions, *Integral Transforms Spec. Funct.* **22** (2011), 453–466 (see also [1093] above).

[1146] G. J. Tee, A tricky Tripos problem - generalized, 2006 *New Zealand Mathematics Colloquium* (Hamilton, New Zealand; December 4–6, 2006); Preprint 2007.

[1147] A. Terras, Some formulas for the Riemann Zeta function at odd integer argument resulting from Fourier expansions of the Epstein Zeta function, *Acta Arith.* **29** (1976), 181–189.

[1148] A. Terras, *Harmonic Analysis on Symmetric Spaces and Applications*, Vol. I, Springer-Verlag, Berlin, Heidelberg and New York, 1985.

[1149] E. C. Titchmarsh, A series inversion formula, *Proc. London Math. Soc.* (*Ser.* 2) **26** (1926), 1–11.

[1150] E. C. Titchmarsh, On an inequality satisfied by the Zeta-function of Riemann, *Proc. London Math. Soc.* (*Ser.* 2) **28** (1927), 70–80.

[1151] E. C. Titchmarsh, *The Theory of the Riemann Zeta-Function*, Clarendon (Oxford University) Press, Oxford, London and New York, 1951; Second edition (Revised by D. R. Heath-Brown), Clarendon (Oxford University) Press, Oxford, London and New York, 1986.

[1152] E. C. Titchmarsh, *The Zeta-Function of Riemann*, Stechert-Hafner Service Agency, New York and London, 1964.

[1153] P. G. Todorov, Une formule simple explicite des nombres de Bernoulli généralisés *C. R. Acad. Sci. Paris Sér. I Math.* **301** (1985), 665–666.

[1154] P. G. Todorov, Explicit and recurrence formulas for generalized Euler numbers, *Funct. Approx.* **22** (1993), 113–117.

[1155] Ž. Tomovski, New double inequalities for Mathieu type series, *Univ. Beograd. Publ. Elektrotehn. Fak. Ser. Mat.* **15** (2004), 80–84.

[1156] Ž. Tomovski, Integral representations of generalized Mathieu series via Mittag-Leffler type functions, *Fract. Calc. Appl. Anal.* **10** (2007), 127–138.

[1157] Ž. Tomovski, New integral and series representations of the generalized Mathieu series, *Appl. Anal. Discrete Math.* **2** (2008), 205–212.

[1158] Ž. Tomovski, On Hankel transforms of generalized Mathieu series via Mittag-Leffler type functions, *Fract. Calc. Appl. Anal.* **12** (2009), 97–107.

[1159] Ž. Tomovski and R. Hilfer, Some bounds for alternating Mathieu type series, *J. Math. Inequal.* **2** (2008), 17–26.

[1160] Ž. Tomovski and D. Leskovski, Refinements and sharpness of some inequalities for Mathieu type series, in *Proceedings of the Mathematical Conference held in Honour of*

Eighty-Five Years of Professor Blagoj Sazdov Popov's Life (Ohrid, Macedonia; September 4–7), *Math. Maced.* **6** (2008), 67–79.

[1161] Ž. Tomovski and T. K. Pogány, New upper bounds for Mathieu-type series, *Banach J. Math. Anal.* **3** (2) (2009), 9–15.

[1162] L. Tornheim, Harmonic double series, *Amer. J. Math.* **72** (1950), 303–314.

[1163] S. Tričkovič, M. V. Vidanovič, and M. S. Stankovič, Series involving the product of a trigonometric integral and a trigonometric function, *Integral Transforms Spec. Funct.* **18** (2007), 751–763.

[1164] F. G. Tricomi, Sulla somma delle inverse delle terze e quinte potenze dei numeri naturali, *Atti Accad. Naz. Lincei Rend. Cl. Sci. Fis. Mat. Natur.* **47** (8) (1969), 16–18.

[1165] S. Y. Trimble, J. Wells and F. T. Wright, Superadditive functions and a statistical application, *SIAM. J. Math. Anal.* **20** (1989), 1255–1259.

[1166] H. Tsumura, On a p-adic interpolation of the generalized Euler numbers and its applications, *Tokyo J. Math.* **10** (1987), 281–293.

[1167] H. Tsumura, On the values of a q-analogue of the p-adic L-functions, *Mem. Fac. Sci. Kyushu Univ. Ser. A* **44** (1990), 49–60.

[1168] H. Tsumura, A note on q-analogues of the Dirichlet series and q-Bernoulli numbers, *J. Number Theory* **39** (1991), 251–256.

[1169] H. Tsumura, On evaluation of the Dirichlet series at positive integers by q-calculation, *J. Number Theory* **48** (1994), 383–391.

[1170] H. Tsumura, A note on q-analogues of Dirichlet series, *Proc. Japan Acad. Ser. A Math. Sci.* **75** (1999), 23–25.

[1171] K. Uchimura, An identity for the divisor generating function arising from sorting theory, *J. Combin. Theory Ser. A* **31** (1981), 131–135.

[1172] K. Uchimura, A generalization of identities for the divisor generating function, *Utilitas Math.* **25** (1984), 377–379.

[1173] K. Uchimura, Divisor generating functions and insertions into a heap, *Discrete Appl. Math.* **18** (1987), 73–81.

[1174] K. Ueno and M. Nishizawa, Quantum Groups and zeta functions, in *Quantum Groups* (Karpacz, 1994), pp. 115–126, PWN, Warsaw, 1995.

[1175] K. Ueno and M. Nishizawa, Quantum Groups and zeta functions, arXiv:hep-th/9408143v1, (1994).

[1176] K. Ueno and M. Nishizawa, The multiple Gamma functions and the multiple q-Gamma functions, *Publ. RIMS Kyoto Univ.* **33** (1997), 813–838.

[1177] R. S. Underwood, An expression for the summation $\Sigma_{m=1}^{n} m^p$, *Amer. Math. Monthly* **35** (1928), 424–428.

[1178] V. R. A. Uppuluri, On a stronger version of Wallis' formula, *Pacific J. Math.* **19** (1966), 183–187.

[1179] G. Vacca, A new series for the Eulerian constant $\gamma = \cdot 577 \cdots$, *Quart. J. Pure Appl. Math.* **41** (1910), 363–364.

[1180] A. van der Poorten, Some wonderful formulae $\cdots$ footnotes to Apéry's proof of the irrationality $\zeta(3)$, *Séminaire Delange-Pisot-Poitou (Théorie des nombres)* **29** (1978/1979), 1–7.

[1181] A. J. van der Poorten, Some wonderful formulas $\cdots$ an introduction to polylogarithms, in *Proceedings of the Queen's Number Theory Conference* (Kingston, Ontario; 1979), pp. 269–286, *Queen's Papers in Pure and Applied Mathematics* **54**, Queen's University, Kingston, Ontario, 1980.

[1182] A. van der Poorten, A proof that Euler missed $\cdots$ Apéry's proof of the irrationality of $\zeta(3)$, *Math. Intelligencer* **1** (1979), 195–203.

[1183] H. van Engen, Concerning Gamma function expansions, *Tôhoku Math. J.* **45** (1938), 124–129.

[1184] H. van Haeringen, *Completely Monotonic and Related Functions*, Report 93–108, Faculty of Technical Mathematics and Informatics, Delft University of Technology, Delft, The Netherlands, 1993; see also *J. Math. Anal. Appl.* **204** (1996), 389–408.

[1185] L. van Hamme, Advanced Problem 6407, *Amer. Math. Monthly* **89** (1982), 703–704.

[1186] H. S. Vandiver, On generalizations of the numbers of Bernoulli and Euler, *Proc. Nat. Acad. Sci. U. S. A.* **23** (1937), 555–559.

[1187] H. S. Vandiver, Simple explicit expressions for generalized Bernoulli numbers of the first order, *Duke Math. J.* **8** (1941), 575–584.

[1188] H. S. Vandiver, An arithmetical theory of the Bernoulli numbers, *Trans. Amer. Math. Soc.* **51** (1942), 502–531.

[1189] I. Vardi, Integrals, an introduction to analytic number theory, *Amer. Math. Monthly* **95** (1988), 308–315.

[1190] I. Vardi, Determinants of Laplacians and multiple Gamma functions, *SIAM J. Math. Anal.* **19** (1988), 493–507.

[1191] I. Vardi, *Computational Recreations in Mathematica*, Addison-Wesley, Redwood City, 1991.

[1192] V. Varlamov, Convolution of Rayleigh functions with respect to the Bessel index, *J. Math. Anal. Appl.* **306** (2005), 413–424.

[1193] V. Varlamov, Convolutions of Rayleigh functions and their application to semi-linear equations in circular domains, *J. Math. Anal. Appl.* **327** (2007), 1461–1478.

[1194] D. P. Verma, A note on Euler's constant, *Math. Student* **29** (1961), 140–141.

[1195] D. P. Verma and A. Kaur, Summation of some series involving Riemann Zeta function, *Indian J. Math. (N. S.)* **25** (1983), 181–184.

[1196] J. A. M. Vermaseren, Harmonic sums, Mellin transforms and integrals, *Internat. J. Modern Phys. A* **14** (1999), 2037–2076.

[1197] H. A. Verrill, Congruences related to modular forms, *Internat. J. Number Theory* **6** (2010), 1367–1390.

[1198] R. Vidūnas, Expressions for values of the Gamma function, *Kyushu J. Math.* **59** (2005), 267–283.

[1199] M.-F. Vignéras, L'équation fonctionnelle de la fonction zêta de Selberg du groupe moudulaire $PSL(2,Z)$, in *"Journées Arithmétiques de Luminy"* (Colloq. Internat. CNRS, Centre Univ. Luminy, Luminy, 1978), pp. 235–249, *Astérisque* **61**, Soc. Math. France, Paris, 1979.

[1200] A. Voros, The Hadamard factorization of the Selberg Zeta function on a compact Riemann surface, *Phys. Lett. B* **180** (1986), 245–246.

[1201] A. Voros, Special functions, spectral functions and the Selberg Zeta function *Comm. Math. Phys.* **110** (1987), 439–465.

[1202] W. R. Wade, *An Introduction to Analysis*, Third edition, Pearson Prentice-Hall, Englewood Cliffs, New Jersey, 2004.

[1203] A. Walfisz, *Weylsche Exponentialsummen in der Neueren Zahlentheorie*, pp. 114–115, B. G. Teubner, Leipzig, 1963.

[1204] P. L. Walker, The zeros of Euler's Psi function and its derivatives, *J. Math. Anal. Appl.* **332** (2007), 607–616.

[1205] W. Walter, Old and new approaches to Euler's trigonometric expansions, *Amer. Math. Monthly* **89** (1982), 225–230.

[1206] M.-J. Wang, A remark on Andrews-Askey integral, *J. Math. Anal. Appl.* **341** (2008), 1487–1494.

[1207] W.-P. Wang, C.-Z. Jia and T.-M. Wang, Some results on the Apostol-Bernoulli and Apostol-Euler polynomials, *Comput. Math. Appl.* **55** (2008), 1322–1332.
[1208] W.-P. Wang and W.-W. Wang, Some results on power sums and Apostol-type polynomials, *Integral Transforms Spec. Funct.* **21** (2010), 307–318.
[1209] M. Ward, A calculus of sequences, *Amer. J. Math.* **58** (1936), 255–266.
[1210] L. C. Washington, p-adic L-functions and sums of powers, *J. Number Theory* **69** (1998), 50–61.
[1211] G. N. Watson, Ramanujan's integrals and Gauss's sums, *Quart. J. Math. Oxford Ser.* **7** (1936), 175–183.
[1212] G. N. Watson, An expansion related to Stirling's formula, derived by the method of steepest descents, *Quart. J. Pure Appl. Math.* **48** (1920), 1–18.
[1213] G. N. Watson, A note on Spence's logarithmic transcendant, *Quart. J. Math. Oxford Ser.* **8** (1937), 39–42.
[1214] G. N. Watson, A note on Lerch's functions, *Quart. J. Math. Oxford Ser.* **8** (1937), 43–47.
[1215] G. N. Watson, *A Treatise on the Theory of Bessel Functions*, Second edition, Cambridge University Press, Cambridge, London and New York, 1944.
[1216] G. Wechsung, Über Kummers funktionalgleichung für den Pentalogarithmus, *Jahresber. Deutsch. Math.-Verein.* **68** (1966), 140–143.
[1217] G. Wechsung, Über die unmöglichkeit des vorkommens von funktionalgleichungen gewisser struktur für Ploylogarithmen, *Aequationes Math.* **5** (1970), 54–62.
[1218] A. Weil, Sur une formule classique, *J. Math. Soc. Japan* **20** (1968), 400–402.
[1219] W. I. Weisberger, Conformal invariants for determinants of Laplacians on Riemann surfaces, *Comm. Math. Phys.* **112** (1987), 633–638.
[1220] W. I. Weisberger, Normalization of the path integral measure and the coupling constants for bosonic strings, *Nuclear Phys. B* **284** (1987), 171–200.
[1221] E. W. Weisstein, [http://mathworld.wolfram.com/HarmonicNumber.html].
[1222] D. Wells, *The Penguin Dictionary of Curious and Interesting Numbers*, p. 28, Penguin Books, Middlesex, England, 1986.
[1223] J. G. Wendel, Note on the gamma function, *Amer. Math. Monthly* **55** (1948), 563–564.
[1224] D. J. Whitehouse and M. J. Phillips, Two-dimensional discrete properties of random surfaces, *Philos. Trans. Roy. Soc. London Ser. A* **305** (1982), 441–468.
[1225] E. T. Whittaker and G. N. Watson, *A Course of Modern Analysis: An Introduction to the General Theory of Infinite Processes and of Analytic Functions; With an Account of the Principal Transcendental Functions*, Fourth edition, Cambridge University Press, Cambridge, London and New York, 1963.
[1226] D. V. Widder, *The Laplace Transform*, Princeton University Press, Princeton, 1941.
[1227] T. Wieting, A Khinchin sequence, *Proc. Amer. Math. Soc.* **136** (2008), 815–824.
[1228] H. S. Wilf, Problem 10588 *Amer. Math. Monthly* **104** (1997), 456.
[1229] G. T. Williams, A new method of evaluating $\zeta(2n)$, *Amer. Math. Monthly* **60** (1953), 19–25.
[1230] K. P. Williams, The asymptotic form of the function $\psi(x)$. *Bull. Amer. Math. Soc.* **19** (1912/1913), 472–479.
[1231] K. S. Williams, On $\Sigma_{n=1}^{\infty}(1/n^{2k})$, *Math. Mag.* **44** (1971), 273–276.
[1232] K. S. Williams and N.-Y. Zhang, Special values of the Lerch Zeta function and the evaluation of certain integrals, *Proc. Amer. Math. Soc.* **119** (1993), 35–49.
[1233] J. R. Wilton, A proof of Burnside's formula for $\log\Gamma(x+1)$ and certain allied properties of Riemann's ζ-function, *Messenger Math.* **52** (1922/1923), 90–93.
[1234] J. R. Wilton, A note on the coefficients in the expansion of $\zeta(s,x)$ in power of $s-1$, *Quart. J. Pure Appl. Math.* **50** (1927), 329–332.

[1235] J. W. Wrench Jr. and D. Shanks, Questions concerning Khintchine's constant and the efficient computation of regular continued fractions, *Math. Comput.* **20** (1966), 444–448.

[1236] T.-C. Wu, C.-H. Chang and H. M. Srivastava, A unified presentation of identities involving Weierstrass-type functions, *Appl. Math. Lett.* **23** (2010), 864–870.

[1237] T.-C. Wu, S.-H. Leu, S.-T. Tu and H. M. Srivastava, A certain class of infinite sums associated with Digamma functions, *Appl. Math. Comput.* **105** (1999), 1–9.

[1238] T.-C. Wu, S.-T. Tu and H. M. Srivastava, Some combinatorial series identities associated with the Digamma function and harmonic numbers, *Appl. Math. Lett.* **13** (3) (2000), 101–106.

[1239] B. Xu and W. Zhang, Small divisor problem for an analytic q-difference equation, *J. Math. Anal. Appl.* **342** (2008), 694–703.

[1240] A. M. Yaglom and I. M. Yaglom, An elementary derivation of the formulas of Wallis, Leibniz and Euler for the number π, *Uspehi Mat. Nauk* (*N. S.*) **8** (5) (57) (1953), 181–187 (in Russian).

[1241] A. M. Yaglom and I. M. Yaglom, *Challenging Mathematical Problems with Elementary Solutions*, Vol. II. *Problems from Various Branches of Mathematics* (Translated from the Russian by J. McCawley Jr.), Dover Publications, New York, 1987.

[1242] K. Yamagata, A contribution to the theory of non-isothermal Laminar flow of fluids inside a straight tube of circular cross section, *Mem. Fac. Engrg. Kyushu Imp. Univ.* **8** (1940), 365–449.

[1243] S.-L. Yang, An identity of symmetry for the Bernoulli polynomials, *Discrete Math.* **308** (2008), 550–554.

[1244] C.-E Yen, M.-L. Lin and K. Nishimoto, An integral form for a generalized Zeta function, *J. Fract. Calc.* **22** (2002), 99–102.

[1245] D. Zagier, Polylogarithms, Dedekind zeta functions, and the algebraic K-theory of fields, in *Arithmetic Algebraic Geometry* (Texel, 1989) (B. van der Geer, F. Oort and J. Steenbrink, Editors), pp. 391–430, Progress in Mathematics, Vol. 89, Birkhäuser, Basel, Boston and Berlin, 1991.

[1246] D. Zagier, Values of Zeta functions and their applications, in *Proceedings of the First European Congress of Mathematics*, Vol. II (Paris, 1992) (A. Joseph, F. Mignot, F. Murat, B. Prum and R. Rentschler, Editors) pp. 497–512, Progress in Mathematics, Vol. 120, Birkhäuser, Basel, Boston and Berlin, 1994.

[1247] D. Zagier, Periods of modular forms, traces of Hecke operators, and multiple zeta values, in *Research into Automorphic Forms and L Functions* (Kyoto, 1992), Sūrikaisekikenkyūsho Kōkyūroku **843** (1993), 162–170 (in Japanese).

[1248] G. B. M. Zerr, Summation of series, *Amer. Math. Monthly* **5** (1898), 128–135.

[1249] N.-Y. Zhang, Euler's constant and some sums related to the Riemann Zeta function, *Theory Practice Math.* (*Beijing*) **4** (1990), 62–70 (in Chinese)

[1250] N.-Y. Zhang and K. S. Williams, Some infinite series involving the Riemann Zeta function, in *Analysis, Geometry and Groups*: *A Riemann Legacy Volume*, Parts I and II (H. M. Srivastava and Th. M. Rassias, Editors), Part II, pp. 691–712, Hadronic Press, Palm Harbor, Florida, 1993.

[1251] N.-Y. Zhang and K. S. Williams, Some series representations of $\zeta(2n+1)$, *Rocky Mountain J. Math.* **23** (1993), 1581–1592.

[1252] N.-Y. Zhang and K. S. Williams, Application of the Hurwitz Zeta function to the evaluation of certain integrals, *Canad. Math. Bull.* **36** (1993), 373–384.

[1253] N.-Y. Zhang and K. S. Williams, Values of the Riemann Zeta function and integrals involving $\log\left(2\sinh\frac{\theta}{2}\right)$ and $\log\left(2\sin\frac{\theta}{2}\right)$, *Pacific J. Math.* **168** (1995), 271–289.

[1254] R. Zhang, On asymptotics of q-Gamma functions, *J. Math. Anal. Appl.* **339** (2008), 1313–1321.

[1255] Z. Zhang and Q. Hu, On the bilateral series ${}_5\psi_5$, *J. Math. Anal. Appl.* **337** (2008), 1002–1009.

[1256] Z. Zhang and X. Wang, A generalization of Calkin's identity, *Discrete Math.* **308** (2008), 3992–3997.

[1257] Z.-Z. Zhang and J.-Y. Jin, Some identities involving generalized Genocchi polynomials and generalized Fibonacci-Lucas sequences, *Fibonacci Quart.* **36** (1998), 329–334.

[1258] Z.-Z. Zhang and H.-Q. Yang, Several identities for the generalized Apostol-Bernoulli polynomials, *Comput. Math. Appl.* **56** (2008), 2993–2999.

[1259] G. Zhao and H. Feng, A new q-analogue of the sum of cubes, *Discrete Math.* **307** (2007), 2861–2865.

[1260] J.-Q. Zhao, Analyitic continuation of multiple Zeta functions, *Proc. Amer. Math. Soc.* **128** (2000), 1275–1283.

[1261] J.-Q. Zhao, q-multiple Zeta functions and q-multiple Polylogarithms, arXiv:math.QA/0304448 v2, **6** (2003) (see also [1262]).

[1262] Multiple q-Zeta functions and multiple q-Polylogarithms, *Ramanujan J.* **14** (2007), 189–221 (see also [1261]).

[1263] J.-Q. Zhao, Bernoulli numbers, Wolstenholme's theorem, and p^5 variations of Lucas' theorem, *J. Number Theory* **123** (2007), 18–26.

[1264] D.-Y. Zheng, Further summation formulae related to generalized harmonic numbers, *J. Math. Anal. Appl.* **335** (2007), 692–706.

[1265] X. Zhou and T. Cai, A generalization of a curious congruence on harmonic sums, *Proc. Amer. Math. Soc.* **135** (2007), 1329–1333.

[1266] S. Zimering, On a Mercerian theorem and its application to the equiconvergence of Cesàro and Riesz transforms, *Publ. Inst. Math. (Beograd) (N. S.)* **1** (1961), 83–91.

[1267] I. J. Zucker, On the series $\sum_{k=1}^{\infty} \binom{2k}{k}^{-1} k^{-n}$ and related sums, *J. Number Theory* **20** (1985), 92–102.

[1268] I. J. Zucker and M. M. Robertson, Some properties of Dirichlet L-series, *J. Phys. A: Math. Gen.* **9** (1976), 1207–1214.

[1269] I. J. Zucker and M. M. Robertson, A systematic approach to the evaluation of $\Sigma_{(m,n\neq 0,0)} \left(am^2 + bmn + cn^2\right)^{-s}$, *J. Phys. A: Math. Gen.* **9** (1976), 1215–1225.

www.ingramcontent.com/pod-product-compliance
Ingram Content Group UK Ltd.
Pitfield, Milton Keynes, MK11 3LW, UK
UKHW020128290926
15471UKWH00003B/362

9 780323 165266